Power Systems

More information about this series at http://www.springer.com/series/4622

Magdi S. Mahmoud · Fouad M. AL-Sunni

Control and Optimization of Distributed Generation Systems

Springer

Magdi S. Mahmoud
Department of Systems Engineering
King Fahd University of Petroleum
and Minerals
Dhahran
Saudi Arabia

Fouad M. AL-Sunni
Department of Systems Engineering
King Fahd University of Petroleum
and Minerals
Dhahran
Saudi Arabia

ISSN 1612-1287 ISSN 1860-4676 (electronic)
Power Systems
ISBN 978-3-319-16909-5 ISBN 978-3-319-16910-1 (eBook)
DOI 10.1007/978-3-319-16910-1

Library of Congress Control Number: 2015934050

Mathematics Subject Classification: 93A10, 93A13, 93A14, 93A15, 93A30

Springer Cham Heidelberg New York Dordrecht London

Printed on acid-free paper

Springer International Publishing AG Switzerland is part of Springer Science+Business Media (www.springer.com)

In the Name of

GOD

the All-Compassionate, the All-Merciful.
"And of knowledge, you (mankind)
have been given only a little."

To Our Families

Magdi S. Mahmoud, Fouad M. AL-Sunni
Dhahran-Saudi Arabia, 2013

Preface

Distributed generation (DG) of electricity based on renewable energy sources such as wind and solar is gaining more and more attention all over the world because of ever growing concerns on energy cost, energy security, and environmental issues. Although DG has a great potential for economic and environmental benefits, how to establish efficient and reliable control over a large number of DG units is one of the fundamental problems to be solved in the near future. One promising solution to this problem is the *microgrid*, which interconnects a group of DG units and loads at a distribution voltage level in a local area such as a university or a residential community. Microgrids are small-scale power systems that facilitate the integration of distributed generators and can operate in both grid-connected and islanded modes. In normal operation, the microgrid is connected to the main grid, and its frequency is dictated by the nominal frequency of the main grid. However, the microgrid may disconnect from the main grid and go to the islanded operation due to preplanned or unplanned events. Islanding process results in active power unbalance between generation and consumption units which, in turn, may cause frequency instability.

Microgrids are established based on localized control and can operate in either a grid-connected mode or an islanded mode, which significantly reduces the complexity of DG unit control. In order to avoid high capital expenditure and low reliability in microgrid operation, decentralized control is indispensable.

This book is essentially written for senior and first-year graduate students interested in studying distributed energy systems and future power systems. The different chapters and sections are organized to treat three broad avenues, namely *architectures and integration, modeling and analysis, and communication and control*. The main perspective of these sections is to capture the main sources for expanding the present electric power grid. Our goal is to capture the spectrum of this exponential transformation, and at the same time present the plethora of open problems that this transformation poses for our control theory colleagues.

The layout of the different chapters is as follows:

- The architectures and integration avenue opens the book with visionary ideas on sustainable architectures for power system operation and control under significant penetration of highly variable renewable energy resources presented in Chap. 1.
- This is followed by a detailed discussion on the distribution generation plants and their dynamic models in Chap. 2.
- Chapter 3 addresses the problem of supply–demand in electric power grid.
- Chapter 4 illustrates several practical control methods for microgrids.
- Chapter 5 focuses on a technical survey of computational intelligence methods.
- Chapter 6 provides a system of systems approach to the modeling, analysis, and control/optimization of microgrids.
- Chapter 7 delves into novel networked control techniques for wide-area oscillation tracking in large-scale power systems.
- This is further elaborated in Chap. 8 to highlight the importance of distributed control and processing as a major tool for wide-area monitoring research.
- In Chap. 9, a hierarchical structuring and control approach is introduced to model the dynamic mechanisms of cascading failures in geographically dispersed grids.
- The book in concluded by an Appendix to complement the math analysis of the book.

Looked at in this light, this book deals with the basic concept, generation technologies, impacts, operation, control and management aspects, and economic viability and market participation issues of microgrid and active distribution networks in a broad perspective.

Each chapter includes examples/simulation cases and a list of appropriate problems at the end.

Throughout this book, the following terminologies, conventions, and notations have been adopted. All of them are quite standard in the scientific media and only vary in form or character. Matrices, if their dimensions are not explicitly stated, are assumed to be compatible for algebraic operations. In symmetric block matrices or complex matrix expressions, we use the symbol $\bullet$ to represent a term that is induced by symmetry.

Acknowledgments

The book is mainly targeted at senior undergraduate or graduate-level students, at academic and industrial researchers working in the field of electrical energy systems, and also at engineers developing actual solutions for distributed generation systems deploying communication networks. One can view this book as a good basis to teach a class on networked-control systems for distributed generation systems.

The idea of writing the book arose and developed during extensive discussions within KFUPM, where we took full advantage of the supportive scientific environment. It is our great pleasure to acknowledge the financial funding afforded by DSR through project no. **IN141019** and for providing overall support for research activities at KFUPM.

During the past years, we had the privilege of teaching various senior and graduate courses at KFUPM (Saudi Arabia). The course notes, updated and organized, were instrumental in generating different chapters of this book and valuable comments and/or suggestions by graduate students were greatly helpful, particularly by those who attended the courses **SCE 507**, **SCE 527**, **SCE 614**, and **SCE 691** offered at the Systems Engineering Department over the period 2007–2013.

Heartfelt thanks are owed to our colleagues from all over the world who have helped us, one way or another, in grasping the fundamental concepts of this exciting field and in writing this book. Moreover, we deeply appreciate the efforts of Dr. Muhammad Sabih, Azhar M. Memon, Nezar M. Alyazidi, Mohammad S. Rahman, Azhar S. Hussain, and Gulam D. Khan for their unfailing help in preparing portions of the manuscript and performing effective numerous simulations.

In writing this volume, we took the approach of referring within the text to papers and/or books which we believe taught us some concepts, ideas, and methods. We further complemented this by adding some remarks, observations, and notes within and at the end of each chapter to shed some light on other related results. We apologize in advance if we have committed injustice and assure all of the colleagues that any mistake was definitely unintentional. We also want to express our deep gratitude to all the researchers in the field who have made their results and publications easily available over the World Wide Web. Without this help, collecting the material discussed in the present book alone would have been too big a challenge to embark on.

Most of all, however, we would like to express our deepest gratitude to our parents who taught us the value of the written word and to all the members of our families. Without their constant love, incredible patience, and (mostly) enthusiastic support this volume would not have been finished.

We would appreciate any comments, questions, criticisms, or corrections that readers may take the trouble of communicating to me at msmahmoud@kfupm.edu.sa, alsunni@kfupm.edu.sa, magdisadekmahmoud@gmail.com.

Dhahran, Saudi Arabia
March 2015

Magdi S. Mahmoud
Fouad M. AL-Sunni

Contents

Part I Modeling and Analysis

1 Introduction 3
1.1 Distributed Generation 3
1.1.1 Technology 3
1.1.2 Value of Distributed Generation 5
1.1.3 Applications and Issues 5
1.1.4 Distributed Resources 6
1.1.5 Distributed Capacity 7
1.1.6 Factors of DG Growth 7
1.1.7 Impacts on Transmission System Operation 9
1.1.8 General Structure 10
1.1.9 Integrating Distributed Energy Resources 12
1.2 Supply–Demand in Electric Power Grid 12
1.2.1 Understanding the Grid 12
1.2.2 Reliability Concepts 13
1.2.3 Electric Power Dynamic Demand 14
1.2.4 The Need for Spinning Reserve 14
1.2.5 Local Load Control 15
1.2.6 Ancillary Services 15
1.2.7 Implementation Issues 15
1.3 Overview of Microgrids 16
1.3.1 Control Tasks 18
1.3.2 A Classification 19
1.3.3 Control Objectives and Methods 21
1.3.4 Microsource Control 23
1.3.5 Control and Protection Requirements 25
1.3.6 Reliable and Economical Operation 26
1.4 Smart Grid 26
1.4.1 Efficiency and Reliability 27
1.4.2 Environmental Benefits 28

1.4.3 Benefits to Consumers 28
1.4.4 Security 29
1.5 Technical Aspects 29
1.5.1 Two-Way Communications 29
1.5.2 Control and Monitoring Techniques 30
1.5.3 Advanced Components 30
1.5.4 Energy Storage 31
1.5.5 Robust Energy Management 33
1.6 DC Microgrids 35
1.6.1 PV Sources Control 36
1.6.2 Storage Control 37
1.6.3 Grid Connection Control 38
1.6.4 DC Load Control 38
1.6.5 Power Balancing Principle 38
1.7 Outline of the Book 39
1.7.1 Methodology 39
1.7.2 Chapter Organization 40
References 41

2 Distributed Generation Plants 47
2.1 Combined Heat and Power Plants 47
2.1.1 Introduction 47
2.1.2 Microcogeneration Systems 48
2.1.3 Internal Combustion Engines 49
2.1.4 Stirling Engines 49
2.1.5 Microturbines 51
2.1.6 Fuel Cells 53
2.2 Renewable Energy Generation 55
2.2.1 Wind Power Plants 55
2.2.2 Small-Scale Hydrogeneration 57
2.3 Solar Photovoltaic Generation 58
2.3.1 Technology Basics 59
2.3.2 Grid-Connected Solar Systems 60
2.3.3 Future Trends 61
2.4 Small Wind Turbine Systems 62
2.4.1 Types of Wind Turbine Systems 63
2.4.2 Wind Turbine Fundamentals 66
2.4.3 Control Loops 67
2.4.4 Generator Side Control 68
2.4.5 Boost Converter Control 68
2.4.6 Rectifier Control 70
2.5 Storage Technologies 70
2.5.1 Classification of Electrical Energy Storage 71
2.5.2 Mechanical Storage Systems 72

2.5.3 Batteries 72
2.5.4 Flywheels 73
2.5.5 Superconducting Magnetic Energy Storage 73
2.5.6 Supercapacitors 74
2.6 Inverter Interfaces 74
2.6.1 Voltage Source Inverters 74
2.6.2 Inverter Realization for Microsources 76
2.6.3 Inverter Realization 76
2.6.4 Unbalanced AC Voltages 77
2.7 Conclusions 78
2.8 Suggested Problems 78
References 84

Part II Architectures and Integration

3 Control Methods for Microgrids 89
3.1 Introduction 89
3.2 Microgrids 90
3.2.1 Definition and Applications 90
3.2.2 Control Functions 91
3.2.3 Components and Formation 92
3.2.4 Overview of Modeling 94
3.2.5 Modes of Operation 103
3.3 Control Approaches 104
3.3.1 Control of Grid-Connected Mode 104
3.3.2 Power Flow Control by Current Regulation 105
3.3.3 Power Flow Control by Voltage Regulation 105
3.3.4 Agent-Based Control 106
3.3.5 Distributed Control 107
3.3.6 H_∞ Control 109
3.3.7 Autonomous/Islanded Mode 109
3.3.8 PQ and VSI Control 110
3.3.9 Autonomous Control 112
3.3.10 New Q–$\dot{V}$ Droop Control 112
3.3.11 Control Design Based on Transfer Function 113
3.3.12 Microgrid Control in both Modes 114
3.4 System of Systems 116
3.4.1 Introduction 116
3.4.2 SoS Control 116
3.4.3 Decentralized Control 117
3.4.4 Multilevel Control 120
3.4.5 Networked Control Systems 122
3.4.6 Comparative Analysis 125

3.5 Modeling and Analysis of Inverter-Based Microgrids....... 126
3.5.1 Introduction 128
3.5.2 Microgrid Model in Autonomous Mode.......... 129
3.5.3 State-Space Model of a Voltage Source Inverter 132
3.5.4 Combined Model of All the Inverters 140
3.5.5 Network Model 141
3.5.6 Load Model............................ 143
3.5.7 Complete Microgrid Model 144
3.5.8 Sensitivity Analysis......................... 145
3.6 Conclusions 147
3.7 Suggested Problems 148
References.. 152

4 Optimal Energy Management 159
4.1 Introduction 159
4.2 A Microgrid Model.............................. 160
4.3 Microgrid and Load Forecasting 162
4.3.1 Proposed NNE 162
4.3.2 Microgrid Environment Forecasting............. 166
4.4 Multiobjective Energy Management.................. 167
4.4.1 Battery Scheduling 169
4.4.2 Fuzzy-Logic-Based Expert System 170
4.5 Simulation Results I 174
4.5.1 RE Power Generation and Load Forecasting....... 174
4.5.2 Multiobjective Intelligent Energy Management 177
4.6 Optimal Energy Cost Management 181
4.7 Microgrid Overview 183
4.7.1 PV Sources Control........................ 184
4.7.2 Storage Control 185
4.7.3 Grid Connection Control 186
4.7.4 DC Load Control 186
4.7.5 Power Balancing Principle 187
4.8 Supervision Control Design 188
4.8.1 Human–Machine Interface Layer............... 189
4.8.2 Prediction Layer 189
4.8.3 Energy Management Layer................... 190
4.8.4 Operation Layer 192
4.9 Simulation Results II........................... 193
4.9.1 Optimization Results 194
4.9.2 Powers Flow Simulation I 196
4.9.3 Powers Flow Simulation II................... 199
4.9.4 Comparison and Discussion 200

4.10 Conclusions 201
4.11 Suggested Problems 202
References. 203

5 A System of Systems Framework for Microgrids 209
5.1 Introduction 209
5.2 Microgrid Subsystems 210
5.2.1 Microgrid Central Controller 211
5.2.2 Microsource and Load Controllers. 212
5.2.3 Microturbines 213
5.2.4 Microsources and Fuel Cells. 213
5.2.5 Storage Devices 214
5.3 The Concept of SoS 214
5.3.1 Micrograds as SoS. 215
5.3.2 Grid-Connected Operation 216
5.3.3 Grid-Islanded Operation. 216
5.3.4 Operation of the Microgrid Under the SoS Framework. 219
5.4 Modeling of Microgrid 220
5.4.1 Microturbine Model 220
5.4.2 PV Solar Cell Model. 222
5.4.3 Wind Turbine Model. 223
5.5 Microgrid Control Architecture. 225
5.5.1 Introduction 226
5.5.2 Hierarchical Control 228
5.5.3 Consensus Control 229
5.5.4 Centralized and Decentralized Control 230
5.5.5 More on Decentralized Control. 234
5.6 Application to Islanded Microgrid. 235
5.6.1 Two-Level Control Strategy 237
5.6.2 Local Subsystem Control 237
5.6.3 Global Corrective Control 238
5.6.4 Simulation Results 239
5.7 Conclusions 241
5.8 Suggested Problems 242
References. 245

Part III Communication and Control

6 Networked Control of Microgrid System of Systems. 251
6.1 Introduction 251
6.2 Microgrid as SoS 253
6.3 Microgrid Islanded System Modeling 256

6.4 Networked Control System. 258
6.5 Closed-Loop Stability Results 261
6.6 Illustrative Example. 265
6.7 Microalternator and Photovoltaic Systems 269
6.7.1 Introduction 269
6.7.2 Modeling of the Microalternator–PV System 272
6.7.3 Microalternator 272
6.7.4 Photovoltaic System 275
6.7.5 Networked Control System Modeling 286
6.7.6 Simulation Results 298
6.8 Conclusions 301
6.9 Suggested Problems 304
References. 307

7 Decentralized Voltage Control Methods 309
7.1 Introduction 309
7.2 Control Strategy I 310
7.2.1 Frequency Control 311
7.2.2 Voltage Control 312
7.2.3 Overcurrent Limiters 313
7.2.4 Islanding Detection Approach. 314
7.3 Small-Signal Dynamic Analysis 315
7.3.1 Dynamics of Grid-Connected Mode 316
7.3.2 Dynamics of Autonomous Mode-Case 1 318
7.3.3 Dynamics of Autonomous Mode-Case 2 320
7.4 Time-Domain Simulation Results 321
7.4.1 Grid-Connected Mode 321
7.4.2 Ride-Through Capability of DG Unit. 322
7.4.3 Transition Capability from Grid-Connected to Islanded Mode 323
7.4.4 Autonomous Mode 327
7.5 Robust Control Strategy for Multi-Microgrids. 328
7.5.1 Introduction 329
7.5.2 System Description 330
7.5.3 Power Management. 330
7.5.4 Mathematical Model 335
7.6 Control Strategy II 337
7.6.1 Design Requirements. 339
7.6.2 Existence Conditions 339
7.6.3 Real Stability Radius Constraints 341
7.6.4 Controller Design Procedure. 341
7.6.5 A Decentralized Controller Scheme. 343
7.6.6 Properties of the Closed-Loop System 343
7.6.7 Other Robustness Measures 347

7.7 Decentralized Inverter Control . . . 348
7.7.1 System Model . . . 349
7.7.2 Power Sharing Control Strategy . . . 352
7.7.3 Controller Design . . . 353
7.7.4 Decentralized Information Acquisition . . . 353
7.7.5 Stability Analysis Without Communication Delay . . . 354
7.7.6 Model of Individual Inverter. . . . 355
7.7.7 Combined Inverter Model . . . 359
7.7.8 Network Model . . . 361
7.7.9 Microgrid Model. . . . 363
7.7.10 System Stability Evaluation . . . 363
7.7.11 Stability Analysis with Communication Delay . . . 363
7.7.12 Simulation Results . . . 365
7.8 Conclusions . . . 368
7.9 Suggested Problems . . . 369
References. . . . 374

8 Advanced Control Approaches . . . 379
8.1 Introduction . . . 379
8.2 Distributed Control Architecture . . . 379
8.2.1 Integrated Wind/Solar/RO System Modeling. . . . 381
8.2.2 Water Desalination System Description . . . 384
8.2.3 Short-Term Supervisory Predictive Control. . . . 386
8.2.4 Supervisory Control System Design I . . . 387
8.2.5 Simulation Results I . . . 389
8.2.6 Integration for Long-Term Operation. . . . 393
8.2.7 Supervisory Control System Design II . . . 394
8.2.8 Simulation Results II. . . . 396
8.2.9 Distributed Energy Systems . . . 400
8.2.10 Distributed Frequency Control . . . 402
8.2.11 Simulation Results III . . . 405
8.3 Multilevel Control of Droop-Controlled Microgrids . . . 408
8.3.1 A Generalized Multilevel Structure . . . 409
8.3.2 Multilevel Control of AC Microgrids. . . . 411
8.3.3 Inner Control Loops . . . 412
8.3.4 Primary Control . . . 412
8.3.5 Secondary Control . . . 414
8.3.6 Tertiary Control . . . 415
8.3.7 Simulation Results IV . . . 416
8.4 Multilevel Control of DC Microgrids. . . . 421
8.4.1 Primary Control . . . 422
8.4.2 Secondary Control . . . 423
8.4.3 Tertiary Control . . . 424
8.4.4 Simulation Results V. . . . 425

8.5 Enhanced Compensation Technique 427
8.5.1 Microgrid Multilevel Control Scheme 428
8.5.2 DG Local Control 430
8.5.3 Fundamental Positive Sequence Powers Controllers 431
8.5.4 Voltage and Current Controllers 431
8.5.5 Virtual Impedance Loop 431
8.5.6 Compensation Effort Controller 434
8.5.7 Secondary Controller 435
8.5.8 Simulation Results VI 436
8.6 Distributed Cooperative Control 440
8.6.1 Microgrid Control Levels 442
8.6.2 Large-Signal Inverter-Based Model 443
8.6.3 Cooperative Secondary Voltage Control 447
8.6.4 Feedback Linearization and Tracking Synchronization 447
8.6.5 Required Sparse Communication Topology 451
8.6.6 Simulation Results VII 452
8.7 Conclusions 455
8.8 Suggested Problems 457
References 462

9 Real-Time Implementation 467
9.1 Neural-Network-Based Secondary Control 467
9.1.1 Introduction 467
9.1.2 An Autonomous Microgrid 470
9.1.3 Primary Control 471
9.1.4 Distributed Secondary Control 474
9.1.5 Neural-Network-Based Distributed Secondary Control 475
9.1.6 Differential Evolution 477
9.1.7 NN Training 478
9.1.8 Simulation Results 481
9.2 Optimal Control for Autonomous Microgrid 487
9.2.1 Introduction 487
9.2.2 Autonomous Microgrid Controller 488
9.2.3 Power Controller 489
9.2.4 Voltage Controller 490
9.2.5 Current Controller 491
9.2.6 Coupling Inductance and Filter 491
9.2.7 Lines Model 492
9.2.8 Load Model 492

9.3 Problem Formulation . . . 492
9.3.1 Results and Discussions . . . 493
9.3.2 Nonlinear Time Domain Simulation . . . 495
9.3.3 Experimental Results . . . 499
9.4 Distributed Control for Autonomous Microgrid . . . 504
9.4.1 Introduction . . . 504
9.4.2 Real-Time Digital Simulator . . . 505
9.4.3 Description of RTDS Hardware . . . 506
9.4.4 Description of RTDS Software . . . 508
9.4.5 Distributed Control Scheme . . . 508
9.4.6 RTDS Implementation . . . 510
9.4.7 Simulation Results . . . 514
9.4.8 Comparison of RTDS and MATLAB Results . . . 515
9.4.9 Load Sharing During Faults . . . 518
9.5 Experimental Verification of Inverter-Based Microgrid . . . 521
9.5.1 Modeling Results . . . 523
9.5.2 Experimental Results . . . 524
9.6 Conclusions . . . 528
References . . . 528

10 Appendix . . . 531
10.1 Important Facts in Linear Algebra . . . 531
10.1.1 Basic Notions . . . 531
10.1.2 Inner Product and Orthogonality . . . 534
10.1.3 Kronecker Product and Stack of Matrices . . . 535
10.2 Linear Transformations and Matrix Groups . . . 536
10.3 Elements of Graph Theory . . . 540
10.3.1 Basic Results . . . 540
10.3.2 Laplacian Spectrum of Graphs . . . 541
10.3.3 Properties of Adjacency Matrix . . . 541
10.4 Matrix Algebra . . . 543
10.4.1 Inverse of Block Matrices . . . 544
10.4.2 Matrix Inversion Lemma . . . 545
10.4.3 Range, Kernel, Rank and Eigenvectors . . . 546
10.4.4 Symmetric and Skew-Symmetric Matrices . . . 549
10.5 Singular Value Decomposition . . . 551
10.5.1 Geometric Interpretation . . . 553
10.5.2 Example A.1 . . . 554
10.5.3 Some Properties of the SVD . . . 555
10.5.4 The QR Decomposition . . . 557
10.6 Useful Formulae . . . 558
10.6.1 Ackermann's Formula for Eigenvalue Assignment . . . 558

10.6.2 Parseval Formula 559
10.6.3 Frobenius Formula 560
10.7 Inequalities. 560
10.7.1 Inequality 1 561
10.7.2 Inequality 2 561
10.7.3 Inequality 3 562
10.7.4 Inequality 4 (Schur Complements) 562
10.7.5 Inequality 5 564
10.8 Lemmas. 565
10.9 Linear Matrix Inequalities 568
10.9.1 Basics 568
10.9.2 Some Standard Problems 569
10.9.3 The S-procedure 571
10.10 Lyapunov Map and Lyapunov Equation. 572
10.11 Persistence of Excitation and Sufficiently Rich Inputs 573
10.12 Notes and References 576
References. 576

Index 577

About the Authors

Magdi Sadek Mahmoud obtained B.Sc. (Honors) in Communication Engineering, M.Sc. in Electronic Engineering, and Ph.D. in Systems Engineering, all from Cairo University in 1968, 1972, and 1974, respectively. He has been a professor of engineering since 1984. He is now a Distinguished University Professor at KFUPM, Saudi Arabia. He was on the faculty at different universities worldwide including Egypt (CU, AUC), Kuwait (KU), UAE (UAEU), UK (UMIST), USA (Pitt, Case Western), Singapore (Nanyang Technological), and Australia (Adelaide). He lectured in Venezuela (Caracas), Germany (Hanover), UK (Kent), USA (Uo Texas at SA), Canada (Montreal, Alberta), and China (BIT, Yanshan). He is the principal author of 34 books, inclusive book-chapters, and the author/co-author of more than 500 peer-reviewed papers. He is the recipient of two national, one regional, and four university prizes for outstanding research in engineering and applied mathematics. He is a fellow of the IEE, a senior member of the IEEE, the CEI (UK), and a registered consultant engineer of information engineering and systems (Egypt). He is currently actively engaged in teaching and research in the development of modern methodologies to distributed control and filtering, networked-control systems, triggering mechanisms in dynamical systems, fault-tolerant systems, and information technology. He is a fellow of the IEE, a senior member of the IEEE, the CEI (UK), and a registered consultant engineer of information engineering and systems Egypt.

Fouad M. AL-Sunni is professor of control and automation in the Department of Systems Engineering. His areas of research include optimal control, robust control, and Identification of dynamic systems. He has published in the areas of fuzzy logic, neural networks, optimal control, identification, time-delay systems, model predictive control, and control and engineering education. He is a senior member of the IEEE, a senior member of IIE, a member of INFORMS, and a member of ISA.

Notation and Symbols

I^+	The set of positive integers
$\Re$	The set of real numbers
$\Re_+$	The set of nonnegative real numbers
$\Re^n$	The set of all n-dimensional real vectors
$\Re^{n \times m}$	The set of $n \times m$-dimensional real matrices
$\mathscr{C}^-$	The open right-half complex plane
$\mathscr{C}^+$	The closed right-half complex plane
$\in$	Belong to or element of
$\subset$	Subset of
$\bigcup$	Union
$\bigcap$	Intersection
$\gg$	Much greater than
$\ll$	Much less than
A^t	The transpose of matrix A
A^{-1}	The inverse of matrix A
I	An identity matrix of arbitrary order
I_s	The identity matrix of dimension $s \times s$
e_j	The jth column of matrix I
x^t or A^t	The transpose of vector x or matrix A
$\lambda(A)$	An eigenvalue of matrix A
$\rho(A)$	The spectral radius of matrix A
$\lambda_j(A)$	The jth eigenvalue of matrix A
$\lambda_m(A)$	The minimum eigenvalue of matrix A where $\lambda(A)$ are real
$\lambda_M(A)$	The maximum eigenvalue of matrix A where $\lambda(A)$ are real
$A^\dagger$	The Moore–Penrose-inverse of matrix A
$P > 0$	Matrix P is real symmetric and positive-definite
$P \geq 0$	Matrix P is real symmetric and positive semi-definite
$P < 0$	Matrix P is real symmetric and negative-definite
$P \leq 0$	Matrix P is real symmetric and negative semi-definite
$A(i,j), A_{ij}$	The ij-th element of matrix A

$det(A)$	The determinant of matrix A
$trace(A)$	The trace of matrix A
$rank(A)$	The rank of matrix A
$\mathscr{L}_2(-\infty, \infty)$	Space of time domain square integrable functions
$\mathscr{L}_2[0, \infty)$	Subspace of $\mathscr{L}_2(-\infty, \infty)$ with functions zero for $t<0$
$\mathscr{L}_2(-\infty, 0]$	Subspace of $\mathscr{L}_2(-\infty, \infty)$ with functions zero for $t>0$
$\mathscr{L}_2(j\Re)$	Square integrable functions on $\mathscr{C}_0$ including at ∞
$\mathscr{H}_2$	Subspace of $\mathscr{L}_2(j\Re)$ with functions analytic in $Re(s)>0$
$\mathscr{L}_\infty(j\Re)$	Subspace of functions bounded on $Re(s)=0$ including at ∞
$\mathscr{H}_\infty$	The set of $\mathscr{L}_\infty(j\Re)$ functions analytic in $Re(s)>0$
$\lvert a\rvert$	The absolute value of scalar a
$\lvert\lvert x\rvert\rvert$	The Euclidean norm of vector x
$\lvert\lvert A\rvert\rvert$	The induced Euclidean norm of matrix A
$\lvert\lvert x\rvert\rvert_p$	The ℓ_p norm of vector x
$\lvert\lvert A\rvert\rvert_p$	The induced ℓ_p norm of matrix A
$Im(A)$	The image of operator/matrix A
$Ker(A)$	The kernel of operator/matrix A
$max\ D$	The maximum element of set D
$min\ D$	The minimum element of set D
$sup\ D$	The smallest number that is larger than or equal to each element of set D
$inf\ D$	The largest number that is smaller than or equal to each element of set D
$arg\ max\ D$	The index of maximum element of ordered set S
$arg\ min\ D$	The index of minimum element of ordered set S
B_r	The ball centered at the origin with radius r
R_r	The sphere centered at the origin with radius r
$\mathscr{N}$	The fixed index set $\{1, 2, \ldots, N\}$
$[a, b)$	The real number set $\{t \in \Re : a \leq t < b\}$
$[a, b]$	The real number set $\{t \in \Re : a \leq t \leq b\}$
S	The set of modes $\{1, 2, \ \ldots, \ s\}$
iff	If and only if
$\otimes$	The Kronecker product
O(.)	Order of (.)
$diag(\ldots)A$	Diagonal matrix with given diagonal elements
$spec(A)$	The set of eigenvalues of matrix A (spectrum)
$min-poly(A)(s)$	The minimal polynomial of matrix A

Abbreviations

AC	Alternating Current
ARE	Algebraic Riccati Equation
CAN	Control Area Network
CB	Circuit Breaker
CC	Central Controller
CCGT	Combined Cycle Gas Turbine
CHP	Combined Heat and Power
CO	Carbon Monoxide
DA	Distribution Automation
DC	Direct Current
DCS	Distributed Control System
DEC	Decentralized Control
DER	Distributed Energy Resource
DFIG	Doubly Fed Induction Generator
DG	Distributed Generation
DGS	Distributed Generation Systems
DMS	Distribution Management Systems
DNO	Distribution Network Operator
DO	Digital Output
DSM	Demand Side Management
GT	Gas Turbine
HC	Hierarchical Control
IC	Internal Combustion
IED	Intelligent Electronic Device
I/O	Input/Output
LMI	Linear Matrix Inequality
LQC	Linear Quadratic Control
LV	Low Voltage
MC	Microsource Controller
MT	Microturbine
MV	Medium Voltage

PEI	Power Electronic Interface
PF	Power-Frequency
PI	Proportional-Integral
PID	Proportional-Integral-Derivative
PMSG	Permanent Magnet Synchronous Generator
PQ	Power Quality
PV	Photovoltaic
PX	Power Exchange
RES	Renewable Energy Source
SCADA	Supervisory Control and Data Acquisition
SYGN	Synchronous Generator
SISO	Single-Input Single-Output
SVD	Singular Value Decomposition
TDS	Time-Delay System
TES	Thermal Energy Storage

Part I
Modeling and Analysis

Chapter 1
Introduction

1.1 Distributed Generation

Strictly speaking, the purpose of distributed generation (DG) is to provide a source of active electric power and, accordingly, it does not need to be able to provide reactive power. As a new approach in the electricity industry there is no generally accepted definition of DG yet; see [2, 16] for further details. In the literature, a large number of terms and definitions are used in relation to DG. The following is selected for convenience [1]:

Definition 1.1 DG is typically from a few kilowatts up to 400 MW.

Of course, the limit 400 MW varies from one country to another. Associated with this is the location of the DG unit as stated below:

Definition 1.2 The location of DG is defined as the installation and operation of electric power generation units connected directly to the distribution network or connected to the network on the customer site of the meter.

The motivation for using this definition is that the connection of generation units to the transmission network is done traditionally by the industry.

1.1.1 Technology

Current practice indicates that the available technology for DG varies widely (see Table 1.1). The following are important observations:

1. First, many of the technologies utilize renewable energy resources. According to the International Energy Agency, *renewable energy resources are defined as resources that are generally not subject to depletion, such as the heat and light from the sun, the force of wind, organic matter (biomass), falling water, ocean energy, and geothermal heat.* Although renewable energy resources can

M.S. Mahmoud and F.M. AL-Sunni, *Control and Optimization of Distributed Generation Systems*, Power Systems, DOI 10.1007/978-3-319-16910-1_1

Table 1.1 Technologies for distributed generation

Technology	Typical available size per module
Combined cycle gas turbine	35–400 MW
Internal combustion engines	5 kW–10 MW
Combustion turbine	1250 MW
Microturbines	25 kW–1 MW
Small hydro	1100 MW
Micro hydro	25 kW–1 MW
Wind turbine	200 W–3 MW
Photovoltaic arrays	20 W–100 kW
Solar thermal, central receiver	1–100 MW[a]
Solar thermal, Lutz system	1080 MW
Biomass (based on gasification)	100 kW–20 MW
Fuel cells, phosacid	200 kW–2 MW
Fuel cells, molten carbonate	250 kW–2 MW
Fuel cells, proton exchange	1–250 kW
Fuel cells, solid oxide	250 kW–5 MW
Geothermal	5–100 MW
Ocean energy	100 kW–1 MW
Stirling engine	210 kW
Battery storage	500 kW–5 MW

[a] *Source* [1]

be described as abundant, availability of the different resources varies significantly between areas and countries, as well as technology efficiency to harvest the renewable energy resources.

2. Secondly, technologies such as micro-hydro units, photovoltaic (PV) arrays, wind turbines (WTs), diesel engines, solar thermal systems, fuel cells, and battery storage consist of a number of small modules, which are assembled in factories. These modules can be installed in a very short time at the final power station location.
3. Another important aspect is the combined production of heat and power (CHP). Combined cycle gas turbines (CCGTs), internal combustion (IC) engines, combustion turbines, biomass gasification, geothermal, sterling engines as well as fuel cells are suitable for a CHP. The combined local production of heat and power has the advantage of high efficiency, if the heat is used locally.

1.1.2 Value of Distributed Generation

It is needless to stress that clean local energy adds tremendous value by providing cost-effective power to ratepayers, utilities, and communities. In this regard, local renewables:

- Avoid the expensive and inefficient long-distance transmission of power.
- Strengthen local economies and communities.
- Enhance energy security by providing for a more robust and resilient grid.
- Boost nation's energy independence.

1.1.3 Applications and Issues

The existing grid has a small number of producers, long distribution ways, and high maintenance cost; it is also difficult to achieve load balancing. Moreover, the depleting fossil fuels and the adverse effect on the environment by its consumption has gained international interest in reducing the excess use of nonrenewable energy resources and many nations are keeping tabs on CO_2 emissions [4].

The main concerns with the existing centralized power system grid are summarized below [37, 43]:

- Increasing demand and lack of high reliability
- Limitations of centralized power system planning
- Risks of volatile bulk power markets
- Security threats
- Limited power quality
- Environmental effect (Release of CO_2, nuclear waste, etc.).

All the above issues urge the need to incorporate DG units into existing power systems [48]. The concept of DG that belongs to the early 1990s has multiple advantages for both source and consumers [22]. In the literature, there exist various definitions of DG which are summarized in [63]. DG is defined as, "Generation of electricity by facilities that are sufficiently smaller than central generating plants so as to allow interconnection at nearly any point in the power system" [88].

DG units are the emerging micro-generating technologies such as microturbines (MTs), fuel cells, IC engines. They also make use of renewable energy sources (RESs) such as PV arrays and WTs. DG units have low emission rates and are environment friendly and economical. The introduction of DG units should reduce the pressure on the central power grid principally, but technically speaking, penetration of DG into the power grid creates a new class of concerns different from those found in traditional power sources. Random applications of DG units will cause as many problems as it may solve [47]. Some of the problems are discussed below:

- First of all, DG units operate close to the distribution voltage level of 480 V as they are geographically located near the loads and provide a direct current (DC) or variable frequency alternating current (AC) output, and hence require power electronic devices in order to interface with the power-grid/load. The power electronic interface (PEI) leads to development of new control strategies [44, 54].

- The output of renewable energy systems fluctuates with conditions of weather, which is also a debatable issue when DG units are connected to the power grid [56].
- The existing power grid follows a multi-level flow of power from transmission to distribution network; any change in power flow causes problems because the DG unit behavior is different from conventional load [4].
- Finally, the initial energy balance for a new load is taken care of by the power stored in the generator inertia and the micro generating units are inertialess. This lack of inertia is the major problem leading to power imbalances between generation and load. There are also a number of barriers in the form of technical, business, and regulatory issues when it comes to connecting DGs to the electrical grid [44].

To overcome these issues and to utilize the potential of DG, the concept of Microgrid (MG) was introduced in [44]. Using power electronic devices along with distributed energy resources (DERs), integration of DG into the utility grid is possible. Power electronic devices improve the flexibility and adaptability of the system by converting the power from source into a fixed frequency AC power. They also provide various ancillary services to the grid [9, 10].

1.1.4 Distributed Resources

Seeking generality, we can state that DG is an electric power source connected directly to the distribution network or to the customer site of the meter. Note that the distinction between distribution and transmission networks is based on the legal definition. In most competitive markets, the legal definition for transmission networks is usually part of the electricity market regulation. Anything that is not defined as transmission network in the legislation can be regarded as distribution network.

According to [57], distributed resources are "demand- and supply-side resources that can be deployed throughout an electric distribution system (as distinguished from the transmission system) to meet the energy and reliability needs of the customers served by that system. Distributed resources can be installed on either the customer side or the utility side of the meter" [57].

Distributed resources consist of two aspects:

1. DG, located within the distribution system or on the customer side of the meter, and
2. demand-side resources, such as load management systems, to move electricity use from peak to off peak periods, and energy efficiency options, e.g., to reduce peak electricity demand, to increase the efficiency of buildings or drives for industrial applications, or to reduce the overall electricity demand. An important aspect of the concept of distributed resources is that the demand-side resources are not only based on local generation within the electrical system on the customer's side of

the meter, but also on means that reduce customer demand. That will influence the electricity supply from the distribution network.

1.1.5 Distributed Capacity

The term distributed capacity is less known than the terms DG or distributed resources, probably because it is even more difficult to clearly define the term.

Distributed capacity includes all aspects of distributed resources, plus the requirements for transmission/distribution capacity. For better distinction between distributed capacity and DG, the following example can be used: one aim of installing DG is to reduce the peak demand. However, DG does not include any reserve capacity, hence the transmission/distribution network usually has to be able to cover at least some of the generation usually supplied by DG. Hence, transmission/distribution lines will be over-dimensioned and the load factor will be worse than without DG. As transmission/distribution systems are regarded as monopolies, the transmission/distribution operator will usually be able to recover the costs for the over-dimensioned system and the poor load factor via higher transmission tariffs.

Distributed capacity now includes all aspects of DG and distributed resources plus reserve capacity, e.g., stand-by generators or load management, to minimize the requirements for over-dimensioning of transmission/distribution system.

1.1.6 Factors of DG Growth

The primary drivers behind the growth of DG and the current focus on its integration into electric power system operation and planning can be classified into three main categories, namely environmental, commercial, and national/regulatory. These drivers are discussed briefly below.

1. **Environmental factors**:
    - The use of renewable energy and CHP to limit greenhouse gas (GHG) emissions is one of the main drivers for DG. In this regard, it is important to point out that integration of renewable sources of electrical energy into power systems is a somewhat different question from that of the integration of DG into power systems. Integration of DG includes some of the issues related to integration of renewable sources but clearly does not deal with integration of transmission connected renewable sources such as large on shore and offshore wind farms.
    - Another important factor for DG from the environmental perspective is the avoidance of construction of new transmission lines and large power plants for which there is increasing public opposition. There is, however, opposition also from some environmental lobby groups to onshore wind farms on grounds of noise and visual "pollution". There is therefore a balance to be struck between

the need for sustainable energy solutions on the one hand and the need to maintain scenic beauty of the environment. Some argue that environmentally benign technologies such as wind that do not emit any GHG and have no long-term waste management problems should be favored.
Technological developments in generator technology are already delivering cost-effective small- to medium-sized generation technologies for domestic applications such as micro-CHP. In times when a premium is placed on land use, these technologies are likely to prove popular.

2. **Commercial factors**:
   - One of the acknowledged consequences of the introduction of competition and choice in electricity is the increased risk faced by all players in the electricity supply chain from generators through transmission and distribution businesses to retailers. It is well known that the capital outlay required to establish new power stations can be very high. The uncertainties associated with a competitive market environment may favor generation projects with a small capacity whose financial risk is commensurately small.
   - The presence of DG close to load centers can have a beneficial impact on power quality (PQ) and supply reliability. One area of improvement is voltage profile improvements, reduction in the number of customer minutes lost, especially if DG is allowed and able to stay on when there are network outages (islanding).
3. **National/regulatory factors**:
   - In recent times, there has been increasing concern among energy policy makers regarding energy security. There is a recognition that modern societies have become dependent on energy resources to the extent that, should there be a disruption in its supply the consequences would be too ghastly to contemplate in political, economic, and social terms. Because of this the EU energy policy focuses on energy security and sustainability. In the context of energy security and sustainability, DG is an attractive proposition in many respects. Some of the more important ones are for example:
     + It is distributed around the network close to customers; failure of one power station will have limited impact on the whole system compared to failure of one large power plant or bulk electricity transmission facility.
     + Diverse technologies and primary energy sources by diversifying the energy sources especially utilizing renewable sources, there is sense of control over the nation's future energy needs. There is increasing concern that the bulk of fossil fuel-based energy supplies come from regions of the world where control of these resources could be potentially unpredictable, thus posing an unacceptable risk.
   - There is an abiding faith among the proponents of reform of electricity supply industries that introduction of competition in generation and customer choice will deliver low energy prices and better service quality. One of the prerequisites for effective competition to occur is that there must be many players in the market. DG clearly advances this cause by providing many small generators

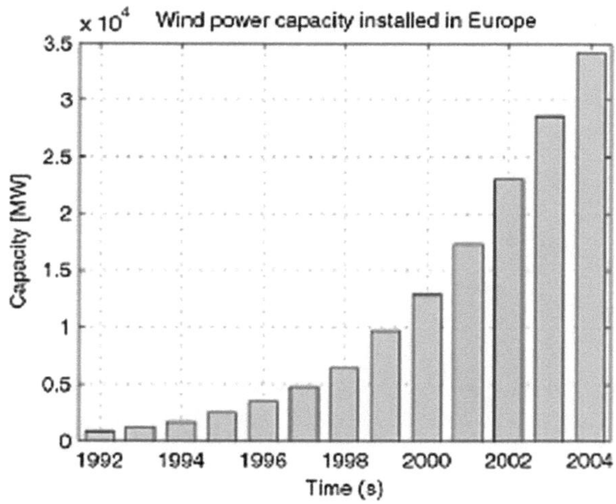

Fig. 1.1 Installed wind energy in Europe

that could potentially trade in the energy market and, and where appropriate market arrangements exist, also trade in ancillary services.

Among the RESs, hydropower and wind energy have the largest utilization nowadays. In countries with hydropower potential, small hydro turbines are used at the distribution level to sustain the utility network in dispersed or remote locations. The wind power potential in many countries around the world has led to a large interest and fast development of WT technology in the past decade [23]. A total amount of nearly 35-GW wind power was installed in Europe by the end of 2004, as shown in Fig. 1.1.

Another renewable energy technology that gains acceptance as a way of maintaining and improving living standards without harming the environment is the PV technology. As shown in Fig. 1.2, the number of PV installations has had an exponential growth, mainly due to the governments and utility companies that support programs that focus on grid-connected PV systems [35, 73].

1.1.7 Impacts on Transmission System Operation

The technical impacts of DG in system operation can be evaluated at different levels:

(a) steady-state operation;
(b) protection coordination;
(c) dynamic behavior;
(d) provision of ancillary services.

Evaluations of the impacts in the transmission network involve a large number of studies, considering different typical scenarios of operation (for instance peak load hours, valley hours, and mid-consumption levels). In power systems where there is a

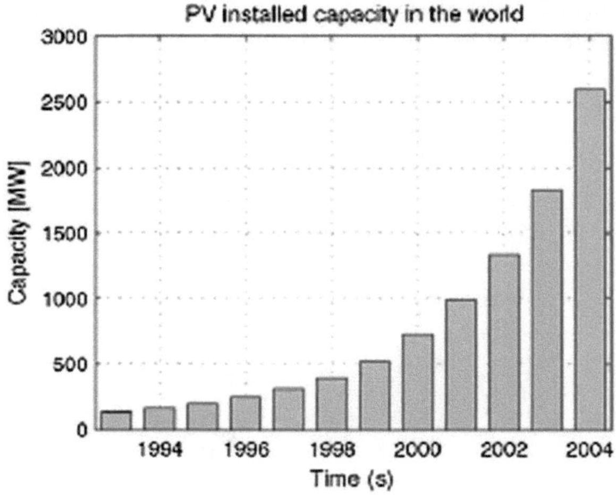

Fig. 1.2 Installed PV energy in the world

considerable contribution from hydropower, the number of these scenarios increases since it will be necessary to evaluate impacts for winter (wet or dry) and summer periods. The intermittent nature of some DG plants (such as wind parks) can also lead to the need to consider additional scenarios with different levels of DG production. These impacts can be analyzed by comparing the solutions obtained for each scenario before and after the introduction of the DG levels for a given study. Two types of analyses can be conducted:

+ Global evaluation of system performance.

+ Analysis of the electrical behavior in specific network areas, due to specifics of the energy conversion systems.

1.1.8 General Structure

A general structure for distributed systems is illustrated in Fig. 1.3. The input power is transformed into electricity by means of a power conversion unit whose configuration is closely related to the input power nature. The electricity produced can be delivered to the local loads or to the utility network, depending on where the generation system is connected.

One important part of the distributed system is its control. The control tasks can be divided into two major parts.

1. **Input-side controller**, with the main property to extract the maximum power from the input source. Naturally, protection of the input-side converter is also considered in this controller.

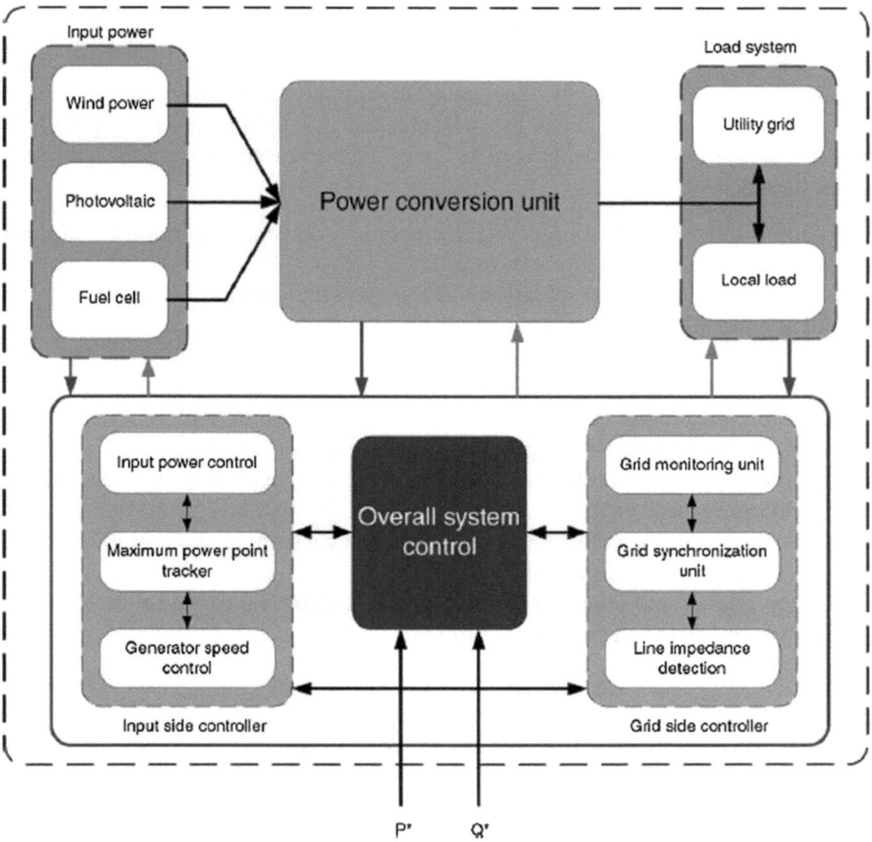

Fig. 1.3 General structure of distributed generation system

2. **Grid-side controller**, which can have the following tasks:
   - control of active power generated to the grid;
   - control of reactive power transfer between the DPGS and the grid;
   - control of DC-link voltage;
   - ensure high quality of the injected power;
   - grid synchronization.

The above items for the grid-side converter are the basic features this converter should have. Additionally, ancillary services like local voltage and frequency regulation, voltage harmonic compensation, or active filtering might be requested by the grid operator.

1.1.9 Integrating Distributed Energy Resources

In order to successfully integrate renewable DERs, many technical challenges must be effectively handled to ensure that the present levels of reliability are not significantly affected, and the potential benefits of DG are fully harnessed. In this sense, the main issues include [30, 42, 51]:

- Schedule and dispatch of units under supply and demand uncertainty, and determination of appropriate levels of reserves.
- Reliable and economical operation of MGs with high penetration levels of intermittent generation in stand-alone mode of operation.
- Design of appropriate demand-side management (DSM) schemes to allow customers to react to the grid's needs.
- Design of new market models that allow competitive participation of intermittent energy sources, and provide appropriate incentives for investment.
- Reengineering of the protection schemes at the distribution level to account for bidirectional power flows.
- Development of new voltage and frequency control techniques to account for the increase in power-electronics-interfaced DG.
- Develop market and control mechanisms that exhibit a plug-and-play feature to allow for seamless integration over time.

1.2 Supply–Demand in Electric Power Grid

This section provides a framework for reducing demand–supply imbalances in the grid, by jointly controlling both the supply-side electric power regulation together with the demand-side energy consumption by residential and commercial consumers demand response. We focus on performance improvements that arise from the complementary dynamics: regulation allows for frequent control updates but suffers from slower dynamics; demand response has faster dynamics but does not allow as frequent control updates.

1.2.1 Understanding the Grid

Unlike water or gas, electricity cannot be stored in large quantities. It must be generated the instant it is used, which requires supply to be kept in constant balance with demand. Furthermore, electricity flows simultaneously over all transmission lines in the interconnected grid system in inverse proportion to the electrical resistance of the lines, so electricity generally cannot be routed over specific lines. This means generation and transmission operations in North America must be monitored and controlled in real time, 24 h a day, to ensure a reliable and continuous supply of

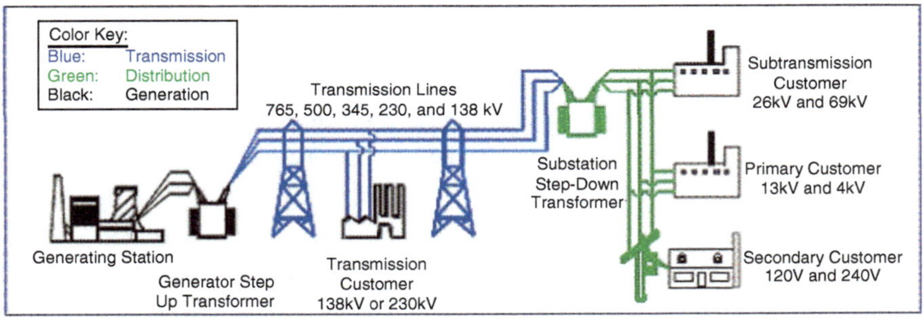

Fig. 1.4 Power generation and transmission components

electricity to homes and businesses. This requires the cooperation and coordination of hundreds of electricity industry participants (Fig. 1.4).

The diagram below depicts the basic elements of the electricity system: how it is created at power generating stations and transported across high-voltage transmission and lower-voltage distribution lines to reach homes and businesses. Transformers at generating stations step up the electric voltage for efficient transport and then step down the voltage at substations to efficiently deliver power to customers.

The generation and transmission components and their associated control systems comprise the bulk-power system.

1.2.2 Reliability Concepts

According to [59], the reliability of the interconnected bulk-power system is defined in terms of two basic and functional aspects:

- **Adequacy**. The ability of the electricity system to supply the aggregate electrical demand and energy requirements of the end-use customers at all times, taking into account scheduled and reasonably expected unscheduled outages of system elements.
- **Operating Reliability**. The ability of the bulk-power system to withstand sudden disturbances, such as electricity short circuits or unanticipated loss of system elements from credible contingencies, while avoiding uncontrolled cascading blackouts or damage to equipment.

Regarding adequacy, system operators can and should take controlled actions or procedures to maintain a continual balance between supply and demand within a balancing area (formerly called a control area). These actions include:

- Interruptible demand: Customer demand that, in accordance with contractual arrangements, can be interrupted by direct control of the system operator or by action of the customer at the direct request of the system operator

- Voltage reductions: Referred to as brownouts because lights will dim as voltage is lowered
- Rotating blackouts: The term is used because each set of distribution feeders is interrupted for a limited time, typically 2030 min, and then those feeders are put back in service and another set is interrupted, and so on, rotating the outages among individual feeders.

All other system disturbances that result in unplanned or uncontrolled interruption of customer demand, regardless of cause, fall under the heading, Operating Reliability. When these interruptions are contained within a localized area, they are considered unplanned interruptions or disturbances. When they spread over a wide area of the grid, they are referred to as cascading blackouts, the uncontrolled successive loss of system elements triggered by an incident at any location. Cascading results in widespread electric service interruption that cannot be restrained from sequentially spreading beyond an area predetermined by studies.

1.2.3 Electric Power Dynamic Demand

Dynamic demand refers to a semi-passive technology for adjusting load demands (LDs) on an electrical power grid. The concept is that by monitoring the frequency of the power grid, as well as their own control parameters, individual, intermittent loads would switch on or off at optimal moments to balance the overall system load with generation, reducing critical power mismatches. As this switching would only advance or delay the appliance operating cycle by a few seconds, it would be unnoticeable to the end user. This is the foundation of dynamic demand control.

1.2.4 The Need for Spinning Reserve

The power utilities are able to predict to a reasonable accuracy (generally to within one or two percent) the demand pattern throughout any particular day. This means that the free market in electricity is able to schedule just enough base load in advance. Any remaining imbalance would then be due either to inaccuracies in the prediction, or unscheduled changes in supply (such as a power station fault) and/or demand. Such imbalances are removed by requesting generators to operate in so-called frequency response mode (also called frequency-control mode), altering their output continuously to keep the frequency near the required value. The grid frequency is a system-wide indicator of overall power imbalance. For example, it will drop if there is too much demand because generators will start to slow down slightly. A generator in frequency-response mode will, under nominal conditions, run at reduced output in order to maintain a buffer of spare capacity. It will then continually alter its output on a second-to-second basis to the needs of the grid with droop speed control.

1.2.5 Local Load Control

In principle, any appliance that operates to a duty cycle (such as industrial or domestic air conditioners, water heaters, heat pumps, and refrigeration) could be used to provide a constant and reliable grid balancing service by timing their duty cycles in response to system load.

Because it is possible to measure grid frequency from any power outlet on the grid, it is possible to design controllers for electrical appliances that detect any frequency imbalance in real time. Dynamic demand-enabled appliances would react to this same signal. When the frequency decreases they would be more likely to switch off, reducing the load on the grid and helping to restore the balance. When the frequency increases past the standard, they would be more likely to switch on, using up the excess power. Obviously, the controller must also ensure that at no point does the appliance stray out of its acceptable operating range. As line frequency is directly related to the speed of rotation of generators on the system, millions of such devices acting together would act like a huge, fast-reacting peaking power plant.

1.2.6 Ancillary Services

The dynamic controller could also provide other ancillary services, such as aiding blackstart recovery of the ability of a power grid to be brought back to service after a power outage if programmed with that function. Generally, blackstarts are made more difficult because of the large number of reactive loads attempting to draw power simultaneously at start-up when voltages are low. This causes huge overloads that trip local breakers delaying full system recovery. The dynamic controller could have these loads “wait their turn”, as it were, until full power has been restored.

Another vital balancing service is fast reserve, which is the use of standby plant to replace possible lost generation (e.g., due to a failed power generator or lost power line). By shedding load quickly while the running generators spin up, then switching back in to bring the frequency back to standard, dynamic controllers could spare the high cost of fast reserve generators. Also, the fast response speed of this method would avoid possible brownouts occurring.

1.2.7 Implementation Issues

Dynamic demand devices have the potential to save considerable amounts of energy by the services they provide. But before dynamic demand control can be widely incorporated regulation must be put in place to mandate installation on at least new appliances or an effective market mechanism must be created to reward installation of the technology fairly. One method contemplated is to enable the electricity meter that measures the electricity consumption also measure the grid frequency, and switch to a higher tariff if the frequency drops below a certain level. The monthly electricity bill will then say that so many hours (and so many kilowatt hours) were on the Regular tariff and a few hours on the Short Supply tariff. Those consumers without smart demand management have to pay the extra cost, but those who install smart technologies that adapt to the short supply periods will save money.

1.3 Overview of Microgrids

A microgrid is an interconnection of distributed energy sources (DES), such as micro-turbines, windturbines, fuel cells, and photovoltaics integrated with storage devices, such as batteries, flywheels, and power capacitors on low voltage distribution systems [31, 60]. A basic MG architecture is shown in Fig. 1.5. This architecture is commonly known as the Consortium for Electric Reliability Technology Solutions architecture [43, 45].

This MG consists of a group of radial feeders, which could be part of a distribution system or a building's electrical system. There are three sensitive-load feeders (Feeders AC) and one nonsensitive-load feeder (Feeder D). The sensitive-load feeders contain sensitive loads that must be always supplied, thus each feeder must have at least a microsource rated to satisfy the load at that feeder. By contrast, the nonsensitive-load feeder is the feeder that may be shut down if there is a disturbance or PQ problems on the utility; the nonsensitive-load feeder will be left to ride through the disturbance or PQ problems [43, 45].

When there is a problem with the utility supply, Feeders AC can island from the grid using the static switch that can separate in less than a cycle to isolate the sensitive loads from the power grid to minimize disturbance to the sensitive loads. In an islanded operation, a MG will work autonomously, therefore it must have enough local generation to meet the demands of the sensitive loads [43, 45]. Furthermore, a disturbance requiring a feeder to operate individually may also occur. If this latter case is considered in the MG design, each sensitive-load feeder must have enough local generation to supply its own loads while the nonsensitive-load feeder will rely on the utility supply.

An essential part of the Consortium for Electric Reliability Technology Solutions architecture is the energy manager which is responsible to manage system operation through power dispatching and voltage setting to each micro-source controller. Some possible criteria for the MG to fulfill this responsibility are as follows [43]:

1. ensure that the necessary electrical loads and heat are fulfilled by the microsources;
2. ensure that the MG satisfies operational contracts with the utility;
3. minimize emissions and/or system losses; and
4. maximize the operational efficiency of the microsources.

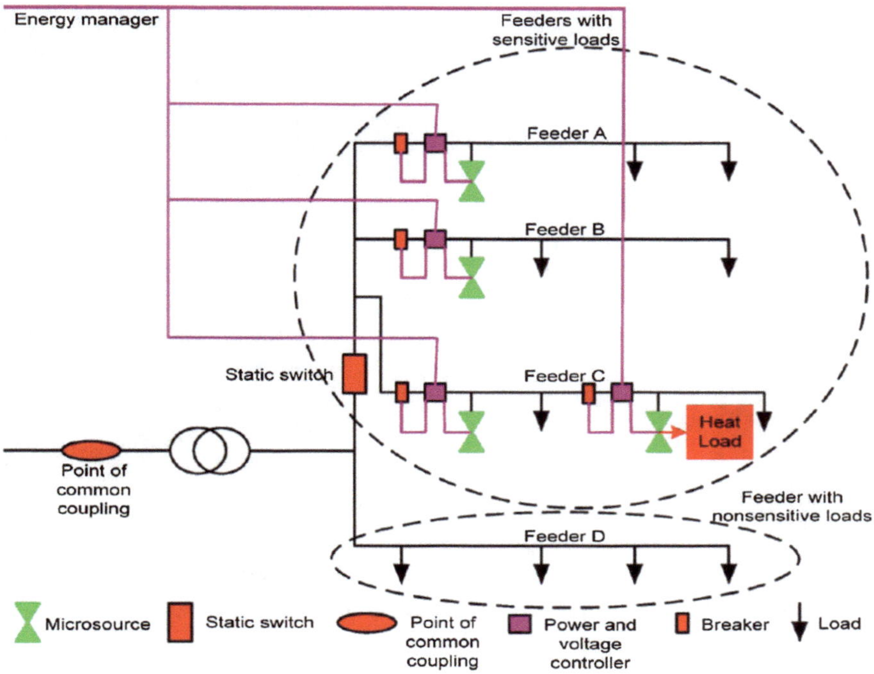

Fig. 1.5 A microgrid layout

Figure 1.6 shows a demonstration system, which has been running successfully on the Greek island of Kythnos since April 2001 [78].

Power system efficiency may increase up to 90 % if CHP is applied in the MG to utilize heat for local uses [46]. In addition, the efficiency increase can also result from loss reduction in transmission lines related to local power generation for local uses. Moreover, the local uses of local power generation will reduce energy or power density of transmission lines so that transmission line congestion can be reduced and investments on transmission line upgrade can be delayed [6, 58]. Environmental friendliness can be achieved due to the current trend to increase the RES participations in MGs; this participation increase will significantly reduce GHG emissions [15, 17, 40, 61, 64]. Furthermore, RES can also ensure energy sustainability due to the nature of their availability; RES can gradually substitute fossil fuels that have limited sources. Ideally, energy can be harvested at no cost, besides installation, operational, and maintenance costs, from RES. However, renewable energy is naturally intermittent so that an energy storage system is required to optimize energy utilization [5, 11, 12, 28, 67, 86, 87].

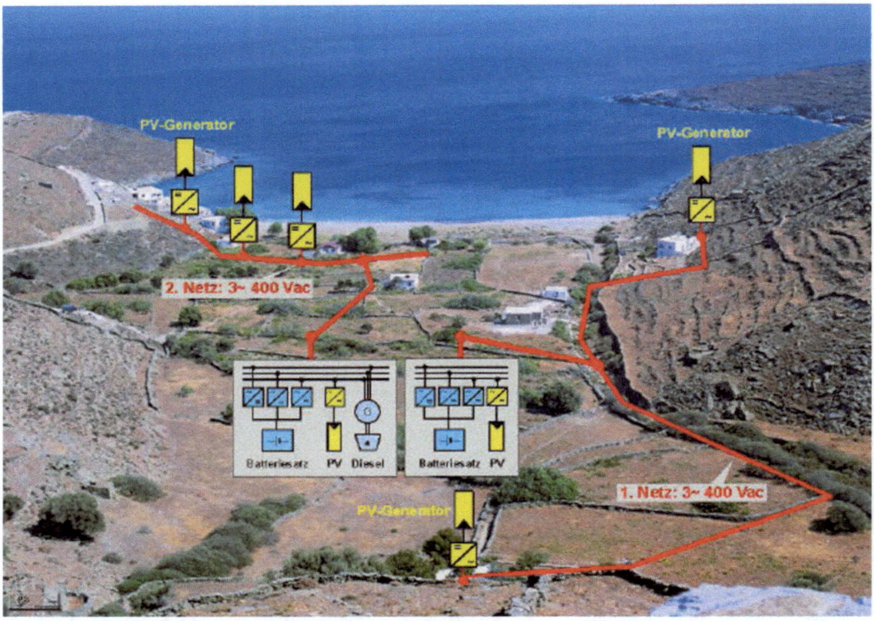

Fig. 1.6 Typical microgrid on Kythnos Island [78]

1.3.1 Control Tasks

The main tasks of MG controllers are to guarantee that [43]:

1. microsources work properly at the predefined operating point or slightly different from the predefined operating point but still satisfy the operating limits;
2. active and reactive powers are transferred according to necessity of the MGs and/or the distribution system;
3. disconnection and reconnection processes are conducted seamlessly;
4. market participation is optimized by optimizing production of local microsources and power exchanges (PXs) with the utility;
5. heat utilization for local installation is optimized;
6. sensitive loads, such as medical equipment and computer servers are supplied uninterruptedly;
7. in case of general failure, the MG is able to operate through black-start; and
8. energy storage systems can support the MG and increase the system reliability and efficiency.

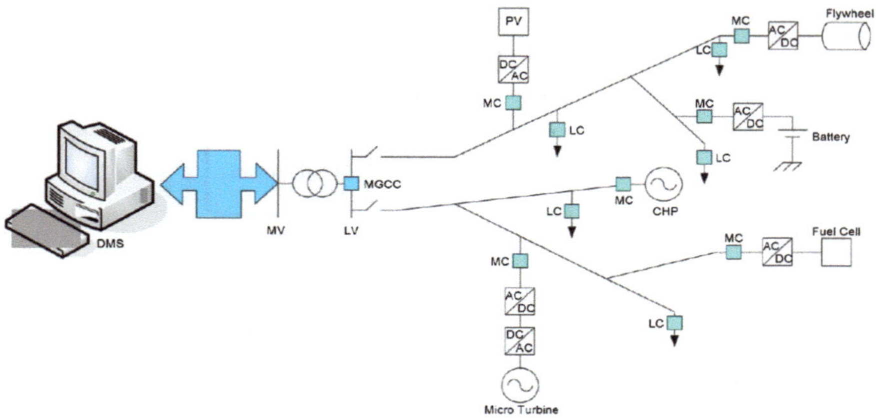

Fig. 1.7 Multilevel control of microgrid

Broadly, the MG controls can be classified as:

- *local controls*, which are the basic category of MG controls. This is achieved by controlling operating points of the microsources and their PEIs without communication systems.
- *centralized controls*, which include microsource controllers (MCs) and load controllers (LCs); microgrid central controllers (MGCCs); and distribution management system (DMS), see Fig. 1.7, and
- *decentralized controls* (DECs), which aimed to maximize autonomy of the microsources and loads.

1.3.2 A Classification

MGs can be broadly classified in many ways as follows:

1. In terms of power type (whether AC or DC) [14], Fig. 1.8 illustrates the following MG power types: DC microgrid [85], high-frequency AC microgrid [13, 84] (which is used to solve power-quality problems caused by the presence of lots of energy conversion devices and to reduce the impact of fluctuations in renewable energy in the MG), line-frequency AC (LFAC) microgrid (the most researched since the concept of MG was formulated [14]), and hybrid DC- and AC-coupled MGs.
2. In terms of the application used for [21], they can be classified into three categories subject to their application: (a) utility MGs (a district of the city operates as a MG), (b) commercial and industrial MGs, and (c) isolated MGs.
3. In terms of system structure, they can be classified into two structures, depending on the number of power processing stages: (a) single-stage power-conversion

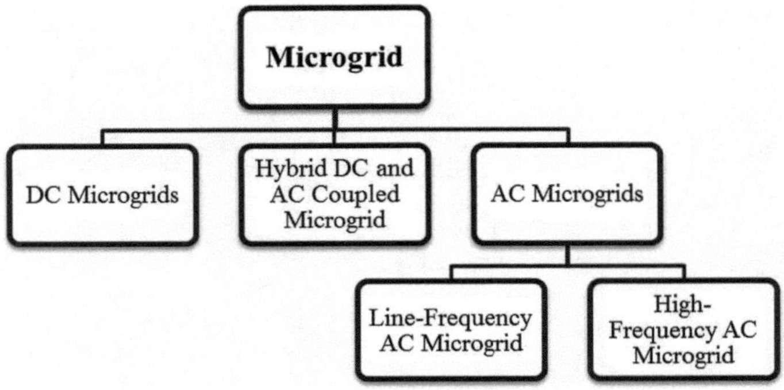

Fig. 1.8 Classification of microgrids based on power type

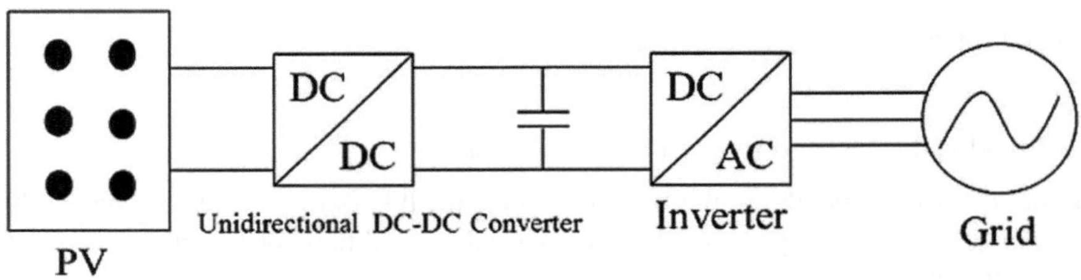

Fig. 1.9 Two-stage conversion systems for PV

systems [14, 41, 69] and (b) two-stage power-conversion systems [9, 36, 66, 75] (which is the most common configuration for all electronically coupled DER). Two-stage conversion systems have two converters, i.e., one on the PV side, which is used to extract the maximum power from the PV, and another is connected to the grid side and controlled to follow grid requirements (refer to Fig. 1.9).

4. In terms of supervisory control, they can be either centralized or decentralized [14]; in centralized control, the central controller (CC) in the MG sends the required set-points to the LC through a two-way communication link, but the technique has low reliability and is redundant [80]; the decentralized technique is a multiagent system providing flexibility to the system and communication between two agents through a communication language such as Java-Jade [14, 20, 29].
5. In terms of the DERs connection to the MG, i.e., either electronically coupled (converters are used) or conventional rotating distributed generators [14, 39]. MGs may also be single- or three-phase systems or connected to low- or medium-voltage distribution networks [62]; their mode of operation is either islanded or grid-connected; each mode has its own operation requirements and requires a distinct control strategy [81].

1.3.3 Control Objectives and Methods

Control of a MG during islanded mode is particularly critical [14, 39, 70]. System control used for the islanded operation includes voltage regulation (VR), frequency regulation (FR), and load-sharing (LS) optimization. Since there is no synchronous machine (SM) in most MGs to achieve demand and supply balancing, the inverters in principle should be responsible for balancing. The use of a voltage source inverter (VSI) to provide a reference for voltage and frequency enables MGs to operate in islanded mode [25, 50].

The terms "grid-forming" and "islanded operation mode" are used interchangeably [14, 39, 70]. The grid-forming mode is customerly used to regulate voltage and frequency and to achieve LS in MGs (Figs. 1.4 to 1.9).

1. *Control Strategies*: There are two main control strategies proposed for MGs during islanded mode [50, 55]:
   (a) single-master operation (SMO) and
   (b) multi-master operation (MMO).
   Both use a VSI to provide a reference for voltage and frequency [25], and a convenient secondary load-frequency control must be considered to maintain the frequency between the specified limits and to run the DER economically [50, 52].

   - SMO [50, 55]: This approach has one inverter acting as VSI (the master) and others as followers (the slaves). When the main power supply is lost, the slaves take voltage reference from the master and operate in the PQ mode. The LCs receive set-points from the MGCC to maintain generation of active and reactive powers at the specified values. The part within the thick dashed lines in Fig. 1.10 illustrates the SMO scheme.
   - MMO [50, 55]: Fig. 1.10 also illustrates MMO approach, in which several inverters act as VSI (the master). The VSI can be connected to storage devices or to DERs. Other inverters with PQ control may also coexist. The generation profile can be modified by the MGCC, which can define new set-points for the LCs.

2. *Wire and Wireless Interconnections*: Instantaneous LS in MGs is achieved through two main control schemes that are classified according to their control-wire interconnections [14]. One scheme is the *active LS technique*, which has parallel-connected MG converters, including masterslave, centralized LS, average LS, current limitation control, and circular chain control (3C).
   The other control scheme for parallel inverters is based mainly on a droop method [39, 70]. This technique adjusts the output voltage and frequency in functions of active power (P) and reactive power (Q) delivered by the inverter. The droop method uses only local power measurements; hence, better reliability and flexibility is achieved in the physical location of the units.

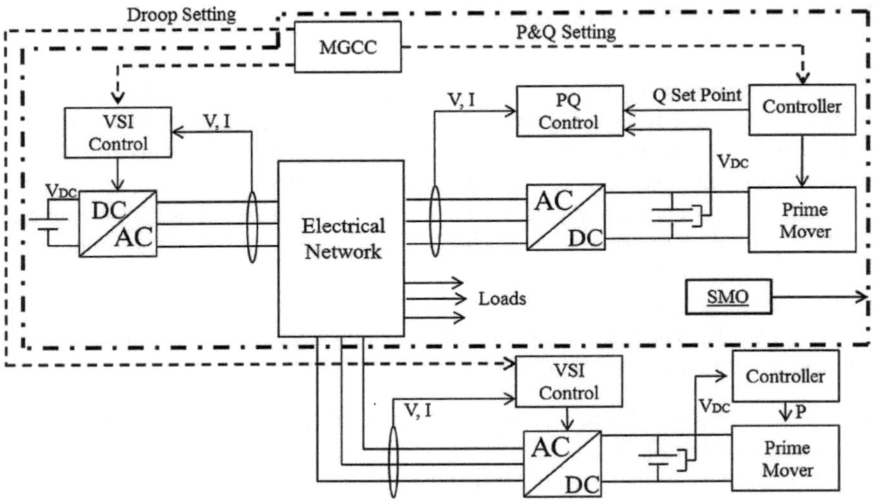

Fig. 1.10 MMO control including SMO

3. *Control Layers*: The MG control system has to ensure that all the control functions are accomplished (e.g., supply of electrical and/or thermal energy, continuous feeding of the critical loads, energy market participation, autoreconnection after failure, etc.). The control objectives can be achieved through either centralized or decentralized control and through three control layers, as shown in Fig. 1.11. The control levels are as follows [14, 39]:

   (1) distribution network operator (DNO) and market operator (MO);
   (2) MGCC;
   (3) LC, which can be either the source controller or the microload controller.

   The DNO is necessary when there is more than one MG in the distribution system. In addition, for the market environment of a specific area, one MO or more is responsible for market management of the MG. Both the DNO and the MO are part of the main grid; they do not belong to the MG. The second level is the MGCC, which is the main integrator of the DER clusters in a MG. It is responsible for stabilizing the voltage and frequency at the point of common coupling (PCC) and also for the active and reactive powers dispatched from each DER [81]. The LCs are the lower layer of control (sometimes called peer-to-peer, as in [53]). They control the DERs and some of the local load and seek to balance active and reactive powers. They have a certain intelligence level and, thus, can make decisions locally in a decentralized operation model.

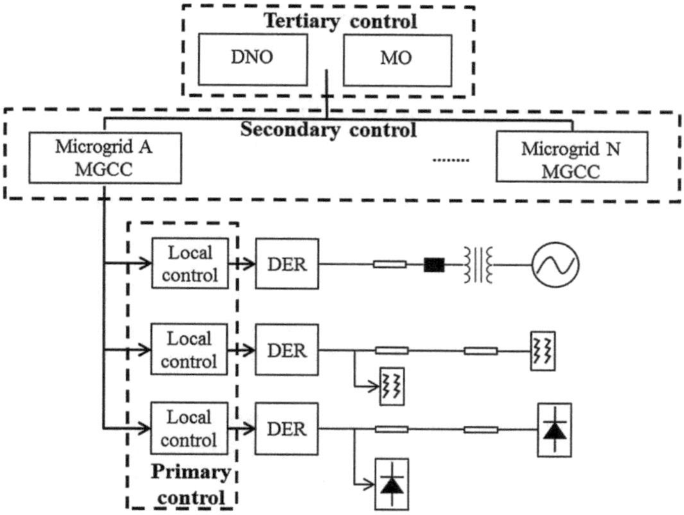

Fig. 1.11 A hierarchical microgrid control system (multiagent controller)

1.3.4 Microsource Control

Microsource controls need to ensure that: New microsources can be added to the system without modification of existing equipment, set-points can be independently chosen, the MG can connect to or isolate itself from the grid in a rapid and seamless fashion, and reactive and active power can be independently controlled and can meet the dynamic needs of the loads.

Each MC must autonomously respond effectively to system changes without requiring data from the loads, the static switch, or other sources. The same applies to the static switch. The basic controller uses a feedback control on power and voltage. The controls use real-time values for P, Q, frequency, and the AC voltage to generate the desired voltage magnitude and angle at the inverter terminals using droop concepts. Operating the inverter as a voltage behind an impedance results in real power being proportional to the phase angle across the coupling inductor, and the injected reactive power is proportional to the magnitude of the inverters output voltage.

- **Voltage–versus–reactive-power (Q) droop**

Integration of large numbers of microsources into a MG is not possible with basic unity power-factor controls. VR is necessary for local reliability and stability. Without local voltage control, systems with high penetrations of microsources could experience voltage and/or reactive-power oscillations. Voltage control must also ensure that there are no large circulating reactive currents between sources. With small errors in voltage set-points, the circulating current can exceed the ratings of the microsources. This situation requires a voltage versus reactive power droop controller so that, as the reactive power generated by the microsource becomes more capacitive, the local voltage set-point is reduced. Conversely, as Q becomes more inductive, the voltage set-point is increased.

- **Power–versus–frequency droop** When the MG is connected to the grid, loads receive power both from the grid and from local microsources, depending on the customers' situation. If the grid power is lost because of IEEE 1547 events, voltage sags, faults, blackouts, etc., the MG can autonomously transfer to island operation. When regulating the output power, each source has a constant negative-slope droop on the P, ω plane where P is output power and ω is frequency in radians per second. Figure 1.12 shows that the slope is chosen by allowing the frequency to drop by a given amount, $\Delta\omega$, as the power spans from 0 to P_{max}, dashed line. Figure 1.12 also shows the power set-points P_{o1} and P_{o2} for two units. This is the amount of power injected by each source when connected to the grid, at system frequency. If the system transfers to island when importing from the grid, then the generation needs to increase the power to balance it in the island. The new operating point will be at a frequency that is lower than the nominal value. In this case, both sources have increased their power output, with Unit 2 reaching its maximum power point. If the system transfers to island when exporting power to the grid, then the new frequency will be higher, corresponding to a lower power output from the sources, with Unit 1 at its zero-power point.

  The characteristics shown in Fig. 1.12 are steady-state characteristics. They have a fixed slope in the region where the unit is operating within its power range. The

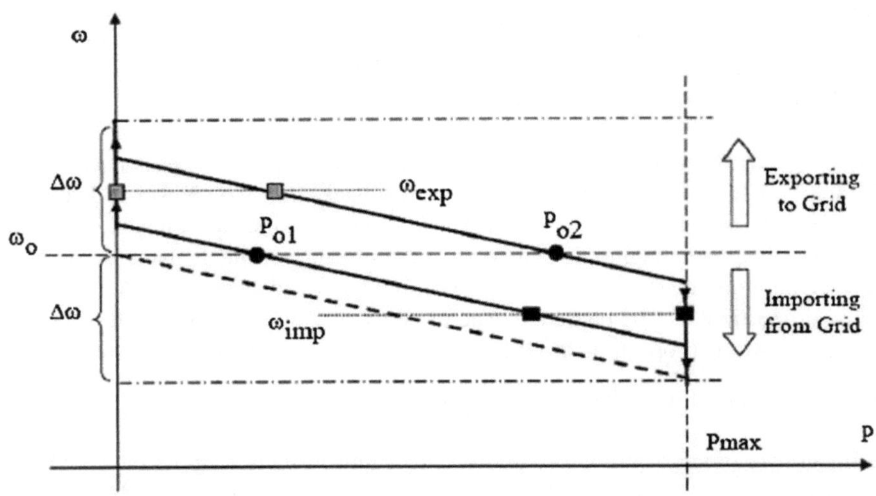

Fig. 1.12 Power-versus-frequency droop

slope becomes vertical as soon as any limit is reached. The droop is the locus where the steady-state points are constrained to come to rest, but during dynamics the trajectory will deviate from the characteristic.

1.3.5 Control and Protection Requirements

MGs, and integration of DER units in general, introduce a number of operational challenges that need to be addressed in the design of control and protection systems in order to ensure that the present levels of reliability are not significantly affected and the potential benefits of DG are fully harnessed. Some of these challenges arise from invalid assumptions typically applied to conventional distribution systems, while others are the result of stability issues formerly observed only at a transmission system level.

The most relevant challenges in MG protection and control include:

- **Bidirectional power flows**: While distribution feeders were initially designed for unidirectional power flow, integration of DG units at low voltage (LV) levels can cause reverse power flows and lead to complications in protection coordination, undesirable power flow patterns, fault current distribution, and voltage control.
- **Stability issues**: Local oscillations may emerge from the interaction of the control systems of DG units, requiring a thorough small-disturbance stability analysis. Moreover, transient stability analyses are required to ensure seamless transition between the grid-connected and stand-alone modes of operation in a MG.
- **Modeling**: Prevalence of three-phase balanced conditions, primarily inductive transmission lines, and constant-power loads are typically valid assumptions when modeling conventional power systems at a transmission level; however, these do not necessarily hold valid for MGs, and consequently models need to be revised.
- **Low inertia**: Unlike bulk-power systems where high number of synchronous generators (SYGNs) ensures a relatively large inertia, MGs might show a low-inertia characteristic, especially if there is a significant share of power electronic-interfaced DG units. Although such an interface can enhance the system dynamic performance, the low inertia in the system can lead to severe frequency deviations in stand-alone operation if a proper control mechanism is not implemented.
- **Uncertainty**: The economical and reliable operation of MGs requires a certain level of coordination among different DERs. This coordination becomes more challenging in isolated MGs, where the critical demand–supply balance and typically higher component failure rates require solving a strongly coupled problem over an extended horizon, taking into account the uncertainty of parameters such as load profile and weather forecast. This uncertainty is higher than those in bulk-power systems, due to the reduced number of loads and highly correlated variations of available energy resources (limited averaging effect).

1.3.6 Reliable and Economical Operation

The MGs control system must be able to ensure the reliable and economical operation of the MG, while overcoming the aforementioned challenges. In particular, desirable features of the control system include:

- **Output control**: Output voltages and currents of the various DER units must track their reference values and ensure oscillations are properly damped.
- **Power balance**: DER units in the MG must be able to accommodate sudden active power imbalances, either excess or shortage, keeping frequency and voltage deviations within acceptable ranges.
- **DSM**: Where applicable, proper DSM mechanisms must be designed to incorporate the ability to control a portion of the load. Additionally, for the electrification of remote communities with abundant local renewable resources, the active participation of the local community may be beneficial to design cost-effective DSM strategies that enhance load-frequency control [3, 82].
- **Economic dispatch**: An appropriate dispatch of DER units participating in the operation of a MG can significantly reduce the operating costs, or increase the profit. Reliability considerations must also be taken into account in the dispatch of units, especially in stand-alone operation.
- **Transition between modes of operation**: A desirable feature of MGs is the ability to work in both grid-connected and stand-alone modes of operation, including a smooth transition between them. Different control strategies might be defined for each mode of operation and, therefore, a high-speed islanding detection algorithm is very important to adjust the control strategy accordingly [38].

1.4 Smart Grid

Concomitant with these advances, researchers have also come to recognize the urgent need for new systems-level knowledge of power and energy systems for sustaining the advancement of this emerging field. This, in turn, has led to a natural demand for the two lifelines of system theory in smart grid (SG) research, namely control and optimization. Almost every facet of making a power system smart or self-regulated boils down to using control theory in some form or other. Relevant examples include modeling, identification, estimation, robustness, optimal control, and decision-making over networks. Over the past 2 years, for instance, several research workshops, conference tutorial sessions, and national meetings have been organized to discuss the strong potential of control and optimization in SG applications starting from small residential-level energy management, smart metering, and power markets to much broader-scale problems such as wide-area monitoring and control.

The main foundations of the emerging SG are

- DER enabled primarily by intermittent, nondispatchable RESs such as wind and solar, and independent MGs and
- Demand response, the concept of controlling loads via cyber-based communication and control and economic signals.

In what follows, we benefit from the survey material contained in [83]. While SG communication technologies offer dynamic information and provide real-time signals to utilities, they inevitably introduce delays in the energy real-time market. In this article, a dynamic, discrete-time model of the wholesale energy market that captures these interactions is derived. Beginning with a framework that includes optimal power flow and real-time pricing, this model is shown to capture the dynamic interactions between generation, demand, and locational marginal price near the equilibrium of the optimal dispatch. It is shown that the resulting dynamic real-time market has stability properties that are dependent on the delay due to the measurement and communication. Numerical studies are reported to illustrate the dynamic model, and a suitable communication topology is suggested.

The current grid system is quickly becoming obsolete. This grid system will not be able to meet our future electricity demands. New efficient technology must be introduced to solve this problem. One solution to this problem is SG. SGs will be able to efficiently handle our increasing energy demands and reduce the environmental impact by incorporating renewable resources. In this chapter, we will discuss why SGs are vital to our future, the different types of new technology they are comprised of, the current advancements, and research that is being conducted.

1.4.1 Efficiency and Reliability

Our current grid system is unable to efficiently supply the energy needed by our country. Heavily populated areas in the United States are often plagued with blackouts and congested transmission lines, also known as bottlenecks. The U.S. Department of Energy reports that these power disturbances and outages cost the country "from USD 25–180 million" every year. With an ever-increasing population and advancements in technology, there is a greater demand for a more resourceful and reliable power grid. The energy consumption rate has risen from 10 % in 1940 to 40 % in 2003 [26].

These increasing failures of our current grid come at a time when the demand for electricity is the highest. In the past decade, our society has been increasingly digitalized. We are more dependent on electricity than ever before. There is a rapidly growing market of technologies that rely on electricity. Just as communication has evolved with technologies such as the Internet and wireless cell phones, our power grid must also evolve. Our current grid will have to evolve quickly to combat the changes that are taking place in our increasingly digital society. SGs will be able to monitor energy usage in real time and predict outages and equipment failures. A SG will be able to restore itself after a blackout or a weather-related outage.

1.4.2 Environmental Benefits

Our aging grid relies heavily on dwindling fossil energy resources. The cost of natural gas, coal, and oil is rising steadily. Oil, specifically, has seen a sharp price increase. Prices for oil have increased by 800 % from 1998 to 2008 [7]. Using these fossil fuels to supply energy to our country also contributes to our growing climate change problem. A cleaner alternative to fossil fuels is needed, and one can be implemented by using our readily available renewable resources. These resources reduce the amount of pollution being expelled into the atmosphere. Wind and solar energy production sites are being placed in remote locations and even offshore. Transporting electricity over long distances, thus far, has proved to be inefficient. SGs will be able to efficiently transport energy over vast distances, and therefore utilize the energy from distant renewable resources.

Recently, the current energy and environmental issues have been addressed in [32]. In this energy plan, USD 3.4 billion of federal stimulus funds is called to be invested to modernize the U.S. current grid system by specifically using SGs [32]. This investment will serve as a down-payment on our future grid system. This plan is also an advocate of smart metering and has made it a point to invest in smart meter technology for American homes.

1.4.3 Benefits to Consumers

SGs will also help consumers save money. Consumers will be able to monitor their home usage by using smart meters. This will encourage consumers to use less energy and will reduce the amount of overall energy needed by the grid. Using less energy at times of peak demand saves money for the consumer. This is because energy produced in periods of high demand costs more to produce than energy produced in times of low demand. Consumers will actively help balance supply and demand and increase reliability by changing the way they use and purchase electricity. Allowing consumers to see the real-time price of electricity will make them more conscious of their usage. Reducing the demand for energy needed from the grid also reduces the amount of pollution being created. Many jobs will be created during both the production (i.e., planning and construction) and post-production (i.e., maintenance, development, etc.) stages. SGs will be compatible with electric vehicles. SGs will be able to use electric cars as energy storage by drawing power from the charging cars when demand is peaking. Thousands of electric cars charging in the grid will decrease the amount of backup power needed and help control peak load leveling. Consumers will be able charge their cars during off-peak times and therefore receive the cheapest energy possible.

1.4.4 Security

Our current grid is susceptible to attacks and natural disasters. SGs will use a large number of smaller, widely distributed plants instead of a few high-producing plants. This decentralization of the grid will help it become more secure. If such an attack or disaster occurs, the SG will be able to restore itself quickly. This is referred to as a self-healing network. SGs will able to isolate problems on the line and reroute the power supply. This will be done by using intelligent switches. These intelligent networks will be able to combat the domino effect, which is a risk for our current grids. They will be able to stop, start, or reroute the energy. This will ensure that the greatest number of customers receive the energy that they require. The SG is a fiscally sound proposition because it will be interconnected to Canada and Mexico, and therefore increase and improve the economic relations with these countries.

1.5 Technical Aspects

SGs will rely on several different technologies. These new technologies will combine with the existing grid to create a more efficient and intelligent grid system. Most of these technologies are already available and are being used in other areas. SGs will use integrated two-way communications, superior control and monitoring techniques, advanced components, energy storage, improved interfaces, and decision support to operate.

1.5.1 Two-Way Communications

SGs will rely greatly on two-way communications. These communications include advanced metering infrastructure (AMI). AMI is a vital component for the implementation of the SG. It is a system that involves two-way communication with smart meters and other smart devices. AMI will benefit both consumers and electric companies, and will allow utilities to provide real-time pricing information to consumers. AMI will also allow utilities to react more quickly to potential power issues. Consumers will be able to reduce their usage when prices are high. AMI will allow electric companies to communicate directly with the smart devices.

Electric companies will be able to reduce the power being used by these devices during peak periods, depending on the preference of the consumer. These devices will control appliances such as washers, dryers, refrigerators, and thermostats. This intelligent system will work without direct supervision by the consumer. This will drastically reduce the cost of energy used by the consumer and reduce the amount of energy needed by the power grid. In addition to lower costs and a reduction in the total amount of energy needed by each customer, AMIs will result in longer feeders, lower voltages, and smaller wire sizes.

1.5.2 Control and Monitoring Techniques

Monitoring and controlling power usage are two of the greatest benefits of the SG. The SG will utilize phasor measurement units (PMUs). These phasors test the voltage and current several times a second at a given location [83]. Wide area measurement systems (WAMS) consist of many PMUs. WAMS offer a way to monitor electricity on a real-timescale. It is widely believed that the Northeast blackout of 2003 could have been confined to a smaller area if WAMS had been in place. SGs will use distribution automation (DA) to remotely control operations that were previously performed by grid managers on-site. This allows for faster and more flexible power adjustments. Automated networks will be able to reroute electricity during outages to services that need it the most. These services include hospitals, police stations, and fire departments. They will do this by communicating with each other. They will be able to reconfigure the system automatically, known as self-healing, to restore power to customers. Another benefit of DA is that it allows for the grid to focus on a system of feeders that are connected, instead of one feeder. The SG will use a system of distribution lines that are connected to multiple substations. This increases the amount of reliable power that the grid can supply.

1.5.3 Advanced Components

SGs will rely on numerous technologies to operate. They will build on our existing technology and add new high-tech components. One of these components is high-temperature superconducting cables. These cables are able to transfer a far greater amount of power than our current cables. These superconducting cables can not only transfer more power than our existing cables, but they can conduct electricity with little to no resistance. Superconducting cables also make it possible for power to be transferred over longer distances than before. Our current cables can transfer electricity over long distances but tend to be inefficient and have high line losses. The superconducting cables will allow the transfer of power over these distances with minimal line loss. This will allow for more renewable resources to be incorporated into our grid. Most of these renewable resources are located in remote areas. One example of this is wind energy. Most wind energy is located away from metropolitan areas in the western area of the United States. By integrating more RESs into our grid, we not only increase the power we are able to supply but also reduce our usage of fossil fuels and our impact on the environment. Another benefit of superconducting cables is the reliability that they offer by using superconductor-based fault current limiters. Superconductor-based fault current limiters limit the amount of current flowing through the system and allow for the continual, uninterrupted operation of the electrical system [83]. Fault current limiters function similar to surge protectors. The excessive current produced during a fault would quickly be controlled by the fault current limiter and would thus be stopped from causing any future damage. The fault current limiter would still allow for normal current to pass through the cables. In addition to superconducting cables, high-voltage direct-current (HVDC) cables will be used to transfer electricity over long distances. HVDC systems are more cost efficient and stable than AC systems when transferring power over long distances. However, AC systems are cheaper for transferring electricity over short distances. HVDC will use fault current limiters as well to protect against blackouts and the damage that they can cause. HVDC systems will work with flexible alternating current transmission system devices. These power flow devices will help monitor the voltage on the grid in real time. They will be able to monitor the power being transferred over the superconducting cables to the power plants. They will also be able to increase power transfer levels. Using HVDC and flexible alternating current transmission system will increase transmission distance, reduce line loss, protect against blackouts, utilize RESs, and increase stability along the grid.

1.5.4 Energy Storage

Energy storage is another vital component of the SG. Energy storage will be placed throughout the grid. This is known as distributing energy resources. This will make it easier for the grid to supply emergency power during a demand increase and to release congestion along the grid. This will also diminish the use of expensive backup plants that utilities use during peak load periods [24]. Energy storage is also a fundamental component of renewable energy. Due to the intermittent nature of renewable energy resources, storage of this energy is greatly needed. The SG will be able to store wind and solar energy and save it for later use during peak energy periods. Using this energy storage will reduce the capacity needed by transmission lines [8]. Using energy storage will increase the reliability and efficiency of the power grid while reducing the impact that it has on the environment. SGs will incorporate fuel cells into the grid system. A fuel cell is a device capable of generating an electrical current by converting the chemical energy of a fuel directly into electrical energy [79]. Fuel cells do not contain active materials required by the electrochemical conversion inside them. These active materials are supplied from the outside. Fuel cells also have a very small carbon footprint [65]. Fuel cells can be recharged easily by adding more solution, and they never run down like traditional batteries. Fuel cells will be able to reduce peak load problems because of their high energy density and reliability. Combined with electrolyzers, fuel cells will be able to store energy. This is done by converting electrical energy into a fuel, usually hydrogen. That fuel is then used to power fuel cells to create energy when needed. Hydrogen fuel cells are particularly useful because they are not affected by temperature, they can withstand long periods of storage, and they can be cycled many times [74]. Hydrogen fuel cells have not been used yet for energy storage because of their cost. Polymer electrolyte membrane (PEM) fuel cells may provide a cost-efficient solution in the near future. PEM fuel cells are commonly used in automobiles but are being researched as an energy storage solution for the power grid. PEM fuel cells will use hydrogen, oxygen, and water to produce electricity and create water as a by-product of the reaction. New developments in self-pressuring PEM electrolyzers and reversible PEM fuel stacks have reduced the cost of hydrogen fuel cells greatly. Fuel cells are useful for storing energy from renewable resources, such as solar and wind energy, because they do not produce consistent amounts of electricity. Hydrogen fuel cells will assist in load management as well by providing energy during peak periods. Hydrogen fuel cells are already being used as power storage on the Indonesian island of Sumatra by Hutchinson Telecommunications. Hutchinson uses these cells to provide backup for their telecom network when power from the grid is lost. These cells, developed by IdaTech, are either fueled directly by hydrogen or by using a liquid fuel consisting of methanol and water that is converted into hydrogen gas to power the system [34].

Ultracapacitors will also be used for storage. Ultracapacitors are DC energy sources that are able to recharge and discharge quickly and can handle very high currents. Unlike batteries, ultracapacitors store energy electrostatically. This process is drastically faster than the chemical process used in traditional batteries. They can be recharged many times but can only store a limited amount of power. Ultracapacitors will be able to store energy for short periods of time and reduce load leveling problems. They will work along with batteries to maintain load levels along the grid. They will be used as an uninterrupted power supply to supply emergency power. Two major drawbacks of using a system of ultracapacitors for wide-scale energy storage are their high cost and low energy storage density. The grid system of the future will encompass many different types of batteries for energy storage. Battery storage has been notorious for its costly nature, but recently new breakthroughs have been made to reduce these costs. Many different types of batteries will be used together to provide storage for the grid. These batteries include lead acid, sodium sulfur, and lithium ion batteries.

Lead acid and sodium sulfur batteries have been the most commonly used batteries for grid storage. Lead acid batteries are becoming increasingly outdated due to their low energy density and their large size. New battery technologies, such as lead-carbon batteries, have proven to be three to four times denser than normal lead acid batteries

[8]. Sodium sulfur batteries offer a better alternative to lead acid batteries. Sodium sulfur batteries store and create energy through the chemical reaction of sodium polysulphide and sodium. This process can be repeated thousands of times. The drawback to sodium sulfur batteries is that the reactants have to be stored separately at very high temperatures. If the batteries are allowed to cool down and discharge they become inoperable. Sodium sulfur batteries are currently being used in Nagoya, Japan for grid storage. This plant can create about 300 MW of extra power when needed for up to 6 h [49]. General Electric plans to construct a large sodium sulfur battery factory in the United States in 2009. This factory intends to have a capacity of 900 mWh/year [8]. Lithium-ion batteries are also being researched as prospective energy storage options. Lithium-ion batteries have a very high energy density and are highly efficient. However, lithium-ion batteries have thus far proven to be costly in comparison with lead acid and sodium sulfur batteries.

In addition to these technologies, the SG will incorporate new ways of generating energy and storing it for later use. One such technology is compressed air. Compressed air energy storage takes off-peak energy from power plants to pump air into an underground air storage tank where it is stored under pressure. This air is then released during periods of peak demand. This released air powers a WT and creates energy. One such compressed air plant in Hurntorf, Germany, is capable of producing up to 300 MW of reserve power for up to 3 h [49]. Another energy storage technology incorporated is pumped hydrogen. This technology has been around for many years and is the most widespread energy storage system in use on power networks [68]. Pumped hydrogen works by using two reservoirs of water, one being at a higher elevation than the other. During off-peak hours, water is pumped into the higher reservoir. When energy is needed, water is released from the top reservoir and flows downward through turbines to the lower reservoir. Pumped-storage hydroelectricity can be used to reduce peak load problems. Another way of storing energy is flywheels. Flywheels take excess energy and convert it into kinetic energy by running it through a motor that spins the flywheel. By slowing down the generator, flywheels give back the energy. Flywheels can absorb energy quickly and release it just as fast. Flywheels need very little upkeep and do not contain toxic chemicals that some batteries do [24].

1.5.5 Robust Energy Management

Robust energy management algorithms should have the ability to operate correctly in the presence of unreliable communication capabilities, and often in the absence of a central control mechanism. Effective distributed control algorithms could be embedded in distributed controllers to properly allocate electrical power among connected buses autonomously. By selecting the incremental cost of each generation unit as the consensus variable, the incremental cost consensus (ICC) algorithm is able to solve the conventional centralized economic dispatch problem in a distributed manner. However, the communication time-delay may cause instability of the system and should be considered during the design process. The mathematical formulation of the ICC algorithm with time-delay is presented in this chapter. Several case studies are also presented to show the system characteristics of the ICC algorithm with time-delay.

Unlike the conventional power system, the SG will rely heavily on an integrated communication and control network. This type of system is known as a large-scale networked control system (LSNCS) in which the components of the system, such as sensors, controllers, and actuators, are connected by the communication network [27]. A SG can be viewed as an LSNCS. The conventional centralized control scheme, however, may encounter severe challenges on applying LSNCS. The control center is required to have a high level of connectivity, which may also impose a substantial computational burden. The centralized control scheme is also more sensitive to failure than distributed control schemes [18]. Moreover, the plug-and- play technologies will make the topology of the SG vary over time [33]. The power system is inherently distributed, so one of the possible control approaches is to apply distributed control algorithms to power system problems.

Figure 1.13 is a layout of a future power system. In this multi-agent system (MAS), effective distributed control algorithms could be embedded in distributed controllers

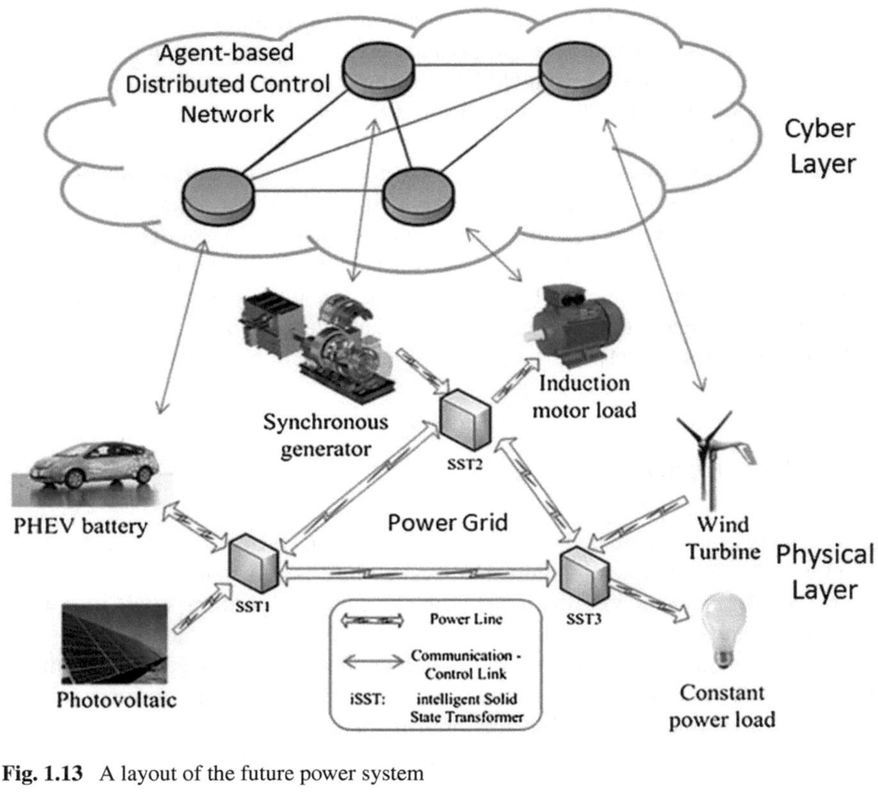

Fig. 1.13 A layout of the future power system

to properly allocate electrical power among connected buses autonomously. The agent-based distributed control for a SG also has been discussed in [19, 76]. A fundamental problem in distributed control systems (DCSs) is that of having all of the distributed controllers reach a consensus.

1.6 DC Microgrids

A local PV power generation combined with storage and grid connection is depicted in Fig. 1.14. The MG feeds directly a DC load through their dedicated converters. Concerning the DC bus, the following considerations are assumed: for the demand side, 90 % of a tertiary building's electrical load is possible to be DC fed in an efficient manner; for the grid side, the power factor can be controlled at 1 [71].

In order to interact with the load environment (user, metadata, SG), a supervision system is added, whose main role is to balance instantaneous power in power system following an energy management algorithm. The system operation must keep power balance while respecting constraints of certain elements. Figure 1.15 shows the power flow in the MG, using unidirectional powers whose sign convention is always positive.

The system operates respecting the available storage level and taking into account the grid connection. In case of insufficient PV energy toward the load, the system security is ensured thanks to the grid connection and by storage, as well as the load

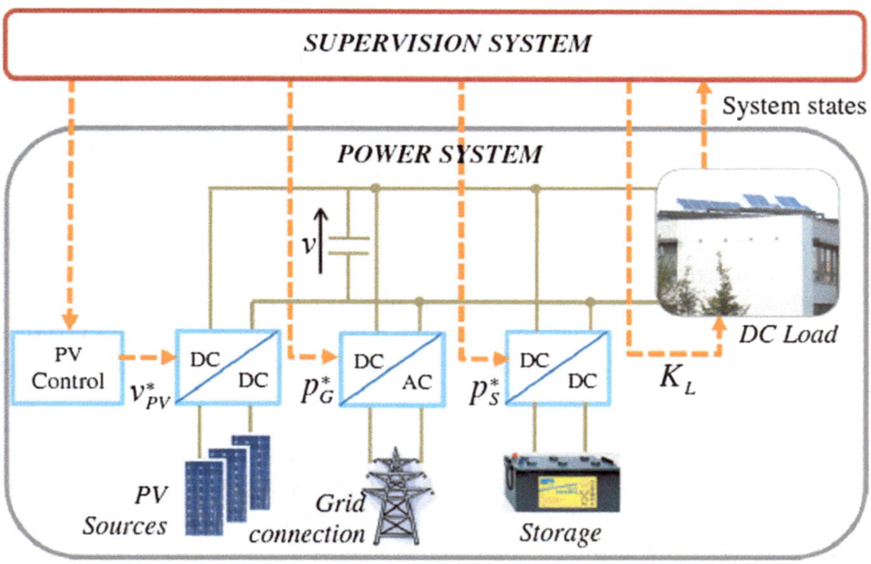

Fig. 1.14 DC microgrid structure

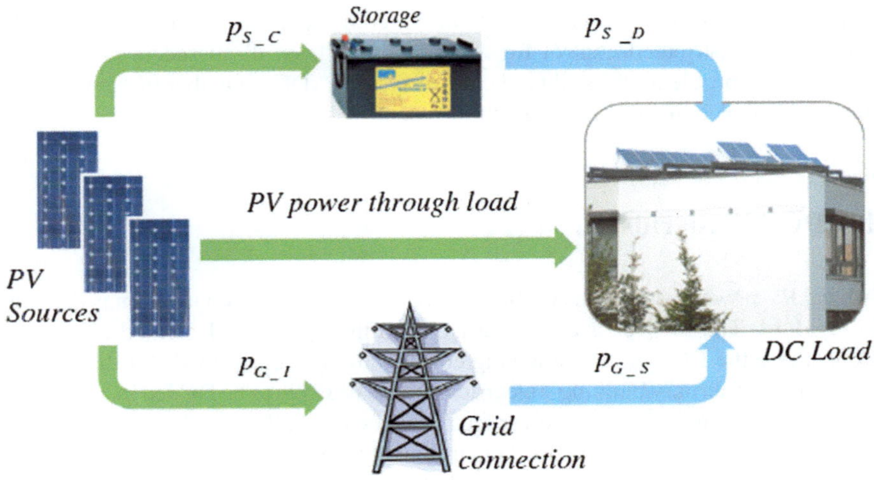

Fig. 1.15 Representation of power flow

shedding program. In case of excess PV power, the storage could be charged and the grid connection gives the possibility to trade it back.

1.6.1 PV Sources Control

It is known that PV system is controlled either by a maximum power point tracking (MPPT) method or by an algorithm to output a limited power $p_{PV_{lim}}$ to protect the storage from overcharging or to maintain grid injection power within imposed constraint. In the following, the chosen MPPT method is the well-known perturb and observe (P&O) [72], according to which PV sources produce an MPPT power $p_{PV_{MPPT}}$. The PV production could be limited thanks to the limited power reference $p^{*}_{PV_{lim}}$ calculated by a supervision algorithm [72]. The PV power p_{PV} control strategy is shown in Fig. 1.16. The P&O algorithm and the PV production limiting algorithm give at the same time corresponding voltage reference $v^{*}_{PV_{MPPT}}$ and $v^{*}_{PV_{lim}}$ to operate PV system. The maximum of these two references is taken as the PV voltage control reference v^{*}_{PV}, which represents the minimum power. Following v^{*}_{PV}, the PV system is operated by voltage and current double closed-loop control via automatic voltage regulator and automatic current regulator.

The output of control, duty cycle of pulse-width modulation (PWM) signal, is given to the power electronic devices to control the PV system. During the MPPT operation, if a limited power within the MPPT ability is given, the proportional-integral (PI) controller would increase the $v^{*}_{PV_{lim}}$. For $v^{*}_{PV_{lim}}$, $>$ $v^{*}_{PV_{MPPT}}$ the $v^{*}_{PV_{lim}}$ reference is taken and the MPPT algorithm is stopped.

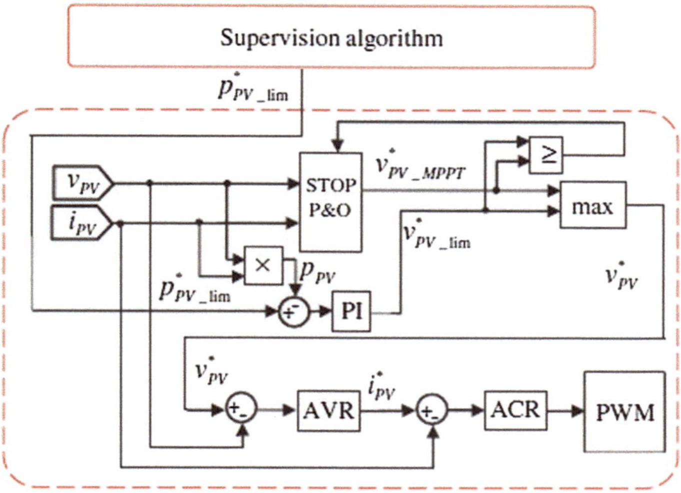

Fig. 1.16 PV system control strategy

By constrained power closed-loop control, the PI controller controls the PV power at the limited level. In case of low solar irradiance, the PV output power ability is less than the limited power reference, so the PI controller will decrease $v^*_{PV_{lim}}$ until the lower limit, and $v^*_{PV_{MPPT}}$ is taken to control the PV system. So, the limited power control does not affect MPPT algorithm and MPPT power is produced.

1.6.2 Storage Control

In general, lead–acid batteries are selected as storage for the DC microgrid applied to building, because of relatively low cost and mature technology [77]. The storage is operated by current closed-loop control, and the storage power is controlled by supervision system which calculates the corresponding power reference. The storage state of charge (SOC) must be respected to its upper and lower limitations, SOC_{max} and SOC_{min} respectively, to protect the storage from overcharging and overdischarging, as described by (1.1). The soc is calculated by (1.2), with SOC_0 as initial SOC at t_0, CREF as storage nominal capacity (Ah) and vS as storage voltage. When the soc limit is not reached, the PV production should not be limited, as in (1.3).

$$SOC_{min} \leq SOC(t) \leq SOC_{max} \tag{1.1}$$

$$SOC(t) = SOC_0 + \frac{1}{3600 \times v_S \times C_{REF}} \int_{t_0}^{t} (p_{S_C} - p_{S_D}(t))\, dt \tag{1.2}$$

$$p_{PV_{lim}}(t) = p_{PV_{MPPT}} \; if \; SOC(t) < SOC_{max} \tag{1.3}$$

1.6.3 Grid Connection Control

Following [72], the grid connection is controlled by current closed-loop control. The grid power is controlled by supervision system which calculates the corresponding power reference. Furthermore, as messages transmitted by the SG via supervision system, limits for grid supply power $p_{G_{Slim}}$ and grid injection power $p_{G_{Ilim}}$ could be imposed. By these two limitations, grid problems, such as performing peak shaving, avoiding undesired injection or downscaling injection fluctuations caused by intermittent PV productions, can be improved.

During MG operation, the grid power should be controlled to satisfy:

$$0 \leq p_{G_I}(t) \leq p_{G_{Ilim}} \quad (1.4)$$

$$0 \leq p_{G_S}(t) \leq p_{G_{Slim}} \quad (1.5)$$

1.6.4 DC Load Control

Obviously, the load power demand p_{G_D} must be satisfied; nevertheless, in case of insufficient storage and grid access limits, p_{G_D} cannot be fully met, and the load must be partially shed, forming load power p_L, to keep power balance. The load limited power $p_{L_{lim}}$ is controlled by the load coefficient K_L defined as:

$$K_L = \frac{p_{L_{lim}}}{p_{L_{max}}} \quad (1.6)$$

where $p_{L_{max}}$ is a constant as contractual subscribed maximum power. It is supposed that the load could not be supplied with a power that exceeds this limit. Given the constant denominator $p_{L_{max}}$, therefore $K_L \in [0,\ 1]$ changes according to the available power for supplying the load. If $p_{L_D} > p_{L_{max}}$, the load would be shed within the limit level. These operations should be controlled by the supervision system. Temporarily, partial load shedding could be a solution to reduce utility grid mismatching, or to obtain less energy consumption, within the limit agreed by end user.

1.6.5 Power Balancing Principle

According to the powers sign convention, *powers always positive*, the physical law of power balancing is described as

$$\begin{aligned} &p_L(t) + p_{G_I} + p_{S_C} = p_{G_S} + p_{S_D} + p_{PV} \\ &p_L(t) = \min(p_{L_D}(t)\ ,\ p_{L_{lim}}(t);\ ),\quad p_{PV}(t) = \min(p_{PV_{MPPT}}(t)\ ,\ p_{PV_{lim}}(t)) \end{aligned} \quad (1.7)$$

Fluctuations in the DC bus voltage, which is noted as ν, arise due to the difference in load consumption and PV generation. The required power reference p^* for power balancing is calculated by regulating ν with a proportional controller:

$$p^*(t) = p_{PV}(t) - p_L(t) - C_p(\nu^*(t) - \nu(t)) \quad (1.8)$$

where ν^* is the DC bus voltage control reference and C_p is the proportional gain. For stabilizing the DC bus voltage, power balance in the system is performed by adjusting storage and grid power. Thus, p^* is shared by the storage and the grid as

$$\begin{aligned} p^*(t) &= p_G^*(t) + p_S^*(t) \\ p_G(t) &= p_{G_I}(t) - p_{G_S}(t), \quad p_S(t) = p_{S_C}(t) - p_{S_D}(t) \end{aligned} \quad (1.9)$$

After computing p^*, the grid power reference p_G^* and storage power reference p_S^* are computed according to

$$p_S^*(t) = K_D(t)\, p^*(t), \quad p_G^*(t) = p^*(t) - p_S^*(t) \quad (1.10)$$

The power balancing of typical DC microgrid operation can work with arbitrary K_D value. To improve efficiency and reduce energy cost, optimization calculations using specifically metadata (power predictions, power limits, real-time grid energy tariffs) are done by the supervision that outputs the distribution coefficient, $K_D(t)$, whose time depending values represent the predicted optimized power flows.

1.7 Outline of the Book

1.7.1 Methodology

Throughout the book, our methodology in each chapter/section is composed of five steps:

- **Mathematical Modeling**
  in which we discuss the main ingredients of the state-space model under consideration.
- **Definitions and/or Assumptions**
  where we state the definitions and/or constraints on the model variables to pave the way for subsequent analysis.
- **Analysis and Examples**
  this signifies the core of the respective sections and subsections, which contain some solved examples for illustration.
- **Results**
  which are provided mostly in the form of theorems, lemmas, and corollaries.

- **Remarks**
  which are given to shed some light on the relevance of the developed results vis-a-vis published work.

In the sequel, theorems (lemmas, corollaries) are keyed to chapters and stated in *italic* font with **bold titles**, for example, **Theorem** 3.4 means Theorem 4 in Chap. 3 and so on. For convenience, we have grouped the reference in one major bibliography cited toward the end of the book. Relevant notes and research issues are offered at the end of each chapter for the purpose of stimulating the reader. We hope that this way of articulating the information will attract the attention of a wide spectrum of readership.

This book aims at providing a rigorous framework for studying analysis, control, and design problems of distributed generation systems (DGS) while addressing the dominating sources of difficulties due to packet dropouts; communication bandwidth constraints; parametric uncertainty and time-delays. The primary objective is three-fold:

1. to review the existing literature on DER from a contemporary perspective,
2. to examine presents modeling, trends, approaches, and applications of DER, and
3. to provide future possibilities, focusing on efficient and reliable design methods.

Looked at in this light, the main features of the book are:

1. *It provides an overall assessment of DER over the past several years*;
2. *It addresses several issues that arise at the interaction of control and communication of DGS*;
3. *It presents key concepts with their proofs followed by efficient computational methods*;
4. *It gives some technical questions and end-of-chapter problems*;
5. *It treats representative industrial applications.*

1.7.2 Chapter Organization

The book is primarily intended for researchers and engineers in the energy systems, control and communication community. It can also serve as complementary reading for elective courses on power control systems at the postgraduate level. The material of the book is divided into chapters and an appendix:

This chapter is an introductory chapter in which the different concepts and general ideas pertaining to DGS are presented.

Chapter 2 is devoted to review of a complete diagnostic profile of the main ingredients of DG plants. The emphasis is on aspects of electricity/heat generation and utilization.

Chapter 3 gives a comprehensive view of control methods for MG systems by scanning the different conventional and recently developed approaches. It pays equal attention to the analysis and modeling aspects.

Chapter 4 addresses the optimal energy management with emphasis on load forecasting and source scheduling. It includes also some aspects of optimal energy cost management.

Chapter 5 outlines a new paradigm for modeling the MGs as a system of systems (SoS) and its impact on microsources. It further emphasizes of the features of this paradigm in terms of components, trends, control methods, and communication profiles for distributed control.

Chapter 6 establishes a detailed characterization of the networked control problem of microgrid system of systems. It includes realistic models of microalternator and PV systems and demonstrates the control design procedure.

Chapter 7 provides pertinent results on decentralized voltage control methods of MGs. It presents two-distinct control strategies. In one strategy, it includes small signal dynamic analysis and grid operational modes. In the other strategy, power management and real stability radius constraint are discussed. Decentralized inverter control is also dealt with.

Chapter 8 is dedicated to advanced control design methods including distributed control architectures, multilevel control approach, and distributed cooperative control approach. Several case studies and simulation experiments were considered.

Chapter 9 contains the setup and real-time implementation of optimal control and distributed control methods for autonomous MG.

Chapter 10 is an appendix containing some relevant mathematical lemmas and basic algebraic inequalities that are used throughout the book.

Several illustrative numerical simulation examples are provided within the individual chapters. At the end of each chapter, a set of suggested problems is provided.

References

1. Ackermann T, Andersson G, Soder L (2001) Distributed generation: a definition. Electr Power Syst Res 57:195–204
2. Ahn SJ, Park JW, Chung IY, Moon SI, Kang SH, Nam SR (2010) Power-sharing method of multiple distributed generators considering control modes and configurations of a microgrid. IEEE Trans Power Deliv 25(3):2007–2016
3. Alvial-Palavicino C, Garrido-Echeverra N, Jimnez-Estvez G, Reyes L, Palma-Behnke R (2011) A methodology for community engagement in the introduction of renewable based smart microgrid. Energy Sustain Deliv 15(3):314–323
4. Arulampalam A et al (2004) Control of power electronic interfaces in distributed generation microgrids. Int J Electron 91(9):503–523
5. Bando S, Sasaki Y, Asano H, Tagami S (2008) Balancing control method of amicrogrid with intermittent renewable energy generators and small battery storage. In: Proceedings of the IEEE PES general meeting conversion and delivery of electrical energy in the 21st century, pp 1–6
6. Basu AK, Chowdhury SP, Chowdhury S, Ray D, Crossley PA (2008) Reliability study of a micro-grid power system. In: Proceedings of the 43rd international university power engineering conference, pp 1–4
7. Battaglini A, Lilliestam J, Bals C, Haas A (2010) The super smart grid. In: Proceedings of the European climate forum (2008). Web. 14 April 2010

8. Batteries and ultra-capacitors for the smart grid: market opportunities 2009–2010, Rep. NanoMarkets. (August 2009). Web. 12 May 2010. http://www.flextech.org/documents/nanomarketsreports/SmartGridES.pdf
9. Blaabjerg F, Chen Z, Kjaer SB (2004) Power electronics as efficient interface in dispersed power generation systems. IEEE Trans Power Electron 19(5):1184–1194
10. Carrasco JM et al (2006) Power-electronic systems for the grid integration of renewable energy sources: a survey. IEEE Trans Ind Electron 53(4):1002–1016
11. Carrasco JM, Franquelo LG, Bialasiewicz JT, Galvan E, Guisado RCP, Prats MA M (2006) Power-electronic systems for the grid integration of renewable energy sources: a survey. IEEE Trans Ind Electron 54(4):1002–1016
12. Cavalho AJ (2001) Energy storage technologies for utility scale intermittent renewable energy systems. J Solar Energy Eng 123:387–389
13. Chakraborty S, Weiss MD, Simes MG (2007) Distributed intelligent energy management system for a single-phase high-frequency AC microgrid. IEEE Trans Ind Electron 54(1):97–109
14. Chen Z (2012) A review of power electronics based microgrids. J Power Electron 12(1): 181–192
15. Chicco G, Mancarella P (2009) Distributed multi-generation: a comprehensive view. Renew Sustain Energy Rev 13:535–551
16. CIRED (1999) Dispersed generation; Preliminary report of CIRED (International conference on electricity distribution), Working Group WG04, Brussels, Belgium, June 1999
17. Costa PM, Matos MA, Lopes JAP (2008) Regulation of microgeneration and microgrids. Energy Policy 36(10):3893–3904
18. D'Andrea R, Dullerud GE (2003) Distributed control design for spatially interconnected systems. IEEE Trans Autom Control 48(9):1478–1495
19. Dimeas AL, Hatziargyriou ND (2007) Agent-based control for microgrids. In: Proceedings of the IEEE power engineering society general meeting, pp 24–28 June 2007
20. Dimeas AL, Hatziargyriou ND (2005) Operation of a multiagent system for microgrid control. IEEE Trans Power Syst 20(3):1447–1455
21. Driesen J, Katiraei F (2008) Design for distributed energy resources. IEEE Power Energy Mag 6(3):30–40
22. El-Khattam W, Salama MMA (2004) Distributed generation technologies, definitions and benefits. Electr Power Syst Res 71(2):119–128
23. EWEA, Online documentation. October 2005, http://www.iea-pvps.org
24. Fehrenbacher K (2010) FAQ: Energy Storage for the Smart Grid. Earth2Tech, The GigaOM Network (13 May 2009). Web. 21 April 2010. http://earth2tech.com/2009/05/13/faq-energy-storage-for-the-smart-grid/
25. Georgakis D, Papathanassiou S, Hatziargyriou N, Engler A, Hardt C (2004) Operation of a prototype microgrid system based on microsources quipped with fast-acting power electronics interfaces. In: Proceedings of the IEEE 35th annual PESC/IEEE power electronic specifications conference, vol 4, pp 2521–2526
26. Grid 2030: A National Vision for Electricitys Second 100 years, U.S. Department of Energy, Office of Electric Transmission and Distribution (2003). Web. 14 April 2010
27. Gupta RA, Chow M-Y (2010) Networked control system: overview and research trends. IEEE Trans Ind Electron 57(7):2527–2535
28. Hall PJ, Bain EJ (2008) Energy-storage technologies and electricity generation. Energy Policy 36(12):4352–4355
29. Hatziargyriou ADN, Hatzivasiliadis S, Jimeno J, Oyarzabal J (2008) DB3: Decentralized control concepts. EU More Microgrid. School of Electrical and Computer Engineering, Greece
30. Hatziargyriou N, Asano H, Iravani R, Marnay C (2007) Microgrids. IEEE Power Energy Mag 5(4):78–94
31. Hatziargyriou N, MICROGRIDS large scale integration of micro-generation to low voltage grids. http://microgrids.eu/micro2000/presentations/16.pdf
32. Holland S (2010) Obama Announces 3.4 Billion in Grants for Smart Grid. Reuters. Business & Financial News, Breaking U.S. & International News, Reuters.com. (27 October 2009). Web. 21 April 2010. http://www.reuters.com/article/idUSTRE59Q1AC20091027

33. Huang AQ, Crow ML, Heydt GT (2011) The future renewable electric energy delivery and management (FREEDM) system: the energy internet. Proc IEEE 99(1):133–148
34. IdaTech, Cascadiant deploy fuel cells with hutchison telecom in indonesia. Renewable Energy Focus. Elsevier Ltd. (15 April 2010). Web. 12 May 2010. http://www.renewableenergyfocus.com/view/8758/idatechcascadiant-deploy-fuel-cells-with-hutchison-telecom-in-indonesia/
35. IEA-PVPS, Cumulative Installed PV Power, October 2005. http://www.iea-pvps.org
36. Inoue S, Akagi H (2006) A bi-directional isolated DC/DC converter as a core circuit of the next-generation medium-voltage power conversion system. In: Proceedings of the 37th IEEE PESC, pp 1–7
37. Jiayi H, Chuanwen J, Rong X (2008) A review on distributed energy resources and MicroGrid. Renew Sustain Energy Rev 12(9):2472–2483
38. Karimi H (2007) Islanding detection and control of an islanded electronically-coupled distributed generation unit, Ph.D. University Toronto, Department of Electrical and Computer Engineering
39. Katiraei F, Iravani R, Hatziargyriou N, Dimeas A (2008) Microgrids management. IEEE Power Energy Mag 6(3):54–65
40. Kaygusuz K (2009) Energy and environmental issues relating to greenhouse gas emissions for sustainable development in Turkey. Renew Sustain Energy Rev 13:253–270
41. Kjaer SB, Pedersen JK, Blaabjerg F (2005) A review of single-phase grid-connected inverters for photovoltaic modules. IEEE Trans Ind Appl 41(5):1292–1306
42. Kroposki B, Lasseter R, Ise T, Morozumi S, Papatlianassiou S, Hatziargyriou N (2008) Making microgrids work. IEEE Power Energy Mag 6(3):40–53
43. Lasseter RH, Akhil A, Marnay C, Stephens J, Dagle J, Guttromson R (2002) The CERTS microgrid concept. White Paper on Integration of distributed energy resources, prepared for Transmission Reliability Program, Office of Power Technologies, U.S. Department of Energy, April 2002. http://certs.lbl.gov/pdf/50829-app.pdf
44. Lasseter RH (2001) Microgrids distributed power generation. In: IEEE power engineering society winter meeting, Columbus, Ohio, vol 1, pp 146–149 February 2001
45. Lasseter RH (2002) MicroGrids. In: Proceedings of the IEEE PES winter meeting, vol 1, pp. 305–308
46. Lasseter RH (2008) Extended CERTS technology. In: Proceedings of the IEEE PES general meeting conversion and delivery of electrical energy in the 21st century, pp 1–5
47. Lasseter RH, Paigi P (2004) Microgrid: a conceptual solution. In: IEEE 35th annual power electronics specialists conference, PESC 04, vol 6
48. Lasseter RH (2007) Microgrids and distributed generation. J Energy Eng 133(3):144–149
49. Lindley D (2010) Smart Grids: The Energy Storage Problem: Nature News, Nature Publishing Group. (6 January 2010). Web. 12 May 2010. http://www.nature.com/news/2010/100106/full/463018a.html
50. Lopes JAP, Moreira CL, Madureira AG (2006) Defining control strategies for microgrids islanded operation. IEEE Trans Power Syst 21(2):916–924
51. Lopes JP, Hatziargyriou N, Mutale J, Djapic P, Jenkins N (2007) Integrating distributed generation into electric power systems: a review of drivers, challenges and opportunities. Electr Power Syst Res 77(9):1189–1203
52. Madureira A, Moreira C, Lopes JP (2005) Secondary load frequency control for microgrids in islanded operation. In: Proceedings of the ICREPQ, Cordoba, Spain, pp 1–5
53. Mao M, Liuchen C, Ding M (2008) Integration and intelligent control of micro-grids with multi-energy generations: a review. In: Proceedings of the IEEE ICSET, pp 777–780
54. Meiqin M, Chang L, Ming D (2008) Integration and intelligent control of microgrids with multi-energy generations: a review. In: Proceedings of the IEEE international conference on sustainable energy technologies, pp 777–780
55. Moreira C, Resende F, Lopes JAP (2007) Using low voltage microgrids for service restoration. IEEE Trans Power Syst 22(1):395–403
56. Morozumi S (2007) Micro-grid demonstration projects in Japan. In: IEEE power conversion conference-Nagoya, PCC'07

57. Moskovitz D, Profits and progress through distributed resources, published by the Regulatory Assistance Project, Maine, USA. http://www.rapmaine.org/distribution.html
58. Nigim, KA, Lee W-J (2007) Microgrid integration opportunities and challenges. In: Proceedings of the IEEE PES general meeting, pp 1–6
59. North American Electric Reliability Corporation, Technical report, August 2013
60. Ogren P, Fiorelli E, Leonard NE (2004) Cooperative control of mobile sensor networks: Adaptive gradient climbing in a distributed environment. IEEE Trans Autom Control 49(8):1292–1302
61. Omer AM (2008) Energy, environment and sustainable development. Renew Sustain Energy Rev 12:2265–2300
62. Peng FZ, Li YW, Tolbert LM (2009) Control and protection of power electronics interfaced distributed generation systems in a customer driven microgrid. In: Proceedings of the IEEE PES general meeting, pp 1–8
63. Pepermans G et al (2005) Distributed generation: definition, benefits and issues. Energy Policy 33(6):787–798
64. Pereira OS, Reis TM, De Araujo RGB, Gongalves FF (2006) Renewable energy as a tool to assure continuity of low emissions in the Brazilian electric power sector. In: Proceedings of the IEEE EIC climate change technology
65. Plug in to Materials Trends for Smart Grid Applications, NanoMarkets. Web. 28 April 2010. http://www.nanomarkets.net/perspectives/articles.cfm?articleID=183&referrer=AZOMDOTCOM
66. Ponnaluri S, Linhofer GO, Steinke JK, Steimer PK (2005) Comparison of single and two stage topologies for interface of BESS or fuel cell system using the ABB standard power electronics building blocks. In: Proceddings of the European conference on power electronic applications, pp 1–9
67. Price A, Thijssen G, Symons P (2008) Electricity storage, a solution in network operation. Distributech Europe. http://energystorage.org/pubs/2000/Distributech2000TechPaper.pdf
68. Pumped Hydroelectricity Storage Association, ESA (April 2009). Web. 28 April 2010. http://www.electricitystorage.org/site/technologies/pumpedhydro/
69. Rahim NA, Omar AM (2002) Three-phase single-stage high-voltage DC converter. In: Proceedings of the institution of electrical engineers: generation, transmission and distribution, vol 149(5), pp 505–509, September 2002
70. Rocabert J, Luna A, Blaabjerg F, Rodriguez P (2012) Control of power converters in AC microgrids. IEEE Trans Power Electron 27(11):4734–4749
71. Sechilariu M, Wang BC, Locment F (2013) Building-integrated microgrid: advanced local energy management for forthcoming smart power grid communication. Energy Build 59(1):236–243
72. Sechilariu M, Wang BC, Locment F (2014) Supervision control for optimal energy cost management in DC microgrid: design and simulation. Electr Power Energy Syst 58:140–149
73. Shahidehpour M, Schwartz F (2004) Dont let the sun go down on PV. IEEE Power Energy Mag 2(3):40–48
74. Smith W (2000) The role of fuel cells in energy storage. J Power Sources 86(12):74–83
75. Song YJ, Chung SK, Enjeti PN (2004) A current-fed HF link direct DC/AC converter with active harmonic filter for fuel cell power systems. In: Conference record IEEE 39th IAS annual meeting, pp 123–128
76. Suryanarayanan S, Mitra J, Biswas S (2010) A conceptual framework of a hierarchically networked agent-based microgrid architecture. Transmission and distribution conference and exposition. IEEE PES, 19–22 April 2010
77. Tan X, Li Q, Wang H (2013) Advances and trends of energy storage technology in microgrid. Int J Electr Power Energy Syst 44(1):179–191
78. The future of village power supply inremote areas. http://www.iset.uni-kassel.de/
79. The Smart Grid: An Introduction, U.S. Department of Energy (2008). Web. 14 April 2010. http://www.oe.energy.gov/DocumentsandMedia/DOESGBookSinglePages.pdf

80. Tsikalakis AG, Hatziargyriou ND (2011) Centralized control for optimizing microgrids operation. In: Proceedings of the IEEE power energy society general meeting, pp 1–8
81. Ustun TS, Ozansoy C, Zayegh A (2011) Recent developments in microgrids and example cases around the world—a review. Renew Sustain Energy Rev 15(8):4030–4041
82. Wang C, Nehrir M (2008) Power management of a stand-alone wind/ photovoltaic/fuel cell energy system. IEEE Trans Energy Convers 23(3):957–967
83. Xiao Y (ed) (2012) Communication and networking in smart grids. CRC Press, Taylor & Francis Group, Boston
84. Xiaofei L, Xin A, Yonggang W (2011) Study of single-phase HFAC microgrid based on Matlab/Simulink. In: Proceedings of the 4th international conference DRPT, pp 1104–1108
85. Xu X, Zha X (2007) Overview of the researches on distributed generation and microgrid. In: Proceedings of the IPEC, pp 966–971
86. Youli S, Litifu Z, Nagasaka K (2009) Efficiency of microgrid with storage battery in reliability, economy and environment assessments. Int J Electr Power Eng 3(3):154–162
87. Yunwei L, Vilathgamuwa DM, Poh Chiang L (2004) Design, analysis, and real-time testing of a controller for multibus microgrid system. IEEE Trans Power Electron 19(5):1195–1204
88. Zareipour H, Bhattacharya K, Canizares CA (2004) Distributed generation: current status and challenges. In: Proceedings of 36th annual North American power symposium (NAPS)

Chapter 2
Distributed Generation Plants

2.1 Combined Heat and Power Plants

Combined heat and power (CHP) or cogeneration systems are most promising as distributed energy resources (DERs) for microgrid (MG) applications. Their main advantage is energy-efficient power generation by judicious utilization of waste heat. Unlike fossil-fueled power plants, CHP systems capture and use the by-product heat locally for domestic and industrial/process heating purposes. Heat produced at moderate temperatures (100180 8C) can also be used in absorption chillers for cooling. Simultaneous production of electricity, heat, and cooling is known as trigeneration or polygeneration.

2.1.1 Introduction

By capturing excess heat, the CHP system allows better usage of energy than conventional generation, potentially reaching an efficiency of more than 80 compared with that of about 35 % for conventional power plants. It is most efficient when the heat is utilized locally. Overall efficiency is reduced if heat is to be transported over long distances using heavily insulated pipes, which are both expensive and inefficient. On the other hand, electricity can be transmitted over much longer distances for lesser energy loss. Thus, CHP plants can be located somewhat remotely from their electrical loads, but they must always be located close to the heat loads for better performance. CHP plants are commonly employed in district heating systems of big towns, hospitals, prisons, oil refineries, paper mills, and industrial plants with large heat loads.

Use of CHP plants has been found to lead to 35 % reduction in primary energy use compared to conventional power generation and heat-only boilers, 30 % reduction in emission with respect to coal-fired power plants and 10 % reduction in emission with respect to combined cycle gas turbine (CCGT) plants.

M.S. Mahmoud and F.M. AL-Sunni, *Control and Optimization of Distributed Generation Systems*, Power Systems, DOI 10.1007/978-3-319-16910-1_2

2.1.2 Microcogeneration Systems

Microcogeneration systems are usually installed in smaller premises like homes or small commercial buildings. They differ from larger CHP units not only in terms of their energy-producing capacities but also in matters of parameter-driven operation. Most large industrial CHP units generate electricity as the primary product with heat as secondary while micro-CHP systems generate heat as the primary commodity with electricity as a by-product. Thus, energy generation of micro-CHP systems is mainly dictated by the heat demand of the end users. Because of this operating model and the fluctuating electrical demand of the structures they operate in (like homes and small commercial buildings), micro-CHP systems often generate more electricity than is demanded.

Micro-CHP sets are basically microturbines coupled to single-shaft, high-speed (50,000–100,000 rpm) permanent magnet synchronous machines (SMs) with airfoil or magnetic bearings. They are provided with power electronic interfaces (PEIs) for connection to the electrical loads. They also have their own heat recovery systems for low and medium temperature heat extraction. Micro-CHP sets are reliable, robust, and cheap. They are available in the range of 10,100 kW capacity. The primary fuel is natural gas, propane, or liquid fuel, which permits clean combustion with low particulates. Biofuelled microturbines are also being considered as a possibility. During operation of a CHP set, the pressure of incoming air is raised after passing through the centrifugal compressor. Temperature of the compressed air is increased on passing through the heat exchanger. When the hot compressed air enters the combustion chamber, it is mixed with fuel and burnt. The high temperature combustion gases are expanded in the turbine to produce mechanical power, which in turn drives the permanent magnet SM to produce electrical power at high frequency. High-frequency output voltage is converted into DC using a rectifier and the DC voltage is reconverted into AC of 50/60 Hz of frequency as per necessity using an inverter interface.

By using micro-CHP plants, MGs can secure the following major advantages:

- Since transportation of electricity is far easier and more cost-effective than heat, it is much more suitable to place micro-CHP plants near heat loads than electrical loads. MG permits this energy optimal placement of CHP plants to achieve full utilization of heat. In case of necessity, fuel cells can also be used in the CHP plants for better utilization of the generated heat.
- The scale of heat generation for individual units is small. Therefore, micro-CHP plants have greater flexibility in matching several small heat loads. Technically, a MG can be designed with a judicious mix of waste and non-waste heat-producing generators so as to optimize the combined generation of heat and electricity. In spite of the above-mentioned flexibility, chances are still there of a mismatch in generating a proper mix of heat and electricity. Hence, attention must be paid to enhance this flexibility.

Micro-CHP systems are primarily based on the following technologies:
Internal Combustion (IC) Engines
Stirling Engines
Microturbines
Fuel cells

2.1.3 Internal Combustion Engines

In IC engines, fuel is burnt in air in a combustion chamber with or without oxidizers. Combustion creates high-temperature and high-pressure gases that are allowed to expand and act on movable bodies like pistons or rotors. IC engines are different from external combustion engines like steam engines and Stirling engines. The external combustion engines use the combustion process to heat a separate working fluid which then works by acting on the movable parts. IC engines include intermittent combustion engines (e.g., reciprocating engine, Wankel engine, and Bourkes engine) and the continuous combustion engines (e.g., Jet engines, rockets, and gas turbines (GTs)). The commonly used fuels are diesel, gasoline, and petroleum gas. Propane gas is also sometimes used as fuel. With some modifications to the fuel delivery components, most IC engines designed for gasoline can run on natural gas or liquefied petroleum gases. Liquid and gaseous biofuels, like ethanol and biodiesel, may also be used. Depending on the type of fuel, the IC engines are provided with spark ignition or compression ignition systems in their cylinders to initiate the fuel combustion process.

2.1.4 Stirling Engines

A stirling engine is a closed-cycle piston heat engine where the working gas is permanently contained within the cylinder. It is traditionally classified as an external combustion engine, though heat can also be supplied by noncombustible sources like solar, geothermal, chemical, and nuclear energy. Stirling engines use an external heat source and an external heat sink. Each is maintained within a limited temperature range and has sufficiently large temperature difference between them.

Stirling engines contain a fixed quantity of air, hydrogen, or helium gas as working fluid. Under normal operation, the engine is completely sealed and no gas can enter or leave the engine. Thus, no valves are required to control the intake and exhaust of gases unlike other piston engines. A Stirling engine cycles through four main processes, viz.

(i) cooling,
(ii) compression,
(iii) heating and
(iv) expansion,

accomplished by the movement of the working gas back and forth between hot and cold heat exchangers. The hot heat exchanger is kept in thermal contact with an external (primary) heat source like a fuel burner, while the cold heat exchanger is kept in thermal contact with an external heat sink like radiators. A change in gas temperature causes a corresponding change in its pressure, and the motion of the piston causes the gas to be alternately expanded and compressed. When the gas is heated, it expands in the sealed chamber and acts on the power piston to produce a power stroke. When the gas is cooled its pressure drops and then less work has to be done by the piston to compress the gas on the return stroke. This work difference yields the net power output.

Stirling engines are basically of three categories:

(i) Alpha Stirling,
(ii) Beta Stirling, and
(iii) Gamma Stirling.

Stirling engines can be economically used as energy producers for CHP applications that utilize a heat source in conjunction with a secondary heating application, such as an industrial process. For such operations, the Stirling engines take the advantage of the temperature differential between the primary heat source and the heating application. The primary heat source enters the Stirling engine heater and produces mechanical power, and the waste heat from the engines heater is used to supply the secondary heating applications. The mechanical power produced by the engine is used to generate electricity as a secondary product. The overall process is very efficient and cost effective.

Stirling engines have several advantages over reciprocating engines. They can achieve high energy conversion efficiency of 80 %, though this is limited by non-ideal properties of the working gas and engine materials (such as friction, thermal conductivity, tensile strength, creep, and melting point). In contrast to IC engines, they are usually more energy efficient, quieter, and more reliable with lower maintenance requirements. But for the same power rating, a Stirling engine has larger size and higher capital cost than an IC engine. Hence, Stirling engines are being used only for those applications where the primary objective is to minimize the capital cost per kilowatt hour but not per kilowatt. The wider application of Stirling engine entirely depends on appropriate cost–benefit analysis. However, with growing concerns of rising energy costs, energy shortages, and environmental pollution caused by combustible fuels, the advantages of Stirling engines have become increasingly significant. They are being employed in various applications like water pumping and electrical generation from plentiful energy sources that are incompatible with the IC engine, such as solar energy, agricultural waste, and domestic refuse, as well as in CHP systems.

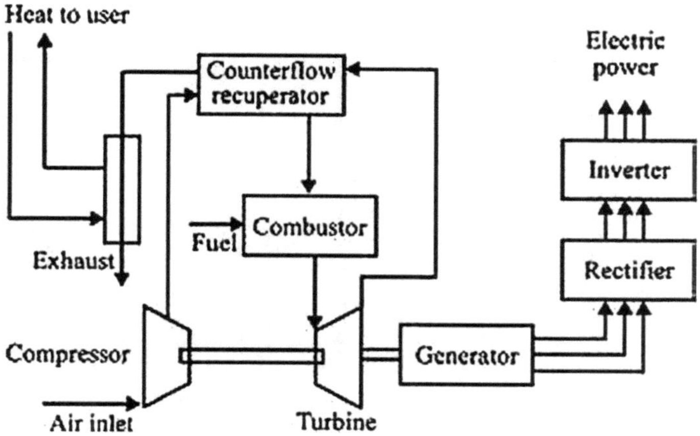

Fig. 2.1 Single-shaft microturbine

2.1.5 Microturbines

Microturbines are widely popular as generating units in distributed generation DG systems and as energy producers in CHP systems. At present they hold the maximum prospect to be used as microsources for MGs. Microturbines are small and simple-cycle GTs. The outputs of the microturbines range typically from around 25 to 300 kW. Performance improvement techniques used in microturbines include recuperation, low NO_x emission technologies, and the use of advanced materials such as ceramic for the hot section parts. Microturbines are available as single-shaft or split-shaft units. Single-shaft unit is a high-speed SM with the compressor and turbine mounted on the same shaft. For these machines, the turbine speed ranges from 50,000 to 120,000 rpm. By contrast, the split-shaft design uses a power turbine rotating at 3,000 rpm and a conventional generator connected via a gearbox for speed multiplication. Figure 2.1 shows a single-shaft microturbine system, while Fig. 2.2 shows a split-shaft one. Unlike traditional backup generators, microturbines are designed to operate for extended periods of time and require little maintenance. They can supply customers' base load requirements or can be used for standby, peak shaving, and cogeneration applications. They can run on most commercially available fuels, such as natural gas, propane, diesel, and kerosene, as well as on biofuels.

Microturbines have the following features:

1. Size. They are relatively smaller in size compared to other DERs.
2. Fuel-to-electricity conversion. They can reach the range of 2530 if the waste heat recovery is used for CHP applications, energy efficiency levels are greater than 80 %.
3. NO_x emissions. These are lower than 7 ppm for natural gas machines.

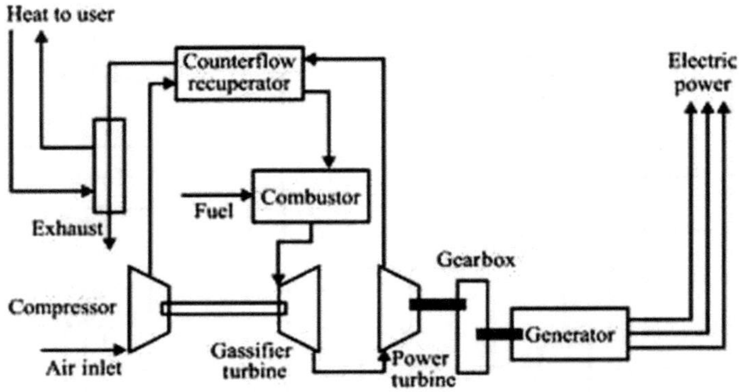

Fig. 2.2 Single-shaft microturbine

4. Operational life. They are designed for 11,000 h of operation between major overhauls with a service life of at least 45,000 h.
5. Economy of operation system costs are lower than 500 per kW. Cost of electricity is competitive with alternatives including grid power for market applications.
6. Fuel flexibility. It is capable of using alternative fuels, like natural gas, diesel, ethanol and landfill gas, and other biomass-derived liquids and gases.
7. Noise level. It has reduced level of noise and vibrations.
8. Installation. It has simpler installation procedure.

The research literature shows that there is an extensive thrust on the application of microturbine as DG systems. Research areas include simulation, offline/real-time studies and development of inverter interfaces for microturbine applications. Several research papers are available on the development of a single-stage axial flow microturbine for power generation, on the study of the facilities of the technology through relevant test results, on the development of active filters and adaptive control mechanisms for microturbines in hybrid power systems, etc. Studies also include development of dynamic models for microturbines to analyze their performance in islanded and grid-connected modes of operation and for cogeneration applications.

Most microturbines use permanent magnet synchronous generator (PMSG) or asynchronous generator for power generation. Ample research has been conducted on PMSG-coupled microturbines. However, very little has been reported on development and load-following performance analysis of microturbine models with synchronous generator (SYGN) in islanded and grid-connected modes. This area needs to be extensively investigated to resolve the technical issues for integrated operation of a microturbine with the main utility grid.

The main advantage of coupling an SG with a split-shaft microturbine is that it eliminates the use of the power converter. In this case, the generator is connected to the turbine via a gearbox to generate conventional 50/60 Hz power. Thus, the need for rectifiers and power converter units is completely eliminated. Moreover, the use of high-speed PMSG has disadvantages such as thermal stress, demagnetization phenomena, centrifugal forces, rotor losses because of fringing effects, and high cost. The disadvantage of coupling induction (asynchronous) generators is that though they are cheaper and robust, their speed is load dependent and they cannot be connected to the grid without the use of expensive power converter systems. The use of PEIs for power conversion introduces harmonics in the system to reduce the output power quality (PQ). These harmonics are eliminated if an SG is used with a gearbox. Also, there are less chances of failure as the gearbox is a much simpler mechanical equipment as compared to complex power electronic devices. However, the main drawback of using a gearbox is that it consumes a fraction of generated power, thus reducing the efficiency of the system. Some manufacturing companies like Ingersoll-Rand Energy Systems, Ballard, Bowman, and Elliott use synchronous machines with their microturbines for both stand-alone and grid-connected operations.

2.1.6 Fuel Cells

A fuel cell converts chemical energy of a fuel directly into electrical energy. It consists of two electrodes (an anode and a cathode) and an electrolyte, retained in a matrix. The operation is similar to that of a storage battery except that the reactants and products are not stored, but are continuously fed to the cell. During operation, the hydrogen-rich fuel and oxidant (usually air) are separately supplied to the electrodes. Fuel is fed to the anode and oxidant to the cathode, and the two streams are separated by an electrode electrolyte system. Electrochemical oxidation and reduction take place at the electrodes to produce electricity. Heat and water are produced as by-products. Figure 2.3 shows the basic construction of a proton exchange membrane fuel cell.

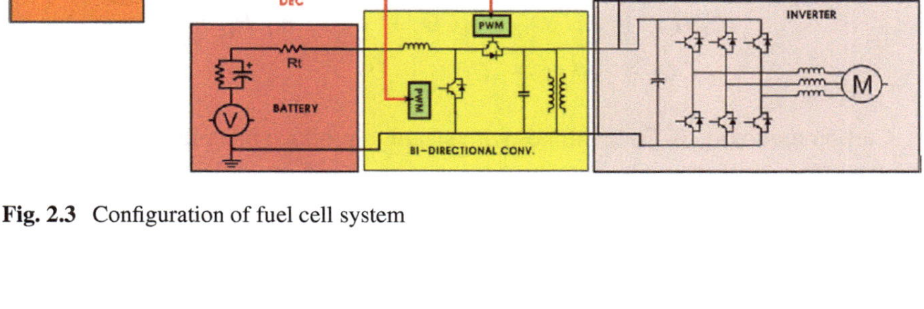

Fig. 2.3 Configuration of fuel cell system

Three major subsystems exist in a typical pure hydrogen fuel cell system:

- Air supply with thermal management in which a reasonable flow rate and pressure are regulated to avoid oxygen starvation, water flooding of the membrane, and excessive auxiliary power consumption.
- Fuel supply, and
- Anode recirculation which is used to reduce hydrogen waste, maintain the pressure difference between anode and cathode to a minimum, and run the fuel in the anode in order to obtain better water management.

A fuel cell system with dynamic response is important for vehicular applications because fluctuations in power demand. A fuel cell and its subsystems usually do not operate at the optimal steady states designed by the manufacturer. Therefore, it is important to construct a precise PEM fuel cell dynamic model and adopt a robust control technique to satisfy the aforementioned power fluctuation requirements.

Fuel cells have several advantages over conventional generators. Due to higher efficiency and lower fuel oxidation temperature, fuel cells emit less CO_2 and NO_x per kilowatt of power generated. Thus, they provide an eco-friendly energy source. As there are no moving parts, they are almost free from noise and vibration, robust and low maintenance. This makes them suitable for urban or suburban locations. Unlike gas and steam turbines, fuel cell efficiency increases at part-load conditions. Moreover, they can use a variety of fuels like natural gas, propane, landfill gas, anaerobic digester gas, diesel, naphtha, methanol, and hydrogen. This versatility ensures that this technology will not become obsolete due to the unavailability of fuels.

A single fuel cell produces output voltage less than 1 V. Therefore, to produce higher voltages, fuel cells are stacked on top of each other and are series connected forming a fuel cell system. Electrical efficiencies of fuel cells lie between 36 % and 60 %, depending on the type and system configuration. By using conventional heat recovery equipment, overall efficiency can be enhanced to about 85 %. Steam reforming of liquid hydrocarbons ($C_n H_m$) is a potential way of providing hydrogen-rich fuel for fuel cells. This is preferred because storage of hydrogen is quite hazardous and expensive. Reformers provide a running stream of hydrogen without having to use bulky pressurized hydrogen tanks or hydrogen vehicles for distribution. The endothermic reaction that occurs in the reforming process in the presence of a catalyst is

$$\begin{aligned} C_n H_m \; + \; n H_2 O \rightarrow \; n CO \; + \; \left(\frac{m}{2} + n\right) \; H_2 O \\ CO \; + \; H_2 O \rightarrow \; CO_2 \; + \; H_2 \end{aligned} \tag{2.1}$$

Carbon monoxide (CO) combines with steam to produce more hydrogen through the water gas shift reaction. Extensive research is going on to design reformer fuel cell system in spite of the following challenges:

- Steam reforming can utilize liquid hydrocarbon fuels like ethanol and biodiesel, but these fuels may not be available in sufficiently large quantities to provide a continuous stream of hydrogen.
- As the reforming reaction takes place at high temperatures, fuel cells have high start-up time and require costly temperature-resistant materials.
- The catalyst is very expensive and the sulfur compounds present in the fuel may poison certain catalysts, making it difficult to run this type of system on ordinary gasoline.
- CO produced in the reaction may poison the fuel cell membrane and may degrade its performance. In that case, complicated CO removal systems must be incorporated into the system.
- Thermodynamic efficiency of the process depends on the purity of the hydrogen product. Normally, thermodynamic efficiency lies between 70 and 85 %.

2.2 Renewable Energy Generation

Renewable power generation can help countries meet their sustainable development goals through provision of access to clean, secure, reliable, and affordable energy. Renewable energy has gone mainstream, accounting for the majority of capacity additions in power generation today [4, 29, 31]. Tens of gigawatts of wind, hydropower, and solar photovoltaic (PV) capacity are installed worldwide every year in a renewable energy market that is worth more than a hundred billion USD annually. Other renewable power technology markets are also emerging. Recent years have seen dramatic reductions in renewable energy technologies costs as a result of R&D and accelerated deployment. Yet policy-makers are often not aware of the latest cost data.

2.2.1 Wind Power Plants

Wind turbines (WTs) operate on a simple principle. The energy in the wind turns two or three propeller-like blades around a rotor. The rotor is connected to the main shaft, which spins a generator to create electricity. WTs are mounted on a tower to capture the most energy. At 100 feet (30 m) or more above ground, they can take advantage of faster and less turbulent wind. WTs can be used to produce electricity for a single home or building, or they can be connected to an electricity grid (shown here) for more widespread electricity distribution.

We have been harnessing the wind's energy for hundreds of years. From old Holland to farms in the United States, windmills have been used for pumping water or grinding grain. Today, the windmill's modern equivalent WT can use the wind's energy to generate electricity.

WTs, like windmills, are mounted on a tower to capture the most energy. At 100 feet (30 m) or more aboveground, they can take advantage of the faster and less turbulent wind. Turbines catch the wind's energy with their propeller-like blades. Usually, two or three blades are mounted on a shaft to form a rotor.

A blade acts much like an airplane wing. When the wind blows, a pocket of low-pressure air forms on the downwind side of the blade. The low-pressure air pocket then pulls the blade toward it, causing the rotor to turn. This is called lift. The force of the lift is actually much stronger than the wind's force against the front side of the blade, which is called drag. The combination of lift and drag causes the rotor to spin like a propeller, and the turning shaft spins a generator to make electricity.

WTs can be used as stand-alone applications, or they can be connected to a utility power grid or even combined with a PV (solar cell) system. For utility-scale (megawatt-sized) sources of wind energy, a large number of WTs are usually built close together to form a wind plant, also referred to as a wind farm. Several electricity providers today use wind plants to supply power to their customers.

Stand-alone WTs are typically used for water pumping or communications. However, homeowners, farmers, and ranchers in windy areas can also use WTs as a way to cut their electric bills.

Small wind systems also have potential as DERs. DERs refer to a variety of small, modular power-generating technologies that can be combined to improve the operation of the electricity delivery system.

Although all WTs operate on similar principles, several varieties are in use today. These include horizontal axis turbines and vertical axis turbines.

Horizontal Axis Turbines

Horizontal axis turbines are the most common turbine configuration used today. They consist of a tall tower, atop which sits a fan-like rotor that faces into or away from the wind, a generator, a controller, and other components. Most horizontal axis turbines built today are two- or three-bladed.

Horizontal axis turbines sit high atop towers to take advantage of the stronger and less turbulent wind at 100 feet (30 m) or more above ground. Each blade acts like an airplane wing, so when wind blows, a pocket of low-pressure air forms on the downwind side of the blade. The low-pressure air pocket then pulls the blade toward it, which causes the rotor to turn. This is called lift. The force of the lift is actually much stronger than the wind's force against the front side of the blade, which is called drag. The combination of lift and drag causes the rotor to spin like a propeller, and the turning shaft spins a generator to make electricity.

Vertical Axis Turbines

Vertical axis turbines are of two types: *Savonius* and *Darrieus*. Neither type is in wide use.

The *Darrieus* turbine was invented in France in the 1920s. Often described as looking like an eggbeater, it has vertical blades that rotate into and out of the wind.

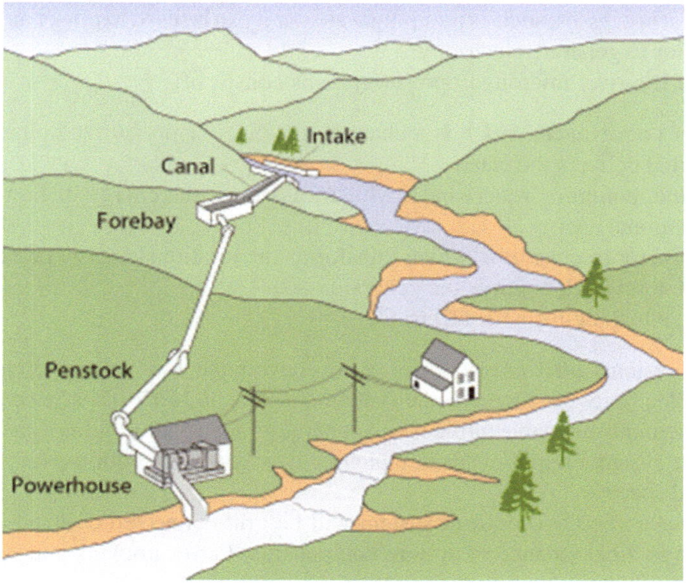

Fig. 2.4 Schematic of run-of-the-river microhydropower systems

Using aerodynamic lift, it can capture more energy than drag devices. The Giromill and cycloturbine are variants on the Darrieus turbine.

The *Savonius* turbine is S-shaped if viewed from above. This drag-type turbine turns relatively slowly but yields a high torque. It is useful for grinding grain, pumping water, and many other tasks, but its slow rotational speeds are not good for generating electricity.

In addition, windmills are still used for a variety of purposes. Windmills have been used by humans since at least 200 B.C. for grinding grain and pumping water. By the 1900s, windmills were used on farms and ranches in the United States to pump water and, later, to produce electricity. Windmills have more blades than modern WTs, and they rely on drag to rotate the blades.

2.2.2 Small-Scale Hydrogeneration

All hydropower systems use the energy of flowing water to produce electricity or mechanical energy. Although there are, several ways to harness moving water to produce energy, "run-of-the-river systems," which do not require large storage reservoirs, are most often used for microhydropower systems (Fig. 2.4).

For run-of-the-river microhydropower systems, a portion of a river's water is diverted to a water conveyance channel, pipeline, or pressurized pipeline (called a penstock) that delivers it to a turbine or waterwheel. The moving water rotates the wheel or turbine, which spins a shaft. The motion of the shaft can be used for mechanical processes, such as pumping water, or it can be used to power an alternator or generator to generate electricity.

Run-of-the-river microhydropower systems consist of:

- A water conveyance, which is a channel, pipeline, or pressurized pipeline (penstock) that delivers the water
- A turbine, pump, or waterwheel, which transforms the energy of flowing water into rotational energy
- An alternator or generator, which transforms the rotational energy into electricity
- A regulator, which controls the generator
- Wiring, which delivers the electricity.

Many systems also use an inverter to convert the low-voltage direct current (LVDC) electricity produced by the system into 120 or 240 V of AC electricity.

Commercially available turbines and generators are usually sold as a package. Do-it-yourself systems require careful matching of a generator with the turbine horsepower and speed.

Whether a microhydropower system will be grid-connected or stand-alone will determine its final balance of system components. For example, some stand-alone systems use batteries to store the electricity generated by the system. However, because hydropower resources tend to be more seasonal in nature than wind or solar resources, batteries may not always be practical. If batteries are used, they should be located as close to the turbine as possible because it is difficult to transmit low-voltage (LV) power over long distances. Dams or diversion structures are rarely used in microhydropower projects. They are an added expense and require professional assistance from a civil engineer. In addition, dams increase the potential for environmental and maintenance problems.

2.3 Solar Photovoltaic Generation

The modern form of the PV, also called solar cells, was invented in 1954 at Bell Telephone Laboratories. Today, PV is one of the fastest growing renewable energy technologies and it is expected that it will play a major role in the future global electricity generation mix [16].

PV technology offers a number of significant benefits, including:

- Solar power is a renewable resource that is available everywhere in the world.
- Solar PV technologies are small and highly modular and can be used virtually anywhere, unlike many other electricity generation technologies.
- Unlike conventional power plants using coal, nuclear, oil and gas; solar PV has no fuel costs and relatively low operation and maintenance (O&M) costs. PV can therefore offer a price hedge against volatile fossil fuel prices.
- PV, although variable, has a high coincidence with peak electricity demand driven by cooling in summer and year round in hot countries.

2.3.1 Technology Basics

Photo of a large silicon solar array on a roof with a blue sky and trees in background. A large silicon solar array installed on the roof of a commercial building. Photo of a traditional-looking home with blue solar tiles integrated into the brown roof. Thin-film solar tiles installed on the roof of a home in Ohio. Photo of a long, blue solar array in a field of grass. A large solar array in Germany.

Solar cells, also called PV cells by scientists, convert sunlight directly into electricity. PV gets its name from the process of converting light (photons) to electricity (voltage), which is called the PV effect. The PV effect was discovered in 1954, when scientists at Bell Telephone discovered that silicon (an element found in sand) created an electric charge when exposed to sunlight. Soon solar cells were being used to power space satellites and smaller items like calculators and watches. Today, thousands of people power their homes and businesses with individual solar PV systems. Utility companies are also using PV technology for large power stations. Solar panels used to power homes and businesses are typically made from solar cells combined into modules that hold about 40 cells. A typical home will use about 10 to 20 solar panels to power the home. The panels are mounted at a fixed angle facing south, or they can be mounted on a tracking device that follows the sun, allowing them to capture the most sunlight. Many solar panels combined together to create one system is called a solar array. For large electric utility or industrial applications, hundreds of solar arrays are interconnected to form a large utility-scale PV system.

Traditional solar cells are made from silicon, are usually flat plate, and generally are the most efficient. Second-generation solar cells are called thin-film solar cells because they are made from amorphous silicon or non-silicon materials such as cadmium telluride. Thin-film solar cells use layers of semiconductor materials only a few micrometers thick. Because of their flexibility, thin-film solar cells can double as rooftop shingles and tiles, building facades, or the glazing for skylights.

Third-generation solar cells are being made from a variety of new materials besides silicon, including solar inks using conventional printing press technologies, solar dyes, and conductive plastics. Some new solar cells use plastic lenses or mirrors to concentrate sunlight onto a very small piece of high efficiency PV material. The PV material is more expensive, but because so little is needed, these systems are becoming cost effective for use by utilities and industry. However, because the lenses must be pointed at the sun, the use of concentrating collectors is limited to the sunniest parts of the country.

PV cell technologies are usually classified into three generations, depending on the basic material used and the level of commercial maturity:

- **First-generation PV systems** (fully commercial) use the wafer-based crystalline silicon (c-Si) technology, either single crystalline (sc-Si) or multicrystalline (mc-Si).
- **Second-generation PV systems** (early market deployment) are based on thin-film PV technologies and generally include three main families:

Fig. 2.5 A large silicon solar array installed on the roof of a commercial building

1. Amorphous (a-Si) and micromorph silicon (a-S_i/c-S_i);
2. Cadmium–Telluride (C_dT_e); and
3. Copper–Indium–Selenide (CIS) and Copper–Indium–Gallium–Diselenide (CIGS).

- **Third-generation PV systems** include technologies, such as concentrating PV (CPV) and organic PV cells that are still under demonstration or have not yet been widely commercialized, as well as novel concepts under development (Fig. 2.5).

2.3.2 Grid-Connected Solar Systems

Solar panels that harness the sun's power to generate electricity provide clean power for homes, communities and businesses, and help cut global carbon emissions.

Solar PV modules generate electricity from sunlight, which can be fed into the mains electricity supply of a building or sold to the public electricity grid. Reducing the need for fossil fuel generation, the growing grid-connected solar PV sector across the globe is helping create jobs, enabling families and businesses to save money, and cut greenhouse emissions.

PV modules use semiconductor materials to generate DC electricity from sunlight. A large area is needed to collect as much sunlight as possible, so the semiconductor is either made into thin, flat, crystalline cells, or deposited as a very thin continuous layer onto a support material. The cells are wired together and sealed into a weather-proof module, with electrical connectors added. Modern modules for grid connection

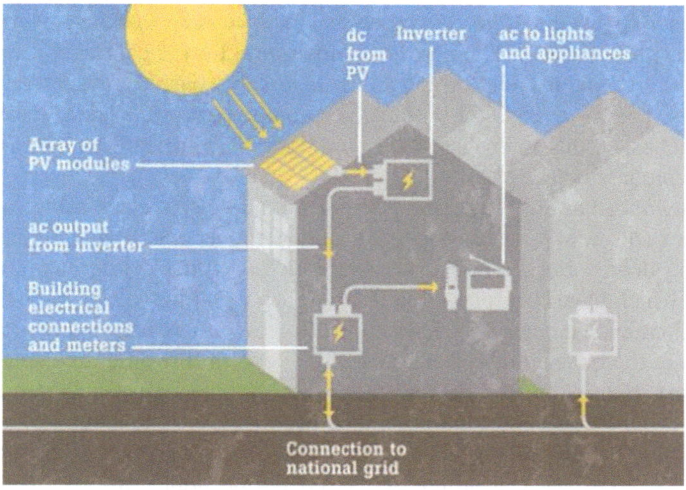

Fig. 2.6 PV connection system

usually have between 48 and 72 cells and produce DC voltages of typically 25–40 volts, with a rated output (see box) of between 150 and 250 Wp.

In order to supply electricity into a mains electricity system, the DC output from the module must be converted to AC at the correct voltage and frequency. An electronic inverter is used to do this. Generally, a number of PV modules are connected in series to provide a higher DC voltage to the inverter input, and sometimes several of these series strings are connected in parallel, so that a single inverter can be used for 50 or more modules. Modern inverters are very efficient (typically 97 %), and use electronic control systems to ensure that the PV array keeps working at its optimum voltage. They also incorporate safety systems as required in the country of use (Fig. 2.6).

PV modules are specified by their watt-peak (Wp) rating, which is the power generated at a solar radiation level of 1000 W/m^2, equivalent to bright sun in the tropics. They still work fine with less solar radiation. The voltage produced by a PV module is largely determined by the semiconductor material and the number of cells, and varies only slightly with the amount of solar radiation. The electrical current and the power generated are proportional to the amount of solar radiation.

2.3.3 *Future Trends*

Grid-connected systems do not usually include batteries for storage, because the mains grid can accept or provide power as needed. However, if rechargeable batteries are included, a grid-connected PV system can be used as a stand-alone AC supply in the event of a power cut, to allow essential loads to keep working. Ashden winner Deng solar provided a 9.2 kWp grid backup system for the central courts in Accra, Ghana, which maintains lighting and thus enables court business to keep going during power cuts. The Aryavart Gramin Bank has provided PV grid backup systems for its rural branches, so that their IT systems and cash machines still work during power cuts and voltage fluctuations.

Looking ahead at the future, the price of PV modules is decreasing rapidly. For crystalline cells, new ways of processing silicon and increased volume manufacture are driving down prices. The market share of thin-film PV is growing rapidly as materials which have been proved in the laboratory go into volume production, and these promise even greater price reductions. However, there is less potential for price reduction in the balance of system, and these costs will soon dominate the overall system cost. Because of the decreasing prices, the rapid growth in the market for grid-connected PV is expected to continue even if government support is reduced. The market will really take off when electricity from PV becomes cheaper than other grid sources. When PV feeds directly into a building supply, this grid parity price is the consumer purchase price, currently 10–20 US cents per kWh. A recent roadmap by the International Energy Agency suggests that this point may be reached in sunny countries by 2020. For systems connecting directly to the national grid, the grid parity price is less than 5 US cents per kWh. The roadmap suggests that even this point could be reached by 2030, and that PV could then be supplying about 5 % of global electricity.

2.4 Small Wind Turbine Systems

In the past two decades, we have witnessed a dramatic increase in the penetration of the WT power generation worldwide. In turn, this growing trend has a major impact on stimulating the research in the power processing field aiming at optimizing the energy extraction from the wind and the energy injection into the grid [41]. Particular attention has been recently focused on DG through small WTs (power unit 200 kW) because of: lower impact on the landscape, lower noise level, grid codes, and national laws imposing simpler grid connection and higher feed-in tariffs and capability to work in island mode for isolated communities [34]. In general, it can be observed that the maximum WT size increases in the case of wind diesel system with battery, and the maximum WT size increases in case of grid connection. The field of small generation was dominated by the use of asynchronous generators directly connected to the grid/load and more recently by PMSG with a diode rectifier, boost converter and inverter [45]. The use of an high number of pole pairs allow the PMSG to operate at low speed without decreasing the efficiency, thus allowing to avoid the gearbox [24]. The use of a diode bridge reduces cost and control algorithm complexity [12].

Moreover, the well-known six-pulse DC voltage waveform allow to implement a simple estimator of the rotor speed. The generator low frequency harmonics (5th and 7th) can be reduced, thanks to the DC inductor that unfortunately has a negative effect on the power factor. Moreover, the extracted power decreases as the wind speed increases due to the major effect of diode commutation and at low speed due to possible discontinuous operation of the DC/DC converter. Hence, for power levels in the order of tens of kW, these generators will use a back-to-back converter leading to a 5, 15 % more power. Besides the choice of the power converter configuration, the control issues in PMSG-based small WT systems are many [14, 21], as shown Fig. 2.8. Simultaneously, in the literature, a lot of attention is committed to DSP and FPGA control system implementation of generator and converters for WT applications [5, 20, 44].

On the generator side two main control issues are present:

- the tracking of the maximum power without using an anemometer, and
- the estimation of the rotor position and speed since the use of a sensor is not always practical.

In case of variable wind conditions that are the most common at low altitude, control strategies involving maximum power point tracking (MPPT) algorithms based on an operational seeking algorithm or a wind speed estimator can be implemented [1, 13, 19, 24, 42, 46].

2.4.1 Types of Wind Turbine Systems

Several types of generators can be adopted in wind power turbines: DC and AC types, parallel and compound DC generators, permanent magnets or electrical field excited, asynchronous or synchronous generators. The right choice depends on the primary source, the type of load and the speed of the turbine. Besides, systems differ with respect to their applications, whether they are stand-alone or connected to the grid [33]. The most adopted WT generator for medium power systems is the doubly fed induction generator (DFIG) that could be used even in stand-alone operation [11, 17, 25]. However, typically, in small WTS for stand-alone applications, the choice is between variable-speed asynchronous and synchronous generators [40]. Variable-speed asynchronous generators can easily operate in parallel with large power systems, since the utility grid controls voltage and frequency, while static and reactive compensating capacitors can be used for correction of the power factor and harmonic reduction. Abrupt speed changes due to load or primary source changes are easily absorbed by the solid rotor of the asynchronous generator, and current surges can be effectively damped without demagnetization issues [39].

Over the last years, the most common solution for wind generator is based on permanent magnet SMs directly connected to the turbine [8, 23, 32, 43]. This solution is particularly favorable as it reduces the frequency and the costs of maintenance operations. In fact these generators are self-excited, allowing operation at high-power factor and high efficiency. Moreover, multipolar PMSG does not need the gearbox to adapt the rotor speed to the blade speed. This advantage becomes crucial for WT installation in harsh environment characterized by low temperature (Fig. 2.7).

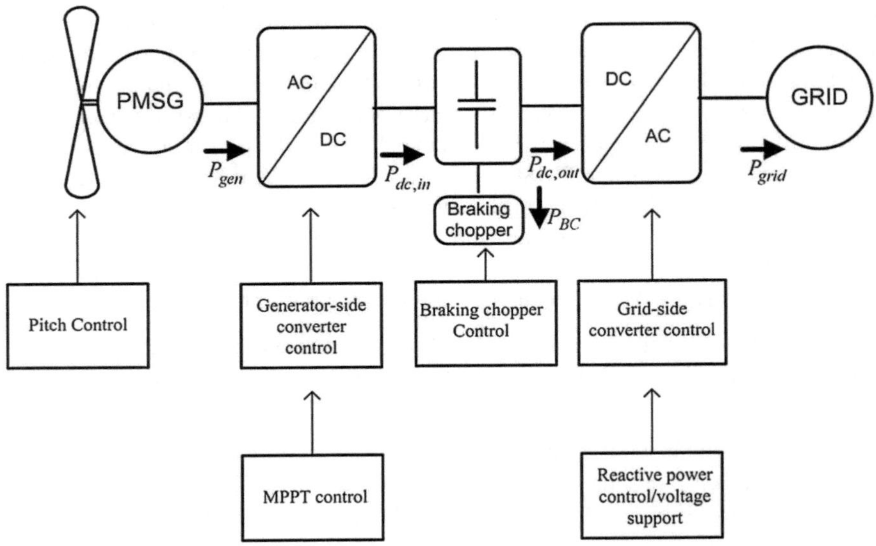

Fig. 2.7 Control issues in PMSG-based small wind turbine systems

Figure 2.8a shows a PMSG connected to the grid through three converters: a diode bridge rectifier, a boost converter, and an inverter. The DC/DC boost converter is used to control the generator [3, 18].

Figure 2.8b shows a PMSG connected to the grid through a bidirectional converter. It consists of two VSI-PWM converters connected by a storage capacitor: the converter connected to the generator is used as a rectifier, while the converter connected to the grid is used as an inverter. To achieve full control of the output, the DC-link voltage must be boosted to a level higher than the amplitude of the grid voltage. The power flow of the grid-side converter is controlled in order to keep the DC-link voltage constant, while the control of the generator side is set to suit the magnetization demand and the reference speed or torque. A technical advantage of this topology is the capacitor decoupling between the grid converter and the generator converter. Besides affording some protection, this decoupling offers separate control of the two converters, allowing compensation of asymmetry both on the generator side and on the grid side, independently.

The two above-mentioned topologies give rise to different efficiencies in the power processing, as shown by Fig. 2.9, where the mechanical power and the power injected into the grid are shown in case of a diode bridge rectifier plus boost converter (a) and back-to-back converter (b).

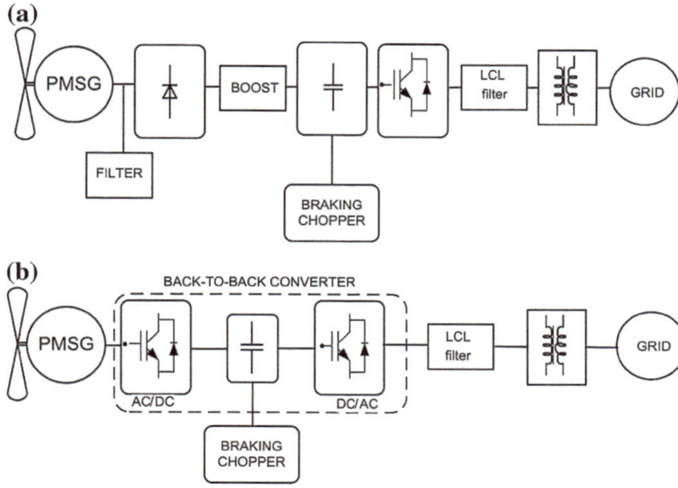

Fig. 2.8 Wind turbine system. **a** Diode bridge rectifier and boost converter. **b** Back-to-back converter

Fig. 2.9 Comparison between the mechanical power and the grid power. **a** Diode bridge rectifier. **b** Back-to-back converter

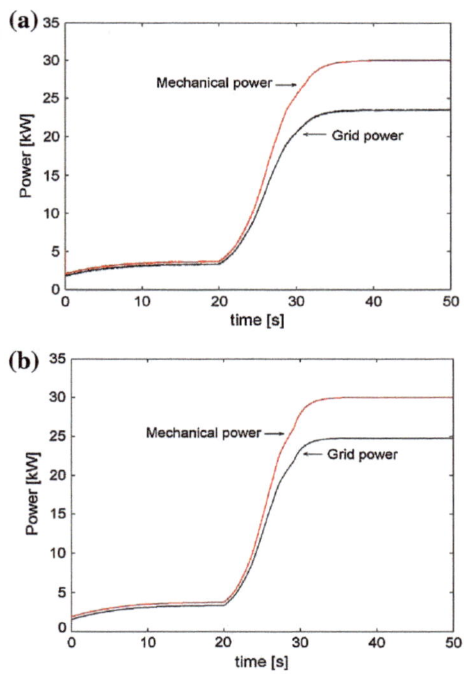

2.4.2 Wind Turbine Fundamentals

This section makes use of the tutorial paper [35]. Wind is recognized worldwide as a cost-effective, environmentally friendly solution to energy shortages. According to the World Wind Energy Association (WWEA), the global installed capacity of WTs grew at an average rate of 27 % per year over the years 2005–2009. Although a WT can be built in either a vertical axis or horizontal axis configuration, focus will be on horizontal axis wind turbines (HAWTs) because they dominate the utility-scale WT market. At the utility scale, HAWTs have aerodynamic and practical advantages [9, 22]. The generating capacity of commercially available HAWTs ranges from less than 1 kW to several megawatts.

The main components of a HAWT that are visible from the ground are the tower, nacelle, and rotor, as shown in Fig. 2.10. The airfoil-shaped blades capture the kinetic energy of the wind and transform it into the rotational kinetic energy of the WTs rotor. The rotor drives the low-speed shaft, which in turn drives the gearbox. The gearbox steps up the rotational speed and drives the generator by means of the high-speed shaft. The gearbox, high-speed shaft, and generator are housed in the nacelle, along with part of the low-speed shaft. Direct drive configurations without gearboxes are being developed to eliminate costly gearbox failures.

WTs may be variable or fixed speed. Variable-speed turbines tend to operate closer to their maximum aerodynamic efficiency for a higher fraction of the time but require electrical power processing so that the generated electricity can be fed into the electrical grid at the proper frequency. Variable-speed turbines are more

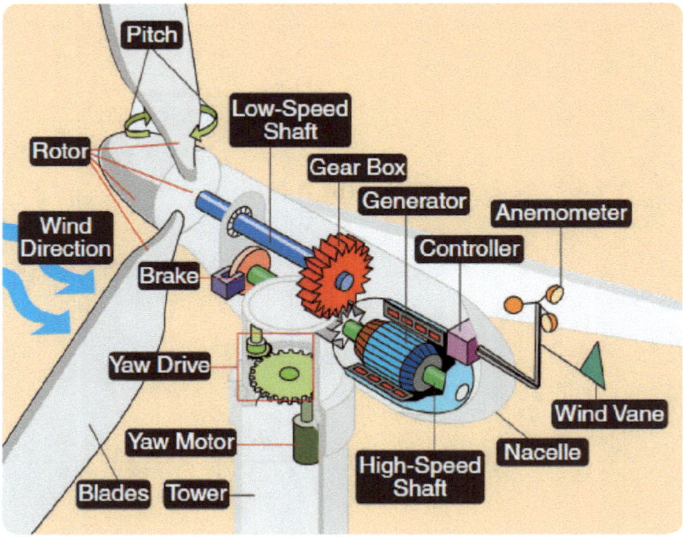

Fig. 2.10 Comparison between the mechanical power and the grid power. **a** Diode bridge rectifier. **b** The main components of a HAWT

cost-effective and thus more popular than constant-speed turbines at the utility scale because of improvements in generator and power electronics technologies. Variable-speed operation can also reduce turbine loads, since sudden increases in wind energy due to gusts can be absorbed by an increase in rotor speed rather than by component bending.

The goals and strategies of WT control are affected by the turbine configuration. A HAWT can be upwind, with the rotor on the upwind side of the tower, or downwind, with the rotor on the downwind side of the tower. This choice affects the turbine dynamics and thus the structural design. A WT can also be variable pitch or fixed pitch, meaning that the blades may or may not be able to rotate about their longitudinal axes. Variable-pitch turbines might allow all or part of their blades to rotate along the pitch axis. Fixed pitch machines are less expensive to build, but the ability of variable-pitch turbines to mitigate loads and affect the aerodynamic torque has driven their dominance in modern utility-scale turbine markets.

2.4.3 Control Loops

The main loops of WT control systems are shown in Fig. 2.11. Since the wind speed varies across the rotor plane, wind speed point measurements convey only a small part of the information about the wind inflow. Rotor speed is the only measurement required for the baseline generator torque and blade pitch controllers.

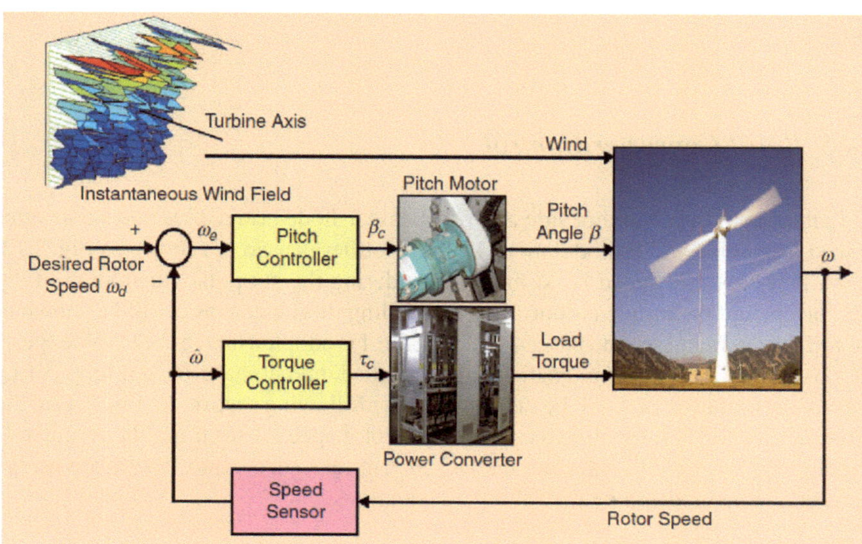

Fig. 2.11 Wind turbine control feedback loops

2.4.4 Generator Side Control

In this section, the generator model and its control is presented. The generator is modeled by the following voltage equations in the rotor reference frame (dq-axes) [7]:

$$\begin{aligned} v_{sd} &= -R_s i_{sd} - \frac{\lambda_{sd}}{dt} + \omega_r \lambda_{sq} \\ v_{sq} &= -R_s i_{sq} - \frac{\lambda_{sq}}{dt} + \omega_r \lambda_{sd} \end{aligned} \tag{2.2}$$

where λ_{sd} and λ_{sq} are d and q flux linkages

$$\lambda_{sd} = L_s i_{sd} + \psi_{PM}, \qquad \lambda_{sq} = L_s i_{sq} \tag{2.3}$$

with ψ_{PM} is the flux of the permanent magnets. Let m_s be the number of phases and n_p be the pole pairs, then the electromagnetic torque is

$$T_e = \frac{m_s n_p}{2} (\lambda_{sq} i_{sd} - \lambda_{sd} i_{sq}) \tag{2.4}$$

Using (2.3), we have

$$T_e = -\frac{m_s n_p}{2} \psi_{PM} i_{sq} = -K_c i_{sq} \tag{2.5}$$

where K_c is the torque constant representing the proportional coefficient between T_e and i_{sq}.

2.4.5 Boost Converter Control

The schemes of the generator side control in case a diode bridge plus boost converter is used or a pulse-width modulation (PWM) rectifier is used are shown in Fig. 2.12. In the first case, the speed is estimated considering the six pulse fluctuation in the DC current and the torque is controlled controlling this current as it will be shown in the following. In the second case, a vector control is adopted hence the knowledge of the position of the electromagnetic force is needed, and a phase-locked loop (PLL) is used to estimate it as it will be discussed in the following section. In Fig. 2.12a, the controller for the first topology is shown. The rotor speed determines the torque reference, which is used to calculate the DC current reference value. The diode bridge rectifier can be modeled by the following equations:

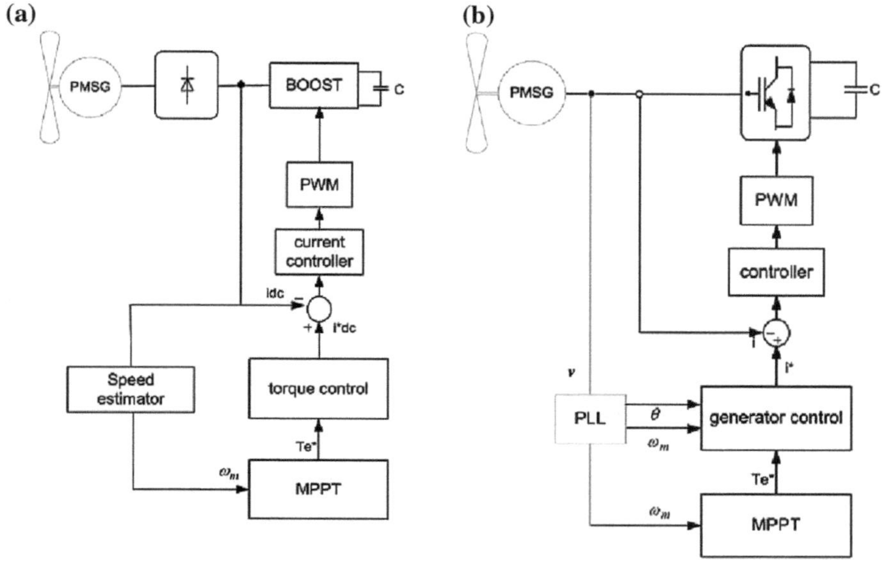

Fig. 2.12 Generator side converter control. **a** Diode bridge rectifier and boost converter. **b** Back-to-back converter

$$v_{dc} = v_a S_a + v_b S_b + v_c S_c \tag{2.6}$$

$$\begin{aligned} i_a &= S_a i_{dc}, \\ i_b &= S_b i_{dc}, \\ i_c &= S_c i_{dc}, \end{aligned} \tag{2.7}$$

In effect, (2.7) and (2.7) represent the rectifier **DC** voltage and the rectifier **AC** currents based on the switching functions

$$\begin{aligned} S_a &= \sum_{k=1}^{\infty} \hat{S_{ak}} sin(k\omega t), \\ S_b &= \sum_{k=1}^{\infty} \hat{S_{bk}} sin\Big(k(\omega t - \frac{2\pi}{3})\Big), \\ S_c &= \sum_{k=1}^{\infty} \hat{S_{ck}} sin\Big(k(\omega t - \frac{4\pi}{3})\Big), \end{aligned} \tag{2.8}$$

For $k = 1$, it yields

$$S_{a1} = \frac{2\sqrt{3}}{\pi} sin(\omega t), \tag{2.9}$$

It is easy to show that the relation existing between I_1, the fundamental component of the **AC** current and I_{ds}, the **DC** current is

$$I_1 \approx 1.1 I_{dc}$$

Further, it follows that $I_{sq} = K'' \; I_1$, where K'' represents the proportional coefficient between I_{sq} and I_1. This eventually leads to the relation between the torque T_e and the **DC** current as

$$T_e = 1.1 K_c K'' I_{dc} \tag{2.10}$$

2.4.6 Rectifier Control

With reference to Fig. 2.12b, the torque control can be achieved through the control of the current i_{sq}. The generator currents can be controlled as follows:

$$\begin{aligned} v_{sd}^* &= v_{sd} \; - \; \omega_r L_s i_{sq} \\ v_{sq}^* &= v_{sq} \; + \; \omega_r L_s i_{sd} \; + \; \omega_r \Psi_{PM} \end{aligned} \tag{2.11}$$

A good selection of the control variables v_{sd}^* and v_{sq}^* leads to a decoupled control of the two current components

$$\begin{aligned} i_{sd} &= \frac{-v_{sd}^*}{sL_s - R_s} \\ i_{sq} &= \frac{-v_{sq}^*}{sL_s - R_s} \end{aligned} \tag{2.12}$$

It follows that the stator q axis current component is used to control the generator torque, but a freedom degree remains to set DC that can be used for flux weakening [41].

2.5 Storage Technologies

Storage is important in the MG both because peak loads are expensive to serve with purchased power and because MG generation sources may not be able to respond to load changes as needed [15]. Load changes are usually caused by short-lived events, such as fast transients resulting from starting of motors or turning on/off of equipment, or from slower changes that exceed the ramping capability of generation available at any given time. All the storage systems mentioned in the sections below require power electronics to convert the stored power to standard, 60-Hz, AC, utility-grade power. These systems can be designed to switch into operation in subcycle time frames, so they are ideal for tracking fast load changes or immediately providing backup if utility power is lost.

2.5.1 Classification of Electrical Energy Storage

Electrical energy storage (EES) has played three main roles.

1. First, EES reduces electricity costs by storing electricity obtained at off-peak times when its price is lower, for use at peak times instead of electricity bought then at higher prices.
2. Secondly, in order to improve the reliability of the power supply, EES systems support users when power network failures occur due to natural disasters, for example.
3. Their third role is to maintain and improve PQ, frequency and voltage.

A widely used approach for classifying EES systems is the determination according to the form of energy used. In Fig. 2.13, EES systems are classified into mechanical, electrochemical, chemical, electrical, and thermal energy storage (TES) systems. Hydrogen and synthetic natural gas are secondary energy carriers and can be used to store electrical energy via electrolysis of water to produce hydrogen and, in an additional step, methane if required. In fuel cells electricity is generated by oxidizing hydrogen or methane. This combined electrolysis-fuel cell process is an electrochemical EES. However, both gases are multipurpose energy carriers. For example, electricity can be generated in a gas or steam turbine. Consequently,

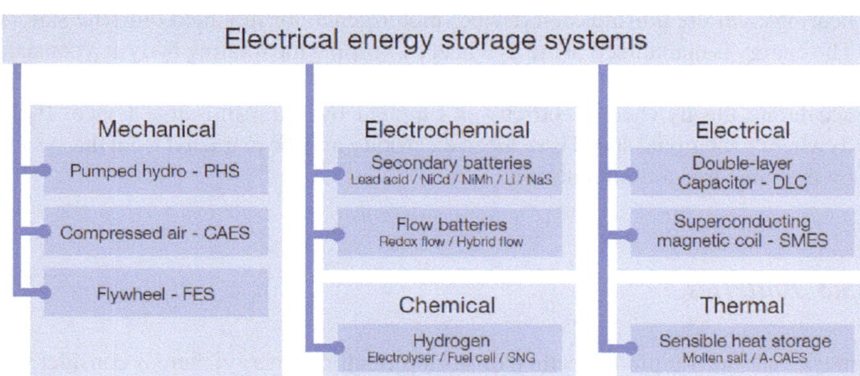

Fig. 2.13 Electrical energy storage

they are classified as chemical energy storage systems. In Fig. 2.13 TES systems are included as well, although in most cases electricity is not the direct input to such storage systems. But with the help of TES the energy from renewable energy sources (RESs) can be buffered and thus electricity can be produced on demand. Examples are hot molten salts in concentrated solar power plants and the storage of heat in compressed air plants using an adiabatic process to gain efficiency.

2.5.2 Mechanical Storage Systems

The most common mechanical storage systems are pumped hydroelectric power plants (pumped hydro storage), compressed air energy storage, and flywheel energy storage.

- Pumped hydro storage power plants represent nearly 99 % of worldwide installed electrical storage capacity, which is about 3 % of global generation capacity. Conventional pumped hydro storage systems use two water reservoirs at different elevations to pump water during off-peak hours from the lower to the upper reservoir (charging). When required, the water flows back from the upper to the lower reservoir, powering a turbine with a generator to produce electricity (discharging).
- In compressed air (compressed gas) energy storage, air is used as storage medium due to its availability. Electricity is used to compress air and store it in either an underground structure or an aboveground system of vessels or pipes. When needed the compressed air is mixed with natural gas, burned and expanded in a modified GT. Typical underground storage options are caverns, aquifers or abandoned mines.
- In flywheel energy storage rotational energy is stored in an accelerated rotor, a massive rotating cylinder. The main components of a flywheel are the rotating body/cylinder (comprised of a rim attached to a shaft) in a compartment, the bearings and the transmission device (motor/generator mounted onto the stator). The energy is maintained in the flywheel by keeping the rotating body at a constant speed. An increase in the speed results in a higher amount of energy stored. To accelerate the flywheel electricity is supplied by a transmission device. If the flywheel's rotational speed is reduced electricity may be extracted from the system by the same transmission device.

2.5.3 Batteries

Batteries are the traditional method of storing electrical energy; there is considerable operational experience with battery systems. Lead–acid batteries, available in almost any size, are used in many applications that require backup power. Batteries using other chemistries are now also available commercially. Recent improvements have increased energy storage density and extended batteries lifetimes. Discharge rates are determined by the batterys design and the chemical reactions used for energy storage.

Batteries store energy in chemical form and are charged/discharged with DC current. This DC current is converted to standard, 60-Hz, AC electrical power by means of power electronics. Most commercial uninterruptible power supplies rely on batteries.

2.5.4 *Flywheels*

Many improvements have been made to flywheel systems in recent years. These systems now incorporate composite rotors, magnetic bearings, and advanced power electronics. Flywheels store energy in high-speed (up to 100,000-rpm) rotating wheel-like rotors or disks connected to motor/generators. High-speed rotation is important because the amount of power stored in the flywheel is proportional to the square of the rotational speed. The flywheel is charged by taking utility power and converting it to drive the flywheel motor, which increases flywheel speed. During a "discharge," power is drawn from the flywheel by the generator, which slows the rotor speed. Because the output of the flywheel generator is variable, an inverter is used to convert power to standard, 60-Hz or 50-Hz AC power.

Flywheel systems come in a wide range of sizes with differing discharge rates for different amounts of time. The flywheel stores a fixed amount of energy (kWh). It can be discharged at high power (kW) for a short time or at a slower rate for a longer period. Flywheels contain no hazardous materials and are not affected by temperature extremes as batteries are, but flywheels are costly and cannot store energy indefinitely.

2.5.5 *Superconducting Magnetic Energy Storage*

Superconductors allow the passage of electrical current without losses. Electrical energy is stored as a circulating current in a superconducting coil of wire. This circulating current establishes a magnetic field in which the energy is stored. The major energy loss in this system results from the need to cool the coil to very low temperatures. PEIs charge and discharge the superconducting coil. Most commercial systems are somewhat larger than 250 kW in capacity. Superconductor storage technology could be adapted to larger MG applications.

2.5.6 Supercapacitors

Supercapacitors are very-high-capacity electrolytic devices that store energy in the form of electrostatic charge. They are composed of two electrodes with a very thin separator. Energy storage capacity increases as the surface area of the electrodes increases. Energy is stored as a DC field in the supercapacitor, and the system uses power electronics to both charge and discharge the capacitors. Supercapacitors can have very high discharge rates and could handle fast load changes in a MG.

2.6 Inverter Interfaces

There are two basic classes of microsources:

- DC sources, such as fuel cells, PV cells, and battery storage; and
- variable high-frequency AC sources such as a microturbines, whose output is converted to DC.

In both cases, the DC voltage that is generated needs to be interfaced to the AC network and its loads. Power electronics provide this interface as a converter periodically switches the DC voltage polarity on the AC side to create an AC waveform of desired magnitude and phase.

A basic understanding of power electronics requires understanding of the creation of waveforms of different magnitudes, phases, and frequencies using circuits containing rapidly acting switches and energy storage elements. Power electronic devices are designed to operate like switches, but, because these devices are made of semiconducting materials like silicon, they can function much more rapidly than mechanical switches. Energy storage elements, such as inductors and capacitors, filter the sharp-edged waveforms created by the switching.

2.6.1 Voltage Source Inverters

A typical circuit and switching sequence is needed that can convert the DC voltage from a microsource to three-phase AC voltage. Consider the circuit in Fig. 2.14a.

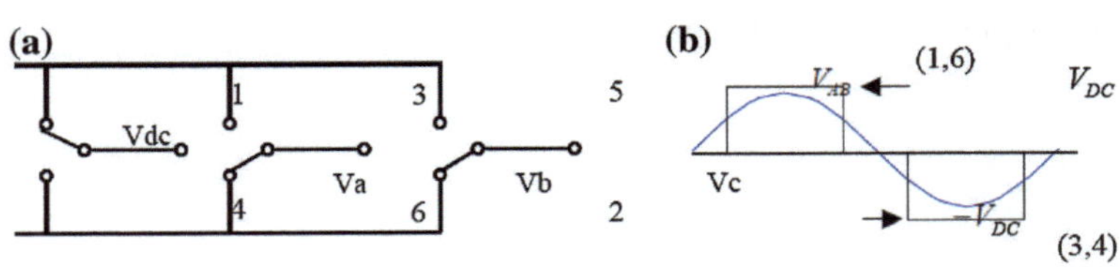

Fig. 2.14 Voltage synthesis. **a** Circuit. **b** Line-to-line voltage

On the left is a DC voltage that is provided by the microsource and connected to a three-phase AC system using double-pole switches. In reality, these switches are power electronic devices with bidirectional current flow capability and rapid switching speeds. In the current positions (1,6,2), the three-phase line-to-line voltage is: $V_{AB} = V_{DC}, V_{BC} = 0\, and\, V_{CA} = V_{DC}$. This strategy allows the converter to synthesize three AC square waves of voltage at the correct phase to each other. Figure 2.14b shows the VAB synthesized square wave. The positive voltage is achieved by the switch positions shown (1,6); negative voltage requires the opposite (3,4) position.

Although a square wave allows full control of rms voltage output and phase, it would require a great deal of filtering to provide the loads with the required sinusoidal waveforms. Power electronic switching devices have the ability to switch much more rapidly than the fundamental frequency, which means that PWM is an option. For example, for the positive voltage section, Fig. 2.14b, the inverter can rapidly switch from the (1,6) position to the (1,3) or (4,6) position, which provides zero voltage between phases $A\&B$. The same can be done on the negative side. This allows the instantaneous average output to be held closer to the desired fundamental output. A converter incorporating PWM considerably less filtering to achieve the required PQ.

In general, PWM is limited by the switching frequency of the power electronic device and the techniques of the controller. Typical switching frequencies are at least 30 times faster than the fundamental frequency. During each switching period, the inverter control selects the times of conduction or duty cycles that create the desired voltage for that period. The resulting voltage is made of pulses of different widths (hence the name PWM). The pattern and number of pulses are designed to provide the required voltage magnitude, and the pulses are placed to minimize the harmonic content. Such a pattern is shown in Fig. 2.15 for all three phases. The arrows indicate the switch position shown in Fig. 2.14a.

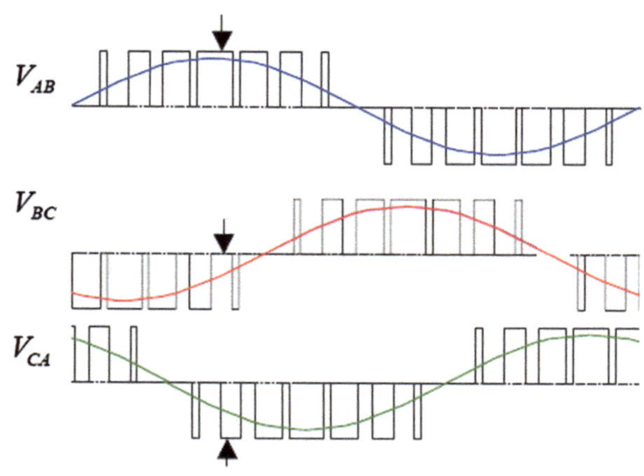

Fig. 2.15 Three-phase PWM voltage

2.6.2 Inverter Realization for Microsources

2.6.3 Inverter Realization

To realize an actual voltage-sourced inverter (VSI) with PWM, two key issues must be addressed: implementation of switches and connection to the customers AC system. The switches shown in Fig. 2.14 are implemented using insulated gate bipolar junction transistor (IGBT) and diodes. The advantages of IGBTs include their simple gate drives derived from voltage control requirements and their ample voltage/current ratings up to 3,000 V and 1,200 A. Their switching times are less than one microsecond. IGBTs with reverse diodes are shown in place of switches in Fig. 2.16.

Microsources such as microturbines, fuel cells, and PV systems seldom exceed 200 kw and in many cases are smaller than 100 kw. This low level of power along with the need to utilize waste heat means that these sources are placed at the customers site rather than at the utilities substation. AC systems are usually 480 volts or less with a four-wire configuration to accommodate single-phase loads. This requirement can be met using the three-legged inverter shown in Fig. 2.14 with the addition of a star-delta transformer the center point of the star can be used to provide the extra wire (see Fig. 2.16a). Another possibility is to add a fourth leg to provide the extra connection point (see Fig. 2.16b). The operation is similar to one discussed in the section on VSIs above except that the DC voltage is always switched between the neutral and single-phase. This allows for direct creation of phase-to-neutral voltages rather than line-to-line voltage provided by a three-legged inverter.

The differences between the two circuits in Fig. 2.16 include:

- the use of magnetic versus extra IGBTs and diodes, and
- the handling of fault current and dependents between DC and AC voltage levels.

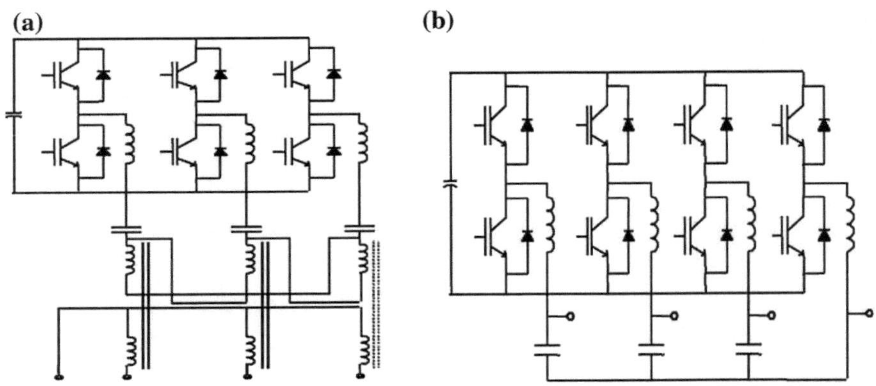

Fig. 2.16 PWM inverters with four-wire interface, **a** Using star-delta transformer, **b** direct connection using four-legged inverter

During a fault, fault current will be flowing in the four-legged inverter; in transformer-coupled systems, the fault current will circulate in the delta winding. For the transformer-coupled system the DC voltage becomes a free variable because of the turns ratio. In the direct-coupled system the DC voltage needs to be 10–20 percent larger than the required peak AC voltage.

2.6.4 Unbalanced AC Voltages

Microsources with AC voltages of 480 V or less and a four-wire configuration to accommodate single-phase loads have unbalanced voltages because of asymmetries in the wiring and the presence of unbalanced loads. When these microsources are connected to the AC system using the inverters discussed above, there will be uneven phase currents. However, microsources may be intolerant of voltage imbalances. Field experience with ONSI fuel cells, for example, has shown that imbalances trip the current protection in the inverter because it assumes balanced currents. (This is the norm for adjustable speed drive protection)

Both inverters shown in Fig. 2.16 can be represented as shown in Fig. 2.17 to demonstrate interactions with the AC system. The VSI creates three AC voltages, $V_{a,b,c}^{inverter}$ that are coupled to the AC system through three inductors, X. The current that flows in each inductor is dependent on the inverter and AC system voltages. If the AC system voltages are balanced and the inverter creates balanced voltages, the currents are equal. When the AC voltages are unbalanced, the currents also become unbalanced. The inverter has the flexibility to create unbalanced voltages between phases that can:

- Rebalance the output currents,
- Correct the systems voltage imbalance,
- Regulate the positive sequence AC voltage but not correct for the imbalance, and
- Remove the negative sequence AC voltage component that results from a voltage dip.

In general, VSIs have the flexibility to deal with most unbalanced situations seen in the field, but this flexibility is not currently used. The Honeywell Parallon 75 uses the system shown in Fig. 2.16a. It can control power and reactive power flow.

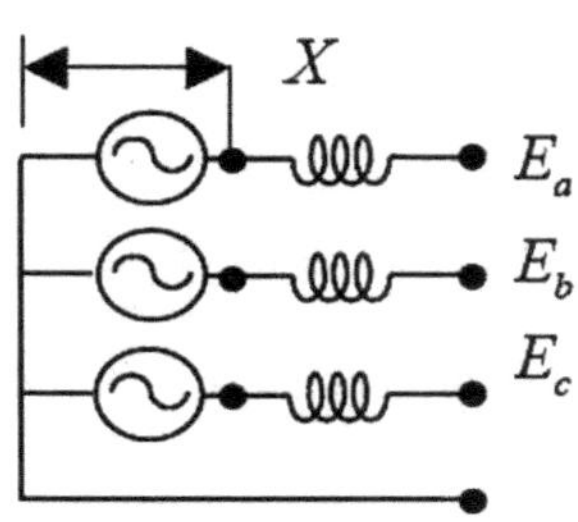

$$\hat{I} = \frac{\hat{V} - \hat{E}}{X}$$

Fig. 2.17 Basic four-wire source

In situations where AC voltages are unbalanced, the currents are also unbalanced and will trip at 20 % overcurrent in the delta winding. The Capstone 330 uses the system in Fig. 2.16b. In island mode, the unit provides balanced three-phase system voltages. The currents are a function of the load imbalance, and the system trips when the neutral line has a power flow above one-third of the systems power rating. For highly unbalanced loads, the total output could be less than one-third the rating of the microturbine.

2.7 Conclusions

This chapter paves the way to discuss numerous concepts pertaining to DG plants. The first section aims to provide a complete overview of the main control issues in small WT systems. The overall system is analyzed both on wind source-side and on grid-side, focusing on the main power converters configurations. Looking at the source-side, it results that the most critical issues are: sensorless operation of the generator and power limitation. Sensorless operation can be based on a PLL, but special care is required in the design since when the converter starts to switch it produces a voltage at the point of connection of the PMSG and the PLL gets synchronized with it, as a consequence an error in the rotor position occurs and it should be estimated and properly compensated. About power limitation, it is highlighted that two systems for limiting the power in excess, one mechanical (the pitch controller) and the other one electrical (the braking chopper), may interact. In fact, in small power WT systems their time constants can be similar and the controller should be proper designed.

2.8 Suggested Problems

In the following set of problems, reference is made to the generic prototype of wireless networked control system (WNCS) depicted.

Problem 2.1 MG dynamics are studied in the island mode. The MG consists of a single inverter-interface DG and a single load connected to the point of common coupling (PCC). The DG model involves a constant voltage source, a power electronic converter, a transformer and a filter. In Fig. 2.18, the load is modeled as an RLC circuit. Investigate a possible control strategy subsequent to an islanding event.

Problem 2.2 Consider Fig. 2.19 which gives comprehensive MG model including both the SYGN-based DGs as well as power electronics-interfaced DGs with the associated controllers. Develop a small signal model of such a model and examine the eigenvalues and participation factors, low frequency electromechanical oscillations and high-frequency modes. Carry out a sensitivity analysis for different operating points.

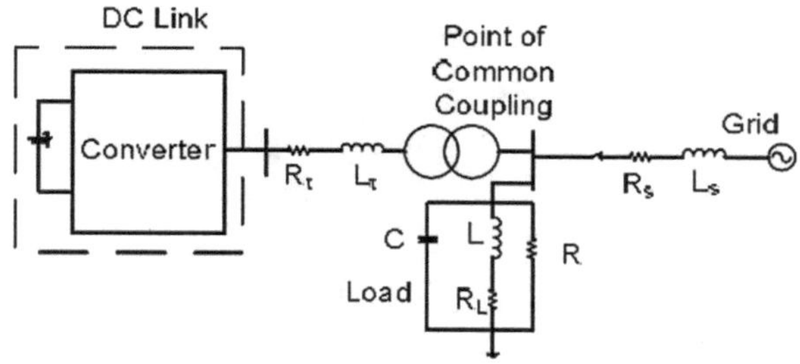

Fig. 2.18 A dynamic model of microgrid

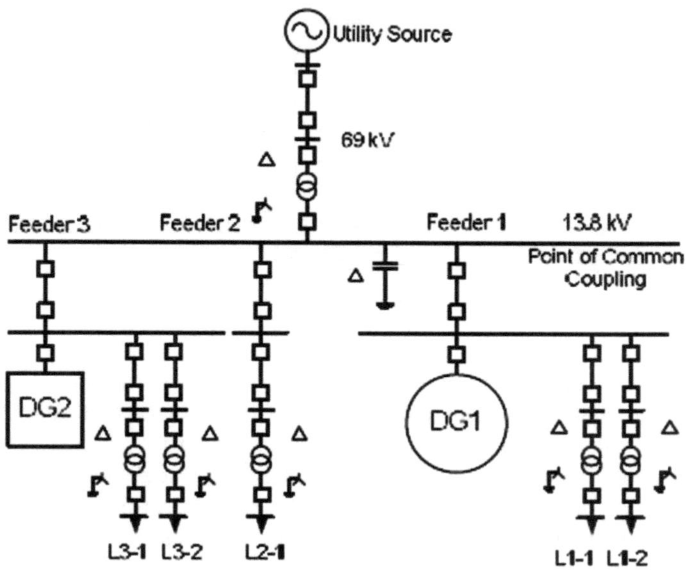

Fig. 2.19 Microgrid circuit diagram

Fig. 2.20 Microgrid frequency control diagram

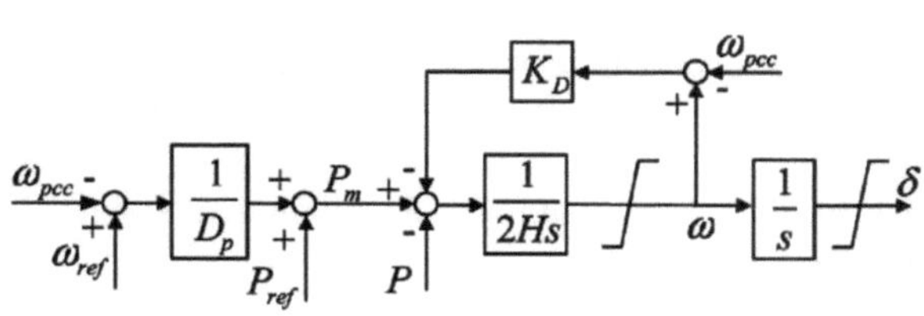

Next consider Fig. 2.20 for a prescribed frequency control of MGs including an oscillator generates reference frequency. Investigate the droop mechanism for the frequency control.

Use the following data for numerical simulation:

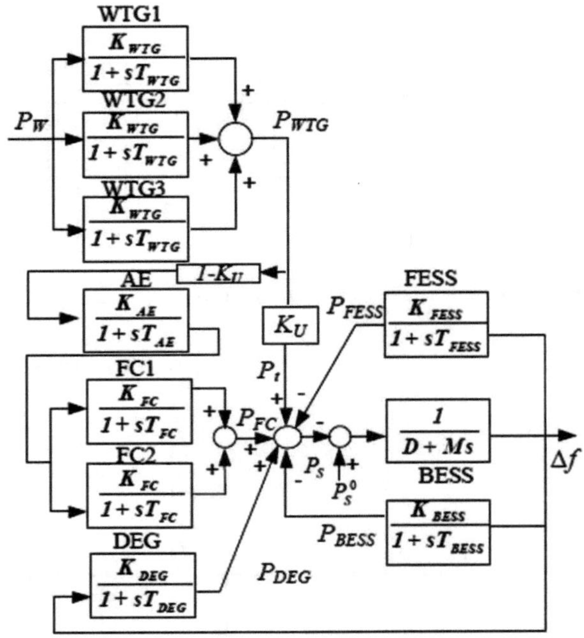

Fig. 2.21 A microgrid dynamic model

Problem 2.3 Dynamics of several DGs such as PV units, WTs, fuel cells, flywheel, battery storage, and diesel generator are approximated by a first order linearized model with a time constant and a gain factor as shown in Fig. 2.21. The studied system is operated in the island mode while the network is neglected. Battery storage, flywheel, and the diesel generator adjust their settings according to the changes in MG frequency. Conduct a small signal stability analysis for such a system and explore the impact of power-frequency (PF) droops (Table 2.1).

Problem 2.4 Consider the detailed MG model presented in Fig. 2.22. It must be noted that each DG supplies its load which can be unbalanced or nonlinear, where the DGs are converter-based. A common balanced load is also supplied by the DGs. Figure 2.23 shows the model used for the converter. Discuss a possible control strategy for load sharing among DGs while maintaining PQ in both the grid-connected and island modes.

Problem 2.5 The power droop control has a long history of use for the SYGN control in power system. It has been recently used for parallel inverter control and especially in inverter-dominated MG. Let E_i and V be the amplitudes of the ith inverter output voltage, and PCC voltage respectively. φ_i is the power angle. R_i and X_i are the resistance and inductance of the line impedance. Consider Fig. 2.24, derive expressions for the inverter output active P_i and reactive power Q_i. Use sensitivity

Table 2.1 System parameters

Parameters	Physical values	Per unit
Rated capacity	0.5	4.3656
Rated voltage (ll, rms)	0.1	9.1206
Rated frequency	0.9	7.3677
L_t	0.35 mH	0.1
R_t	0.026 Ω	0.02
L	2.76 mH	0.8
C	2.55 mF	1.25
R	1.3 Ω	1
L_{eq}	0.345 mH	0.1
R_{eq}	0.013 Ω	0.01
H	0.5 s	–
D_P	–	0.05
K_D	–	20
K_Q	–	10
T_1	0.159 s	–
T_2	0.016 s	–
K_V	–	5

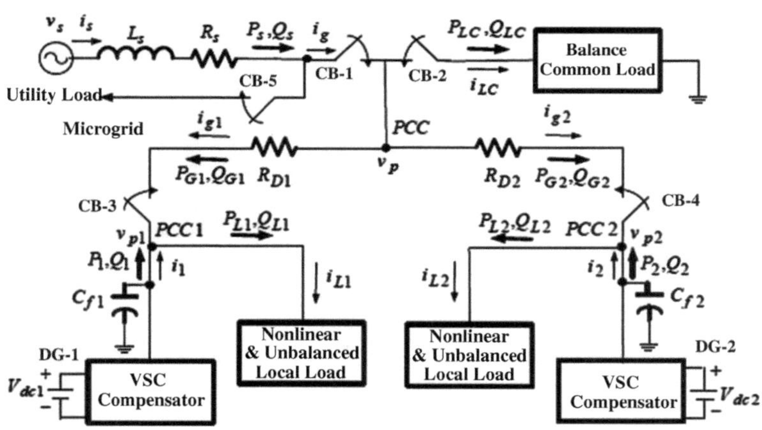

Fig. 2.22 A detailed microgrid model with two distributed generation systems

analysis theory, provide a demonstration that clarifies the basic principle of the power droop control.

Problem 2.6 The proposed system configuration, as shown in Fig. 2.25, incorporates a separate controller for each phase. Same controller may be used for all the three phases. DG is considered as DC source for simulation.

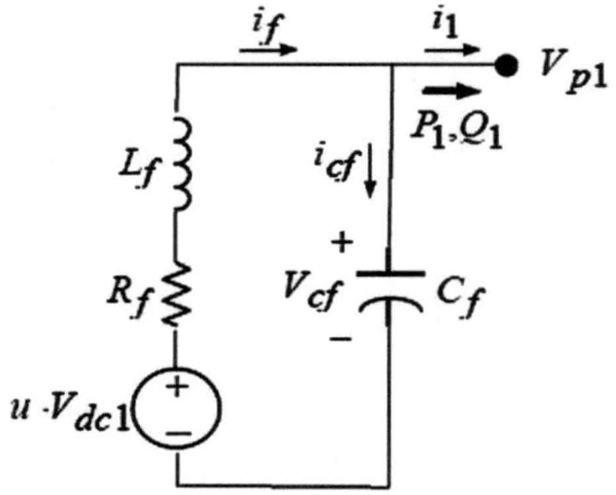

Fig. 2.23 One phase of convertor model

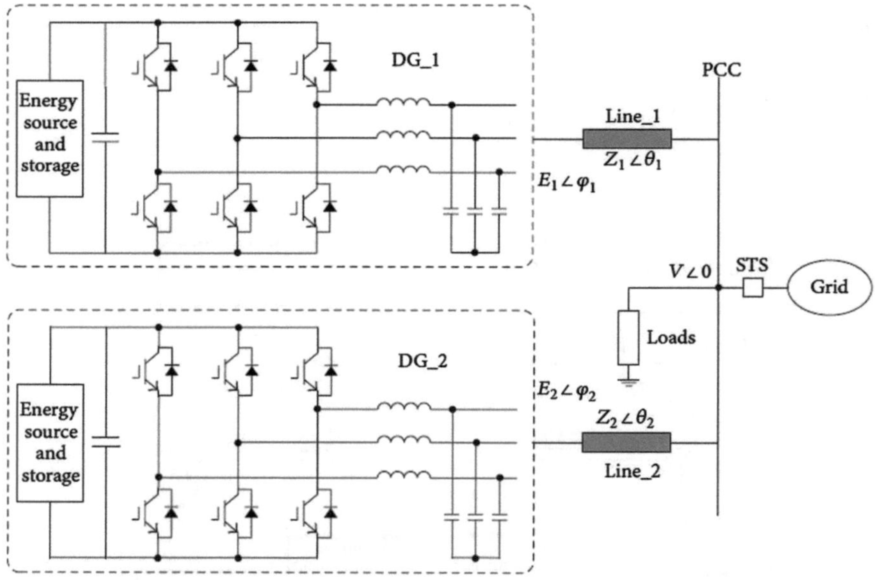

Fig. 2.24 A schematic of an inverter-based microgrid

When the switch is closed, the system is said to be in grid mode. In this mode, the microsources act as constant power sources, which are controlled to inject the demanded power into the network. The interfaced DG needs to be properly controlled in order to provide the required active and reactive power to ensure the system stability for which the active and reactive power control (PQ control) technique is designed. When the switch is opened, the system is said to be in islanded mode. When a fault occurs or for maintenance purpose main grid need to be disconnected from the MG.

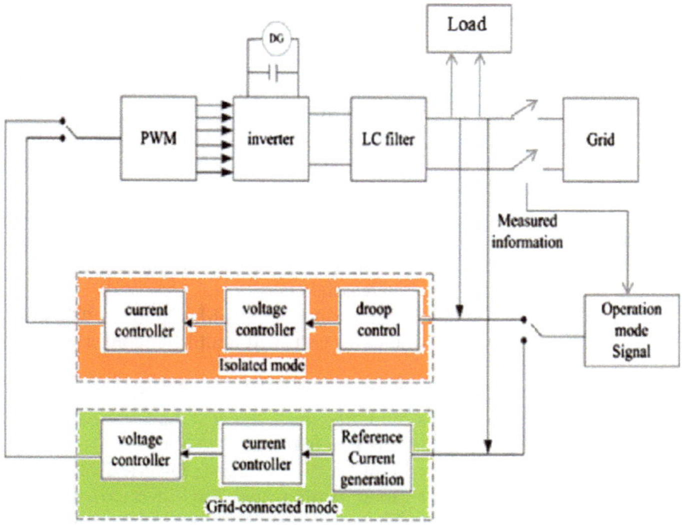

Fig. 2.25 Grid-connected system

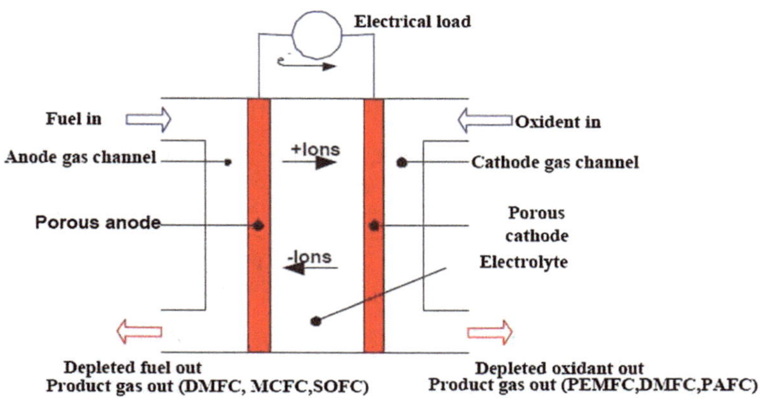

Fig. 2.26 Fuel cell system

MG thus has to provide an uninterrupted power to the loads. Give a representative analysis of the system.

Problem 2.7 The proposed system configuration, as shown in Fig. 2.26, incorporates a particular fuel cell system.

Develop an appropriate dynamic model.

References

1. Abo-Khalil AG, Lee D-C (2008) MPPT control of wind generation systems based on estimated wind speed using SVR. IEEE Trans Ind Electron 55(3):1489–1490
2. Alepuz S, Calle A, Busquets-Monge S, Kouro S, Wu B (2013) Use of stored energy in PMSG rotor inertia for low-voltage ride-through in back-to-back NPC converter-based wind power systems. IEEE Trans Ind Electron 60(5):1787–1796
3. Aliprantis DC, Papathanassiou SA, Papadopoulos MP, Kladas AG (2000) Modeling and control of a variable-speed wind turbine equipped with permanent synchronous generator. In: Proceedings of the ICEM 2000, Espoo, Finland, August 28–30, pp 558–562
4. Ashden: Sustainable solutions, http://www.ashden.org/
5. Atalik T, Deniz M, Koc E, Gercek C, Gultekin B, Ermis M, Cadirc I (2012) Multi-DSP and -FPGA based fully-digital control system for cascaded multilevel converters used in FACTS. IEEE Trans Ind Inf 8(3):511–527
6. Bernal-Perez S, A-Villalba S, Blasco-Gimenez R, Rodrguez-DDerle J (2013) Efficiency and fault ride-through performance of a diode-rectifier- and VSC-inverter-based HVDC link for offshore wind farms. IEEE Trans Ind Electron 60(6):2401–2409
7. Bernal-Perez S, A-Villalba S, Blasco-Gimenez R, Rodrguez-DDerle J (2013) Efficiency and fault ride-through performance of a diode-rectifier- and VSC-inverter-based HVDC link for offshore wind farms. IEEE Trans Ind Electron 60(6):2401–2409
8. Boldea I (2006) Variable speed generator. Taylor & Francis Group, Boca Raton
9. Burton T, Sharpe D, Jenkins N, Bossanyi E (2001) Wind energy handbook. Wiley, West Sussex
10. Cardenas R, Pena R, Wheeler P, Clare J, Asher G (2009) Control of the reactive power supplied by a WECS based on an induction generator fed by a matrix converter. IEEE Trans Ind Electron 56(2):429–438
11. Cardenas R, Pena R, Tobar G, Clare J, Wheeler P, Asher G (2009) Stability analysis of a wind energy conversion system based on a doubly fed induction generator fed by a matrix converter. IEEE Trans Ind Electron 56(10):4194–4206
12. Carranza O, Figures E, Carcera G, Ortega R, Velasco D (2011) Low power wind generation system based on variable speed permanent magnet synchronous generators. In: Proceedings of the international symposium on industrial electronics, ISIE 11, Gdansk, Poland, June 27–30, pp 1063–1068
13. Chen WL, Li ZC, Lin YS, Huang BX (2011) Control and performance identification for small vertical axis wind turbines. In: Proceedings of the IET conference renewable power generation, RPG 11, Edinburgh, UK, September 6–8, p 113
14. Esp JM, Castell J (2013) Wind turbine generation system with optimized DC-link design and control. IEEE Trans Ind Electron 60(3):919–929
15. Integration of distributed energy resources: the CERTS microgrid concept. Consultant report no. P500-03-089, October 2003
16. International Renewable Energy Agency: Working Papers, http://www.irena.org/
17. Iwanski G, Koczara W (2008) DFIG-based power generation system with UPS function for variable-speed applications. IEEE Trans Ind Electron 55(8):3047–3054
18. Kanellos FD, Papathanassiou SA, Hatziargyriou ND (2000) Dynamic analysis of a variable speed wind turbine equipped with a voltage source AC/DC/AC converter interface and a reactive current loop. In: Proceedings of the 10th mediterranean electro-technical conference (MEleCon00), vol 3, pp 986–989
19. Kazmi SMR, Goto H, Guo H-J, Ichinokura O (2011) A novel algorithm for fast and efficient speed-sensorless maximum power point tracking in wind energy conversion systems. IEEE Trans Ind Electron 58(1):36–39
20. Kazmierkowski MP, Jasinski M, Wrona G (2011) DSP-based control of grid-connected power converters operating under grid distortions. IEEE Trans Ind Inf 7(2):204–211
21. Li R, Xu D (2013) Parallel operation of full power converters in permanent—magnet direct-drive wind power generation system. IEEE Trans Ind Electron 60(4):1619–1629

22. Manwell JF, McGowan JG, Rogers AL (2002) Wind energy explained: theory, design, and application. Wiley, West Sussex
23. Milivojevic N, Krishnamurthy M, Emadi A, Stamenkovic I (2011) Theory and implementation of a simple digital control strategy for brushless DC generators. IEEE Trans Power Electron 26(11):3345–3356
24. Mirecki A, Roboam X, Richardeau F (2007) Architecture complexity and energy efficiency of small wind turbines. IEEE Trans Ind Electron 54(1):660–670
25. Morren J, de Haan SWH (2005) Ride through of wind turbines with doubly-fed induction generator during a voltage dip. IEEE Trans Energy Convers 20(2):435–441
26. Nagliero A, Mastromauro RA, Monopoli VG, Liserre M, DellAquila A (2011) Analysis of a universal inverter working in grid-connected, stand-alone and micro-grid. In: Proceedings of the ISIE 2010, Bari, Italy, pp 650–657
27. Nagliero A, Mastromauro RA, Ricchiuto D, Liserre M (2011) Harmonic control strategy for universal operation of wind turbine systems. In: Proceedings of the power engineering 2011, Malaga, Spain, pp 1–5
28. Nagliero A, Mastromauro RA, Ricchiuto D, Liserre M, Nitti M (2011) Gain-scheduling-based droop control for universal operation of small wind turbine systems. In: Proceedings of the ISIE 2011, Gdansk, Poland, pp 1459–1464
29. National Renewable Energy Lab, http://www.nrel.gov/
30. Nguyen TH, Lee D-C (2013) Advanced fault ride-through technique for PMSG wind turbine systems using line-side converter as STATCOM. IEEE Trans Ind Electron 60(7):2842–2850
31. Office of Energy Efficiency and Renewable Energy, http://energy.gov/eere/energybasics
32. Oliveira DS, Reis MM, Silva C, Colado Barreto L, Antunes F, Soares BL (2010) A three-phase high-frequency semi-controlled rectifier for PM WECS. IEEE Trans Power Electron 25(3):677–685
33. Orlando NA, Liserre M, Monopoli VG, Mastromauro RA, DellAquila A (2008) Comparison of power converter topologies for permanent magnet small wind turbine system. In: Proceedings of the international symposium on industrial electronics, pp 2359–2364, June 2008
34. Ozgener O (2006) A small wind turbine system (SWTS) application and its performance analysis. Energy Convers Manag 47:1326–1337
35. Pao LY, Johnson KE (2011) Control of wind turbines: approaches, challenges and recent developments. IEEE Control Syst Mag 11(4):44–62
36. Papathanassiou A, Papadopoulos MP (1999) Dynamic behavior of variable speed wind turbines under stochastic wind. IEEE Trans Energy Convers, 1617–1623
37. Ricchiuto D, Liserre M, Mastromauro R, DellAquila A, Pigazo A (2011) Fractional-order based droop control of an universal wind-turbine system. In: Proceedings of the EPE 2011, p 110
38. Rizo M, Bueno E, DellAquila A, Liserre M, Mastromauro RA (2011) Generalized controller for small wind turbines working grid-connected and stand-alone. In: Proceedings of the ICCEP 2011, Ischia, pp 51–55
39. Simoes MG, Chakraborty S, Wood R (2003) Induction generators for small wind energy. IEEE Power Electron Soc Newsl 18(3):19–23
40. Teodorescu R, Blaabjerg F (2004) Flexible control of small wind turbines with grid failure detection operating in stand. Alone and grid connected mode. IEEE Trans Power Electron 19(5):1323–1332
41. Thongam JS, Bouchard P, Beguanne R, Okou AF, Merabet AF (2011) Control of variable speed wind energy conversion system using a wind speed sensorless optimum speed MPPT control method. In: Proceedings of the 37th annual conference on IEEE industrial electronics society, IECON11, Melbourne, Australia, November 7–10, pp 855–860
42. Wang Q, Chang L (2004) An intelligent maximum power extraction algorithm for inverter-based variable speed wind turbine systems. IEEE Trans Power Electron 19(5):1242–1249
43. Wang J, Xu D, Wu B, Luo Z (2011) A low-cost rectifier topology for variable-speed high-power PMSG wind turbines. IEEE Trans Power Electron 26(8):2192–2200

44. Wang G, Yang R, Xu D (2013) DSP-based control of sensorless IPMSM drives for wide-speed-range operation. IEEE Trans Ind Electron 60(2):720–727
45. Wilamowski B, Irwin D (2011) The industrial electronics handbook power electron and motor drives. 2nd edn. CRC Press, Boca Raton
46. Xia Y, Ahmed KH, Williams BW (2013) Wind turbine power coefficient analysis of a new maximum power point tracking technique. IEEE Trans Ind Electron 60(3):1122–1132

Part II
Architectures and Integration

Chapter 3
Control Methods for Microgrids

3.1 Introduction

The burden on the transmission network is increasing at an unexpected pace due to the increasing demand for power. Since updates to the transmission network are economically challenging, microgrids (MGs) have evolved to become an economically viable alternative. In MGs, generating units are commissioned within the scope of the conventional distribution network so that power can directly flow from the generators to the load without having to pass through the transmission network.break The other advantage of using such an architecture is that loads can be served even if the transmission network is down due to a fault, increasing the overall reliability of the system. An MG is generally known as a system consisting of small distributed generating stations along with the loads which is capable of going into islanded operation at times of need [11, 13, 14, 26, 28, 47, 48].

Among the many benefits of having an MG, one is that it facilitates distributed generation (DG) and high penetration of renewable energy sources (RESs) [5, 7, 76]. They increase power quality (PQ) and reliability of electric supply. An MG having RESs will help to alleviate some of the environmental issues related to burning fossil fuels. There is extensive literature on the various challenges posed by MGs. Despite having some benefits of MG architecture in the grid environment, there are also some challenges related to this. Implementation is an issue. MG protection is also considered one of the most important challenges facing the implementation of MGs. Once an MG is formed, it is important to assure that the loads, lines, and DGs on the island are protected because conventional unidirectional power flow protection method is no longer viable [67]. Solid regulatory base is another issue related to MGs. It is known that the energy-related industry's established policies and 'solid regulatory base' in place, which became important for the growth in market. Government organizations should ensure that these regulatory policies should include guidelines and schemes to implement MG technologies.

M.S. Mahmoud and F.M. AL-Sunni, *Control and Optimization of Distributed Generation Systems*, Power Systems, DOI 10.1007/978-3-319-16910-1_3

Control of the voltage and frequency during islanded operation of DGs is also a major challenge. A method for intentionally islanding a single DG to feed a local load was proposed in [123]. A much more complex and challenging task is to operate more than one DG on the island. With more than one DG on the island, it is necessary to regulate the voltage during MG operation, which could be achieved by using a voltage versus reactive power droop controller [63]. There needs to be an algorithm that should complete the resynchronization process once the grid is restored. A supervisory control mechanism will monitor the overall process and provide information to the local controller (LOC) to respond accordingly.

The concept of power grid is based on the technology introduced around 120 years ago. It is facing a lot of issues in keeping up with modern challenges. One of the main challenges is to guarantee electricity supply to customers and maintaining long-term energy security. Therefore, increased reliability/efficiency is much needed in today's world where the demand of electricity is ever-growing. MGs incorporate various DG units into the utility grid and solve many problems of existing power systems. It is also the vital building block of the future smart grid (SG) [63].

On another front, the concept of system of systems (SoS) is gaining rapid interest in the field of research and can also possibly lead to a new branch of engineering known as SoS Engineering. This branch is related to systems engineering and deals with the optimization of a network of interacting systems. Any large-scale integrated system or any complex system can be viewed as an example of SoS [1, 58]. An MG can be viewed as a SoS.

In this chapter, motivation toward development of MG and an overview is presented on the two key aspects, modeling and control, of MG. Recent developments in these two key aspects are presented. A better control strategy, by viewing MG as a special case of SoS, is discussed.

3.2 Microgrids

The notion of control is central to micro grids. In fact, what distinguishes a micro grid from a distribution system with distributed energy resources (DERs) is exactly the capability of their control, so that they appear to the upstream network as a control led, coordinated unit.

3.2.1 Definition and Applications

An MG can be defined as, "A network of low voltage power generating units, storage devices and loads capable of supplying a local area such as suburban area, an industry or any commercial area with electric power and heat." The components of an MG are interfaced through quick response power electronics and present itself as a single entity and therefore can be connected to the traditional power grid or also operated in stand-alone mode as a self-sustained power system [63].

As stated in [61], "The heart of the MG concept is the notion of a flexible, yet controllable interface between the MG and the wider power system". MG acts as a *good citizen*, that is, ideal conventional load behavior toward the grid which is less troublesome than distributed generation system (DGS). It also has environmental benefits because it uses RESs.

Different countries around the world adopt various topologies and structure based on their priorities on functionality offered by MG. The research on MG is more active in the US, Canada, Europe, and Japan. Several demonstration projects and laboratory facilities are developed and a lot of research is in progress concerning various issues in the MG [43, 115]. Various objectives which can be achieved by the use of MG are listed below; ride-through capability provided by energy-storage is a common objective of MG.

- Reliability of power supply
- Reduction of environmental impact of electric supply
- Reduction of investment in plant, equipment and cost
- Increase of energy efficiency Stable
- Ensure diversity of energy supply
- Power supply to a remote site
- Ride-through capability provided by energy storage

The future *SG* is expected to be a well-organized plug-and-play integration of MGs connected via dedicated highways for exchange of command, data, and power. The emerging standards, research, development, and demonstration are also discussed in [34].

3.2.2 Control Functions

Before analyzing the technical implementation of control, a general overview of the main control functionalities in an MG is presented. These functionalities can be distinguished in three groups, as shown in Fig. 3.1. The lower level is closely related to the individual components and local control (microsources, storage, loads, and electronic interfaces), the medium level to the overall MG control, and the upper level to the interface to the upstream network. More specifically:

1. **Upstream network interface** The core interaction with the upstream network is related to market participation, more specifically the MG actions to import or export energy following the decisions of the ESCO. Owing to the relatively small size of an MG, the ESCO can manage a larger number of MGS, in order to maximize its profit and provide ancillary services to the upstream network. The operation of multi-MGs is discussed in Chap. 5.

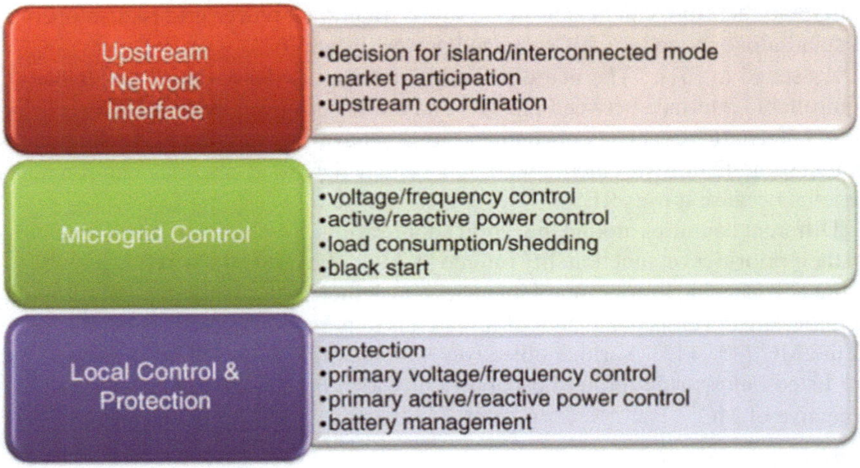

Fig. 3.1 Microgrid system functionalities

2. **Internal MG control** This level includes all the functionalities within the MG that require the collaboration of more than two actors. Functions within this level are:
   - load and RES forecast,
   - load shedding/management,
   - unit commitment/dispatch,
   - secondary voltage/frequency control,
   - secondary active/reactive power control,
   - security monitoring,
   - black start.
3. **Local control** This level includes all the functionalities that are local and performed by a single DG, storage or controllable load, that is:
   - protection functions,
   - primary voltage/frequency control,
   - primary active/reactive power control,
   - battery management.

It should be noted that these functionalities are relevant to the normal state of operation. They might need to change in critical or emergency states.

3.2.3 Components and Formation

A generalized structure of MG is shown in Fig. 3.2. The MG can be connected to the utility grid through single point of common coupling (PCC). The isolating device is used to isolate the MG from the utility grid.

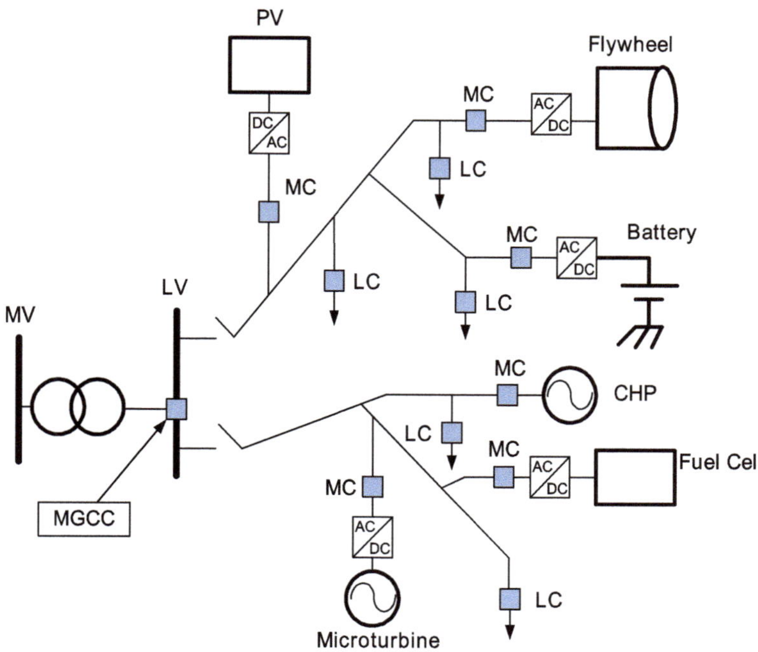

Fig. 3.2 Generalized microgrid structure

The DG unit is responsible for generation of electricity. It consists of rotating type and inverter type generating devices. Rotating type includes IC engines, gas turbines (GTs), microalternator etc. whereas the inverter type includes Photovoltaics (PVs), fuel cells, wind turbines (WTs), etc. Both rotating and inverter type requires power electronic converters for their interface. The power range of DG unit components is small-scale ranging from 4 to 10,000 KW [69].

Energy storage unit is essential to balance the flow of power at the onset of islanding mode of operation. It is also used to control the flow of power to and from the main grid. They help in improving the quality of power (QoP) and assist in voltage control. Batteries, flywheels, supercapacitors, superconducting magnetic energy storage, etc., can be used to store the energy. All these devices again require power electronic devices for their interface [69].

There has to be a *control system* for the safe operation of MG in various modes of its operation. This system can be based on a central controller (CC) or distributed controller. The selection of controller depends mainly on the operation mode of MG and its requirements [10]. Various control strategies are discussed in this paper in other sections.

The purpose of MG is not obtained until the customer is served with nominal voltage and frequency by a stable system during all the modes of operation [10].

3.2.4 Overview of Modeling

An MG integration of various units. Basically, it consists of DG unit, energy-storage unit, controller unit, and conventional load. The DG unit again comprises various micro-generating devices. Therefore, MG modeling varies from one configuration to other depending on the components used. Various approaches for the modeling and control of MG can be found in the literature [25]. We will discuss the different models available in the literature.

A small signal dynamic analysis of an autonomous hybrid system is performed in [70]. The configuration of the system is shown in Fig. 3.3.

The dynamics of all the DG units are approximated by a first order linear model with a time constant and a gain factor while the network is neglected [62, 70, 110]. The transfer functions of various components are obtained and time domain analysis is performed by considering various components at each time. The transfer functions of various components are given as follows and Fig. 3.4 shows the configuration in one of the cases:

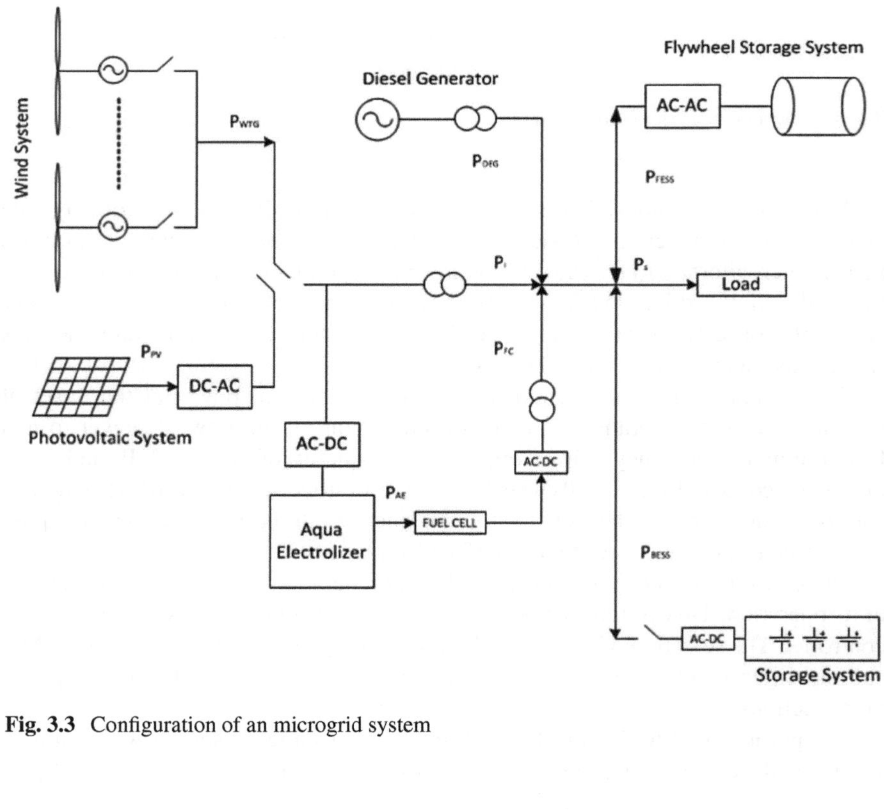

Fig. 3.3 Configuration of an microgrid system

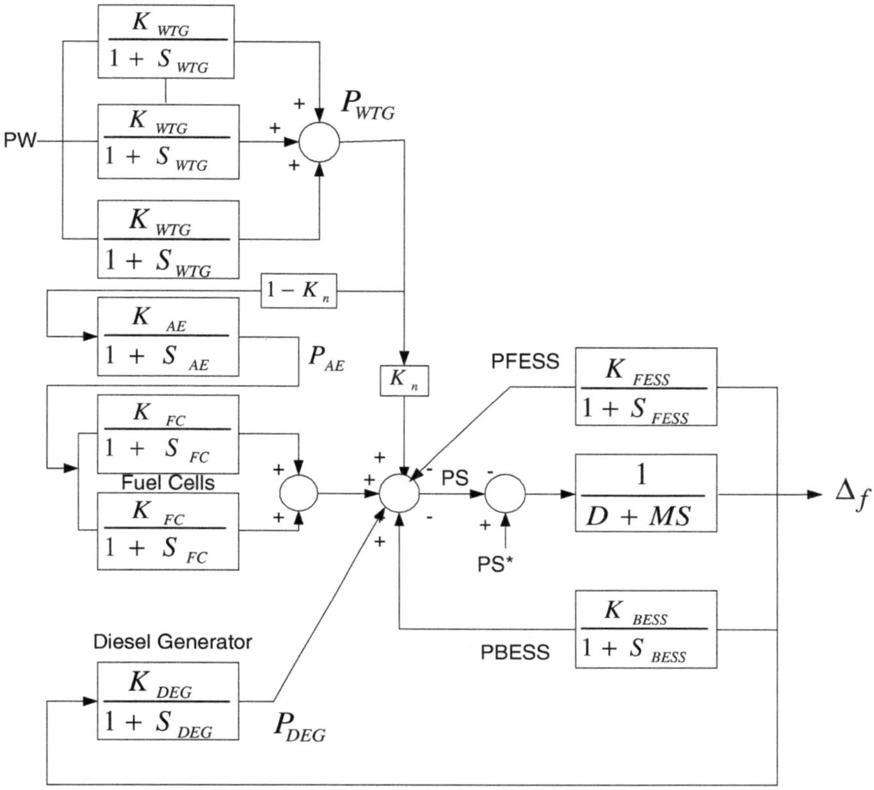

Fig. 3.4 Block diagram of an microgrid

$$Wind\ \ Turbine : G_{WTG}(s) = \frac{K_{WTG}}{1 + sT_{WTG}}$$

$$PV\ \ SystemG_{PVS}(s) = \frac{K_{PV}}{1 + sT_{PV}}$$

$$Fuel\ \ CellG_{FC}(s) = \frac{K_{FC}}{1 + sT_{FC}}$$

$$Diesel\ \ Engine\ \ GeneratorG_{DEG}(s) = \frac{K_{DEG}}{1 + sT_{DEG}}$$

$$Aqua\ \ ElectrolyserG_{AE}(s) = \frac{K_{AE}}{1 + sT_{AE}}$$

$$Storage\ \ SystemG_{SS}(s) = \frac{K_{sto}}{1 + sT_{sto}}$$

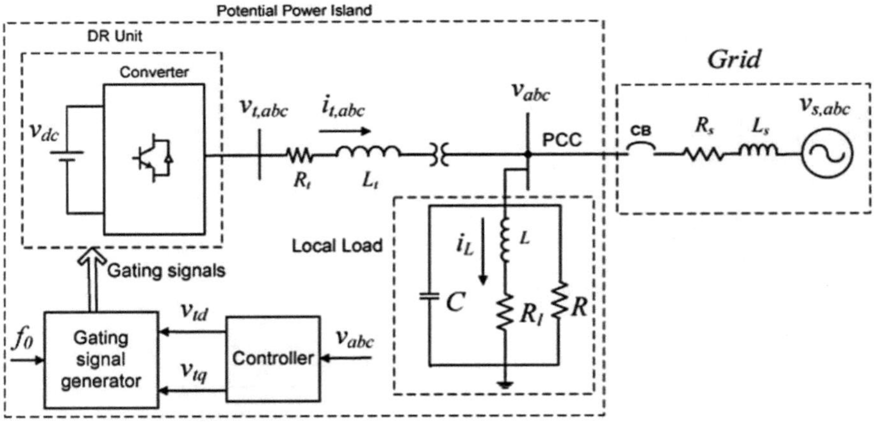

Fig. 3.5 Model of MG in [52, 53, 104]

Since an MG is a power generating unit, it can be represented by a DC source. This concept of modeling an MG with an RLC load in islanded mode is proposed in [36, 52]. As shown in Fig. 3.5, MG is represented by a DC source connected to a voltage-sourced converter (VSC). The MG is connected to the grid by means of an R-L filter, step-up transformer, and a circuit breaker (CB). The CB is open when the MG is islanded. The load which is passive RLC type is connected on the high voltage side of the transformer. A control system is used to control the VSC.

The dynamic model of Fig. 3.5 can be represented by the following nonlinear equations:

$$
\begin{aligned}
\frac{di_{td}}{dt} &= \omega i_{tq} - \frac{R_t}{L_t} i_{td} - \frac{v_d}{L_t} + \frac{v_{td}}{L_t} \\
\frac{dv_d}{dt} &= \frac{1}{C} i_{td} - \frac{1}{RC} v_d - \frac{i_{Ld}}{C} \\
\frac{i_{Ld}}{dt} &= \omega i_{Lq} + \frac{v_q}{L} - \frac{R_l}{L} i_{Ld} \\
\frac{i_{tq}}{dt} &= -\omega i_{td} - \frac{R_t}{L_t} i_{tq} + \frac{v_{tq}}{L_t} \\
\frac{i_{Lq}}{dt} &= -\omega i_{Ld} - \frac{R_l}{L} i_{Lq} \\
\omega C v_d &= i_{tq} - i_{Lq}
\end{aligned}
$$

After performing the linearization whose details can be seen in [52, 53, 104], the state-space matrices are obtained as mentioned below. The overall test system was simulated in the MATLAB® Simulink® and *ATPDraw* environment.

$$A = \begin{bmatrix} -\frac{R_t}{L_t} & \omega_0 & 0 & -\frac{1}{L_t} \\ \omega_0 & -\frac{R_l}{L} & -2\omega_0 & \frac{R_l C \omega_0}{L} - \frac{\omega_0}{R} \\ 0 & \omega_0 & -\frac{R_l}{L} & \frac{1}{L} - \omega_0^2 C \\ \frac{1}{C} & 0 & -\frac{1}{C} & -\frac{1}{RC} \end{bmatrix}$$
$$B^T = \left[\frac{1}{L_t}\ 0\ 0\ 0 \right] \quad C = \left[0\ 0\ 0\ 1 \right]$$
$$X^T = \left[i_{td}\ i_{tq}\ i_{Ld}\ v_d \right] \tag{3.1}$$

This concept of representing MG as a combination of DC source with VSC is also presented in [4, 88]. This paper models the islanded operation of MG consisting of two parallel DG units. Again, the local load is a passive RLC network located at the PCC. The schematic diagram of such an arrangement is shown in Fig. 3.6. The MG structure is used for the application of decentralized control (DEC) and hence there is a separate controller for each DG unit.

By applying KVL and KCL and further application of frame transformation, gives the below dynamic equations governing the MG

$$\begin{aligned} \frac{dV_{dq}}{dt} &= -\frac{1}{RC}V_{dq} - \frac{1}{C}i_{t,dq1} - \frac{1}{C}I_{L,dq} + \frac{1}{C}I_{t,dq2} \\ \frac{I_{t,dq1}}{dt} + j\omega_0 I_{t,dq1} &= -\frac{1}{L_{t1}}V_{dq} - \frac{R_{t1}}{L_{t1}}I_{t,dq1} + \frac{1}{L_{t1}}V_{t,dq1} \\ \frac{I_{L,dq}}{dt} + j\omega_0 I_{L,dq} &= \frac{1}{L}V_{dq} - \frac{R_l}{L}i_{L,dq} \\ \frac{i_{t,dq2}}{dt} + j\omega_0 I_{t,dq2} &= -\frac{1}{L_{t2}} - \frac{R_t 2}{L_t 2}i_{t,dq2} + \frac{1}{L_t 2}v_{t,dq2} \end{aligned}$$

The model is then represented in state-space which is simulated using MATLAB SimPowerSystems toolbox.

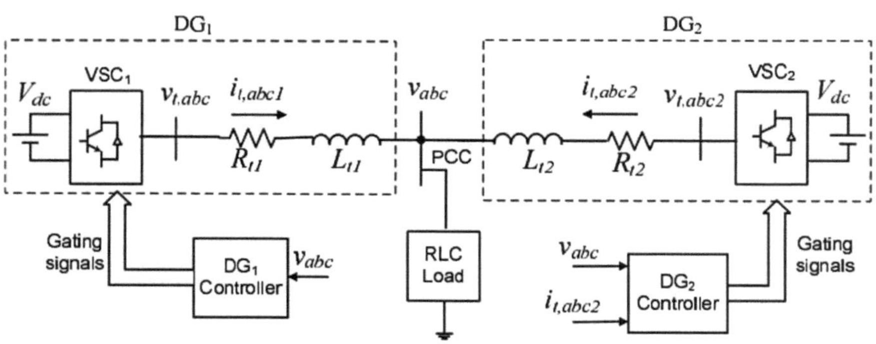

Fig. 3.6 Radial configuration of DGs used in [4]

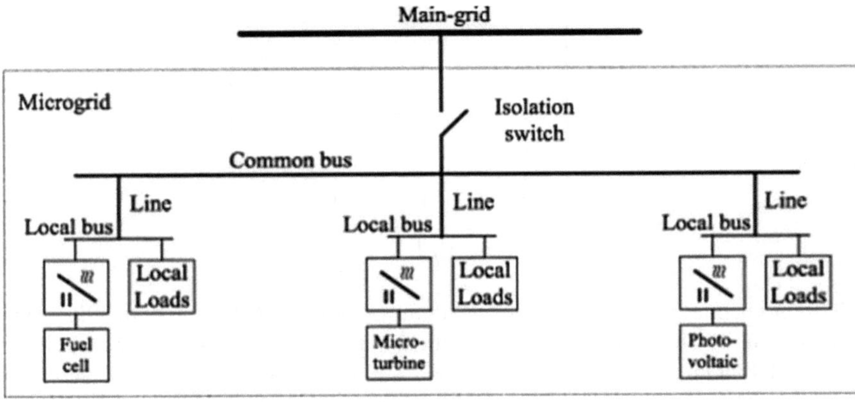

Fig. 3.7 Inverter-based MG structure in [103]

An MG consisting of only inverter-based DGs is modeled in [41, 103]. A typical structure of such MG is shown in Fig. 3.7. The modeling approach considered the full dynamic model of the complete network rather than algebraic equations.

The approach of modeling was divided into three modules, namely inverter, network, and loads. The inverter model comprises the dynamics of controller, output filter, and coupling inductor. The state equations of network and load are represented on one of the inverters' reference frame which is assumed to be a common reference. Then using the transformation technique [116], all the other inverters are transformed to this common frame. Each submodule is modeled in state-space form and combined together on this common reference frame. Block diagram of state-space model of the MG is shown in Fig. 3.8.

By combining individual submodules, the overall state-space model of MG is given below. The subscript represents states of inverter, network and load. Detailed derivation and information on the state-space matrices can be found in [103].

$$\begin{bmatrix} \Delta\dot{x}_{Inv} \\ \Delta\dot{i}_{networkdq} \\ \Delta\dot{i}_{loaddq} \end{bmatrix} = A_{mg} \begin{bmatrix} \Delta x_{Inv} \\ \Delta i_{networkdq} \\ \Delta i_{loaddq} \end{bmatrix} \tag{3.2}$$

A small signal dynamic model of MG which includes synchronous generator-based DG and power electronically interfaced DG is presented in [57]. Figure 3.9 shows the single line diagram of the MG. DG1 is a synchronous machine (SM) (diesel or GT generator) with excitation and governor control system, whereas DG2 is a dispatchable source (microturbine or wind generator, etc.) equipped with a VSC. The system parameters are given in [56].

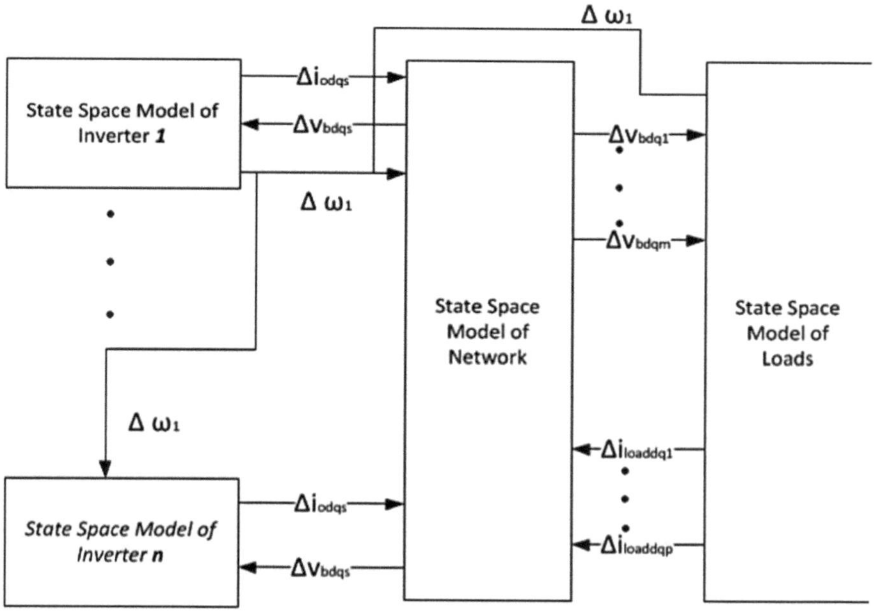

Fig. 3.8 State-space model of MG in [103]

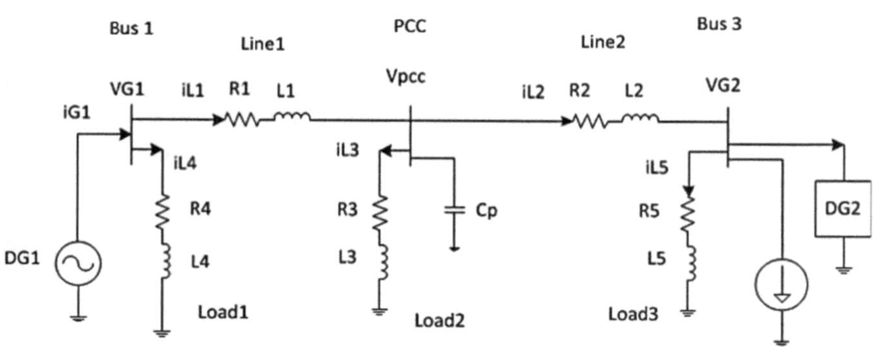

Fig. 3.9 Small signal of synchronous generator model [57]

To obtain the linearized mathematical model of the above system the following steps are followed:

- The ordinary differential equations (ODE) of DG units including network components are developed in their respective local *dq0* reference frames.
- The obtained equations are transformed to the global *dq0* frame of MG.
- Linearized about a nominal operating point and arranged in the state-space form.

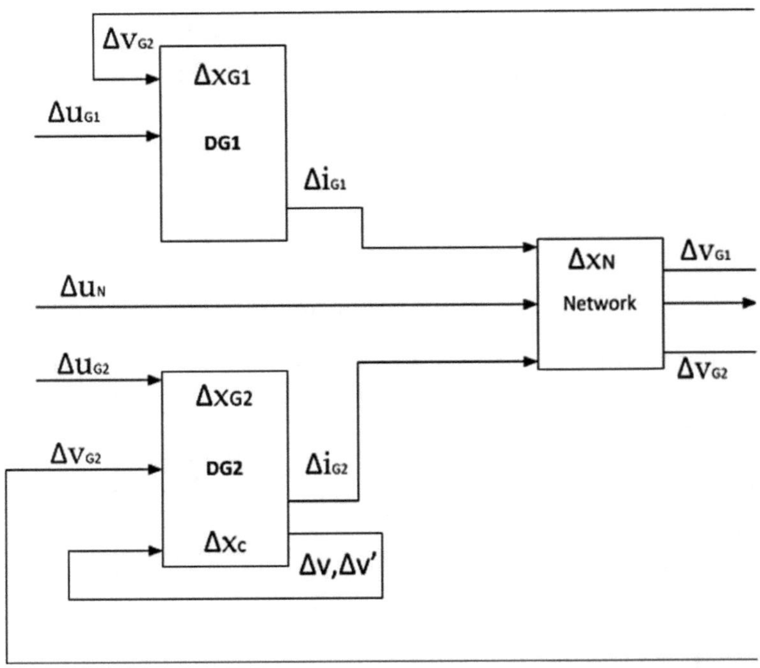

Fig. 3.10 Block diagram of MG model [57]

The dynamic model of DG1 in its local reference *dq0* frame is obtained from [60] and the dynamic model of DG2 can be found in [54, 109]. The electrical network modeling is carried out on the basis outlined in [96] and can be found in [54]. The overall system model is represented by the block diagram in Fig. 3.10. The small signal model was validated in the *PSCAD/EMTDC* environment.

The same MG structure is taken into account in [56] and the stability analysis for various transient conditions such as energizing load, transition from grid-connected mode to islanded mode, and vice versa is performed. An operational architecture developed within EU R&D MGs projects [33, 79] is adopted in [78]. This concept is shown in Fig. 3.11. It is a multi-level type control and management scheme supported by a communication infrastructure.

The head of this multilevel control system is microgrid central controller (MGCC) installed at the MV/LV substation and centrally controls the MG. Load controllers (LCs) and microsource controller (MC) form the second level of hierarchy and exchange information with the MGCC. LC acts as an interface to controllable loads and MC controls the active and reactive power of each microsource. Both LC and MC receive their set-points from MGCC.

The dynamic modeling of each DG components was picked from different literature. The dynamic model of solid oxide fuel cell (SOFC) is described in detail with the values of each parameters in [93, 127]. A GT was used for the primary

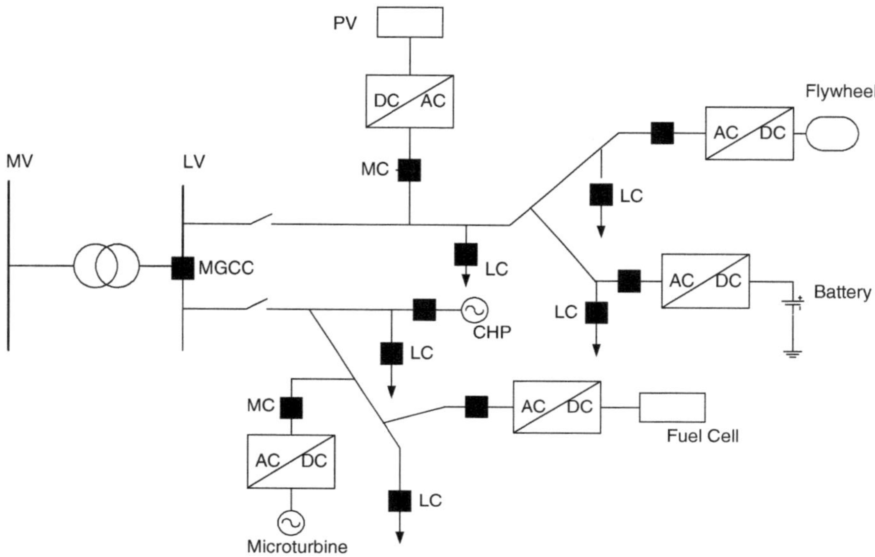

Fig. 3.11 MG structure in [78]

unit of microturbine. The dynamic model of GT is adopted from [127]. A fifth-order induction generator connected directly to the network serves as a wind generator. This model was available in MATLAB Simulink toolboxes. An empirical model for the PV generator based on experimental results was adopted from [42].

Flywheels and batteries are used for the modeling of storage devices. They were modeled as a constant DC voltage sources and were coupled to the electrical network using power electronic interface (PEI).

The inverter modeling can be derived as per two control strategies, PQ inverter control modeling [37] and voltage source inverter (VSI) control model [29, 77]. Inverters are modeled based only on their control functions for the purpose of analyzing the dynamic behavior of MG [8, 12, 27, 56, 66]. Two types of loads were considered, one is constant impedance load and other is motor load.

A low voltage (LV) test network was built in MATLAB/Simulink SimPowerSystems environment. The implementation of this network is shown in Fig. 3.12 whose detailed description can be found in [94, 95].

A low voltage MG with three unbalanced phases was proposed in [114]; Fig. 3.13 shows MATLAB Simulink implementation of the structure used. Inverter droop controls were used to interface DGs and loads were modeled as constant power. Simulations were carried out for both grid connected and isolated modes. This model has the advantage of modeling small unbalanced networks but lacks the analytic details required for stability analysis.

A new method to form the system matrices of large MGs in islanded mode is discussed in [22]. The MG under consideration has DG's power electronically interfaced

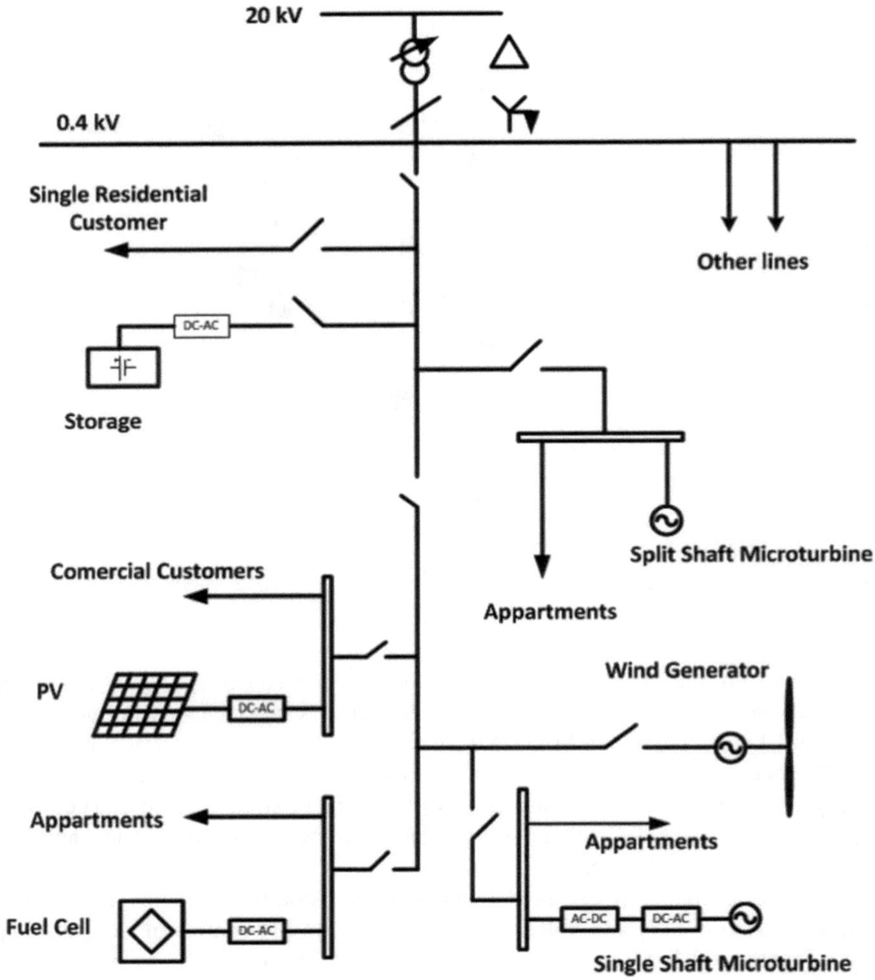

Fig. 3.12 Implementation of MG on LV network

and hence the dynamics are similar to that shown in [103]. There were two types of DG, one was *PQ* regulated and the other was *Vf* regulated which are introduced in [57].

The proposed modeling approach is based on four defined complex vectors. These vectors allow for complex-valued system matrices to be formed in a quite automated way. Moreover, a convenient partition of the system matrices is proposed, which in turn allows fast and easy modifications. Additionally, a multivariable methodology is proposed to simultaneously determine the control system gains in an optimal sense.

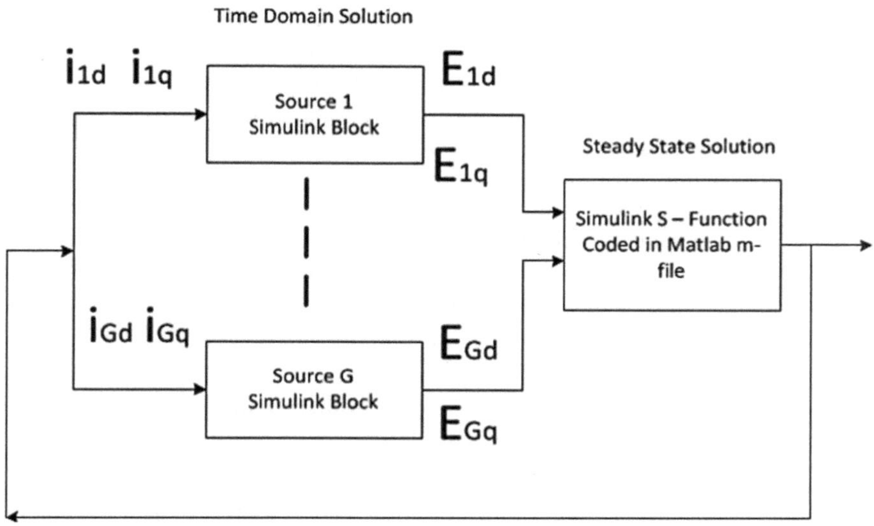

Fig. 3.13 MG network and MATLAB/Simulink structure used in [114]

An alternative approach is based on the "hub model" for MGs [2] in which the couplings between an integrated electricity and natural gas system to yield optimal operation are modeled by energy hubs. It turns out that this concept serves as interface between the loads and the transmission infrastructures and supports the application of distributed control schemes. Similarly, hybrid modeling control techniques are applied to a two-generator power system connected to the grid and the plant consists of a solar field and a secondary power source formed by an electrolyzer, a hydrogen tank, and a fuel cell stack. It is shown that the system has essentially hybrid dynamics, as it can operate in four distinct modes, depending on the power circuit configuration and the fuel cell stack state [21]. With focus on bulk power flow of MGs, research investigations are reported in [101, 118] in which the optimal design of an electrical MG and sizing of its components is sought to balance capital investment with expected operational cost while meeting performance requirements. In [121], a comprehensive review of current control technology is presented with emphasis on challenges of MG controls. The impact of frequency and voltage regulation on the optimal design of an autonomous military MG, comprised of a solar panel and vehicles as power sources, with each vehicle incorporating a battery and generator, is developed in [30, 31].

3.2.5 Modes of Operation

MG can operate autonomously and can also be connected to the utility/main grid. In case a fault occurs while operating in grid connected mode, MG has the ability to disconnect itself from grid and operate independently supplying its local load [51]. Therefore, the MG modes of operation can be classified into grid connected, islanded, transition between grid-connected mode to the islanded mode, and vice versa [56]. In any mode of operation, the heat generated by some of the microsources can be used to supply the heat demand of the local load.

When functioning in parallel with the utility (main) grid, it acts as a model or good citizen and the voltage and frequency are controlled by the main grid. Depending on the load of the main grid, it will either supply or absorb power and act as either controllable load or controllable source. If any fault or disturbance occurs in the main grid, MG has the ability to disconnect and operate autonomously. This ability of MG increases the QoP to its local customer by providing local voltage control. In this mode of operation, the points to be noted are

- The frequency and voltage magnitude are controlled by utility grid;
- DG units supply the total or a part of the load.

Islanding of MG can be due to unplanned faulty events discussed in [56] and can also be due to planned actions like maintenance, etc. The MG controls the voltage and frequency in autonomous mode by continuously adjusting the output active and reactive power. This is a very common mode of operation. In this mode, it supplies a local load which is closely located geographically. The local load can be a small village, a university, an industry or a commercial building, etc. The main issues that the MG should address in this mode is the management of voltage and frequency, QoP, balancing between load and supply, communication among its components, etc. In this mode of operation, the points to be noted are

- The DG units control the frequency and voltage magnitude;
- It supplies active and reactive power to the load.

3.3 Control Approaches

The control strategies for MG depends on the mode of its operation. The aim of control technique should be to stabilize the operation of MG. When designing a controller, operation mode of MG plays a vital role. Therefore, after modelling the key aspect of the MG is control. In this section we will discuss the various control paradigms.

3.3.1 Control of Grid-Connected Mode

In grid connected mode, MG acts as a controllable load/source. It should not actively regulate the voltage at the PCC. Its main function is to satisfy its load requirements with good citizen behavior toward main grid. The balance between generation and demand, control of the parameters of the system is taken care by the utility grid. The voltage and frequency reference of the MG is also set by the main grid. Therefore the main task of a DG unit is to control the output real power (P) and reactive power (Q) [55, 73]. The P, Q generated by a DG can be controlled either by current-based or by voltage-based power flow control [85, 89, 99, 100].

3.3.2 Power Flow Control by Current Regulation

The control scheme for *power flow control through current regulation* is illustrated in Fig. 3.14. It is desired to control both the real and reactive power. The real power control loop is used to obtain the synchronous frame d-axis reference current and reactive power control loop is used to obtain the q-axis current. The synchronous d-q frame current can then be controlled in a closed-loop manner [55].

3.3.3 Power Flow Control by Voltage Regulation

The other method to control the power flow is based on the output voltage of DG. Therefore, it is known as *power flow control through voltage regulation*. It can be shown that real power (P) flow is proportional to the voltage phase angle (δ) and reactive power (Q) flow is proportional to voltage difference ($V_l - V_g$), where V_l is DG voltage and V_g is PCC voltage and δ is the phase angle difference between these two voltages. Therefore, the flow of P can be regulated using δ and flow of Q can be regulated using $V_l - V_g$. This scheme is illustrated in Fig. 3.15. To improve the accuracy of reactive power control, integral control can be included into the reactive power controller [17, 73, 86, 122].

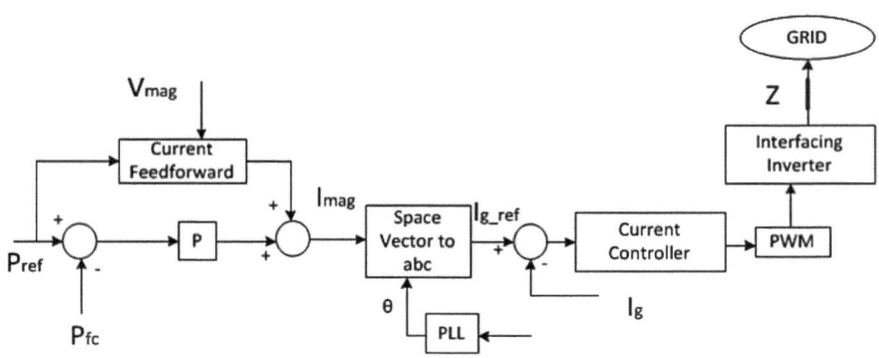

Fig. 3.14 PQ power control through output current regulation [99]

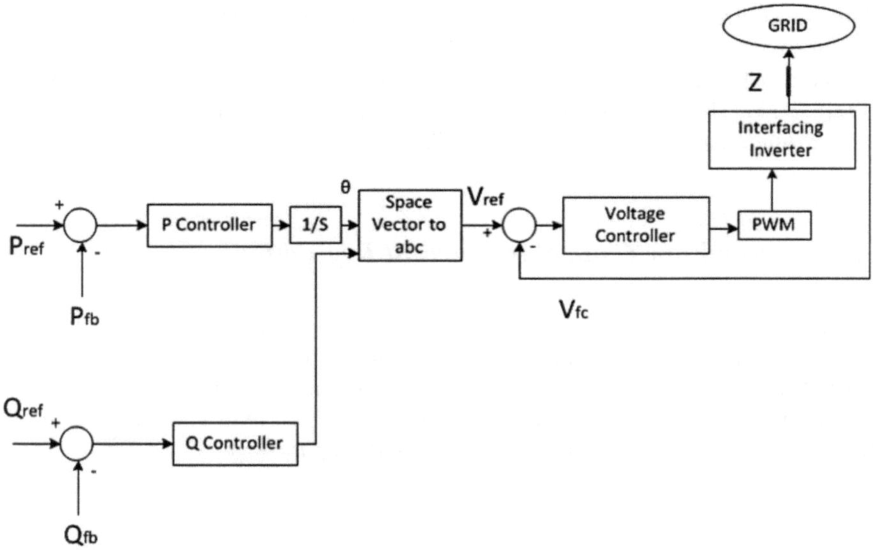

Fig. 3.15 PQ power control through output voltage regulation [99]

Power control through voltage regulation is more sensitive than current regulation to the line impedance between the DG and the PCC.

3.3.4 Agent-Based Control

MG management system was developed using agent-based technology in [92]. MG agents were developed on Java Agent Development Framework. The proposed system has several functionalities like supervisory control and data acquisition (SCADA) system, selling bids managing system, load shifting system, etc. The software architecture of the management system is shown in Fig. 3.16.

MG agent platform consists of the following components:

1. Microgrid central controller: It includes pulling agent, database agent, control agent, shifting agent, curtailment agent
2. Microsource controller: It includes generator agent, schedule agent, bid agent
3. Load controller: It includes load agent, status agent, switch agent

The effectiveness and applicability of the introduced software have been evaluated on a laboratory environment.

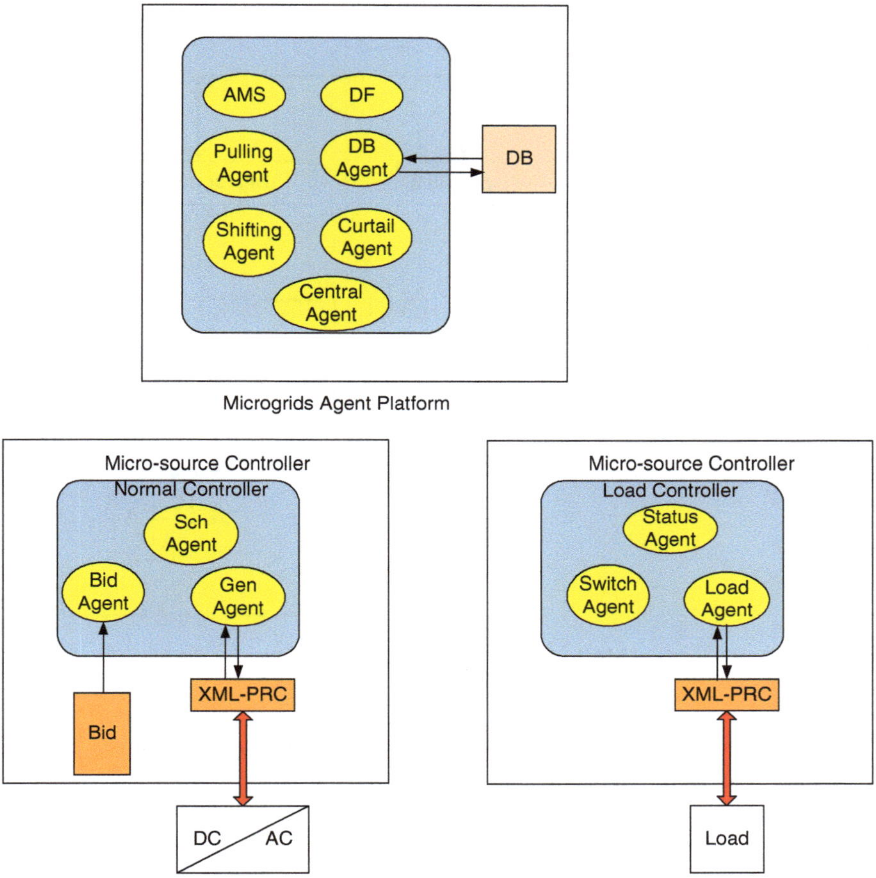

Fig. 3.16 Architecture of management system

3.3.5 Distributed Control

A distributed control approach based on multi-agent system (MAS) for MGs is proposed in [23], where advantages of MAS technology is utilized for controlling MGs. As shown in Fig. 3.17, a fully decentralized approach is adopted with three distinguished control levels.

Distribution network operator (DNO) and market operator (MO) are at medium voltage (MV) level and do not belong to MG. DNO refers to the operational functions of the system and is responsible for technical operation of one or more MGs, whereas one or more MOs are responsible for market functions of the area.

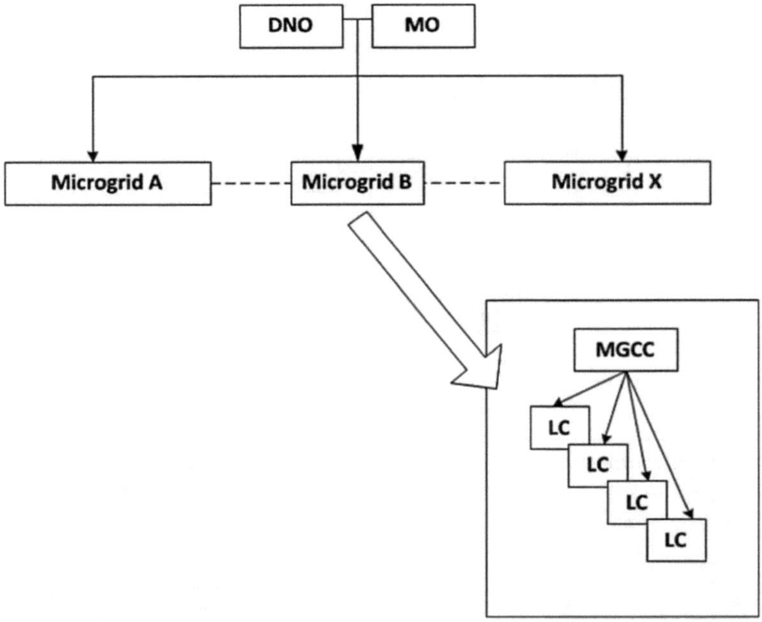

Fig. 3.17 Control levels of MAS environment

MGCC is the main interface between DNO/MO and the MG. Its main function is to optimize the operation of MG and coordinate the LOC.

At the lower level, LCs control the DG, production, storage, and some of the local loads.

Using MAS technology, a model of the system is obtained in detail where every agent uses the the exact piece of information it needs, leaving the technical details for the agents that are below it in the organization chart. The paper proposes three types of agents. *Control agent* which controls physical units of the system directly. *Management agents* which manage the MG and take the decision. *Ancillary agents* which perform tasks like communication and storage of data. The proposed MAS platform is depicted in Fig. 3.18.

In [23], internal operation of the MG and its participation in energy market was focused particularly. The algorithm is proposed so that every DER or controllable load decides what is best for it.

Fig. 3.18 Types of MAS agents

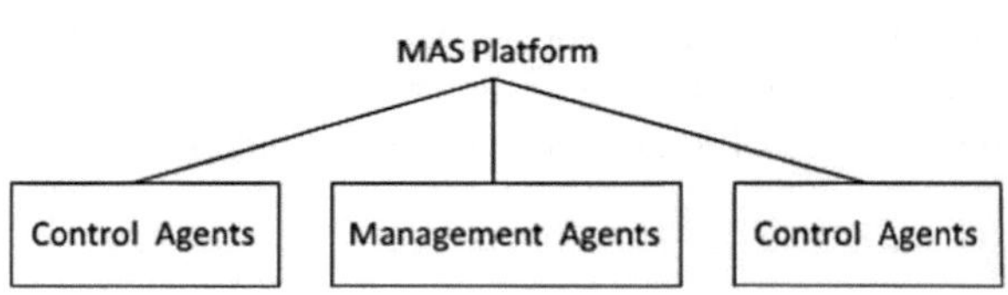

3.3.6 H_∞ Control

A new power balancing method based on H_∞ control theory is proposed in [90]. The power fluctuations were considered as the disturbances added to the MG. Figure 3.19 shows the block diagram of MG structure.

Where G_{ge} and G_{bt} are the first order transfer function representation of GT [120] and K_{ge} and K_{bt} are controller gains for GT and battery, respectively. F is the low/high pass filter gain.

Using the Robust Control Toolbox of MATLAB, the controller gains K_{ge} and K_{bt} were determined as standard controllers. Figure 3.20 shows the block diagram of these controller designs, where W with subscripts 1, 2, and 3 are weighing functions for tracking performance, gain margin of the MG system, and robustness for power fluctuations respectively.

In [91], technical challenges and stability of DGs when connected into the distribution system are detailed. Since high penetration of the DGs can be considered as a MG the same technical challenges can be assumed correct for a grid connected MG.

3.3.7 Autonomous/Islanded Mode

In the autonomous or islanded mode of operation, MG supplies its local load and is not connected to the utility grid. The main challenges in this mode are

1. Voltage and frequency control
2. Balance between supply and demand
3. Power quality
4. Issues relating to microsources
5. Communication among MG components

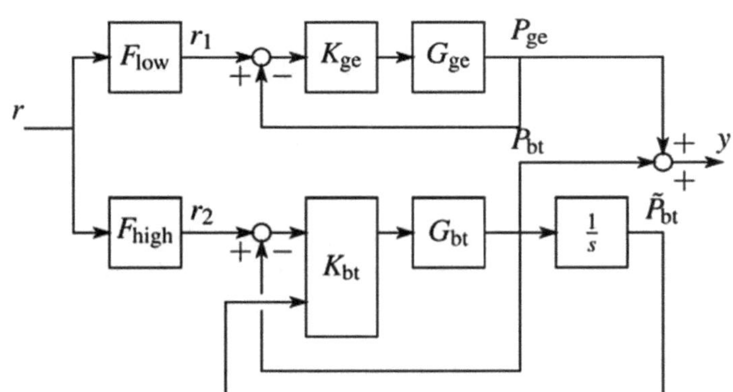

Fig. 3.19 Structure of MG [90]

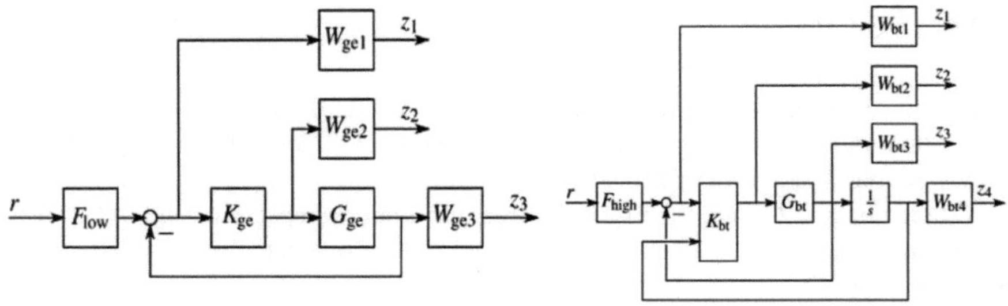

Fig. 3.20 Design of H_∞ controllers [90]

A lot of research has been done on control of MG in autonomous/islanded operation [97], which will be discussed in this section. The two main control strategies *PQ* and *VSI* control are discussed first, detailed description and explanation of these two controls can be found in [8].

When connected to grid, all the DGs can operate in PQ control mode, because the voltage and frequency is dictated by the utility grid but at least one DG has to follow VSI control in islanded mode since the voltage and frequency reference are set by the MG.

3.3.8 PQ and VSI Control

The aim of *PQ control* is to provide constant active and reactive power at the desired power factor [22, 78]. The reference values of power are defined by a LOC or centrally from the MGCC. This scheme can be implemented as a current controlled voltage source or voltage controlled current source as discussed earlier in Sect. 3.3.3. Current or voltage components in direct (I_d or V_d) and in quadrature (I_q or V_d) with inverter terminal voltage are computed based on the method given in [37].

Figure 3.21 shows the control block for this strategy using current control. The direct component (I_d) of the current is used for the control of active power and the quadrature component (I_q) is used for the reactive power control.

VSI matches the behavior of a SM controlling the voltage and frequency on the AC system [14, 37, 66]. It acts as a voltage source whose output voltage's magnitude and frequency are controlled through *Vf* droop characteristics which are shown in Fig. 3.22. Hence, this method is also known as *droop method.*

In VSI control, voltage is related to reactive power (*V–Q*), whereas frequency/ phase shift is related to active power (*f–P*) as shown in the below equations.

$$f = f_0 - K_p \Delta P$$
$$V = V_0 - K_q \Delta Q$$

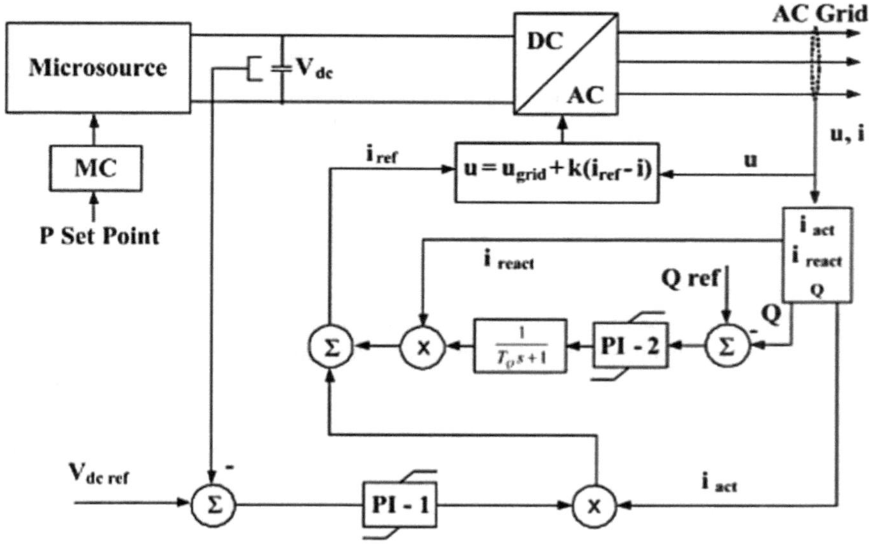

Fig. 3.21 PQ control scheme [78]

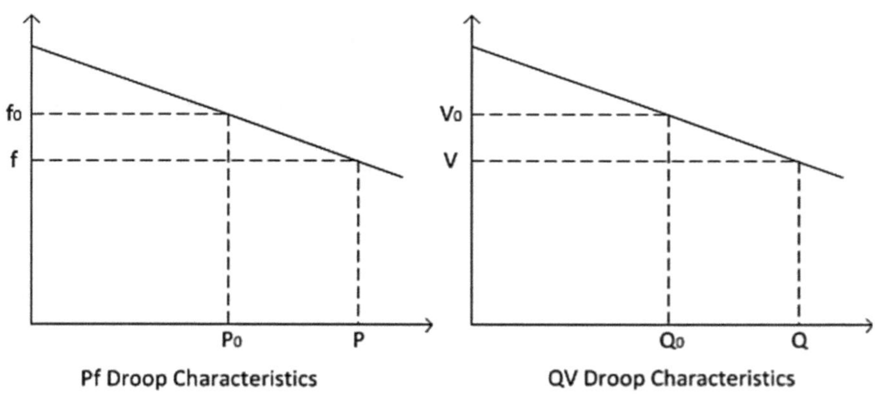

Fig. 3.22 V and f characteristics

where K_p and K_q are respective slopes of droop characteristics and f_0 and V_0 are the idle values of frequency and voltage [29]. The microsources will change their output by ΔP (or ΔQ) when the frequency (or voltage) changes by Δf (or Δv) from the nominal values f_0 (or v_0).

A VSI model is shown in Fig. 3.23 [29, 77]. The active and reactive powers are computed using the VSI terminal voltage. The output voltage frequency f is determined by the active power droops and magnitude V is determined by reactive

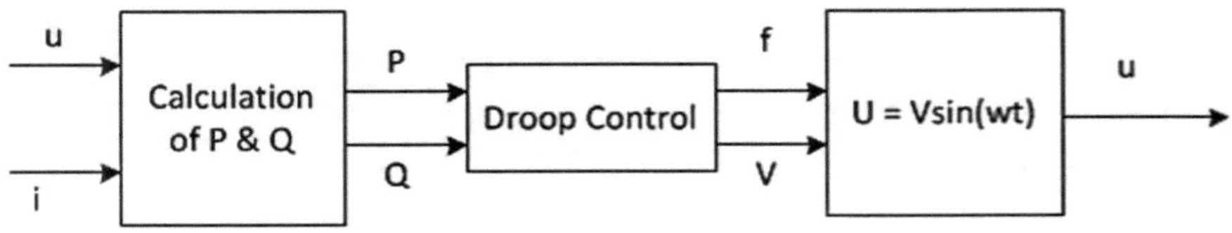

Fig. 3.23 Model of a VSI

power droops. The output voltage are the reference signals which control the VSI switching sequence.

The key point to be observed here is VSI uses the local measurements at its terminals and reacts to any system disturbances quickly and hence does not require any communication infrastructure [14, 66]. But there will be a communication infrastructure within the DG for the optimal management [78]. A complete review of *Vf* control strategies can be found in [9] and validation of *Vf* control in both grid and islanded mode is performed in [10].

3.3.9 Autonomous Control

The concept of utilizing *VQ* and *Pf* droops for controlling the MG is also proposed in [64, 65, 102]. They propose an autonomous control for the peer-to-peer and plug-and-play model of the MG components. The concept of peer-to-peer allows the continuous operation of MG even with loss of any component/DG because there is no master controller or central storage unit. The concept of plug-and-play ensures that any component can be added at any point in the system without re-engineering the controls.

MC based on droop characteristics is shown in Fig. 3.24. P, Q, and V are calculated and then the droops are implemented in two separate blocks. Then the controller generates the voltage at the desired magnitude and frequency at the inverter terminals.

3.3.10 New Q–$\dot{V}$ Droop Control

In islanded mode, the reactive power sharing is highly dependent on impedance of power line. Due to the different distances among DERs interface converters (DICs), the equivalent transmission line impedance could be unequal [29]. *Pf* and *QV* droop characteristics are used in DICs for power sharing operations [68, 102]. The *Pf* droop control provides an accurate real power sharing among the DICs but the problem arises in *QV* droop control. Because of this unequal impedance load sharing performance of *QV* control can be affected. Various control methods addressing this issue have been proposed [38, 72, 108] but with some constraints.

Therefore, a new droop control method for the islanded operation of MG is proposed in [71] to overcome the effect of line impedance on the reactive power flow,

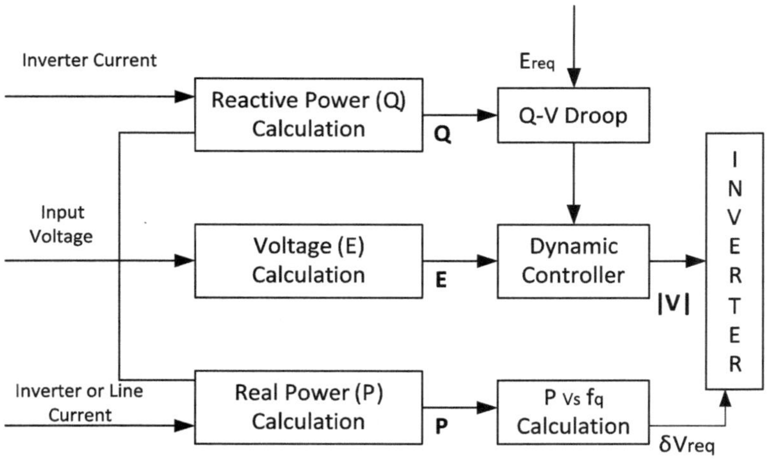

Fig. 3.24 Microsource controller using droops

this method is known as Q–$\dot{V}$ droop control method where $\dot{V}$ represents the rate of change in voltage. By regulating the voltage with $\dot{V}$, the reactive power sharing can be made independent of the line impedance. The operation principle of the proposed method is shown in Fig. 3.25.

3.3.11 Control Design Based on Transfer Function

Controller based on the transfer function of the plant is designed in [52, 104]; this is adopted from the classical feedback control approach presented in [35].

For the islanded mode of operation, Fig. 3.26 shows the structure of controller. Reference angle is provided by a three-phase PLL. The q component of load voltage is set to zero and d component is regulated to the desired peak value. Regulation of v_d is done by comparing with reference signal and the error is applied to the controller. The controller then provides inputs to the gating signal generator of the VSC (Fig. 3.5). More details can be found in [52, 104].

Instead of using frequency droops, an internal oscillator is used to design a multi-variable controller in [53]. The function of this oscillator is to control the frequency in an open loop way. The robust servomechanism controller was designed using the parameter optimization methods [19, 20] along with non-conservative robustness constraint [107].

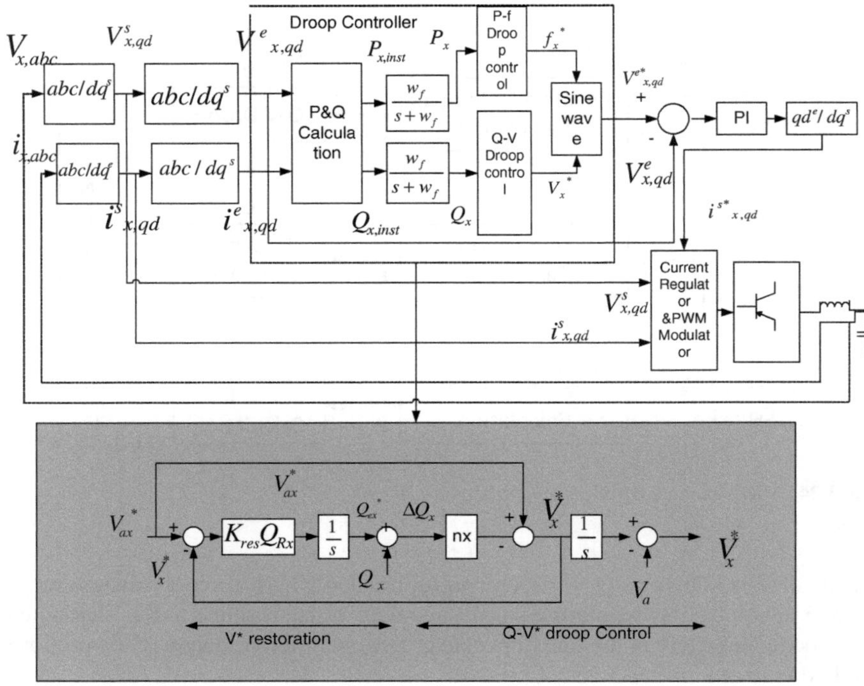

Fig. 3.25 Block diagram of DIC with $Q\dot{V}$ droop control [71]

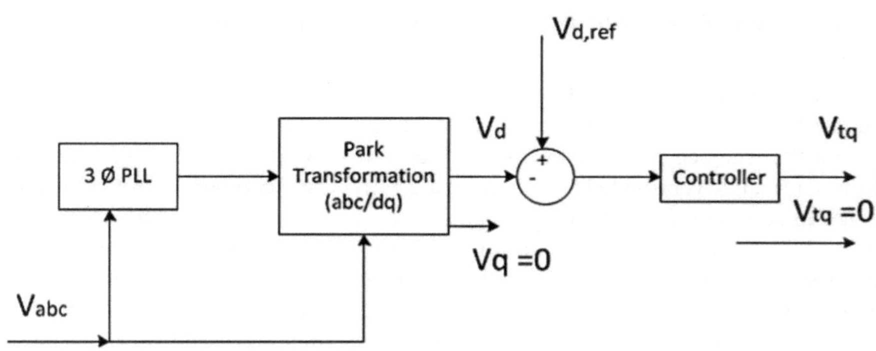

Fig. 3.26 Control strategy of [52, 104]

3.3.12 Microgrid Control in Both Modes

A novel control strategy which can be applied to MG in both the modes is introduced in [36]. This scheme of control allows a DG to control its real and reactive power components in grid-connected mode and to control the voltage and frequency in autonomous mode.

The proposed VSC frequency control is similar to that of a SM and is shown by the block diagram in Fig. 3.27. In grid-connected mode, the frequency at PCC (ω_{pcc}) is equal to the grid frequency and hence has no impact on the system dynamics and only the output real power reference has to be set. In autonomous mode, it is the same as VSC frequency determined by the droop characteristics.

The block diagram in Fig. 3.28 shows the proposed voltage control scheme. During grid-connected mode, output reactive power of VSC at PCC is set while in autonomous mode, the DG has to supply the load with reactive power, which is achieved by setting E_2 to zero and only E_1 remains effective.

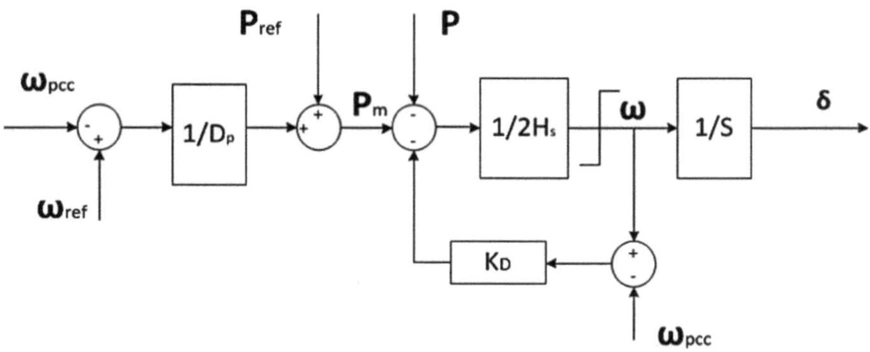

Fig. 3.27 Block diagram for the *frequency* control [36]

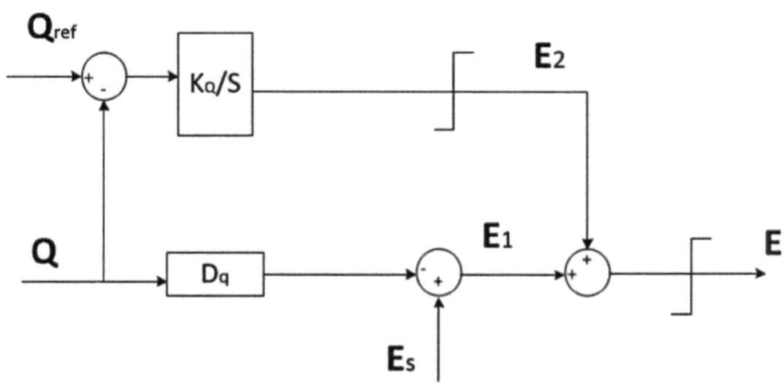

Fig. 3.28 Block diagram for the *Voltage* control [36]

3.4 System of Systems

To understand the concept of SoS or cyber-physical system (CPS), let us consider an airplane which is an example of large-scale complex system. Various parts of airplane are operated by different systems but the plane flies only when all its systems operate collectively and does not fly if they operate individually. Therefore, a SoS is the large-scale integration of many systems that combine their capabilities together to form a more complex system offering more functionalities than the individual sum of the constituent systems [84].

3.4.1 Introduction

SoS is inherently multidisciplinary. The synthesis of multi systems requires the study of their interdependency because each effects the other. This will result in different problems ranging from modeling to control. Therefore almost all the key issues of systems engineering have to be revised. The methodology of developing models, tools, and control of SoS is typically referred to as SOS engineering. SoS methodology finds large number of applications in the defense sector but recently it is also being used in auto transportation, space exploration, search and rescue, and many other non-defense areas [46].

For SoS engineering branch, all aspects of systems engineering have to be revised. But the main problem arises in the two key aspects of modeling and control. The challenge in modeling point of view lies in the indirect effects, the cause and effect need not be directly related, any change *here* can produce effect *over there* because of their interdependency. It is clear that these systems are very large and possess unpredictable behavior which is difficult to model. SoS models available so far are still immature and there should be focus on additional development. These models are quite complex and need a multidisciplinary approach [46, 80–83].

Control is perhaps the most critical challenge facing SoS designers. Due to the difficulty or impossibility of developing a comprehensive SoS model, either analytically or through simulation, SoS control remains an open problem and is, of course, different for each application domain. Moreover, real-time control which is required in almost all application domains of interdependent systems poses an especially difficult problem [49, 50].

3.4.2 SoS Control

The key issue of SoS, which is control, faces the main challenge of developing a comprehensive SoS model, analytically or by simulation. Availability of a proper model is necessary to design a controller. If a proper mathematical model is available then there are several available control strategies. Also, control strategy for each system is not only dependent on its own sensory information but also on the communication links among its neighboring systems or components; this is another difficulty which rises from the control point of view. Control of SoS, which is different for each application domain, is still an open research area. In this section we will discuss several potential control strategies.

3.4.3 Decentralized Control

Another control lacking real-time consideration is DEC [46]. In this scheme of control, SoS is assumed to be having multiple input and output variables. The control design aims at assigning proper inputs for proper controller which can observe a set of outputs. Thus there are multiple controllers, each one controls a particular operation of SoS. As it can be seen in Fig. 3.29, this scheme avoids storage of data.

For the islanded operation of an MG, a robust DEC strategy is proposed in [4]. The proposed model of MG is shown in Fig. 3.6. It is shown that the MG can be represented by an interconnected composite system consisting of two subsystems [18] and each subsystem can be controlled using the LOCs.

The dynamic model of the MG is decomposed into two subsystems as follows:

- Master subsystem
- Slave subsystem

Since the two subsystems are controllable and observable, it is shown that the composite system is stabilizable by using only LOCs, i.e., DEC strategy can be applied.

For the master subsystem, an H_∞ controller was designed to meet the robust characteristics [126]. This control strategy fulfills the voltage and frequency requirements of the load. The configuration of H_∞ control is shown in Fig. 3.30. MATLAB LMI toolbox is used to synthesize H_∞ controller [112].

A simple PI controller was designed for the slave subsystem using the conventional dq current control method [113, 119]. This is depicted in Fig. 3.31. The overall model and its controllers were simulated in MATLAB/SimPowerSystems toolbox.

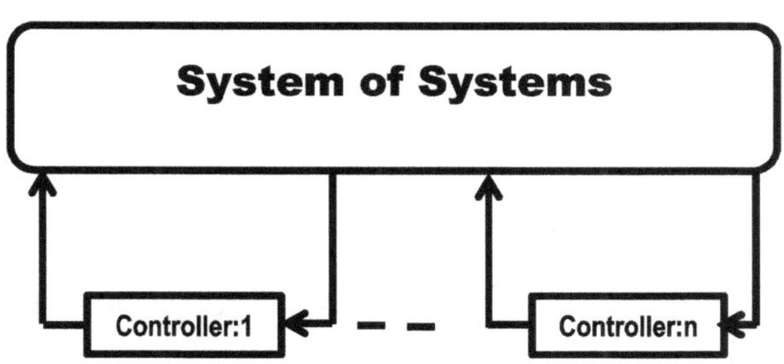

Fig. 3.29 Decentralized SoS control scheme

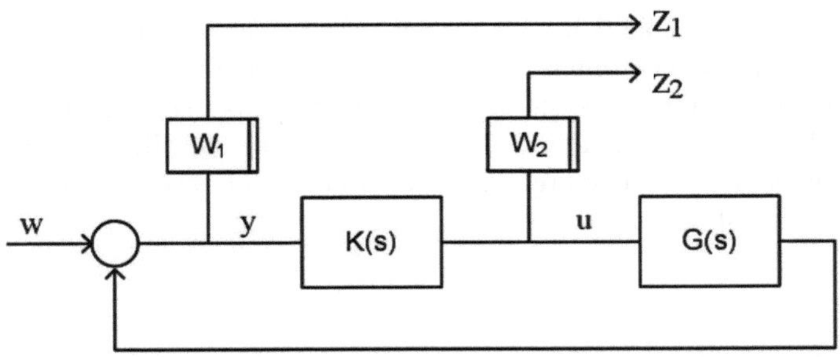

Fig. 3.30 H_∞ Control for Master subsystem [4]

A fundamental concepts of power management system (PMS) and robust DEC strategy for the islanded MG is proposed in [32]. The schematic diagram of proposed control is illustrated in Fig. 3.32. It consists of a PMS, LOC for each DER and MG frequency control and synchronization scheme.

A low bandwidth communication system is used to supply the instantaneous values of real/reactive power of each DER unit and load to the PMS. The PMS determines the set-points of real, reactive power and voltage for the PC buses and

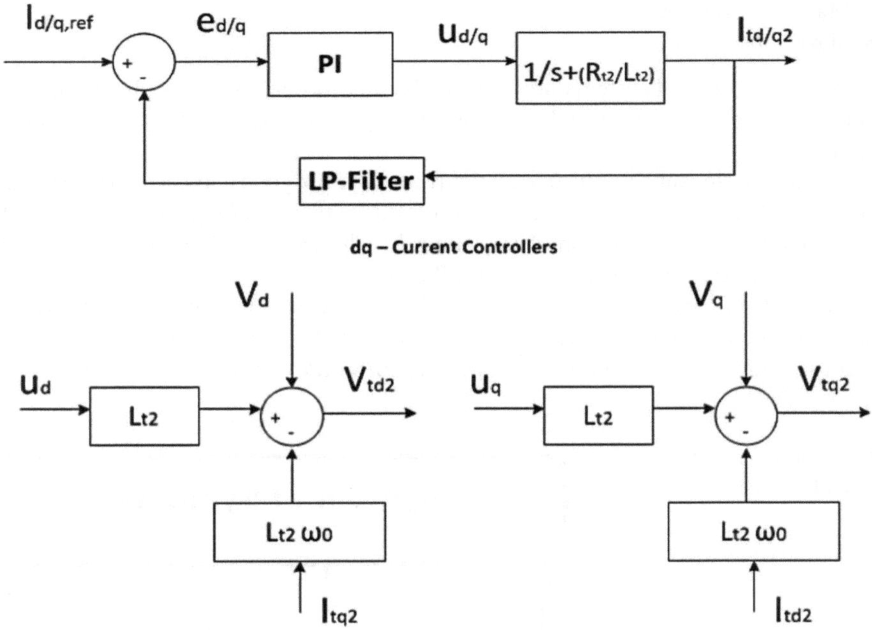

Fig. 3.31 dq Current control method for slave subsystem [4]

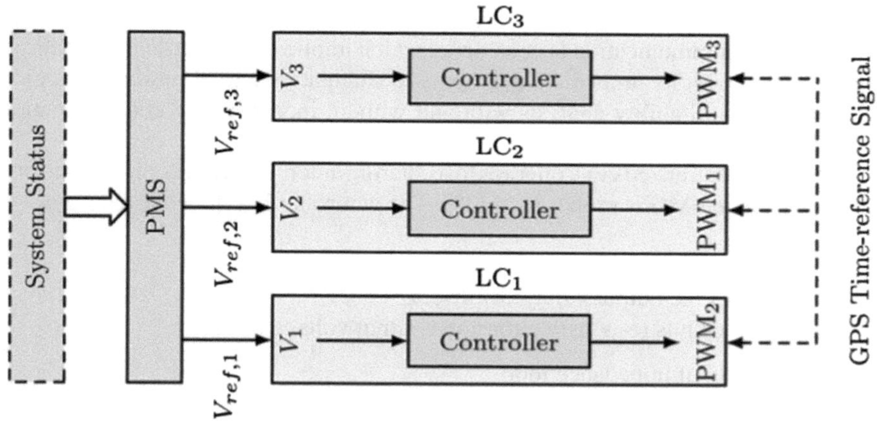

Fig. 3.32 PMS and control scheme in [32]

transmits to LCs that measure the magnitude of voltage at its PC bus providing voltage tracking based on the received reference set-point.

A decentralized inverter control based on wireless communication is proposed in [74]. Wireless communication is used to enhance the stability of droop-based decentralized inverter control. A wireless network is developed so that each inverter can communicate with a certain set of inverters. Figure 3.33 shows the block diagram on inverter control.

Droop-based inverter control scheme is adopted [87, 103]. The controller of each individual inverter consists of three parts; the power controller, voltage controller, and current controller. Only the stability of power controller is considered whereas

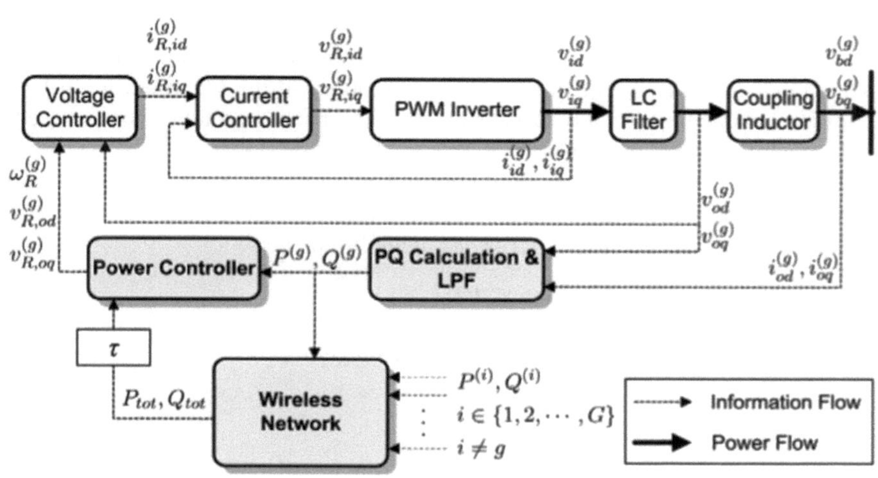

Fig. 3.33 Inverter Control scheme in [74]

the voltage and current controllers are based on traditional PI controllers. A fully decentralized communication is considered which implies any inverter only needs to communicate with its immediate neighbors to calculate the total power generation of all DG units. Stability analysis with and without incorporating communication delay is presented.

A novel decentralized controller for load sharing among parallel connected inverters in an islanded MG is proposed in [39]. The controller has three nested loops

1. Inner loop
   - Regulates the output voltage of inverter
   - Voltage gain is responsible for good output voltage tracking
2. Resistive output impedance loop
   - Reduces the impact of line impedance unbalance
   - Used to fix the output impedance of the inverter in terms of magnitude and phase
   - The output impedance presented to harmonic components can be fixed
3. P/Q sharing outer loop
   - Used to obtain proper P/Q sharing
   - Droop/boost control scheme is used.

3.4.4 Multilevel Control

As discussed earlier, SoS is integration of large-scale systems and large-scale systems can be decomposed into subsystems. *MultiLevel control* assumes SoS is characterized by *N* finite set of subsystems coordinated by system coordinator as shown in Fig. 3.34. By employing any optimal control method the subsystems can be optimized and repeatedly perform the modeling, the interactions between the coordinator and subsystems can be converged to an optimal solution. In the literature, multi-level control is obtained by the classical steady-state approach but a lot of issues have to be dealt while its implementation in real time [46, 80–83].

In what follows, a control scheme comprised of three main levels, namely primary control, secondary control, and tertiary control is shown in Fig. 3.35 [40, 117].

The key points related to *primary control* are listed below

1. Used to share load between converters
2. Improves the system performance and stability
3. Regulates the output frequency and voltage magnitude

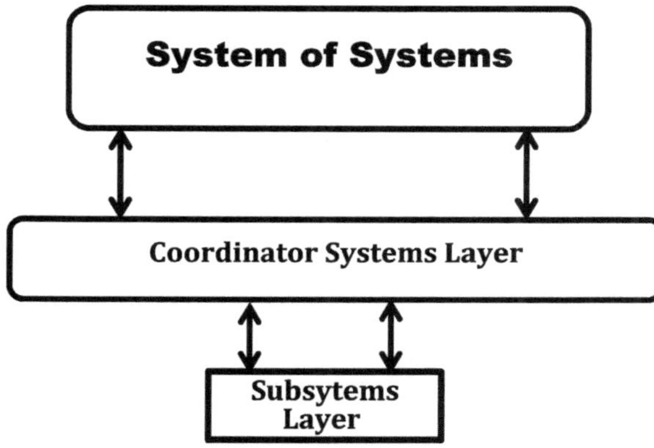

Fig. 3.34 Multilevel SoS control

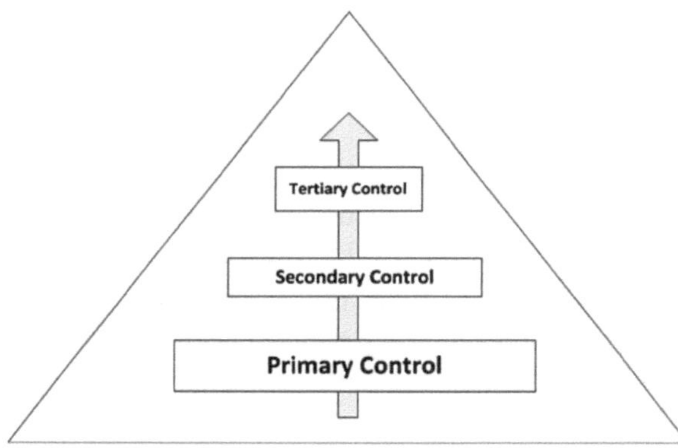

Fig. 3.35 Structure of multilevel control scheme

4. *Droop-control* method is often used
5. Can also include virtual impedance control loop to provide proper output impedance

The key points related to *secondary control* are listed below:

1. Restores the f and V to nominal values whenever load change occurs
2. Removes any steady-state error introduced by the droop control
3. During transition from islanded to grid-connected mode, this control can perform synchronization to the main grid before interconnection
4. Makes use of low bandwidth communication
5. More global responsibilities.

The primary control loop makes use of only local output voltage and current to perform calculations of droop control method, whereas the secondary control level consists of an external centralized controller to correct the errors produced by the primary control. Both controls are depicted in Fig. 3.36.

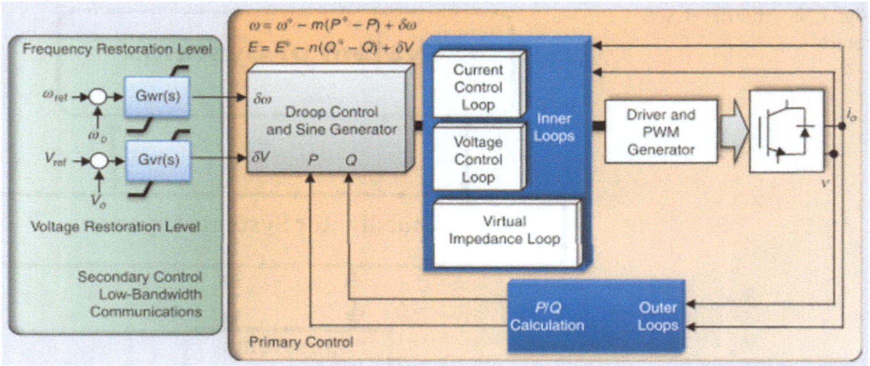

Fig. 3.36 Primary and secondary control

The tertiary control level comes into play mainly when the MG interacts with the utility grid. The key points related to *Tertiary control* are listed below:

1. Controls the power flow between MG and the utility grid;
2. Sends the frequency and voltage references to the secondary control;
3. Can perform islanding detection or voltage harmonic reduction;
4. Can also improve the QoP at PCC.

The proposed multilevel control scheme allows the system to integrate more and more MGs and with this scheme of control MGs can operate in both grid connected and islanded mode.

3.4.5 Networked Control Systems

In modern control systems, we find more and more application of networks owing to impressive advancements in network technology. One such example is *networked control system* (NCS). In NCS, the feedback channel is closed using a real-time communication network and all the data among the components of system are exchanged through this communication network [124]. In [45], NCS is properly defined as, "Network Control Systems (NCS) are spatially distributed systems in which communication between sensors, actuators, and controllers occurs through a shared band limited digital communication network." This definition explains that the components of NCS are distributed and may operate asynchronously to reach some overall objective [6].

One of the main issues in NCS is the transmission delays and packet dropouts, therefore the challenge in NCS for SoS is to develop an SoS distributed control system (DCS) which can overcome these issues. As mentioned in [59], these communication infractions can be compensated by

1. Adjusting control power and controlling distances between systems (power control);
2. Trading off modulation, coding, and antenna diversity versus throughput (adaptive modulation coding);
3. The (non-wireless) intra-feedback (on-board hardware) loop of the autonomous control within S_i is lower latency than the inter-wireless distributed control loop between S_i and S_j or the inter-wireless SoS controller and the S_i controller.

Another way to check on these communications is to design a wireless network control system (WNCS) taking account of all the aspects of the ad hoc network [16]. In this design, the distributed control will generate two components at each sampling period, one is LOC, which is classical or modern control and the other is correction component of the controller, which compensates for the ad hoc network quality of service (QoS) parameters. As shown in Fig. 3.37, with the combination of a LOC, correction component, and adaptive sampler the stability and robustness will be enhanced.

In the previous section, a multilevel control scheme for MG is discussed. The primary and tertiary control levels are decentralized and centralized, respectively, because one aims at the control of DG and other at global optimization of MG. Conventionally, secondary control is implemented in the MGCC but recently a new distributed control scheme for the secondary known as NCS is proposed in [111], which is shown in Fig. 3.38.

This strategy is proposed for power electronically based MGs. The primary and secondary control are implemented in DG unit. The primary control which is generally droop control is already discussed in Sect. 3.3. The secondary control has frequency, voltage, and reactive power controls in a distributed manner. The secondary

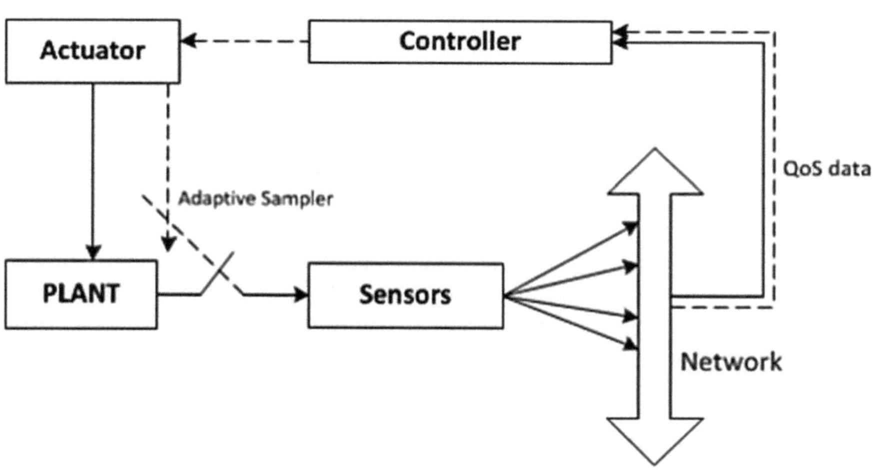

Fig. 3.37 NCS for SoS with sampler

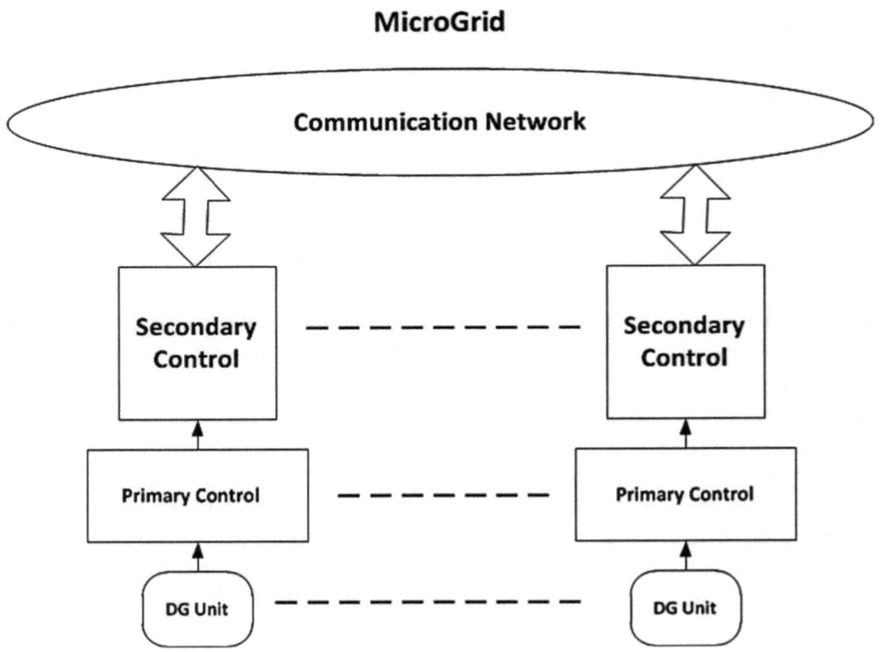

Fig. 3.38 Networked controlled MG [111]

control gathers all the measurements from DG units using the communication system, averages them, and generates the proper control signal for the primary control level. The schematic diagram of this proposed control scheme can be seen in Fig. 3.39 and the detailed explanation can be found in [111].

In another approach, a real-time network is used for the control of parallel multi-inverter system [125]. MG makes use of this type of inverter connections to deliver energy.

The considered system is shown in Fig. 3.40. It consists of a CC, communication network, and inverters with their LOCs. The control strategy is as follows: LOCs send the voltage and current measurements to the CC via network frame. A closed-loop control inside the CC produces the satisfactory pulse-width modulation (PWM) duty ratio for each inverter module and sends it back to the LOC via another network frame. This is a centralized control strategy where CC has all the central information.

A proportional-integral-derivative (PID) controller is used to achieve better inverter performance [98] and D-partition method is used to determine the stability region of PI controller [75]. It was found that network-induced delay brings about a considerable effect on closed-loop control of a single inverter. The practical model implementing the NCS is shown in Fig. 3.41.

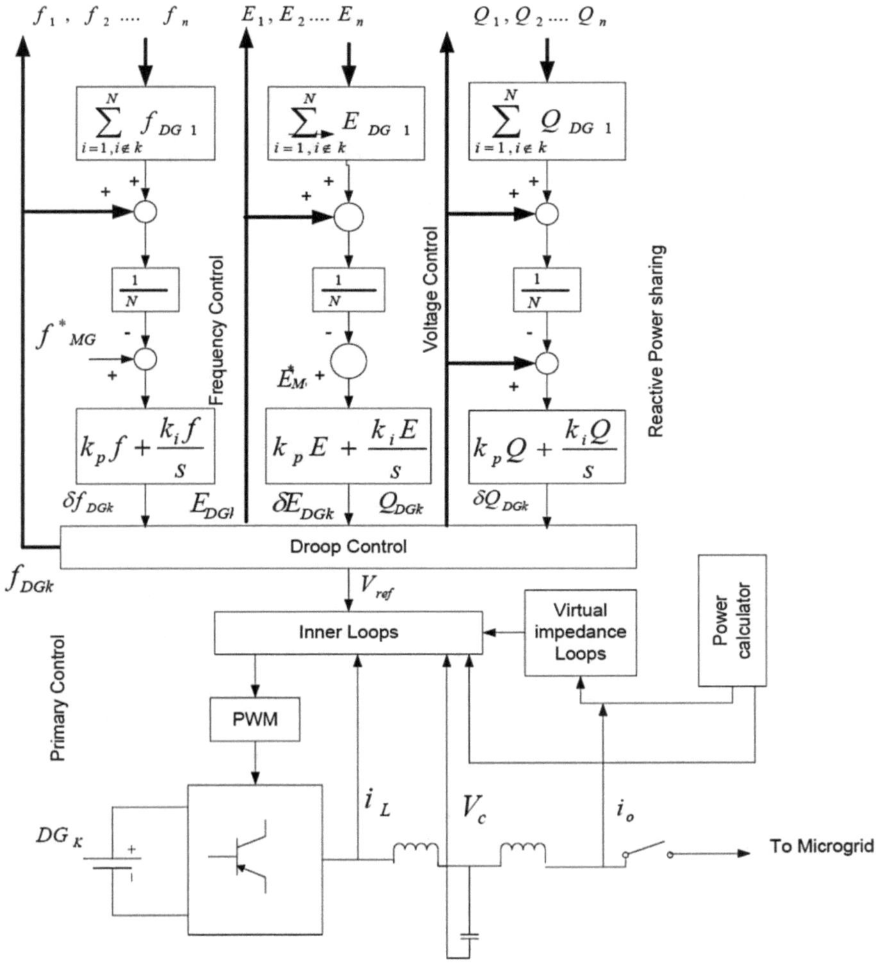

Fig. 3.39 Distributed secondary control [111]

3.4.6 Comparative Analysis

After discussing the control techniques, it is worth performing the comparative analysing of the control techniques most commonly used. In this section, we classify the control techniques considering vital aspects for the purpose of simplification and better understanding. Control strategies for MG are vast and detailed comparison of each techniques with another is out of the scope of this paper.

MG control (depending on architecture) can be generally classified into two main streams, namely centralized, distributed (or decentralized). Multi-level control is also most widely used, but again depending on architecture of control levels, even this

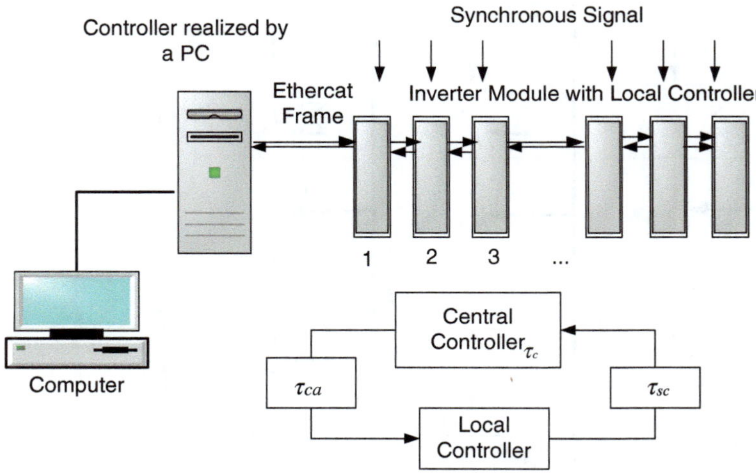

Fig. 3.40 Control of parallel multi-inverter system [125]

control techniques falls in the former mentioned categories. For instance in multi-level control, the secondary control level can be single (centralized) [40] or it can be implemented in a distributed way [111]. Figure 3.42 shows the performance of the controller when implemented in centralized and decentralized manner. The time delay in the communication network is taken in account here. It is obvious from the figure that, the control implemented in distributed way is better.

Depending on the power sharing, the methods available in the literature can be broadly divided into two groups, one using communication medium and another is non-communication based. Because of their inherent advantages, only communication-less control which is also known as Droop-based control is used. It was initially proposed in [14] and since then there were many variations performed keeping the basic idea of droop control [3, 71, 86]. One such comparison is shown in Fig. 3.43. A modified droop control technique, designed to improve the reactive power sharing among the DG units, is proposed and analyzed with the conventional one [71].

3.5 Modeling and Analysis of Inverter-Based Microgrids

The analysis of the small-signal stability of conventional power systems is well established, but for inverter-based MGs there is a need to establish how circuit and control features give rise to particular oscillatory modes and which of these have poor damping. This section develops the modeling and analysis of autonomous operation of inverter-based MGs. Each submodule is modeled in state-space form and all are combined together on a common reference frame. The model captures the detail of the control loops of the inverter but not the switching action. Some inverter modes are

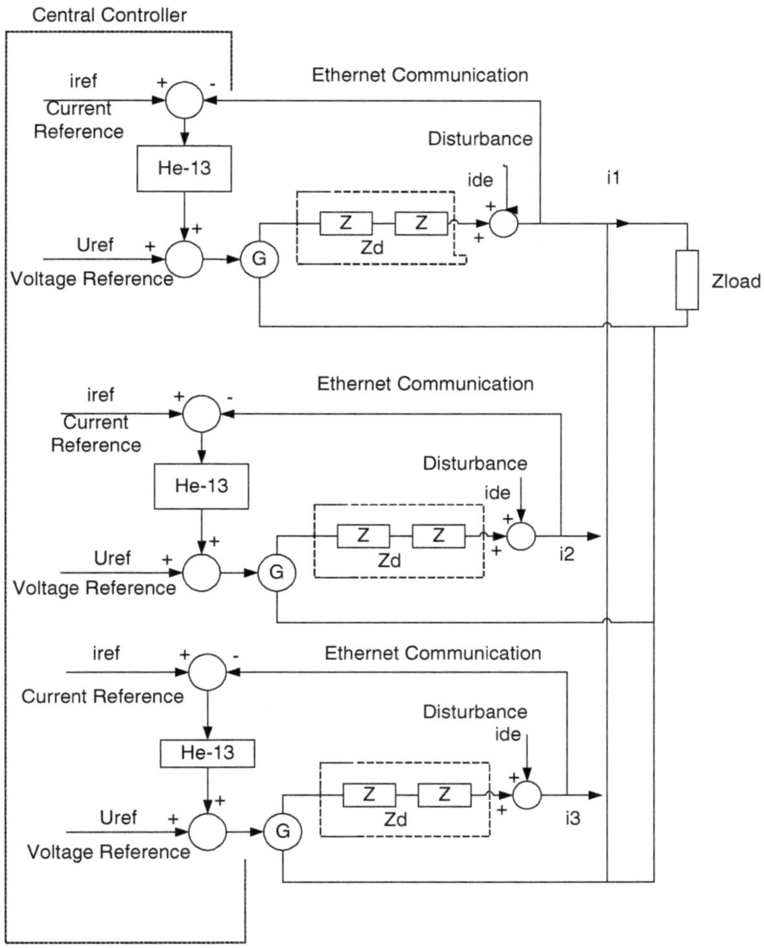

Fig. 3.41 NCS controlled parallel multi-inverter model [125]

found at relatively high frequency and so a full dynamic model of the network (rather than an algebraic impedance model) is used. The complete model is linearized around an operating point and the resulting system matrix is used to derive the eigenvalues. The eigenvalues (termed "modes") indicate the frequency and damping of oscillatory components in the transient response. A sensitivity analysis is also presented which helps in identifying the origin of each of the modes and identifies possible feedback signals for design of controllers to improve the system stability. With experience, it is possible to simplify the model (reduce the order) if particular modes are not of interest as is the case with SM models. Experimental results from an MG of three 10-kW inverters are used to verify the results obtained from the model.

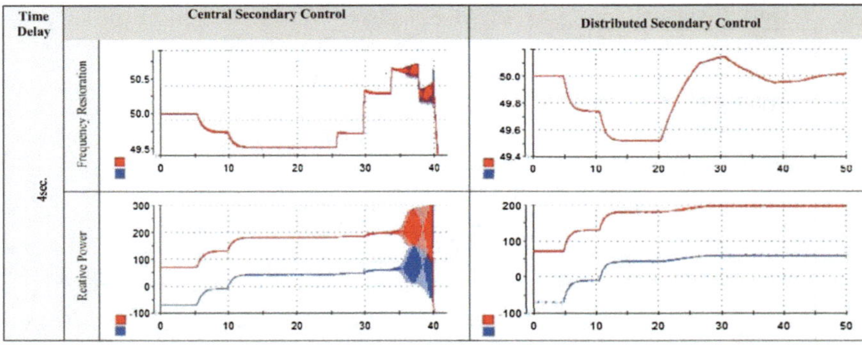

Fig. 3.42 Performance analysis of centralized and distributed controllers

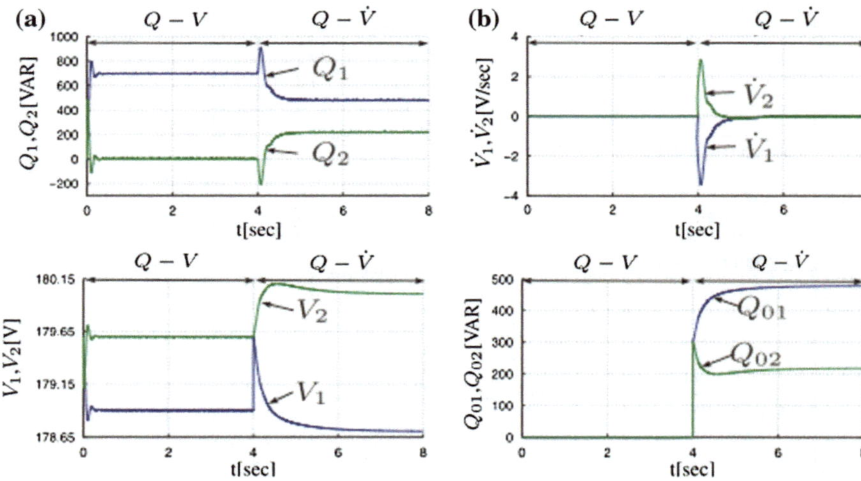

Fig. 3.43 Performance analysis of primary controllers

3.5.1 Introduction

Recent innovations in small-scale distributed power generation systems combined with technological advancements in power electronic systems led to concepts of future network technologies such as MGs. These small autonomous regions of power systems can offer increased reliability and efficiency and can help integrate renewable energy and other forms of DG [57]. Many forms of DG such as fuel-cells, photovoltaic, and microturbines are interfaced to the network through power electronic converters [3, 61]. These interface devices make the sources more flexible in their operation and control compared to conventional electrical machines. However, due to their negligible physical inertia they also make the system potentially susceptible to oscillation resulting from network disturbances.

An MG can be operated either in grid connected mode or in stand-alone mode. In grid connected mode, most of the system-level dynamics are dictated by the main grid due to the relatively small size of microsources. In stand-alone mode, the system dynamics are dictated by microsources themselves, their power regulation control and, to an unusual degree, by the network itself.

One of the important concerns in the reliable operation of an MG is small-signal stability. In conventional power systems, stability analysis is well established and for the different frequency ranges (or time horizons) of possible concern there are models which include the appropriate features. The features have been established on the basis of decades of experience so that there are standard models of synchronous machines, governors, and excitation systems of varying orders that are known to capture the important modes for particular classes of problem. This does not yet exist for MGs and may be difficult to achieve because of the range of power technologies that might be deployed. However, we can begin by developing full-order models of inverters and the inverter equivalents of governors and exciters. Examination of these models applied to various systems will develop a body of experience that allows reduced order models to be selected for some problems.

Previous dynamic analysis of stand-alone systems has been carried out by assuming an ideal inverter as in [15]. This means that the closed-loop inner controllers that track voltage and current references are assumed to track perfectly, accurately, and quickly. They therefore do not have any effect on the small-signal stability. This assumption is based on the fact that the closed-loop bandwidth of the inverter is well above the bandwidth of power sharing controllers that set the voltage and current references. This is a relatively safe assumption for low power inverters with a high switching frequency, but cause important dynamics to be omitted for large inverters where low switching frequency limits the control bandwidth of the innermost control loop. The modeling approach presented in [70] concentrates on stability issues for an individual inverter connected to a stiff AC bus. This is valuable in illuminating inverter properties but needs extension to cover the interaction of inverters with each other and with network dynamics before it can indicate the nature of stability issues in MGs.

3.5.2 Microgrid Model in Autonomous Mode

A typical characteristic of an MG is that it can be operated either in grid connected or in islanded (autonomous) mode. Normally, when an MG is operated in grid connected mode the microsources act as constant power sources which means that they are controlled to inject the demanded power into the network. In autonomous mode the microsources are controlled to supply all the power needed by the local loads while maintaining the voltage and frequency within the allowed limits. Autonomous operation of an MG might be initiated for either of the following two reasons. First, because of preplanned (intentional) islanding due to maintenance or economical reasons. Depending on the market situation the owner of an MG can chose between autonomous and grid connected modes [24, 58, 62]. Second, because of unplanned (unintentional) islanding due to the failure of the main grid caused by a network fault.

Autonomous operation is realized by opening the isolating switch (shown in Fig. 3.44) which disconnects the MG from the main grid. Once the MG is isolated the microsources feeding the system are responsible for maintaining the voltage and frequency while sharing the power. During autonomous operation it is important to avoid overloading of inverters and to ensure that the changes in load are taken by inverters in a well-controlled manner. Control techniques based on a communication link, such as the master–slave approach [106], can be adapted in systems where microsources are connected to a common bus or located in close proximity. However, a communication link makes the system more expensive and less reliable. Also, in a typical MG microsources can be located far away from each other making a communication link less attractive. Control techniques based on local measurements, which do not require expensive communication facilities, have been proposed [14]. See also [66–74, 78–92].

In this section, the latter method is adopted. The modeling approach presented in this paper divides the whole system into three major submodules; inverter, network, and loads (Fig. 3.45). Each inverter is modeled on its individual reference frame whose rotation frequency is set by its local power sharing controller. The inverter model includes the power sharing controller dynamics, output filter dynamics, coupling inductor dynamics, and voltage and current controller dynamics. These last two elements introduce high frequency dynamics which are apparent at peak and light load conditions and during large changes in load. The small-signal flow among the submodules shown in Fig. 3.45 will be explained in the following sections.

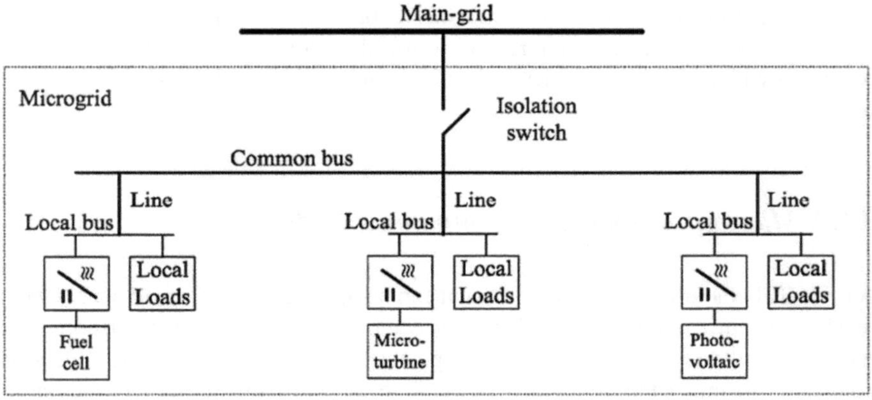

Fig. 3.44 Typical structure of inverter-based microgrid

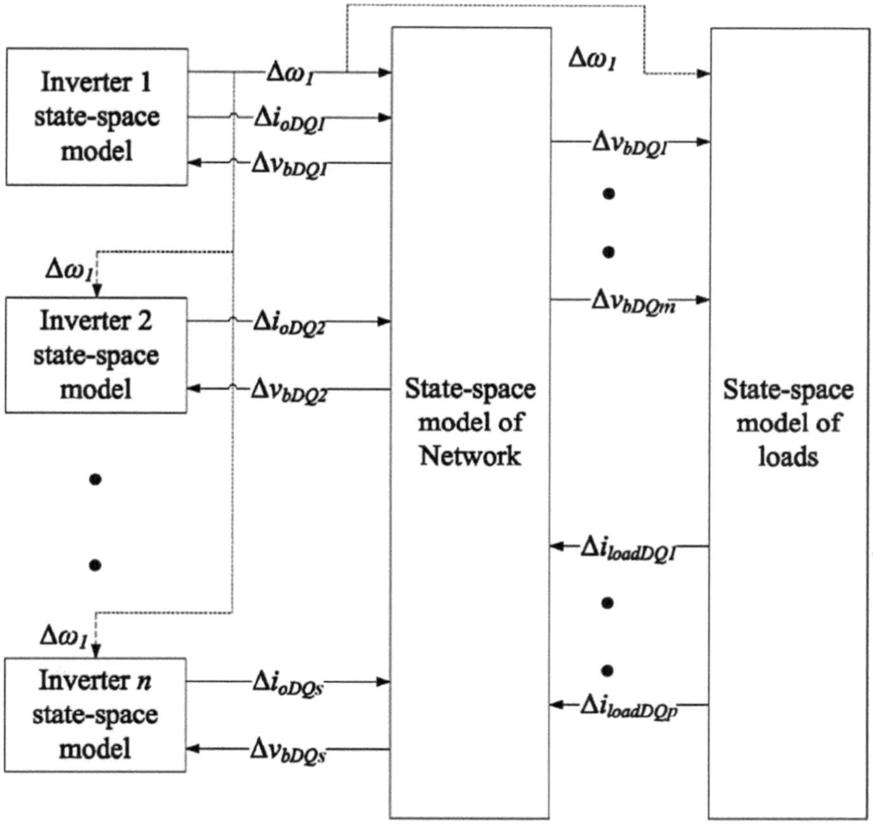

Fig. 3.45 Block diagram of complete small-signal state-space model of an microgrid

Network dynamics are generally neglected in small-signal modeling of conventional power systems. The reason behind this is that the time constants of rotating machines and their controls are much larger than those of the network. In the case of MGs, the microsources are connected through inverters whose response times are very small and network dynamics would influence the system stability. Previous work [15] on small-signal modeling of parallel connected inverters was carried out without considering the network dynamics.

Here, the state equations of the network and the loads are represented on the reference frame of one of the individual inverters. This reference frame is considered as the common reference frame. All the other inverters are translated to this common reference frame using the transformation technique [99] depicted in Fig. 3.46 and defined in (3.3). Here, the axis set $(D - Q)$ is the common reference frame rotating

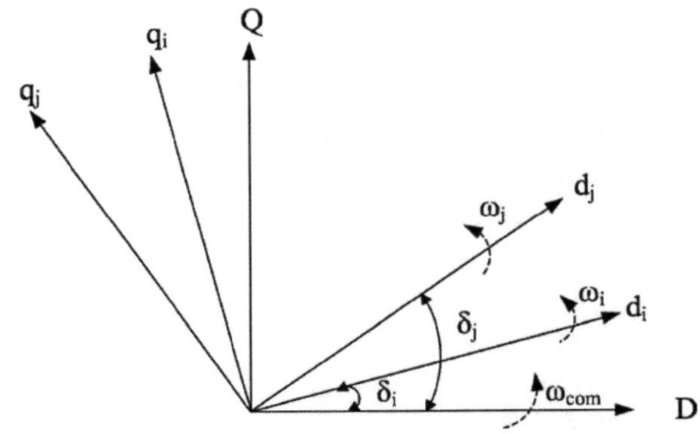

Fig. 3.46 Reference frame transformation

at a frequency W_{com}, whereas axes $(d-q)_i$ and $(d-q)_j$ are the reference frames of ith and jth inverters rotating at W_i and W_j, respectively

$$f_{DQ} = [T_i][f_{dq}] \tag{3.3}$$

$$T_i = \begin{bmatrix} \cos(\delta_i) & -\sin(\delta_i) \\ \sin(\delta_i) & \cos(\delta_i) \end{bmatrix}. \tag{3.4}$$

In (3.3) and (3.4), δ_i is the angle of the reference frame of ith inverter with respect to the common reference frame. In the following sections the internal modeling of all the three modules is discussed in detail. It is to be noted that in the equations of the following sections the three-phase voltages and currents are represented as vectors in $d-q$ frame, whereas the other variables such as real and reactive powers and angles are scalars.

3.5.3 State-Space Model of a Voltage Source Inverter

VSI is commonly used to interface distributed generators to the network. Figure 3.47 shows the block diagram of an inverter connected to the MG. The power processing section consists of a three-leg inverter, an output *LC* filter, and coupling inductor. Assuming an ideal source from the DG side, the DC bus dynamics can be neglected.

With the realization of high switching frequencies (4–10 kHz), the switching process of the inverter may also be neglected. The controller of a DG inverter can be divided into three different parts. First, is an external power control loop which sets the magnitude and frequency (and hence phase) for the fundamental component of the inverter output voltage according to the droop characteristics set for the real and reactive powers (harmonic power sharing has been treated as an additional function with a different control topology [86]). The second and third parts of the control system are the voltage and current controllers, which are designed to reject

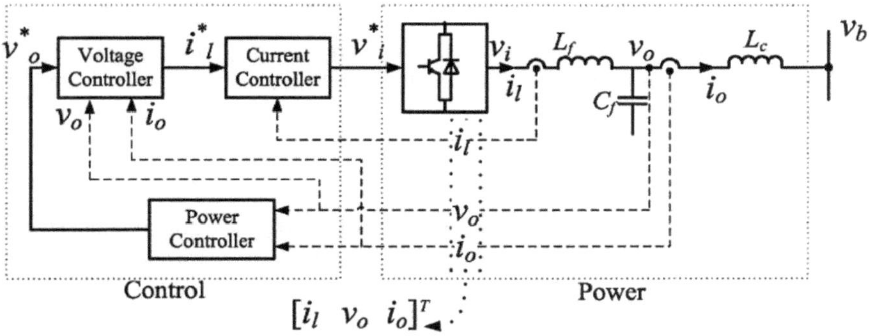

Fig. 3.47 DG inverter block diagram

high frequency disturbances and provide sufficient damping for the output LC filter [86, 95–106].

In this section, a state-space model is presented for all of the subsystems: control loops, output filter, and coupling inductor. The model is constructed in a rotational reference frame set by the external power controller of the particular individual inverter.

1. *Power controller*: The basic idea behind the droop control is to mimic the governor of a synchronous generator (SYGN). In a conventional power system, SYGNs will share any increase in the load by decreasing the frequency according to their governor droop characteristic. This principle is implemented in inverters by decreasing the reference frequency when there is an increase in the load. Similarly, reactive power is shared by introducing a droop characteristic in the voltage magnitude.
   As shown in Fig. 3.48, instantaneous active and reactive power components $\tilde{p}$ and $\tilde{q}$ are calculated from the measured output voltage and output current as

$$\begin{aligned}\tilde{p} &= v_{od}i_{od} + v_{oq}i_{oq}\\ \tilde{q} &= v_{od}i_{oq} + v_{oq}i_{od}.\end{aligned} \tag{3.5}$$

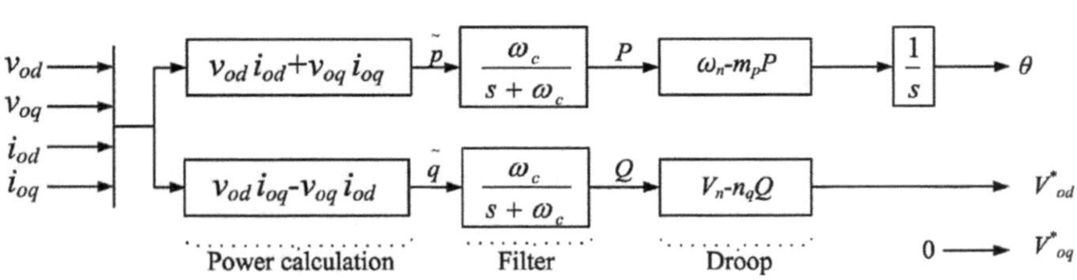

Fig. 3.48 Power controller

The instantaneous power components are passed through low-pass filters, shown in (3.6), to obtain the real and reactive powers P and Q corresponding to the fundamental component. W_c represents the cut-off frequency of low-pass filters

$$P = \frac{W_c}{s + W_c}\tilde{p}, \qquad Q = \frac{W_c}{s + W_c}\tilde{q}. \tag{3.6}$$

The real power sharing between inverters is obtained by introducing an artificial droop in the inverter frequency as in (3.7). The frequency W is set according to the droop gain (m_p) and phase is set by integrating the frequency. This mimics to governor and inertia characteristics of conventional generators and provides a degree of negative feedback. For instance, if the power drawn from a generator increases, then the rotation of its voltage slows and its angle retards. In the following equations W_n will represent the nominal frequency set-point, whereas α is the angle of the inverter reference frame seen from a reference frame rotating at W_n. From (3.7) it can be seen that the angle of the inverter voltage, α, changes in response to the real power flow in the required negative sense and with a gain set by the droop

$$\begin{aligned} w &= w_n - m_p P \\ \dot{\theta} &= w, \quad \theta = w_n t - \int m_p P dt \\ \alpha &= -\int m_p P dt, \quad \dot{\alpha} = -m_p P. \end{aligned} \tag{3.7}$$

To share the reactive power among multiple inverters, a droop is introduced in the voltage magnitude as given in (3.8). Here, V_n stands for the nominal set-point of d-axis output voltage. The control strategy is chosen such that the output voltage magnitude reference is aligned to the d-axis of the inverter reference frame, and the q-axis reference is set to zero.

$$v_{od}^* = V_n - n_q Q, \quad v_{oq}^* = 0. \tag{3.8}$$

The droop gains m_p and n_q are calculated using (3.9) for the given range of frequency and voltage magnitude

$$m_p = \frac{w_{\max} - w_{\min}}{P_{\max}}, \quad n_q = \frac{V_{od\max} - V_{od\min}}{Q_{\max}}. \tag{3.9}$$

As discussed earlier, to construct the complete model on a common reference frame, the reference frame of one of the inverters is taken as the common frame. To translate the variables from an individual inverter reference frame onto the common frame, we define an angle δ for each inverter, given in (3.10). It should be noted that δ represents the angle between an individual inverter reference frame and the common reference frame, as shown in Fig. 3.46

$$\delta = \int (W - W_{com}). \tag{3.10}$$

Now, to allow a simpler system representation, the d and q axis components of voltages and currents in the following equations are combined to form vectors as in (3.11):

$$\begin{aligned} v^*_{odq} &= \left[v^*_{od}\ v^*_{oq}\right]^T, i_{ldq} = \left[i_{ld}\ i_{lq}\right]^T, \\ v_{odq} &= \left[v_{od}\ v_{oq}\right]^T, i_{odq} = \left[i_{od}\ i_{oq}\right]^T. \end{aligned} \tag{3.11}$$

By linearizing and rearranging the equations above, the small-signal power controller model can be written in a state-space form as in (3.12). The outputs of the power controller are the small-signal variation of output voltage reference Δv^*_0 and the frequency Δw. Matrices of (3.12) are defined in (3.13). An additional input signal Δw_{com}, which is the frequency deviation of the common reference frame, is also included in the model. It facilitates the connection of an individual inverter model to the common reference frame.

$$\begin{aligned} \begin{bmatrix} \Delta\dot{\delta} \\ \Delta\dot{P} \\ \Delta\dot{Q} \end{bmatrix} &= A_P \begin{bmatrix} \Delta\delta \\ \Delta P \\ \Delta Q \end{bmatrix} + B_P \begin{bmatrix} \Delta i_{ldq} \\ \Delta v_{odq} \\ \Delta i_{odq} \end{bmatrix} + B_{Pwcom}[\Delta w_{com}] \\ \begin{bmatrix} \Delta w \\ \Delta v^*_{odq} \end{bmatrix} &= \begin{bmatrix} C_{Pw} \\ C_{Pv} \end{bmatrix} \begin{bmatrix} \Delta\delta \\ \Delta P \\ \Delta Q \end{bmatrix} \end{aligned} \tag{3.12}$$

$$\begin{aligned} A_P &= \begin{bmatrix} 0 & -m_p & 0 \\ 0 & -w_c & 0 \\ 0 & 0 & -w_c \end{bmatrix} B_{Pxcom} = \begin{bmatrix} -1 \\ 0 \\ 0 \end{bmatrix} \\ B_P &= \begin{bmatrix} 0 & 0 & 0 & 0 & 0 & 0 \\ 0 & 0 & w_c I_{od} & w_c I_{oq} & w_c V_{od} & w_c I_{oq} \\ 0 & 0 & w_c I_{oq} & -w_c I_{od} & -w_c V_{oq} & w_c V_{od} \end{bmatrix} \\ C_{Pw} &= \left[0\ -m_p\ 0\right] C_{pv} = \begin{bmatrix} 0 & 0 & -n_q \\ 0 & 0 & 0 \end{bmatrix}. \end{aligned} \tag{3.13}$$

2. *Voltage controller:* Fig. 3.49 shows the voltage controller block diagram including all feedback and feedforward terms. Output voltage control is achieved with a standard PI controller. The corresponding state equations are

$$\frac{d\phi_d}{dt} = v^*_{od} - v_{od}, \quad \frac{d\phi_q}{dt} = v^*_{oq} - v_{oq} \tag{3.14}$$

Fig. 3.49 Voltage controller

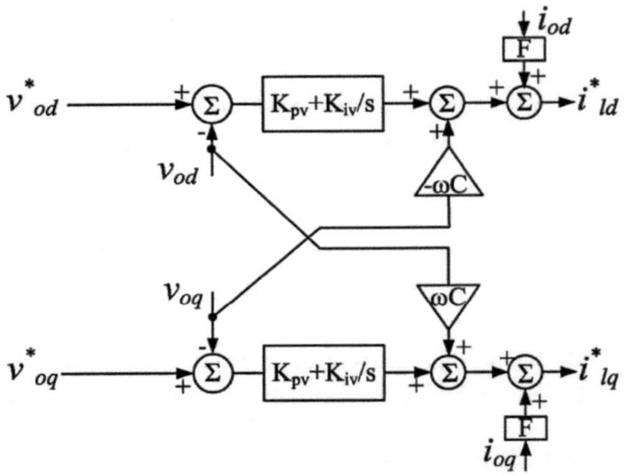

along with the algebraic equations

$$i_{ld}^{*} = Fi_{od} - w_n C_f v_{oq} + K_{pv}(V_{od}^{*} - v_{od}) + K_{iv}\phi_d \tag{3.15}$$
$$i_{lq}^{*} = Fi_{oq} - w_n C_f v_{od} + K_{pv}(V_{oq}^{*} - v_{oq}) + K_{iv}\phi_q. \tag{3.16}$$

Note that (3.17) and (3.20) represent the linearized small-signal state-space form of the voltage controller. Here, the input to the subsystem is split into two terms: the reference input and the feedback inputs

$$[\Delta\dot{\phi}_{dq}] = [0][\Delta\phi_{dq}] + B_{V1}[\Delta v_{odq}^{*}] + B_{v2}\begin{bmatrix}\Delta i_{ldq}\\ \Delta v_{odq}\\ \Delta i_{odq}\end{bmatrix}. \tag{3.17}$$

wherein (3.17)

$$\Delta\phi_{dq} = \begin{bmatrix}\Delta\phi_d & \Delta\phi_q\end{bmatrix}^T \tag{3.18}$$

$$B_{v1} = \begin{bmatrix}1 & 0\\ 0 & 1\end{bmatrix}, B_{v2} = \begin{bmatrix}0 & 0 & -1 & 0 & 0 & 0\\ 0 & 0 & 0 & -1 & 0 & 0\end{bmatrix} \tag{3.19}$$

$$\Delta i_{ldq}^{*} = C_V[\Delta\phi_{dq}] + D_{V1}\Delta v_{odq}^{*} + D_{v2}\begin{bmatrix}\Delta i_{ldq}\\ \Delta v_{odq}\\ \Delta i_{odq}\end{bmatrix} \tag{3.20}$$

$$C_V = \begin{bmatrix}K_{iv} & 0\\ 0 & K_{iv}\end{bmatrix}, D_{V1} = \begin{bmatrix}K_{pv} & 0\\ 0 & K_{pv}\end{bmatrix}$$

$$D_{V2} = \begin{bmatrix}0 & 0 & -K_{pv} & -w_n C_f & F & 0\\ 0 & 0 & w_n C_f & -K_{pv} & 0 & F\end{bmatrix}. \tag{3.21}$$

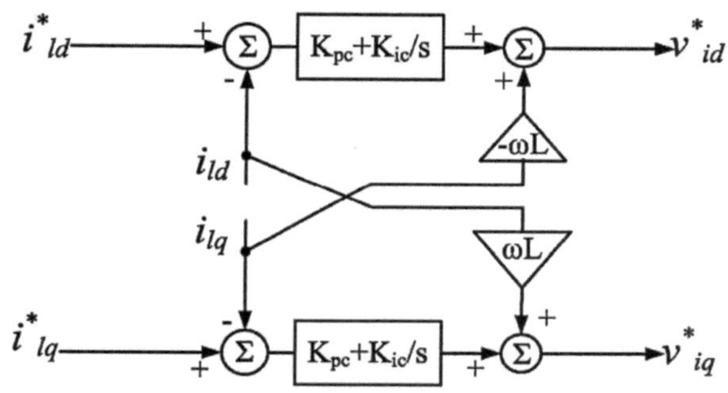

Fig. 3.50 Current controller

3. *Current controller:* Figure 3.50 shows the current controller structure. Output filter inductor current control is achieved with a standard PI controller. The corresponding state equations are

$$\frac{d\gamma_d}{dt} = i_{ld}^* - i_{ld} \frac{d\gamma_q}{dt} = i_{lq}^* - i_{lq} \tag{3.22}$$

along with the algebraic equations

$$v_{id}^* = -w_n L_f i_{lq} + K_{pc}(i_d^* - i_{ld}) + K_{ic}\gamma_d \tag{3.23}$$
$$v_{iq}^* = w_n L_f i_{ld} + K_{pc}(i_{lq}^* - i_{lq}) + K_{ic}\gamma_q. \tag{3.24}$$

Similarly, (3.25) and (3.28) represent the linearized small-signal state-space form of current controller

$$[\Delta\dot{\gamma}_{dq}] = [0][\Delta\gamma_{dq}] + B_{C1}[\Delta i_{ldq}^*] + B_{C2}\begin{bmatrix}\Delta i_{ldq}\\ \Delta v_{odq}\\ \Delta i_{odq}\end{bmatrix} \tag{3.25}$$

where

$$\Delta\gamma_{dq} = \left[\Delta\gamma_d \ \Delta\gamma_q\right]^T \tag{3.26}$$

$$B_{C1} = \begin{bmatrix}1\,0\\0\,1\end{bmatrix}, B_{C2} = \begin{bmatrix}-1 & 0 & 0\,0\,0\,0\\ 0 & -1 & 0\,0\,0\,0\end{bmatrix} \tag{3.27}$$

$$\Delta v_{idq}^* = C_C[\Delta\gamma_{dq}] + D_{C1}[\Delta i_{ldq}^*] + D_{C2}\begin{bmatrix}\Delta i_{ldq}\\ \Delta v_{odq}\\ \Delta i_{odq}\end{bmatrix} \tag{3.28}$$

$$C_C = \begin{bmatrix}K_{ic} & 0\\ 0 & K_{ic}\end{bmatrix}, D_{C1} = \begin{bmatrix}K_{pc} & 0\\ 0 & K_{pc}\end{bmatrix}$$

$$D_{C2} = \begin{bmatrix}-K_{pc} & -w_n L_f & 0\,0\,0\,0\\ w_n L_f & -K_{pc} & 0\,0\,0\,0\end{bmatrix}. \tag{3.29}$$

4. *Output LC filter and coupling inductance:* Output *LC* filter and the coupling inductance small-signal model can be represented with the following state equations by assuming that the inverter produces the demanded voltage ($v_i = v_i^*$):

$$\frac{di_{ld}}{dt} = \frac{-r_f}{L_f} i_{ld} + w i_{lq} + \frac{1}{L_f} v_{id} - \frac{1}{L_f} v_{od} \tag{3.30}$$

$$\frac{di_{lq}}{dt} = \frac{-r_f}{L_f} i_{lq} - w i_{ld} + \frac{1}{L_f} v_{iq} - \frac{1}{L_f} v_{oq} \tag{3.31}$$

$$\frac{dv_{od}}{dt} = w v_{oq} + \frac{1}{C_f} i_{ld} - \frac{1}{C_f} i_{od} \tag{3.32}$$

$$\frac{dv_{oq}}{dt} = -w v_{od} + \frac{1}{C_f} i_{lq} - \frac{1}{C_f} i_{oq} \tag{3.33}$$

$$\frac{di_{od}}{dt} = \frac{-r_c}{L_c} i_{od} + w i_{oq} + \frac{1}{L_c} v_{od} - \frac{1}{L_c} v_{bd} \tag{3.34}$$

$$\frac{di_{oq}}{dt} = \frac{-r_c}{L_c} i_{oq} - w i_{od} + \frac{1}{L_c} v_{oq} - \frac{1}{L_c} v_{bq}. \tag{3.35}$$

The following equations represent the linearized small-signal state-space form of the LC filter and coupling inductance. Frequency W_0 is the system steady-state frequency at the given operating point

$$\begin{bmatrix} \Delta \dot{i}_{ldq} \\ \Delta v_{odq} \\ \Delta i_{odq} \end{bmatrix} = A_{LCL} \begin{bmatrix} \Delta i_{ldq} \\ \Delta v_{odq} \\ \Delta i_{odq} \end{bmatrix} + B_{LCL1}[\Delta v_{idq}] + B_{LCL2}[\Delta v_{bdq}] + B_{LCL3}[\Delta w] \tag{3.36}$$

$$A_{LCL} = \begin{bmatrix} \frac{-r_{L_f}}{L_f} & w_0 & \frac{-1}{L_f} & 0 & 0 & 0 \\ -w_0 & \frac{-r_{L_f}}{L_f} & 0 & \frac{-1}{L_f} & 0 & 0 \\ \frac{1}{C_f} & 0 & 0 & w_0 & -\frac{1}{C_f} & 0 \\ 0 & \frac{1}{C_f} & -w_0 & 0 & 0 & -\frac{1}{C_f} \\ 0 & 0 & \frac{1}{L_c} & 0 & \frac{-r_{L_c}}{L_c} & w_0 \\ 0 & 0 & 0 & \frac{1}{L_c} & -w_0 & \frac{-r_{L_c}}{L_c} \end{bmatrix}$$

$$B_{LCL1} = \begin{bmatrix} \frac{1}{L_f} & 0 \\ 0 & \frac{1}{L_f} \\ 0 & 0 \\ 0 & 0 \\ 0 & 0 \\ 0 & 0 \end{bmatrix} \quad B_{LCL2} = \begin{bmatrix} 0 & 0 \\ 0 & 0 \\ 0 & 0 \\ 0 & 0 \\ -\frac{1}{L_c} & 0 \\ 0 & -\frac{1}{L_c} \end{bmatrix}$$

$$B_{LCL3} = \left[I_{lq} \; -I_{ld} \; V_{oq} \; -V_{od} \; I_{oq} \; -I_{od} \right]^T. \tag{3.37}$$

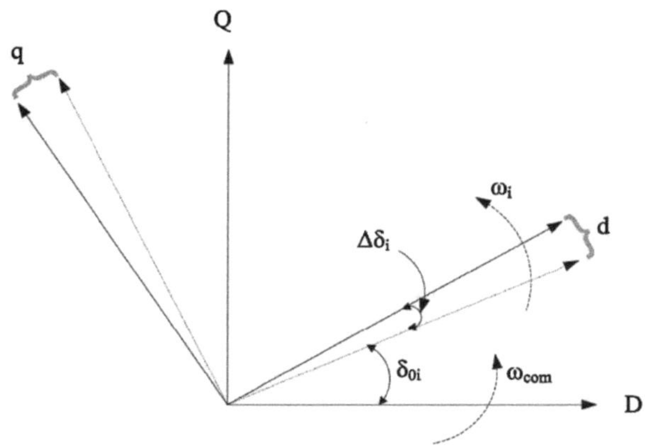

Fig. 3.51 Small-signal equivalent of reference frame transformation

5. *Complete model of an individual inverter:* To connect an inverter to the whole system the output variables need to be converted into the common reference frame. In this case the output variables of an inverter are the output currents represented as a vector Δi_{odq}. Using the transformation technique introduced in (3.3) and (3.4), the small-signal output current Δi_{oDQ} on the common reference frame can be obtained, as in (3.38). The small-signal equivalent of the reference transformation is shown in Fig. 3.51

$$\Delta i_{oDQ} = T_s \Delta i_{odq} + T_C \Delta \delta. \tag{3.38}$$

where

$$T_s = \begin{bmatrix} \cos(\delta_0) & -\sin(\delta_0) \\ \sin(\delta_0) & \cos(\delta_0) \end{bmatrix}$$
$$T_C = \begin{bmatrix} -I_{od}\sin(\delta_0) - I_{oq}\cos(\delta_0) \\ I_{od}\cos(\delta_0) - I_{oq}\sin(\delta_0) \end{bmatrix}. \tag{3.39}$$

Similarly, the input signal to the inverter model is the bus voltage which is expressed on the common reference frame. The bus voltage can be converted into the individual inverter reference frame using reverse transformation, given by

$$\Delta v_{bdq} = T_S^{-1} \Delta v_{bDQ} + T_V^{-1} \Delta \delta \tag{3.40}$$

where

$$T_V^{-1} = \begin{bmatrix} -V_{bD}\sin(\delta_0) + V_{bQ}\cos(\delta_0) \\ -V_{bD}\cos(\delta_0) - I_{bQ}\sin(\delta_0) \end{bmatrix}. \tag{3.41}$$

It is to be noted that the inverter whose reference frame is taken as the common reference frame has to provide its reference frequency Δw_{com} to all the submodules of the model as shown in (3.10). Also, care should be taken to apply this modification in the power controller model shown in (3.12).
A complete state-space small-signal model of the inverter can be obtained by combining the state-space models of the power controller, voltage controller, current controller, and output LCL filter, given by (3.12), (3.17), (3.20), (3.25), (3.28), (3.36), (3.38) and (3.40). There are totally 13 states, three inputs, and two outputs in each individual inverter model (except the inverter whose reference frame is the common reference frame, which has three outputs)

$$[\Delta \dot{x}_{invi}] = A_{INVi}[\Delta x_{invi}] + B_{INVi}[\Delta v_{bDQi}] + B_{iwcom}[\Delta w_{com}] \tag{3.42}$$

$$\begin{bmatrix} \Delta w_i \\ \Delta i_{oDQi} \end{bmatrix} = \begin{bmatrix} C_{INVw_i} \\ C_{INVc_i} \end{bmatrix} [\Delta x_{invi}]. \tag{3.43}$$

In (3.42) and (3.43)

$$\Delta x_{invi} = [\Delta \delta_i \ \Delta P_i \ \Delta Q_i \quad \Delta \phi_{dqi} \ \Delta \gamma_{dqi} \quad \Delta i_{ldqi} \ \Delta v_{odqi} \ \Delta i_{odqi}]^T \tag{3.44}$$

and (3.45)–(3.48), shown below

$$A_{INVi} = \begin{bmatrix} A_{Pi} & 0 & 0 & B_{Pi} \\ B_{V1i}C_{Pvi} & 0 & 0 & B_{V2i} \\ B_{C1i}D_{V1i}C_{Pvi} & & & \\ B_{LCL1i}D_{C1i}D_{V1i}C_{Pvi} + & B_{C1i}C_{Vi} & 0 & B_{C1i}D_{V2i} + B_{C2i} \\ B_{LCL2i}[T_{Vi}^{-1} 00] & B_{LCLli}D_{Cli}C_{Vi} & B_{LCLli}C_{Ci} & A_{LCLi} + \\ B_{LCL3i}C_{Pw_i} & & & B_{LCLli}(D_{C1i}D_{V2i} + D_{C2i}) \end{bmatrix}_{13\times 13} \tag{3.45}$$

$$B_{INVi} = \begin{bmatrix} 0 \\ 0 \\ 0 \\ B_{LCL2}T_S^{-1} \end{bmatrix}_{13\times 2} \qquad B_{iwcom} = \begin{bmatrix} B_{Pwcom} \\ 0 \\ 0 \\ 0 \end{bmatrix}_{13\times 1} \tag{3.46}$$

$$C_{INVwi} = \begin{cases} [\, C_{pw} \ 0 \ 0 \ 0 \,]_{1\times 13} & i = 1 \\ [\, 0 \ 0 \ 0 \ 0 \,]_{1\times 13} & i \neq 1 \end{cases} \tag{3.47}$$

$$C_{INVci} = \left[[\, T_C \ 0 \ 0 \,] \ 0 \ 0 \ [\, 0 \ 0 \ T_s \,] \right]_{2\times 13} \tag{3.48}$$

3.5.4 Combined Model of All the Inverters

The small-signal modeling of an individual DG inverter on a common reference frame was discussed. In an MG there can be several inverters acting as sources and connected remotely from each other. The modeling approach in this work is to form a submodel of all the individual DG inverters and combine them with the network and individual load models.

Let us consider a system with "s" number of DG inverters where the reference frame of inverter number 1 is taken as the common reference frame. Then a combined small-signal model of all the inverter units together is obtained, as shown in

$$\Delta x_{INV} = A_{INV}[\Delta x_{INV}] + B_{INV}[\Delta v_{bDQ}] \tag{3.49}$$

$$\Delta i_{oDQ} = C_{INVc}[\Delta x_{INV}]. \tag{3.50}$$

In (3.49) and (3.50)

$$[\Delta x_{INV}] = \begin{bmatrix} \Delta x_{inv1} & \Delta x_{inv2} & \ldots & \Delta x_{invs} \end{bmatrix}^T \tag{3.51}$$

and (3.52)–(3.54) shown below

$$A_{INV} = \begin{bmatrix} A_{INV1} + B_{1wcom}C_{INVw1} & 0 & 0 & . \\ 0 & A_{INV2} + B_{2wcom}C_{INVw1} & 0 & . \\ . & . & . & . \\ . & . & . & A_{INVs} + B_{swcom}C_{INVw1} \end{bmatrix}_{13s\times 13s} \tag{3.52}$$

$$B_{INV} = \begin{bmatrix} B_{INV1} \\ B_{INV2} \\ . \\ . \\ B_{INVs} \end{bmatrix}_{13s\times 2m}$$

$$[\Delta v_{bDQ}] = \begin{bmatrix} \Delta v_{bDQ1} & \Delta v_{bDQ2} & \ldots & \Delta v_{bDQm} \end{bmatrix} \tag{3.53}$$

$$C_{INVc} = \begin{bmatrix} [C_{INVc1}] & 0 & 0 & . \\ 0 & [C_{INVc2}] & 0 & . \\ . & . & . & . \\ . & . & . & [C_{INVcs}] \end{bmatrix}_{2s\times 13s} \tag{3.54}$$

3.5.5 Network Model

An example network of n lines and m nodes with s inverters and p load points is shown in Fig. 3.52. On a common reference frame the state equations of line current of ith line connected between nodes j and k are

$$\frac{di_{lineDi}}{dt} = \frac{-r_{linei}}{L_{linei}} i_{lineDi} + w i_{lineQi} + \frac{1}{L_{linei}} v_{bDj} - \frac{1}{L_{linei}} v_{bDk} \tag{3.55}$$

$$\frac{di_{lineQi}}{dt} = \frac{-r_{linei}}{L_{linei}} i_{lineQi} - w i_{lineDi} + \frac{1}{L_{linei}} v_{bQj} - \frac{1}{L_{linei}} v_{bQk}. \tag{3.56}$$

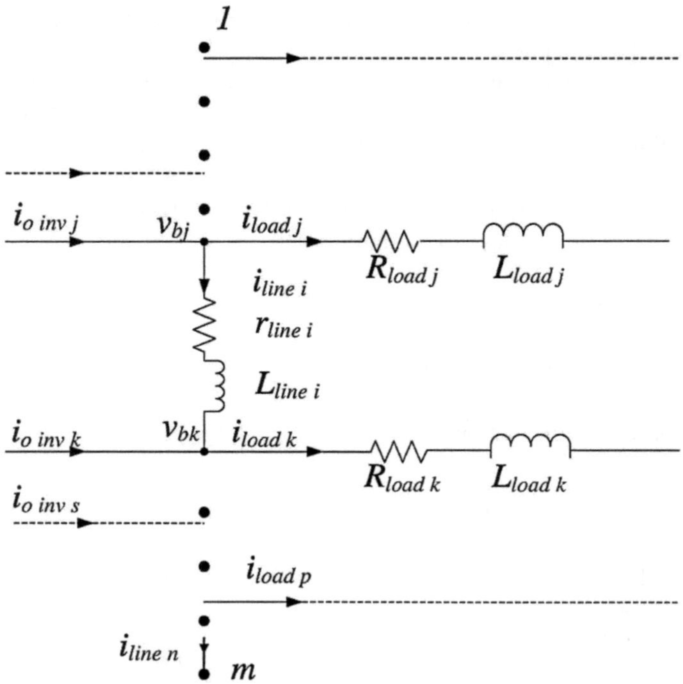

Fig. 3.52 Network representation

Hence, the small-signal state-space model of a network with n lines is given by

$$[\Delta i_{lineDQ}] = A_{NET}[\Delta i_{lineDQ}] + B_{1NET}[\Delta v_{bDQ}] + B_{2NET}\Delta w \tag{3.57}$$

In (3.57)

$$\Delta i_{lineDQ} = \begin{bmatrix} \Delta i_{lineDQ1} & \Delta i_{lineDQ2} & \ldots & \Delta i_{lineDQn} \end{bmatrix}^T \tag{3.58}$$

$$\Delta v_{bDQ} = \begin{bmatrix} \Delta v_{bDQ1} & \Delta v_{bDQ2} & \ldots & \Delta v_{bDQm} \end{bmatrix}^T \tag{3.59}$$

$$\Delta w = \Delta w_{com} \tag{3.60}$$

$$A_{NET} = \begin{bmatrix} A_{NET1} & 0 & \ldots & 0 \\ 0 & A_{NET2} & \ldots & 0 \\ \ldots & \ldots & \ldots & \ldots \\ 0 & 0 & \ldots & A_{NETn} \end{bmatrix}_{2n\times 2n} \tag{3.61}$$

$$B_{1NET} = \begin{bmatrix} B_{1NET1} \\ B_{1NET2} \\ \ldots \\ \ldots \\ B_{1NETn} \end{bmatrix}_{2n\times(2m)}, \quad B_{2NET} = \begin{bmatrix} B_{2NET1} \\ B_{2NET2} \\ \ldots \\ \ldots \\ B_{2NETn} \end{bmatrix}_{2n\times 1} \tag{3.62}$$

where you get (3.63) as follows:

$$A_{NETi} = \begin{bmatrix} \frac{-r_{linei}}{L_{linei}} & w_0 \\ -w_0 & \frac{-r_{linei}}{L_{linei}} \end{bmatrix} B_{NETi} = \begin{bmatrix} I_{lineQi} \\ -I_{lineDi} \end{bmatrix}$$

$$B_{1NETi} = \begin{bmatrix} \cdots & \frac{1}{L_{linei}} & 0 & \cdots & \frac{-1}{L_{linei}} & 0 & \cdots \\ \cdots & 0 & \frac{1}{L_{linei}} & \cdots & 0 & \frac{-1}{L_{linei}} & \cdots \end{bmatrix}_{2\times(2m)}. \tag{3.63}$$

3.5.6 Load Model

Although many types of load can exist in MGs, a general RL load is considered in this paper. The state equations of the RL load connected at ith node are

$$\frac{di_{loadDi}}{dt} = \frac{-R_{loadi}}{L_{loadi}} i_{loadDi} + w i_{loadQi} + \frac{1}{L_{loadi}} v_{bDi} \tag{3.64}$$

$$\frac{di_{loadQi}}{dt} = \frac{-R_{loadi}}{L_{loadi}} i_{loadQi} - w i_{loadDi} + \frac{1}{L_{loadi}} v_{bQi}. \tag{3.65}$$

Hence, for a network with p load points the small-signal state-space model of loads is given by

$$[\Delta i_{loadDQ}] = A_{load}[\Delta i_{loadDQ}] + B_{1LOAD}[\Delta v_{bDQ}] + B_{2LOAD}\Delta w. \tag{3.66}$$

where in (3.66)

$$[\Delta i_{loadDQ}] = \begin{bmatrix} \Delta i_{loadDQ1} & \Delta i_{loadDQ2} & \ldots & \Delta i_{loadDQp} \end{bmatrix}^T \tag{3.67}$$

$$A_{load} = \begin{bmatrix} A_{load1} & 0 & \ldots & 0 \\ 0 & A_{load2} & \ldots & 0 \\ \ldots & \ldots & \ldots & \ldots \\ 0 & 0 & \ldots & A_{loadp} \end{bmatrix}_{2p\times 2p} \tag{3.68}$$

$$B_{1LOAD} = \begin{bmatrix} B_{1LOAD1} \\ B_{1LOAD2} \\ \ldots \\ \ldots \\ B_{1LOADp} \end{bmatrix}_{2p\times(2m)}$$

$$B_{2LOAD} = \begin{bmatrix} B_{2LOAD1} \\ B_{2LOAD2} \\ \ldots \\ \ldots \\ B_{2LOADp} \end{bmatrix}_{2p\times(1)} \tag{3.69}$$

where

$$A_{loadi} = \begin{bmatrix} \frac{-R_{loadi}}{L_{loadi}} & w_0 \\ -w_0 & \frac{-R_{loadi}}{L_{loadi}} \end{bmatrix} B_{2LOADi} = \begin{bmatrix} I_{loadQi} \\ -I_{loadDi} \end{bmatrix}$$

$$B_{1LOADi} = \begin{bmatrix} \dots & \frac{1}{L_{loadi}} & 0 & \dots \\ \dots & 0 & \frac{1}{L_{loadi}} & \dots \end{bmatrix}_{2\times(2m)}. \quad (3.70)$$

3.5.7 Complete Microgrid Model

It can be seen in (3.42), (3.57), and (3.66) that the node voltages are treated as inputs to each subsystem. To ensure the node voltage is well-defined (and that the numerical solution is well conditioned) a virtual resistor r_N is assumed between each node and ground. The resistance of virtual resistor is chosen sufficiently large such that its introduction would have minimum influence on the dynamic stability of the system. Hence, the voltage of ith node is given by

$$v_{bDi} = r_N(i_{oDi} - i_{loadDi} + i_{lineDi,j}) \quad (3.71)$$

$$v_{bQi} = r_N(i_{oQi} - i_{loadQi} + i_{lineQi,j}). \quad (3.72)$$

In symbolic form, for a network with m nodes

$$[\Delta v_{bDQ}] = R_N(M_{INV}[\Delta i_{oDQ}] + M_{load}[\Delta i_{loadDQ}] + M_{NET}[\Delta i_{lineDQ}]). \quad (3.73)$$

In (3.73), matrix R_N is of size $2m \times 2m$, whose diagonal elements are equal to r_N. The mapping matrix M_{INV} is of size $2m \times 2s$, which maps the inverter connection points onto network nodes. For example, if ith inverter is connected at jth node, the element $M_{INV}(j, i)$ will be 1 and all the other elements in that row will be 0. Similarly, M_{load} is of size $2m \times 2p$ maps load connection points onto the network nodes with -1. Matrix M_{NET} of size $2m \times 2n$, maps the connecting lines onto the network nodes. Here, care should be taken to put either $+1$ or -1 based on whether the given line current is entering or leaving the node.

Now, the complete MG small-signal state-space model and hence the system state matrix can be obtained by using the individual subsystem models given by (3.42), (3.43), (3.57), (3.66), and (3.73) to yield

$$\begin{bmatrix} \Delta x_{INV} \\ \Delta i_{lineDQ} \\ \Delta i_{loadDQ} \end{bmatrix} = A_{mg} \begin{bmatrix} \Delta x_{INV} \\ \Delta i_{lineDQ} \\ \Delta i_{loadDQ} \end{bmatrix}. \quad (3.74)$$

The complete system state matrix A_{mg} is given in (3.75). The small-signal flow among all the submodules is shown in Fig. 3.45 and Eq. (3.75).

3.5.8 Sensitivity Analysis

The eigenvalue concept of control theory has been extensively used to determine the stability of conventional power systems. Eigenvalues, termed modes, are the solution of the characteristic equation of a system's linearized state matrix [3]. Eigenvalues reveal the different frequency components in the system and their available damping.

Further information on the origin of different frequency components can be obtained by observing the participation of different state variables in a particular mode [3]. This can be achieved with a sensitivity analysis conducted on the system state matrix. The sensitivity factor Ph, given by (3.76), is the measure of the association between the state variables and the modes and is equal to the sensitivity of the eigenvalue λ_i to the diagonal element a_{kk} of the system state matrix. Sensitivity factors can be calculated using left and right eigenvectors (Figs. 3.53, 3.54 and 3.55)

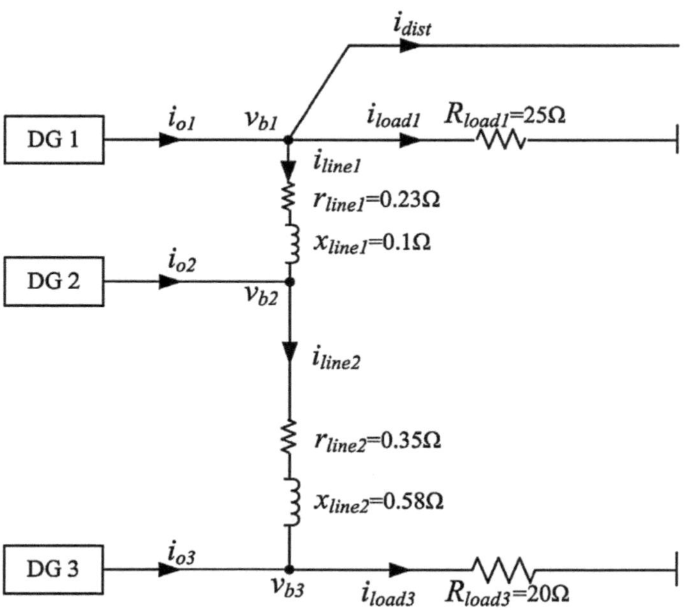

Fig. 3.53 Test system

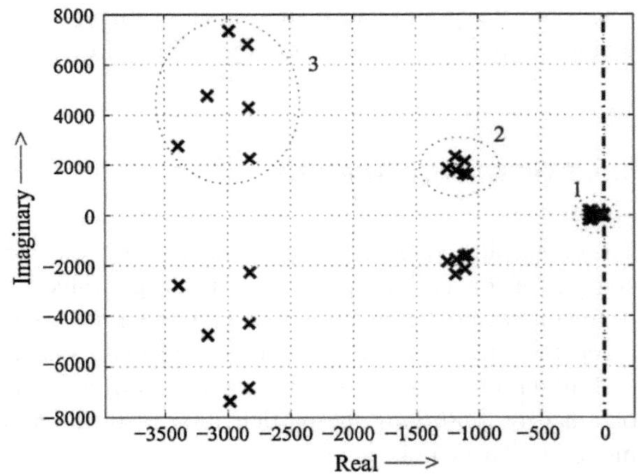

Fig. 3.54 Eigenvalue spectrum of the system indicating various modes

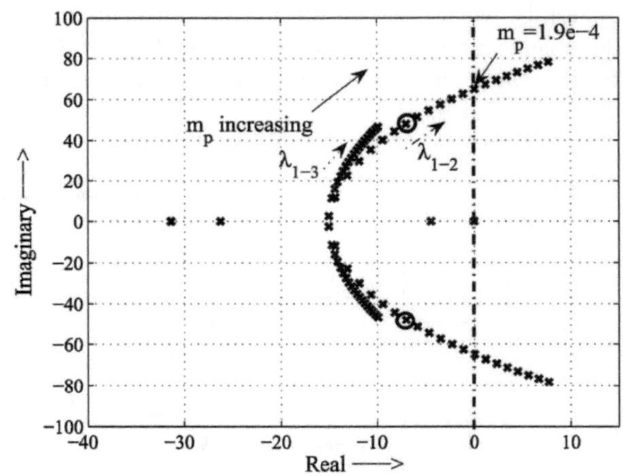

Fig. 3.55 Trace of low-frequency modes as a function of real power droop gain: $1.578e^{-5}$ (0.05 % droop) $\le m_p \le 3.14e^{-4}$ (1 % droop)

$$A_{mg} = \begin{bmatrix} A_{mg1} & B_{INV}R_N M_{NET} & B_{INV}R_N M_{load} \\ A_{mg2} & A_{NET} + B_{1NET}R_N M_{NET} & B_{1NET}R_N M_{load} \\ A_{mg3} & B_{1LOAD}R_N M_{NET} & A_{load} + B_{1LOAD}R_N M_{load} \end{bmatrix}. \quad (3.75)$$

$$A_{mg1} = A_{INV} + B_{INV}R_N M_{INV} C_{INVc}$$

$$A_{mg2} = B_{1NET}R_N M_{INV} C_{INVc} + B_{2NET} C_{INVw}$$

$$A_{mg3} = B_{1LOAD}R_N M_{INV} C_{INVc} + B_{2LOAD} C_{INVw}$$

$$rp_{ki} = \frac{\partial \lambda_i}{\partial a_{kk}} \quad (3.76)$$

3.6 Conclusions

The role of the MG in penetration of DGs in the present utility network is discussed. Modeling of MG is a key aspect and the recent developments in the modeling of MG are presented in both grid-connected and autonomous mode. The control techniques of MG available in the literature for various modes of operation are also discussed. The MG can be viewed as a special case of SoS. It can be concluded that using NCS, a better control of MG can be obtained.

In this section, a small-signal state-space model of an MG is presented. The model includes inverter low frequency dynamics, high frequency dynamics, network dynamics, and load dynamics. All the submodules are individually modeled and are then combined on a common reference frame to obtain the complete model of the MG.

The model was analyzed in terms of the system eigenvalues and their sensitivity to different states. With the help of this analysis the relation between different modes and system parameters was established. It was observed that the dominant low-frequency modes are highly sensitive to the network configuration and the parameters of the power sharing controller of the microsources. The high frequency modes are largely sensitive to the inverter inner loop controllers, network dynamics, and load dynamics.

Results obtained from the model were verified experimentally on a prototype MG. It was observed that the model successfully predicts the complete MG dynamics both in the low and high frequency range (Tables 3.1 and 3.2).

Small signal modeling has had a long history of use in conventional power systems. The inverter models (and the inclusion of network dynamics) illustrated in this paper allow MGs to be designed to achieve the stability margin required of reliable power systems.

Table 3.1 Test system parameters

Inverter parameters (10 kVA rating)			
Parameter	Value	Parameter	Value
f_s	8 kHz	m_p	9.4e–5
L_f	1.35 mH	n_q	1.3e–3
Cf	50 μF	K_{pv}	0.05
r_f	0.1 Ω	K_{iv}	390
L_c	0.35 mH	K_{pc}	10.5
r_{Lc}	0.03 Ω	K_{ic}	16e3
W_c	31.41	F	0.75
Network and Load parameters (see Fig. 3.10)			

Table 3.2 Electrical MG system parameters

Quantity	Value
Resistance of VSC filter	$R_t = 1.5\,\text{m}\Omega$
Inductance of VSC filter	$L_t = 300\,\mu\text{H}$
VSC rated power	2.5 MW
VSC terminal voltage	600 V line–line
Switching frequency	$f_{sw} = 1985\,\text{Hz}$
DC bus voltage	$V_{dc} = 1500\,\text{V}$
Load nominal resistance	$R = 76\,\Omega$
Load nominal inductance	$L = 111.9\,\text{mH}$
Load nominal capacitance	$C = 62.86\,\mu\text{F}$
Load quality factor	$Q = R\sqrt{\frac{C}{L}} = 1.8$
Load resonant frequency	$f_{res} = \frac{1}{2\pi\sqrt{LC}} = 60\,\text{Hz}$
Inductor quality factor	$q_\ell = \frac{L\Omega_0}{R_\ell} = 120$
System nominal frequency	$f_0 = 60\,\text{Hz}$
Transformer voltage ratio	$0.6/13.8\,\text{kV}$ (Y/Δ)
Transformer rating	2.5 MVA

3.7 Suggested Problems

Problem 3.1 The objective here is to design a linear controller and investigate its robust performance for an islanded operation of a DG unit and its local load. The islanded system and its controller is depicted in Fig. 3.56 while the electrical parameters are presented in Fig. 3.2. In this system, the DG unit utilizes a VSC as the interface medium. Develop an appropriate controller to guarantee frequency control and to regulate the island voltage. Demonstrate the robust stability and robust performance criteria with respect to fast transient response and zero steady-state error.

Problem 3.2 The structure of a variable speed wind energy conversion system based on a doubly fed induction generator (DFIG) is given in Fig. 3.57. Seeking to develop an efficient wind energy conversion system, it is desired to study the robust control of a WT derived by a DFIG by developing appropriate mathematical model and designing a control strategy that improves the performance of the WT and assures an optimum energy capture while guaranteeing the stability of the controlled system.

Problem 3.3 The objective here is to examine the effectiveness of grid-oriented converter control of small WT using the configuration in Fig. 3.58. It is observed that the current control has been operated by two PI controllers, hence, it is possible to neglect the dynamics of the inner control loops for the voltage loop design. In order to achieve high current tracking, the control algorithm accounts for the grid-side inductance.

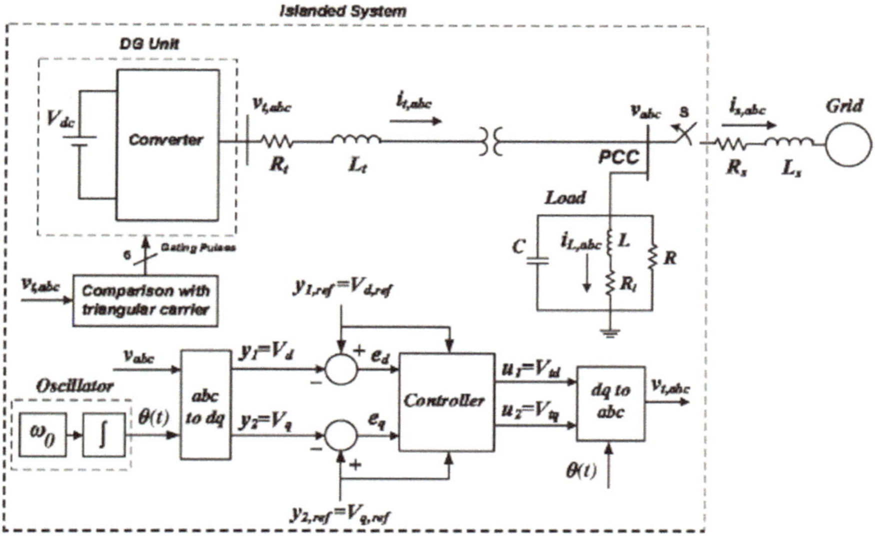

Fig. 3.56 An islanded system and its controller

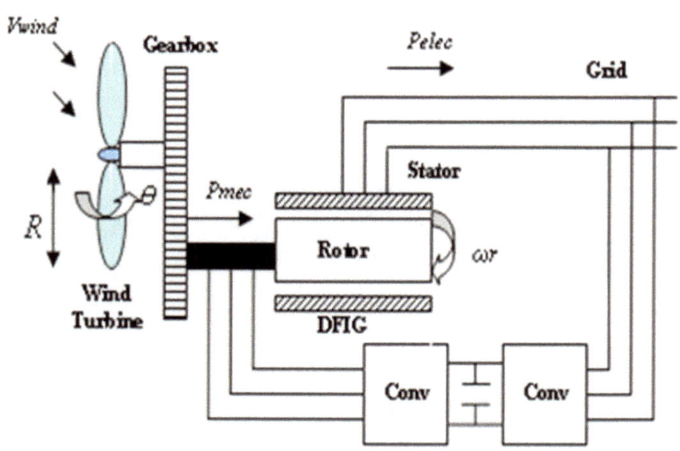

Fig. 3.57 Variable speed wind turbine

A suggested scenario is to develop a model of the grid-side system in the d-q framework. Then design a grid-side converter control that leads to decoupling of the dynamics of the d and q axes. Use the technical optimum design method to tune the coefficient of the PI controllers. Finally provide a simulation of the converter voltage, grid voltage, and grid current for a 10 % voltage sag.

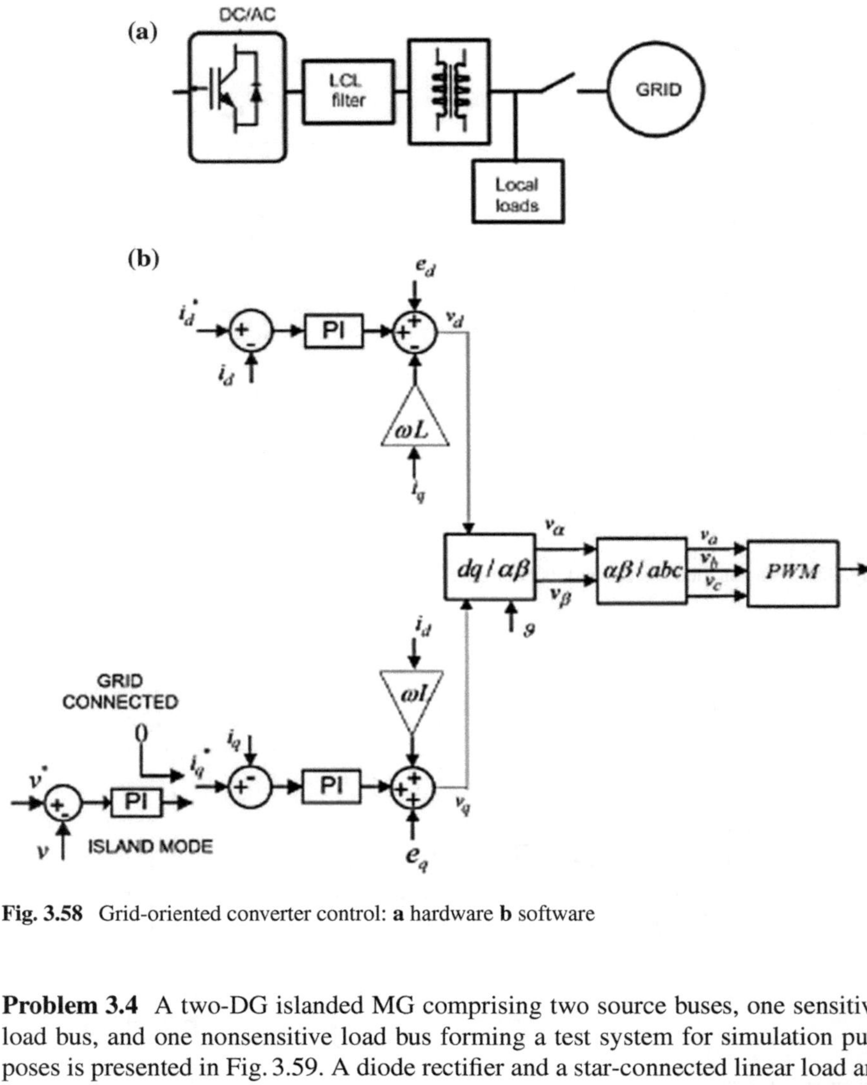

Fig. 3.58 Grid-oriented converter control: **a** hardware **b** software

Problem 3.4 A two-DG islanded MG comprising two source buses, one sensitive load bus, and one nonsensitive load bus forming a test system for simulation purposes is presented in Fig. 3.59. A diode rectifier and a star-connected linear load are connected to sensitive load bus. Henceforth, it is assumed that one phase of nonlinear load is disconnected to generate unbalanced voltage distortion. The switching frequency of the DG inverters is 10 kHz. The DC link voltage $V_{dc} = 650\,\text{V}$, the LC filter inductance $L = 1.8\,\text{mH}$, the LC filter capacitance $C = 25 = \mu\text{F}$, the nonlinear loads tie lines $Z = R_t + j\omega L_t,\ R_t = 0.1\,\Omega,\ L_t = 1.8\,\text{mH}$, the linear load $Z = R_\ell + j\omega L_\ell,\ R_\ell = 50\,\Omega,\ L_\ell = 20\,\text{mH}$, the impedance of the distribution lines $Z_1,\ Z_2,\ Z_3$ has a restive part of $R_z = 0.1\,\Omega$ and an inductive part of $L_z = 1.8\,\text{mH}$ and finally the nonlinear loads have $C_{NL}/R_{NL1} = 235\,\mu\text{F}/50\,\Omega,\ \ R_{NL2} = 50\,\Omega/L_{NL} = 0.084\,\text{mH}$

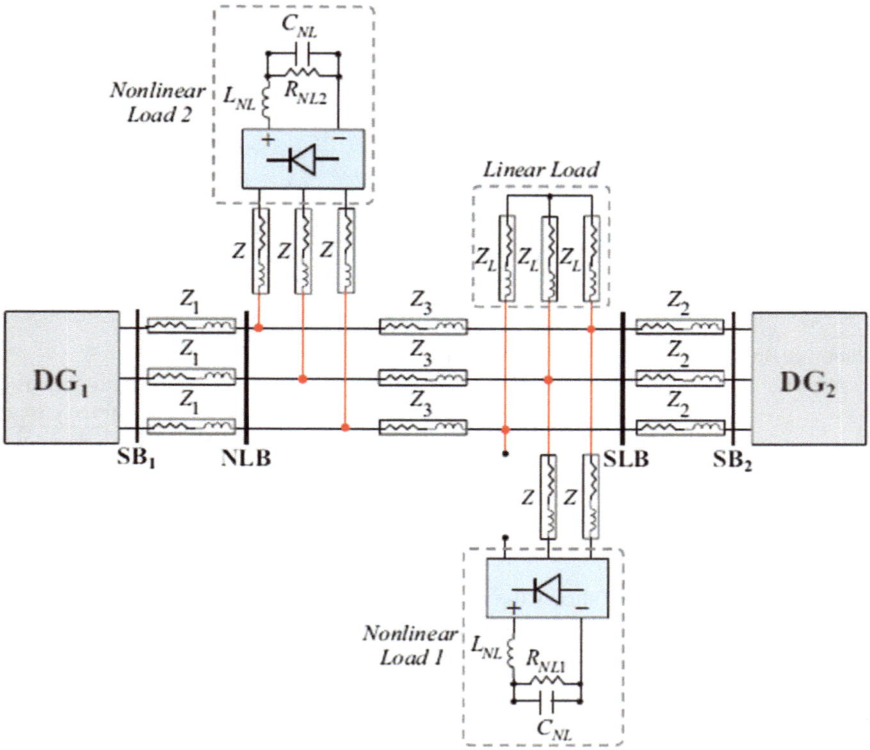

Fig. 3.59 A test system

It is desired to use MATLAB/Simulink to perform simulations along the following steps:

1. In the interval ($0 \leq t < 2s$), DGs operate only with fundamental positive sequence virtual impedance and secondary control is not acting.
2. In the interval ($2 \leq t < 4s$), Virtual resistances for fundamental negative sequence and harmonic components are added.
3. In the interval ($4 \leq t < 7s$), Secondary control is activated. The reference values of unbalance factor and harmonic distortion indices are 0.2 %.

The parameters of the DGs local controller and the secondary PI controller are left to the designer to select.

Problem 3.5 In the technical literature, a **virtual power plant** (VPP) is a cluster of DER which is collectively operated by a central control entity; a VPP can replace a conventional power plant, while providing higher efficiency and more flexibility. Despite the fact that the MG and the VPP appear to be similar concepts, there are certainly some differences. The task here is to undertake a careful examination of these differences.

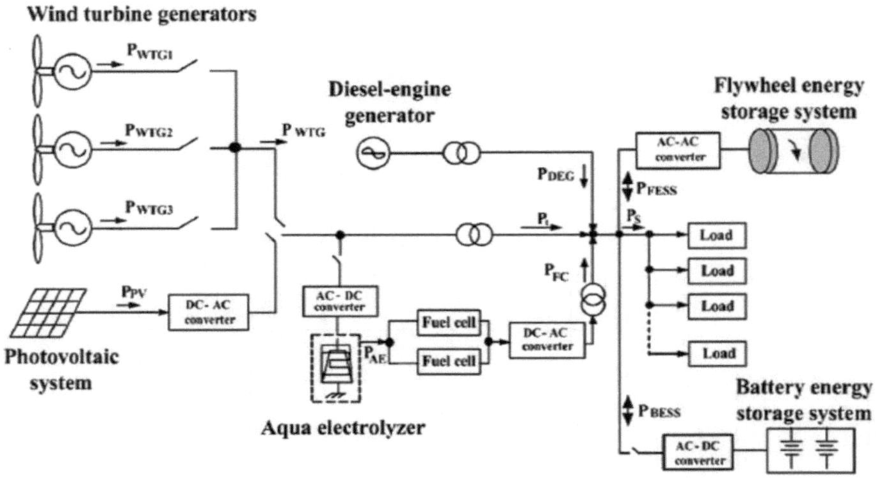

Fig. 3.60 Schematic of a hybrid power generation/energy storage system

Problem 3.6 Figure 3.60 shows the schematic of a hybrid power generation/energy storage system comprised of: three wind turbine generators, a photovoltaic system, two fuel cells, and a standby diesel engine generator.

The objective is to conduct a small-signal stability analysis coupled with time-domain and/or frequency-domain simulations to evaluate the performance of the system using typical data values.

References

1. Alex G, Sauser B, Boardman J (2008) System-of-systems engineering management: a review of modern history and a path forward. IEEE Syst J 2(4):484–499
2. Arnold M, Negenborn RR, Andersson G, Schutter BD (2009) Multi-area predictive control for combined electricity and natural gas systems. In: Proceedings of the European control conference. Budapest, Hungary, pp 1408–1413, August 2009
3. Arulampalam A, Barnes M, Engler A, Goodwin A, Jenkins N (2004) Control of power electronic interlaces in distributed generation microgrids. Int J Electron 91(9):503–523
4. Babazadeh M, Karimi H (2011) Robust decentralized control for islanded operation of a microgrid. In: Proceedings of IEEE power and energy society general meeting
5. Bae I, Kim J (2008) Reliability evaluation of customers in a microgrid. IEEE Trans Power Syst 23(3):1416–1422
6. Baillieul J, Antsaklis PJ (2007) Control and communication challenges in networked real-time systems. Proc IEEE 95(1):9–28
7. Barker P, Herman D (2002) Technical and economic feasibility of microgrid-based power systems. In: Proceedings of the seventh EPRI distributed resources conference and exhibition Dallas, TX, 20–22 March 2002
8. Barsali S et al (2002) Control techniques of dispersed generators to improve the continuity of electricity supply. In: Proceedings of IEEE power engineering society winter meeting, vol 2

9. Baudoin S, Vechiu I, Camblong H (2012) A review of voltage and frequency control strategies for islanded microgrid. In: Proceedings of the 16th international IEEE conference system theory, control and computing (ICSTCC)
10. Bhaskara SN, Rasheduzzaman M, Chowdhury (2012) Laboratory-based microgrid setup for validating frequency and voltage control in islanded and grid-connected modes. In: IEEE green technologies conference
11. Blaabjerg F, Chen Z, Kjaer SB (2004) Power electronics as efficient interface in dispersed power generation systems. IEEE Trans Power Electron 19(5):1184–1194
12. Caldon R et al (2003) Analysis of dynamic performance of dispersed generation connected through inverters to distribution networks. In: Proceedings of the 17th international conference on electricity distribution—CIRED, Barcelona, May 2003
13. Carrasco JM et al (2006) Power-electronic systems for the grid integration of renewable energy sources: a survey. IEEE Trans Ind Electron 53(4):1002–1016
14. Chandorkar MC, Divan DM, Adapa R (1993) Control of parallel connected inverters in standalone ac supply systems. IEEE Trans Ind Appl 29(1):136–143
15. Coelho EAA, Cortizo P, Gracia PFD (2002) Small signal stability for parallel-connected inverters in stand-alone ac supply systems. IEEE Trans Ind Appl 38(2):533–542
16. Colandairaj J, Irwin GW, Scanlon WG (2007) Wireless networked control systems with QoS-based sampling. Proc IET Control Theory Appl 1(1):430–438
17. Dai M et al (2008) Power flow control of a single distributed generation unit. IEEE Trans Power Electron 23(1):343–352
18. Davison E (1976) The robust decentralized control of a general servomechanism problem. IEEE Trans Autom Control 21(1):14–24
19. Davison EJ, Ferguson I (1981) The design of controllers for the multivariable robust servomechanism problem using parameter optimization methods. IEEE Trans Autom Control 26(1):93–110
20. Davison DE, Davison EJ (2002) Optimal transient response shaping of the servomechanism problem. J Optim Theory Appl 115(3):491–515
21. del Real AJ, Arce A, Bordons C (2007) Hybrid model predictive control of a two-generator power plant integrating photovoltaic panels and a fuel cell. In: Proceedings of the 46th IEEE conference on decision and control, New Orleans, LA, USA, pp 5447–5452, 12–14 December 2007
22. Diaz G et al (2009) Complex-valued state matrices for simple representation of large autonomous microgrids supplied by ⟨formula type⟩. IEEE Trans Power Syst 24(4): 1720–1730
23. Dimeas AL, Hatziargyriou ND (2003) Operation of a multiagent system for microgrid control. IEEE Trans Power Electron 20(3):1447–1455
24. Dimeas AL, Hatziargyriou ND (2005) Operation of a multiagent system for microgrid control. IEEE Trans Power Syst 20(3):1447–1455
25. Dobakhshari A, Salehi S, Azizi S, Ranjbar AM (2011) Control of microgrids: aspects and prospects. In: Proceedings of IEEE international conference on networking, sensing and control (ICNSC)
26. El-Khattam W, Salama MMA (2004) Distributed generation technologies, definitions and benefits. Electric Power Syst Res 71(2):119–128
27. El-Sharkh MY et al (2004) Analysis of active and reactive power control of a stand-alone PEM fuel cell power plant. IEEE Trans Power Syst 19(4):2022–2028
28. Engler A (2004) Applicability of droops in low voltage grids. Int J Distrib Energy Resour 1(1):3–15
29. Engler A (2005) Applicability of droops in low voltage grids. Int J Distrib Energy Resour 1(1):1–6
30. Ersal T, Ahn C, Hiskens IAI, Peng H, Stefanopoulou A, Stein JL (2012) On the effect of DC source voltage on inverter-based frequency and voltage regulation in a military microgrid. In: Proceedings of the American control conference, Fairmont Queen Elizabeth, Montral, Canada, 27–29 June, pp 2965–2971

31. Ersal T, Ahn C, Peters DL, Whitefoot JW, Mechtenberg AR, Hiskens IA, Peng H, Stefanopoulou AG, Papalambros PY, Stein JL (2013) Coupling between component sizing and regulation capability in microgrids. IEEE Trans Smart Grid. doi:10.1109/TSG.2013.2260363
32. Etemadi AH, Davison EJ, Iravani R (2012) A decentralized robust control strategy for multi-DER microgrids part I: fundamental concepts. IEEE Trans Power Syst 1(1):271–278
33. European Research Project MicroGrids. http://microgrids.power.ece.ntua.gr/
34. Farhangi H (2010) The path of the smart grid. IEEE Power Energy Mag 8(1):18–28
35. Franklin GF, Powell JD, Emami-Naeini A (1986) Feedback control of dynamics systems. Prentice Hall, New York
36. Gao F, Iravani MR (2008) A control strategy for a distributed generation unit in grid-connected and autonomous modes of operation. IEEE Trans Power Deliv 23(2):850–859
37. Georgakis D et al (2004) Operation of a prototype microgrid system based on micro-sources quipped with fast-acting power electronics interfaces. In: Proceedings of IEEE 35th annual power electronics specialists conference (PESC 04), vol 4
38. Guerrero JM et al (2005) Output impedance design of parallel-connected UPS inverters with wireless load-sharing control. IEEE Trans Ind Electron 52(4):1126–1135
39. Guerrero JM et al (2007) Decentralized control for parallel operation of distributed generation inverters using resistive output impedance. IEEE Trans Ind Electron 54(2):994–1004
40. Guerrero JM et al (2011) Hierarchical control of droop-controlled AC and DC microgrids—a general approach toward standardization. IEEE Trans Ind Electron 58(1):158–172
41. Hassan MA, Abido MA (2011) Optimal design of microgrids in autonomous and grid-connected modes using particle swarm optimization. IEEE Trans Power Electron 26(3): 755–769
42. Hatziagyriou N et al (2004) Modeling of micro-sources for security studies. CIGRE session, Paris
43. Hatziargyriou N et al (2007) Microgrids: an overview of ongoing research, development, and demonstration projects. IEEE Power Energy Mag
44. Hernandez-Aramburo CA, Green TC, Mugniot N (2005) Fuel consumption minimization of a microgrid. IEEE Trans Ind Appl 41(3):673–681
45. Hespanha JP, Naghshtabrizi P, Xu Y (2007) A survey of recent results in networked control systems. Proc IEEE 95(1):138–162
46. Hipel KW et al (2007) The future of systems, man, and cybernetics: application domains and research methods. IEEE Trans Syst, Man, Cybern, Part C: Appl Rev 37(5):726–743
47. IEEE Standards Coordinating Committee 21 (2011) IEEE Std. 1547.4-2011: IEEE guide for design, operation, and integration of distributed resource island systems with electric power systems, July 2011
48. Illindala MS, Piagi P, Zhang H, Venkataramanan G, Lasseter RH (2004) Hardware development of a laboratory-scale microgrid phase 2: operation and control of a two-inverter microgrid. National Renewable Energy Report, March 2004
49. Jamshidi M (2008) System of systems engineering-principles and application. Taylor & Francis CRC Publishers, Boca Raton
50. Jamshidi M (2009) System of systems engineering-innovations for the 21 century. Wiley, New York
51. Kanellos FD, Tsouchnikas AI, Hatziargyriou ND (2005) Micro-grid simulation during grid-connected and islanded modes of operation. In: Proceedings of international conference on power systems transients, vol 6, pp 1–6
52. Karimi H, Nikkhajoei H, Iravani R (2008) Control of an electronically-coupled distributed resource unit subsequent to an islanding event. IEEE Trans Power Deliv 23(1):493–501
53. Karimi H, Davison EJ, Iravani R (2010) Multivariable servomechanism controller for autonomous operation of a distributed generation unit: design and performance evaluation. IEEE Trans Power Syst 25(2):853–865
54. Katiraei F (2005) Dynamic analysis and control of distributed energy resources in a micro-grid. Ph.D. thesis Dissertation, University of Toronto

55. Katiraei F, Iravani MR (2006) Power management strategies for a microgrid with multiple distributed generation units. IEEE Trans Power Electron 21(4):1821–1831
56. Katiraei F, Iravani MR, Lehn PW (2005) Micro-grid autonomous operation during and subsequent to islanding process. IEEE Trans Power Deliv 20(1):248–257
57. Katiraei F, Iravani MR, Lehn PW (2007) Small-signal dynamic model of a micro-grid including conventional and electronically interfaced distributed resources. IET Gener, Transm Distrib 1(3):369–378
58. Keating C, Rogers R, Unal R, Dryer D, Sousa-Poza A, Safford R, Peterson W, Rabadi G (2003) System of systems engineering. Eng Manag J 15:36–43
59. Kelley B, Jamshidi M, Akopian D (2009) MRI development of a test-bed swarm of autonomous unmanned vehicles ad-hoc networked control via secure 4G wireless. NSF MRI Proposal
60. Krause PC, Wasynczuk O, Sudhoff SD (2002) Analysis of electric machinery and drive systems. Wiley-IEEE, New York
61. Lasseter B (2001) Microgrids [distributed power generation]. In: IEEE power engineering society winter meeting, Columbus, Ohio, vol 1, pp 146–149
62. Lassetter RH (2002) Integration of distributed energy resources: the CERTS microgrid concept. CERT Report, April 2002
63. Lasseter RH (2002) MicroGrids. IEEE Power Eng Soc Winter Meet 1:305–308
64. Lasseter RH (2007) Microgrids and distributed generation. J Energy Eng 133(3):144–149
65. Lasseter RH, Paigi P (2004) Microgrid: a conceptual solution. In: IEEE 35th annual power electronics specialists conference, PESC 04, vol 6
66. Lasseter RH, Piagi P (2000) Providing premium power through distributed resources. In: Proceedings of the 33rd annual Hawaii international conference on system sciences
67. Lasseter R, Akhil A, Marnay C, Stephens J, Dagle J, Guttromson R, Sakis Meliopoulous A, Yinger R, Eto J (2002) Integration of distributed energy resources—the CERTS microgrid concept. Report LBNL-50829, April 2002
68. Lasseter RH et al (2011) CERTS microgrid laboratory test bed. IEEE Trans Power Deliv 26(1):325–332
69. Lasseter RH et al, White paper on integration of distributed energy resources—the CERTS microgrid concept. http://certs.lbl.gov/pdf/50829-app.pdf
70. Lee DJ, Wang L (2008) Small-signal stability analysis of an autonomous hybrid renewable energy power generation/energy storage system part I: time-domain simulations. IEEE Trans Energy Convers 23(1):311–320
71. Lee C-T, Chu C-C, Cheng P-T (2013) A new droop control method for the autonomous operation of distributed energy resource interface converters. IEEE Trans Power Electron 28(4):1980–1993
72. Li YW, Kao C-N (2009) An accurate power control strategy for power-electronics-interfaced distributed generation units operating in a low-voltage multi-bus microgrid. IEEE Trans Power Electron 24(12):2977–2988
73. Li YW et al (2004) Design, analysis and real-time testing of controllers for multi-bus microgrid system. IEEE Trans Power Electron 19:1195–1204
74. Liang H, Choi BJ, Zhuang W, Shen X (2013) Stability enhancement of decentralized inverter control through wireless communications in microgrids. IEEE Trans Smart Grid 4(1):321–331
75. Lin OL, Dan-ying G, Wei-dong Z (2007) Optimal tuning method of PID controller based on gain margin and phase margin. IET Control Theory Appl 24(5):837–840
76. Llria A, Curea O, Jimenez J, Camblong H (2011) Survey on microgrids: unplanned islanding and related inverter control techniques. Int J Renew Energy 36:2052–2061
77. Lopes JAP, Moreira CL, Resende FO (2005) Microgrids black start and islanded operation. In: Proceedings of the 15th power systems computation conference (PSCC), Liege
78. Lopes JAP, Moreira CL, Madureira AG (2006) Defining control strategies for microgrids islanded operation. IEEE Trans Power Syst 21(2):916–924
79. Lopes JAP et al, Management of microgrids, presented at JIEEC2003. http://microgrids.power.ece.ntua.gr/documents/Microgrids_management_jieec2003.pdf

80. Mahmoud MS (2004) Resilient control of uncertain dynamical systems. Springer, Heidelberg
81. Mahmoud MS (2010) Decentralized control and filtering in interconnected dynamical systems. CRC Press, New York
82. Mahmoud MS (2011) Decentralized systems with design constraints. Springer, London
83. Mahmoud MS (2012) Distributed control and filtering for industrial systems. IET Press, London
84. Maier MW (1998) Architecting principles for systems-of-systems. Syst Eng 1(4):267–284
85. Marwali MN, Keyhani A (2004) Control of distributed generation systems—part I: voltages and current control. IEEE Trans Power Electron 19(6):1541–1550
86. Marwali MN, Jung J-W, Keyhani A (2004) Control of distributed generation systems—part II: load sharing control. IEEE Trans Power Electron 19(6):1551–1561
87. Mohamed YARI, El-Saadany EF (2008) Adaptive decentralized droop controller to preserve power sharing stability of paralleled inverters in distributed generation microgrids. IEEE Trans Power Electron 23(6):2806–2816
88. Moradi R, Karimi H, Ghartemani MK (2010) Robust decentralized control for islanded operation of two radially connected DG systems. In: Proceedings of IEEE international symposium on industrial electronics (ISIE)
89. Morozumi S (2007) Micro-grid demonstration projects in Japan. In: IEEE power conversion conference-Nagoya, PCC'07
90. Nagahara M et al (2012) H-infinity control of microgrids involving gas turbine engines and batteries. In: Proceedings of 51th IEEE conference on decision and control
91. Nazari MH, Ilic M (2009) Technical challenges in modernizing distribution electric power systems with large number of distributed generators. In: Proceedings of IEEE PowerTech
92. Oyarzabal J et al (2005) Agent based micro grid management system. In: Proceedings of IEEE international conference on future power systems
93. Padulles I, Ault GW, McDonald JR (2000) An integrated SOFC plant dynamic model for power systems simulation. J Power Sources 86(1):495–500
94. Papathanassiou S, Hatziargyriou N, Strunz K (2005) A benchmark low voltage microgrid network. In: Proceedings of the CIGRE symposium: power systems with dispersed generation
95. Papathanassiou S et al, Study-Case LV Network. http://microgrids.power.ece.ntua.gr/documents/Study-Case%20LV-Network.pdf
96. Parniani M, Iravani MR (1995) Computer analysis of small-signal stability of power systems including network dynamics. IEE Proc C: Gener, Transm Distrib 142(6):613–617
97. Pedrasa MA, Spooner T (2006) A survey of techniques used to control microgrid generation and storage during island operation. In: Proceedings of the Australian universities power engineering conference
98. Peng I (2004) Research on control technique for PWM inverters based on state-space theory. Ph.D. Dissertation, Hua Zhong University of Science and Technology, Wuhan, P. R. China, September 2004
99. Peng FZ, Li YW, Tolbert LM (2009) Control and protection of power electronics interfaced distributed generation systems in a customer-driven microgrid. In: Proceedings of power and energy society general meeting (PES'09)
100. Pepermans G et al (2005) Distributed generation: definition, benefits and issues. Energy Policy 33(6):787–798
101. Peters DL, Mechtenberg AR, Whitefoot JW, Papalambros PY (2011) Model predictive control of a microgrid with plug-in vehicles: error modeling and the role of prediction horizon. In: Proceedings of ASME dynamic systems and controls conference, Arlington, VA, October 31–November 2, DSCC2011-6030
102. Piagi P, Lasseter RH (2006) Autonomous control of microgrids. In: Proceedings of IEEE power engineering society general meeting
103. Pogaku N, Prodanovic M, Green TC (2007) Modeling, analysis and testing of autonomous operation of an inverter-based microgrid. IEEE Trans Power Electron 22(2):613–625
104. Popov M et al (2010) Modeling, control and islanding detection of microgrids with passive loads. In: Proceedings of the 14th international power electronics and motion control conference (EPE/PEMC)

105. Prodanovic M (2004) Power quality and control aspects of parallel connected inverters in distributed generation. Ph.D. Dissertation, Imperial College, University of London, London, UK
106. Prodanovic M, Green T, Mansir H (2000) A survey of control methods for parallel three-phase inverters connection. Proc Inst Elect Eng 475:472–477
107. Qiu L, Davison EJ (1986) New perturbation bounds for the robust stability of linear state space models. In: Proceedings of the 25th IEEE conference on decision and control, vol 25
108. Sao CK, Lehn PW (2005) Autonomous load sharing of voltage source converters. IEEE Trans Power Deliv 20(2):1009–1016
109. Schauder C, Mehta H (1993) Vector analysis and control of advanced static VAR compensators. IEE Proc C: Gener, Transm Distrib 140(4):299–306
110. Senjyu T et al (2005) A hybrid power system using alternative energy facilities in isolated island. IEEE Trans Energy Convers 20(2):406–414
111. Shafiee Q, Vasquez JC, Guerrero JM (2012) Distributed secondary control for islanded MicroGrids—a networked control systems approach. In: IECON 2012—38th annual conference on IEEE industrial electronics society. IEEE
112. Skogestad S, Postlethwaite I (2007) Multivariable feedback control: analysis and design, vol 2. Wiley, New York
113. Song HS, Kwanghee N (1999) Dual current control scheme for PWM converter under unbalanced input voltage conditions. IEEE Trans Ind Electron 46(5):953–959
114. Soultanis NL, Papathanasiou SA, Hatziargyriou ND (2007) A stability algorithm for the dynamic analysis of inverter dominated unbalanced LV microgrids. IEEE Trans Power Syst 22(1):294–304
115. US Department of Energy Electricity Distribution Programme, Advanced distribution technologies and operating concepts-microgrids. http://www.electricdistribution.ctc.com/MicroGrids.htm
116. Uudrill JM (1968) Dynamic stability calculations for an arbitrary number of interconnected synchronous machines. IEEE Trans Power Appar Syst 3:835–844
117. Vasquez JC et al (2010) Hierarchical control of intelligent microgrids. Ind Electron Mag 4(4):23–29
118. Whitefoot JW, Mechtenberg AR, Peters DL, Papalambros PL (2011) Optimal component sizing and forward-looking dispatch of an electrical microgrid for energy storage planning. In: Proceedings of the ASME 2011 international design engineering technical conferences and computers and information in engineering conference. Paper number DETC2011-48513
119. Yazdani A, Iravani R (2006) A unified dynamic model and control for the voltage-sourced converter under unbalanced grid conditions. IEEE Trans Power Deliv 21(3):1620–1629
120. Yee SK, Milanovic JV, Hughes FM (2008) Overview and comparative analysis of gas turbine models for system stability studies. IEEE Trans Power Syst 23(1):108–118
121. Zamora R, Srivastava A (2010) Controls for microgrids with storage: review, challenges, and research needs. Renew Sustain Energy Rev 14(7):2009–2018
122. Zareipour H, Bhattacharya K, Canizares CA (2004) Distributed generation: current status and challenges. In: Proceedings of the 36th annual North American power symposium (NAPS)
123. Zeineldin H, Marei MI, El-Saadany EF, Salama MMA (2004) Safe controlled islanding of inverter based distributed generation. In: IEEE 35th annual power electronics specialists conference, PESC 04, vol 4, pp 2515–2520, June 2004
124. Zhang W, Branicky MS, Phillips SM (2001) Stability of networked control systems. IEEE Control Syst 21(1):84–99
125. Zheng W, Ma H, He X (2012) Modeling, analysis, and implementation of real time network controlled parallel multi-inverter systems. In: Proceedings of the 7th international power electronics and motion control conference (IPEMC), vol 2
126. Zhou K et al (1997) Essentials of robust control. Prentice Hall, Englewood Cliffs
127. Zhu Y, Tomsovic K (2002) Development of models for analyzing the load-following performance of microturbines and fuel cells. Electr Power Syst Res 62(1):1–11

Chapter 4
Optimal Energy Management

4.1 Introduction

The increasing incidence of distributed generation (DG) and active distributed networks, changing regularities, and needs to improve the power system reliability and clean power support is providing development of a new power system perception commonly referred to as the smart grid (SG). In this regard, the microgrid (MG) can be considered as one of the most promising concepts. A MG is essentially an active distribution network defined as an integrated power delivery system. A MG consists of a low-voltage (LV) network composed of loads, renewable energy sources (RESs), and DG units operating as a single controllable load connected to the main grid. Compared to the conventional power plants, a MG is characterized by specific operation and constraints depending on several critical stochastic parameters. So far, MGs have been mostly established as test-bed platforms in some developed countries such as Japan, Canada, and the U.S. In fact, such concept cannot be widely implemented in practice without a prior successful establishment of an energy manager to achieve optimal and reliable control of the MG for a given site to minimize its operation cost while consolidating its reliability and environment-friendly features [1, 5–7, 12, 17, 23–27, 31].

As an essential element in the new era of smart power, several approaches have been reported in the literature in relation to MG intelligent energy management applicable within the SG system [22, 56]. A fuel consumption minimization approach has been proposed in [47] based on power sharing approach, in which the optimized cost function includes a penalty function for heat generation excess. The authors promoted the importance of a MG communication infrastructure allowing the coordination with RES forecasting as a part of the central control. However, the proposed approach does not take into account the prediction of RE power generation and the management of storage devices. In [76], mesh-adaptive direct-search-based optimization algorithm is applied for MG online energy management; however, this approach does not provide information regarding the forecasting of the RESs or load demand (LD). Moreover, the MG energy management problem has been reduced to

M.S. Mahmoud and F.M. AL-Sunni, *Control and Optimization of Distributed Generation Systems*, Power Systems, DOI 10.1007/978-3-319-16910-1_4

a generation scheduling without taking into consideration the storage facility management as the battery is controlled based on conventional opportunity charging approach. A mathematical framework for environmental economic dispatch of MG energy management is proposed in [25], where advantages of intelligent methods over traditional computational techniques for optimization are discussed. However, the proposed formulations do not take into consideration the energy storage management. An energy management system for a MG using multilayered perceptron artificial neural networks (ANNs) is proposed in [14], where hourly decisions of DG dispatch for cost optimization are projected by the neural network. The training set is generated using a separate optimal power flow software package. Unfortunately, complexity of the problem, eventual noisy sets in the training data, and the generalization performances of the multilayered perceptron network may result in serious reliability issues in the MG operation. A one-day-ahead multiperiod optimization model under supply demand balancing constraints for a MG is proposed in [73] to determine the amount of power that a MG should exchange with a wholesale energy market so that its total benefit is maximized. However, environmental aspects and RE generation (REG) forecasting are not considered in their investigations. For most of the MG energy management relevant literature, forecasting values of REG and LD are not treated in practice but simply estimated or assumed as available data [22, 77], which implies that forecasting uncertainties are not considered during the optimization process. Therefore, the energy management process is regarded based on a deterministic approach which naturally affects the result accuracy and, therefore, the overall performances of the MG. In this context, an intelligent online energy management approach was proposed in [15], where the authors took into consideration the solar power generation (SPG) forecasting and the battery management. However, the proposed battery scheduling is considered in a deterministic approach achieved based on an HF where uncertainty related to the forecasted parameters cannot be considered.

In this chapter, a generalized formulation for MG energy management approach is investigated and developed using artificial intelligence techniques.

4.2 A Microgrid Model

A MG normally operates in a grid-connected mode, as power can be imported or exported from and to the main grid. During disturbances, such as power quality (PQ) events, IEEE 1547 events, or faults, the generation and corresponding loads can be autonomously separated from the distribution system to isolate the MG's load from the disturbances (and thereby maintaining high level of service) without harming the transmission grid's integrity via the sectionalizing circuit breaker (SCB). In terms of power flow control, a DG unit can be either a dispatchable or nondispatchable unit [59]. The power output of a dispatchable unit is controlled externally over operation set-points provided by the MIEM as a part of the central controller (CC) as the fuel cell and the microturbine that are controlled over fuel-in-flow adjustments and governor

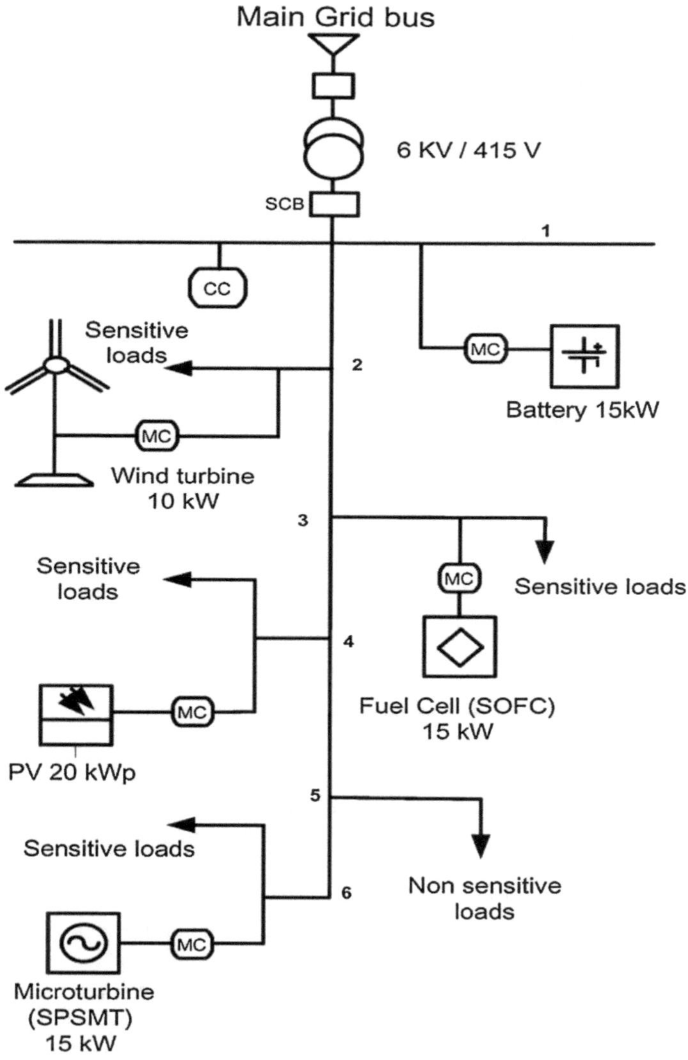

Fig. 4.1 Architecture of the microgrid

speed control. On the other hand, the RESs are typical nondispatchable DG units, and their generated power depends on the actual weather conditions (wind speed, temperature, and irradiance) and is controlled in a way to generate their maximum power [18]. The developed MG consists of six buses (Fig. 4.1) and is composed of a 15-kW battery connected to bus 1, a 10-kW rated wind electric generation system located at bus 2, a 15-kW rated solid oxide fuel cell (SOFC) connected to bus 3, a 20-kWp rated photovoltaic (PV) array connected to bus 4, and a 15-kW rated single-shaft microturbine connected to bus 6. In the MG simulation model, inverters play a vital role by interfacing DC microsources with the AC power side. Two kinds of control strategies are mainly used to operate an inverter within a MG. The PQ inverter control type is used to supply a given active and reactive power set point. This type of inverter is used to interface the single-shaft microturbine, the SOFC, the PV array, and the battery during grid-connected operation. The voltage source inverter (VSI) control type is used to "feed" the load with predefined values of voltage and frequency. The VSI is used to interface the battery to the MG representing the reference bus during islanding mode. When analyzing the dynamic behavior of a MG, inverters are modeled based only on their control functions; therefore, fast switching transients, harmonics, and inverter losses are neglected. Further details of the MG's simulation model description are available in [18, 58–84].

4.3 Microgrid and Load Forecasting

In this section, the applicability of ANNs for MG environment short-term forecasting is investigated. Although ANNs are providing a relevant methodology for solving several types of nonlinear problems, they are still considered as an unstable learning model [34]. In fact, the changes on training data and the architecture of the network which incorporate the number of hidden layers, the number of neurons, and the initial connection weights have considerable effect on the network training and predicting performances. Moreover, the fitting to the regularities of the data and the fitting to the noise sets during the training of a single neural network restrain it to be trapped in a local optimum solution with low forecasting accuracy. On the other hand, there is no systematic investigation for those issues; mostly, researchers have adopted trial and error methodology to deal with these inconsistencies [44]. In this paper, a robust NNE based on regularized negative correlation learning (NCL) (RNCL) is proposed for an enhanced learning model performance by improving the generalization ability and reducing the variance and the performance instability with a relative simple implementation. Indeed, the major feature of the proposed machine learning is redundancy; based on combination of different learning machines, each neural network could perform the task on its own, leading to better generalization performance. The MG's environment forecasting tasks include the following: 24-h-ahead SPG forecasting of a 20-kWp rated PV array, 1-h-ahead WPG forecasting of a 10-kW rated wind electric generation system, and 1-h-ahead LD forecasting of a building in Tokyo University of Agriculture and Technology (TUAT).

4.3.1 Proposed NNE

Generalized machine learning for MG environment forecasting is proposed based on RNCL for a heterogeneous NNE composed by an aggregation of the following: multilayered perceptron neural network (MLPNN), radial basis function neural network (RBFNN), and recurrent neural network (RNN). The aggregation of all predictors constitutes the network ensemble $\{f_b(x)\}_{b=1}^{B}$, where the forecasting result of all the networks is aggregated based on averaging to get the output of the complete predictor ensemble, as shown in the following equation:

$$f_B(x_n) = \frac{1}{B}\sum_{b=1}^{B} f_b(x_n). \tag{4.1}$$

The developed NNEs are associated to a training set D (composed of data recorded during fiscal years 2007 and 2008). However, using the same training data set for all the component networks (to estimate the same function) results in high correlated outputs that affect the efficiency and the robustness of the learning machine [91]. One way to mitigate the potential colinearity problem is to add a negative colinearity penalty to the error function of each predictor [10, 16], where the training error cost function of each single predictor f_b is defined as

$$e_b = \sum_{n=1}^{N}(f_b(x_n) - y_n)^2 + \lambda p_b \tag{4.2}$$

where p_b is the correlation penalty term with a weighting parameter λ expressed as

$$p_b = \sum_{n=1}^{N}\left\{(f_b(x_n) - f_B(x_n))\sum_{i\neq b}(f_i(x_n) - f_B(x_n))\right\} \tag{4.3}$$

$$= -\sum_{n=1}^{N}(f_b(x_n) - f_B(x_n))^2. \tag{4.4}$$

The goal is minimizing p_b to negatively correlate the predictor's error inside the overall ensemble. A null value of e_b results in an independently trained predictors, while for $\lambda = 1$, the error function of the NNE is equal to averaging of all individual networks expressed as

$$\begin{aligned} e_B &= \sum_{b=1}^{M}\sum_{n=1}^{N}\{(f_b(x_n) - y_n)^2 - (f_b(x_n) - f_B(x_n))^2\} \\ &= \sum_{n=1}^{N}(f_B(x_n) - y_n)^2. \end{aligned} \tag{4.5}$$

To improve the NCL generalization performances in relation to over-fitting issues, particularly for the case of nontrivial data as microgrid LD, wind speed, and solar radiation [4], RNCL has been proposed for homogeneous NNE composed of RBFNN that is applicable to any nonlinear estimator [21]. In this paper, the learning approach is extended to a heterogeneous NNE providing a more diversified output while guaranteeing interaction between the ensemble networks. In fact, the ensemble training objectives are decomposed into a set of subobjectives while each subobjective is implemented by an individual neural network. In comparison with the RNCL algorithm, a regularization term $\sum_{b=1}^{B} \alpha_b w_b^T w_b$ is incorporated into the ensemble error function

$$e_B = \sum_{n=1}^{N} (f_B(x_n) - y_n)^2 + \sum_{b=1}^{B} \alpha_b w_b^T w_b \tag{4.6}$$

where $w_b = [w_{b,1}, \ldots, w_{b,nb}]$ is the weight vector of a neural network b in the NNE and n_b is its total number of weights.

The regularization term is decomposed into B parts as for the training of each network in the ensemble, where the error function of a single network in the ensemble is defined as

$$\begin{aligned} e_B = & \frac{1}{B} \sum_{n=1}^{N} (f_b(x_n) - y_n)^2 \\ & - \frac{1}{B} \sum_{n=1}^{B} (f_b(x_n) - f_B(x_n))^2 \alpha_b w_b^T w_b. \end{aligned} \tag{4.7}$$

The error function (4.7) imposes a regulation term to each neural network to optimize the regularization parameter α_b instead of the correlation parameter λ. Scaled conjugate gradient algorithm [75, 86, 87, 89, 92, 95] is implemented for the training process of the neural networks, where the partial derivative of (4.7) with respect to the neural network weights is expressed as

$$\begin{aligned} \frac{\partial e_b}{\partial w_{b,i}} = & \frac{2}{B} \sum_{n=1}^{N} (f_B(x_n) - y_n) \frac{\partial f_b(x_n)}{\partial w_{b,i}} \\ & - \frac{2}{B} \sum_{n=1}^{N} (f_b(x_n) - f_B(x_n)) \left(\frac{M-1}{M} \right) \frac{\partial f_b(x_n)}{\partial w_{b,i}} \\ & + \alpha_b w_b^T w_b. \end{aligned} \tag{4.8}$$

For a training dataset $D : \{x_n, y_n\}_{n=1}^{N}$, the optimal regularization parameter α_b^{opt} is calculated based on Bayesian inference [36, 80]. Assuming that the targets are sampled with independent noise sample e_n with zero-mean Gaussian distribution and β^{-1} variance where the weight vector related to each network has a zero-mean Gaussian distribution and μ_b^{-1} variance, with α_b^{-1}/β^{-1}, then the target standard probabilistic formulation is expressed as

$$y_n = f_B(x_n) + e_n. \tag{4.9}$$

For the hyperparameters μ and β, the weight vector w is obtained by maximizing the posterior $P(w|D)$ which, according to the Bayesian theorem, is expressed as

$$P(w|D) = \frac{P(D|w, \beta)P(w|\mu)}{P(D|\mu, \beta)} \tag{4.10}$$

where the prior of weight vector $w = [w_1^T, \ldots, w_B^T]^T$ is expressed as

$$P(w|\mu) = \prod_{b=1}^{B} (\frac{\mu_i}{2\pi})^{\frac{n_i}{2}} e^{(-\frac{1}{2}\mu_b w_b^T w_b)}. \tag{4.11}$$

As the noise en follows a Gaussian distribution with zero mean and variance β^{-1}, the likelihood $P(D|w, \beta)$ is expressed as

$$P(D|w, \beta) = \prod_{n=1}^{N} (\frac{\beta}{2\pi})^{\frac{1}{2}} e^{(-\frac{\beta}{2} e_{n2})}. \tag{4.12}$$

After omitting the constants and normalization factors, the posterior $P(w|D)$ can be approximated as

$$P(w|D) \propto e^{(-\frac{\beta}{2}\sum_{n=1}^{N} e_{n2})} e^{(-\sum_{b=1}^{B} \frac{\mu_b}{2} w_b^T w_b)}. \tag{4.13}$$

Finally, the optimal regularization parameter α_b^{opt} is calculated by maximizing the posterior of $P(\mu, \beta|D)$ while the marginal likelihood is expressed in (4.14); further calculation details are available in [21, 91, 97–99, 102–104, 108–110]

$$P(D|\mu, \beta) = \frac{P(D|w, \beta)P(w|\mu)}{P(w|D)} \tag{4.14}$$

Figure 4.2 shows the flowchart of the RNCL-based algorithm for the developed NNE composed of B predictors. The original training dataset $D = \{(x_1, d_1), \ldots, (x_N, d_N)\}$ is normalized (rescaled) using across-channel normalization [55]. The learning is conducted in an iterative mode, for the first epoch initial regulation parameters $\{\alpha_b\}_{b=1}^{B}$ are set for the training of the networks. Each single predictor training is executed for a desired number of iterations j_{Max}; once all the predictors are trained ($b = B$), the NNE average error is calculated to update the regularization parameter α_b^{opt} and start a new epoch. The process is repeated until a suitable convergence criterion is reached (goal).

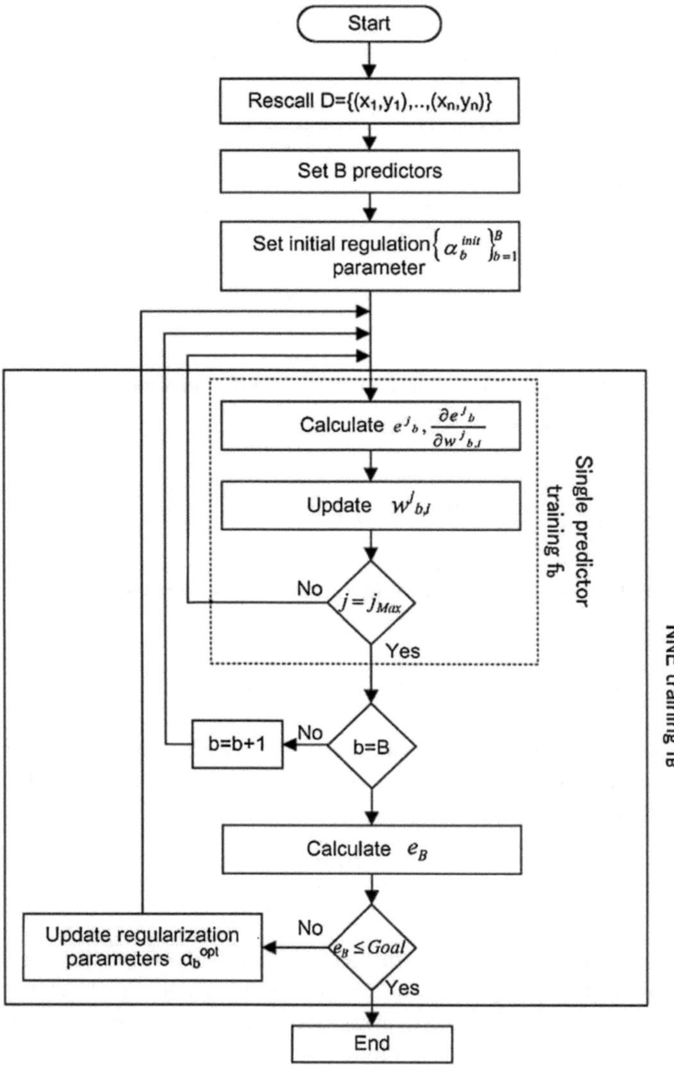

Fig. 4.2 NNE bagging training flowchart

4.3.2 Microgrid Environment Forecasting

Table 4.1 illustrates the input layer's variables associated to the proposed NNEs, where TUAT refers to the data recorded inside the TUAT and JMA refers to the data collected from Japan Meteorological Agency, measured in the local area of the TUAT (Koganei city campus).

Table 4.1 ANN inputs

	Input	Description	Source
Load demand	Hour (k)	time (4 digit)	operator
	Temperature (k)	one hour ago (average)	TUAT
	Pressure(k)	one hour ago (average)	JMA
	Humidity (k)	one hour ago (average)	JMA
	Day type	working / holiday day	operator
	Load demand (k)	one hour ago (average)	TUAT
	Load demand (k-1)	two hour ago (average)	TUAT
	Load demand (k-2)	three hour ago (average)	TUAT
WPG	Max wind speed (k)	one hour ago (peak)	TUAT
	Wind speed (k)	one hour ago (average)	TUAT
	Wind speed (k-1)	two hour ago (average)	TUAT
	Wind speed (k-2)	three hour ago (average)	TUAT
	Humidity (k)	one hour ago (average)	JMA
	Pressure(k)	one hour ago (average)	JMA
	Wind direction (k)	one hour ago (4 digit)	TUAT
SPG	Max temperature (d)	Last day (daily average)	TUAT
	Min temperature (d)	Last day (daily average)	TUAT
	Temperature (d)	Last day (daily average)	TUAT
	Pressure(d)	Last day (daily average)	JMA
	Precipitation (d)	Last day (daily average)	JMA
	Radiation$_{k-1,..k-24}$ (d)	Last day 24 input	TUAT

4.4 Multiobjective Energy Management

Decision problems with multiple conflicting objectives (goals or criteria) are known as multiple criteria decision making or MO [28, 57, 66]. Such subject covers both discrete problems (with a finite set of alternatives, also called actions or solutions) and continuous problems. In other words, the MO is a methodology aimed to find the best solution among different objectives, usually conflicting objectives, where the problem can be handled as follows:

$$Minimize\{f_1(P), \ldots, f_k(P)\}(k \geq 2) \tag{4.15}$$

$$\begin{aligned} &Subject\ to: \\ &h_i(P) = 0, i = 1, 2, \ldots, q \\ &g_j(P) \leq 0, j = 1, 2, \ldots, p \\ &p_i^{\min} \leq P_i \leq P_i^{\max} \forall i = 1, \ldots, N. \end{aligned} \tag{4.16}$$

The conflicting objective functions $f_i : R^n \rightarrow R$ have to be minimized simultaneously; the decision vector (variable) $P = (P_1, P_2, \ldots, P_N)^T$ belongs to the nonempty feasible region $S \subset R^n$. A decision vector is regarded as optimal if none of its components can be improved without deterioration to at least one of the other components. More precisely, a decision vector $P^{'} \in S$ is called Pareto optimal if there is no other $P \in S$ such that $f_i(P) \leq f_i(P^{'}) \forall i = 1, \ldots, k$ and $f_j(P) < f_j(P^{'})$ for at least one index j.

For MG energy management, the solution of the optimal multiobjective process amounts to finding an output power generation vector $P = (P_1, P_2, \ldots, P_N)$ providing the generation set-points of each controllable DG to guarantee the minimum cost and gaseous emissions during the MGs operation while satisfying the load balance within the power generation rating limit of each DG. Therefore, the general model described in (4.15) is expressed in the following to define the proposed microgrid MO energy management involving two objective functions:

$$f_1(P) = \sum_{i=1}^{N} FC_i(P_i) + P_b \dot{P}_B - P_s \dot{P}_S + M_i(P_i) \tag{4.17}$$

$$f_2(P) = \sum_{i=1}^{N} \sum_{j=1}^{M} (EF_{ij} P_i) + GEF\dot{P}_B - (GEF - MGEF)\dot{P}_S \tag{4.18}$$

where P_i is the power output related to the DG unit i in kilowatt hours, FC_i is the fuel consumption cost related to the DG unit i in dollars per kilowatt hours, P_b is the actual hour electricity buying price in dollars per kilowatt hours, P_B is the amount of purchased electricity, P_s is the actual hour electricity selling price in dollars per kilowatt hours, P_S is the sold electricity in kilowatt hours, M_i is the maintenance cost related to the DG unit i in dollars per kilowatt hours, EF_{ij} is the emission factor related to the DG unit i in pounds per kilowatt hours where j is the emission type (NO_x, CO_2, and SO_2), MGEF is the microgrid average emission factor in pounds per kilowatt hours, and, finally, GEF is the main grid average emission factor in pounds per kilowatt hours.

The fuel consumption FC_i for the controllable DG (fuel cell and microturbine) is expressed in (4.19); more details can be found in [13, 29]

$$FC_i(P_i) = C_i \frac{P_i}{\eta_i} \tag{4.19}$$

where C_i is the fuel price and η_i is the efficiency rate related to the DG unit i.

The objective functions (4.17) and (4.18) are subjected to an equality constraint for active power and load balance (4.20) and restriction limits of the output related to each DG unit (4.21)

$$\sum_{i=1}^{N} P_i + P_{PV} + P_{WT} + P_{batt} + P_B - L - P_S = 0 \tag{4.20}$$

$$P_i^{\min} \leq P_i \leq P_i^{\max} \quad \forall i = 1, 2, \ldots, N. \tag{4.21}$$

The cost objective function expressed in (4.17) depends on three parameters: The first one is the power generation cost related to each DG based on the fuel price and the consumption rate of each unit described in (4.19), the second one is the energy trade between the main grid and the MG, and the third one is the maintenance cost assumed to be proportional to the power generation of each DG unit [8]. On the other hand, the emission objective function expressed in (4.18) depends on the individual emission rate of each DG during autonomous generation and on the main grid average emission rate (fossil, coal, and nuclear) in case of a MG's generation lack [29, 48]. Furthermore, selling power to the grid is equivalent to reducing the overall emission (MG + main grid) as the MG emits less than conventional power plants; in this case, the emission saving is approximated as the difference between the microgrid average emission (MGEF) and the main grid average emission (GEF).

4.4.1 Battery Scheduling

A 15-kWh lead–acid battery is used as storage facility; such batteries are relatively cheap but suffer from deep depth of discharge (DOD) as it dramatically decreases its lifetime by reducing the expected average cycles (Fig. 4.3). In this chapter, the lead–acid battery efficiency is assumed to be 90 % for both charging and discharging efficiencies. Based on the previous assumptions, (4.20) is modified depending on the following two cases:

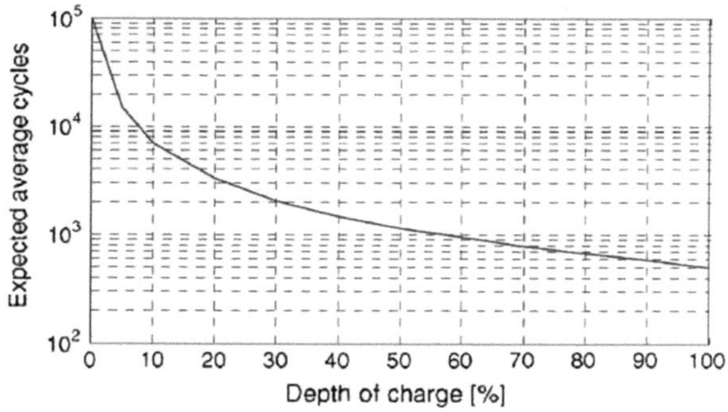

Fig. 4.3 Cycle expectancy for a lead–acid battery

1. charging

$$P_{batt} = 0.9.(\sum P_i + P_{PV} + P_{WT} + P_B - L - P_S) \quad (4.22)$$

2. discharging

$$P_{batt,k} = \frac{1}{0.9}.(L + P_S - \sum P_i - P_{PV} - P_{WT} - P_B). \quad (4.23)$$

4.4.2 Fuzzy-Logic-Based Expert System

The battery scheduling as a part of optimal online energy management can be regarded as a decision-making process. In fact, one way to solve this problem is to design an expert system that is able to decide, referring to the forecasted entities, whether the battery should be charged or discharged and at which convenient rates. The whole process takes into account the safe operation of the battery, the minimization of the operation costs, and the gaseous emissions. Such approach cannot be implemented using a deterministic analysis or a linear-programming-based MO due to the sizeable scenario combinations and their stochastic nature. In this paper, fuzzy logic (FL)-based expert system is applied for MG battery scheduling along with the multiobjective (cost and emission) linear programming optimization. Indeed, fuzzy set theory can encompass such subjective decision-making process due to its ability to define human reasoning that can handle uncertainties regarding to the MG exogenous environment and the uncertainty of the forecasted parameters. Moreover, such approach can be easily extended to fit the plug and play concept of MGs irrespectively of the generation rating and the architecture of its components.

1. *Fuzzification process*: The fuzzy-based expert system applied to the battery scheduling is based on seven measured/forecasted parameters listed as follows: the actual time (T), the battery state of charge (BSOC), electricity price (EP), LD, REG, actual solar daily generation (ASDG) which is equal to $\sum_1^{24}$ SPGi, and tomorrow's solar daily availability (TSDA) which is directly assigned in linguistic terms based on Japan Meteorological Agency weather forecasting referred to as sunny, mostly sunny, and cloudy. The input membership functions forward a degree of membership related to its respective input patterns to the fuzzy inference engine.
   Figure 4.4a shows the time membership function that takes as input the actual time and assigns it to a degree of a membership for each fuzzy set in the graph, where Mr stands for morning, Md stands for midday, and N stands for night. Figure 4.4b shows the BSOC membership functions, where the input is the actual state of charge (SOC) of the battery assigned into four membership sets listed as VL (very low), L (low), M (medium), and H (high). Figure 4.4c shows the EP membership functions, where the input is the actual EP (in dollars per kilowatt hours) assigned into three membership sets listed as L (low), M (medium), and H (high).

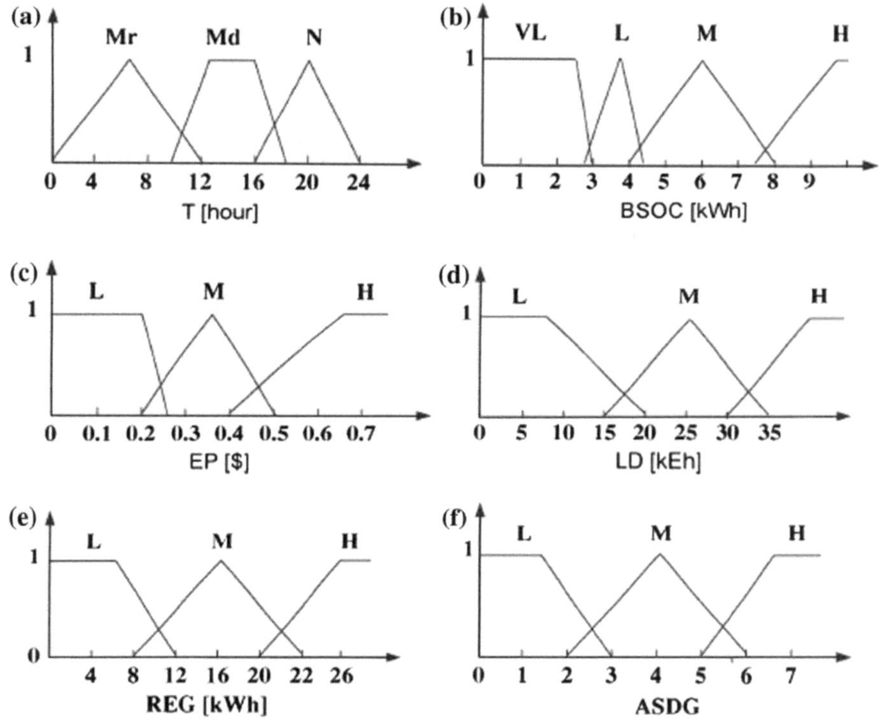

Fig. 4.4 Input membership functions

Figure 4.4d shows the LD membership functions, where the input is the next-hour forecasted LD (in kilowatt hours) assigned into three membership sets listed as L (low), M (medium), and H (high). Figure 4.4e shows the REG membership function, where the input is the next-hour forecasting of the SPG and WPG (in kilowatt hours) assigned into three membership sets listed as L (low), M (medium), and H (high). Finally, Fig. 4.4f shows the ASDG membership function, where the input is the average of the 24-h-ahead SPG forecasting (in kilowatt hours) assigned into three membership sets listed as $\{L(low), M(medium), H(high)\}$.

The parameters related to the input membership functions are listed in Table 4.2, where each variable coordinate is defined based on the membership trapezoidal function model shown in Fig. 4.5.

2. *Inference engine*: Once the degrees of membership functions of each fuzzy set have been determined for a particular input, they are forwarded to the inference engine that defines which rules should be evaluated; a typical fuzzy rule is expressed in this form.

   If (ASDG is "H") and (TSDA is "L") and (T is "Md") and (BSOC is "H") and (LD is "H") and (EP is "M") and (REG is "L") Then (Discharge "M").

   The implemented fuzzy rules summarize an efficient battery scheduling based on

Table 4.2 Parameters relative to the input membership function sets

	T			BSOC				EP		
	Mr	*Md*	*N*	*VL*	*L*	*M*	*H*	*L*	*M*	*H*
x1	0	10	16	-	2.8	4	7.4	-	0.2	0.4
x2	6	12	20	-	3.6	6	9.5	-	0.35	0.65
x3	6	16	20	2.5	3.6	6	-	0.2	0.35	-
x4	12	18	24	3	4.3	8	-	0.25	0.5	-

	LD			REG			ASDG		
	L	*M*	*H*	*L*	*M*	*H*	*L*	*M*	*H*
x1	-	15	30	-	8	20	-	2	5
x2	-	25	40	-	10	26	-	4	6.5
x3	7.4	25	-	6	10	-	1.3	4	-
x4	20	35	-	12	22	-	3	6	-

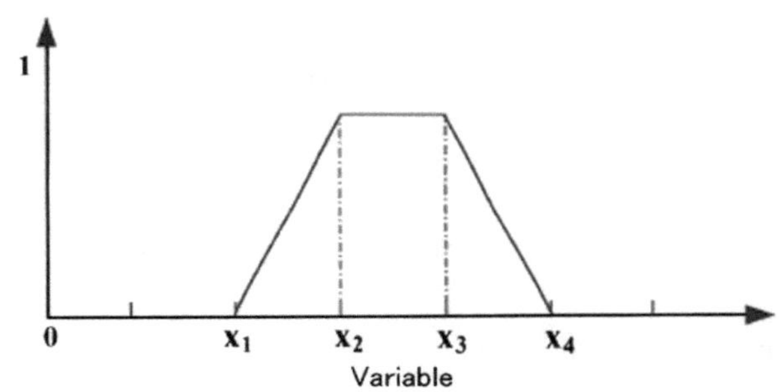

Fig. 4.5 Typical membership function

human expertise, and the goal is to decide the most appropriate action whether to charge or discharge the battery at a specific rate while mainly taking into consideration the availability of the REG sources, the LD, and the EP. The rule ensemble with charging consequent mainly tends to reach the full BSOC with the slowest possible rate where the priority for charging is associated to the low LD period and, eventually, high generation potential of the renewable sources (i.e., high REG). In case of potential lack of renewable energies (low REG or ASDG), the battery scheduling tends to accelerate the charging process during low LD period (morning) so that the full BSOC is reached antecedent to the midday load peak demand; thereby, the dispatchable DG can be fully dedicated to meet the load balance and therefore avoid expensive energy purchase from the main grid, particularly in case of SPG unavailability. On the other hand, the rule ensemble with discharging consequent mainly tends to meet the LD balance with the highest priority to avoid expensive energy purchases or trespass the DOD limit. At the end of the day, BSOC is managed with respect to the actual and next-day SPG availabilities (ASDA and TSDA). Taking the case of an actual sunny day followed by a cloudy day, the scheduling strategy will tend to store an important quantity of energy during the actual day which obviously increases the actual operation cost; however, the overall cost for the two days will noticeably decrease as important energy purchases will be avoided on the next day during load peak hours. The evaluated rules are selected based on nonzero memberships of the input values. For instance, the previously referred rule would only be selected if all of the inputs had a membership value other than zero. Once this rule (and any others meeting these criteria) is found, the output degrees of membership would be asserted according to the membership's values of the inputs.

3. *Defuzzification*: Once the degrees of membership have been calculated via the inference engine, the next step is the defuzzification into an output signal of charging or discharging rate. This is done much like the fuzzification process but in a reverse process. In the case where multiple rules have been asserted (and, hence, multiple degrees of membership for the outputs), the center of mass of the weighted outputs is used. This output value is then multiplied by a normalizing value to return it to the level of real-world outputs as a charging (Chg) or a discharging (Dsc) rate. However, in case of possible linguistic conflicting output rules (i.e., charging and discharging in the same time), then the rule with the higher membership output is retained due to the fact that the two outputs are mutually exclusive. Finally, the output corresponding to each input pattern of the fuzzy expert system is forwarded to the multiobjective algorithm and substituted in (4.22) or (4.23) where the energy stored or restored to/from the battery is expressed as

$$P_{batt} = (Chg + Dsc)BSOC. \tag{4.24}$$

Compared to MG's conventional formulations for battery scheduling, the proposed FL-based expert system approach tends to be less "conservative." Indeed, conventional binary logic-based methodologies handle decision making related to the management of the battery based on parameters that can only have one of two states, i.e., "true" or "false", symbolized by 1 and 0. On the other hand, FL uses notions such as "partially true" and "partially false" or "completely true" and "completely false". Such feature outcomes a more tolerant decision-making process capable of adaptation as situations evolve as well as taking consideration of potential uncertainties regarding to the forecasted variables. Figure 4.6 summarizes the proposed microgrid MIEM control strategy based on the FL battery scheduling. Local measurement of the microgrid LD, WPG, and SPG as well as weather conditions is prerequisite to forecast the RE power generation and the microgrids LD. The forecasted entities are forwarded along with the BSOC, time, and cost factors (electricity and fuel prices) to the FL-based decision-making system and the MO linear programming to calculate the optimal operation set-points of the battery and the controllable DG sources.

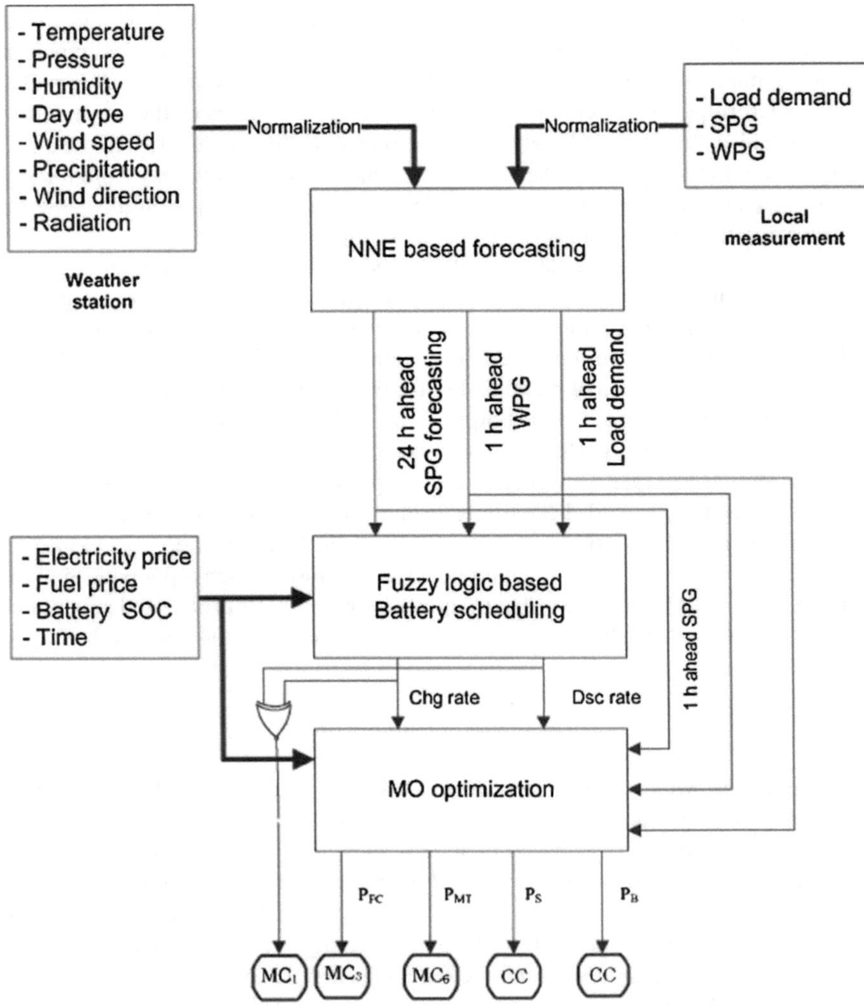

Fig. 4.6 MIEM control structure based on FL battery scheduling

4.5 Simulation Results I

4.5.1 RE Power Generation and Load Forecasting

For the validation of the proposed NNE-based forecasting tasks, statistical error forecasting performance is investigated basing on the mean average deviation (MAD), and the mean absolute percentage error (MAPE). For each forecasting task, the NNE

Table 4.3 Statistical error forecasting performances

			RBFNN	MLPNN	RNN	NNE
Winter	Load demand	MAPE	3.20	3.97	3.39	3.04
		MAD	0.21	0.25	0.23	0.21
	SPG	MAPE	3.93	3.51	3.98	2.52
		MAD	0.24	0.18	0.21	0.15
	WPG	MAPE	5.4251	5.02	6.05	4.68
		MAD	0.44	0.41	0.49	0.37
Spring	Load demand	MAPE	3.13	4.02	3.46	3.08
		MAD	0.23	0.26	0.24	0.21
	SPG	MAPE	5.24	5.17	6.26	4.15
		MAD	0.42	0.45	0.38	0.30
	WPG	MAPE	6.11	6.04	6.48	5.12
		MAD	0.49	0.48	0.51	0.38
Summer	Load demand	MAPE	3.23	4.07	3.39	3.21
		MAD	0.21	0.26	0.22	0.21
	SPG	MAPE	6.63	5.76	7.21	4.61
		MAD	0.44	0.37	0.40	0.28
	WPG	MAPE	6.53	6.13	6.62	5.13
		MAD	0.47	0.48	0.52	0.38
Fall	Load demand	MAPE	3.13	3.91	3.29	3.11
		MAD	0.17	0.22	0. 18	0.17
	SPG	MAPE	5.94	4.03	5.82	3.61
		MAD	0.26	0.17	0.23	0.15
	WPG	MAPE	6.58	6.43	6.98	5.03
		MAD	0.52	0.59	0.60	0.41

forecasting performances are compared to conventional single ANNs consisting of an MLPNN, an RBFNN, and an RNN.

Table 4.3 summarizes the entire statistical performances of the proposed forecasting tasks. The forecasting results are compared based on MAD and MAPE criteria for each of the developed ANNs, and testing datasets are composed of four seasons with different weather conditions. Table 4.3 shows clearly that the proposed NNE has the lowest forecasting error including different testing datasets covering winter, spring, summer, and fall. On the other hand, RBFNN, RNN, and MLPNN achieved also an acceptable forecasting accuracy. Among the conventional neural networks, the RBFNN presented the best overall results, while the MLPNN achieved the lowest forecasting accuracy. Indeed, the RBFNN can overcome several limitations of MLPNN and RNN such as a highly nonlinear weight update and slow convergence rate.

Figure 4.7 shows a daily 2D SPG prediction error graph in kilowatt hours of the developed ANNs, where it can be observed that the proposed NNE achieved the lowest forecasting error. Figure 4.8 shows a comparison of the LD forecasting projected by the developed ANNs during two days; the first one is a holiday with

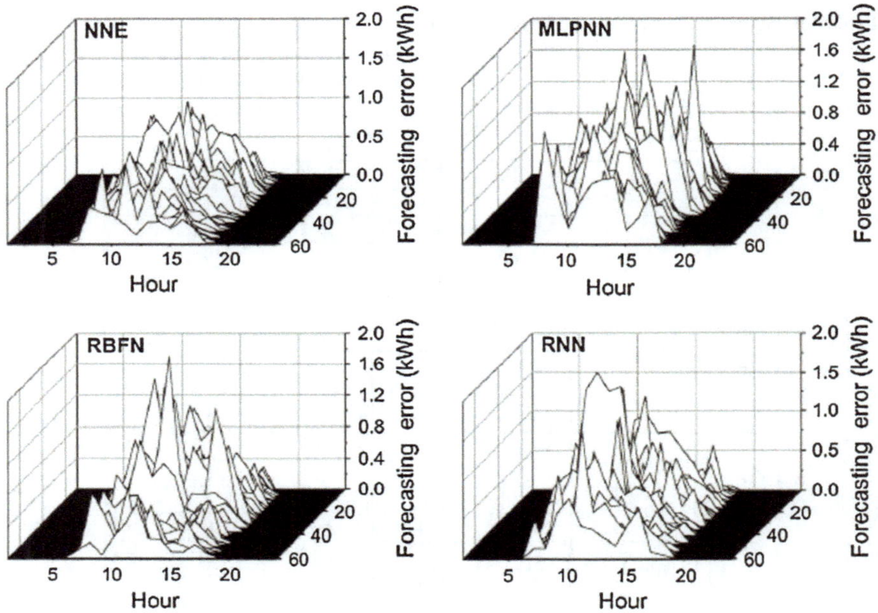

Fig. 4.7 SPG forecasting error of the proposed ANNs

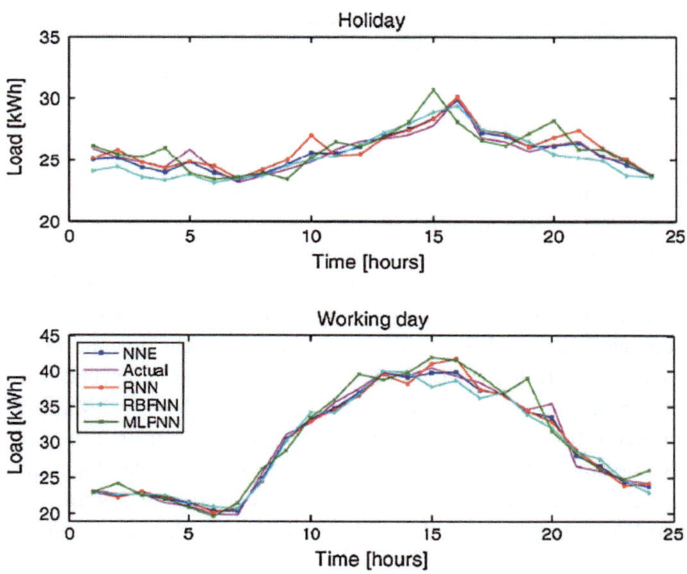

Fig. 4.8 LD forecasting for a working day and a holiday

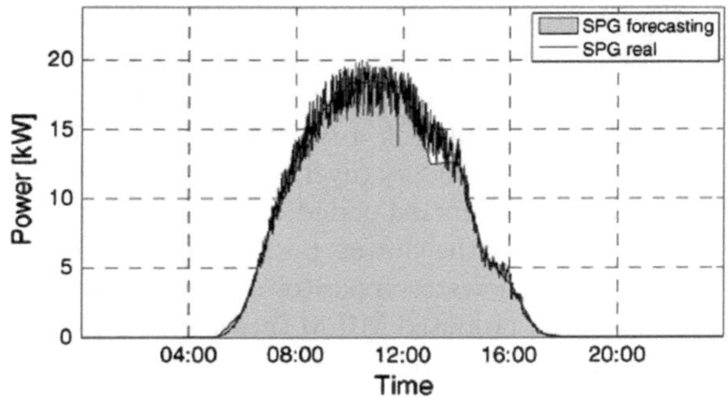

Fig. 4.9 Comparison of the forecasted and the real generated power of the PV array

a low LD peak of 31 kWh, while the second one is a working day with a high LD peak reaching 41 kWh. It can be observed that the LD characteristics during the two days are very different. In fact, for a holiday day type, the LD is almost flat, while the working day is characterized by low (6 A.M.) and high peak (3 P.M.) demands. Because of the important dynamic dissimilarities of the two daily LD characteristics, it can be seen that the NNE accomplished a visible error attenuation improvement resulting in stable forecasting performances

Figure 4.9 shows one-day measurement of the real power generated from the PV array that is used for the MG real-time simulation model. The measured real SPG similitude with the 24-h-ahead SPG forecasting used to predict the MIEM of the microgrid corroborated the results obtained based on NNE forecasting.

4.5.2 Multiobjective Intelligent Energy Management

To validate the MIEM, three approaches were compared: The first two approaches consist in a MIEM based on FL expert system and HF battery scheduling, while the third one consists in MO of the MG where the battery scheduling is managed using conventional opportunity charging. The developed HF is composed of six scenarios, and each scenario represents a scheduling strategy of one-day typical operation. The choice of a specific scenario depends on the actual and the next-day REG. Dealing with each scenario is involving the time, the actual SOC of the battery, the actual load, and the actual RE power generations where the time variable is correlated to the EP (as the EP peak usually occurs during a specific time period). Each scenario of the battery scheduling flowchart establishes operation set-points for charging or discharging the battery during operation hours, which comes to define the power stored or restored (P_{batt}). After the execution of replaces the conventional equality constraint (4.21) while running the MO process; further details can be found in [15, 18, 19].

Figure 4.10 shows the battery scheduling for one-day operation using the proposed FL-based expert system. The full BSOC is reached before the peak of LD occurrence, then followed by a gradual battery discharge (peak shaving). The delineated hourly charging rates resulted in a smoother charging dynamic; therefore, there are less solicitations on the battery (overheating) and the DG controllable sources. Similarly, during load peak demand period, the proposed approach achieved a successful load peak shaving with the slowest possible discharging rates.

Figure 4.11 shows a comparison of the emission and operation cost during one-day operation. The proposed MIEM (based on FL battery scheduling) achieved important saving on operation cost as well as the gaseous emission level compared to a conventional control operation (without MIEM). It can be observed that the proposed methodology achieved considerable cost saving during LD peak that coincides with nonavailability of the PV generator and consequent rise of the EPs.

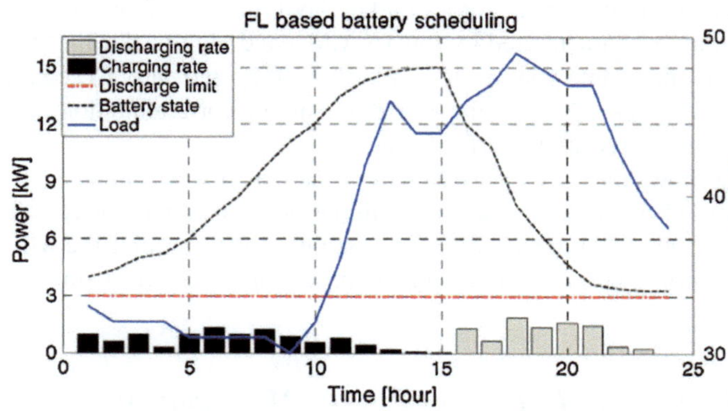

Fig. 4.10 Battery scheduling process for one-day operation

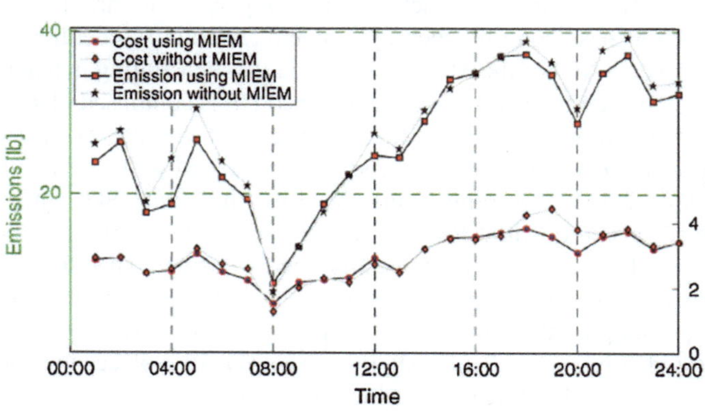

Fig. 4.11 Cost and emission comparison of the microgrid operation

The simulation results cover two successive sunny and cloudy days and one week using the HF, the FL, and opportunity battery management (no MIEM). Table 4.4 summarizes the simulation results that include the operation cost, CO_2, SO_2, and NO_x emissions, and the battery initial and final states of storage (beginning and end of the day). It can be seen that the MIEM (using intelligent battery scheduling) achieved a visible saving on the operation cost and the gaseous emissions. The proposed MIEM based on FL battery scheduling accomplished 1.35 and 5.76 % of cost saving and 2.96 and 6.1 % of lower emissions compared to the HF and the conventional opportunity charging approaches, respectively, during the whole referred simulation period. For the first-day simulation, the cost saving of the proposed MIEM approach compared to the opportunity charging was equal to 4.63 % compared to 8.15 % for the next day as a result of the intelligent battery scheduling that takes into account the next-day solar energy availability. In fact, during the first simulation day, the management strategy adapts the charge, discharge, and the end-of-the-day storage so that the saved energy can be restored during the next day characterized by a poor SPG potential; therefore, more important operation cost savings are achieved during the second day compared to the first one.

In addition to the saving on the operation cost and emissions, indirect saving is also incited using the proposed intelligent management strategy that consists in less maintenance and replacement cost for the lead–acid battery. In fact, the lead–acid battery lifetime and operation performance are affected (therefore, maintenance cost and overall MG operation cost) by fast charging rates that can lead to overcharge and/or important temperature rise. Moreover, the life cycle of the battery depends closely on the DOD in each cycle; the HF and FL-based scheduling approaches achieved DOD averages equal to 64 and 61 %, respectively, compared to 80 % for the opportunity battery management. According to Fig. 4.3, the FL and HF battery scheduling approaches assure approximately 820 and 790 expected average cycles, respectively, compared to 515 cycles for the conventional opportunity charging approach.

Table 4.4 Operation cost and emissions of the microgrid

		Cost [$]	CO_2 [lb]	SO_2 [10^{-2}lb]	NO_x [10^{-2}lb]	Battery [kWh]
Sunny	HF	70.83	646.56	0. 33	8.47	3.48/6.9
	FL	70.02	639.81	0.32	8.14	3.16/7.05
	No MIEM	73.42	673.49	0. 35	8.71	3.24/3.31
Cloudy	HF	73.84	688.53	0. 35	9.33	6.9/3.06
	FL	72.94	680.84	0.34	9.27	7.05/3.16
	No MIEM	79.42	701.35	0. 36	9.42	3.31/3.49
One Week	HF	379.74	2.68×10^3	1.61	38.86	3.78/5.71
	FL	374.36	2.59×10^3	1.59	37.21	3.04 /5.20
	No MIEM	396.15	2.78×10^3	1.65	40.87	4.14/4.09

Assuming that the battery accomplishes a charging and discharging cycle once a day, then the battery needs to be approximately replaced in 27 and 26 months for the case of FL- and HF-based approaches, respectively, whereas it needs to be replaced in only 17 months for an opportunity charging approach. For a replacement cost of the batteries equal to \$180/kWh (\$2700 for the whole system), extra daily maintenance costs equal to \$3.28, \$3.4, and \$5.4 for the FL, the HF, and the opportunity charging battery management, respectively, need to be considered as part of the overall operation cost. It is clear that the developed MIEM approach based on FL expert system for battery scheduling achieved the lowest operation cost and emissions while paradoxically requiring a lower average DOD of the battery compared to the two other approaches. This is due to the efficiency of the battery scheduling as for avoiding important purchase of power from the main grid during the electricity demand peaks. In fact, during load peak hours, the EPs can dramatically increase in a deregulated electricity market, resulting in an important increase of the overall operation costs.

The proposed MIEM predictive control system is implemented on the MG simulation model developed under MATLAB® Simulink® environment to validate the proposed approach over 24-h real-condition operation including experimental data measured in TUAT.

Figure 4.12 shows the profile of power exchanged with the main grid in real condition based on the MG simulation model for a MIEM based on FL and HF battery scheduling approaches. It is expected that the equality constraint on load balance (4.20) may be partially violated in real-condition operation as the forecasted inputs of the MIEM are hourly averaged values without considering the forecasting errors. However, the proposed MIEM with FL expert system for battery scheduling presented noticeably limited power import from the main utility compared to the HF-based approach. As a result, considering a MG's long-term operation, important minimization on operation cost and gaseous emissions can be achieved using the proposed approach.

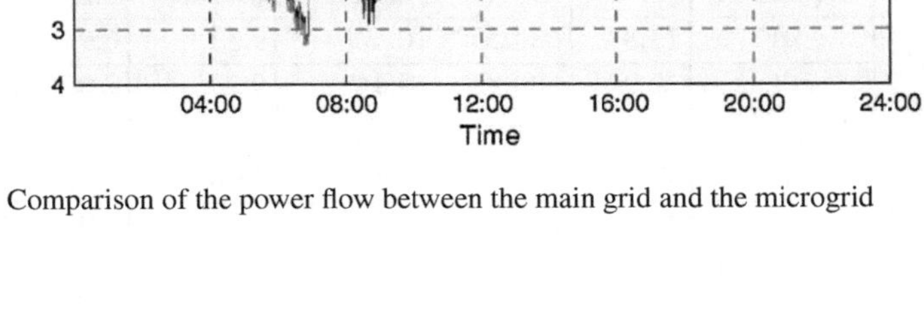

Fig. 4.12 Comparison of the power flow between the main grid and the microgrid

4.6 Optimal Energy Cost Management

Aiming to avoid grid voltage fluctuations [65, 107], or even blackout, at any time instant, the electric grid must balance power between the production and the consumption with a small margin of error. The grid capacity is built to satisfy the peak consumption. If the peak consumption can be shifted during a day, referred to as "peak shaving", the power adjustment, often ensured by excess capacities working in stand-by mode, could be largely reduced. To build a more robust utility grid, strategies and means of power management are being developed, as well as information on grid needs and availability [67], which could assist in power balancing by avoiding undesired injection and performing load shaving during peak hours. For this, the SG is being created to facilitate information exchange. SG is electric networks that employ innovative and intelligent monitoring, control communication, and self-healing technologies to deliver better services for power producers and distributors, flexible choices for end-users, and reliability and security of power supply [85, 93]. SG is expected mainly for the following aspects: bidirectional power distribution; bidirectional communication, and reduction mismatching between supply and demand.

The concept of MG is proposed for better RE penetration into the utility grid and helps energy management to respond to some grid issues, such as peak shaving, and reduces energy cost [3, 64]. MGs are considered as one of the possible approaches helping to develop the SG [61]. By aggregating loads and multisource, renewable and traditional, MG can operate in both off-grid and grid-connected configuration. It is generally considered that MG controls onsite generation and power demand to meet the objectives of providing local power, ancillary services, and injecting power into the utility grid if required [42]. Concerning MG approach, several main advantages can be given: improving RE penetration level, facilitating the SG implementation, better energy supply for remote areas, power balancing at local level with self-supplying possibility, and maintaining load supply during islanding operation or off-grid mode [94]. Thus, the MG controller becomes essential for balancing power and energy management, and facilitates the sources pooling during islanding.

Depending on the usage of AC or DC bus for coupling different elements within microgrid, AC microgrid, DC microgrid, and hybrid AC/DC microgrids structures exist [42]. At present, the DC grid is not ubiquitous [96, 105], but more HVDC transmission lines are being built in MW level, while low voltage DC grid is being adopted starting with data centers, for the reason of more efficiency, less cost, less occupied space, lower lifetime cost and more reliability [2, 100]. See also [108–110].

In [42] presents a three-level hierarchical control (HC) is presented according to ISA-95 and applied to AC or DC microgrids. This general approach of HC for MGs is conceived for a large-scale power system, upstream in the utility grid hierarchy. Imitating the behavior of a grid synchronous generator control, the proposed HC strategy aims at balancing power between multi inverters coupled on the same bus without communication, while controlling the power at the point of common coupling (PCC) at the same time. The proposed HC is considered as a part of the central control and does not take into account the prediction of the power generation and the energy optimization.

In [100], support for autonomous DC microgrid applications is proposed by integrating the device-level service-oriented architecture paradigm into the international standard IEC 61850 applications. In order to create self-manageable MG with semantic-enabled plug and play process for distributed energy resources (DERs), this solution provides generic middle-ware platform required for vertical communication. However, the proposed solution applied to the real MG power systems requires additional control and regulation policy.

A high-level energy management supervision, by means of multiagent systems, is presented in [60]. In this work, the authors focus on two-level architecture for multiple interconnected MGs aiming to manage DER in order to match the buyers and sellers in the energy markets.

Generalized formulation for intelligent energy management of a MG is proposed in [20] using multiobjective optimization to minimize the operation cost and the environmental impact. An ANN ensemble is developed to predict renewable energy generation (REG) and LD. In addition, a battery scheduling is proposed as a part of an optimal online energy management, seen as a decision-making process. However, SG data exchanges online or dynamic energy pricing are not considered.

To increase penetration of small PV production into the grid, a local HC with energy management is proposed in [90]. The system is presented as multilayer control structure, each layer with a different function, and is based on an optimal power flow management with predictions, which considers batteries aging and day-ahead approach into the optimization process. However, the exchange data with the SG, such as limitations of the grid capacity, is not taken into account. Moreover, due to uncertainty of prediction and lack of grid information, the grid power could be out of control.

Concerning the energy management two main approaches are considered: rule-based and optimization-based approaches. Rule-based approach manages the system according to prefixed rules, such as simple rule base, multiagent system [60] and fuzzy logic approaches [15, 20]. Optimization-based approach manages the system by mathematical optimization, carried out with objective function and constraints. The optimization methods include the artificial intelligence joint with linear programming [20], linear programming [15] or dynamic programming [9, 90], and genetic algorithms [11].

To sum up the results presented in these works, the rule-based system is simple and robust, but does not guarantees the optimal performance with given operating conditions. Moreover, rules become complex when facing different scenarios. The optimization gives an optimal solution within given constraints and operation condition. However, optimization is usually treated as separate problem from the power balancing strategy, and optimization is not implemented in real-time operation. Since optimization are preformed based on prediction, errors between prediction and real condition could result in degraded real-time operation by violating certain constraints, or even result in failure. Thus, optimization and power balancing should be designed together and coupled with adequate interface.

Hence, in contrast of the above cited works, in this study the goal is to design and develop a MG local supervision control which can respond to all following requirements:

- to design optimization and power balancing together and couple them through adequate interface;
- to implement optimization in real-time operation with resistance to uncertainties;
- to process prediction data used for optimization;
- to exchange data with the SG and with the end-user;
- to adapt various data in MG operation, such as real-time market pricing and user demand.

Thus, this section focuses on supervisory control of DC microgrid which is supposed to manage the power flow in MG and power flow exchanged with the utility grid, with the objective of making full use of each source while respecting their constraints on capacity and power.

4.7 Microgrid Overview

The MG presented in Fig. 4.13 is suggested for local PV power generation combined with storage and grid connection, which feed directly a DC load through their dedicated converters. Concerning the DC bus, following considerations are assumed: for the demand side, 90 % of a tertiary building's electrical load is possible to be DC fed in efficiency manner; for the grid side, the power factor can be controlled at 1 [93].

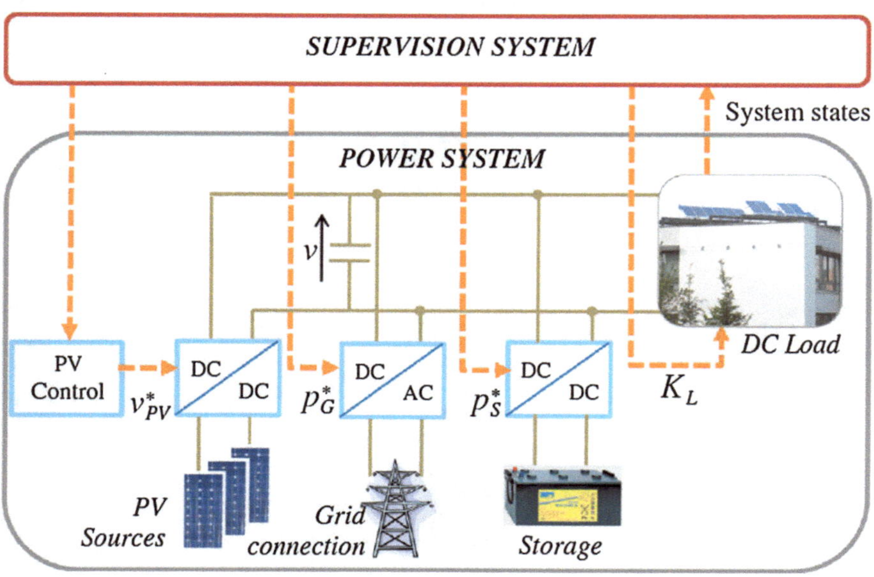

Fig. 4.13 DC microgrid overview

Aiming to interact with the load environment (user, metadata, SG), a supervision system is added, whose main role is to balance instantaneous power in power system following an energy management algorithm. The system operation must keep power balance while respecting constraints of certain elements. Figure 4.14 shows the powers flow in the MG, using unidirectional powers whose sign convention is always positive.

The system operates respecting the available storage level and taking into account the grid connection. In case of insufficient PV energy toward the load, the system security is ensured thanks to the grid connection and by storage, as well as the load shedding program. If any excess PV power, the storage could be charged and the grid connection gives the possibility to trade it back (Fig. 4.14).

4.7.1 PV Sources Control

PV system is controlled either by a maximum power point tracking (MPPT) method or by an algorithm to output a limited power $p_{PV_{\lim}}$ to protect the storage from overcharging or to maintain grid injection power within imposed constraint. In this study, the chosen MPPT method is the very well known Perturb & Observe (P&O) [50, 52], according to which PV sources produce an MPPT power $p_{PV_{MPPT}}$. The PV production could be limited thanks to the limited power reference $p^*_{PV_{\lim}}$ calculated by the supervision with the algorithm described later.

The PV power p_{PV} control strategy is shown in Fig. 4.15. The P&O algorithm and the PV production limiting algorithm give at the same time corresponding voltage reference $v^*_{PV_{MPPT}}$ and $v^*_{PV_{\lim}}$ to operate PV system. The maximum of these two

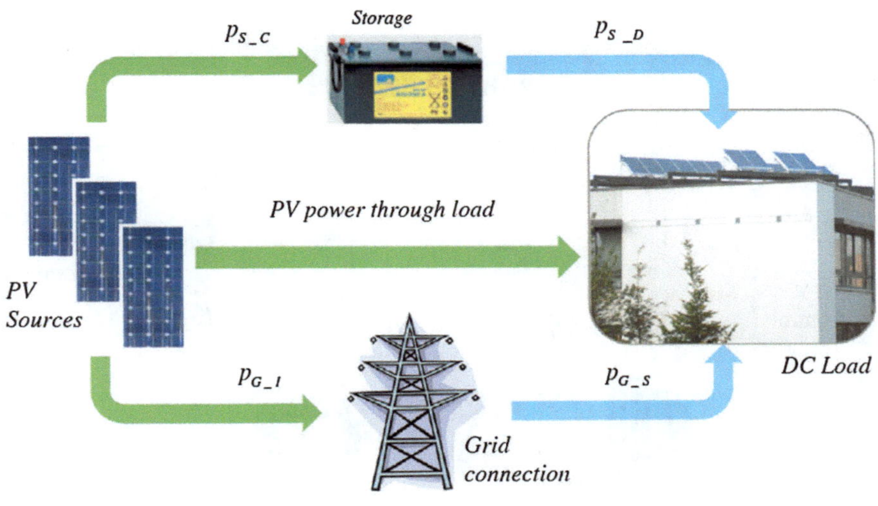

Fig. 4.14 Powers flow representation

references is taken as the PV voltage control reference v^*_{PV}, which represents the minimum power. Following v^*_{PV}, the PV system is operated by voltage and current double closed-loop control via automatic voltage regulator and automatic current regulator.

The output of control, duty cycle of pulse-width modulation (PWM) signal, is given to the power electronic devices to control the PV system. During the MPPT operation, if limited power within the MPPT ability is given, the proportional integral (PI) controller would increase the $v^*_{PV_{\lim}}$. For $v^*_{PV_{\lim}} > v^*_{PV_{MPPT}}$ the $v^*_{PV_{\lim}}$ reference is taken and the MPPT algorithm is stopped. By constrained power closed-loop control, the PI controller controls the PV power at the limited level. In case of low solar irradiance, the PV output power ability is less than the limited power reference, so the PI controller will decrease $v^*_{PV_{\lim}}$ until the lower limit, and $v^*_{PV_{MPPT}}$ is taken to control the PV system. So, the limited power control does not affect MPPT algorithm and MPPT power is produced. Experiment validation of the control is provided in [106].

4.7.2 Storage Control

Lead–acid batteries are selected as storage for the DC microgrid applied to building, because of relatively low cost and mature technology [101]. The storage is operated by current closed-loop control, and the storage power is controlled by a supervision system which calculates the corresponding power reference. The storage SOC must be respected to its upper and lower limitations, $SOC_{\max}$ and $SOC_{\min}$ respectively, to protect the storage from overcharging and over discharging,

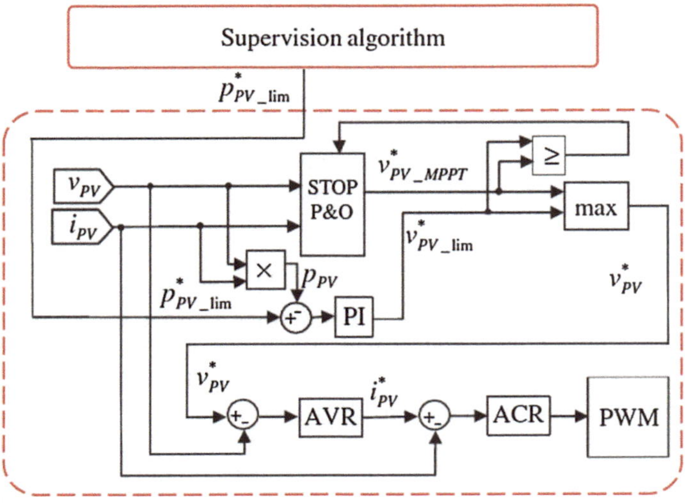

Fig. 4.15 PV system control strategy

as described by (4.25). The soc is calculated by (4.26), with SOC_0 as initial soc at t_0, C_{REF} as storage nominal capacity (Ah) and v_S as storage voltage. When the soc limit is not reached, the PV production should not be limited, as in (4.27).

$$SOC_{\min} \leq soc(t) \leq SOC_{\max} \tag{4.25}$$

$$soc(t) = SOC_0 + \frac{1}{3600\dot{v}_S\dot{C}_{REF}} \int_{t_0}^{t} (p_{S_C}(t) - p_{S_D}(t))dt \tag{4.26}$$

$$p_{PV_{\lim}}(t) = p_{PV_{MPPT}}(t) \quad if\ soc(t) < SOC_{\max} \tag{4.27}$$

4.7.3 Grid Connection Control

The grid connection is controlled by current closed-loop control. The grid power is controlled by supervision system which calculates the corresponding power reference. Furthermore, as messages transmitted by the SG via supervision system, limits for grid supply power $p_{G\_S\_\lim}$ and grid injection power $p_{G\_I\_\lim}$ could be imposed. By these two limitations, grid problems, such as performing peak shaving, avoiding undesired injection or downscaling injection fluctuations caused by intermittent PV productions, can be improved. During MG operation, the grid power should be controlled to satisfy (4.28) and (4.29).

$$0 \leq p_{G\_I}(t) \leq p_{G\_I\_\lim} \tag{4.28}$$

$$0 \leq p_{G\_S}(t) \leq p_{G\_S\_\lim} \tag{4.29}$$

4.7.4 DC Load Control

The load power demand p_{L_D} should be satisfied; nevertheless, in case of insufficient storage and grid access limits, p_{L_D} cannot be fully met, and the load must be partially shed, forming load power p_L, to keep power balance. The load limited power $p_{L\_\lim}$ is controlled by the load coefficient K_L defined by (4.30):

$$K_L = p_{L\_\lim}/p_{L\_\max} \tag{4.30}$$

where $p_{L\_\max}$ is a constant as contractual subscribed maximum power. It is supposed that the load could not be supplied with a power that exceeds this limit. Given the constant denominator $p_{L\_\max}$, therefore $K_L \in [0, 1]$ changes according to the available power for supplying the load. If $p_{L\_D} > p_{L\_\lim}$, the load would be shed within the limit level. These operations should be controlled by the supervision system. Temporarily load partial shedding could be a solution to reduce utility grid mismatching, or to obtain less energy consumption, within the limit agreed by end-user.

4.7.5 Power Balancing Principle

According to the powers sign convention (powers always positive), the physical law of power balancing is described by:

$$p_L(t) + p_{G\_I}(t) + p_{S\_C}(t) = p_{G\_S}(t) + p_{S\_D}(t) + p_{PV}(t) \tag{4.31}$$

with $p_L(t) = \min(p_{L\_D}(t), p_{L\_\lim}(t))$ and $p_{PV}(t) = \min(p_{PV\_MPPT}(t), p_{PV\_\lim}(t))$.

Fluctuations in the DC bus voltage, which is noted v, are caused by the difference between load consumption and PV generation. The required power reference p^* for power balancing is calculated by regulating v with a proportional controller as in Eq. (4.32):

$$p^*(t) = p_{PV}(t) - p_L(t) - C_P(v^*(t) - v(t)) \tag{4.32}$$

where v^* is the DC bus voltage control reference and C_P is the proportional gain. For stabilizing the DC bus voltage, power balance in the system is performed by adjusting storage and grid power. Thus, p^* is shared by the storage and the grid as in (4.33)

$$p^*(t) = p_G^*(t) + p_S^*(t) \tag{4.33}$$

with $p_G(t) = p_{G\_I}(t) - p_{G\_S}(t)$ and $p_S(t) = p_{S\_C}(t) - p_{S\_D}(t)$.

After calculating p^*, the grid power reference p_G^* and storage power reference p_S^* are calculated according to a distribution coefficient $K_D(t)$, as described in (4.34) and (4.35).

$$p_S^*(t) = K_D(t)p^*(t) \tag{4.34}$$

$$p_G^*(t) = p^*(t) - p_S^*(t) \tag{4.35}$$

The power balancing of studied DC microgrid operation can work with any K_D value. To improve efficiency and reduce energy cost, optimization calculations using specifically metadata (power predictions, power limits, real-time grid energy tariffs) are done by the supervision which outputs the distribution coefficient, $K_D(t)$, whose time depending values represent the predicted optimized powers flow.

4.8 Supervision Control Design

The supervision system, proposed in Fig. 4.16, is designed in four-layer structure, which consists of human-machine interface (HMI), prediction layer that predicts load consumption and PV production, energy management layer that optimizes the predicted powers flow, and operation layer that balance instantaneous power, based on unique interface parameter $K_D(t)$, in power system.

The supervision control takes into consideration prediction data and various constraints, such as grid power limits and storage capacity, which can be fully respected for both optimization and real operation. Each layer provides an independent function and thus, the structure is flexible and can be implemented in several microcontrollers or computers so that real-time power balancing control and complicated optimization can be executed at the same time without affecting each other. The multilayer structure simplifies implementation of such complex control strategy. Thus, the control of the power balancing is separated from energy management layer, yet they are linked through one interface parameter $K_D(t)$. On one hand, the energy management layer is able to optimize the MG energy cost through predictive data and thus obtain the predicted optimized power flow which is then translated into $K_D(t)$ sequence. On the other hand, the power balancing control in operation layer is an independent function that can work with any $K_D(t)$ value. The distribution coefficient $K_D(t)$ is a single interface parameter yet represents the power flow from different sources. Hence, the communication of $K_D(t)$ does not need high-speed communication between layers.

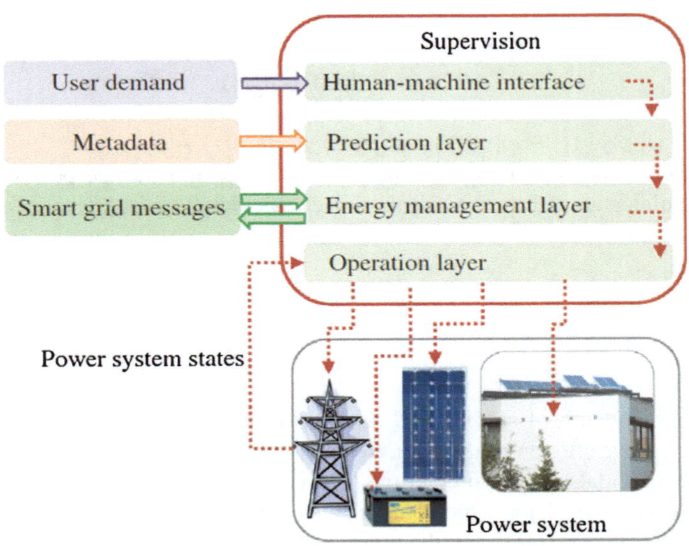

Fig. 4.16 Design principle of supervision

4.8.1 Human–Machine Interface Layer

The HMI design is presented in Fig. 4.17. User can specify the lowest limit of the load coefficient, $K_{L\_\lim}$. Aiming to maintain power balance, if the system requires $K_L < K_{L\_\lim}$, the operation is assumed with the necessary K_L value, but the user will be notified.

4.8.2 Prediction Layer

The prediction layer design is presented in Fig. 4.18. It calculates possible PV power and load power evolutions for the next day. Based on solar irradiance and temperature forecasting, and on PV model, built with parameters identification or PV solar irradiance mapping [51, 69], the PV predictable power $p_{PV\_prediction}$ could be calculated with as less error as possible.

The load power supply can be predicted by statistical data and/or by information of load scheduling from building management system and with respect of user demand [4, 88]. The load scheduling, that represents the operating program of building facilities, building energy needs linked to weather, and the partial and full load shedding program are supposed known, and they are implemented and updated continuously by the building management system. According to the value of $K_{L\_\lim}$ chosen by the user, the load prediction power $p_{L\_prediction}$ can be calculated by this layer.

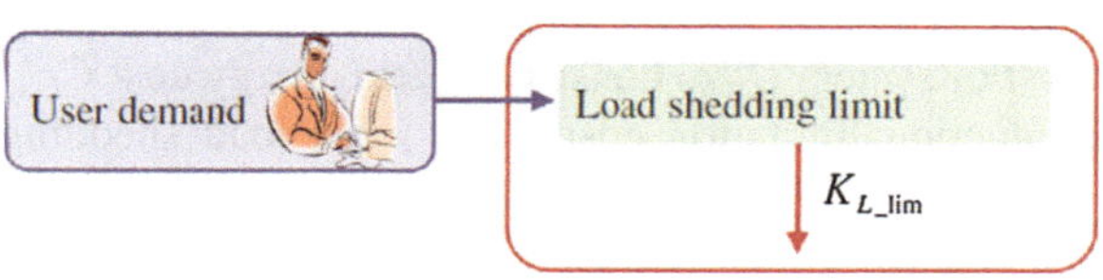

Fig. 4.17 HMI layer design

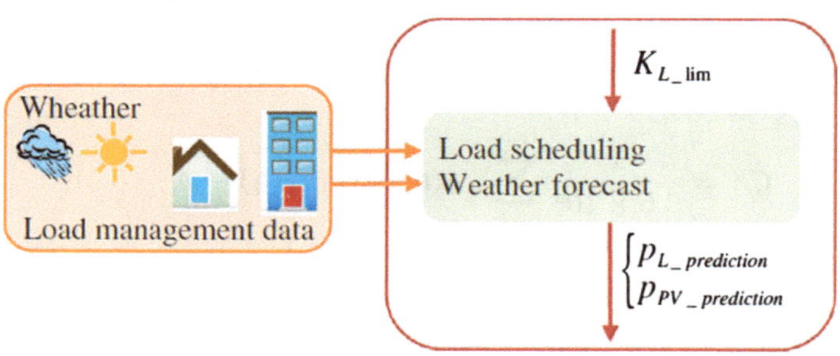

Fig. 4.18 Prediction layer design

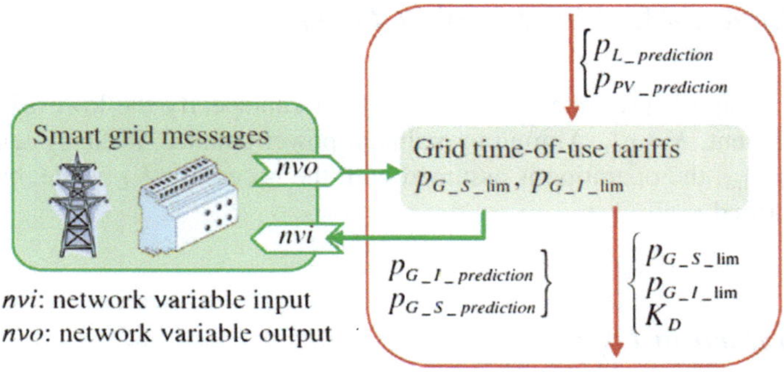

Fig. 4.19 Energy management layer design

4.8.3 Energy Management Layer

Energy management layer (Fig. 4.19) interacts with the prediction layer and controls the operation layer by calculating the distribution coefficient K_D following an optimization method. The optimization goal is to obtain the best power distribution between the grid and the storage, so to reduce energy cost, grid power peak consumption, load shedding, and limiting PV production at the same time. In this study, the SG message is supposed to provide real-time grid energy tariffs and grid power limits, which assist in reducing peak supply and avoid undesired injection. Furthermore, thanks to the SG connection, this layer is able to inform the grid operator of the grid supply power prediction $p_{G\_S\_prediction}$ and grid injection power prediction $p_{G\_I\_prediction}$.

The energy cost of system C_{total} consists of grid energy cost C_G, storage energy cost C_S, PV production limitation cost C_{PVL}, and load shedding cost C_{LS}, as in (4.36).

$$C_{total} = C_G + C_S + C_{PVL} + C_{LS} \tag{4.36}$$

By calculating the energy cost for each time duration Δt, C_G is defined by (4.37). According to this definition, the grid power could be bought or sold.

$$\begin{aligned} C_G &= \frac{1}{3.6 \times 10^6} \sum_{t_i - t_0}^{t_F} [c_G(t_i)\dot{\Delta} t \dot{(} p_{G\_I}(t_i) + p_{G\_S}(t_i))] \\ t_i &= \{t_0, t_0 + \Delta t, t_0 + 2\Delta t, \ldots, t_F\} \end{aligned} \tag{4.37}$$

This study takes into account the same price for energy purchased or sold, and the grid energy tariff is defined by

$$c_G(t) = \begin{cases} c_{NH} = 0.1\varepsilon = kWh & for\ t \in normal\ hours\ (NH) \\ c_{PH} = 0.7\varepsilon = kWh & for\ t \in peak\ hours\ (PH) \end{cases} \tag{4.38}$$

Storage aging should be considered to give an energy tariff of storage using. For this study, the storage energy cost CS and an arbitrary storage energy tariff are defined in Eq. (4.39).

$$C_S = \frac{1}{3.6 \times 10^6} \sum_{t_i - t_0}^{t_F} [c_S(t_i)\dot{\Delta t}(p_{S\_C}(t_i) + p_{S\_D}(t_i))]$$
$$with\ t_i = \{t_0, t_0 + \Delta t, t_0 + 2\Delta t, \ldots, t_F\}\ and\ c_S(t) = 0.05\varepsilon/kWh \quad (4.39)$$

The cost of PV system shedding and an arbitrary PV shedding tariff are defined in (4.40).

$$C_{PVL} = \frac{1}{3.6 \times 10^6} \sum_{t_i - t_0}^{t_F} [C_{PVL}(t_i)\dot{\Delta t}(p_{PV\_MPPT}(t_i) - p_{PV\_\lim}(t_i))]$$
$$t_i = \{t_0, t_0 + \Delta t, t_0 + 2\Delta t, \ldots, t_F\}, \quad c_{PVL}(t) = 1\varepsilon/kWh \quad (4.40)$$

The cost of load shedding and an arbitrary load shedding tariff are defined by

$$C_{LS} = \frac{1}{3.6 \times 10^6} \sum_{t_i - t_0}^{t_F} [C_{LS}(t_i)\dot{\Delta t}(p_{L\_D}(t_i) - p_{L\_\lim}(t_i))]$$
$$t_i = \{t_0, t_0 + \Delta t, t_0 + 2\Delta t, \ldots, t_F\}\ and\ c_{LS}(t) = 1\varepsilon/kWh \quad (4.41)$$

Aiming to limit the power grid fluctuations, grid power changing rate limits are introduced as:

$$\begin{cases} p_G(t_i) - p_G(t_{i-1}) \le Limit \\ p_G(t_i) - p_G(t_{i-1}) \ge -Limit \end{cases} \quad (4.42)$$
$$t_i = \{t_0, t_0 + \Delta t, t_0 + 2\Delta t, \ldots, t_F\}$$

As PV energy grid injection benefits incentive tariffs, energy grid injection by power grid charged storage is forbidden. Thus, the following limits are imposed in order to ensure storage energy charge and grid injection only from PV production.

$$\begin{aligned} &p_G(t_i) \ge 0,\ p_S(t_i) \ge 0 \quad if\ p_{PV}(t_i) - p_{L\_D}(t_i) \ge 0 \\ &p_G(t_i) \le 0,\ p_S(t_i) \le 0 \quad if\ p_{PV}(t_i) - p_{L\_D}(t_i) < 0 \\ &t_i = \{t_0, t_0 + \Delta t, t_0 + 2\Delta t, \ldots, t_F\} \end{aligned} \quad (4.43)$$

Finally, by considering the discrete time instant ti, from t_0 to t_F, with the time interval Δt, the optimization problem can be completely mathematically expressed by (4.44):

$$Minimize C_{total} = C_G + C_S + C_{PVL} + C_{LS}$$

with respect to :

$$\begin{cases} p_L(t_i) + p_{G\_I}(t_i) + p_{S\_C}(t_i) = p_{G\_S}(t_i) + p_{S\_D}(t_i) + p_{PV}(t_i) \\ SOC_{\min} \leq soc(t_i) \leq SOC_{\max} \\ soc(t_i) = SOC_0 + \frac{1}{3600 \dot{v}_S C_{REF}} \sum_{t_i=t_0}^{t_F} (p_{S\_C}(t_i) - p_{S\_D}(t_i))\Delta t \\ 0 \leq p_{G\_I}(t_i) \leq p_{G\_I\_\lim} \\ 0 \leq p_{G\_S}(t_i) \leq p_{G\_S\_\lim} \\ p_G(t_i) - p_G(t_{i-1}) \leq Limit \\ p_G(t_i) - p_G(t_{i-1}) \geq -Limit \\ p_G(t_i) \geq 0,\ p_S(t_i) \geq 0\ if\ p_{PV}(t_i) - p_{L\_D}(t_i) \geq 0 \\ p_G(t_i) < 0,\ p_S(t_i) < 0\ if\ p_{PV}(t_i) - p_{L\_D}(t_i) < 0 \\ p_{PV\_MPPT}(t) - p_{PV\_\lim}(t) = 0\ if\ soc(t) < SOC_{\max} \end{cases} \tag{4.44}$$

$$t_i = \{t_0, t_0 + \Delta t, t_0 + 2\Delta t, \ldots, t_F\}$$

Energy optimization are usually solved by linear programming [15] or dynamic programming [33, 90] technique. Dynamic programming can solve nonlinear problem, while linear programming solves problems satisfying linear forms. Linear programming can be more efficiently solved with less time and memories. It can be verified that, mathematical formulation given by Eq. (4.44) follows standard linear programming form, except for the last constraint which does not take a linear form. However, this constraint can be easily linearized by introducing for each time point ti, a variable array which is integer; this is why the optimization formulation is a mixed integer linear program [32].

In this study, the optimization problem is solved using the IBM ILOG CPLEX solver [53], which is a powerful tool for solving different types of optimization problems. However, any other mixed integer linear programming solver also can be used (LP_SOLVE, GUROBI,. . .). In addition, in order to express our problem in the syntax of the solver and to call the solving algorithm of the solver, a procedure written in C++ is used. This procedure output to a file the optimal powers flow, which is the evolution of $p_{S\_D}(t)$, $p_{S\_C}(t)$, $p_{G\_S}(t)$, $p_{G\_I}(t)$. The estimated optimum powers flow is then translated into a control parameter that is the optimum distribution coefficient $K_D(t)$. Taking into account (4.33) and (4.34), the optimum K_D values are calculated by (4.45):

$$K_D(t) = \frac{p_{S\_C}(t) - p_{S\_D}(t)}{p_{S\_C}(t) - p_{S\_D}(t) + p_{G\_I}(t) - p_{G\_S}(t)} \tag{4.45}$$

4.8.4 Operation Layer

According to the energy management layer output ($K_D(t)$ and grid power limits), the operation layer presented in Fig. 4.20 aims at balancing power in the power system while meeting all constraints.

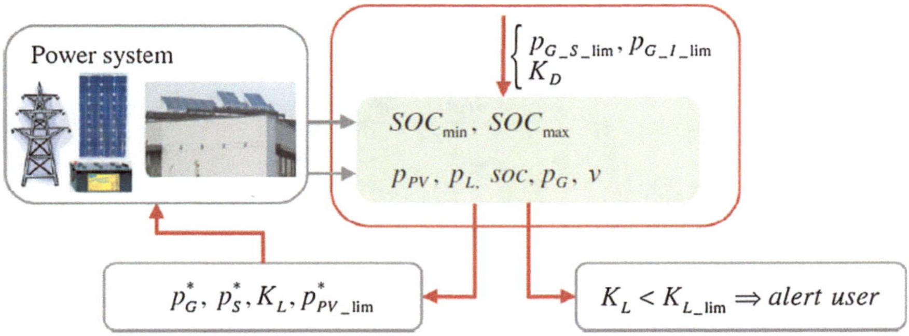

Fig. 4.20 Operation layer design

If there is not sufficient power to supply the load (insufficient PV production when the grid power is limited and the storage is empty), the load shedding is performed. In such cases, the load power has to be limited to $p_{L\_\lim}$, which is calculated as in (4.46).

$$p_{L\_\lim} = P_{PV} + p_{G\_S\_\lim} \tag{4.46}$$

The system compares the actual load power p_L with the load power limit $p_{L\_\lim}$; if $p_L > p_{L\_\lim}$ the load would be shed within the limit level. When the storage is full and the grid injection limit does not permit absorbing all excess of PV production, the PV limited production is performed by calculating $p^*_{PV\_\lim}$:

$$p^*_{PV\_\lim} = P_L + p_{G\_I\_\lim} \tag{4.47}$$

To sum up the above described power balancing, and taking into consideration the limits of storage and grid given by (4.25), (4.27)–(4.29), the overall algorithm of the operation layer is shown in Fig. 4.21.

Experiment validation of the technical feasibility of the operation layer for a constant value of K_D is provided in [105]. As the prediction layer requires available and reliable forecast data, for the PV sources geolocation and for few hours ahead, in this study, the optimized DC MG is tested by simulation.

4.9 Simulation Results II

The MG control is simulated for an operation on April 23, 2011 in Compiegne, France. The objective of this study is more to validate a comprehensive approach rather than purely numerical results. For this reason, we do not give the numerical values of various system components studied, yet powers values are based on our experimental multisource system platform.

The simulation results of the DC microgrid were obtained under the MATLAB-Simulink. All the implemented automatic controls, described earlier, are working

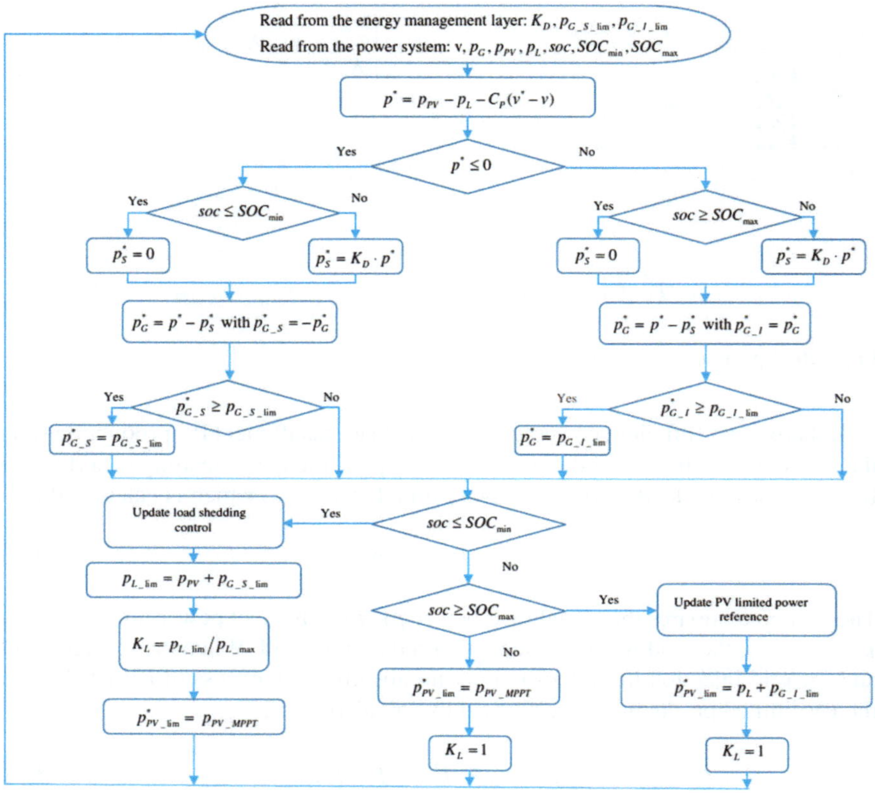

Fig. 4.21 Flowchart of operation layer control

satisfactorily. The proportional controller, whose proportional gain C_P is used in Eq. (4.32), provides a wide control bandwidth adapting to simulation step. However, synthesis for parameter tuning is not realized in this work.

4.9.1 Optimization Results

Nowadays services can be found providing worldwide PV power forecast according to location and weather information. However, public data of solar irradiance (W/m^2) forecasting are not yet precise enough for a specified location. This is why is very difficult to calculate the PV power prediction following real solar irradiance forecasting. The measured PV MPPT power, whose values have been recorded by our experimental device, is shown in Fig. 4.22a: the green curve shows the real-time PV power evolution, while the gray bars are the hourly average PV power.

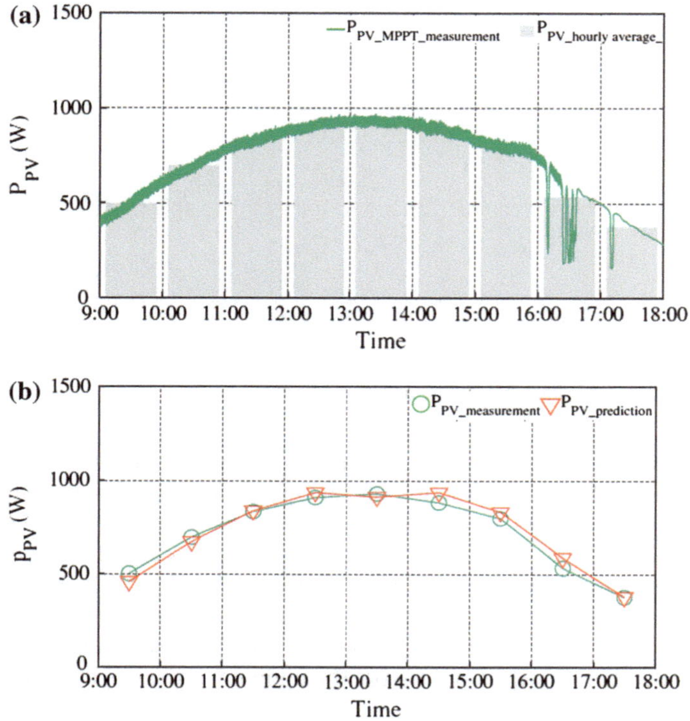

Fig. 4.22 PV MPPT power evolution (**a**), hourly PV power measurement and prediction (**b**)

As this section aims at demonstrating feasibility of the designed supervision, to overcome the lack of real solar irradiance forecasting, the PV power prediction data are calculated from the real measurement data. PV power hourly prediction data given in Fig. 4.22b are assumed having random ±10 % error with the hourly average of the measurement data.

Load prediction data are supposed to be given by load management system, which implies additional uncertainties. In this study, a simple arbitrary load power evolution is considered. The difference of load power and load power prediction is shown in Fig. 4.23. In order to perform an optimized operation for the next day, prediction layer is supposed to give to the energy management layer the forecasts of the PV power and the load power hourly evolutions.

Peak hours during the day are assumed 11:00–13:00 and 16:00–18:00. Grid and storage constraints, as arbitrary values are imposed for the system operation. To mitigate grid power strong fluctuations, grid power changing limit is imposed as 20 W/s. Arbitrary soc limits are considered as 45 and 55 %, while $SOC_0 = 50\,\%$ and $C_{REF} = 130$ Ah. Considering the storage capacity of our experimental platform, whose parameter values are used in the simulation study, these soc limits are selected to show the system behavior with relevant storage events (full, empty) in a day

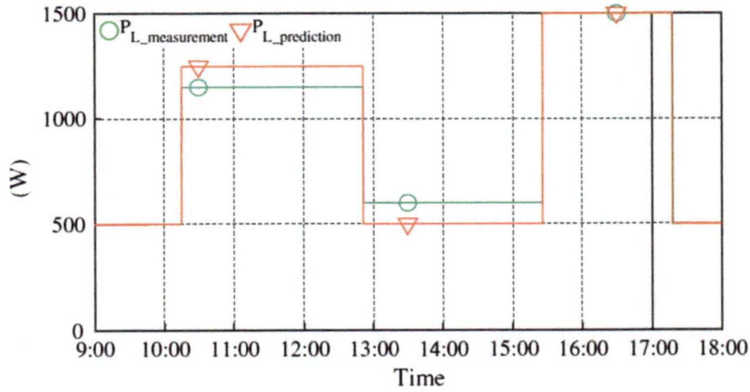

Fig. 4.23 Load power measurement and day-ahead prediction

run. Grid power injection limit is imposed as 700 W, and supply limit as 600 W. Based on the prediction information, the energy management layer calculates the optimization problem by CPLEX and gives the optimum powers flow evolution, as presented in Fig. 4.24a. Corresponding K_D(t) sequence is calculated by Eq. (4.45) from the optimum powers flow evolution, as presented in Fig. 4.24b. For performing a day optimization during 9 h, the data resolution is chosen at 10 s/point, i.e. 3240 points each power curve. The optimization program execution time is within 10 s for a computer with CORE i5 processor. The optimum cost estimated by CPLEX is 0.517 ε.

Figure 4.24 shows the optimization calculation results. It can be observed that for peak hours the PV production is not sufficient to supply the load; so, the storage and the grid have to supply the load for the remain part of power. The considered optimization problem is formulated to minimize the global energy cost for the whole period from 9:00 to 18:00, while respecting all constraints. So, as for the peak hours the grid energy tariff is very high and largely superior to storage energy tariff, it seems normal that the storage has to be proposed to supply the load. During the peak hours the soc decreases continually, as shown in Fig. 4.24b. Hence, the storage is used to supply the load as much as possible with respect to its soc lower limit. However, as the storage energy is not enough, the grid power is also used to supply the load. This is why, during peak hours, storage power and grid power are proposed to share the necessary power to supply the load in an optimized manner while respecting all constraints. This sharing power is proposed in an intermittent manner by the used solver.

4.9.2 Powers Flow Simulation I

The calculated $K_D(t)$ optimum evolution is given to the operation layer to run the power system following the conditions given by 23rd of April 2011 (meteorological and load). The operation powers flow, as real situation, simulated by MATLAB,

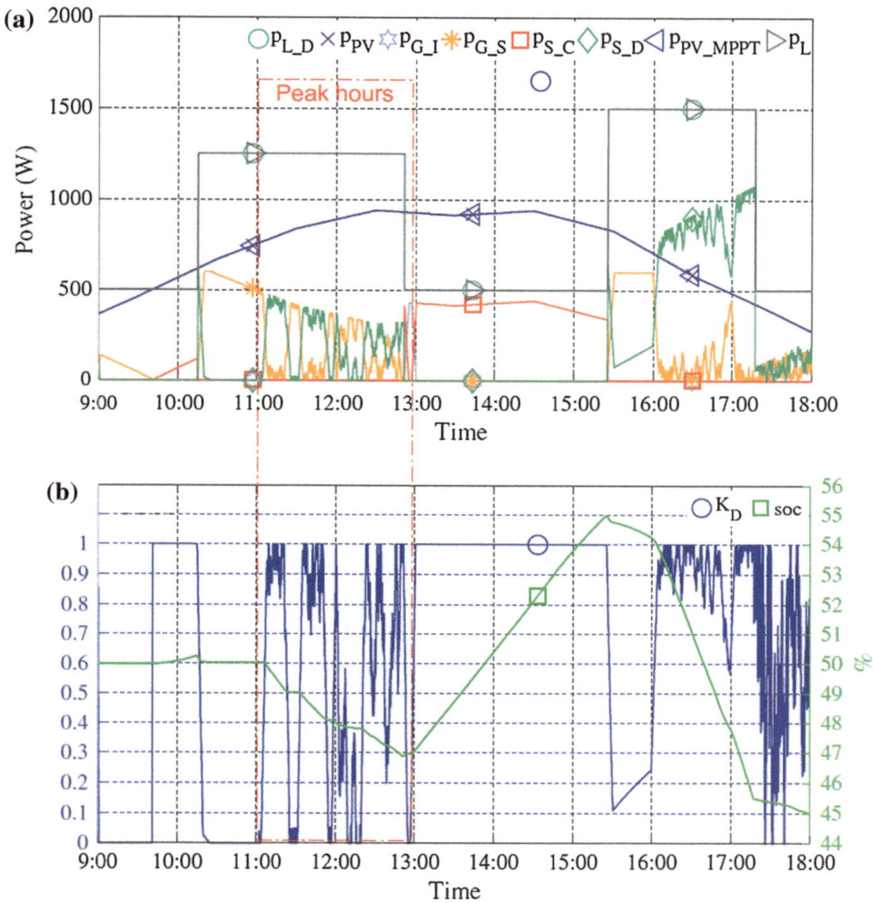

Fig. 4.24 Optimized powers flow (**a**), optimized $K_D(t)$ and soc evolution (**b**)

is shown in Fig. 4.25a. During this day operation, grid and storage share power for supplying energy or for receiving energy at the same time. In the first off-peak hours (9:00–11:00), grid mainly supplies the load for reserving storage for peak hour supply.

During the first peak hours (11:00–13:00), the load is supplied by storage and grid, the sharing proportion is determined by optimization calculation which aims also to reserve storage for supplying during second period of peak hours.

Just before 13:00 the surplus of the PV production is injected into the grid in order to make the maximum profit. Aiming to reduce the energy cost by avoiding grid to supply during peak hours, in the second peak hour period (16:00–18:00), the storage is mainly used for supplying the load. During 13:00–15:00 with the excess PV production, the storage is charged for supplying in the second peak hour. Grid power injection limit and supply limit are respected. Short time load shedding can be seen in

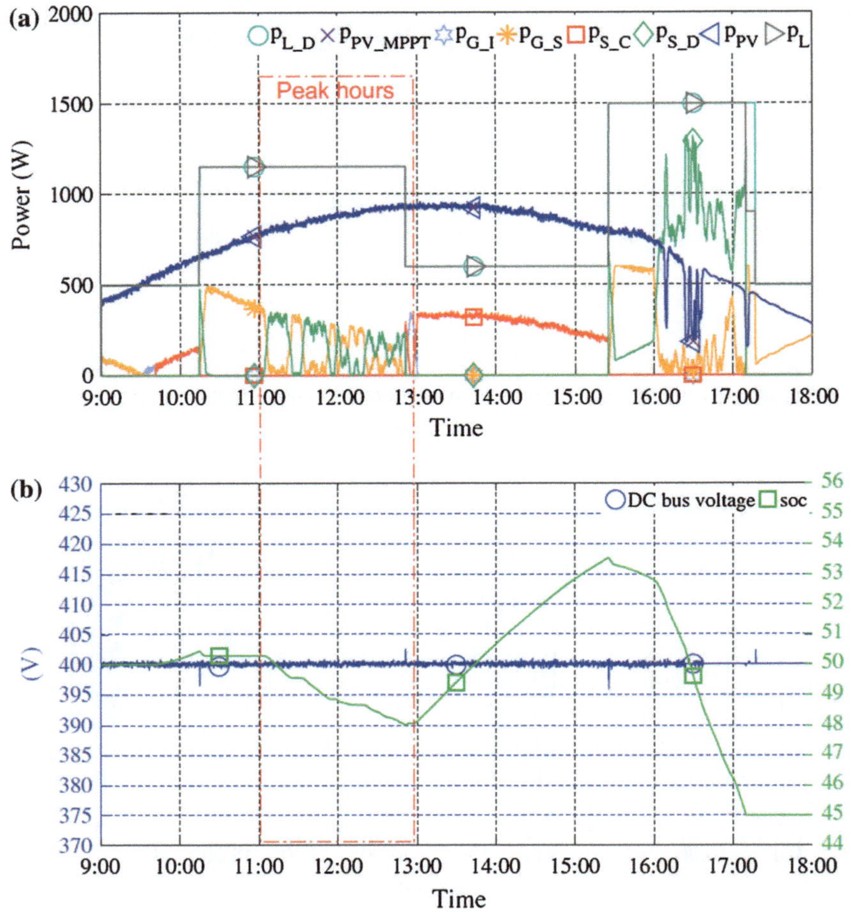

Fig. 4.25 Simulated powers flow (**a**), DC bus voltage and soc evolution (**b**) for optimum $K_D(t)$

the operation after 17:00, when the battery is empty. The load shedding is performed based on instantaneous power information. To avoid load shedding fluctuations in PV power fluctuating circumstances, it is also possible to impose duration for load shedding in CPLEX optimization, and optimized load shedding information could be given to operation layer to override the operation layer load shedding control. The energy cost is 0.512 , which is close to the optimization cost. The soc evolution with the optimum $K_D(t)$ and the DC bus voltage are illustrated in Fig. 4.25b. The DC bus voltage fluctuations are negligible compared to the value of 400 V, signifying the power is well balanced.

By comparing the results presented in Figs. 4.24 and 4.25, it can be seen that the simulated powers flow is slightly different from the optimization due to the uncertainties of solar irradiance prediction and load power prediction. During the solar fluctuations between 16:00 and 17:00, the storage provided more powers. However, in this case, the storage is still able to be the main load supply during the second period of peak hours, but a slight load shedding occurs when the soc reaches its low limit.

4.9.3 Powers Flow Simulation II

In order to further analyze, a simulation case for a constant K_D is presented in Fig. 4.26. It is chosen as a constant value $K_D = 0.5885$ which is the average value of optimum $K_D(t)$ evolution shown in Fig. 4.24b. In this case, the obtained energy

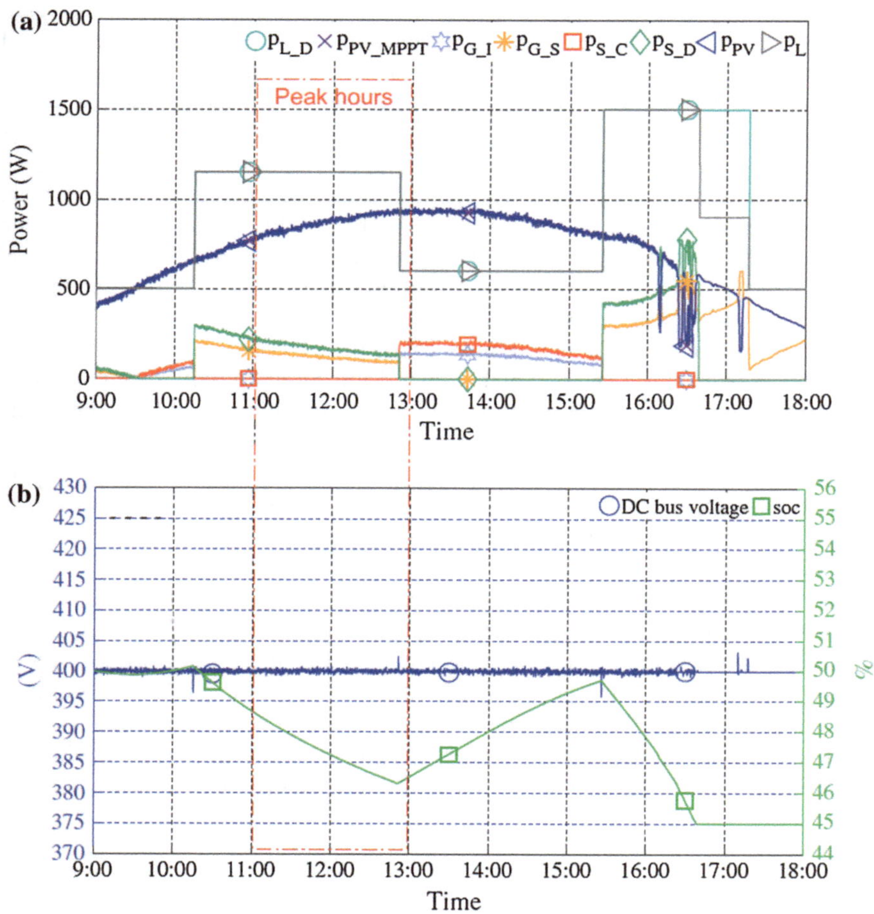

Fig. 4.26 Simulated powers flow (**a**), DC bus voltage and soc evolution (**b**) for constant $K_D = 0.5885$

cost, 0.652ε, the difference with optimization is larger compared with using optimal $K_D(t)$, and longer load shedding can be seen during this operation. The soc evolution, illustrated in Fig. 4.26b, is very different from the optimum soc evolution shown in Fig. 4.25b. Even optimization effect is affected, the power balancing is robust. Regarding the DC bus voltage illustrated in Fig. 4.26b, it can be seen that the DC bus voltage remains stable with very slight fluctuations, signifying the power is well balanced.

4.9.4 Comparison and Discussion

Table 4.1 shows the energy cost of the MG C_{total} given by Eq. (4.36) and occurrences of load shedding for these three cases: optimized operation by energy management layer with ±10 % uncertainties prediction data, simulated operation in case of a real PV production with the calculated optimum $K_D(t)$, and simulated operation with constant $K_D = 0.5885$. It can be seen that simulated cost for the optimum $K_D(t)$ is close to the prediction cost, and the error is within 1 %, which is from the uncertainties of prediction.

For the constant K_D case, the cost is 26 % more than the prediction case; moreover, longer load shedding can be seen in this case. Even with some uncertainties, the optimized $K_D(t)$ operates the MG with respect of the utility grid requirements and storage capacity. This comparison validates the presented simulation case for the proposed supervision of the proposed DC microgrid.

As previously mentioned, the communication of $K_D(t)$ does not need high-speed communication between layers. So, the supervision control provides the possibility of reperforming optimization and updating the $K_D(t)$ sequence during the actual operation without interrupting the power balancing. Thus, hourly or more frequent optimization that updates $K_D(t)$ sequence, with latest prediction and power system status, is expected to give better energy performance of the supervision control.

However, the limit of the supervision control is that optimizing effectiveness is affected on the prediction precision. Prediction uncertainties do not influence power balancing but the optimal energy cost is affected. Future research should focus on enhancing optimization performance, especially facing low prediction precision. An optimization technique that able to optimize power flow with consideration on uncertainties of the prediction combined with a rule-based algorithm in operation layer that corrects $K_D(t)$ in real time with respect of power system status can be developed as one solution. Besides, a second storage can be installed as backup for correcting the errors between optimized power and real operation (Table 4.5).

Table 4.5 Comparison of different cases

Case operation	K_D	$C_{total}cost(\varepsilon)$	Load shedding duration
Optimization	$Optimum K_D(t)$, Fig. 4.24b	0.517	*No shedding*
Simulation	$Optimum K_D(t)$, Fig. 4.24b	0.512	7 min
Simulation	$K_D = 0.5885$	0.652	37 min

4.10 Conclusions

In this chapter, a neural network based distributed secondary control (DSC) scheme for an autonomous smart MG system has been proposed in the first portion. In this scheme, the controller is designed to act dynamically to load changes and the associated optimized gains have been evaluated using differential evolution (DE) optimization procedure. The performance comparison between the proposed controller with fixed-gain controller is also performed. The proposed controller is speedier than the traditional one. The adaptability and robustness of the controller is also demonstrated with respect to the load sharing and voltage/frequency regulation with the help of MATLAB simulations. The ensuing results have emphasized that the proposed controller has been able to restore the output voltage and frequency to their nominal values, by eliminating the transients, whenever there is a change in load. Proper load sharing among the generating units have been also achieved.

In the second portion, MG intelligent online energy management under cost and emission minimization has been investigated; the problem formulation includes optimal battery scheduling, taking into account the uncertainty of the MG exogenous variables and forecasted entities.

In order to validate the NNE forecasting module, the forecasting reliability of the machine learning was evaluated in terms of forecasting error. The experimental results showed that the NNE achieved a higher forecasting accuracy than conventional MLPNN, RBFNN, and RNN. In fact, the NNE can improve the generalization and noise tolerance of the learning machine effectively through aggregating numbers of neural networks with different models by improving the generalization ability and reducing the variance and the performance instability. The forecasted entities are crucial data for the implementation of the online intelligent energy management.

The MIEM for MG was validated by comparing three approaches for battery scheduling jointly with the MO for dispatchable DG sources. The simulation results proved the effectiveness of the proposed MIEM using FL-based expert system. The proposed intelligent energy management approach achieved clearly important minimization on the operation cost and emission level of the MG. Moreover, the fuzzy expert system for battery scheduling decreased considerably the battery maintenance cost by extending the operation lifetime through lower depth of battery discharge.

The proposed approach is finally implemented on a MG simulation model to validate the proposed methodology. Simulation during 24-h real-condition operation including experimental data has been performed. The results of the suggested MIEM compared with HF-based battery management corroborated the results obtained based on the proposed control system.

Facing the advent of SG context, a MG control combining power balancing, optimization, and SG interaction is proposed through a multilayer supervision control structure. The research issue of implementing optimization in real-time operation is particularly addressed.

Based on PV sources, storage, power grid connection, and DC load, the MG aims at self-supply with limited access to grid. Taking into account forecast of PV production and load power demand, the four-layer supervision system performs optimization and implements optimization in instantaneous power balancing through simple interface. It handles also constraints such as storage capability, grid power limitations, energy grid tariffs, grid peak hours. The optimization is based on mixed integer linear programming, solved by CPLEX. Simulation results, even with uncertainties of prediction and arbitrary energy tariffs, taken as assumptions in this study, show that the proposed supervision design is able to perform efficiency and cost-effective powers flow in real-time operation with respect to constraints such as grid power limits and storage capacity. Load shedding and PV power limiting ensures power balancing in any case. The simulation shows that the optimization gives better energy performance while minimizing load shedding and PV production limitation and the operation layer respects all constrains of power system elements. On the other hand, the optimization efficiency is based on prediction precision, which may limit the final performance. The designed operation layer can work with any K_D value, so the prediction errors and nonoptimum K_D does not affect the power balance. However, if uncertainties are higher than $\pm 10\,\%$, the energy cost and the load shedding duration could be severely affected. The developed simulation will permit in further work to design an additional supervision layer aiming to mitigate the differences between the optimized power flow and the real one.

4.11 Suggested Problems

Problem 4.1 The adaptive neurofuzzy inference system is a hybrid approach composed of an ANN and a fuzzy inference system. It has a network-type structure and maps the input–output dataset using the parameters of fuzzy membership functions. Consider the development of two PV simulators: the first is based on single-diode simulator and the second is using two-diode PV model that will add two additional unknown parameters to optimize. Make a comparison between the two simulators to investigate their efficiencies.

Problem 4.2 Different optimization techniques can be used instead of DE to estimate the model parameters of PV. It is desired to conduct a systematic study to establish which optimizing technique has superior performance. Consider employing different objective functions to test the efficiency of the evolutionary techniques.

Problem 4.3 It is desired to build up an architecture based an adaptive neurofuzzy inference system with MPPT controller to be interfaced with the power grid through inverter. The objective here is to investigate the effects of changing environmental conditions on power grid.

Problem 4.4 It is observed from the foregoing sections that the MG only refers to a building scale and involves only a few sources. It is desired here to investigate generalizing the idea of parameterizing power balancing and interfacing with optimization, as well as SG interaction to be used as solution for advanced energy management for other MGs to optimize local power flow and improve future PV penetration.

Problem 4.5 For urban areas, extensive effort is going at increasing rate to construct integrated photovoltaic (COIPV) primarily for self-feeding of buildings equipped with PV array and storage. With an aim of elimination of multiple energy conversions, a DC network distribution is frequently considered through a multilevel supervision able to exchange messages with the SG and metadata. The objective here is to get benefits from the materials provided in this chapter to investigate the impact of predictions and uncertainties on power sources and power load request and to critically examine the efficiency of energy conversion.

References

1. Akagi H, Kanazawa Y, Nabae A (1984) Instantaneous reactive power compensators comprising switching devices without energy storage components. IEEE Trans Ind Appl 3:625–630
2. AlLee G, Tschudi W (2012) Edison Redux: 380 Vdc brings reliability and efficiency to sustainable data centers. IEEE Power Energy Mag 10:50–59
3. Alvarez E, Campos AM, Arboleya P, Gutirrez AJ (2012) Microgrid management with a quick response optimization algorithm for active power. Int J Electr Power Energy Syst 43(1): 465–473
4. Amjady N, Keynia F, Zareipour H (2010) Short-term load forecast of microgrids by a new bi-level prediction strategy. IEEE Trans Smart Grid 1(3):286–294
5. Arulampalam A et al (2004) Control of power electronic interfaces in distributed generation microgrids. Int J Electron 91(9):503–523
6. Back T, Fogel DB, Michalewicz Z (1997) Handbook of evolutionary computation. IOP Publishing Ltd, Bristol
7. Bakken BH, Grande OS (1998) Automatic generation control in a deregulated power system. IEEE Trans Power Electron 13(4):1401–1406
8. Barbir F, Gmez T (1996) Efficiency and economics of proton exchange membrane PEM fuel cell. Int J Hydrog Energy 21(10):891–901
9. Bo G, Mills JK, Dong S (2011) Energy management control of microturbine-powered plug-in hybrid electric vehicles using the telemetry equivalent consumption minimization strategy. IEEE Trans Veh Technol 60:4238–4248
10. Brown G, Yao X (2001) On the effectiveness of negative correlation learning. In: Proceedings of the1st UK workshop on computational intelligence, Edinburgh, UK, pp 57–62

11. Bustos C, Watts D, Ren H (2012) MicroGrid operation and design optimization with synthetic wind and solar resources. IEEE Trans Lat Am 10:1550–1562
12. Campanari S, Macchi E (2004) Technical and tariff scenarios effect on microturbine trigenerative applications. J Eng Gas Turbines Power 126(3):581–589
13. Campanari S, Boncompagni L, Macchi E (2004) Microturbines and trigeneration: optimization strategies and multiple engine configuration effects. J Eng Gas Turbines Power 12(1): 92–101
14. Celli G, Pilo F, Pisano G, Soma GG (2005) Optimal participation of a microgrid to the energy market with an intelligent EMS. In: Proceedings of IEEE international power engineering conference, Singapore, pp 663–668
15. Chakraborty S, Weiss MD, Simes MG (2007) Distributed intelligent energy management system for a single-phase high-frequency AC microgrid. IEEE Trans Ind Electron 54(1): 97–109
16. Chan ZSH, Kasabov N (2005) Fast neural network ensemble learning via negative-correlation data correction. IEEE Trans Neural Netw 16(6):1707–1710
17. Chandorkar MC, Divan DM, Adapa R (1993) Control of parallel connected inverters in standalone ac supply systems. IEEE Trans Ind Appl 29(1):136–143
18. Chaouachi A, Kamel RM, Nagasaka K (2010) Microgrid efficiency enhancement based on neuro-fuzzy MPPT control for photovoltaic generator. In: Proceedings of IEEE photovoltaic specialists conference, Honolulu, HI, pp 2889–2894
19. Chaouachi A, Kamel RM, Nagasaka K (2010) Multiobjective online management of a microgrid. In: Proceedings of renewable energy conference, Yokohama, Japan, pp 81–84
20. Chaouachi A, Kamel RM, Andoulsi R, Nagasaka K (2013) Multiobjective intelligent energy management for a microgrid. IEEE Trans Ind Electron 60(1):1688–1699
21. Chen H, Yao X (2009) Regularized negative correlation learning for neural network ensembles. IEEE Trans Neural Netw 20(12):1962–1979
22. Chen SX, Gooi HB (2011) Jump and shift method for multi-objective optimization. IEEE Trans Ind Electron 58(10):4538–4548
23. Chiou J-P, Wang F (1998) A hybrid method of differential evolution with application to optimal control problems of a bio-process system. In: Proceedings IEEE international conference world congress on computational intelligence and evolutionary computation
24. Coelho EAA, Cabaleiro Cortizo P, Garcia PFD (2002) Small-signal stability for parallel-connected inverters in stand-alone AC supply systems. IEEE Trans Ind Appl 38(2):533–542
25. Colson CM, Nehrir MH, Pourmousavi SA (2010) Towards real-time microgrid power management using computational intelligence methods. In: Proceedings of IEEE power energy society general meeting, Minneapolis, MN, p 18
26. Demuth H, Beale M (2000) Neural network toolbox for use with Matlab. The MathWorks Inc, User Guide, Natick, USA
27. Dobakhshari L, Salehi A, Azizi S, Ranjbar AM (2011) Control of microgrids: aspects and prospects. In: IEEE international conference networking, sensing and control (ICNSC)
28. Ehrgott M, Gandibleux X (2003) Multiple criteria optimization: state of the art annotated bibliographic surveys. Kluwer Academic Publishers, Boston
29. Emission Rates for New DG Technologies, the regulatory assistance project. http://www.raponline.org/docs/RAP_DGEmissions_2001_05.pdf
30. Farhangi H (2010) The path of the smart grid. IEEE Power Energy Mag 8(1):18–28
31. Firestone R, Marnay C (2005) Energy manager design for microgrids. CERTS, California Energy Commission, Sacramento, CA, LBNL-54447
32. Franco JF, Rider MJ, Lavorato M, Romero R (2013) A mixed-integer LP model for the optimal allocation of voltage regulators and capacitors in radial distribution systems. Int J Electr Power Energy Syst 48:123–130
33. Fuselli D, De Angelis F, Boaro M, Squartini S, Wei Q, Liu D (2013) Action dependent heuristic dynamic programming for home energy resource scheduling. Int J Electr Power Energy Syst 48:148–160

34. Gao Z, Ming F, Hongling Z (2005) Bagging neural networks for predicting water consumption. J Commun Comput 2(3):19–25
35. Georgilakis PS (2011) Integration of distributed generation in the power system. Wiley-IEEE Press, New Jersey
36. Gestel TV, Suykens JAK, Lanckriet G, Lambrechts A, Moor BD, Vandewalle J (2002) Bayesian framework for least-squares support vector machine classifiers, Gaussian processes, and Kernel Fisher discriminant analysis. Neural Comput 14(5):1115–1147
37. Guerrero JM et al (2009) Control strategy for flexible microgrid based on parallel line-interactive UPS systems. IEEE Trans Ind Electron 56(3):726–736
38. Guerrero JM et al (2011) Hierarchical control of droop-controlled AC and DC microgrids—a general approach toward standardization. IEEE Trans Ind Electron 58(1):158–172
39. Guerrero JM, Vasquez JC, Matas J, de Vicuna LG, Castilla M (2011) Hierarchical control of droop-controlled AC and DC microgrids—a general approach toward standardization. IEEE Trans Ind Electron 58(1):158–172
40. Guerrero JM et al (2013) Advanced control architectures for intelligent microgrids—part I: decentralized and hierarchical control. IEEE Trans Ind Electron 60(4):1254–1262
41. Guerrero JM et al (2013) Advanced control architectures for intelligent microgrids—part II: power quality, energy storage, and AC/DC microgrids. IEEE Trans Ind Electron 60(4): 1263–1270
42. Guerrero JM, Chandorkar M, Lee TL, Loh PC (2013) Advanced control architectures for intelligent microgrids part I: decentralized and hierarchical control. IEEE Trans Ind Electron 60(4):1607–1618
43. Hagan MT, Menhaj M (1994) Training feedforward networks with Marquardt algorithm. IEEE Trans Neural Netw 5(6):989–993
44. Hansen LK, Salamon P (1990) Neural network ensembles. IEEE Trans Pattern Anal 12(10):993–1001
45. Hatziargyriou N, Asano H, Iravani MR, Marnay C (2007) Microgrids. IEEE Power Energy Mag 5(4):78–94
46. Haykin S (2004) Neural networks: a comprehensive foundation. Neural Netw 2(3):251–270
47. Hernandez-Aramburo CA, Green TC, Mugniot N (2005) Fuel consumption minimization of a microgrid. IEEE Trans Ind Appl 41(3):673–681
48. Hggmark S, Neimane V, Axelsson U, Holmberg P, Karlsson G, Kauhaniemi K, Olsson M, Liljegren C (2003) Aspects of different distributed generation technologies. CODGUNet WP 3, Vattenfall Utveckling Ab
49. Hornik K, Stinchcombe M, White H (1990) Universal approximation of an unknown mapping and its derivatives using multilayer feedforward networks. Neural Netw 3(5):551–560
50. Houssamo I, Locment F, Sechilariu M (2010) Maximum power tracking for photovoltaic power system: development and experimental comparison of two algorithms. Renew Energy 35(10):2381–2387
51. Houssamo I, Wang BC, Sechilariu M, Locment F, Friedrich G (2012) A simple experimental prediction model of photovoltaic power for DC microgrid. In: Proceedings of IEEE international symposium on industrial electronics, pp 963–98
52. Houssamo I, Locment F, Sechilariu M (2013) Experimental analysis of impact of MPPT methods on energy efficiency for photovoltaic power systems. Int J Electr Power Energy Syst 46:98–107
53. IBM.com, IBM ILOG CPLEX Optimizer, http://www.ibm.com
54. Jian Z et al (2011) Modeling and control of automotive antilock brake systems through PI and neural network arithmetic. In: Proceedings of international conference on electronic and mechanical engineering and information technology (EMEIT), vol 4
55. Kamel RM, Chaouachi A, Nagasaka K (2010) Wind power smoothing using fuzzy logic pitch controller and energy capacitor system for improvement micro-grid performance in islanding mode. Energy 35(5):2119–2129
56. Kanchev H, Lu D, Colas F, Lazarov V, Francois B (2011) Energy management and operational planning of a microgrid with a PV-based active generator for smart grid applications. IEEE Trans Ind Electron 58(10):4583–4592

57. Kashfi F, Hatami S, Pedram M (2011) Multiobjective optimization techniques for VLSI circuits. In: Proceedings international symposium on quality electronic design, pp 1–8
58. Katiraei F, Reza Iravani F, Lehn PW (2005) Micro-grid autonomous operation during and subsequent to islanding process. IEEE Trans Power Deliv 20(1):248–257
59. Katiraei F, Iravani R, Hatziargyriou N, Dimeas A (2008) Microgrids management. IEEE Power Energy Mag 6(3):54–65
60. Kumar Nunna HSVS, Doolla S (2013) Multiagent-based distributed-energy-resource management for intelligent microgrids. IEEE Trans Ind Electron 60(1):1678–1687
61. Lasseter RH (2011) Smart distribution: coupled microgrids. Proc IEEE 99:1074–1082
62. Lasseter RH, Paigi P (2004) Microgrid: a conceptual solution. In: IEEE 35th annual power electronics specialists conference PESC'04, vol 6
63. Lasseter R, Tomsovic K, Piagi P (2000) Scenarios for distributed technology applications with steady state and dynamic models of loads and micro-sources. CERTS. U.S. Department of Energy, Washington, DC
64. Lasseter RH, Eto JH, Schenkman B, Stevens J, Vollkommer H, Klapp D (2010) CERTS microgrid laboratory test bed. IEEE Trans Power Deliv 26(1):2531–2540
65. Lee TL, Hu SH, Chan YH (2013) D-STATCOM with positive-sequence admittance and negative-sequence conductance to mitigate voltage fluctuations in high-level penetration of distributed-generation systems. IEEE Trans Ind Electron 60(4):1417–1428
66. Li H, Zhang Q (2009) Multiobjective optimization problems with complicated Pareto sets, MOEA/D and NSGA-II. IEEE Trans Evol Comput 13(2):284–302
67. Liserre M, Sauter T, Hung JY (2010) Future energy systems, integrating renewable energy sources into the smart power grid through industrial electronics. IEEE Ind Electron Mag 4(1):18–37
68. Locment F, Sechilariu M, Houssamo I (2012) DC load and batteries control limitations for photovoltaic systems: experimental validation. IEEE Trans Power Electron 27(9):4030–4038
69. Lorenz E, Hurka J, Heinemann D, Beyer HG (2009) Irradiance forecasting for the power prediction of grid-connected photovoltaic systems. IEEE J Sel Top Appl Earth Observ Remote Sens 2:2–10
70. MacKay DJC (1992) Bayesian interpolation. Neural Comput 4(3):415–447
71. Mahmoud MS, Hassan MF, Darwish MG (1985) Large scale control systems: theories and techniques. Marcel Dekker Inc, New York
72. Mahmoud MS, Azher Hussain S, Abido MA (2014) Modeling and control of microgrid: an overview. J Frankl Inst 351(5):2822–2859
73. Mashhour E, Moghaddas-Tafreshi SM (2010) Integration of distributed energy resources into low voltage grid: a market-based multi-period optimization model. Electr Power Syst Res 80(4):473–480
74. Mazumder SK, Tahir M, Acharya K (2008) Pseudo-decentralized control-communication optimization framework for microgrid: a case illustration. In: Proceedings of transmission and distribution conference and exposition (T & D). IEEE Press
75. Mller MF (1993) A scaled conjugate gradient algorithm for fast supervised learning. Neural Netw 6(4):525–533
76. Mohamed FA, Koivo HN (2010) System modeling and online optimal management of microgrid using mesh adaptive direct search. Int J Electr Power Energy Syst 32(5):398–407
77. Mohamed FA, Koivo HN (2011) Multiobjective optimization using modified game theory for online management of microgrid. Eur Trans Electr Power 21(1):839–854
78. Mohamed Y-R, Radwan AA (2011) Hierarchical control system for robust microgrid operation and seamless mode transfer in active distribution systems. IEEE Trans Smart Grid 2(2): 352–362
79. Narendra KS, Parthasarathy K (1990) Identification and control of dynamical systems using neural networks. IEEE Trans Neural Netw 1(1):4–27
80. Neal RM (1996) Bayesian learning for neural networks. Springer, New York
81. Ng GW (1997) Application of neural networks to adaptive control of nonlinear systems. Research Studies Press, New York

82. Patterson BT (2012) DC, come home: DC microgrids and the birth of the Enernet. IEEE Power Energy Mag 10:60–69
83. Paudyal S, Canizares CA, Bhattacharya K (2011) Optimal operation of distribution feeders in smart grids. IEEE Trans Ind Electron 58(10):4495–4503
84. Pedrasa MA, Spooner T (2006) A survey of techniques used to control microgrid generation and storage during island operation. In: Proceedings of the Australian universities power engineering conference
85. Peeters E, Belhomme R, Batlle C, Bouffard F, Karkkainen S, Six D (2009) ADDRESS: scenarios and architecture for active demand development in the smart grid of the future. In: Proceedings of CIRED 20th international conference on electricity distribution, pp 1–4
86. Piagi P, Lasseter RH (2006) Control and design of microgrid components. Final report
87. Pogaku N, Prodanovic M, Green TC (2007) Modeling, analysis and testing of autonomous operation of an inverter-based microgrid. IEEE Trans Power Electron 22(2):613–625
88. Ren P, Xiang Z, Qiu Z (2013) Intelligent domestic electricity management system based on analog-distributed hierarchy. Int J Electr Power Energy Syst 46:400–404
89. Riedmiller M, Braun H (1993) A direct adaptive method for faster back-propagation learning: the RPROP algorithm. In: Proceedings of IEEE international conference neural networks
90. Riffonneau Y, Bacha S, Barruel F, Ploix S (2012) Optimal power flow management for grid connected PV systems with batteries. IEEE Trans Sustain Energy 2(3):325–332
91. Rosen B (1996) Ensemble learning using decor related neural networks. Connect Sci 8(3): 373–383
92. Savaghebi M et al (2012) Secondary control for compensation of voltage harmonics and unbalance in microgrids. In: Proceedings of the 3rd IEEE international symposium on power electronics for distributed generation systems (PEDG)
93. Sechilariu M, Wang BC, Locment F (2013) Building-integrated microgrid: advanced local energy management for forthcoming smart power grid communication. Energy Build 59(1):236–243
94. Sechilariu M, Wang BC, Locment F (2013) Building integrated photovoltaic system with energy storage and smart grid communication. IEEE Trans Ind Electron 60(4):1607–1618
95. Shafiee Q, Vasquez JC, Guerrero JM (2012) Distributed secondary control for islanded microgrids—a networked control systems approach. In: Proceedings IEEE 38th annual conference on industrial electronics—IECON'12
96. Shenai K, Shah K (2011) Smart DC micro-grid for efficient utilization of distributed renewable energy. In: Proceedings of the IEEE energy technologies, pp 1–6
97. Simpson-Porco JW, Drfler F, Bullo F (2013) Synchronization and power sharing for droop-controlled inverters in islanded microgrids. Automatica 49(9):2603–2611
98. Soleimani-Mohseni M, Thomas B (2004) Neural networks for self tuning of PI- and PID controllers. In: Proceedings of intelligent systems design and applications, ISDA04, vol 2. Budapest, pp 1955–1962
99. Storn R, Price K (1997) Differential evolutiona simple and efficient heuristic for global optimization over continuous spaces. J Glob Optim 11(4):341–359
100. Sucic S, Havelka JG, Dragicevic T (2014) A device-level service-oriented middleware platform for self-manageable DC microgrid applications utilizing semantic-enabled distributed energy resources. Int J Electr Power Energy Syst 54:576–588
101. Tan X, Li Q, Wang H (2013) Advances and trends of energy storage technology in microgrid. Int J Electr Power Energy Syst 44(1):179–191
102. Vandoorn TL et al (2012) Communication-based secondary control in microgrids with voltage-based droop control. In: Proceedings of transmission and distribution conference and exposition (T&D). IEEE Press
103. Vasquez JC et al (2010) Hierarchical control of intelligent microgrids. IEEE Ind Electron Mag 4(4):23–29
104. Wang GJ, Fong CT, Chang KJ (2001) Neural-network-based self-tuning PI controller for precise motion control of PMAC motors. IEEE Trans Ind Electron 48(2):408–415

105. Wang BC, Sechilariu M, Locment F (2012) Intelligent DC microgrid with smart grid communications: control strategy consideration and design. IEEE Trans Smart Grid 3(4):2148–2156
106. Wang BC, Houssamo I, Sechilariu M, Locment F (2012) A simple PV constrained production control strategy. In: Proceedings of IEEE international symposium on industrial electronics, pp 969–974
107. Wei-Lin H, Chia-Hung L, Chao-Shun C, Hsu CT, Te-Tien K, Cheng-Ta T (2011) Impact of PV generation to voltage variation and power losses of distribution systems. In: Proceedings of the 4th international conference on electric utility deregulation and restructuring and power technologies, pp 1474–1478
108. Yang LG-, Ming-Nguang L (2010) The summary of differential evolution algorithm and its improvements. In: Proceedings of the 3rd international conference advanced computer theory and engineering (ICACTE), vol 3
109. Zhang Y, Ma H (2012) Theoretical and experimental investigation of networked control for parallel operation of inverters. IEEE Trans Ind Electron 59(4):1961–1970
110. Zheng W, Ma H, He X (2012) Modeling, analysis, and implementation of real time network controlled parallel multi-inverter systems. In: Proceedings the 7th international power electronics and motion control conference (IPEMC), vol 2

Chapter 5
A System of Systems Framework for Microgrids

5.1 Introduction

Economic challenges, technological advancements, and environmental impacts are now demanding distributed generation (DG) in place of the conventional centralized generation [40]. Power operation companies are now confronted with unprecedented difficulties in terms of meeting the load requirements, consumer satisfaction, and environmental considerations. Thus, DG has received good attention because of its potential to alleviate pressure from the main transmission system by supplying a few local loads [2]. The waste heat generated from the fuel to electricity conversion is exploited by the distributed generation system (DGS) with the help of microturbines, reciprocating engines, and fuel cells to provide heat and power to the customers. Adding to the system distributed energy sources (DES) like photovoltaic (PV) panels, wind turbines (WTs), energy storage devices such as batteries and capacitors, generators extracting energy from other renewable and controllable loads can provide momentous contributions to future energy generation and distribution. Another noteworthy feature is that the carbon emission is reduced to a large extent satisfying the commitment of many nations concerning decrease of carbon footprints [24]. However, the DG faces technical issues regarding its connection to the intermittent renewable generation and feeble areas of the distribution network. Further, owing to the distinct behavior of the DG unlike the conventional load, alteration in power flow results in problems. To introduce an intelligent electric network, the notion of the *smart grid* (SG) has recently emerged. Among desired characteristics of SG affecting the distribution level are improved reliability and sustainability. These attributes are mainly realized through microgrids (MGs) which facilitate the effective integration of distributed energy resources (DERs) [7, 13, 55, 73, 78].

The MG has paved its way into DG and looks promising for future aspects. It has the ability to respond to changes in the load, while decreasing feeder losses and improving local reliability. Basically designed to cater the heat and power requirements of local customers, it can serve as an uninterruptible power supply for critical loads. The challenges posed by the world today, the environmental and economical

M.S. Mahmoud and F.M. AL-Sunni, *Control and Optimization of Distributed Generation Systems*, Power Systems, DOI 10.1007/978-3-319-16910-1_5

considerations are asking for large-scale, complex, integrated, and well-organized systems. The gap between the conventional technologies and the ever-growing needs of the society need to bridged, and hence development of systems engineering principles needs to undergo a transition. This is where, the extension of systems engineering to heterogeneous, interoperable, and evolutionary SoS becomes imperative. However, system of systems (SoS) is now a widespread notion and has entered several domains including defense, IT, health care, manufacturing, energy and space stations, and exploration to name a few.

Thus, this chapter aims at creating a SoS framework for the MGs for their efficient operation and management. The fact that MGs exhibit similar characteristics to SoS, makes it easier to draft the entire architecture of MGs in a SoS framework.

5.2 Microgrid Subsystems

The concept of MG has received considerable attention owing to its potential to serve as an alternate power source, utilizing unconventional sources, and supplying the most critical loads of the main grid in case of a network failure. MGs are low voltage (LV) networks or distributed energy systems which provide heat and power to a particular area by employing generators and loads. They have the ability to operate independently and isolate themselves from the main grid in case of a fault [16, 17]. If proper control techniques are implemented, they may improve the reliability of electrical energy supply. MG comprises of microturbines, WTs, fuel cells, PV cells, etc., as sources of energy which are interfaced with the help of power electronic converters. All these units are connected to the main grid through a point of common coupling (PCC) and look as a solitary unit to the distribution network. No additional inertia is added to the system from the distributed generators. However, because of this the power balance amid generation and load and the network frequency becomes complicated to maintain, especially when the MG is in islanded mode [77]. The islanded mode is an operating condition in which the MG isolates itself from the main grid in case of a fault. However, the transition from the grid-connected and the islanded mode must be stable [62]. If the MG is consuming or supplying power to the main grid before disconnection, a power imbalance occurs. This is compensated by the energy storage units because of the fact that the microsources have low inertia and slow dynamic response [19, 31].

The architecture of MG is shown in Fig. 5.1. As it can be seen from the figure, the MG consists of microgrid central controller (MGCC) or central energy manager, microsource controllers (MCs), LCs, microturbines, fuel cells, battery storage, loads, renewable generators like the PV panel, and a combined heat and power (CHP) unit. The common coupling point can also be seen which connects the MG with the main grid. We shall now discuss each component to understand their role in the overall operation and performance of the MG.

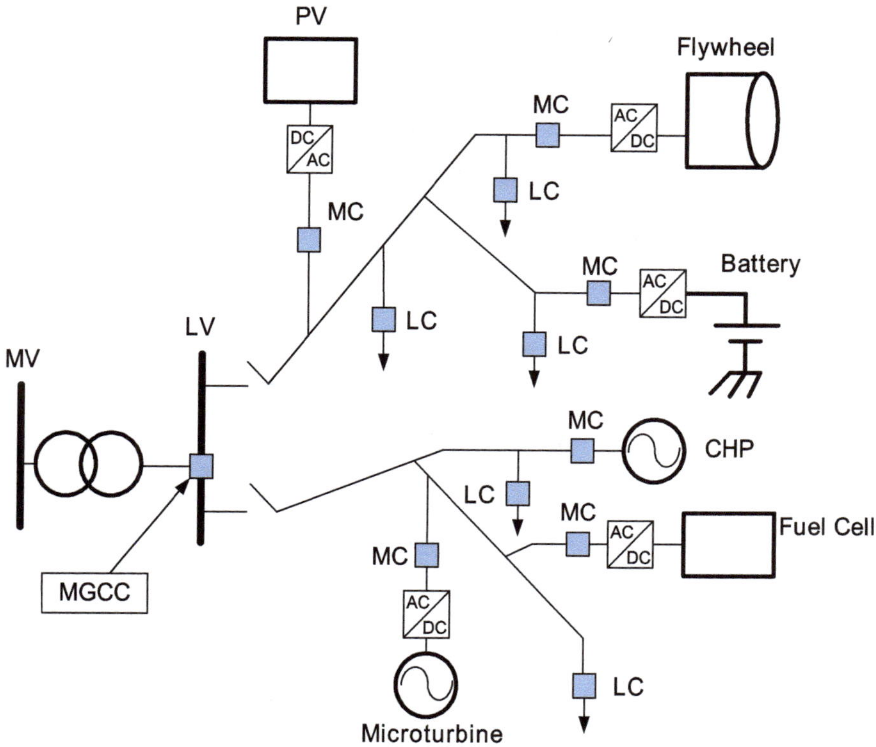

Fig. 5.1 Microgrid architecture

5.2.1 Microgrid Central Controller

The MGCC acts as a connector between the medium voltage (MV) and LV side of the grid. The LV side of the grid is occupied by the MG and its components. The MGCC is responsible for allocating set-points to the load and MCs and take care of the technical and management concerns of the MG. Also, it establishes communication with the distributed management system to perk up the operation of the MV distribution system. Apart from this, the MGCC carries out numerous other functions including forecasting studies, scheduling of microgeneration from an economic point of view, security assessment, and demand-side management (DSM) functions. All these are accomplished by collecting data from the loads and microsources. Additionally, during the islanded mode of operation of the MG, the MGCC acts as a secondary loop to facilitate a change in the output power control of microgenerators. Further, it must also devise interruption strategies and intelligently make use of the storage devices available to ensure reliability and reduce interruption time [44].

MGs with DES has a major feature in that the sources are dispersed over a wide area. These sources are tight to each other and to loads by a distribution network. In

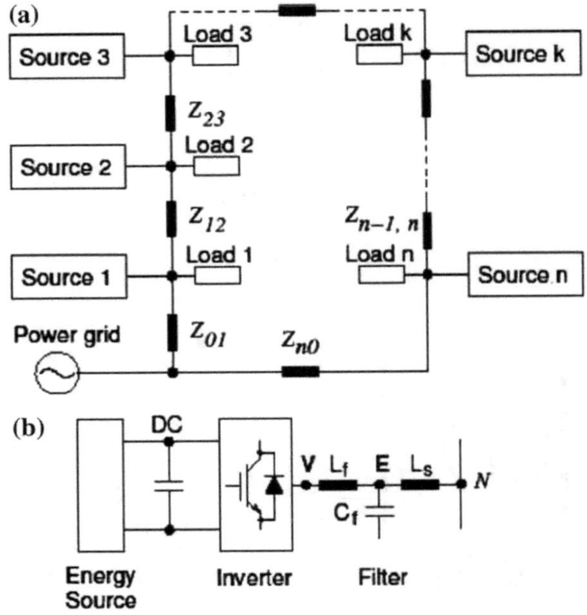

Fig. 5.2 Microgrid with distributed sources and loads

addition, the distributed MG may be coupled to the main power grid at some point as well. Figure 5.2a shows a distributed MG structure connected to the main grid. It is also shown the MG line impedance (Z_{01}, Z_{12}, ..., $Z_{n-1,n}$). The source is connected to the MG distribution network by an inverter interface through a filter, that is, an LCL filter, shown in Fig. 5.2b.

5.2.2 Microsource and Load Controllers

The microsource and load controllers are dedicated controllers for a particular component of the MG. By making use of local measurements of current and voltage, the MC controls its assigned microsource. They are responsible to carry out load tracking and management, load control of the storage devices, and economic scheduling functions. The primary functions of the MC are regulation of power flow on a feeder where the operating points of the loads are varying, regulation of voltage at each microsource to accommodate the changing loads on the system, and most importantly see that every microsource takes its load during islanded operation [39].

5.2.3 Microturbines

Microturbines are single shaft, simple mechanical devices consisting of a generator which is a permanent magnet machine functioning at variable speed typically in the range of (50,000–100,000) rpm. The variable speed generation system is interfaced to the electrical system through power electronics. The microturbines are flexible to operate on different fuels such as natural gas and gasoline. Equipped with good reliability, they are also economical to afford [41].

5.2.4 Microsources and Fuel Cells

The microsources include renewable sources of energy like solar, wind, and hydropower. Mini-hydro generators, WTs, and PV panels are located geographically in a MG. The energy from these renewable sources is harnessed and converted to electricity. The only shortcoming of these sources is their intermittent nature.

Fuel cells are unconventional and produce electricity from hydrogen and oxygen. They release water vapor and the emissions are less. Also, they offer higher efficiencies when compared to the microturbines discussed above. However, fuel cells are currently uneconomical and quite expensive. From the environmental perspective, renewable sources and fuel cells are well suited for DG in place of their conventional counterparts including the combustion engines, see Fig. 5.1.

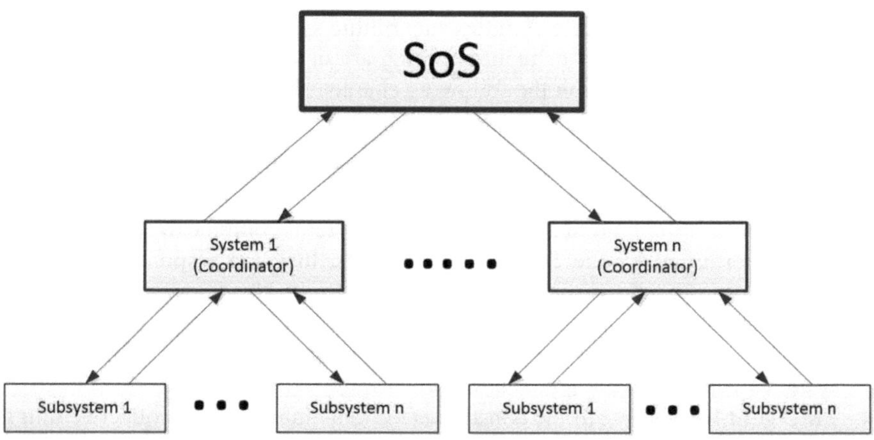

Fig. 5.3 Structure of SoS

5.2.5 Storage Devices

Storage devices have a crucial role to play in the MG. These are back up energy systems which provide energy to the critical loads much like the uninterrupted power supply (UPS). The storage devices include batteries, flywheels, and ultra capacitors. With the advancement in technology, higher capacity storage devices are available which can provide considerable supply of energy.

5.3 The Concept of SoS

The concept of SoS has opened up a new school of thought in systems engineering. SoS has emerged as a hot topic for research over the past few years. Although still in the infant stages, the concept of SoS has managed to achieve widespread acclaim. Being restricted to defense and IT, at one point of time, SoS has now entered a plethora of domains [25, 27]. Many definitions have been proposed in the literature for SoS, the more descriptive one being "systems of systems are large-scale integrated systems that are heterogeneous and independently operable on their own, but are networked together for a common goal" [26], where the goal is cost effectiveness, robustness, and performance. A SoS comprises of numerous heterogeneous subsystems which are independently operable. These subsystems also possess the ability to continue operating even if separated from the SoS. Each constituent subsystem has no power over the other but communicate with each other through a network to effectively carry out tasks and collectively achieve a mission. However, one needs to understand the difference between large-scale complex monolithic systems and actual SoS. There exists certain characteristics or features which are unique to SoS as given in [12]. The SoS is expected to exhibit the following characteristics.

1. **Operational independence**: All the constituent systems within the SoS architecture operate independently and have no interference with other neighbor systems in their functionality.
2. **Managerial independence**: The constituent systems continue to operate on their own unperturbed by the SoS. In other words, they are responsible for their autonomous operation.
3. **Evolutionary development**: The SoS is not designed as a single unit, and is rather flexible which can accommodate numerous new systems or do away with systems which are no longer necessary.
4. **Emergent behavior**: All the constituent systems function as a collective unit to accomplish a common objective, which cannot be achieved by a single component system.
5. **Geographic distribution**: The distribution of the subsystems is sequential to facilitate flow of information among them.

5.3.1 Microgrids as SoS

The question to be asked here is whether the MG falls under the SoS category and does it exhibit the characteristics of a SoS. For this, we need to consider the subsystems within the MG architecture and observe their operation and interaction with the MGCC.

The MG as described above is a complex system comprising of micro sources, loads, and energy storage devices. Most of the elements are nonlinear systems and possess strong cross-coupling between them. However, complexity is not the only thing that would be sufficient for the MG to qualify as a SoS, it must comply with other features of SoS presented by [48] mentioned in the previous section, refer to Fig. 5.3.

The elements of the MG, which in other words may be termed as subsystems include microsources which may be PV and wind generators, micro turbines deriving energy from gas or biofuels, fuel cells, and storage devices constituting batteries or flywheels. All of these subsystems operate independently without interfering with each other. Thus, operational independence is seen in the subsystems of the MG. Also, each subsystem depends on its own source to function and produce energy and is hence responsible for its autonomous operation. For instance, the microturbine runs on biofuel to perform its function of producing energy and has nothing to do with the microsource generators which harness energy from renewable sources such as the wind and solar power. The operation of the storage devices too is distinct and are concerned only with storing energy for use when required. Moreover, the MG supports evolutionary development and is supposed to accommodate new subsystems into its architecture whenever needed unperturbed the complexity of the overall system. A new microsource may be added to the MG which utilizes hydroenergy for example, or a new storage device may be incorporated for better and efficient storage of energy. On the contrary, a subsystem may be discarded from the system or architecture owing to some reason. All these explains the evolutionary characteristic of the MG. Further, the emergent behavior of the MG is evident from the fact that the subsystems are all operating to collectively satisfy the main objective of providing combined heat and electric power to the local community even in the case of a failure of the main grid. This objective cannot be achieved by a single subsystem of the MG and requires a combined effort of all the constituent subsystems. Finally, the MG is geographically distributed and all of its subsystems are spatially distributed in a particular region. A communication network facilitates the flow of information among all subsystems and the MGCC. Necessary data is retrieved from the subsystems to make the overall decision by the central controller (CC). From the above discussion, it can be concluded that the MG architecture satisfies the characteristics and falls perfectly on the lines of a SoS.

5.3.2 Grid-Connected Operation

For grid-connected operation, the MG is required to follow the distribution rules of the network with out any participation in the operation of the main power system. The MG operating based on this approach becomes important for stable operation of the power system. In the case of grid-connected, the MG can draw its power from the main grid or it can supply its power to the main grid and it works like controllable load or source. Figure 5.4 presents an overview of primary, secondary, and tertiary control of the DER units in the grid-connected operation [10].

The primary controller is used to give reliable operation and to response to communication fails. The primary controller is associated by frequency f and voltage U constants and it gives Q_{prim} and P_{prim} control actions to control the deviation of voltage and frequency.

The secondary controller is used to regulate the voltage and frequency based on its response to the load and supply changes. Its control actions P_{sec} and Q_{sec} are given by calculating the average of primary control actions.

The tertiary controller is used to improve the operation of DER units to be in the optimal cost, i.e., as far as the power limits allow to the units of distributed energy resources (DERs) to exchange their power at equal marginal.

5.3.3 Grid-Islanded Operation

The MG units can be controlled based on decentralized approach to balance the MG components energy and the demand. When the grid affected by any an abnormal

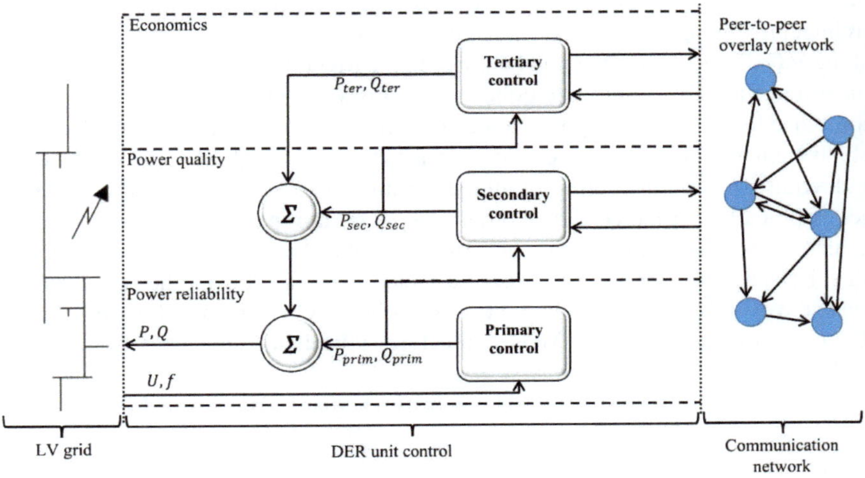

Fig. 5.4 Overview of the primary, secondary, and tertiary control building blocks and their relation with each other, the LV grid and the communication network

operation or conditions the MG should be disconnected and changed to grid-island operation mode. If the MG becomes grid-island mode, it should handle the following considerations [57]:

1. Supply and demand balancing.
2. Acceptable power quality.
3. Voltage and frequency balance.
4. Communication among the MG components.

The control of an isolated MG means balancing the generation and the demand power to keep high performances with acceptable range of frequency and voltages amplitude. In the literature, there are two main strategies of control are used for this purpose: PQ inverter controller to keep the active power constant at a desirable power factor and voltage source inverter (VSI) controller to regulate the frequency and voltage amplitude. These control strategies for isolated MG are discussed in [56] for single master operation and multi-master operation. In [3], the experimental setup results of the VSI controller for the output voltages of the MG sources during various loads. The islanded MG frequency and the generated power control with big load variations connected load is given in [5]. The controller in [80] is automatically controls the frequency drop and restores the frequency to its desired reference.

Proportional plus integral (PI) controller is proposed to control the isolated MG frequency in [22, 29, 36, 37, 42, 56, 63]. The fuzzy controller [66] is used to select optimal PI gains by considering the microsources parameters changing and the particle swarm optimization (PSO) is used for tuning PI controller [9]. Also the power source's reactive current can be used to control the frequency [71], when the frequency depends on the reactive power, also presented in [67]. Instead of frequency control, the voltage balance control presented in [33], a controller applied to VSI to improve voltage transient response. Voltage balance controllers also proposed in [67–69, 71]. An optimal controller is proposed in [70] to control the MG islanded mode frequency and voltage variations, and the laboratory setup for MG frequency and voltage and control is implemented in [6]. $H\infty$ robust control is proposed in [4, 50] to control the operation of the connection of two (DG) microgrid units.

The MG islanded mode has to continue providing energy to the MG loads, so the focus here will be on the islanded mode operation.

The objective of the MG is best served when its constituent subsystems are well coordinated, organized, and are able to communicate effectively among each other. The microsources depend on renewable sources which are intermittent, and hence require meticulous management and strategic utilization of the energy produced in order to meet the requirements of the load. This calls for a integrated framework which allows each subsystem to operate independently and establishes good communication among them to accomplish the desired goal. The emergent nature and geographic distribution of the MG makes it convenient to be architected in a SoS framework. A SoS framework addresses all the management, communication, and control needs of a MG in a systematic manner.

Figure 5.5 shows a SoS framework for the MG. The MGCC is well connected to all the constituent subsystems through a local control network. An electrical network

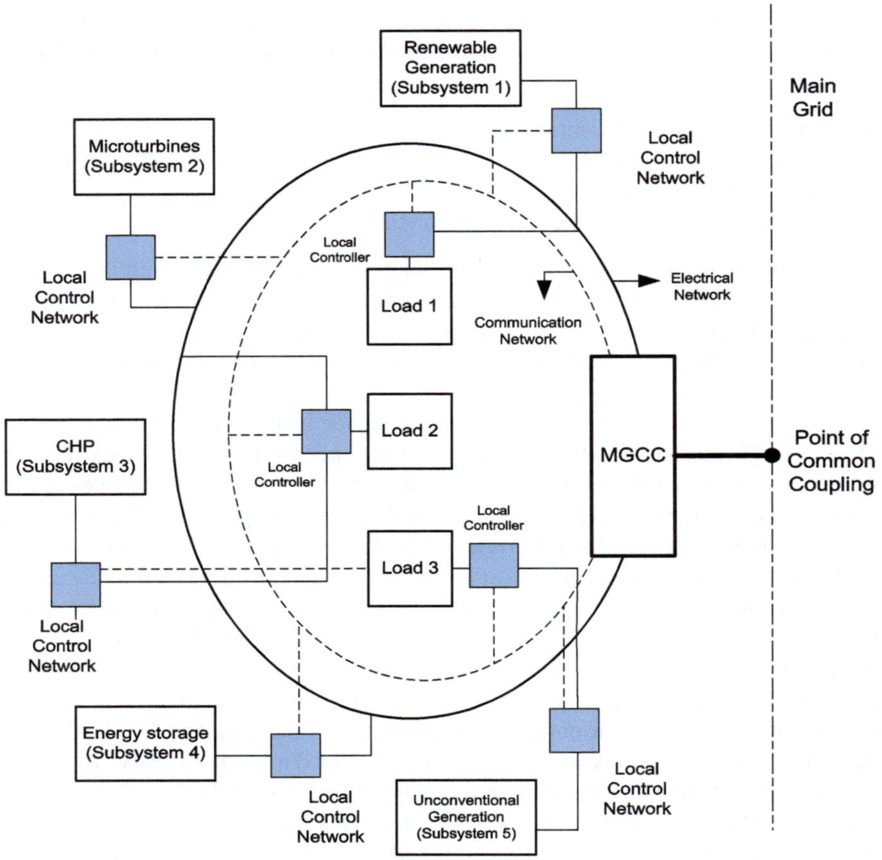

Fig. 5.5 SoS framework for microgrids

and a communication network are used for the transfer of control signals and data collection, respectively. The MGCC is connected to the main grid via a PCC. All the components of MG are classified as subsystems. The first subsystem is the renewable sources of generation which include the PV panel, the hydro generator, and the WT. All these sources are independent systems in the subsystem, thereby making it complex. The second subsystem is microturbines which combines all the single shaft turbines running both on liquid and gas. The third subsystem is the CHP unit which is responsible to supply the local loads with heat. The waste heat emitted from the power generation sources is transported to the required location. The CHP unit is preferably placed close to the local loads because heat cannot be transported like electricity for long distances. The fourth subsystem is the energy storage system which has batteries, ultra large capacitors and flywheels to store the power, and supply the domestic loads when there is an emergency. Finally, the fifth subsystem consists of unconventional generation systems like the fuel cell. Additionally, there are MCs dedicated for every subsystem and load controllers for the local loads which are connected through an electrical as well as a communication network. In other words, all the five subsystems are connected to the MGCC to receive control and communication signals. Collectively, all systems and subsystems constitute the SoS framework.

5.3.4 Operation of the Microgrid Under the SoS Framework

The MG, now enclosed in the SoS framework can carry out all its functions in a coordinated and organized manner maintaining effective communication among its subsystems to strategically use each of them to accomplish its objectives. The framework will better address the operation of all the subsystems both during the connected and islanded or autonomous mode.

Speaking about the MGCC, it needs to perform energy management by looking into the energy storage systems and the energy flow in both connected and islanded operation modes. The MGCC facilitates energy transfer with the main grid to control the active and reactive power flows and keeps a check on the storage system [60, 61]. Normally, the MGCC collects data from the loads and microsources to perform functions like forecasting studies, DSM, and scheduling of microgeneration. With the communication network covering all the subsystems of the MG, it would become convenient for the MGCC to retrieve data and monitor the subsystems. Although the amount of data is not proliferating, the communication network could well support any addition to the system in the future considering its evolutionary feature. Further, since the subsystems are operating independently without mutual interference with each other, it becomes simpler for the MGCC to allocate set-points to local and MCs. Making use of the well-connected electrical network the controllers can take the necessary action including regulating power flow and load tracking. The local and MCs enjoy managerial independence and are responsible for their respective microsource or load. During the islanded mode, the local controllers (LOCs) must ensure that their microsource takes its share of load, so that the needs of the domestic consumers are optimally met. Because of operational independence, the MCs regulate voltage of its particular source to accommodate the changing loads on the system and simultaneously have to communicate with the MGCC to stay informed about the load changes coming from the main grid.

The operational and managerial independence features come handy during the islanded mode, when the MGCC must devise interruption strategies and make efficient use of the storage systems to maintain high reliability and lessen interruption time. All unconventional generation systems and storage systems operating independently can be assigned by the MGCC to supply the critical loads during a blackout, while it forms a secondary loop changing the output power of microgenerators to prepare for the task. This is possible because of the well-organized electrical and communication network ensuring proper and efficient transfer of signals from the MGCC. Moreover, the operational and managerial independence of the local control network makes things easier, since the MGCC does not have to worry about their operation and just needs to communicate necessary information to them as explained above to perform their functions. It is clearly evident how the subsystems under the SoS framework operate independently and communicate with each other to accomplish a desired goal.

5.4 Modeling of Microgrid

There exists several papers available in the literature where detailed modeling of various components of the MG can be found. For instance, [79] presents the modeling of microsources in a MG. Mathematical models are given for the diesel generator, batteries, and WT. Also, detailed models for fuel cells and microturbines having single shaft are available in [81], diesel generator having a prime mover, generator unit and governor is modeled in [65, 72].

In [23, 34], the modeling of almost all significant components is explained including the PV systems, WTs, microturbines, and fuel cells which constitute the microsources. Further, basic models are also described for the power electronic interfaces (PEIs). Similarly [49] deals with the detailed modeling of almost all components at steady state and studies their transient responses for input changes.

5.4.1 Microturbine Model

Figure 5.6 shows the significant blocks forming a microturbine. The purpose of this simplified diagram is to perform the dynamic behavior analysis. P_{dem} is the demanded power, P_{ref} represents the reference power and P_{in} represents the power control variable which is to be applied to the turbine. The microturbine is equipped with recuperation, advanced materials, and technologies enabling low NO_x emissions [76]. Further explanation and detailed modeling of the microturbines can be found in [52].

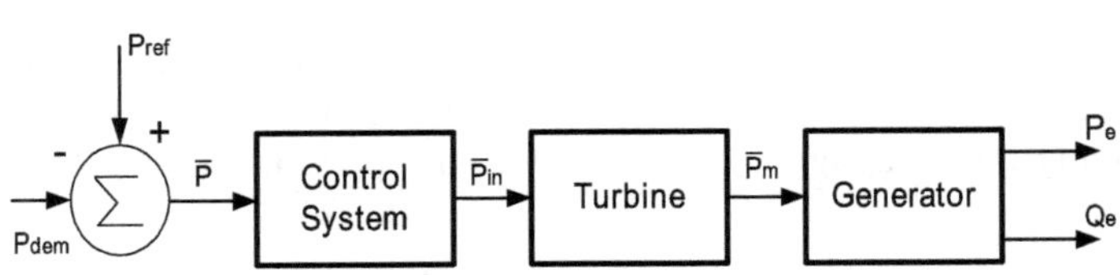

Fig. 5.6 Significant blocks of a microturbine model

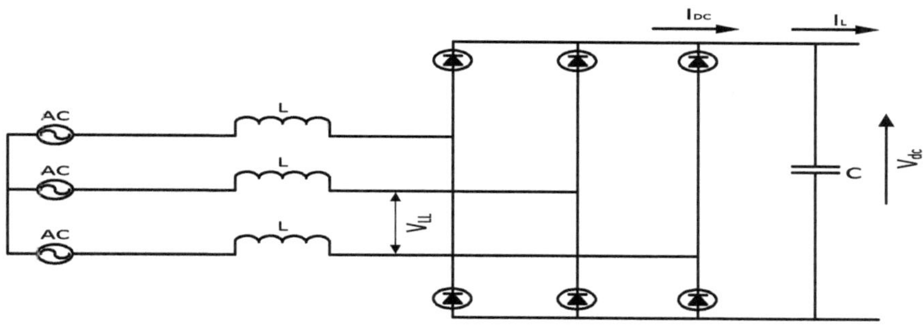

Fig. 5.7 Equivalent circuit of generator and rectifier in a microturbine

Figure 5.7 shows an equivalent circuit representation of the generator and the rectifier of the microturbine. Although the microturbine also consists of a recuperator, it has not been considered in modeling because it is used only for enhancing engine efficiency. Considering the equivalent circuit in Fig. 5.7, the voltage induced in the generator terminal V_{LL} is given as

$$V_{LL} = K_v \omega \sin(\omega t) \tag{5.1}$$

where K_v is the voltage constant and ω is the electrical angular frequency. Now, the output DC voltage from the rectifier is given as

$$V_{dc} = \frac{3}{\pi}|V_{LL}| - \frac{3\omega L}{\pi} I_{dc} \tag{5.2}$$

Substituting (5.1) in (5.2), we have

$$V_{dc} = \frac{3}{\pi} K_v \omega - \frac{3}{\pi} \omega L I_{dc} \tag{5.3}$$

Let us now define a no load voltage (DC) E as

$$E = \frac{3K_v}{\pi} \omega \tag{5.4}$$

Substituting in (5.2), we have

$$E = V_{dc} + \frac{3L}{\pi} \omega I_{dc} \tag{5.5}$$

Under the condition that the system does not suffer from any losses, the input power P_m can be defined as

$$P_m = V_{dc} I_{dc} \tag{5.6}$$

Substituting the value of V_{dc} from (5.3) in the above equation, we have

$$P_m = \frac{3}{\pi} K_v \omega I_{dc} - \frac{3}{\pi} L \omega I_{dc}^2 \tag{5.7}$$

For the same system, the mechanical shaft torque is given as

$$T_m = \frac{P_m}{\omega} = \frac{3}{\pi} K_v I_{dc} - \frac{3}{\pi} L I_{dc}^2 \tag{5.8}$$

And, the representation of mechanical part of the system is

$$\frac{d\omega}{dt} = \frac{1}{J_r}(T_m - T_t) \tag{5.9}$$

where J_r is the inertia of shaft and T_m, T_t are the mechanical and load torques, respectively.

5.4.2 PV Solar Cell Model

Figure 5.8 shows a representation of the solar cell through an electrical single diode model [59]. The inputs to the model are array voltage, ambient temperature, and solar irradiance while the output is array current. Also, a lot of work has been done to design a comprehensive PV model by utilizing power electronic components [12, 75].

Considering the equivalent circuit of the PV cell given in Fig. 5.6, the output current i is given by

$$i = i_{ph} - i_D - i_p \tag{5.10}$$

where i_{ph} is the current generated by light, i_D is the diode current, and i_p is the shunt leakage current through resistance R_p. The open circuit voltage V_{oc} is given by

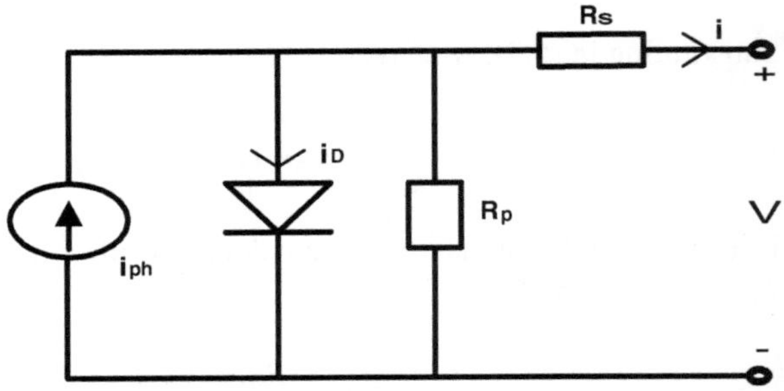

Fig. 5.8 Simple PV cell model

$$V_{oc} = V + iR_s$$

where V is the terminal voltage and R_s is the series resistance. Also, the diode current i_D is given by [51]

$$i_D = i_d\left[\frac{qV_{oc}}{A_{cf}KT} - 1\right] \tag{5.11}$$

where i_d is the saturation current of the diode, q is the electron charge, A_{cf} is the curve fitting constant, K is the Boltzmann constant, and T is the temperature. Now the load current is expressed as

$$\begin{aligned} i &= i_{ph} - i_{os}\left(exp\left[\frac{qV_{oc}}{AKT} - 1\right]\right) - \frac{V_{oc}}{R_p} \\ i_{ph} &= \frac{G}{100}[i_{SR} + K_I(T - 25)] \\ i_{os} &= i_{or}\left(\frac{T}{T_r}\right)^3 exp\left[\frac{qE}{BK}\left(\frac{1}{T_r} - \frac{1}{T}\right)\right] \end{aligned} \tag{5.12}$$

Note that i_{os} is reverse saturation current of the PV cell, A, B are p-n junction ideality factors, G is solar irradiation, E is the band gap for Silicon, i_{SR} is the short circuit current at 25° and 1000 W/m^2, K_I is the short circuit current temperature coefficient at i_{SR}, T_r is the reference temperature and i_{or} is the cell saturation current at T_r. More information about the modeling can be found in [49].

5.4.3 Wind Turbine Model

Figure 5.9 represents a squirrel cage induction generator-based WT. It has a conventional grid-connected to a squirrel cage induction generator, which is connected to a rotor through a gearbox. The function of the gearbox is to adjust the speeds of the rotor and generator, whose ranges are different [1, 29, 30]. There are different subsystems in the WT that can be modeled separately. Out of which are the aerodynamic and mechanical subsystem whose modeling is presented below: The aerodynamic power is given by [34, 35]

$$P_\alpha = \omega_r T_w = \frac{1}{2}\rho \cdot A \cdot C_p(\lambda, \beta) \cdot v_w^3 \tag{5.13}$$

where P_α is the aerodynamic power, ρ is the density of air, β is the pitch angle, A represents the rotor area, T_w represents the aerodynamic torque, ω_r expresses the blade rotating speed, λ is the tip speed ratio, v_w is the wind speed and $C_p(\lambda, \beta)$ is a dimensionless constant. Now, the mechanical subsystem can be simulated using

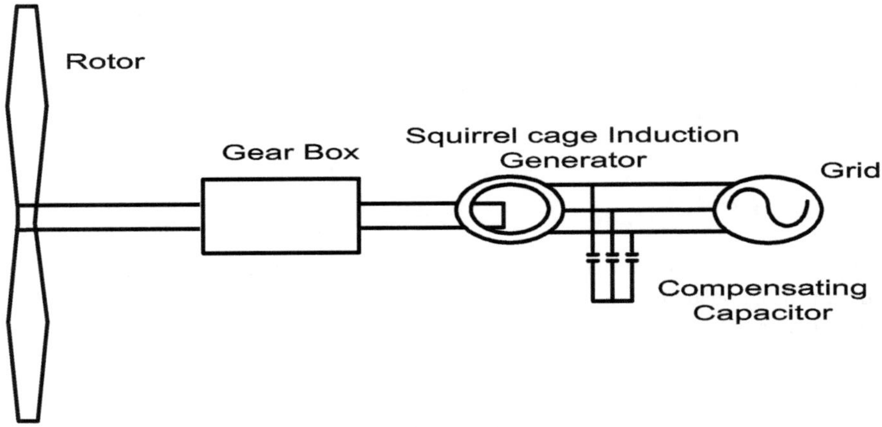

Fig. 5.9 Wind turbine model

three or six elastically connected masses. To represent the low-speed shaft torsional mode, a minimum of two masses are needed. Described below are the three mass equivalent state-space equations of the model

$$\frac{d}{dt}\begin{bmatrix}\theta\\ \omega\end{bmatrix} = \begin{bmatrix}[0]_{3\times3} & [I]_{3\times3}\\ -\frac{1}{2}[H]^{-1}[C] & -\frac{1}{2}[H]^{-1}[D]\end{bmatrix}\begin{bmatrix}\theta\\ \omega\end{bmatrix} + \begin{bmatrix}[0]_{3\times3}\\ \frac{1}{2}[H]^{-1}\end{bmatrix}T \tag{5.14}$$

where $\theta^T = [\theta_R \;\; \theta_{GB} \;\; \theta_G]$ represents the angular position vector, $\omega^T = [\omega_R \;\; \omega_{GB} \;\; \omega_G]$ denotes the angular speed vector and $T^T = [T_w \;\; 0 \;\; T_G]$ expresses the external torque vector including the electromagnetic torque T_G and aerodynamic torque T_w. $[H] = diag(H_R, H_{GB}, H_G)$ is the diagonal inertia matrix and C and D are stiffness and damping matrices, respectively. C also denotes the high- and low-speed shaft elasticities and is given as

$$[C] = \begin{bmatrix} C_{HGB} & -C_{HBG} & 0\\ -C_{HBG} & C_{HBG} + C_{GBG} & -C_{GBG}\\ 0 & -C_{GBG} & C_{GBG}\end{bmatrix}$$

D represents the torque losses and internal friction and is given as

$$[D] = \begin{bmatrix} D_R + d_{HGB} & -d_{HGB} & 0\\ -d_{HGB} & \begin{matrix} D_{GB} + d_{HGB}\\ +d_{GBG}\end{matrix} & -d_{GBG}\\ 0 & -d_{GBG} & D_G + d_{GBG}\end{bmatrix}$$

Note that the subscripts GB and G represent the gearbox and generator, respectively. The induction generator is represented by a fourth order model in an arbitrary reference frame. From [23], using the generator convention for stator currents, we have

$$\begin{aligned} u_{sd} &= -r_s i_{sd} - \omega\psi_{sq} + p\psi_{sd} \\ u_{sq} &= -r_s i_{sq} + \omega\psi_{sd} + p\psi_{sq} \\ u_{rd} &= r_r i_{rd} - (\omega - \omega_r)\psi_{rq} + p\psi_{rd} \\ u_{rq} &= r_r i_{rq} + (\omega - \omega_r)\psi_{rd} + p\psi_{rq} \end{aligned}$$

where $p = \frac{1}{\omega_0}\frac{d}{dt}$ and ω_0 is the base cyclic frequency, ω represents the rotating speed of the arbitrary reference frame, and the subscripts dq represent the dq axis of the frame and s, r denote the stator and the rotor, respectively.

Further, the fluxes ψ are related to the stator and rotor winding currents by the equations given below:

$$\begin{aligned} \psi_{sd} &= -X_s i_{sd} + X_m i_{rd} \\ \psi_{sq} &= -X_s i_{sq} + X_m i_{rq} \\ \psi_{rd} &= -X_m i_{sd} + X_r i_{rd} \\ \psi_{rq} &= -X_m i_{sq} + X_r i_{rq} \end{aligned}$$

where the electromagnetic torque is given by

$$T_e = \psi_{qr} i_{dr} - \psi_{dr} i_{qr}$$

5.5 Microgrid Control Architecture

Although extensive research has been carried out in designing control strategies for MGs, there is still ambiguity regarding the best MG control strategies. Reference [77] presents a survey of the various control techniques developed for MGs. However, since we are looking toward the MG from a SoS perspective, we need to use the control strategies devised for SoS. The first hurdle lies in finding a mathematical model for the framework. The control paradigms given in [28] need a mathematical model for their implementation. Nevertheless, a few mathematical models of the MG are available in the literature which are complex and of high order [14, 15]. Additionally, state-space modeling has also been done for the MG [8, 82].

5.5.1 Introduction

There is no general structure of MG control architecture, since the configuration depends on the type of MG or the existing infrastructure. Before analyzing the MG control and management architecture, lets have a look at todays distribution systems.

Figure 5.10 presents the major parts of the control and management infrastructure of a typical distribution system with increased DG penetration. We can distinguish the distribution management system (DMS) and the automated meter reading (AMR) systems. The DMS is mainly responsible for the monitoring of the main HV/MV and maybe some critical MV/LV substations. The hardware system consists of the main server and several remote terminal units or intelligent electronic devices spread across the distribution system. Usually, the DMS does not control the DGs/RESs (except for some large installations in certain cases) or the loads. Typical control actions are network reconfiguration, by switching operations in the main feeders, and voltage control via capacitor switching or perhaps transformer tap changing (mostly manually). The AMR system is responsible for the collection of electronic meter readings and is used mainly for billing purposes. In Fig. 5.10, we do not consider the existence of the advanced meter infrastructure (AMI), since this is considered next, as part of the MG control system. By AMI, we also mean the capability of

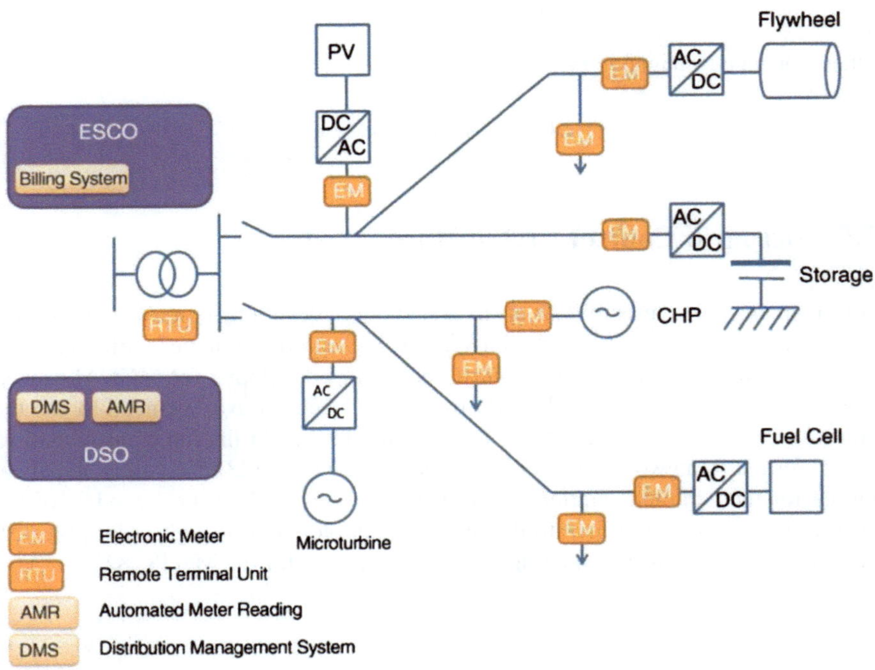

Fig. 5.10 Typical distribution system management structure

controlling some loads locally, either directly via the meter or via the home area network, in which case the electronic meter is the gateway.

As discussed in the forgoing sections the DSO is responsible for managing and controlling the distribution system and is also responsible for collecting the energy metering data, although in some countries meter reading can be handled by an independent entity. The DSO sends the metering data to the supplier/ESCO, who is a market player and is responsible among other things for the billing of customers.

The structure shown in Fig. 5.10 is not sufficient for MG management, since it provides limited control capabilities, especially within a market environment. Thus, it is important to introduce a new control level locally at DG and loads, capable of meeting the goals of

- enabling all relevant actors to advanced market participation
- being scalable in order to allow the integration of a large number of users (scalability)
- allowing integration of the components of different vendors (open architecture)
- ease of installing new components (plug-and-play)
- ease of integrating new functionalities and business cases (expandability).

Using the local control level, a more complicated, hierarchical architecture is introduced in Fig. 5.11. This architecture comprises the following:

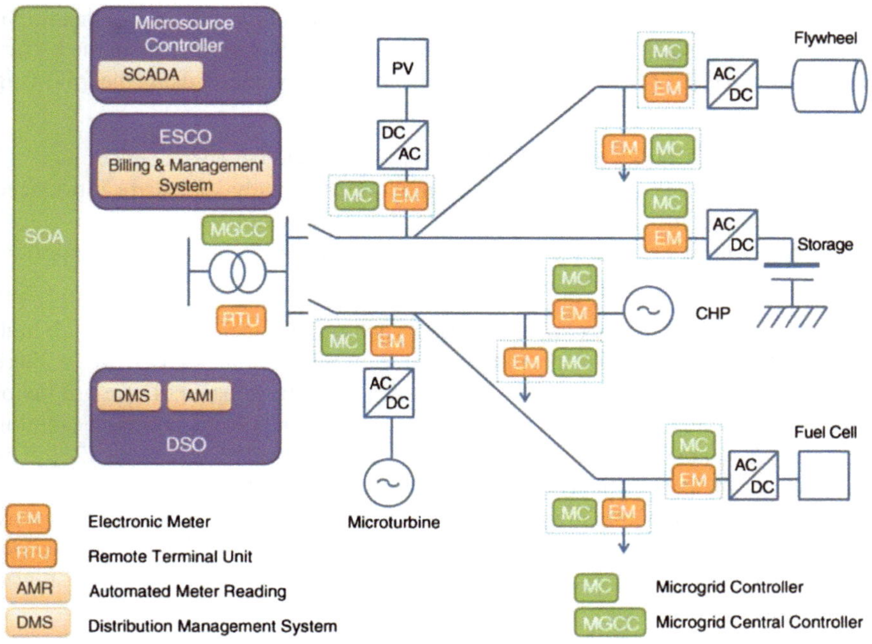

Fig. 5.11 Typical microgrid management structure

1. **The MC** which is responsible for controlling and monitoring DER, such as DGs, storage devices and loads, including electric vehicles. The MC could be a separate hardware device or piece of software installed in either the electronic meter, the DG power electronic interface, or any device in the field with sufficient processing capacity. This is shown as a dashed frame that surrounds both the MC and the EM.
2. **The MGCC** provides the main interface between the MG and other actors such as the DSO or the ESCO, and can assume different roles, ranging from the main responsibility for the maximization of the MG value to simple coordination of the local MCs. It can provide set-points for the MCs or simply monitor or supervise their operation. It is housed in the MV/LV substation and comprises a suite of software routines of various functionalities depending on its role.
3. **The DMS**, discussed previously, is responsible among others for the collaboration between the DSO, the ESCO and the MG operator. The existence of a backbone system, a platform, based on service oriented architecture is assumed for the integration of its functionalities. In some cases, the MGCC software can be integrated in this platform.

5.5.2 Hierarchical Control

Once we obtain the mathematical model, we look to the control strategies that might be applied to the MGs. One interesting control strategy would have been the hierarchical control (HC) [11, 74]. Functionally, the MG can operate by using the following three main HC levels:

- Primary control is the droop control used to share load between converters.
- Secondary control is responsible for removing any steady-state error introduced by the droop control.
- Tertiary control concerning more global responsibilities decides the import or export of energy for the MG.

Such kind of systems operate over large synchronous machines (SMs) with high inertia and inductive networks. However, in power electronics-based MGs, there are no inertias, and the nature of the networks is mainly resistive. There are major differences between both systems that we have to take into account when designing their control schemes. A three-level HC can be constructed following [21]. The primary control deals with the inner control of the DG units by adding virtual inertias and controlling their output impedance. The secondary control is conceived to restore the frequency and amplitude deviations produced by the virtual inertias and output virtual impedance. The tertiary control regulates the power flows between the grid and the MG at the PCC.

In what follows, we look at the basic elements of consensus-based control that can be possibly applied to the MG. This is an important class type of cooperative control paradigm [32]. Our view is to focus on the mutual agreement among the different units of MG as subsystems in a SoS.

5.5.3 Consensus Control

The consensus control problem has been discussed extensively in the literature owing to its applicability in the formation of robots, cooperative control of unmanned aerial vehicles, and communication between sensor networks [18]. Consensus among a group of agents or subsystems means to arrive at a certain agreement concerning a particular value or quantity which is dependent on the states of every agent. A consensus algorithm or protocol is a rule that explains the interaction between an agent and its surrounding subsystems enclosed in the network. However, management of the shared information between the agents is a very critical issue which has to be resolved in order to carry out coordination among the subsystems. To address the consensus problems for dynamic networked agents, a theoretical framework was developed by [53].

A consensus protocol is introduced with x_i representing the information state of the ith agent, where the information state is responsible for the information that has to be coordinated among the agents, the consensus protocol for continuous systems can be formulated as [43, 54, 58, 64].

$$\dot{x}_i(t) = \sum_{j \in J_i(t)} \alpha_{ij}(t)(x_i(t) - x_j(t)) \tag{5.15}$$

where $J_i(t)$ is the group of agents whose information is available to agent i at time t, $\alpha_{ij}(t)$ represents a positive time-varying weighing factor.

Additionally, in [38], consensus among a group of networked agents or systems is obtained by using an optimal control design method. A set of agents are considered and each of them can be dynamically represented as

$$\begin{aligned} \dot{x}_i &= A_i x_i + B_i u_i, \quad x_i \in \Re^n, u_i \in \Re^m \\ y_i &= C_i x_i, \quad y_i \in \Re^q, i = 1, \dots N \end{aligned} \tag{5.16}$$

where x_i represents the state vector, u_i represents the input vector and Y_i represents the output vector of each agent i. $n.m$ and q give the dimension of the corresponding state, input, and output vectors. However, the whole network of agents can dynamically be represented as

$$\dot{X} = AX + BU, \quad Y = CX$$

where X, U, and Y represent the state, input, and output vectors of the whole set of agents. All the respective individual agent vectors are concatenated in one vector as follows:

$$X_{Nn\times 1} = [x_i^T, \ldots, x_N^T]^T, U_{Nm\times 1} = [u_i^T, \ldots, u_N^T]^T$$
$$Y_{Nq\times 1} = [y_i^T, \ldots, y_N^T]^T$$

Therefore, matrices A, B, and C are given as

$$A = diag[A_1, \ldots A_N], \; B = diag[B_1, \ldots B_N]$$
$$C = diag[C_1, \ldots C_N]$$

5.5.4 Centralized and Decentralized Control

The MG structure depicted in Fig. 5.11 can be operated in a *centralized* or *decentralized* way, depending on the responsibilities assumed by the different control levels.

In *centralized control*, the main responsibility for the maximization of the MG value and the optimization of its operation lies with the MGCC. The MGCC using market prices of electricity and gas costs, and based on grid security concerns and ancillary services requests by the DSO, determines the amount of power that the MG should import from the upstream distribution system, optimizing local production or consumption capabilities. The defined optimized operating scenario is realized by controlling the microsources and controllable loads within the MG by sending control signals to the field. In this framework, noncritical, flexible loads can be shed, when profitable. Furthermore, it is necessary to monitor the actual active and reactive power of the components.

In a fully *decentralized approach*, the main responsibility is given to the MCs that compete or collaborate, to optimize their production, in order to satisfy the demand and probably provide the maximum possible export to the grid, taking into account current market prices. This approach is suitable in cases of different ownership of DERs, where several decisions should be taken locally, making centralized control very difficult. Apart from the main objectives and characteristics of the controlled MG, the choice between centralized and decentralized approaches for MG control, depends on the available or affordable resources: Personnel and equipment. The two approaches are presented in Figs. 5.12 and 5.13. In both approaches, some basic functions are centrally available, such as local production and demand forecasting and security monitoring.

In the following, the effectiveness of the two approaches in terms of calculation time, scalability, and accuracy is discussed. At this point, the introduction of algorithm complexity is useful. Algorithm complexity refers to the calculation time that an algorithm needs to finish a task. This calculation usually correlates the execution time needed with one or more arguments; for example, the solution time of the unit

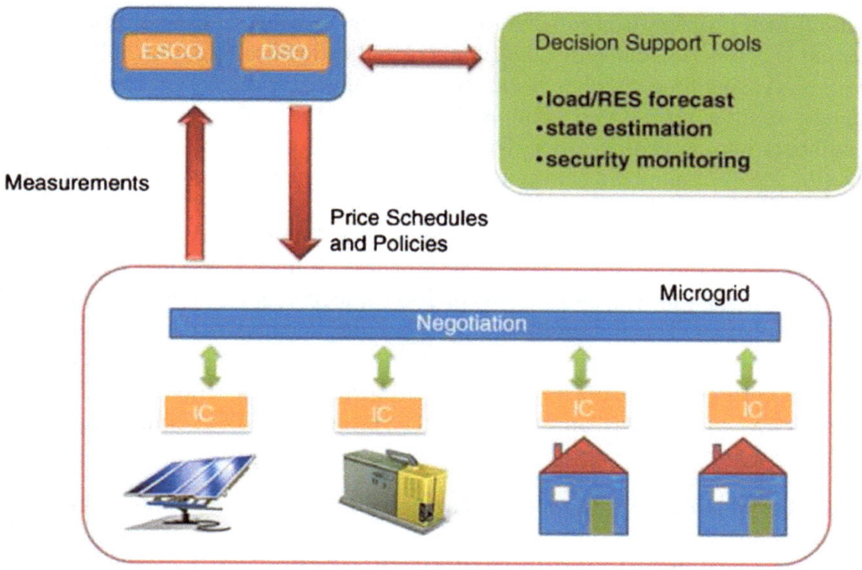

Fig. 5.12 Principle of centralized control

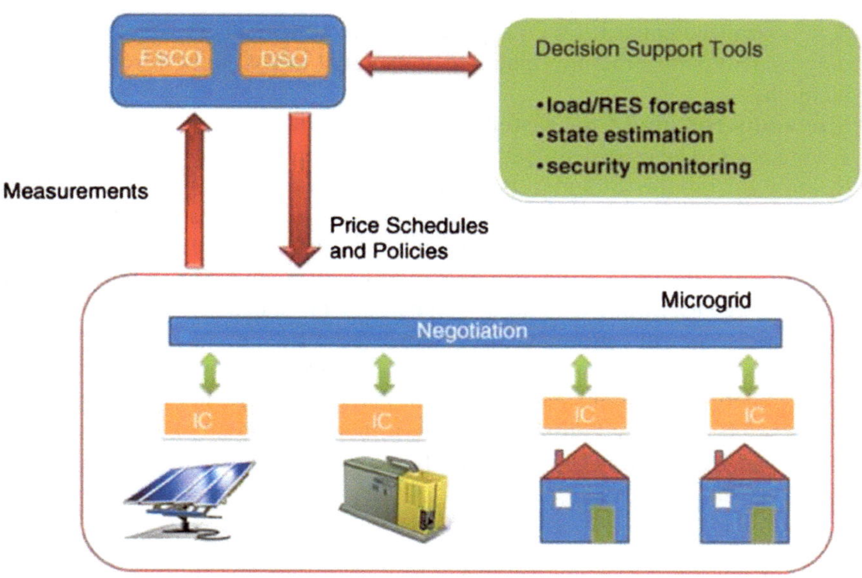

Fig. 5.13 Principle of decentralized control

commitment problem is correlated to the number of the generating units available. Capital O is used to denote the algorithms classification according to their response (e.g., processing time, number of messages exchanged, or memory requirements) to changes in input size. As an example, lets consider the problem of sorting a list of n numbers into ascending order. The algorithm sorts the list by swapping at each iteration a pair of numbers between two positions of the list. The critical task in this algorithm is the number of swaps between numbers.

If the complexity of the sorting algorithm is $\mathbf{O}(n^2)$, this means that the maximum number of swaps is the square of the number of items in the list, so a list of 10 numbers requires at maximum 100 swaps, a list of 100 numbers requires at maximum 10,000 swaps, and so on. This notion is important in order to compare the two approaches not only with respect to the number of nodes (e.g., number of items in the list), but also their correlation with the most time- or effort-consuming action (e.g., number of swaps). Next, the key attributes that affect the performance of the control algorithms for MGs are listed.

- **Number of nodes**: A MG consists of several microsources and controllable loads. The number of DERs affects critically the complexity of the problem and the computational time.
- **Number of message exchanges**: DGs and loads in MGs are usually dispersed, and communication systems at LV usually have limited bandwidth. In several cases the number of messages required to perform a task is of primary importance. A decentralized control (DEC) approach reduces the number of messages, since only a small part of the information is transferred to the higher levels of control hierarchy.
- **Size and structure of the system model**: The structure and complexity of the system need to be considered. Decisions taken by different actors might not only increase the number of nodes, but also impose extra technical and nontechnical constraints. A relevant issue is the level of information, such as what parameter or constraint should be made available for the decision process of the various actors. For example, the state of charge (SOC) of a battery might be important to the neighboring DGs, while the internal temperature or voltage level of a cell and the associated technical constraints, might not be relevant.
- **Accuracy and optimality**: An algorithm may converge to the optimal solution or near the optimal solution. It is self-evident that the convergence and the accuracy of the solutions depend on the accuracy of the models used and of the relevant input data. The question is whether a suboptimal solution is acceptable and, if so, at what cost.

The choice between the *centralized* and *decentralized* approach for MG control, depends on the main objectives and the special characteristics of the controlled MG and the available or affordable resources: Personnel and equipment.

Clearly, *centralized control* is more suitable if the users of the microgrid (DG and load owners) have common goals or a common operational environment seeking cooperation, in order to meet their goals. Such an example is an industrial MG, in which a single owner might exercise full control of all its energy sources and loads, is able to continuously monitor them and aims to operate the system in the most economical way. Considering the general attributes listed earlier, the number of nodes is generally limited and it is relatively easy to install a fast communication system and a set of sensors. Dedicated operating personnel for the operation of the MG might be available. Furthermore, the optimization problem has a limited set of constraints and specific objectives, such as cost minimization. Finally, the requested solution should be as accurate as possible since a suboptimal solution may lead to profit losses.

MGs operating in a market environment might require that competitive actions of the controller of each unit have a certain degree of independence and intelligence. Furthermore, local DER owners might have different objectives: In addition to selling power to the network, they might have other tasks, such as producing heat for local installations, keeping the voltage locally at a certain level, or providing a backup system for local critical loads in case of main system failure. Some MG customers might primarily seek their own energy cost minimization and have diverse needs, although they all might benefit from the common objective of lowering operating costs of their feeder. In a residential MG, for example, one household might have at one particular moment increased electric energy needs, for example, for cooking while another household might need no electricity at all, because all its tenants are absent. Both households would like to sell the extra power produced locally to the grid, but it is unlikely that they would accept remote control of their production. Considering again the general attributes mentioned earlier, we should identify that in this case the number of nodes might increase significantly. A neighborhood might have dozens of households or installed DGs, and if we consider multi-MGs the number is increasing further. In such cases, it is not possible to have a dedicated communication system, but existing infrastructure should be used. Thus, part of the system might not have sufficient bandwidth or the communication could be expensive. An approach that limits the amount of data transfer is essential. The availability of powerful computing facilities or dedicated operators is also highly unlikely.

Furthermore, the optimization problem is becoming extremely complex, as a result of specific characteristics. For example, it is extremely complicated to model the comfort requirements in each household or to include all the special technical constraints of all appliances in a single optimization problem. The decentralized approach suggests that this type of constraint and subproblem should be solved locally in each household or DG. In the general control problem, each household could be presented as a load node that has the ability to shed or shift some load, or a production node with DGs that can offer a certain amount of energy without caring about the type of the engine or the technical constraints. Finally, in this case, a suboptimal solution is probably acceptable, given the high costs of the installation of fast communication networks or powerful processors dedicated to energy optimization.

Another important factor is the openness of the system. The distributed computing technology allows DG manufacturers and loads to provide plug-and-play components by embedding control agents (following some rules) in their devices.

The software should be intelligent enough to monitor the process and follow the best policy. The availability of dedicated personnel responsible for system installation and maintenance, likely in centralized systems, might relax this requirement to some extent. In this case, dedicated personnel could also be available for monitoring the process, which could intervene in an emergency.

The general conclusion is that the centralized approach is suitable for a system with one specific goal and a decentralized one in a system with several goals.

5.5.5 More on Decentralized Control

With reference to Fig. 5.2, the control of the inverter plus filter interfaces is crucial to the operation of the MG. Because of the distributed nature of the system, these interfaces need to be controlled on the basis of local measurements only. The DEC of the individual interfaces should address the following basic issues.

1. Interfaces should share the total load (linear or nonlinear) in a desired way.
2. DEC based on local measurement should guarantee stability on a global scale.
3. Inverter control should prevent any DC voltage offsets on the MG.
4. Inverter control should actively damp oscillations between the output filters.

From the viewpoint of DEC, it is convenient to classify DG architectures into three classes:

- In *highly dispersed networks*, the interconnecting impedances are predominantly inductive, and the voltage magnitude and phase angle at different source interconnects can be very different.
- In *networks spread over a smaller area*, the impedances are still inductive with a significant resistive component. The voltage magnitude does not differ much, but the phase angles can be different for different sources.
- In *very small networks*, the impedances are small and predominantly resistive. Neither magnitude nor phase angle differences are significant at any point.

The main common quantity in all cases is the steady-state frequency which must be the same for all sources. In the grid-connected mode, the MG frequency is decided by the grid. In the islanded mode, the frequency is decided by the MG control.

In each of these classes, if every source is connected to at most two other sources, as shown in Fig. 5.2a, then the MG is radial. Otherwise, it is meshed. If there is a line connecting Source 1 with Source k in Fig. 5.2a, then it is a meshed MG. It is worth noting that the majority of work done on MG decentralized control has been for radial-MG topologies. The DEC of interfaces in meshed topologies is an area that needs further research.

5.6 Application to Islanded Microgrid

Figure 5.14 shows a MG system which consists of the parallel connection of several DG units. All DG units are connected to the PCC where the main grid and the local load are also connected. Each DG unit is coupled to the main grid via a voltage-sourced converter (VSC) [4]. The study system is an islanded MG consisting of two DG units and a passive RLC load which are all connected to the PCC, the MG parameters are given as in Table 5.1 [4].

In Fig. 5.15 a single-line diagram of the three-phase MG system is shown.

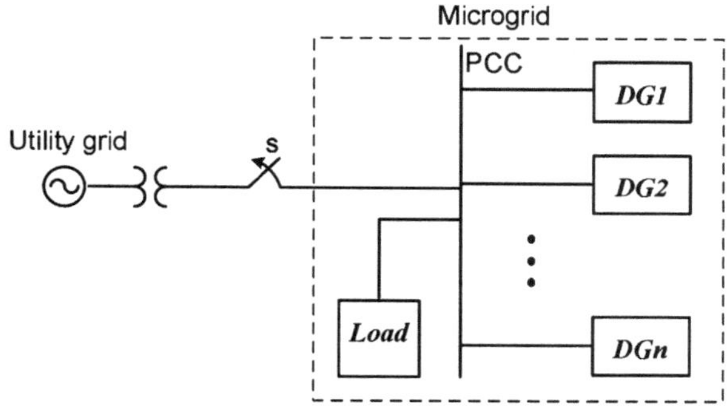

Fig. 5.14 Microgrid system with multiple parallel connected DGs

Table 5.1 Parameters of the microgrid system

Quantity	Value
R_{t1} (series filter resistance in DG1)	1.5 mΩ
L_{t1} (series filter resistance in DG1)	300 μH
R_{t2} (series filter resistance in DG2)	6 mΩ
L_{t2} (series filter resistance in DG2)	900 μH
VSC rated power	$S_{base} = 2.5$ MVA
VSC terminal voltage (line-line)	$V_{base} = 600$ V
f_{sw} (PWM carrier frequency)	1980 Hz
R (load nominal resistance)	76 Ω
L (load nominal inductance)	111.9 mH
C (load nominal capacitance)	62.86 μF
$q = \frac{L\omega_0}{R_L}$ (Inductor quality factor)	120
f_0 (system frequency)	60 Hz
V_{dc} (DC bus voltage)	1500 V
Transformer voltage ratio (Y/Δ)	0.6/13.8 kV

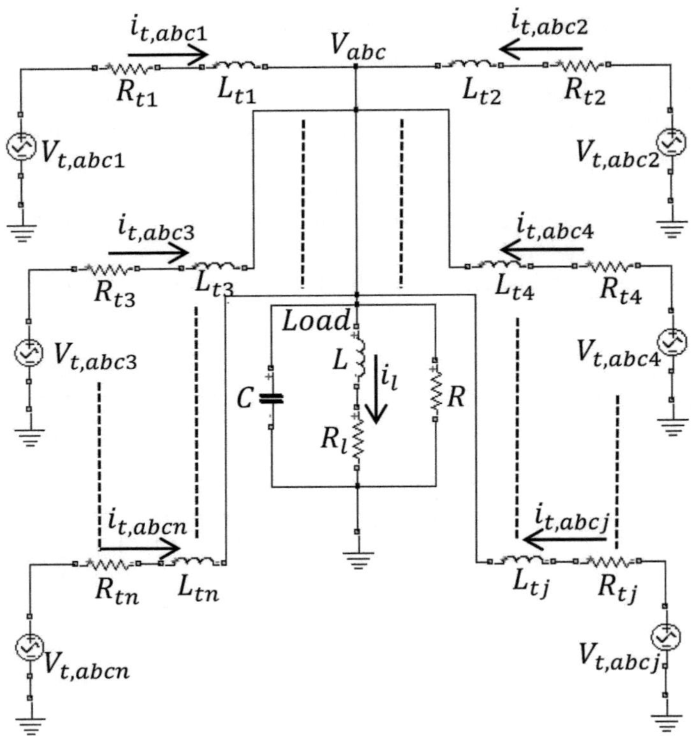

Fig. 5.15 Single-line diagram of the islanded n DG system

By applying the KVL and KCL, the equations of the MG system can be compactly written as

$$\begin{aligned}
\dot{x} &= Ax + Bu \\
y &= Cx \\
x &= [x_1, \quad x_2]^T \\
x_1 &= [V_d, \ V_q, \ I_{td1}, \ I_{tq1}, \ I_{Ld}, \ I_{Lq}, \ I_{td2}, \ I_{tq2}] \\
x_2 &= [\ldots, \ \ldots, \ I_{td_n}, \ I_{tq_n}]^T
\end{aligned}$$

The control vector u and the output vector y are

$$u = \begin{bmatrix} V_{td_1} & V_{tq_1} & V_{td_2} & V_{tq_2} & \cdots & V_{td_n} & V_{tq_n} \end{bmatrix}^T$$

$$y = \begin{bmatrix} V_d & V_q & I_{td_2} & I_{tq_2} & \cdots & I_{td_n} & I_{tq_n} \end{bmatrix}^T$$

where the details of the matrices A, B, and C can be easily obtained.

5.6.1 Two-Level Control Strategy

A DEC approach for an autonomous operation of islanded MG system was proposed in [4], the islanded MG system is controlled by only using the LOCs. This approach is not guaranteed to give satisfactory performance for islanded MG consisting of n interconnected subsystems, because the interconnections among the subsystems eventually introduce a perturbation signals acting in contradiction to the autonomy of local MG subsystems, which cannot be effectively handled by local decentralized controllers. A global control strategy is proposed in this paper to provide corrective signals to account for this interaction effect.

In this section, a two-level coordinating control scheme [45–47] is proposed for the MG model (5).

5.6.2 Local Subsystem Control

It is first shown that, the system (5) described by an interconnection of n subsystems and given by

$$\dot{x_i}(t) = A_i x_i(t) + B_i(t) + \sum_{j=1}^{n} H_{ij} x_j(t), i = 1, 2, \ldots, n \tag{5.17}$$

where

- $x_i : \Re \to \Re^{n_i}$ is the state of the ith subsystem.
- $u_i : \Re \to \Re^{m_i}$ is the control function.
- A_i,B_i: are the isolated MG subsystems dynamics.
- H_{ij}: specifies the interaction between n subsystems.

It is assumed that each subsystem has a quadratic cost function

$$J_i(t_o, x_i(t_o), u_i(t_o)) = \int_{t_o}^{\infty} \{\| x_i(t_o) \|_{Q_i}^2 + \| u_i(t_o) \|_{R_i}^2\} dt, \tag{5.18}$$

where $i = 1, 2, \ldots, n.$, and $0 < Q_i^T = Q_i \in \Re^{n_i \times n_i}$, $0 < R^T = R_i \in \Re^{m_i \times m_i}$.

The subsystems cooperates to minimize global cost function

$$J(t_o, x(t_o), u(t_o)) = \sum_{i=1}^{n} J_i(t_o, x_i(t_o), u_i(t_o)) \tag{5.19}$$

The purpose of the ith decentralized subsystem controllers is to control the isolated MG subsystem as a function only of it is state $x_i(t)$ with a knowledge only of $\{A_i$, B_i, Q_i, and $R_i\}$.

5.6.3 Global Corrective Control

In this subsection, a centralized optimal controller (corrective controller) is proposed to minimize the cost function J based on the requirements of decentralized costs. The decentralized subsystem control $u_i^*(t)$ which minimizes the subsystem cost J_i under the subsystem constraints

$$\dot{x_i}(t) = A_i x_i(t) + B_i(t), \quad i = 1, 2, \ldots, n \tag{5.20}$$

is given by

$$u_i^*(t) = -K_i x_i(t) = -R_i^{-1} B_i^T P_i x_i(t), \quad i = 1, 2, \ldots, n \tag{5.21}$$

P_i is the symmetric positive definite solution of *Riccati equation*

$$A_i^T P_i + P_i A_i - P_i B_i R_i^{-1} B_i^T P_i + Q_i = 0 \tag{5.22}$$

The optimum costs of these controllers

$$J_i^*(t_o, x_i(t_o)) = x_i^T(t_o) P_i x_i(t_o), \quad i = 1, 2, \ldots, n, \tag{5.23}$$

And the overall optimum cost of the system (6) is

$$J(t_o, x(t_o)) = \sum_{i=1}^{n} J_i^*(t_o, x_i(t_o)) \tag{5.24}$$

The system performance form is resulted from the interactions based on using a global corrective controller in addition to the subsystem decentralized optimal controllers $u_i^*(t)$. This global controller which will be applied to any MG is used to account for the interactions based on a knowledge of the states of the other MGs. The information structure of this global system consists of $\{x_i(t), t \geq t_o, i = 1, 2, \ldots, n\}$ and $\{H_{ij}, i, j = 1, 2, \ldots, n\}$ with the a priori information of A_i, B_i, and P_i. For $i = 1, 2, \ldots, n$.

The gain of the global corrective controller [50] will be achieved based on the local subsystem gain matrices K_i with perturbation of the composite matrix $B = diag[B_1\ B_2\ \ldots\ B_n]$, so the input to each MG subsystem will be consists from two components [50]

$$u_i(t) = u_i^*(t) + u_i^g(t) \tag{5.25}$$

where $u_i^*(t) = -K_i x_i(t)$ is the LOC and $u_i^g(t) = -\sum_{j=1}^{n} M_{ij} x_j(t)$ is the required corrective controller.

5.6.4 Simulation Results

To verify the proposed controller design performance the system given in Fig. 5.15 is implemented in the MATLAB®. In the study system, the microgrid $DG1$ is considered as a master DG, and the microgrids $DG2$ to DG_n are considered as slaves, a multilevel control scheme is applied to this isolated grid such as a local subsystem optimal controller is used to control each DG individually and a global corrective controller to give the required control action to each subsystem. The simulation results show a good performance which verified the proposed control design. Figures 5.16, 5.17, 5.18, 5.19 and 5.20 shows the proposed control strategy response

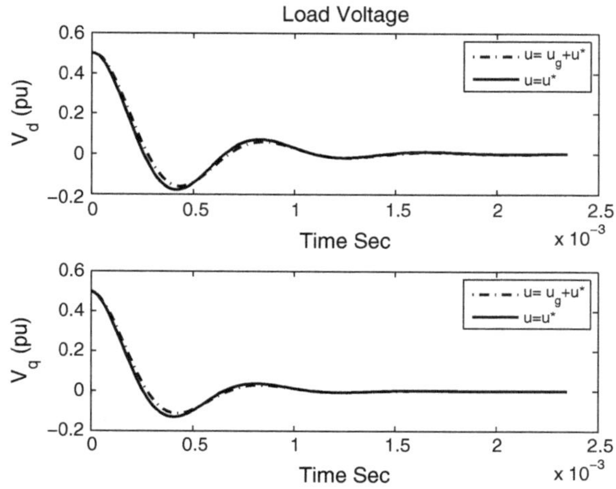

Fig. 5.16 *d* and *q* components of the load voltage

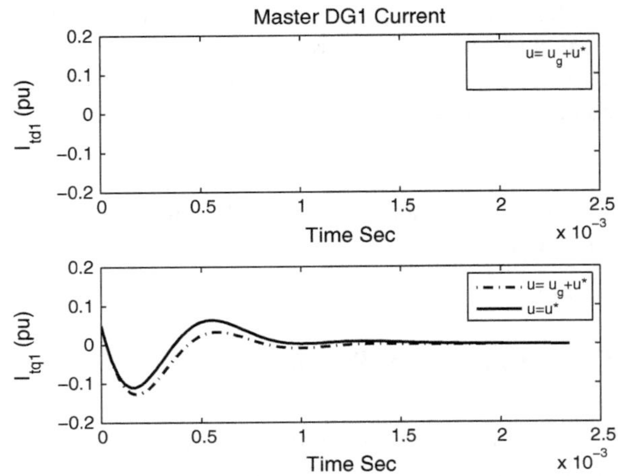

Fig. 5.17 *d* and *q* components of master DG1 current

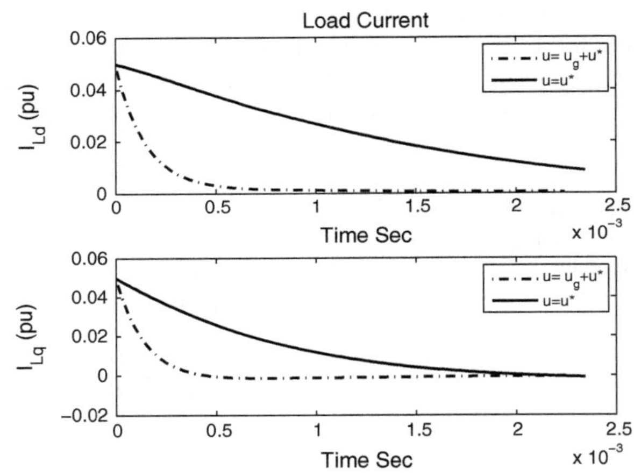

Fig. 5.18 *d* and *q* components of load current

and how it controls the load current and voltage, MG_1, MG_2, and MG_3 currents. This results show that the corrective controller improve the transient response of the subsystem current which means that the subsystem DEC is not enough to control the subsystems currents of the islanded MG.

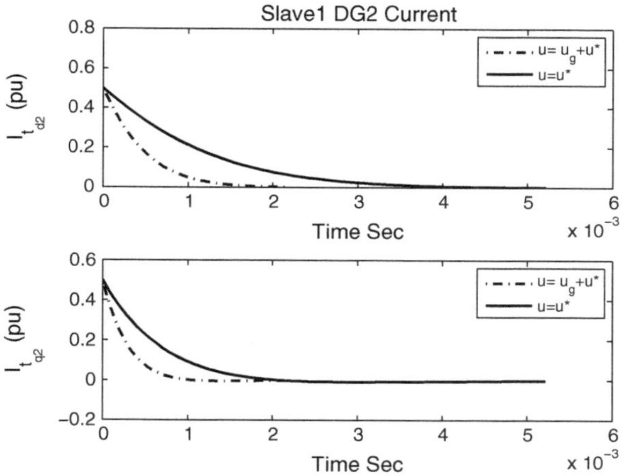

Fig. 5.19 *d* and *q* components of slave1 DG2 current

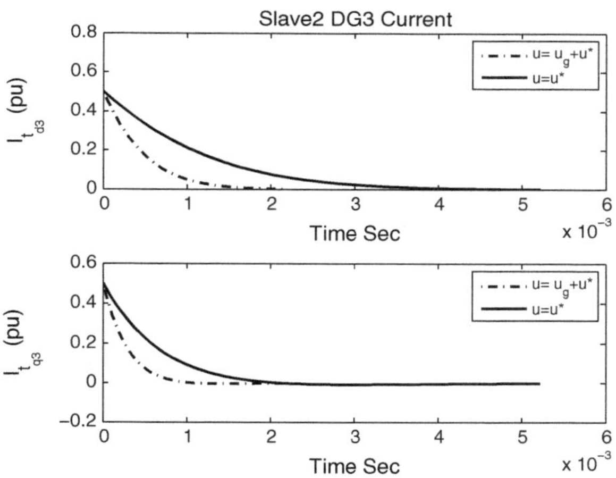

Fig. 5.20 *d* and *q* components of slave1 DG3 current

5.7 Conclusions

The need for DG to combat the concerns faced by the conventional generation and distribution system is reviewed. The concept of MG is presented and its role in providing ancillary services to the main grid is discussed. All the components of MG and their respective role and operation are presented. The SoS notion is introduced and its characteristics are explained. The resemblance of MG features to SoS is highlighted. Further, an SoS framework is proposed for the MG. The framework is designed keeping in consideration the architecture of MG. After which, the operation of the MG under the proposed framework is explained and the benefits of the framework are discussed. Further, modeling of components of the MG is presented including the microturbine, PV solar cell, and the WT. The control strategy of MGs is discussed in terms of the HC, the consensus-based control and the DEC.

5.8 Suggested Problems

Problem 5.1 The review of SoS literature for MG application illustrates the necessity to create an SoS engineering management framework based on the demands of constant technological progress in the complex dynamic environment. Following [20], a detailed view of interaction of different processes within SoS and the functionality of the effective SoS engineering management (SoSEMM) framework is shown in Fig. 5.21. It is a challenging task to apply this framework to DGS with microsources.

Problem 5.2 The block diagram of an inverter-based DG system is presented in Fig. 5.22 along with its droop power controller . The power system includes a primary power source (essentially PV panels), the voltage source converter (VSC), and the power, voltage, and current control loops. The power controller contains the droop technique as presented in Fig. 5.23.

It is desired to build up a linearized mathematical model appropriate for time-domain simulations in the PSCAD/EMTDC software environment. Use convenient set of parameter values.

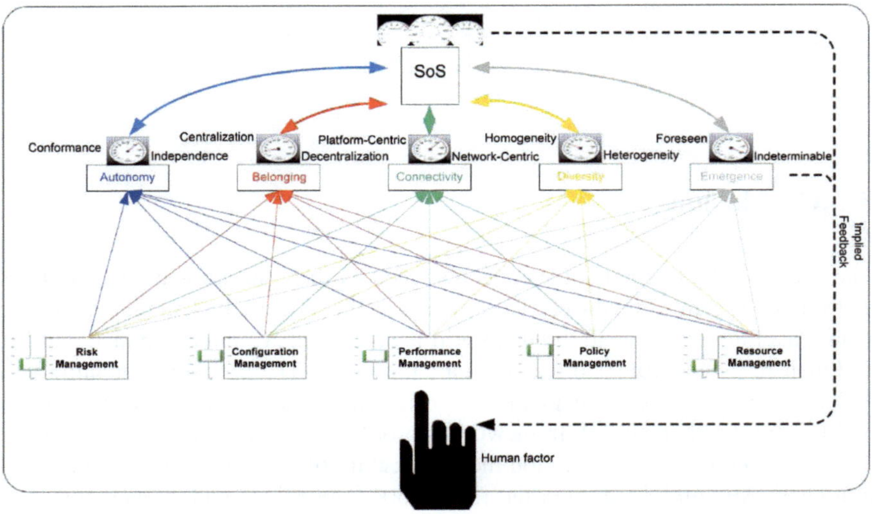

Fig. 5.21 SoS engineering management framework

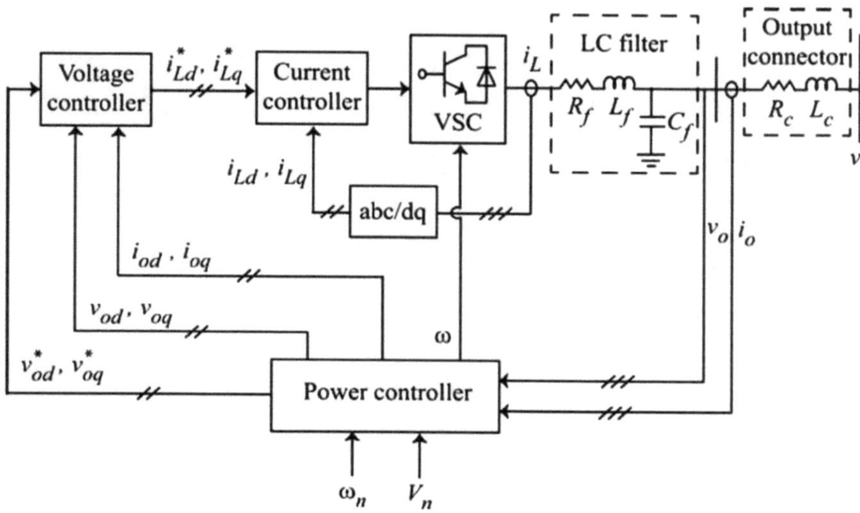

Fig. 5.22 An inverter-based distributed generation system

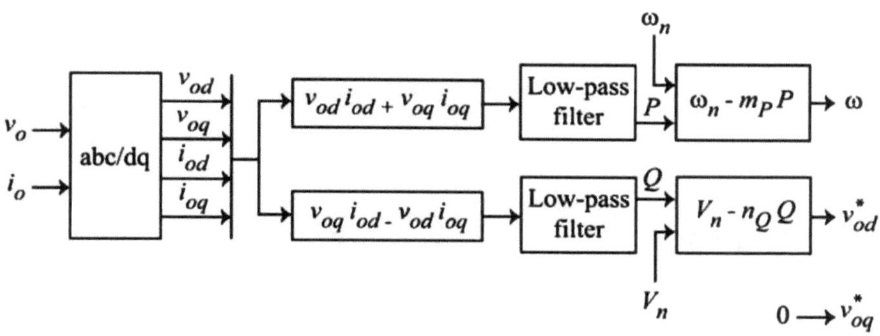

Fig. 5.23 Power controller schematic

Problem 5.3 A multiple DG system is presented in Fig. 5.24 where three DGs are connected at different location of the MG. The four loads that are connected to the MG are shown as $Load_1$, $Load_2$, $Load_3$, and $Load_4$ and the real and reactive power flow for different line sections and load demand (LD) are also shown in Fig. 5.24. Each of the DG controllers needs to measure its local quantities only.

The objective here is to investigate the DG synchronization problem in islanded operation with particular emphasis on method of generating the common angle reference frame. Discuss the DG operation when the generated common signal is lost.

Problem 5.4 Figure 5.25 provides an example of a multilayered MG, in which the lower layer illustrates a typical physical structure of the MG, including buildings, wind generators, solar panel generators, and a wireless access point (AP). These

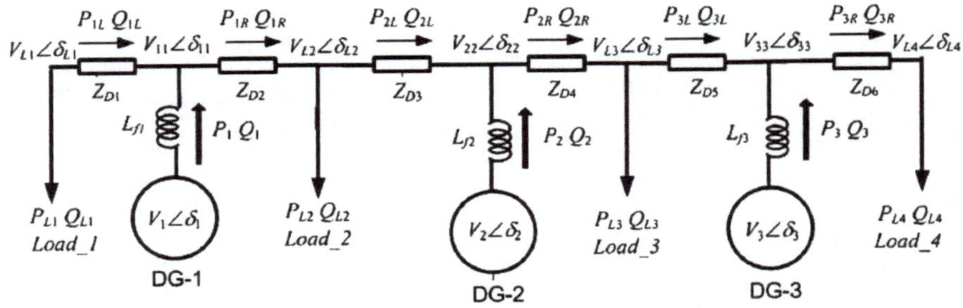

Fig. 5.24 Multiple DG connected to microgrid

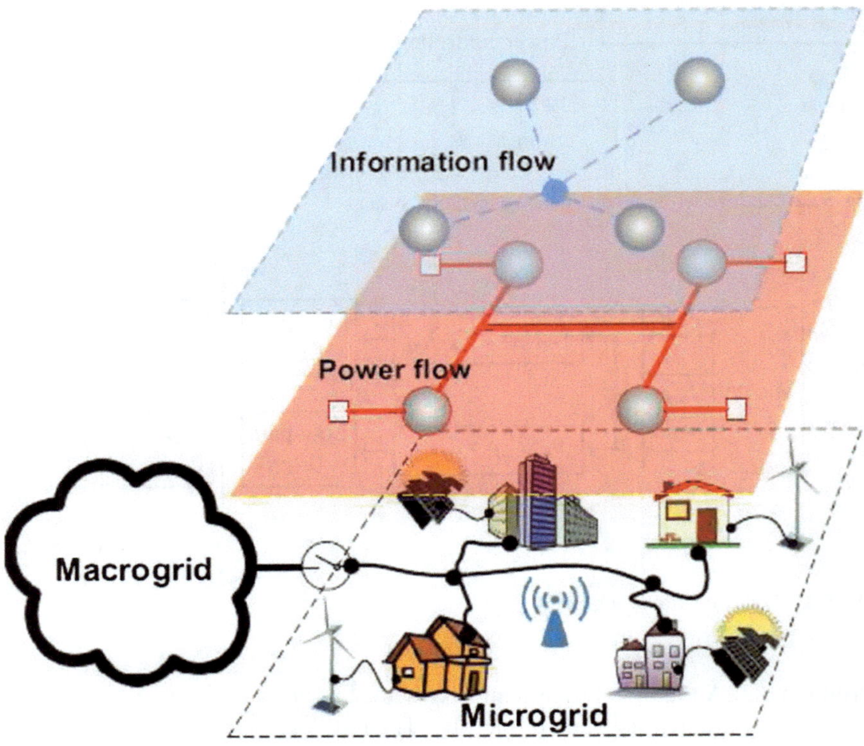

Fig. 5.25 Representative example of multilayered microgrid

buildings and generators exchange power using power-lines. They also exchange information via an AP-based wireless network where the top layer shows the information flow within this MG and the middle layer shows the power flow.

It is of interest to demonstrate possible islanding operation of generations and loads and discuss potential advantage of such intentional action on the power

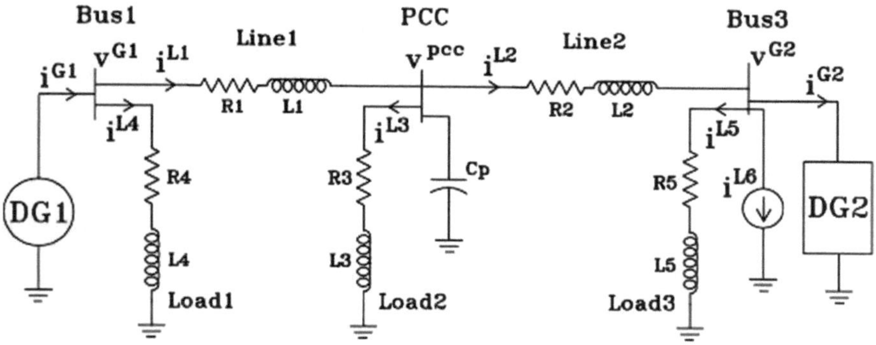

Fig. 5.26 Single-line diagram of a microgrid

system reliability. Discuss effective forecasting and scheduling methods to model such renewable energy sources (RESs) in face of the intermittent and fluctuating nature of wind and solar generation.

Problem 5.5 Figure 5.26 depicts a single-line diagram of typical distribution system during an autonomous MG mode of operation. The task is to investigate the dynamic behavior of the autonomous MG, including the controller of $DG1$ and $DG2$ using state-space analysis.

References

1. Ackermann T (2005) Wind power in power systems. Wiley, Royal Institute of Technology, Stockholm, Sweden
2. Arulampalam A, Barnes M, Engler A, Goodwin A, Jenkins N (2004) Control of power electronic interfaces in distributed generation microgrids. Int J Electron 91:503–523
3. Ashourian MH, Mohd Zin AA, Mokhtar AS, Mirazimi SJ, Muda Z (2011) Controlling and modeling power-electronic interface DERs in islanding mode operation micro grid. In: IEEE symposium on industrial electronics and applications (ISIEA), pp 161–166
4. Babazadeh M, Karimi H (2011) Robust decentralized control for islanded operation of a microgrid. In: IEEE power and energy society general meeting, pp 1–8
5. Barklund E, Pogaku N, Prodanovic M, Hernandez-Aramburo C, Green TC (2008) Energy management in autonomous microgrid using stability-constrained droop control of inverters. IEEE Trans Power Electron 23(5):2346–2352
6. Bhaskara SN, Rasheduzzaman M, Chowdhury BH (2012) Laboratory-based microgrid setup for validating frequency and voltage control in islanded and grid-connected modes. In: IEEE green technologies conference, pp 1–6
7. Bidram A, Davoudi A (2012) Hierarchical structure of microgrids control system. IEEE Trans Smart Grid 3:1963–1976
8. Chen CL, Lai JS, Martin D, Lee YS (2010) State-space modeling, analysis, and implementation of paralleled inverters for microgrid applications. IEEE
9. Das DCH, Roy AK, Sinha N (2011) PSO based frequency controller for wind-solar-diesel hybrid energy generation/energy storage system. In: International conference on energy, automation, and signal (ICEAS), pp 1–6

10. De Brabandere K, Vanthournout K, Driesen J, Deconinck G, Belmans R (2007) Control of microgrids. In: Proceedings of the IEEE power engineering society general meeting, pp 1–7
11. Dimeas AL, Hatziargyriou ND (2005) Operation of a multi-agent system for microgrid control. IEEE Trans Power Syst 20(3)
12. Eltawil MA, Zhao Z (2010) Grid-connected photovoltaic power systems: technical and potential problems—a review. Renew Sustain Energy Rev 14:112–129
13. Elkhatib ME, El-Shatshat R, Salama MMA (2011) Novel coordinated voltage control for smart distributed networks with DG. IEEE Trans Smart Grid 2:598–605
14. Etemadi AH, Iravani R (2012) Eigenvalue and robustness analysis of a decentralized voltage control scheme for an islanded multi-der microgrid. IEEE Trans Power Deliv 27(4)
15. Etemadi AH, Davison EJ, Iravani R (2012) A Decentralized robust control strategy for multi-der microgrids—part I: fundamental concepts. IEEE Trans Power Deliv 27(4)
16. European Research Project Microgrids. http://Microgrids.power.ece.ntua.gr
17. European Research Project More Microgrids. http://Microgrids.power.ece.ntua.gr
18. Fax JA, Murray RM (2004) Information flow and cooperative control of vehicle formations. IEEE Trans Autom Control 49(9):1465–1476
19. Georgakis D, Papathanassiou S (2004) Operation of a prototype microgrid system based on micro-sources equipped with fast acting power electronics interfaces. In: Proceedings of IEEE 35th annual power electronics specialists conference, Germany, vol 4, pp 2521–2526
20. Gorod A, Sauser B, Boardman J (2008) System-of-systems engineering management: a review of modern history and a path forward. IEEE Syst J 2(4):484–499
21. Guerrero JM, Vasquez JC, Matas J, de Vicuna LG, Castilla M (2011) Hierarchical control of droop-controlled AC and DC microgrids: a general approach toward standardization. IEEE Trans Ind Electron 58(1):158–172
22. Haruni AMO, Gargoom A, Haque ME, Negnevitsky M (2010) Dynamic operation and control of a hybrid wind-diesel stand alone power systems. In: The twenty-fifth annual IEEE applied power electronics conference and exposition (APEC), pp 162–169
23. Hatziargyriou N, Kariniotakis G (2004) Modelling of microsources for security studies. In: Proceedings of the 2004 CIGRE SESSION. Paris
24. Hatziargyriou N, Asano H, Iravani R, Marnay C (2007) Microgrids: an overview of ongoing research and development and demonstration projects. Proc IEEE Power Energy Mag 95(7):78–94
25. Henshaw M, Siemieniuch C, Sinclair M, Henson S, Barot V, Jamshidi MO, Delaurentis D, Ncube C, Lim SL, Dogan H (2013) Systems of systems engineering: a research imperative. In: Proceedings of IEEE international conference system science and engineering, Hungary
26. Jamshidi M (2002) System of systems engineering–new challenges for the 21'st century. IEEE Syst Mag
27. Jamshidi M (2008) System of systems engineering: principles and applications. Taylor & Francis CRC Publishers, Boca Raton
28. Jamshidi M (2009) Control of system of systems. In: Proceedings of the 7th IEEE international conference on industrial informatics, pp 1–16
29. Jayalakshmi NS, Gaonkar DN (2011) Performance study of isolated hybrid power system with multiple generation and energy storage units. In: International conference power and energy systems (ICPS), pp 1–5
30. Jenkins N, Allan R, Crossley P, Kirschen D, Strbac G (2000) Embedded generation. The Institution of Electrical Engineers, UK
31. Jones E, Fitzer C, Barnes M (2006) Investigation of microgrids. In: Proceedings of the 3rd IET international conference power electronics, machines and drives, pp 510–514
32. Joordens MA, Jamshidi M (2010) Consensus control for a system of underwater swarm robots. IEEE Syst J 4(1)
33. Karimi H, Nikkhajoei H, Iravani R (2008) Control of an electronically-coupled distributed resource unit subsequent to an islanding event. IEEE Trans Power Deliv 23(1):493–501
34. Kariniotakis G et al (2003) DA1-digital models for microsources. Microgrids Project Deliverable DA1

35. Kariniotakis GN, Soultanis NL, Tsouchnikas AI, Papathanasiou SA, Hatziargyriou ND (2005) Dynamic modeling of microgrids. In: IEEE international conference, future power systems
36. Kasal GK, Singh B (2011) Voltage and frequency controllers for an asynchronous generator-based isolated wind energy conversion system, IEEE Trans Energy Convers 26(2):402–416
37. Katiraei F, Iravani MR (2005) Power management strategies for a microgrid with multiple distributed generation units. IEEE Trans Power Syst
38. Kazerooni ES, Khorasani K (2010) Optimal consensus seeking in a network of multiagent systems: an LMI approach. IEEE Trans Syst, Man, Cybern—Part B: Cybern 40(2):540–547
39. Lasseter RH, Akhil A, Marnay C, Stephens J, Dagle J, Guttronmson R, Meliopoulous AS, Yinger R, Eto J, White paper on integration of distributed energy resources- the CERTS microgrid concept. http://certs.lbl.gov/pdf/50829-app.pdf
40. Lasseter RH (2007) Microgrids and distributed generation. J Energy Eng, Am Soc Civ Eng 2:1–6
41. Lasseter RH, Paigi P (2004) Microgrids: a conceptual solution. In: Proceedings of 35th Annual IEEE power electronics specialists conference, Germany
42. Lee DJ, Wang L (2008) Small-signal stability analysis of an autonomous hybrid renewable energy power generation/energy storage system part I: time-domain simulations. IEEE Trans Energy Convers 23(1):311–320
43. Lin Z, Broucke M, Francis BA (2004) Local control strategies for groups of mobile autonomous agents. IEEE Trans Autom Control 49(4):622–629
44. Lopes JAP, Madureira AG, Moreira CCLM (2012) A view of microgrids WIREs. Energy Environ 2(1):86–103
45. Mahmoud MS (1979) Multilevel systems: information flow in large linear problems. In: Singh MG, Title A (eds) Handbook of large scale systems engineering applications. North Holland, Amsterdam, pp 96–109
46. Mahmoud MS (2010) Decentralized control and filtering in interconnected dynamical systems. CRC Press, New York
47. Mahmoud MS, Selim SZ, Shi P, Baig MH (2012) New results on networked control systems with non-stationary packet dropouts. IET Control Theory Appl 6(15):2442–2452
48. Maier MW (1998) Architecting principles for systems-of-systems. Syst Eng 1(4):267–284
49. Mohamed F (2006) Microgrid modelling and simulation. Helsinki University of Technology, Finland
50. Moradi R, Karimi H, Karimi-Ghartemani M (2010) Robust decentralized control for islanded operation of two radially connected DG systems. In: IEEE international symposium on industrial electronics (ISIE), pp 2272–2277
51. Mukund R (1999) Wind and solar power systems. CRC Press, Florida
52. Nagpal M, Moshref A, Morison GK et al (2001) Experience with testing and modeling of gas turbines. In: Proceedings of the IEEE/PES 2001 winter meeting, Ohio, USA, pp 652–656
53. Olfati-Saber R, Murray RM (2003) Consensus protocols for networks of dynamic agents. In: Proceedings American control conference, pp 951–956
54. Olfati-Saber R, Fax JA, Murray RM (2007) Consensus and cooperation in networked multi-agent systems. Proc IEEE 95(1):215–233
55. Pagani GA, Aiello M (2011) Towards decentralization: a tropological investigation of the medium and low voltage grids. IEEE Trans Smart Grid 2:538–547
56. Peças Lopes JA, Moreira CL, Madureira AG (2005) Defining control strategies for analyzing microgrids islanded operation. In: IEEE Russia power technology, pp 1–7
57. Pedrasa MA, Spooner T (2006) A survey of techniques used to control microgrid generation and storage during island operation. In: Proceedings the Australian universities power engineering conference
58. Pepermans G, Driesen J, Haeseldonckx D, Belmans R, Dhaeseleer W (2005) Distributed generation: definition, benefits and issues. Energy Policy 33(6):787–798
59. Phang JCH, Chan DSH, Phillips JR (1984) Accurate analytical method for the extraction of solar cell model parameters. Electron Lett 20(10):406–408

60. Piagi P, Lasseter RH (2006) Autonomous control of microgrids. In: Proceedings of IEEE power engineering society general meeting (PES)
61. Pogaku N, Prodanovic M, Green TC (2007) Modelling, analysis and testing of autonomous operation of an inverter based microgrid. IEEE Trans Power Electron 22(2):613–625
62. Prodanovic M, Green TC (2006) High-quality power generation through distributed control of a power park microgrid. IEEE Trans Ind Electron 53(5):1471–1482
63. Ray PK, Mohanty SR, Kishor N (2010) Dynamic modeling and control of renewable energy based hybrid system for large band wind speed variation. In: Proceedings of the IEEE PES innovative smart grid technologies conference europe (ISGT Europe), pp 1–10
64. Ren W, Beard RW (2004) Formation feedback control for multiple spacecraft via virtual structures. Proc IEE Control Theory Appl 151(3):357–368
65. Roy S, Malik O, Hope G (1991) An adaptive control scheme for speed control of diesel driven power-plants. IEEE Trans Energy Convers 6(4):605–611
66. Salhi I, Doubabi S, Essounbouli N (2010) Fuzzy control of micro hydro power plants. IET
67. Sao CK, Lehn PW (2006) Intentional islanded operation of converter fed microgrids. In: Proceedings IEEE power engineering society general meeting, pp 6–12
68. Sao C, Lehn PW (2008) Control and power management of converter fed microgrids. IEEE Trans Power Syst 23(3):1088–1098
69. Sao C, Lehn PW (2008) Voltage balancing of converter fed microgrids with single phase loads. In: Proceedings of the IEEE power and energy society general meeting-conversion and delivery of electrical energy in the 21st century, pp 1–7
70. Shabestary SMA, Saeedmanesh M, Rahimi-Kian A, Jalalabadi E (2012) Real-time frequency and voltage control of an islanded mode microgrid. In: The 2nd Iranian conference on smart grids (ICSG), pp 1–6
71. Soto D, Edrington C, Balathandayuthapani S, Ryster S (2012) Voltage balancing of islanded microgrids using a time-domain technique. Electric Power Syst Res 84(1):214–223
72. Stavrakakis G, Kariniotakis G (1995) A general simulation algorithm for the accurate assessment of isolated diesel-wind turbines systems—part I: a general multi-machine power system model. IEEE Trans Energy Convers 10(3):577–583
73. Vandoorn TL, Renders B, Degroote L, Meersman B, Vandevelde L (2011) Active load control in islanded microgrids based on the grid voltage. IEEE Trans Smart Grid 2:139–151
74. Vasquez JC, Guerrero JM, Miret J, Castilla M, Vicuna LGD (2010) Hierarchical control of intelligent microgrids. IEEE Ind Electron Mag 23–29
75. Wai RJ, Wang WH (2008) Grid connected photovoltaic generation system. IEEE Trans Circuits Syst 1: Regul Pap 55: 953–964
76. Watts JH (1999) Microturbines: a new class of gas turbine engines. Gas Turbine News Brief 39(1):5–11
77. Xiao Z, Wu J, Jenkins N (2010) An overview of microgrid control. Intell Autom Soft Comput 16(2):199–212
78. Yang Q, Barria JA, Green TC (2011) Advanced power electronic conversion and control system for universal and flexible power management. IEEE Trans Smart Grid 2:231–243
79. Yubing D, Yulei G, Qingmin L, Hui W (2008) Modelling and simulation of the microsources within a microgrid. In: IEEE international conference electrical machines and systems ICEMS
80. Zhang X, Liu J, Liu T, Zhou L (2010) A novel power distribution strategy for parallel inverters in islanded mode microgrid. In: the twenty-fifth annual IEEE applied power electronics conference and exposition (APEC), pp 2116–2120
81. Zhu Y, Tomsovic K (2002) Development of models for analyzing load-following performance of microturbines and fuel cells. Electric Power Syst Res 62(1):1–11
82. Zhu M, Li H, Li X (2013) Improved state-space model and analysis of islanding inverter-based microgrid. In: IEEE international symposium on industrial electronics (ISIE)

Part III
Communication and Control

Chapter 6
Networked Control of Microgrid System of Systems

6.1 Introduction

Economic challenges, technological advancements, and environmental impacts are now demanding distributed generation (DG) in the place of the conventional centralized generation [18]. Power operation companies are now confronted with unprecedented difficulties in terms of meeting the load requirements, consumer satisfaction, and environmental considerations. Thus, DG has received good attention because of its potential to alleviate pressure from the main transmission system by supplying a few local loads [1]. The waste heat generated from the fuel to electricity conversion is exploited by the distributed generation system (DGS) with the help of microturbines, reciprocating engines, and fuel cells to provide heat and power to the customers. Adding to the system distributed energy sources (DES) like photovoltaic (PV) panels, wind turbines (WTs), energy storage devices such as batteries and capacitors, generators extracting energy from other renewable and controllable loads can provide momentous contributions to future energy generation and distribution. Another noteworthy feature is that the carbon emission is reduced to a large extent satisfying the commitment of many nations concerning decrease of carbon footprints [8]. However, the DG faces technical issues regarding its connection to the intermittent renewable generation and feeble areas of the distribution network. Further, owing to the distinct behavior of the DG unlike the conventional load, alteration in power flow results in problems. To counter the irregular behavior and increasing penetration of the DG, the microgrid (MG) was introduced.

The concept of MG has received considerable attention owing to its potential to serve as an alternate power source, utilizing unconventional sources and supplying the most critical loads of the main grid in case of a network failure. MGs are low voltage networks or distributed energy systems which provide heat and power to a particular area by employing generators and loads. They have the ability to operate independently and isolate themselves from the main grid in case of a fault [4, 5, 26]. If proper control techniques are implemented, they may improve the reliability of electrical energy supply. MG comprises of microturbines, WTs, fuel cells, PV

M.S. Mahmoud and F.M. AL-Sunni, *Control and Optimization of Distributed Generation Systems*, Power Systems, DOI 10.1007/978-3-319-16910-1_6

cells, etc., as sources of energy which are interfaced with the help of power electronic converters. All these units are connected to the main grid through a point of common coupling (PCC) and look as a solitary unit to the distribution network.

No additional inertia is added to the system from the distributed generators. However, because of this the power balance amid generation and load and the network frequency becomes complicated to maintain, especially when the MG is in islanded mode [15]. The MG operates in two modes, namely, the grid connected and the islanded mode.

Grid-connected mode: In the grid-connected mode, the MG is supposed to follow the rules of distribution network without being involved in the operation of main power system. The MG operation based on this approach is significant for the stable operation of power system. During this mode, the MG can draw power from the main grid or can supply its power to the main grid, thus functioning similar to a controllable load or source. By supplying or drawing power, the MG should be able to control the active and reactive power flows and have an eye on the energy storage [7]. However, in this mode due to the small size of distribution units the system dynamics have to be fixed by a wide extent. Another issue is the slow response at the control signals whenever there is a change in output power. Furthermore, due to lack of synchronous machines (SMs) connected to low power grid, virtual inertias have to be incorporated in the control loops of the power electronic interfaces (PEIs) [19].

Islanded mode: The islanded mode is an operating condition in which the MG isolates itself from the main grid in case of a fault. However, the transition from the grid connected and the islanded mode must be stable [6]. If the MG is consuming or supplying power to the main grid before disconnection, a power imbalance occurs. This is compensated by the energy storage units because of the fact that the microsources have low inertia and slow dynamic response [11, 38].

The MG must address the following issues when operating in the islanded mode:

1. Supply and demand balancing.
2. Acceptable power quality.
3. Voltage and frequency balance.
4. Communication among the MG components.

Controlling the islanded MG means balancing the generation and demand power to deliver high performance, while maintaining acceptable ranges of frequency and voltage amplitude. The islanded operation of the MG will be the focus of this paper.

Several control strategies for the MG have been proposed in the literature including PI controllers in [2, 7, 14, 16, 17, 19, 20, 32, 39]. Robust H_∞ control is presented in [3, 28] for the control of two DG units. An optimal controller is presented for controlling the frequency and voltage fluctuations during islanded mode in [33].

It is worth noting that a MG is a complex system comprising of a variety of systems which are nonlinear in nature and possess strong cross-coupling between them. Hence viewing the MG from an intelligent SoS perspective is need of the hour. Moreover an efficient control methodology based on Sos has to be established in order to overcome the challenges posed by the MG. The concept of SoS is now widespread and has entered several domains including defense, IT, health care, manufacturing, energy and space stations and exploration to name a few.

Networked control of system of system (SoS) has been introduced in [38]. A control system consisting of a real-time network in its feedback can be termed as a networked control system (NCS) [38]. The same concept is applied in this paper to the MG operating in islanded mode where three DG units supplying a load are considered as three subsystems. At the primary level, the islanded system is assumed to be equipped with PI controllers. While at the secondary level, the NCS is designed for controlling the interconnected DG units, forming a networked MG system of systems. To the best of the author's knowledge this is the first time, where the concept of SoS, networked control and MG are integrated together.

As mentioned in [38] that the need to design a SoS control system which can tolerate packet loss and delays is one of the prime challenges in SoS networked control, we considered a network which is subjected to bounded random packet losses and the controller stabilizes the system in the presence of packet losses.

6.2 Microgrid as SoS

The concept of SoS has opened up a new school of thought in systems engineering. SoS has emerged as a hot topic for research over the past few years. Although still in the infant stages, the concept of SoS has managed to achieve widespread acclaim. Being restricted to defense and IT, at one point in time, SoS has now entered a plethora of domains [9, 12]. Many definitions have been proposed in the literature for SoS, the more descriptive one being "systems of systems are large-scale integrated systems that are heterogeneous and independently operable on their own, but are networked together for a common goal" [13], where the goal is cost-effectiveness, robustness, and performance. A SoS comprises of numerous heterogeneous subsystems which are independently operable. These subsystems also possess the ability to continue operating even if separated from the SoS. Each constituent subsystem has no power over the other but communicate with each other through a network to effectively carry out tasks and collectively achieve a mission. However, one needs to understand the difference between large-scale complex monolithic systems and actual SoS. There exists certain characteristics or features which are unique to SoS as given in [27]. The SoS is expected to exhibit the following characteristics:

1. **Operational independence**: All the constituent systems within the SoS architecture operate independently and have no interference with other neighbor systems in their functionality.
2. **Managerial independence**: The constituent systems continue to operate on their own unperturbed by the SoS.
3. **Evolutionary development**: The SoS is not designed as a single unit and is rather flexible which can accommodate numerous new systems.

4. **Emergent behavior**: All the constituent systems function as a collective unit to accomplish a common objective, which cannot be achieved by a single component system.
5. **Geographic distribution**: The distribution of the subsystems is sequential to facilitate flow of information among them.

Some key issues on the control of SoS are:

- Typically, in SoS control structure there are sensor feedback of the individual systems in addition to wired or wireless sensor and/or data transmission from each neighboring system sensor to any given system controller/actuator.
- The two general configurations in network control systems are direct structure and hierarchical structure. In the NCS direct structure, see Fig. 6.1, the control network carry the data of the sensors to the loop controllers that will calculate the control signal. These control values are transferred on the control network to the actuators. The network is also used to synchronize the sampling on sensors. As mentioned before, the advantage of such a configuration is the economy in cabling and remote commissioning of sensors and actuators.

The question to be asked here is whether the MG falls under the SoS category and does it exhibit the characteristics of a SoS. For this, we need to consider the subsystems within the MG architecture and observe their operation and interaction among each other.

The MG as described above is a complex system comprising of micro-sources, loads, and energy storage devices. Most of the elements are nonlinear systems and possess strong cross-coupling between them. However, complexity is not the only thing that would be sufficient for the MG to qualify as a SoS, it must comply with other features of SoS presented by [19] as mentioned above.

The elements of the MG, which in other words may be termed as subsystems include microsources which may be PV and wind generators, micro turbines deriving energy from gas or bio-fuels, fuel cells and storage devices constituting batteries or flywheels. All of these subsystems operate independently without interfering with each other. Thus, operational independence is seen in the subsystems of the MG. Also, each subsystem depends on its own source to function and produce energy and is hence responsible for its autonomous operation. For instance, the microturbine runs on biofuel to perform its function of producing energy and has nothing to do with the microsource generators which harness energy from renewable sources such as the wind and solar power. The operation of the storage devices too is distinct and are concerned only with storing energy for use when required. Moreover, the MG supports evolutionary development and is supposed to accommodate new subsystems into its architecture whenever needed unperturbing the complexity of the overall system. A new microsource may be added to the MG which utilizes hydroenergy, for example, or a new storage device may be incorporated for better and efficient storage of energy. On the contrary, a subsystem may be discarded from the system or architecture owing to some reason. All these explains the evolutionary characteristic of the MG. Further, the emergent behavior of the MG is evident from the fact that the

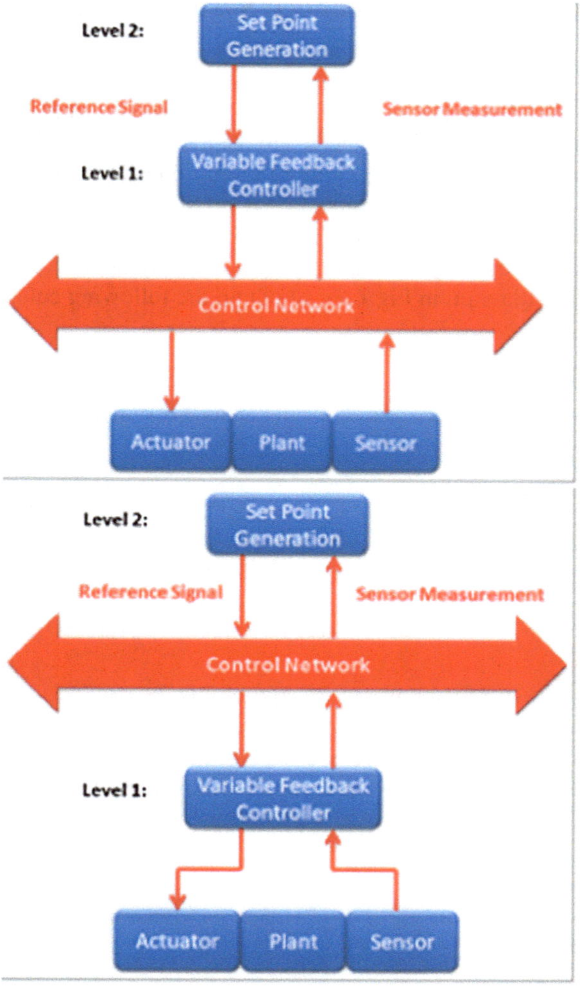

Fig. 6.1 (*Top*) direct structure of a two-layer NCS, (*bottom*) hierarchical structure of a two-layer NCS

subsystems are all operating to collectively satisfy the main objective of providing combined heat and electric power to the local community even in the case of a failure of the main grid. This objective cannot be achieved by a single subsystem of the MG and requires a combined effort of all the constituent subsystems. Finally, the MG is geographically distributed and all of its subsystems are spatially distributed in a particular region. From the above discussion it can be concluded that the MG architecture satisfies the characteristics and falls perfectly on the lines of a SoS.

6.3 Microgrid Islanded System Modeling

The mathematical modeling of the islanded system of MG is presented in this section. The islanded system having n DG units and a local load is shown in Fig. 6.2. All the DGs and load are connected to the main grid via a PCC. A single line diagram of the islanded system consisting of multiple DGS and load is shown in Fig. 6.3.

Each DG unit is modeled by a DC voltage source, a three-phase voltage source converter and a series RL filter. However, we take into consideration three DG units as three subsystems supplying a load. By applying KVL and KCL from the single line diagram in Fig. 6.3, we obtain the following equations:

$$
\begin{aligned}
v_{abc} &= L\frac{di_{L,abc}}{dt} + R_l i_{L,abc} \\
v_{t,abc_1} &= L_{t_1}\frac{di_{t,abc_1}}{dt} + R_{t_1} i_{t,abc_1} + v_{abc} \\
v_{t,abc_2} &= L_{t_2}\frac{di_{t,abc_2}}{dt} + R_{t_2} i_{t,abc_2} + v_{abc} \\
v_{t,abc_3} &= L_{t_3}\frac{di_{t,abc_3}}{dt} + R_{t_3} i_{t,abc_3} + v_{abc} \\
i_{t,abc_1} + i_{t,abc_2} + i_{t,abc_3} &= \frac{1}{R} v_{abc} + i_{L,abc} + C\frac{dv_{abc}}{dt}
\end{aligned}
\tag{6.1}
$$

In (6.1), v_{abc}, v_{t,abc_1}, v_{t,abc_2}, v_{t,abc_3}, $i_{L,abc}$, i_{t,abc_1}, i_{t,abc_2}, i_{t,abc_3} are basically 3×1 vectors which include the phase quantities. Every 3-phase variable x_{abc} in (6.1) can be converted to a $\alpha\beta$ reference frame system under balanced conditions. This conversion is accomplished by applying the following transformation (*abc to* $\alpha\beta$).

$$
x_{\alpha\beta} = x_a e^{j0} + x_b e^{j\frac{2\pi}{3}} + x_c e^{j\frac{4\pi}{3}} \tag{6.2}
$$

where $x_{\alpha\beta} := x_\alpha + jx_\beta$. Hence the dynamic model of the above system in the $\alpha\beta$ frame is

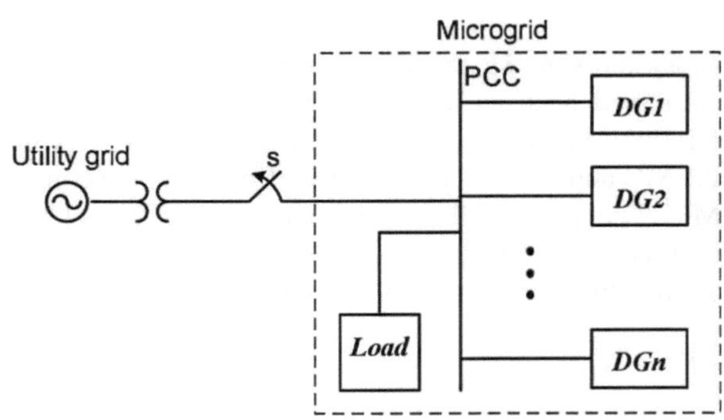

Fig. 6.2 Microgrid with multiple parallel connected DGs

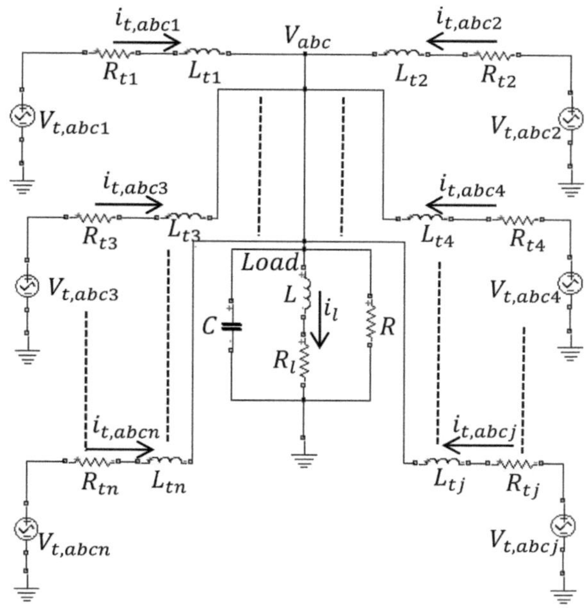

Fig. 6.3 Single line diagram of the islanded system consisting of multiple DGs

$$\begin{aligned}
\frac{dv_{\alpha\beta}}{dt} &= -\frac{1}{RC}v_{\alpha\beta} + \frac{1}{C}i_{t,\alpha\beta_1} - \frac{1}{C}i_{L,\alpha\beta} + \frac{1}{C}i_{t,\alpha\beta_2} + \frac{1}{C}i_{t,\alpha\beta_3} \\
\frac{di_{t,\alpha\beta_1}}{dt} &= -\frac{R_{t_1}}{L_{t_1}}i_{t,\alpha\beta_1} - \frac{1}{L_{t_1}}v_{\alpha\beta} + \frac{1}{L_{t_1}}v_{t,\alpha\beta_1} \\
\frac{di_{L,\alpha\beta}}{dt} &= \frac{1}{L}v_{\alpha\beta} - \frac{R_l}{L}i_{L,\alpha\beta} \\
\frac{di_{t,\alpha\beta_2}}{dt} &= -\frac{R_{t_2}}{L_{t_2}}i_{t,\alpha\beta_2} - \frac{1}{L_{t_2}}v_{\alpha\beta} + \frac{1}{L_{t_2}}v_{t,\alpha\beta_2} \\
\frac{di_{t,\alpha\beta_3}}{dt} &= -\frac{R_{t_3}}{L_{t_3}}i_{t,\alpha\beta_3} - \frac{1}{L_{t_3}}v_{\alpha\beta} + \frac{1}{L_{t_3}}v_{t,\alpha\beta_3}
\end{aligned} \tag{6.3}$$

Now (6.3) is transferred to a rotating frame which has its basis on

$$x_{\alpha\beta} = x_{dq}e^{j\theta} = (x_d + jx_q)e^{j\theta} \tag{6.4}$$

where $\theta(t) := \int_0^t \omega(\zeta)d\zeta + \theta_0$. $\theta(t)$ represents the phase angle of a reference vector $(x_\alpha^{ref} + jx_\beta^{ref})$ in the $\alpha\beta$ frame. The equations for the dq-frame can be written as [3]:

$$\begin{aligned}
\frac{dV_{dq}}{dt} + j\omega_0 V_{dq} &= -\frac{1}{RC}V_{dq} + \frac{1}{C}I_{t,dq_1} - \frac{1}{C}I_{L,dq} + \frac{1}{C}I_{t,dq_2} + \frac{1}{C}I_{t,dq_3}\\
\frac{dI_{t,dq_1}}{dt} + j\omega_0 I_{t,dq_1} &= -\frac{1}{L_{t_1}}V_{dq} - \frac{R_{t_1}}{L_{t_1}}I_{t,dq_1} + \frac{1}{L_{t_1}}V_{t,dq_1}\\
\frac{dI_{L,dq}}{dt} + j\omega_0 I_{L.dq} &= \frac{1}{L}V_{dq} - \frac{R_l}{L}I_{L,dq}\\
\frac{dI_{t,dq_2}}{dt} + j\omega_0 I_{t,dq_2} &= -\frac{1}{L_{t_2}}V_{dq} - \frac{R_{t_2}}{L_{t_2}}I_{t,dq_2} + \frac{1}{L_{t_2}}V_{t,dq_2}\\
\frac{dI_{t,dq_3}}{dt} + j\omega_0 I_{t,dq_3} &= -\frac{1}{L_{t_3}}V_{dq} - \frac{R_{t_3}}{L_{t_3}}I_{t,dq_3} + \frac{1}{L_{t_3}}V_{t,dq_3}
\end{aligned}$$

6.4 Networked Control System

After formulating the above equations, considering the states as $x_p = [V_d, V_q, I_{td_1}, I_{tq_1}, I_{Ld}, I_{Lq}, I_{td_2}, I_{tq_2}, I_{td_3}, I_{tq_3}]$, the control vector as $u_p = [V_{td_1}, V_{tq_1}, V_{td_2}, V_{tq_2}, V_{td_3}, V_{tq_3}]$ and output vector as $y_p = [V_d, V_q, I_{td_2}, I_{tq_2}, I_{td_3}, I_{tq_3}]$, we can obtain the state-space model of the islanded system in the standard form

$$\dot{x}_p(t) = A_p x_p(t) + B_p u_p(t), \quad y_p(t) = C_p x(t) \tag{6.5}$$

The matrices A_p, B_p and C_p are shown below.

$$A_p = \begin{bmatrix}
-\frac{1}{RC} & \omega_0 & \frac{1}{C} & 0 & -\frac{1}{C} & 0 & \frac{1}{C} & 0 & \frac{1}{C} & 0\\
-\omega_0 & -\frac{1}{RC} & 0 & \frac{1}{C} & 0 & -\frac{1}{C} & 0 & \frac{1}{C} & 0 & \frac{1}{C}\\
-\frac{1}{L_{t1}} & 0 & -\frac{R_{t1}}{L_{t1}} & \omega_0 & 0 & 0 & 0 & 0 & 0 & 0\\
0 & -\frac{1}{L_{t1}} & -\omega_0 & -\frac{R_{t1}}{L_{t1}} & 0 & 0 & 0 & 0 & 0 & 0\\
\frac{1}{L} & 0 & 0 & 0 & -\frac{R_l}{L} & \omega_0 & 0 & 0 & 0 & 0\\
0 & \frac{1}{L} & 0 & 0 & -\omega_0 & -\frac{R_l}{L} & 0 & 0 & 0 & 0\\
-\frac{1}{L_{t2}} & 0 & 0 & 0 & 0 & 0 & -\frac{R_{t2}}{L_{t2}} & \omega_0 & 0 & 0\\
0 & -\frac{1}{L_{t2}} & 0 & 0 & 0 & 0 & -\omega_0 & -\frac{R_{t2}}{L_{t2}} & 0 & 0\\
-\frac{1}{L_{t_3}} & 0 & 0 & 0 & 0 & 0 & 0 & 0 & -\frac{R_{t_3}}{L_{t_3}} & \omega_0\\
0 & -\frac{1}{L_{t_3}} & 0 & 0 & 0 & 0 & 0 & 0 & -\omega_0 & -\frac{R_{t_3}}{L_{t_3}}
\end{bmatrix}$$

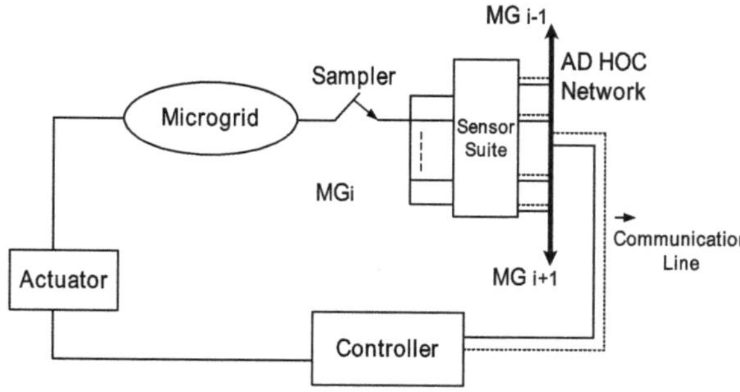

Fig. 6.4 Microgrid SoS structure

$$B_p = \begin{bmatrix} 0 & 0 & 0 & 0 & 0 & 0 \\ 0 & 0 & 0 & 0 & 0 & 0 \\ \frac{1}{L_{t1}} & 0 & 0 & 0 & 0 & 0 \\ 0 & \frac{1}{L_{t1}} & 0 & 0 & 0 & 0 \\ 0 & 0 & 0 & 0 & 0 & 0 \\ 0 & 0 & 0 & 0 & 0 & 0 \\ 0 & 0 & \frac{1}{L_{t2}} & 0 & 0 & 0 \\ 0 & 0 & 0 & \frac{1}{L_{t2}} & 0 & 0 \\ 0 & 0 & 0 & 0 & \frac{1}{L_{t3}} & 0 \\ 0 & 0 & 0 & 0 & 0 & \frac{1}{L_{t3}} \end{bmatrix}; C_p = \begin{bmatrix} 1\,0\,0\,0\,0\,0 \\ 0\,1\,0\,0\,0\,0 \\ 0\,0\,0\,0\,0\,0 \\ 0\,0\,0\,0\,0\,0 \\ 0\,0\,0\,0\,0\,0 \\ 0\,0\,0\,0\,0\,0 \\ 0\,0\,1\,0\,0\,0 \\ 0\,0\,0\,1\,0\,0 \\ 0\,0\,0\,0\,1\,0 \\ 0\,0\,0\,0\,0\,1 \end{bmatrix}^T$$

For all practical purposes, we seek digital control and henceforth using appropriate sampling period $T_s = 0.05$ s we convert model (6.5) into the following discrete-time linear time-invariant model as:

$$x_p(k+1) = Ax_p + Bu_p, \quad y_p = Cx_p \tag{6.6}$$

We now consider the operation of SoS-NCS system under the effect of random communication delays, where the sensor is clock driven and the controller and the actuator are event driven. See Fig. 6.4 where the solid line for wired connections and dotted line for wireless connections. For a more general case, we assume that the measurement after passing through the network exhibits a randomly varying communication delay and is described by

$$y_c(k) = y_p(k - \delta(k)\tau_k^m), \tag{6.7}$$

where τ_k^m is the *communication delay* and $\delta(k)$ is a stochastic binary variable with *Prob*$\{\delta(k) = 1\} = p_k$ where p_k assumes discrete values. When the full state information is not available and the time delay occurs on the actuation side, it is desirable to design the following observer-based controller [25]:

$$\hat{x}(k+1) = A\hat{x} + Bu_p(k) + L(y_c(k) - \hat{y}_c(k))$$
$$\hat{y}_c(k) = C\hat{x}(k - \delta(k)\tau_k^m), \tag{6.8}$$
$$u_c(k) = K\hat{x}(k), \quad u_p = K\hat{x}(k - \alpha(k)\tau_k^a) \tag{6.9}$$

where $\hat{x}(k) \in \Re^n$ is the estimate of the system (6.5), $\hat{y}_c(k) \in \Re^p$ is the observer output, and $L \in \Re^{n\times p}$ and $K \in \Re^{m\times n}$ are the observer and controller gains, respectively, and τ_k^a is the *actuation delay*. The stochastic binary variable $\alpha(k)$, mutually independent of $\delta(k)$, is also a Bernoulli distributed white sequence with $Prob\{\alpha(k) = 1\} = s_k$ where s_k assumes discrete values.

In this section, to reflect the actual operation of MGs, we assume that τ_k^a and τ_k^m are time-varying and have the following bounded condition:

$$\tau_m^- \le \tau_k^m \le \tau_m^+, \quad \tau_a^- \le \tau_k^a \le \tau_a^+ \tag{6.10}$$

Define the estimation error by $e(k) = x_p(k) - \hat{x}(k)$. Then, it yields

$$x_p(k+1) = Ax_p(k) + BKx_p(k - \alpha(k)\tau_k^a) - BKe(k - \alpha(k)\tau_k^a), \tag{6.11}$$
$$e(k+1) = Ae(k) - LCe(k - \delta(k)\tau_k^m) \tag{6.12}$$

In terms of $\xi(k) = [x_p^T(k) \quad e^T(k)]^T$, systems (6.11) and (6.12) can be cast into the form:

$$\xi(k+1) = \mathsf{A}_j\xi(k) + \mathsf{B}_j\xi(k - \tau_k^m) + \mathsf{C}_j\xi(k - \tau_k^a) \tag{6.13}$$

where $\{\mathsf{A}_j, \mathsf{B}_j, \mathsf{C}_j, j = 1, \ldots, 4\}$ and j is an index identifying one of the following pairs $\{(\delta(k) = 1, \alpha(k) = 1), (\delta(k) = 1, \alpha(k) = 0), (\delta(k) = 0, \alpha(k) = 0), (\delta(k) = 0, \alpha(k) = 1)\}$:

$$\mathsf{A}_1 = \begin{bmatrix} A & 0 \\ 0 & A \end{bmatrix}, \ \mathsf{A}_2 = \begin{bmatrix} A + BK & -BK \\ 0 & A \end{bmatrix},$$
$$\mathsf{A}_3 = \begin{bmatrix} A + BK & -BK \\ 0 & A - LC \end{bmatrix}, \ \mathsf{A}_4 = \begin{bmatrix} A & 0 \\ 0 & A - LC \end{bmatrix},$$
$$\mathsf{B}_1 = \begin{bmatrix} BK & -BK \\ 0 & 0 \end{bmatrix}, \ \mathsf{B}_2 = \begin{bmatrix} 0 & 0 \\ 0 & 0 \end{bmatrix},$$
$$\mathsf{B}_3 = \begin{bmatrix} 0 & 0 \\ 0 & 0 \end{bmatrix}, \ \mathsf{B}_4 = \begin{bmatrix} BK & -BK \\ 0 & 0 \end{bmatrix},$$
$$\mathsf{C}_1 = \begin{bmatrix} 0 & 0 \\ 0 & -LC \end{bmatrix}, \ \mathsf{C}_2 = \begin{bmatrix} 0 & 0 \\ 0 & -LC \end{bmatrix},$$
$$\mathsf{C}_3 = \begin{bmatrix} 0 & 0 \\ 0 & 0 \end{bmatrix}, \ \mathsf{C}_4 = \begin{bmatrix} 0 & 0 \\ 0 & 0 \end{bmatrix} \tag{6.14}$$

Remark 6.1 Following [25], we note from (6.14) that

$$\mathsf{A}_j + \mathsf{B}_j + \mathsf{C}_j = \begin{bmatrix} A + BK & -BK \\ 0 & A - LC \end{bmatrix}, j = 1, \ldots, 4 \tag{6.15}$$

The interpretation of this significant result is that $\mathsf{A}_j + \mathsf{B}_j + \mathsf{C}_j$ represents the fundamental matrix of the delayed system (6.13), which must be independent of the mode of operation. This fact will help in simplifying the control design algorithm.

The aim of the remaining part of the paper is to *design an observer-based feedback stabilizing controller in the form of* (6.8) *and* (6.9) *such that the closed-loop system* (6.13) *is exponentially stable in the mean square*. Our approach is based on the concepts of switched time-delay systems (TDSs) [24]. For simplicity in exposition, we introduce

$$\begin{aligned}
\sigma_1(k) &= Prob\{\delta(k) = 1,\ \alpha(k) = 1\},\ \ \hat{\sigma}_1 = \mathrm{E}[\sigma_1] \\
\sigma_2(k) &= Prob\{\delta(k) = 1,\ \alpha(k) = 0\},\ \ \hat{\sigma}_2 = \mathrm{E}[\sigma_2] \\
\sigma_3(k) &= Prob\{\delta(k) = 0, \alpha(k) = 0\},\ \ \hat{\sigma}_3 = \mathrm{E}[\sigma_3] \\
\sigma_4(k) &= Prob\{\delta(k) = 0,\ \alpha(k) = 1\},\ \ \hat{\sigma}_4 = \mathrm{E}[\sigma_4]
\end{aligned} \tag{6.16}$$

where $\mathrm{E}[\sigma_i]$ is the expected value of $\sigma_i,\ i = 1, \ldots, 4$.

6.5 Closed-Loop Stability Results

In this section, we will thoroughly investigate the stability analysis problem for the closed-loop system (6.13). First, let us deal with the stability analysis problem and derive a sufficient condition under which the closed-loop system (6.13) with the given controllers (6.8) and (6.9) is exponentially stable in the mean square. Extending on [23], the following Lyapunov function candidate is constructed to establish the main theorem:

$$V(\xi(k)) = \sum_{i=1}^{5} V_i(\xi(k)) \tag{6.17}$$

$$V_1(\xi(k)) = \sum_{j=1}^{4} \xi^T(k) P \xi(k),\ \ P > 0,\ \ V_2(\xi(k))$$

$$= \sum_{j=1}^{4} \sum_{i=k-\tau_k^m}^{k-1} \xi^T(i) Q_j \xi(i),\ \ Q_j = Q_j^T > 0$$

$$V_3(\xi(k)) = \sum_{j=1}^{4} \sum_{i=k-\tau_k^a}^{k-1} \xi^T(i) Q_j \xi(i), \quad V_4(\xi(k))$$

$$= \sum_{j=1}^{4} \sum_{\ell=-\tau_m^+ +2}^{-\tau_m^- +1} \sum_{i=k+\ell-1}^{k-1} \xi^T(i) Q_j \xi(i)$$

$$V_5(\xi(k)) = \sum_{j=1}^{4} \sum_{\ell=-\tau_a^+ +2}^{-\tau_a^- +1} \sum_{i=k+\ell-1}^{k-1} \xi^T(i) Q_j \xi(i) \tag{6.18}$$

It is not difficult to show that there exist real scalars $\mu > 0$ and $\upsilon > 0$ such that

$$\mu \|\xi\|^2 \leq V(\xi(k)) \leq \upsilon \|\xi(k)\|^2 \tag{6.19}$$

We now present the main results for system (6.13) to be exponentially stable in two theorems: Theorem 6.1concerns the conditions under which the closed-loop stability is guaranteed given that the controller gains are known and Theorem 6.2 provides a way to determine these gains.

Theorem 6.1 *Let the controller and observer gain matrices K and L be given. The closed-loop system (6.13) is exponentially stable if there exist matrices* $0 < P, \; 0 < Q_j^T = Q_j, \; j = 1, \ldots, 4$ *and matrices* $R_i, \; S_i,$ *and* $M_i, \; i = 1, 2,$ *such that the following matrix inequality holds*

$$\Lambda_j = \begin{bmatrix} \Lambda_{1j} & \Lambda_{2j} \\ \bullet & \Lambda_{3j} \end{bmatrix} < 0 \tag{6.20}$$

$$\Lambda_{1j} = \begin{bmatrix} \Psi_j + \Phi_{j1} & -R_1 + S_1^T & -R_2 + S_2^T \\ \bullet & -S_1 - S_1^T - \hat{\sigma}_j Q_j & 0 \\ \bullet & \bullet & -S_2 - S_2^T - \hat{\sigma}_j Q_j \end{bmatrix}$$

$$\Lambda_{2j} = \begin{bmatrix} -R_1 + M_1^T - \Phi_{j2} & -R_2 + M_2^T - \Phi_{j3} \\ -S_1 - M_1^T & 0 \\ 0 & -S_2 - M_2^T \end{bmatrix}$$

$$\Lambda_{3j} = \begin{bmatrix} -M_1 - M_1^T + \Phi_{j4} & \Phi_{j5} \\ \bullet & -M_2 - M_2^T + \Phi_{j6} \end{bmatrix} \tag{6.21}$$

where

$$\begin{aligned}
\Psi_j &= -P + \hat{\sigma}_j(\tau_m^+ - \tau_m^- + \tau_a^+ - \tau_a^- + 2) Q_j + R_1 + R_1^T + R_2 + R_2^T \\
\Phi_{j1} &= (\mathsf{A}_j + \mathsf{B}_j + \mathsf{C}_j)^T \hat{\sigma}_j P (\mathsf{A}_j + \mathsf{B}_j + \mathsf{C}_j) \\
\Phi_{j2} &= (\mathsf{A}_j + \mathsf{B}_j + \mathsf{C}_j)^T \hat{\sigma}_j P \mathsf{B}_j \\
\Phi_{j3} &= (\mathsf{A}_j + \mathsf{B}_j + \mathsf{C}_j)^T \hat{\sigma}_j P \mathsf{C}_j, \; \Phi_{j5} = \mathsf{B}_j^T P \mathsf{C}_j \\
\Phi_{j4} &= \mathsf{B}_j^T \hat{\sigma}_j P \mathsf{B}_j, \; \Phi_{j6} = \mathsf{C}_j^T \hat{\sigma}_j P \mathsf{C}_j
\end{aligned}$$

Proof Follow parallel lines to [25] and therefore it is omitted.

Next, a solution to the problem of the observer-based stabilizing controller design is provided by the following theorem:

Theorem 6.2 *Let the delay bounds τ_m^+, τ_m^-, τ_a^+, τ_a^- be given. Evaluate the quantities $\hat{\sigma}_j$, $j = 1, \ldots, 4$. Then the closed-loop system (6.105) is exponentially stable if there exist matrices $0 < X_1$, X_2, Y_1, Y_2, Z_1, $0 < \Xi_j$, $j = 1, \ldots, 4$ and matrices Π_i, Υ_i and Γ_i, $i = 1, 2$, such that the following matrix inequality holds for $j = 1, \ldots, 4$:*

$$\begin{bmatrix} \widehat{\Lambda}_{1j} & \widehat{\Lambda}_{2j} & \widehat{\Omega}_j \\ \bullet & \Lambda_{3j} & 0 \\ \bullet & \bullet & -\hat{\sigma}_j \widehat{X} \end{bmatrix} < 0 \tag{6.22}$$

$$\widehat{X} = \begin{bmatrix} X_1 & X_2 \\ X_2^T & X_2 \end{bmatrix} \tag{6.23}$$

$$\begin{aligned}
\widehat{\Psi}_j &= -\hat{X} + \hat{\sigma}_j(\tau_m^+ - \tau_m^- + \tau_a^+ - \tau_a^- + 2)\Xi_j + \Pi_1 + \Pi_1^T + \Pi_2 + \Pi_2^T \\
\widehat{\Lambda}_{1j} &= \begin{bmatrix} \widehat{\Psi}_j & -\Pi_1 + \Upsilon_1^T & -\Pi_2 + \Upsilon_2^T \\ \bullet & -\Upsilon_1 - \Upsilon_1^T - \hat{\sigma}_j \Xi_j & 0 \\ \bullet & \bullet & -\Upsilon_2 - \Upsilon_2^T - \hat{\sigma}_j \Xi_j \end{bmatrix} \\
\widehat{\Lambda}_{2j} &= \begin{bmatrix} -\Pi_1 + \Gamma_1^T & -\Pi_2 + \Gamma_2^T \\ -\Upsilon_1 - \Gamma_1^T & 0 \\ 0 & -\Upsilon_2 - \Gamma_2^T \end{bmatrix}, \quad \widehat{\Lambda}_{3j} = \begin{bmatrix} -\Gamma_1 - \Gamma_1^T & 0 \\ \bullet & -\Gamma_2 - \Gamma_2^T \end{bmatrix} \\
\widehat{\Omega}_j &= \begin{bmatrix} \widehat{\Omega}_{1j} & 0 & 0 & -\widehat{\Omega}_{4j} & -\widehat{\Omega}_{5j} \end{bmatrix} \\
\widehat{\Omega}_{1j} &= \begin{bmatrix} X_1 A^T + Y_1^T B^T - Y_1^T B^T Z_1 & X_2 A^T - Y_2^T \\ X_2 A^T & X_2 A^T - Y_2^T \end{bmatrix}, \quad \forall\, j \\
\widehat{\Omega}_{4j} &= \begin{bmatrix} Y_1^T B^T - Y_1^T B^T Z_1 & 0 \\ 0 & 0 \end{bmatrix}, \quad j = 1, 4 \\
\widehat{\Omega}_{5j} &= \begin{bmatrix} 0 & -Y_2^T \\ 0 & -Y_2^T \end{bmatrix}, \quad j = 1, 2, \quad \widehat{\Omega}_{4j} = 0, \quad j = 2, 3, \quad \widehat{\Omega}_{5j} = 0, \quad j = 3, 4
\end{aligned} \tag{6.24}$$

where the gain matrices are given by

$$K = Y_1 X_1^{-1}, \quad L = Y_2 X_2^{-1} C^{\dagger}$$

Proof Define

$$\Omega_j = \left[(\mathsf{A}_j + \mathsf{B}_j + \mathsf{C}_j) \; 0 \; -\mathsf{B}_j \; -\mathsf{C}_j \right]^T$$

then matrix inequality (6.115) can be expressed as

$$\Lambda_j = \widetilde{\Lambda} + \Omega_j P \Omega_j^T < 0 \tag{6.25}$$

$$\widetilde{\Lambda}_j = \begin{bmatrix} \widetilde{\Lambda}_{1j} & \widetilde{\Lambda}_{2j} \\ \bullet & \widetilde{\Lambda}_{3j} \end{bmatrix} < 0$$

$$\widetilde{\Lambda}_{1j} = \begin{bmatrix} \Psi_j & -R_1 + S_1^T & -R_2 + S_2^T \\ \bullet & -S_1 - S_1^T - Q_j & 0 \\ \bullet & \bullet & -S_2 - S_2^T - Q_j \end{bmatrix}$$

$$\widetilde{\Lambda}_{2j} = \begin{bmatrix} -R_1 + M_1^T & -R_2 + M_2^T \\ -S_1 - M_1^T & 0 \\ 0 & -S_2 - M_2^T \end{bmatrix}$$

$$\widetilde{\Lambda}_{3j} = \begin{bmatrix} -M_1 - M_1^T & 0 \\ \bullet & -M_2 - M_2^T \end{bmatrix} \tag{6.26}$$

Setting $\widehat{X} = P^{-1}$, invoking Schur complements, we write matrix Λ_j in (6.137) equivalently as

$$\begin{bmatrix} \widetilde{\Lambda}_{1j} & \widetilde{\Lambda}_{2j} & \Omega_j \\ \bullet & \widetilde{\Lambda}_{3j} & 0 \\ \bullet & \bullet & -\widehat{X} \end{bmatrix} < 0 \tag{6.27}$$

Applying the congruence transformation

$$T_j = diag[\widehat{X},\ \widehat{X},\ \widehat{X},\ \widehat{X},\ \widehat{X},\ I]$$

to matrix inequality in (6.139) and manipulating using (6.135) and

$$\begin{aligned} &\Xi_j = \widehat{X} Q_j \widehat{X},\ \Pi_j = \widehat{X} R_j \widehat{X},\ \Upsilon_j = \widehat{X} S_j \widehat{X}, \\ &\Gamma_j = \widehat{X} M_j \widehat{X},\ Z_1 = X_1^{-1} X_2. \end{aligned}$$

we readily obtain matrix inequality (6.134) subject (6.136).

Some remarks stand out:

Remark 6.2 In our work, the $\widehat{X}$ matrix considered in Theorem 6.3 is distinct from the one in [25] from several aspects. First, the $\widehat{X}$ matrix incorporates two different elements X_1 and X_2 unlike the $\widehat{X}$ matrix in [25] which contains the same element X. This extension results in a distinct LMI term $\widehat{\Omega_j}$ given in (6.24). Second, the expression for calculating the controller and observer gains (K and L) is also different from [25] This is done to present a more realistic approach in determining the controller and observer gains. Further, the computational results of our work would be less sensitive than that of [25] since the feedback gains are independent in our case but they are somehow coupled in [25].

Remark 6.3 The selection of $\widehat{X}$ as given by (6.135) has the advantage of converting the solution of bilinear matrix inequalities to that of seeking the feasibility of linear matrix inequalities and hence avoiding iterative procedures. It should be noted that the LMI (6.134) depends of the particular average dropout patterns under consideration, which is quite useful in illustrating different operating conditions of the communications network.

Remark 6.4 It is noted that the implementation of Theorem 6.2 is on-line in nature as it requires calling random generators to pick up numbers corresponding to the scalars $\hat{\sigma}_1, \ldots, \hat{\sigma}_4$ and to subsequently evaluate the probabilities to compute the state and error trajectories. This represents a salient feature not shared by other methods for networked control design under unreliable communication links.

Remark 6.5 The NCS model considered in this section is distinct from the one presented in [25]. The model represents a typical SoS networked control model consisting of a sensor suite and an ad hoc network for multiple subsystems. Moreover the illustrated example for the proposed control strategy is exclusively a microgrid system of systems.

6.6 Illustrative Example

The example considered here is a 3-subsystem SoS model of the MG with three DG units and a load. The respective state-space matrices are obtained by substituting the parameter values given in Table 6.1. The system is discretized at a sampling time $T = 0.1$ s and the simulations are carried out in MATLAB® and Simulink®. The controller and observer gains can be obtained by using the relation $K = Y_1 X_1^{-1}$ and $L = Y_2 X_2^{-1} C^{\dagger}$ respectively. Where X_1, X_2, Y_1, Y_2 can be found by solving the LMI (6.24). The values of $\bar{\alpha}$ and $\bar{\delta}$ are set at 0.1 each. The measurement delay is varied from $\tau_m^- = 1$ to $\tau_m^+ = 5$ and the actuation delay is varied from $\tau_a^- = 3$ to $\tau_a^+ = 9$. The matrices K and L after solving the LMI (6.24) are

$$K = \begin{bmatrix} K_1 & K_2 \end{bmatrix}$$

$$K_1 = \begin{bmatrix} -0.0119 & 0.0002 & -0.0025 & -0.0002 & 0.0000 \\ 0.0010 & -0.0000 & 0.0001 & 0.0000 & 0.0000 \\ 0.0075 & -0.0001 & 0.0016 & 0.0002 & -0.0000 \\ -0.0042 & 0.0000 & -0.0308 & -0.0002 & -0.0221 \\ -0.2286 & 0.0074 & 0.0514 & 0.0017 & 0.2692 \\ 0.2442 & -0.0064 & -0.0055 & 0.0697 & -0.0026 \end{bmatrix}$$

Table 6.1 Parameters of the microgrid system

Quantity	Value
R_{t1} (series filter resistance in DG1)	1.5 mΩ
L_{t1} (series filter inductance in DG1)	300 μH
R_{t2} (series filter resistance in DG2)	6 mΩ
L_{t2} (series filter inductance in DG2)	900 μH
R_{t3} (series filter resistance in DG3)	9 mΩ
L_{t3} (series filter inductance in DG3)	1200 μH
VSC rated power	$S_{base} = 2.5$ MVA
VSC terminal voltage (line-line)	$V_{base} = 600$ V
f_{sw} (PWM carrier frequency)	1980 Hz
R (Load nominal resistance)	76 Ω
L (Load nominal inductance)	111.9 mH
C (Load nominal capacitance)	62.86 μF
$q = \frac{L\omega_0}{R_L}$ (Inductor quality factor)	120
f_0 (System frequency)	60 Hz
V_{dc} (DC bus voltage)	1500 V
Transformer voltage ratio (Y/Δ)	0.6/13.8 kV

Table 6.2 Parameters of the microgrid system

Parameter	Value (p.u)	Parameter	Value (p.u)	Parameter	Value (p.u)
H	3	P_{e1}	0.02	L_{dc}	0.05
M	2.5	P_{e2}	0.1	C_{dcp}	1.1
ω_0	377	P_{e3}	0.15	C_{pf}	0.2
K_A	15	e_{q1}	0.04	R_{eq}	0.1
T_A	0.03	e_{q2}	0.01	R_{eq1}	0.2
x_d	1.3	m_p	100	R_d	0.1
x_q	0.47	$k_p = k_{pf}$	1885	i_{pfq0}	0.67
x_d'	0.3	k_{pd1}	0.08	i_{pfd0}	0.13
T_{do}'	7	k_{pd2}	0.9	r_t	0.1
i_{td1}	0.28	k_{pv}	0.01	r_b	0.15
i_{td2}	0.08	d_c	0.6	x_t	0.2
x_b	0.265				

$$K_2 = \begin{bmatrix} 0.0037 & -0.0002 & 0.0035 & -0.0002 & -0.0010 \\ -0.0001 & 0.0002 & -0.0000 & 0.0011 & -0.0000 \\ -0.0023 & 0.0001 & -0.0022 & 0.0001 & 0.0006 \\ -0.0029 & 0.0088 & 0.0020 & 0.0020 & -0.0154 \\ 0.3100 & -0.0279 & 0.0128 & -0.0097 & -0.0021 \\ 0.3061 & 0.0109 & 0.0021 & 0.0036 & 0.0040 \end{bmatrix}$$

$$L = \begin{bmatrix} -0.0166 & 0.0004 & -0.0010 & -0.0290 & -0.0034 & -0.0001 \\ -0.0301 & -0.0024 & -0.0025 & 0.0022 & 0.0006 & -0.0018 \\ 0.0351 & -0.0012 & 0.0003 & 0.0003 & -0.0008 & -0.0001 \\ 0.0015 & 0.0012 & 0.0012 & 0.0046 & 0.0005 & -0.0010 \\ 0.0071 & 0.0000 & -0.0001 & 0.0002 & -0.0003 & -0.0000 \\ -0.0829 & 0.0014 & 0.0182 & -0.3126 & 0.0903 & -0.0005 \\ 0.0723 & -0.0005 & 0.0042 & 0.0055 & 0.0059 & -0.0005 \\ 0.1060 & 0.0006 & 0.0025 & -0.0901 & 0.0101 & 0.0017 \\ -0.1834 & 0.0007 & -0.0167 & 2.2114 & -0.0582 & -0.0051 \\ 0.0103 & -0.0017 & 0.0032 & -0.0309 & 0.0097 & -0.0047 \end{bmatrix}$$

The plots obtained above demonstrate the effectiveness of the networked control strategy. The controller stabilizes the system in the presence of delays and nonstationary packet dropouts. The d-q components of currents of all the DG units are controlled in addition to the load currents and load voltages of the islanded system (Figs. 6.5, 6.6, 6.7, 6.8, 6.9 and 6.10).

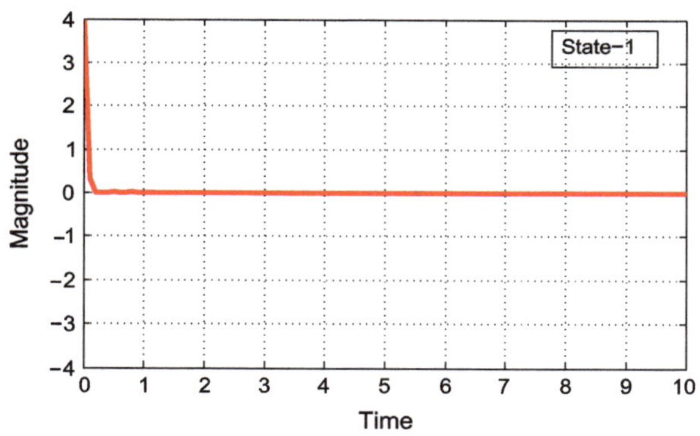

Fig. 6.5 d component of load voltage

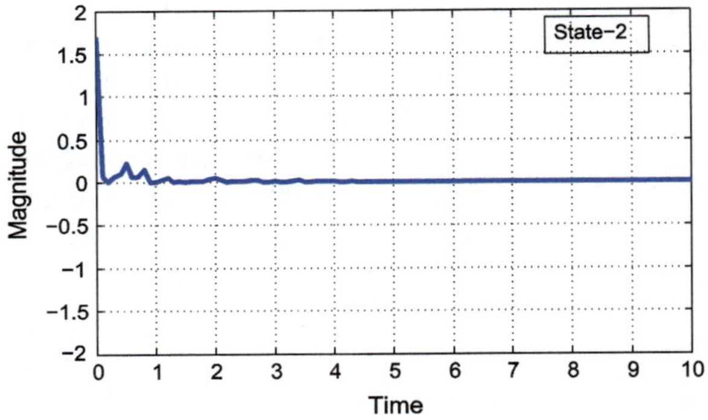

Fig. 6.6 q component of load voltage

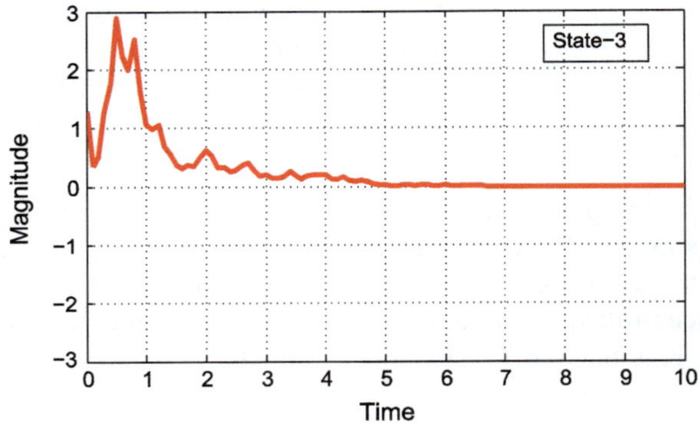

Fig. 6.7 d component of DG1 current

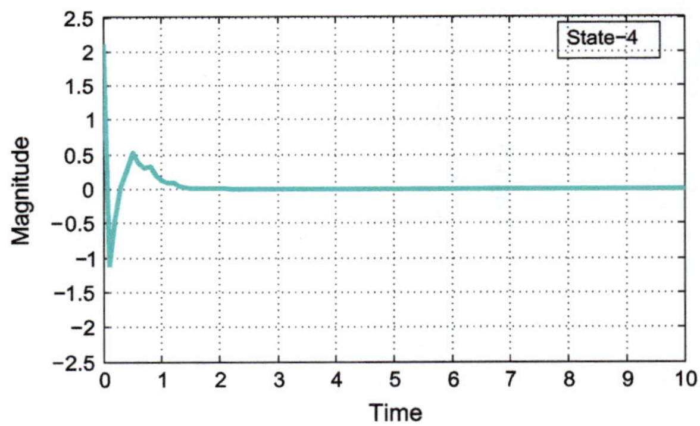

Fig. 6.8 q component of DG1 current

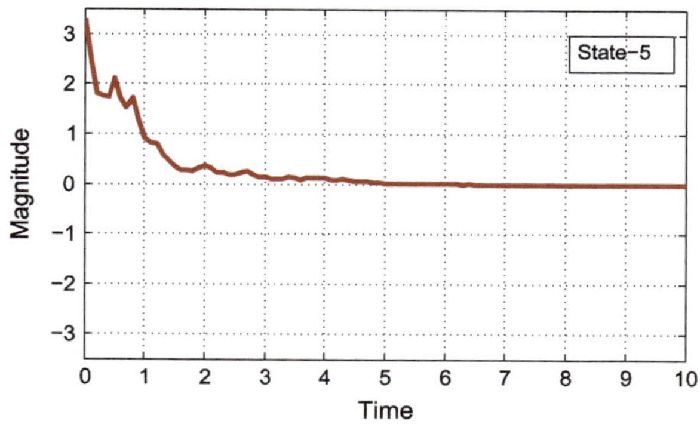

Fig. 6.9 d component of load current

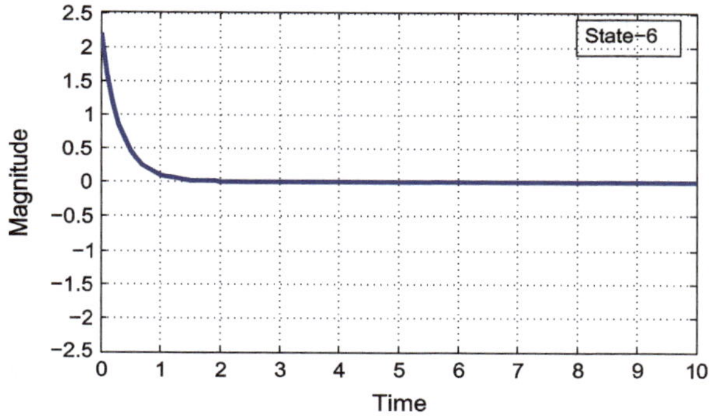

Fig. 6.10 q component of load current

6.7 Microalternator and Photovoltaic Systems

6.7.1 Introduction

Economic challenges, technological advancements and environmental impacts are now demanding DG in place of the conventional centralized generation [18, 29, 30]. Power operation companies are now confronted with unprecedented difficulties in terms of meeting the load requirements, consumer satisfaction, and environmental considerations (Figs. 6.11, 6.12, 6.13 and 6.14). Thus, DG has received good attention because of its potential to alleviate pressure from the main transmission system by supplying a few local loads [1]. The waste heat generated from the fuel to electricity conversion is exploited by the DGS with the help of microturbines, reciprocating engines and fuel cells to provide heat and power to the customers. Adding to the system DES like PV panels, WTs, energy storage devices such as batteries and capacitors, generators extracting energy from other renewable and controllable loads can provide momentous contributions to future energy generation and distribution. Another noteworthy feature is that the carbon emission is reduced to a large extent satisfying the commitment of many nations concerning decrease of carbon footprints

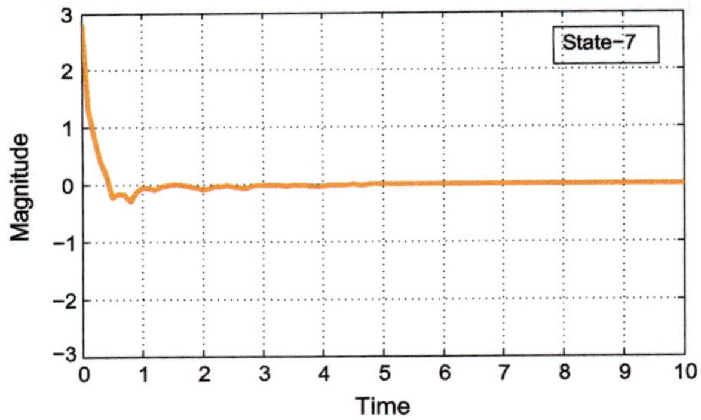

Fig. 6.11 d component of DG2 current

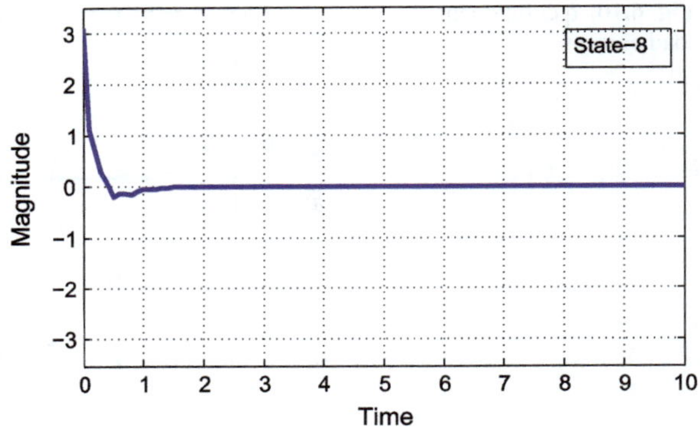

Fig. 6.12 q component of DG2 current

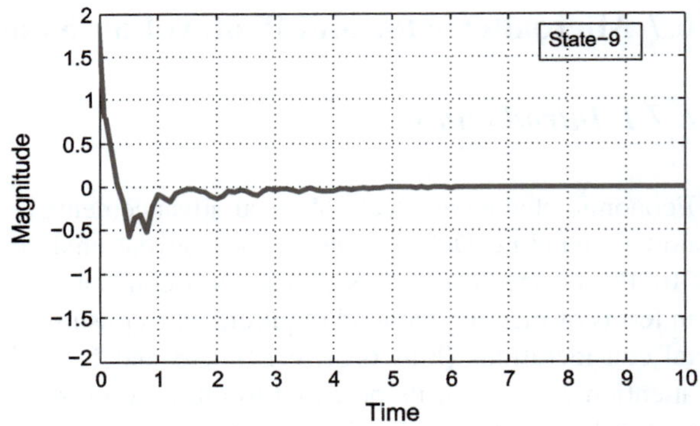

Fig. 6.13 d component of DG3 current

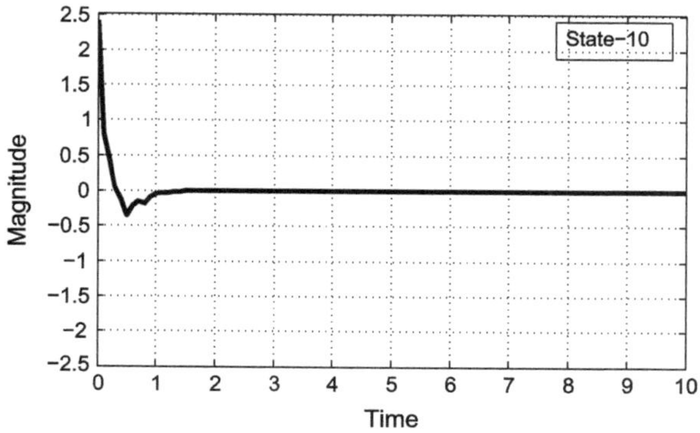

Fig. 6.14 q component of DG3 current

[8]. However, the DG faces technical issues regarding its connection to the intermittent renewable generation and feeble areas of the distribution network. Further, owing to the distinct behavior of the DG unlike the conventional load, alteration in power flow results in problems. To counter the irregular behavior and increasing penetration of the DG, the MG was introduced.

The MG has paved its way into DG and looks promising for future aspects. It has the ability to respond to changes in the load, while decreasing feeder losses and improving local reliability. Basically designed to cater the heat and power requirements of local customers, it can serve as an uninterruptible power supply (UPS) for critical loads. Several control strategies for the MG have been proposed in the literature including PI controllers in [2, 7, 14, 16, 17, 19, 20, 32, 34, 35, 39]. Robust H_∞ control is presented in [3, 28] for the control of two DG units. An optimal controller is presented for controlling the frequency and voltage fluctuations during islanded mode in [33].

The concept of SoS has opened up a new school of thought in systems engineering. SoS has emerged as a hot topic for research over the past few years. Although still in the infant stages, the concept of SoS has managed to achieve widespread acclaim. Being restricted to defense and IT, at one point in time, SoS has now entered a plethora of domains [9, 12].

It is worth noting that a MG is a complex system comprising of a variety of systems which are nonlinear in nature and possess strong cross-coupling between them. Hence viewing the MG from an intelligent SoS perspective is need of the hour. Moreover an efficient control methodology based on SoS has to be established in order to overcome the challenges posed by the MG. The concept of SoS is now widespread and has entered several domains including defense, IT, health care, manufacturing, energy and space stations and exploration to name a few.

Networked control of SoS has been introduced in [38]. A control system consisting of a real-time network in its feedback can be termed as a NCS [38]. As mentioned in [38] that the need to design a SoS control system which can tolerate packet loss and delays is one of the prime challenges in SoS networked control, we considered a network which is subjected to both delays and nonstationary packet dropouts and the controller stabilizes the system in the presence of these communication infractions. The controller design for such a networked control is presented in [39]. An addition is made in the controller design by introducing a distinct $\hat{X}$ matrix in the LMI that is used to obtain the controller gain matrices and a realistic approach is adopted to compute the gains. However, the main objective of this paper is to formulate the DG units of a MG in a SoS framework and stabilize the network control SoS in the presence of communication infractions.

The stability of its constituent systems is a major issue in MGs. The inverters connected to these systems operate under imbalance conditions due to sensitive loads. This leads to switching harmonics, voltage and frequency variations in the MG system and disturbs the stability of the system. Therefore, this paper attempts at proposing a SoS framework for better operation and a network control SoS controller to stabilize the MG system when connected to load and in the presence of delays and packet dropouts. Hence two sets of microalternator and PV systems are considered as DG units, which are eventually connected to a load and the main grid.

6.7.2 Modeling of the Microalternator–PV System

To model the MG system consisting of microalternator and PV system, we consider the separate modeling of microalternator and PV system initially. After modeling both these individual systems, we integrate them into a MG system which is connected to a load and the main grid. Two sets of such systems are considered eventually which form a SoS structure for the MG system.

6.7.3 Microalternator

The swing equation of the alternator can be written as two first order differential equations [37]

$$\frac{d\delta}{dt} = \omega_0(\omega - 1) \tag{6.28}$$

$$\frac{d\omega}{dt} = \frac{1}{2H}(P_m - P_e) \tag{6.29}$$

where δ and ω are rotor angle and rotor speed. P_m and P_e are mechanical power input and electrical power output of the generator (Fig. 6.15).

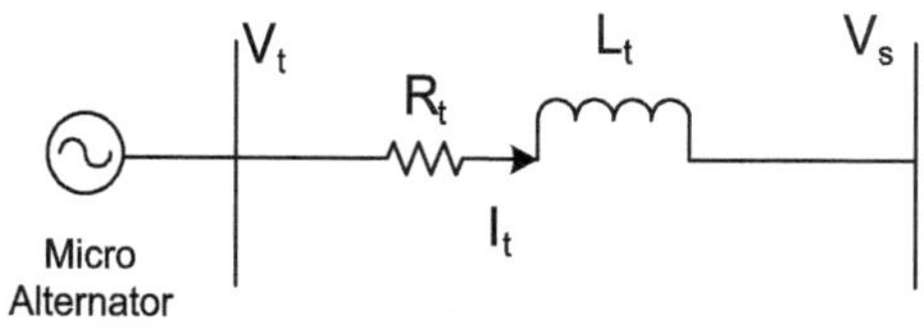

Fig. 6.15 Microalternator connected to grid

The internal voltage e'_q is given by

$$\frac{de'_q}{dt} = \frac{1}{T'_{do}}[E_{fd} - e'_q - (x_d - x'_d)i_{td}] \tag{6.30}$$

where x_d, x'_d and T'_{do} are the d-axis synchronous resistance, transient reactance and open circuit field constants, respectively. e'_q is the voltage behind the transient reactance along q-axis.

An IEEE type ST is used for the voltage regulator excitation.

$$\frac{dE_{fd}}{dt} = \frac{1}{T_A}[K_A(V_{tref} - V_t) - (E_{fd} - E_{fdo})] \tag{6.31}$$

where E_{fd} is the field voltage along d-axis, K_A and T_A are the gain and time constants of the exciter.

Figure 6.1 shows a microalternator connected to the main grid. The terminal voltage of the alternator is given as

$$V_t = V_s + (r_t + jx_t)i_t$$

In d-q terms it becomes

$$\begin{aligned} V_d + jV_q &= V_{sd} + jV_{sq} + (r_t + jx_t)(i_{td} + ji_{tq}) \\ x_q i_{tq} + j(e'_q - x'_d i_{td}) &= (V_{sd} + r_t i_{td} - x_t i_{tq}) \\ &\quad + j(V_{sq} + r_t i_{tq} + x_t i_{td}) \end{aligned}$$

The real part can be written as

$$V_{sd} = (x_q + x_t)i_{tq} - r_t i_{td} \tag{6.32}$$

and the imaginary part can be written as

$$e'_q - V_{sq} = r_t i_{tq} + (x'_d + x_t)i_{td} \tag{6.33}$$

Substituting $x_1 = (x'_d + x_t)$ and $x_2 = (x_q + x_t)$ and solving (6.32) and (6.33), we obtain

$$i_{td} = \frac{-r_t V_{sd} + (e'_q - V_{sq})(x_q + x_t)}{r_t^2 + (x'_d + x_t)(x_q + x_t)} \tag{6.34}$$

$$i_{tq} = \frac{V_{sd}[(x'_d + x_t)(x_q + x_t)] + r_t(e'_q - V_{sq})(x_q + x_t)}{(x_q + x_t)[r_t^2 + (x'_d + x_t)(x_q + x_t)]} \tag{6.35}$$

The terminal voltage of the alternator is given as

$$\begin{aligned} V_t &= (V_d^2 + V_q^2)^{1/2} \\ V_t &= ((x_q i_{tq})^2 + (e'_q - x'_d i_{td})^2)^{1/2} \end{aligned} \tag{6.36}$$

The power output is given as

$$\begin{aligned} P_e &= V_d i_{td} + V_q i_{tq} \\ P_e &= (e'_q i_{tq}) + (x_q - x'_d) i_{td} i_{tq} \end{aligned} \tag{6.37}$$

By substituting (6.34) and (6.35) in (6.36) and (6.37), we can express the terminal voltage and power output of the microalternator in terms of MG voltage components V_{sd} and V_{sq}. The above nonlinear equations are linearized by perturbing the set of equations around a normal operating point. The linearized state equations are

$$\Delta\dot{\delta} = \omega_0 \Delta\omega \tag{6.38}$$

$$\Delta\dot{\omega} = \frac{1}{2H}[-\Delta P_e] \tag{6.39}$$

$$\Delta\dot{e}'_q = \frac{1}{T'_{do}}[\Delta E_{fd} - \Delta e'_q - (x_d - x'_d)\Delta i_{td}] \tag{6.40}$$

$$\Delta\dot{E}_{fd} = \frac{K_A}{T_A}\Delta V_t - \frac{1}{T_E}\Delta E_{fd} \tag{6.41}$$

The generator output current is

$$\Delta i_t = \Delta i_{td} + j\Delta i_{tq} \tag{6.42}$$

From (6.34), we have the d-axis generator output current as

$$i_{td} = \frac{-r_t V_{sd} + (e'_q - V_{sq})x_2}{r_t^2 + x_1 x_2}$$

Differentiating on both side yields

$$\Delta i_{td} = \frac{-r_t}{r_t^2 + x_1 x_2}\Delta V_{sd} + \frac{x_2}{r_t^2 + x_1 x_2}(\Delta e'_q - \Delta V_{sq}) \tag{6.43}$$

Similarly, obtaining the q-axis generator output current from (6.35) and differentiating both sides yields

$$\Delta i_{tq} = \frac{\Delta V_{sd}}{x_1} + \frac{r_t}{x_d' + x_t}(\Delta e_q' - \Delta V_{sq}) \tag{6.44}$$

The terminal voltage is

$$\begin{aligned} V_t^2 &= V_d^2 + V_q^2 \\ \Delta V_t &= \frac{V_{do}}{V_{to}}\Delta V_d + \frac{V_{qo}}{V_{to}}\Delta V_q \\ \Delta V_t &= \frac{V_{do}}{V_{to}}(x_q \Delta i_{tq}) + \frac{V_{qo}}{V_{to}}(\Delta e_q' - x_d' \Delta i_{td}) \end{aligned} \tag{6.45}$$

From (6.37) we have the power output of the alternator as

$$\begin{aligned} P_e &= (e_q' i_{tq}) + (x_q - x_d') i_{td} i_{tq} \\ \Delta P_e &= e_{qo}' \Delta i_{tq} + i_{tqo} \Delta e_q' \\ &\quad + (x_q - x_d')[i_{tdo} \Delta i_{tq} + i_{tqo} \Delta i_{td} \end{aligned} \tag{6.46}$$

Now considering the states as $[\delta, \omega, e_q', E_{fd}]$ and substituting the corresponding values, we have the state matrix A of the microalternator as

$$A = \begin{bmatrix} 0 & \omega_0 & 0 & 0 \\ \frac{-P_{e1}}{2H} & 0 & \frac{-P_{e3}}{2H} & 0 \\ \frac{-(x_d - x_d') i_{td1}}{T_{do}'} & 0 & \frac{-(x_d - x_d') i_{td3}}{T_{do}'} & \frac{1}{T_{do}'} \\ \frac{K_A}{T_A} V_{t1} & 0 & \frac{K_A}{T_A} V_{t3} & \frac{-1}{T_A} \end{bmatrix}$$

6.7.4 Photovoltaic System

A mathematical expression describing the I-V characteristics of a solar cell has been studied extensively. An equivalent model of a solar cell shown in Fig. 6.16

Fig. 6.16 Model of PV cell

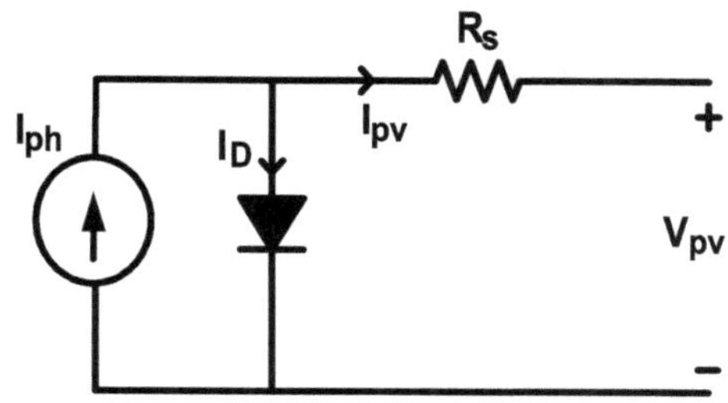

includes a photo diode, a shunt resistor depicting leakage current and a series resistor representing an internal resistance to current flow. The difference between PV current I_{ph} and sum of normal diode current I_D and leakage current in the shunt resistor I_{sh} gives the net output current from the PV cell.

$$I_{pv} = I_{ph} - I_D - \frac{(V_{pv} + I_{pv} R_s)}{R_{sh}} \tag{6.47}$$

The diode current I_D is given as

$$I_D = I_s(e^{\frac{(V_{pv}+R_s I_{pv})}{nV_T}} - 1) \tag{6.48}$$

where I_{pv} is the cell current, V_{pv} is the cell voltage, I_s is the reverse saturation current(depends on temperature), n is the ideality factor, R_s is the series resistance, R_{sh} is the shunt resistance, $V_T(=\frac{kT}{q})$ is the thermal voltage, k is Boltzmann constant, T is working temperature of the cell, and q is the charge of the electron.

The solar irradiation and the working temperature of the cell determines the photo current. At given cell temperature T, the photo current can be expressed as

$$I_{ph} = [I_{sc} + a(T - T_{ref})]G \tag{6.49}$$

where I_{sc} is the short current of the cell at 25 °C and 1 kW/m^2, a is the temperature coefficient of I_{sc}, T_{ref} is the reference temperature of the cell and G is the irradiation in kW/m^2.

The reverse saturation current also depends on temperature and can be expressed as

$$I_s = I_{sref}\left(\frac{T}{T_{ref}}\right)^{\frac{3}{n}} e^{\frac{-qE_g}{nk}(\frac{1}{T} - \frac{1}{T_{ref}})} \tag{6.50}$$

where I_{sref} is the saturation current of the cell, E_g is the energy band gap of the semiconductor used in the solar cell.

The characteristic equation of the cell from the approximate model is given as

$$I_{pv} = I_{sc} - I_s\left[e^{q\left(\frac{V_{pv}+I_{pv}R_s}{nV_T}\right)} - 1\right] \tag{6.51}$$

Because of the fact that the power generated by a solar cell is low, multiple solar cells are connected in series and parallel to generate power in the range of watts. Thus the characteristic equation of the PV array consisting of N_s series and N_p parallel modules can be derived from the PV cell equation mentioned above

$$I_{pv} = N_p I_{ph} - N_p I_s\left[e^{\left(\frac{V_{pv}}{N_s} + \frac{I_{pv}R_s}{N_p}\right)/nV_T} - 1\right] \tag{6.52}$$

The above equation establishes a nonlinear current voltage relationship of a PV array. To solve this equation, it is initially expressed in the form $f(x) = 0$ then the following iterative procedure, based on Newton Raphson algorithm [31] is applied:

Starting from an initial solution x_n, it is updated as

$$x_{n+1} = x_n - \frac{f(x_n)}{f'(x_n)} \quad (6.53)$$

The iterations continue until the error is less than or equal to a tolerance E_s

$$|\frac{x_{n+1} - x_n}{x_n}| \leq E_s \quad (6.54)$$

The I-V characteristics of the PV are related as

$$V_{pv} = N_s \left[ln \left(\frac{N_p I_{ph} - I_{pv}}{N_p I_s} + 1 \right) n V_T - \frac{I_{pv} R_s}{N_p} \right] \quad (6.55)$$

This can be related to the function of the form $f(x) = 0$

$$f(V) = V - N_s \left[n V_T * ln \left(\frac{N_p I_{ph} - I_{pv} - N_p I_s}{N_p I_s} \right) + \frac{I_{pv} R}{N_p} \right] = 0 \quad (6.56)$$

Since $f'(V) = 1$, the recursive formula can be written as

$$V_{n+1} = N_s \left[n V_T * ln \left(\frac{N_p I_{ph} - I_{pv} - N_p I_s}{N_p I_s} \right) + \frac{I_{pv} R}{N_p} \right] \quad (6.57)$$

On the other hand, the power conditioning unit (PCU) consists of devices which are needed to connect the PV array to the MG [10]. The significant components of the PCU are

1. *DC/DC converter*
2. *DC link capacitor*
3. *Inverter*
4. *Output filter circuit*

DC/DC converter model: The primary function of the DC/DC converter is to either increase or decrease the DC output voltage. Out of the various topologies of DC/DC converter, buck and boost converters are the more fundamental ones. Because the PV output voltage has be to stepped up, a boost converter is used in this case. A typical converter configuration is shown in Fig. 6.17. The boost converter steps up the DC voltage level. It consists of an inductor, a diode, and a power electronic switch.

The dynamics of the converter can be expressed as

$$V_{pv} = \dot{L}_{dc} I_{pv} + (1 - d_c) V_{dcp}$$

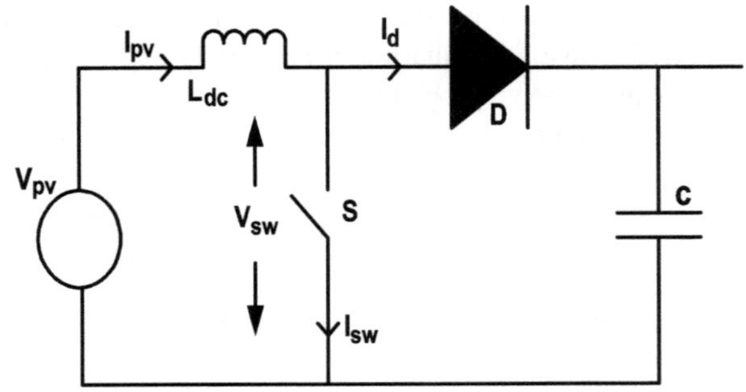

Fig. 6.17 DC/DC converter configuration

which leads to

$$\dot{i}_{pv} = \frac{1}{L_{dc}}(V_{pv} - (1 - d_c)V_{dcp}) \tag{6.58}$$

where L_{dc} is the inductance of the converter and d_c is the duty cycle defined as the ratio of ON period to the switching time period (T).
DC link capacitor model: The DC link capacitor functions as an energy storage and filter for the DC voltage.

By applying KCL at the DC link node, the dynamics of the DC link capacitor can be obtained

$$\frac{dV_{dcp}}{dt} = \frac{1}{C_{dc}}(I_{dc1} - I_{dc2}) \tag{6.59}$$

where $I_{dc1} = (1 - d_c)I_{pv}$ and I_{dc2} is the input current to the inverter, which is derived later in this section. V_{dc} is the voltage across capacitor C_{dc}.

Inverter model: The inverter is responsible for the conversion of the PV array DC output and giving it to the grid at an appropriate frequency. A voltage gain model of a voltage source inverter (VSI) operating in PWM mode is considered as shown in Fig. 6.18.

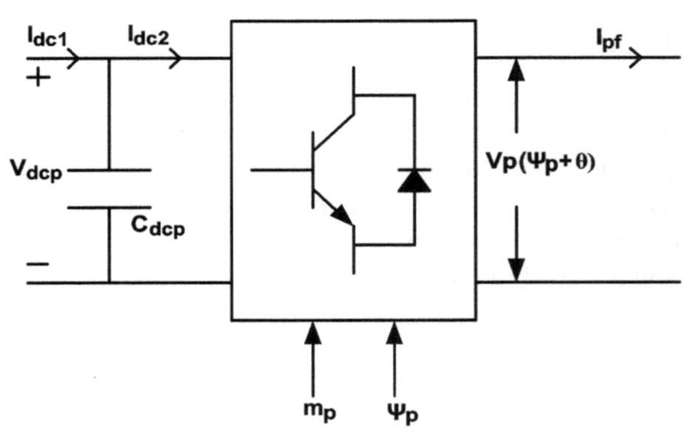

Fig. 6.18 Inverter model

The power on DC side of the inverter is given as

$$P_{dc} = V_{dcp} I_{dc2} \tag{6.60}$$

The instantaneous active power on the AC side of the inverter is expressed as

$$P_{ac} = Re[V_p I_{pf}^*] \tag{6.61}$$

where V_p and I_{pf} are the inverter output voltage and inverter output current, respectively. In d-q terms, V_p and I_{pf} can be expressed as

$$V_p = V_{pd} + jV_{pq}$$
$$I_{pf} = I_{pfd} + jI_{pfq}$$

Substituting the values of V_p and I_{pf} in (6.61), we obtain

$$P_{ac} = V_{pd} I_{pfd} + V_{pq} I_{pfq} \tag{6.62}$$

Equating the AC and DC power

$$V_{dcp} I_{dc2} = V_{pd} I_{pfd} + V_{pq} I_{pfq} \tag{6.63}$$

During its operation in PWM mode, the output voltage of the inverter can be written as

$$V_p = m_p + V_{dcp} \psi_p \tag{6.64}$$

where V_{dcp} is the DC link voltage, m_p is the modulation index and ψ_p is the phase angle of the inverter.

In d-q terms, it becomes

$$V_{pd} = m_p * V_{dcp} * cos(\psi_p + \theta) \tag{6.65}$$
$$V_{pq} = m_p * V_{dcp} * sin(\psi_p + \theta) \tag{6.66}$$

Substituting (6.65) and (6.66) into (6.63), we obtain the expression for I_{dc2}.

$$I_{dc2} = (I_{pfd} m_p cos(\psi_p + \theta) + I_{pfq} m_p sin(\psi_p + \theta)) \tag{6.67}$$

LC filter and coupling inductance model: The purpose of using a low pass filter is attenuation of switching frequency ripple of the output voltage of an inverter. The filter is a T section of an RL circuit shunted by a capacitor. While the inductor blocks high-frequency harmonics, the capacitor stops low-frequency harmonics. Collectively, they block most of the harmonics, thereby reducing ripples from going through the system [22]. By applying KVL around the PV inverter and filter capacitor, we obtain a nonlinear equation as

$$V_p = I_{pf}R_{pf} + L_{pf}\frac{dI_{pf}}{dt} + V_{cp} + (I_{pf} - I_p)R_{pdr} \tag{6.68}$$

where R_{pf} is the filter resistance, L_{pf} is the filter inductance, R_{pdr} is the damping resistance and V_{cp} is the capacitor voltage. The above equation in d-q frame can be written as

$$\begin{aligned}\frac{dI_{pfd}}{dt} &= \frac{-\omega_0 R_{pf}}{L_{pf}}I_{pfd} + \omega_0\omega I_{pfq} + \frac{\omega_0 m_p V_{dcp}cos(\psi_p + \theta)}{L_{pf}} \\ &- \frac{\omega_0 V_{cpd}}{L_{pf}} - \omega_0 R_{pdr} I_{pcd} \end{aligned} \tag{6.69}$$

$$\begin{aligned}\frac{dI_{pfq}}{dt} &= \frac{-\omega_0 R_{pf}}{L_{pf}}I_{pfq} - \omega_0\omega I_{pfd} + \frac{\omega_0 m_p V_{dcp}sin(\psi_p + \theta)}{L_{pf}} \\ &- \frac{\omega_0 V_{cpq}}{L_{pf}} - \omega_0 R_{pdr} I_{pcq} \end{aligned} \tag{6.70}$$

By coupling the transmission line between MG and PV filter capacitor, we obtain a nonlinear equation as

$$V_{cp} = I_p R_p + L_p\frac{dI_p}{dt} + V_s - (I_{pf} - I_p)R_{pdr} \tag{6.71}$$

Writing in d-q terms, we have

$$\frac{dI_{pd}}{dt} = \frac{-\omega_0 R_p}{L_p}I_{pd} + \omega_0\omega I_{pq} + \frac{\omega_0}{L_p}(V_{cpd} - V_{sd}) + \omega_0 R_{pdr} I_{pcd} \tag{6.72}$$

$$\frac{dI_{pq}}{dt} = \frac{-\omega_0 R_p}{L_p}I_{pq} + \omega_0\omega I_{pd} + \frac{\omega_0}{L_p}(V_{cpq} - V_{sq}) + \omega_0 R_{pdr} I_{pcq} \tag{6.73}$$

where I_p is the coupling current, R_p is the coupling resistance and L_p is the coupling inductance. The voltage across the capacitor is given as

$$C_{pf}\frac{dV_{cp}}{dt} = (I_{pf} - I_p) \tag{6.74}$$

where C_{pf} is the filter capacitor.

Next, writing in d-q terms, we get

$$\frac{dV_{cpd}}{dt} = \frac{1}{C_{pf}}(I_{pfd} - I_{pd}) + \omega_0\omega V_{cpq} \tag{6.75}$$

$$\frac{dV_{cpq}}{dt} = \frac{1}{C_{pf}}(I_{pfq} - I_{pq}) + \omega_0\omega V_{cpd} \tag{6.76}$$

Linearized model of the PV system: A linearized model of the PV system includes a small signal model of characteristic equation of the PV array and PCU. From (6.55) the characteristic equation of the PV arrays is

$$V_{pv} = N_s \left[ln \left(\frac{N_p I_{ph} - I_{pv}}{N_p I_s} + 1 \right) n V_T - \frac{I_{pv} R_s}{N_p} \right]$$

Differentiating both sides, we get

$$\Delta V_{pv} = -N_s \left[\frac{n V_T}{N_p I_{ph} - I_{pvo} + N_p I_s} + \frac{R_s}{N_p} \right] \Delta I_{pv} \tag{6.77}$$

which can be written as

$$\Delta V_{pv} = K_{pv} \Delta I_{pv} \tag{6.78}$$

where

$$K_{pv} = -N_s \left[\frac{n V_T}{N_p I_{ph} - I_{pvo} + N_p I_s} + \frac{R_s}{N_p} \right]$$

By linearization each of the components of the PCU, a small signal model is obtained. The linearized state equations are given as [10].

$$\begin{aligned} \Delta \dot{V}_{dcp} = & -\frac{1}{C_{dcp}} [-I_{pfd0} m_p sin(\psi_p + \theta) \Delta \psi_p \\ & + m_p cos(\psi_p + \theta) \Delta I_{pfd} + I_{pfd0} cos(\psi_p + \theta) \Delta m_p \\ & + I_{pfq0} m_p cos(\psi_p + \theta) \Delta \psi_p + m_p sin(\psi_p + \theta) \Delta I_{pfq} \\ & + I_{pfq0} sin(\psi_p + \theta) \Delta m_p - (1 - d_c) \Delta I_{pv}] \end{aligned} \tag{6.79}$$

The PV inverter output filter current

$$\begin{aligned} \Delta \dot{I}_{pfd} = & \frac{-\omega_0 R_{pf}}{L_{pf}} \Delta I_{pfd} + \omega_0 (\Delta I_{pfq} + I_{pfq0} \Delta \omega) \\ & + \frac{\omega_0}{L_{pf}} [m_p cos(\psi_p + \theta) \Delta V_{dcp} - m_p V_{dcp0} sin(\psi_p + \theta) \Delta \psi_p \\ & + V_{dcp0} cos(\psi_p + \theta) \Delta m_p] \\ & - \frac{\omega_0 \Delta V_{cpd}}{L_{pf}} - \frac{\omega_0 R_{pdr}}{L_{pf}} (\Delta I_{pfd} - \Delta I_{pq}) \end{aligned} \tag{6.80}$$

$$\begin{aligned}\Delta \dot{I}_{pfq} &= \frac{-\omega_0 R_{pf}}{L_{pf}}\Delta I_{pfq} - \omega_0(\Delta I_{pfd} + I_{pfd0}\Delta\omega) \\ &+ \frac{\omega_0}{L_{pf}}[m_p sin(\psi_p + \theta)\Delta V_{dcp} + m_p V_{dcp0} cos(\psi_p + \theta)\Delta\psi_p \\ &+ V_{dcp0} sin(\psi_p + \theta)\Delta m_p] \\ &- \frac{\omega_0 \Delta V_{cpq}}{L_{pf}} - \frac{\omega_0 R_{pdr}}{L_{pf}}(\Delta I_{pfq} - \Delta I_{pq}) \end{aligned} \tag{6.81}$$

The PV output coupling inductance current

$$\begin{aligned}\Delta \dot{I}_{pd} &= -\frac{R_p}{L_p}\Delta I_{pd} + \omega_0(\Delta I_{pq} + I_{pq0}\Delta\omega) \\ &+ \frac{1}{L_p}(\Delta V_{cpd} - \Delta V_{sd}) + \frac{\omega_0 R_{pdr}}{L_p}(\Delta I_{pfd} - \Delta I_{pd}) \end{aligned} \tag{6.82}$$

$$\begin{aligned}\Delta \dot{I}_{pq} &= -\frac{R_p}{L_p}\Delta I_{pq} + \omega_0(\Delta I_{pd} + I_{pd0}\Delta\omega) \\ &+ \frac{1}{L_p}(\Delta V_{cpq} - \Delta V_{sd}) + \frac{\omega_0 R_{pdr}}{L_p}(\Delta I_{pfq} - \Delta I_{pq}) \end{aligned} \tag{6.83}$$

The voltage across the filter capacitor

$$\Delta \dot{V}_{cpd} = \omega_0(\Delta V_{cpq} + V_{cpq0}\Delta\omega) + \frac{1}{C_{pf}}(\Delta I_{pfd} - \Delta I_{pd}) \tag{6.84}$$

$$\Delta \dot{V}_{cpq} = \omega_0(\Delta V_{cpd} + V_{cpd0}\Delta\omega) + \frac{1}{C_{pf}}(\Delta I_{pfq} - \Delta I_{pq}) \tag{6.85}$$

Now let us consider a system where the microalternator and the PV system are considered under one system. So we have a MG with two DG units.

The microalternator–PV combination is integrated and a load is connected to the system. The next step is to model the combined system and load.

Figure 6.19 shows a MG system with a microalternator and PV generator along with a load connected to the main grid. All of these are connected through a common bus having voltage V_s. Microalternator and PV system dynamics have already been explained above. Both the systems had components of V_s along d-q axes. Now to get a closed form representation of the combined state model, V_{sd} and V_{sq} have to be expressed in terms of the selected states. For this we apply KCL at the common bus V_s, which gives

$$I_t + I_p = I_b + I_l$$

where I_t, I_p, I_b, and I_l are microalternator output current, PV output current, grid current, and load current, respectively.

Considering the d-q components of these currents, we get

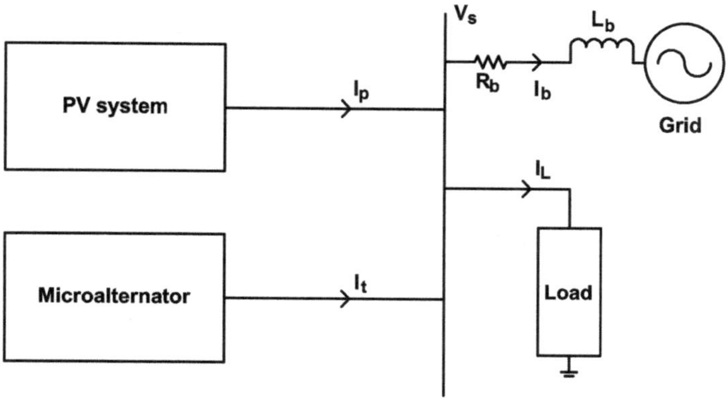

Fig. 6.19 Combined system

$$I_{td} + I_{pd} = I_{bd} + I_{ld} \tag{6.86}$$

$$I_{tq} + I_{pq} = I_{bq} + I_{lq} \tag{6.87}$$

Further, we express the non-state currents ($I_{td}, I_{tq}, I_{bd}, I_{bq}, I_{ld}, I_{lq}$) as functions of V_{sd} and V_{sq}. However, the microalternator output currents I_{td} and I_{tq} are already presented in terms of V_{sd} and V_{sq} in (6.33) and (6.35), respectively.

Load current:

At the MG, the load is modeled as admittance $Y = g - jb$. The load current is $I_l = V_s Y$

$$I_{ld} + jI_{lq} = 4(V_{sd} + jV_{sq})(g - jb)$$

Equating real and imaginary parts, we get

$$I_{ld} = gV_{sd} + bV_{sq} \tag{6.88}$$

$$I_{lq} = gV_{sq} - bV_{sd} \tag{6.89}$$

Grid current:

The main grid current I_b is given as

$$I_b = \frac{V_s - V_b}{r_b + jx_b}$$

$$I_{bd} + jI_{bq} = \frac{V_{sd} + jV_{sq} - (V_b sin\delta + jV_b cos\delta)}{r_b + jx_b}$$

Equating real and imaginary parts

$$I_{bd} = \frac{(V_{sq} - V_b sin\delta)r_b + (V_{sq} - V_b cos\delta)x_b}{r_b^2 + x_b^2} \tag{6.90}$$

$$I_{bq} = \frac{(V_{sq} - V_b cos\delta)r_b + (V_{sq} - V_b sin\delta)x_b}{r_b^2 + x_b^2} \tag{6.91}$$

Now let us substitute the values of I_{td}, I_{bd}, I_{ld} in (6.87) and solve for the bus voltage components V_{sd} and V_{sq}

$$\begin{aligned} I_{pd} + I_{td} &= I_{bd} + I_{ld} \\ I_{pd} &= \frac{-r_t V_{sd} + (e_q' - V_{sq})x_2}{z_1} \\ &= \frac{(V_{sq} - V_b sin\delta)r_b + (V_{sq} - V_b cos\delta)x_b}{z_b} + gV_{sd} + bV_{sq} \end{aligned}$$

where $z_1 = r_t^2 + x_1 x_2$ and $z_b = r_b^2 + x_b^2$.

$$\begin{aligned} &[gz_b z_1 + z_1 r_b + r_t z_b]V_{sd} + [bz_b z_1 + x_b z_1 + x_2 z_b]V_{sq} \\ &= z_b z_1 I_{pd} + z_b e_q' x_2 + V_b cos\delta x_b z_1 + V_b sin\delta r_b z_1 \end{aligned} \tag{6.92}$$

Substitute (6.25), (6.90) and (6.91) in the above equation

$$\begin{aligned} I_{tq} + I_{pq} &= I_{bq} + I_{lq} \\ I_{pq} &= \frac{V_{sd} z_1 - r_t^2 V_{sd} + r_t(e_q' - V_{sq})x_2}{x_2 z_1} \\ &= \frac{(V_{sq} - V_b cos\delta)r_b + (V_{sq} - V_b sin\delta)x_b}{z_b} \end{aligned} \tag{6.93}$$

Simplifying this as above and solving with (6.93), we obtain

$$\begin{aligned} V_{sd} &= \frac{1}{\Theta_1}[z_b z_1 I_{pd} + z_b x_2 e_q' \\ &\quad + V_b z_1 (r_b sin\delta + x_b cos\delta) - \Theta_2 V_{sq}] \end{aligned} \tag{6.94}$$

$$V_{sq} = \Theta_4 I_{pd} + \Theta_5 e_q' + \Theta_6 I_{pq} + \Theta_7 V_b \tag{6.95}$$

where

$$\begin{aligned} \Theta_1 &= [gz_b z_1 + z_1 r_b + r_t z_b] \\ \Theta_2 &= [bz_b z_1 + x_b z_1 + x_2 z_b] \\ \Theta_3 &= [x_b x_2 z_1 + bz_b x_2 z_1 + z_1 z_b - r_t^2 z_b] \end{aligned}$$

$$
\begin{aligned}
Den &= \Theta_2\Theta_3 + \Theta_1^2 x_2 \\
\Theta_4 &= \frac{1}{Den}\Theta_3 z_b z_1 \\
\Theta_5 &= \frac{1}{Den}(\Theta_3 z_b x_2 + z_b x_2 r_t \Theta_1) \\
\Theta_6 &= \frac{1}{Den}(\Theta_1 x_2 z_1 z_b) \\
\Theta_7 &= \frac{1}{Den}[z_1\Theta_3(r_b sin\delta + x_b cos\delta) \\
&+ \Theta_1 x_2 z_1 (r_b cos\delta - x_b sin\delta)]
\end{aligned}
$$

Linearized model of the combined system:

A linearized model of the combined system is obtained by expressing the linearized MG voltage components (ΔV_{sd}, ΔV_{sq}) along d-q axes in terms of the selected states.

Differentiating (6.95) and (6.96), we obtain

$$
\Delta V_{sq} = \Theta_4 \Delta I_{pd} + \Theta_5 \Delta e_q' + \Theta_6 \Delta I_{pq} + \Theta_0 \Delta\delta \tag{6.96}
$$

where

$$
\begin{aligned}
\Theta_0 = V_b\Big[&\frac{1}{Den}(z_1\Theta_3(r_b cos\delta_0 - x_b sin\delta_0) \\
&- \Theta_1 x_2 z_1 (r_b sin\delta_0 + x_b cos\delta_0))\Big]
\end{aligned}
$$

Similarly, we also obtain

$$
\begin{aligned}
\Delta V_{sd} = &\frac{1}{\Theta_1}[z_b z_1 \Delta I_{pd} + z_b x_2 \Delta e_q' \\
&+ V_b z_1 (r_b cos\delta_0 - x_b sin\delta_0)\Delta\delta \\
&- \Theta_2(\Theta_4 \Delta I_{pd} + \Theta_5 \Delta e_q' + \Theta_6 \Delta I_{pq} + \Theta_0 \Delta\delta)]
\end{aligned} \tag{6.97}
$$

Finally, the values of ΔV_{sd} and Δv_{sq} are substituted in the linearized equations of the individual component models to obtain the closed form equation. Selecting the states as $[\delta, \omega, e_q', E_{fd}, I_{pv}, V_{dcp}, I_{pfd}, I_{pfq}, I_{pd}, I_{pq}, V_{cpd}, V_{cpq}]^T$ and the control input as $[m_p, \psi_p]$, we consequently obtain the state matrices for the combined microalternator–PV system.

$$A = \begin{bmatrix} 0 & \omega_0 & 0 & 0 & 0 & 0 & 0 & 0 & 0 & 0 & 0 & 0 \\ -\frac{P_{e1}}{2H} & 0 & -\frac{P_{e3}}{2H} & 0 & 0 & 0 & P_{e1} & P_{e2} & 0 & 0 & 0 & 0 \\ A_{(3,1)} & 0 & A_{(3,3)} & \frac{1}{T'_{do}} & 0 & 0 & e_{q1} & e_{q2} & 0 & 0 & 0 & 0 \\ \frac{K_A}{T_A} V_{t1} & 0 & \frac{K_A}{T_A} V_{t3} & \frac{-1}{T_A} & 0 & 0 & 0 & 0 & 0 & 0 & 0 & 0 \\ 0 & \omega_0 I_{pfq0} & 0 & 0 & \frac{k_{pv}}{L_{dc}} & \frac{(d_c-1)}{L_{dc}} & 0 & 0 & 0 & 0 & 0 & 0 \\ 0 & \omega_0 I_{pfd0} & 0 & 0 & \frac{(1-d_c)}{C_{dcp}} & 0 & m_p k_{pd1} & m_p k_{pd2} & 0 & 0 & 0 & 0 \\ -k_p D_1 & \omega_0 I_{pd0} & -k_p C_1 & 0 & k_{pf} & 0 & -k_{pf} R_{eq} & \omega_0 & k_{pf} R_d & 0 & -k_{pf} & 0 \\ -k_p G & -\omega_0 I_{pq0} & -k_p C_1 & 0 & 0 & 0 & -\omega_0 & -k_{pf} R_{eq} & 0 & k_{pf} R_d & 0 & -k_{pf} \\ 0 & \omega_0 V_{cpq0} & 0 & 0 & 0 & 0 & k_p R_d & 0 & A_{(9,9)} & A_{(9,10)} & k_p & 0 \\ 0 & -\omega_0 V_{cpq0} & 0 & 0 & 0 & 0 & 0 & k_p R_d & A_{(10,9)} & A_{(10,10)} & 0 & k_p \\ 0 & 0 & 0 & 0 & 0 & 0 & \frac{\omega_0}{C_{pf}} & 0 & -\frac{\omega_0}{C_{pf}} & 0 & 0 & \omega_0 \\ 0 & 0 & 0 & 0 & 0 & 0 & 0 & \frac{\omega_0}{C_{pf}} & 0 & -\frac{\omega_0}{C_{pf}} & -\omega_0 & 0 \end{bmatrix}$$

$$B = \begin{bmatrix} 0 & 0 \\ 0 & 0 \\ 0 & 0 \\ 0 & 0 \\ k_{pf} V_{dcp0} cos(\psi_p) & -k_{pf} V_{dcp0} m_p sin(\psi_p) \\ k_{pf} V_{dcp0} sin(\psi_p) & k_{pf} V_{dcp0} m_p cos(\psi_p) \\ 0 & 0 \\ 0 & 0 \\ 0 & 0 \\ 0 & 0 \\ 0 & 0 \\ \frac{(I_{pfd0} cos(\psi_P) + I_{pfq0} sin(\psi_P))}{C_{dcp}} & \frac{-m_p (I_{pfd0} sin(\psi_P) - I_{pfq0} cos(\psi_P))}{C_{dcp}} \end{bmatrix}$$

$$A_{3,1} = \frac{-(x_d - x'_d) i_{td1}}{T'_{do}}, \quad A_{3,3} = \frac{-(x_d - x'_d) i_{td3}}{T'_{do}}$$

$$A_{9,9} = -k_p (R_{eq1} + \Theta_{11}), \quad A_{9,10} = -\omega_0 - k_p \left(\frac{-\Theta_2 \Theta_6}{\Theta_1} \right)$$

$$A_{10,9} = -\omega_0 - k_p \Theta_4, \quad A_{10,10} = -k_p (R_{eq1} + \Theta_6)$$

$$\Theta_{11} = \left(\frac{z_b z_1}{\Theta_1} - \frac{\Theta_2 \Theta_4}{\Theta_1} \right)$$

6.7.5 Networked Control System Modeling

Consider the MG with two sets of microalternator and PV system representing a SoS–NCS with random communication delays, where the sensor is clock driven and the controller and the actuator are event driven (Fig. 6.20). The discrete-time linear time-invariant plant model is as follows:

$$x_p(k+1) = A x_p + B u_p, \quad y_p = C x_p \tag{6.98}$$

where $x_p(k) \in \Re^n$ is the plant's state vector and $u_p(k) \in \Re^m$ and $y_p(k) \in \Re^p$ are the plant's control input and output vectors, respectively. A, B, and C are known as real matrices with appropriate dimensions. For a more general case, we assume

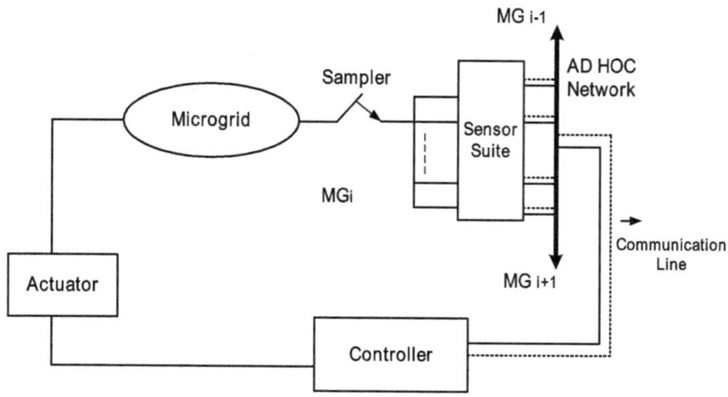

Fig. 6.20 Networked control system

Table 6.3 Pattern of p_k

p_k	q_1	q_2	...	q_{n-1}	q_n
$Prob(p_k = q)$	r_1	r_2	...	r_{n-1}	r_n

that the measurement after passing through the network exhibits a randomly varying communication delay and is described by [39]

$$y_c(k) = \begin{cases} y_p(k - \tau_k^m), \ \delta(k) = 1 \\ y_p(k), \ \delta(k) = 0 \end{cases} \tag{6.99}$$

where τ_k^m stands for *measurement delay*, the occurrence of which satisfies the Bernoulli distribution, and $\delta(k)$ is Bernoulli distributed white sequence exhibiting the occurrence of message (packet) dropouts. It order to capture the current practice of computer communication management that experiences different time-dependent operational modes, we let

$$Prob\{\delta(k) = 1\} = p_k$$

where p_k assumes discrete values, see Table 6.2. Two particular classes can be considered:

Class 1: p_k has the probability mass function where $q_r - q_{r-1}$ = constant for $r = 2, \ldots, n$.

This covers a wide range of cases including

1. If there is no information about the likelihood of different values, we use the uniform discrete distribution, $r_i = 1/n, \ i = 1, 2, \ldots, n$,
2. If it is suspected that p_k follows a symmetric triangle distribution, we use the following function: (i) For n even, $r_i = a + jd, \ j = 0, 1, \ldots, n/2$ and $r_i = a + (n - j)d, \ j = 0, 1, \ldots, n/2 + 1, n/2 + 2, \ldots, n,$ where $na + dn(n-1)/4 = 1$,

(ii) For n odd, $r_i = a + jd, \; j = 0, 1, \ldots, (n-1)/2$ and $r_i = a + (n-j)d,$ $j = 0, 1, \ldots, (n+1)/2, (n+2)/2, \ldots, n,$ where $na + dn(n-1)^2/4 = 1$
3. If it is suspected that r_i is a decreasing linear function, we use $r_i = a - jd,$ $j = 0, 1, \ldots, n$ where $na - dn(n-1)/2 = 1$
4. If it is suspected that r_i is a increasing linear function, we use $r_i = a - (n-j)d, \; j = 0, 1, \ldots, n$ where $na - dn(n-1)/2 = 1$

Class 2: $p_k = X/n, \; n > 0$ and $0 \le X \le n$ is a random variable that follows the binomial distribution $\mathbf{B}(q, n), \; q > 0$, that is

$$Prob(p_k = (ax+b)/n) = \binom{n}{x} q^x (1-q)^{n-x}, \quad b > 0,$$
$$x = 0, 1, 2, \ldots, n, \quad an + b < n$$

Remark 6.6 It is significant to note that the case $Prob\{\delta(k) = 1\} = \bar{\delta},$ where $\bar{\delta}$ is a constant value, is widely used in majority of results on NCS. In this paper, we focus on nonstationary dropouts.

When the full state information is not available and the time delay occurs on the actuation side, it is desirable to design the following observer-based controller [21]:

$$\begin{aligned} &Observer: \\ &\hat{x}(k+1) = A\hat{x} + Bu_p(k) + L(y_c(k) - \hat{y}_c(k)) \\ &\hat{y}_c(k) = \begin{cases} C\hat{x}(k), \; \delta(k) = 0 \\ C\hat{x}(k - \tau_k^m), \; \delta(k) = 1 \end{cases} \end{aligned} \tag{6.100}$$

$$\begin{aligned} &Controller: \\ &u_c(k) = K\hat{x}(k) \\ &u_p = \begin{cases} u_c(k), \; \alpha(k) = 0 \\ u_c(k - \tau_k^a), \; \alpha(k) = 1 \end{cases} \end{aligned} \tag{6.101}$$

where $\hat{x}(k) \in \Re^n$ is the estimate of the system (6.98), $\hat{y}_c(k) \in \Re^p$ is the observer output, and $L \in \Re^{n \times p}$ and $K \in \Re^{m \times n}$ are the observer and controller gains, respectively, and τ_k^a is the *actuation delay*. The stochastic variable $\alpha(k)$, mutually independent of δ, is also a Bernoulli distributed white sequence with

$$Prob\{\alpha(k) = 1\} = s_k$$

where s_k assumes discrete values. By similarity, a particular class is that s_k has some probability mass function as in Table 6.3, where $s_r - s_{r-1} =$ constant for $r = 2, \ldots, n$.

In this chapter, we assume that τ_k^a and τ_k^m are time varying and have the following bounded condition (Table 6.4):

$$\tau_m^- \le \tau_k^m \le \tau_m^+, \quad \tau_a^- \le \tau_k^a \le \tau_a^+ \tag{6.102}$$

Table 6.4 Pattern of s_k

s_k	s_1	s_2	$\cdots$	s_{n-1}	s_n
$Prob(s_k = t)$	t_1	t_2	$\cdots$	t_{n-1}	t_n

Define the estimation error by $e(k) = x_p(k) - \hat{x}(k)$. Then, it yields

$$x_p(k+1) = \begin{cases} Ax_p(k) + BKx_p(k - \tau_k^\alpha) \\ -BKe(k - \tau_k^\alpha), \quad \alpha(k) = 1, \\ (A + BK)x_p(k) - BKe(k), \\ \qquad\qquad \alpha(k) = 0, \end{cases} \tag{6.103}$$

$$\begin{aligned} e(k+1) &= x_p(k+1) - \hat{x}(k+1) \\ &= \begin{cases} Ae(k) - LCe(k - \tau_k^m), & \delta(k) = 1, \\ (A - LC)e(k), & \delta(k) = 0 \end{cases} \end{aligned} \tag{6.104}$$

In terms of $\xi(k) = [x_p^T(k) \quad e^T(k)]^T$, systems (6.103) and (6.104) can be cast into the form:

$$\xi(k+1) = \mathsf{A}_j \xi(k) + \mathsf{B}_j \xi(k - \tau_k^m) + \mathsf{C}_j \xi(k - \tau_k^a) \tag{6.105}$$

where $\{\mathsf{A}_j, \mathsf{B}_j, \mathsf{C}_j, j = 1, \ldots, 4\}$ and j is an index identifying one of the following pairs $\{(\delta(k) = 1, \alpha(k) = 1), (\delta(k) = 1, \alpha(k) = 0), (\delta(k) = 0, \alpha(k) = 0), (\delta(k) = 0, \alpha(k) = 1)\}$:

$$\begin{aligned} \mathsf{A}_1 &= \begin{bmatrix} A & 0 \\ 0 & A \end{bmatrix}, \ \mathsf{A}_2 = \begin{bmatrix} A + BK & -BK \\ 0 & A \end{bmatrix}, \\ \mathsf{A}_3 &= \begin{bmatrix} A + BK & -BK \\ 0 & A - LC \end{bmatrix}, \ \mathsf{A}_4 = \begin{bmatrix} A & 0 \\ 0 & A - LC \end{bmatrix}, \\ \mathsf{B}_1 &= \begin{bmatrix} BK & -BK \\ 0 & 0 \end{bmatrix}, \ \mathsf{B}_2 = \begin{bmatrix} 0 & 0 \\ 0 & 0 \end{bmatrix}, \\ \mathsf{B}_3 &= \begin{bmatrix} 0 & 0 \\ 0 & 0 \end{bmatrix}, \ \mathsf{B}_4 = \begin{bmatrix} BK & -BK \\ 0 & 0 \end{bmatrix}, \\ \mathsf{C}_1 &= \begin{bmatrix} 0 & 0 \\ 0 & -LC \end{bmatrix}, \ \mathsf{C}_2 = \begin{bmatrix} 0 & 0 \\ 0 & -LC \end{bmatrix}, \\ \mathsf{C}_3 &= \begin{bmatrix} 0 & 0 \\ 0 & 0 \end{bmatrix}, \ \mathsf{C}_4 = \begin{bmatrix} 0 & 0 \\ 0 & 0 \end{bmatrix} \end{aligned} \tag{6.106}$$

Remark 6.7 It is remarked for simulation processing that we can express (6.103) and (6.104) in the form

$$x_p(k+1) = s_k[Ax_p(k) + BKx_p(k - \tau_k^{\alpha})$$
$$-BKe(k - \tau_k^{\alpha})]$$
$$+(1 - s_k)[(A + BK)x_p(k) - BKe(k)] \quad (6.107)$$
$$e(k+1) = p_k[Ae(k) - LCe(k - \tau_k^{m})]$$
$$+(1 - p_k)[(A - LC)e(k)] \quad (6.108)$$

where the values of the random variables $p_k,\ s_k$ are generated in the manner discussed earlier.

Remark 6.8 It is important to note from (6.106) that

$$\mathsf{A}_j + \mathsf{B}_j + \mathsf{C}_j = \begin{bmatrix} A + BK & -BK \\ 0 & A - LC \end{bmatrix}, j = 1, \ldots, 4 \quad (6.109)$$

The interpretation of this result is that $\mathsf{A}_j + \mathsf{B}_j + \mathsf{C}_j$ represents the fundamental matrix of the delayed system (6.105), which must be independent of the mode of operation. This will help in simplifying the control design algorithm.

The aim of this section is to design an observer-based feedback stabilizing controller in the form of (6.100) and (6.101) such that the closed-loop system (6.105) is exponentially stable in the mean square. Our approach is based on the concepts of switched TDSs [24]. For simplicity in exposition, we introduce

$$\sigma_1(k) = Prob\{\delta(k) = 1,\ \alpha(k) = 1\},\ \ \hat{\sigma}_1 = \mathrm{E}[\sigma_1]$$
$$\sigma_2(k) = Prob\{\delta(k) = 1,\ \alpha(k) = 0\},\ \ \hat{\sigma}_2 = \mathrm{E}[\sigma_2]$$
$$\sigma_3(k) = Prob\{\delta(k) = 0, \alpha(k) = 0\},\ \ \hat{\sigma}_3 = \mathrm{E}[\sigma_3]$$
$$\sigma_4(k) = Prob\{\delta(k) = 0,\ \alpha(k) = 1\},\ \ \hat{\sigma}_4 = \mathrm{E}[\sigma_4] \quad (6.110)$$

where $\mathrm{E}[\sigma_i]$ is the expected value of $\sigma_i,\ i = 1, \ldots, 4$. Since we assume that $\delta(k)$ and $\alpha(k)$ are independent random variables, then it follows that

$$\hat{\sigma}_1 = \mathrm{E}[p_k]\mathrm{E}[s_k],\ \ \hat{\sigma}_2 = \mathrm{E}[p_k]\mathrm{E}[1 - s_k]$$
$$\hat{\sigma}_3 = \mathrm{E}[1 - p_k]\mathrm{E}[1 - s_k],\ \ \hat{\sigma}_4 = \mathrm{E}[1 - p_k]\mathrm{E}[s_k] \quad (6.111)$$

In what follows, we will thoroughly investigate the stability analysis and controller synthesis problems for the closed-loop system (6.105). First, let us deal with the stability analysis problem and derive a sufficient condition under which the closed-loop system (6.105) with the given controller (6.100) and (6.101) is exponentially stable in the mean square. Extending on [23], the following Lyapunov function candidate is constructed to establish the main theorem:

$$V(\xi(k)) = \sum_{i=1}^{5} V_i(\xi(k)) \quad (6.112)$$

$$V_1(\xi(k)) = \sum_{j=1}^{4} \xi^T(k)P\xi(k), \ P > 0$$

$$V_2(\xi(k)) = \sum_{j=1}^{4} \sum_{i=k-\tau_k^m}^{k-1} \xi^T(i)Q_j\xi(i), \ Q_j = Q_j^T > 0$$

$$V_3(\xi(k)) = \sum_{j=1}^{4} \sum_{i=k-\tau_k^a}^{k-1} \xi^T(i)Q_j\xi(i)$$

$$V_4(\xi(k)) = \sum_{j=1}^{4} \sum_{\ell=-\tau_m^+ +2}^{-\tau_m^- +1} \sum_{i=k+\ell-1}^{k-1} \xi^T(i)Q_j\xi(i)$$

$$V_5(\xi(k)) = \sum_{j=1}^{4} \sum_{\ell=-\tau_a^+ +2}^{-\tau_a^- +1} \sum_{i=k+\ell-1}^{k-1} \xi^T(i)Q_j\xi(i) \quad (6.113)$$

It is not difficult to show that there exist real scalars $\mu > 0$ and $\upsilon > 0$ such that

$$\mu\|\xi\|^2 \leq V(\xi(k)) \leq \upsilon\|\xi(k)\|^2 \quad (6.114)$$

Remark 6.9 By carefully considering Remark 6.8 in view of model (6.105), it is justified to select matrix P to be the same for all operational modes, hence independent of j, while keeping matrix Q_j dependent on mode j.

We now present the analysis result for system (6.105) to be exponentially stable.

Theorem 6.3 *Let the controller and observer gain matrices K and L be given. The closed-loop system (6.105) is exponentially stable if there exist matrices $0 < P$, $0 < Q_j^T = Q_j$, $j = 1, \ldots, 4$ and matrices R_i, S_i, and M_i, $i = 1, 2$, such that the following matrix inequality holds [39]*

$$\Lambda_j = \begin{bmatrix} \Lambda_{1j} & \Lambda_{2j} \\ \bullet & \Lambda_{3j} \end{bmatrix} < 0 \quad (6.115)$$

$$\Lambda_{1j} = \begin{bmatrix} \Psi_j + \Phi_{j1} & -R_1 + S_1^T & -R_2 + S_2^T \\ \bullet & -S_1 - S_1^T - \hat{\sigma}_j Q_j & 0 \\ \bullet & \bullet & -S_2 - S_2^T - \hat{\sigma}_j Q_j \end{bmatrix}$$

$$\Lambda_{2j} = \begin{bmatrix} -R_1 + M_1^T - \Phi_{j2} & -R_2 + M_2^T - \Phi_{j3} \\ -S_1 - M_1^T & 0 \\ 0 & -S_2 - M_2^T \end{bmatrix}$$

$$\Lambda_{3j} = \begin{bmatrix} -M_1 - M_1^T + \Phi_{j4} & \Phi_{j5} \\ \bullet & -M_2 - M_2^T + \Phi_{j6} \end{bmatrix} \quad (6.116)$$

where

$$\begin{aligned}\Psi_j &= -P + \hat{\sigma}_j(\tau_m^+ - \tau_m^- + \tau_a^+ - \tau_a^- + 2)Q_j \\ &\quad + R_1 + R_1^T + R_2 + R_2^T \\ \Phi_{j1} &= (\mathsf{A}_j + \mathsf{B}_j + \mathsf{C}_j)^T \hat{\sigma}_j P(\mathsf{A}_j + \mathsf{B}_j + \mathsf{C}_j) \\ \Phi_{j2} &= (\mathsf{A}_j + \mathsf{B}_j + \mathsf{C}_j)^T \hat{\sigma}_j P\mathsf{B}_j \\ \Phi_{j3} &= (\mathsf{A}_j + \mathsf{B}_j + \mathsf{C}_j)^T \hat{\sigma}_j P\mathsf{C}_j,\ \ \Phi_{j5} = \mathsf{B}_j^T P\mathsf{C}_j \\ \Phi_{j4} &= \mathsf{B}_j^T \hat{\sigma}_j P\mathsf{B}_j,\ \ \Phi_{j6} = \mathsf{C}_j^T \hat{\sigma}_j P\mathsf{C}_j\end{aligned}$$

Proof Defining $y(k) = x(k+1) - x(k)$, one has

$$\xi(k - \tau_k^m) = \xi(k) - \sum_{i=k-\tau_k^m}^{k-1} y(i) \tag{6.117}$$

$$\xi(k - \tau_k^a) = \xi(k) - \sum_{i=k-\tau_k^a}^{k-1} y(i) \tag{6.118}$$

Then the system (6.105) can be transformed into

$$\xi(k+1) = (\mathsf{A}_j + \mathsf{B}_j + \mathsf{C}_j)\xi(k) - \mathsf{B}_j\lambda(k) - \mathsf{C}_j\gamma(k) \tag{6.119}$$

where

$$\lambda(k) = \sum_{i=k-\tau_k^m}^{k-1} y(i),\ \ \gamma(k) = \sum_{i=k-\tau_k^a}^{k-1} y(i).$$

Evaluating the difference of $V_1(\xi(k))$ along the solution of system (6.119), we have

$$\begin{aligned}\mathrm{E}[\Delta V_1(\xi(k))] &= \mathrm{E}[V_1(\xi(k+1))] - V_1(\xi(k)) \\ &= \sum_{j=1}^{4}\Big[\xi^T(k)[\Phi_{j1} - P]\xi(k) - 2\xi^T(k)\Phi_{j2}\lambda(k) \\ &\quad - 2\xi^T(k)\Phi_{j3}\gamma(k) + \lambda^T(k)\Phi_{j4}\lambda(k) \\ &\quad + 2\lambda^T(k)\Phi_{j5}\gamma(k) + \gamma^T(k)\Phi_{j6}\gamma(k)\Big]\end{aligned} \tag{6.120}$$

A straightforward computation gives

$$\begin{aligned}
\mathrm{E}[\Delta V_2(\xi(k)] &= \sum_{j=1}^{4}\hat{\sigma}_j\Bigg[\sum_{i=k+1-\tau_{k+1}^m}^{k}\xi^T(i)Q_j\xi(i) \\
&\quad - \sum_{i=k-\tau_k^m}^{k-1}\xi^T(i)Q_j\xi(i) \\
&= \xi^T(k)Q\xi(k) - \xi(k-\tau_k^m)Q_j\xi(k-\tau_k^m) \\
&\quad + \sum_{i=k+1-\tau_{k+1}^m}^{k-1}\xi^T(i)Q_j\xi(i) - \sum_{i=k+1-\tau_k^m}^{k-1}\xi(i)Q_j\xi(i)\Bigg]
\end{aligned} \tag{6.121}$$

In view of

$$\begin{aligned}
&\sum_{i=k+1-\tau_{k+1}^m}^{k-1}\xi^T(i)Q_j\xi(i) \\
&= \sum_{i=k+1-\tau_{k+1}^m}^{k-\tau_k^m}\xi^T(i)Q_j\xi(i) + \sum_{i=k+1-\tau_k^m}^{k-1}\xi^T(i)Q_j\xi(i) \\
&\le \sum_{i=k+1-\tau_k^m}^{k-1}\xi^T(i)Q_j\xi(i) + \sum_{i=k+1-\tau_m^+}^{k-\tau_m^-}\xi^T(i)Q_j\xi(i)
\end{aligned} \tag{6.122}$$

We readily obtain

$$\begin{aligned}
\mathrm{E}[\Delta V_2(\xi(k))] \le \sum_{j=1}^{4}\hat{\sigma}_j\Bigg[&\xi^T(k)Q_j\xi(k) \\
&- \xi^T(k-\tau_k^m)Q_j\xi(k-\tau_k^m) \\
&+ \sum_{i=k+1-\tau_m^+}^{k-\tau_m^-}\xi^T(i)Q_j\xi(i)\Bigg]
\end{aligned} \tag{6.123}$$

Following parallel procedure, we get

$$\begin{aligned}
\mathrm{E}[\Delta V_3(\xi(k))] &\le \sum_{j=1}^{4}\hat{\sigma}_j\Bigg[\xi^T(k)Q_j\xi(k) \\
&\quad - \xi^T(k-\tau_k^a)Q_j\xi(k-\tau_k^a) + \sum_{i=k+1-\tau_a^+}^{k-\tau_a^-}\xi^T(i)Q_j\xi(i)\Bigg]
\end{aligned} \tag{6.124}$$

Finally

$$\begin{aligned}
\mathrm{E}[\Delta V_4(\xi(k))] &= \sum_{j=1}^{4} \hat{\sigma}_j \Bigg[\sum_{\ell=-\tau^{+}{}_{m}+2}^{-\tau^{-}{}_{m}+1} [\xi^T(k) Q_j \xi(k) \\
&\quad - \xi^T(k+\ell-1) Q_j \xi(k+\ell-1)] \Bigg] \\
&= \sum_{j=1}^{4} \hat{\sigma}_j \Bigg[(\tau^{+}{}_{m} - \tau^{-}{}_{m}) \xi^T(k) Q_j \xi(k) \\
&\quad - \sum_{i=k+1-\tau^{+}{}_{m}}^{k-\tau^{-}{}_{m}} \xi^T(i) Q_j \xi(i) \Bigg] \qquad (6.125)
\end{aligned}$$

$$\begin{aligned}
\mathrm{E}[\Delta V_5(\xi(k))] &= \sum_{j=1}^{4} \hat{\sigma}_j \Bigg[(\tau^{+}{}_{a} - \tau^{-}{}_{a}) \xi^T(k) Q_j \xi(k) \\
&\quad - \sum_{i=k+1-\tau^{+}{}_{a}}^{k-\tau^{-}{}_{a}} \xi^T(i) Q_j \xi(i) \Bigg] \qquad (6.126)
\end{aligned}$$

It follows from (6.117) and (6.118) that:

$$\xi(k) - \xi(k - \tau_k^m) - \lambda(k) = 0 \qquad (6.127)$$

$$\xi(k) - \xi(k - \tau_k^a) - \gamma(k) = 0 \qquad (6.128)$$

Therefore, for any appropriately dimensioned matrices R_i, S_i and M_i, $i = 1, 2$, we have the following equations:

$$2[\xi^T(k) R_1 + \xi^T(k - \tau_k^m) S_1 + \lambda^T(k) M_1][\xi(k) - \xi(k - \tau_k^m) - \tau(k)] = 0 \qquad (6.129)$$

$$2[\xi^T(k) R_2 + \xi^T(k - \tau_k^a) S_2 + \gamma^T(k) M_2][\xi(k) - \xi(k - \tau_k^a) - \gamma(k)] = 0 \qquad (6.130)$$

On combining (6.120)–(6.130), we reach

$$\begin{aligned}
\mathrm{E}[\Delta V(\xi(k))] \le \sum_{j=1}^{4} \Bigg[& \xi^T(k) \Psi_j \xi(k) + \sum_{j=1}^{4} \xi^T(k)(-2R_1 + 2S_1^T) \xi(k - \tau_k^m) \\
& + \xi^T(k)(-2R_2 + 2S_2^T) \xi(k - \tau_k^a) \\
& + \xi^T(k)(-2R_1 + 2M_1^T - 2\Phi_{j2}) \lambda(k) \\
& + \xi^T(k)(-2R_2 + 2M_2^T - 2\Phi_{j3}) \gamma(k) \\
& + \xi^T(k - \tau_k^m)(-S_1 - S_1^T - \hat{\sigma}_j Q_j) \xi(k - \tau_k^m) \\
& + \xi^T(k - \tau_k^m)(-2S_1 - 2M_1^T) \lambda(k)
\end{aligned}$$

$$
\begin{aligned}
&+ \xi^T(k-\tau_k^a)(-S_2 - S_2^T - \hat{\sigma}_j Q_j)\xi(k-\tau_k^a) \\
&+ \xi^T(k-\tau_k^a)(-2S_2 - 2M_2^T)\gamma(k) \\
&+ \lambda^T(k)(-M_1 - M_1^T + \Phi_{j4})\lambda(k) \\
&+ \gamma^T(k)(-M_2 - M_2^T + \Phi_{j5})\gamma(k) \\
&+ \lambda^T(k)\Phi_{j6}\gamma(k)\Big] = \sum_{j=1}^{4}\Big[\zeta^T(k)\widetilde{\Lambda}_j\zeta(k)\Big]
\end{aligned} \tag{6.131}
$$

where

$$
\begin{aligned}
\zeta(k) &= \big[\,\zeta_1^T\ \zeta_2^T\,\big]^T,\ \zeta_2 = \big[\,\lambda^T(k)\ \gamma^T(k)\,\big]^T \\
\zeta_1 &= \big[\,\xi^T(k)\ \xi^T(k-\tau_k^m)\ \xi^T(k-\tau_k^a)\,\big]^T
\end{aligned} \tag{6.132}
$$

and $\widetilde{\Lambda}_j$ corresponds to Λ_j in (6.116) by Schur complements. If $\Lambda_j < 0,\ j = 1, \ldots, 4$ holds, then

$$
\begin{aligned}
\mathrm{E}[V(\xi(k+1)) - V(\xi(k))] &= \sum_{j=1}^{4}\Big[\zeta^T(k)\widetilde{\Lambda}_j\zeta(k)\Big] \\
&\le \sum_{j=1}^{4}\Big[-\widetilde{\Lambda}_{min}(\widetilde{\Lambda}_j)\zeta^T(k)\zeta(k)\Big] \\
&< -\sum_{j=1}^{4}\Big[\beta_j\zeta^T(k)\zeta(k)\Big]
\end{aligned} \tag{6.133}
$$

where

$$
0 < \beta_j < min\big[\lambda_{min}(\Lambda_j), max\{\lambda_{max}(P),\ \lambda_{max}(Q_j)\}\big]
$$

Inequality (6.133) implies that $\mathrm{E}[V(\xi(k+1)) - V(\xi(k))] < -\phi V(\xi(k)),\ 0 < \phi < 1$. In the manner of [36], we get

$$
||\xi(k)||^2 \le \frac{\upsilon}{\kappa}||\xi(0)||^2(1-\phi)^k + \frac{\lambda}{\mu\phi}
$$

Therefore, it can be verified that the closed-loop system (6.105) is exponentially stable. This completes the proof.

A solution to the problem of the observer-based stabilizing controller design is provided by the following theorem:

Theorem 6.4 *Let the delay bounds $\tau_m^+,\ \tau_m^-,\ \tau_a^+,\ \tau_a^-$ be given. Evaluate the quantities $\hat{\sigma}_j,\ j=1,\ldots,4$. Then the closed-loop system (6.105) is exponentially stable if there exist matrices $0 < X_1,\ X_2,\ Y_1,\ Y_2,\ Z_1,\ 0 < \Xi_j,\ j=1,\ldots,4$ and matrices $\Pi_i,\ \Upsilon_i$ and $\Gamma_i,\ i=1,2$, such that the following matrix inequality holds for $j=1,\ldots,4$:*

$$\begin{bmatrix} \widehat{\Lambda}_{1j} & \widehat{\Lambda}_{2j} & \widehat{\Omega}_j \\ \bullet & \Lambda_{3j} & 0 \\ \bullet & \bullet & -\hat{\sigma}_j \widehat{X} \end{bmatrix} < 0 \tag{6.134}$$

$$\widehat{X} = \begin{bmatrix} X_1 & X_2 \\ X_2^T & X_2 \end{bmatrix} \tag{6.135}$$

$$\widehat{\Psi}_j = -\hat{X} + \hat{\sigma}_j(\tau_m^+ - \tau_m^- + \tau_a^+ - \tau_a^- + 2)\Xi_j + \Pi_1 + \Pi_1^T + \Pi_2 + \Pi_2^T$$

$$\widehat{\Lambda}_{1j} = \begin{bmatrix} \widehat{\Psi}_j & -\Pi_1 + \Upsilon_1^T & -\Pi_2 + \Upsilon_2^T \\ \bullet & -\Upsilon_1 - \Upsilon_1^T - \hat{\sigma}_j \Xi_j & 0 \\ \bullet & \bullet & -\Upsilon_2 - \Upsilon_2^T - \hat{\sigma}_j \Xi_j \end{bmatrix}$$

$$\widehat{\Lambda}_{2j} = \begin{bmatrix} -\Pi_1 + \Gamma_1^T & -\Pi_2 + \Gamma_2^T \\ -\Upsilon_1 - \Gamma_1^T & 0 \\ 0 & -\Upsilon_2 - \Gamma_2^T \end{bmatrix}$$

$$\widehat{\Lambda}_{3j} = \begin{bmatrix} -\Gamma_1 - \Gamma_1^T & 0 \\ \bullet & -\Gamma_2 - \Gamma_2^T \end{bmatrix}$$

$$\widehat{\Omega}_j = \begin{bmatrix} \widehat{\Omega}_{1j} & 0 & 0 & -\widehat{\Omega}_{4j} & -\widehat{\Omega}_{5j} \end{bmatrix}$$

$$\widehat{\Omega}_{1j} = \begin{bmatrix} X_1 A^T + Y_1^T B^T - Y_1^T B^T Z_1 & X_2 A^T - Y_2^T \\ X_2 A^T & X_2 A^T - Y_2^T \end{bmatrix}, \ \forall\, j$$

$$\widehat{\Omega}_{4j} = \begin{bmatrix} Y_1^T B^T - Y_1^T B^T Z_1 & 0 \\ 0 & 0 \end{bmatrix}, \quad j = 1, 4$$

$$\widehat{\Omega}_{5j} = \begin{bmatrix} 0 & -Y_2^T \\ 0 & -Y_2^T \end{bmatrix}, \quad j = 1, 2$$

$$\widehat{\Omega}_{4j} = 0, \ j = 2, 3, \quad \widehat{\Omega}_{5j} = 0, \quad j = 3, 4 \tag{6.136}$$

where the gain matrices are given by

$$K = Y_1 X_1^{-1}, \quad L = Y_2 X_2^{-1} C^{\dagger}.$$

Proof Define

$$\Omega_j = \begin{bmatrix} (\mathsf{A}_j + \mathsf{B}_j + \mathsf{C}_j) & 0 & -\mathsf{B}_j & -\mathsf{C}_j \end{bmatrix}^T$$

then matrix inequality (6.115) can be expressed as

$$\Lambda_j = \widetilde{\Lambda} + \Omega_j P \Omega_j^T < 0 \quad (6.137)$$

$$\widetilde{\Lambda}_j = \begin{bmatrix} \widetilde{\Lambda}_{1j} & \widetilde{\Lambda}_{2j} \\ \bullet & \widetilde{\Lambda}_{3j} \end{bmatrix} < 0$$

$$\widetilde{\Lambda}_{1j} = \begin{bmatrix} \Psi_j & -R_1 + S_1^T & -R_2 + S_2^T \\ \bullet & -S_1 - S_1^T - Q_j & 0 \\ \bullet & \bullet & -S_2 - S_2^T - Q_j \end{bmatrix}$$

$$\widetilde{\Lambda}_{2j} = \begin{bmatrix} -R_1 + M_1^T & -R_2 + M_2^T \\ -S_1 - M_1^T & 0 \\ 0 & -S_2 - M_2^T \end{bmatrix}$$

$$\widetilde{\Lambda}_{3j} = \begin{bmatrix} -M_1 - M_1^T & 0 \\ \bullet & -M_2 - M_2^T \end{bmatrix} \quad (6.138)$$

Setting $\widehat{X} = P^{-1}$, invoking Schur complements, we write matrix Λ_j in (6.137) equivalently as

$$\begin{bmatrix} \widetilde{\Lambda}_{1j} & \widetilde{\Lambda}_{2j} & \Omega_j \\ \bullet & \widetilde{\Lambda}_{3j} & 0 \\ \bullet & \bullet & -\widehat{X} \end{bmatrix} < 0 \quad (6.139)$$

Applying the congruence transformation

$$T_j = diag[\widehat{X},\ \widehat{X},\ \widehat{X},\ \widehat{X},\ \widehat{X},\ I]$$

to matrix inequality in (6.139) and manipulating using (6.135) and

$$\Xi_j = \widehat{X} Q_j \widehat{X},\ \Pi_j = \widehat{X} R_j \widehat{X},\ \Upsilon_j = \widehat{X} S_j \widehat{X},$$
$$\Gamma_j = \widehat{X} M_j \widehat{X},\ Z_1 = X_1^{-1} X_2.$$

we readily obtain matrix inequality (6.134) subject (6.136).

Remark 6.10 The selection of $\widehat{X}$ as given by (6.135) has the advantage of converting the solution of bilinear matrix inequalities to that of seeking the feasibility of linear matrix inequalities and hence avoiding iterative procedures. It should be noted that the LMI (6.134) depends on the average dropout patterns identified by (6.111), which is quite useful in illustrating different operating conditions of the communications network.

Remark 6.11 It is remarked that the implementation of Theorem 6.2 is on-line in nature as it requires calling random generators to pick up numbers corresponding to the scalars $\hat{\sigma}_1, \ldots, \hat{\sigma}_4$ and to evaluate the probabilities in model (6.107) and (6.108) to compute the state and error trajectories. This represents a salient feature not shared by other methods for networked control design under unreliable communication links.

Remark 6.12 In this section, the $\widehat{X}$ matrix considered in Theorem 6.2 is distinct from the one in [39]. The $\widehat{X}$ matrix incorporates two different elements X_1 and X_2 unlike the $\widehat{X}$ matrix in [39] which contains the same element X. This extension results in a distinct LMI term $\widehat{\Omega_j}$ given in (6.24). Further, the expression for calculating the controller and observer gains (K and L) is also different from [39]. This is done to present a more realistic approach in determining the controller and observer gains.

Remark 6.13 The NCS model considered in this paper is distinct from the one presented in [39]. The model represents a typical SoS networked control model consisting of a sensor suite and an ad hoc network for multiple subsystems. Moreover the illustrated example for the proposed control strategy is exclusively a microgrid system of systems.

6.7.6 Simulation Results

In this section, the MG system which is modeled as a NCS of system is simulated. The output feedback controller design is implemented which stabilizes the system in the presence of packet dropouts and delays. The parameter values given in Table 6.1 are substituted in the state matrices A and B presented earlier in this paper. Both the sets of PV–microalternator system are similar in nature but have different parameter values.

The controller and observer gains can be obtained by using the relation $K = Y_1 X_1^{-1}$ and $L = Y_2 X_2^{-1} C^{\dagger}$, respectively. Where X_1, X_2, Y_1, Y_2 can be found by solving the LMI (6.134). The values of $\bar{\alpha}$ and $\bar{\delta}$ are set at 0.1 each. The measurement delay is varied from 1 to 5 and the actuation delay is varied from 3 to 9. The system is discretized at a sampling time of 0.01 s. The gain matrices obtained for both sets are shown below. The figures shown afterward represent the state response of both sets of microalternator–PV system. Each state represents a typical dynamic of the system.

Set 1:

$$K = \begin{bmatrix} K_1 & K_2 \end{bmatrix}$$

$$K_1 = \begin{bmatrix} -0.0003 & -0.0002 & 0.0001 & -0.0002 & 0.0000 & 0.0001 \\ -0.0001 & -0.0016 & -0.0002 & 0.0004 & -0.0003 & -0.0001 \end{bmatrix}$$

$$K_2 = \begin{bmatrix} -0.0112 & 0.0001 & 0.0000 & -0.0006 & -0.0003 & 0.0000 \\ -0.0003 & -0.0001 & -0.0004 & -0.0001 & -0.0002 & -0.0067 \end{bmatrix}$$

$$L = \begin{bmatrix} -0.0076 & -0.0172 \\ -0.0172 & -0.0013 \\ 0.0041 & 0.0005 \\ -0.0244 & -0.0002 \\ 0.0121 & -0.0042 \\ 0.0061 & 0.0004 \\ -0.0069 & -0.0170 \\ -0.0489 & 0.0010 \\ -0.0039 & -0.0009 \\ -0.0002 & 0.0005 \\ 0.0156 & 0.0008 \\ -0.0015 & 0.0221 \end{bmatrix}$$

Set 2:

$$K = \begin{bmatrix} K_1 & K_2 \end{bmatrix}$$

$$K_1 = \begin{bmatrix} -0.0007 & -0.0100 & 0.0028 & 0.0004 & 0.0003 & -0.0419 \\ -0.0005 & -0.0003 & 0.0000 & 0.0020 & -0.0002 & -0.0090 \end{bmatrix}$$

$$K_2 = \begin{bmatrix} 0.0000 & -0.0001 & 0.0016 & 0.0021 & 0.0023 & 0.0000 \\ 0.0000 & 0.0000 & -0.0012 & 0.0011 & 0.0000 & -0.0047 \end{bmatrix}$$

$$L = \begin{bmatrix} -0.0152 & -0.0036 \\ -0.0030 & -0.0003 \\ 0.0016 & -0.0025 \\ 0.0000 & 0.0005 \\ -0.0116 & 0.0010 \\ 0.0008 & 0.0002 \\ -0.0003 & -0.0001 \\ 0.0011 & -0.0126 \\ 0.0000 & 0.0003 \\ 0.0000 & -0.0124 \\ 0.0166 & -0.0008 \\ -0.0003 & 0.0003 \end{bmatrix}$$

It can be seen that the output feedback controller stabilizes the system in the presence of delays and packet dropouts. The effectiveness of the proposed methodology is demonstrated by the plots (Figs. 6.21, 6.22 and 6.23).

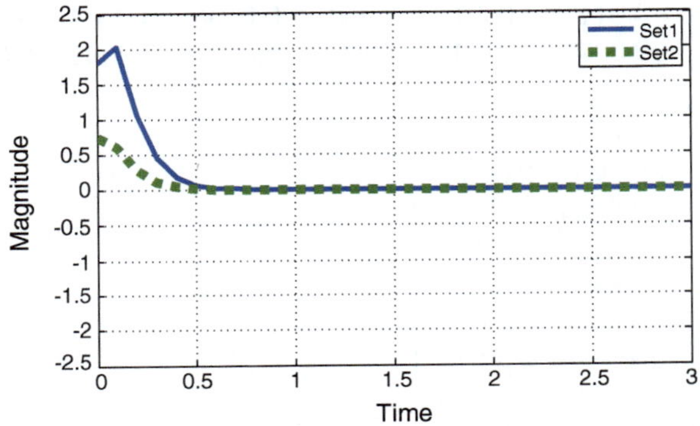

Fig. 6.21 Rotor angle of the microalternator

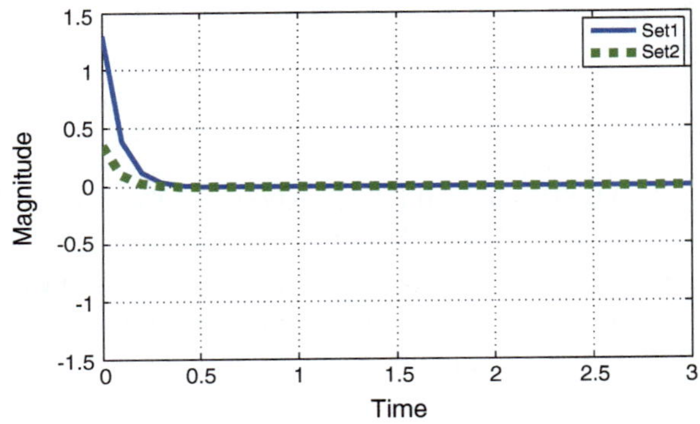

Fig. 6.22 Rotor speed of the microalternator

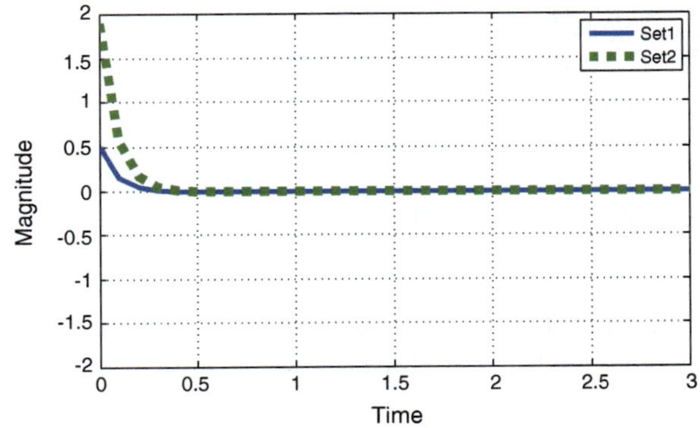

Fig. 6.23 Internal voltage of microalternator along q-axis

6.8 Conclusions

The networked control of microgrid system of systems is presented in this paper. The MG operating in islanded mode and consisting of three DG units supplying a load is devised into a SoS framework where the generation units are considered as three subsystems of the SoS. The SoS networked control system modeling is presented which is subjected to both measurement and actuation delays and nonstationary packet dropouts. Also an observer-based controller is introduced which stabilizes the system in the presence of delays and nonstationary dropouts (Figs. 6.24, 6.25, 6.26, 6.27, 6.28 and 6.29). Simulation results are incorporated to demonstrate the effectiveness of the proposed networked control strategy for the microgrid system of systems.

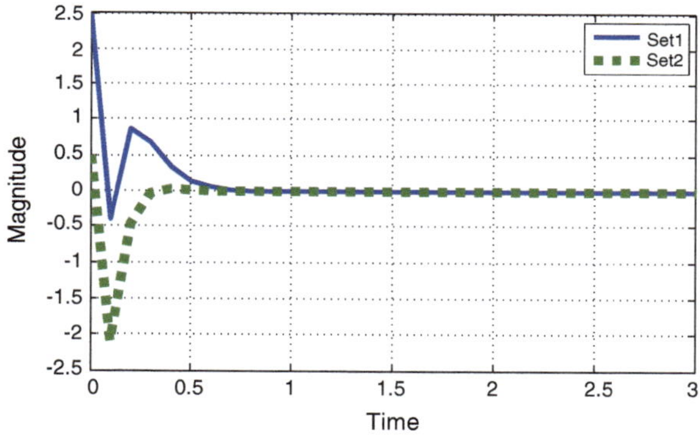

Fig. 6.24 Field voltage of microalternator along d-axis

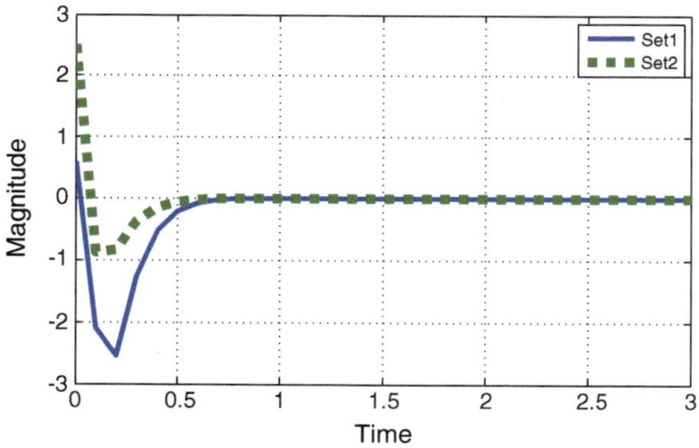

Fig. 6.25 Photovoltaic cell current

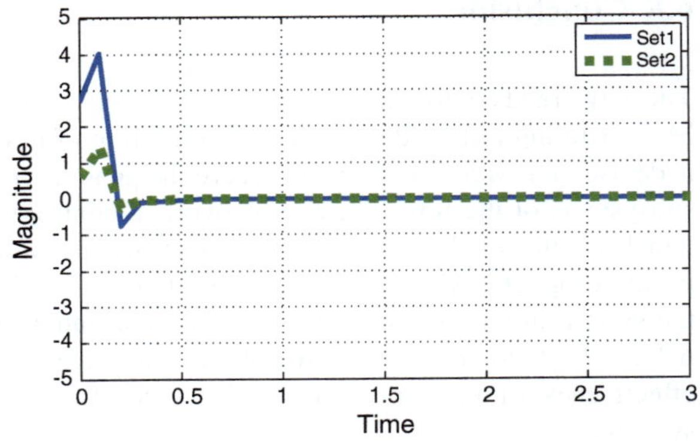

Fig. 6.26 Voltage across DC link capacitor

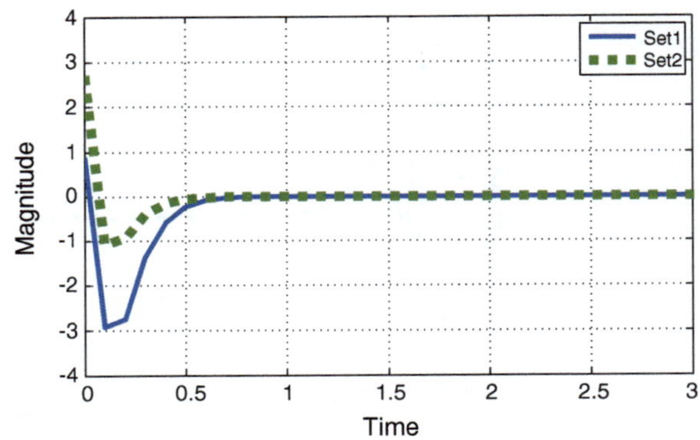

Fig. 6.27 Inverter output current along d-axis

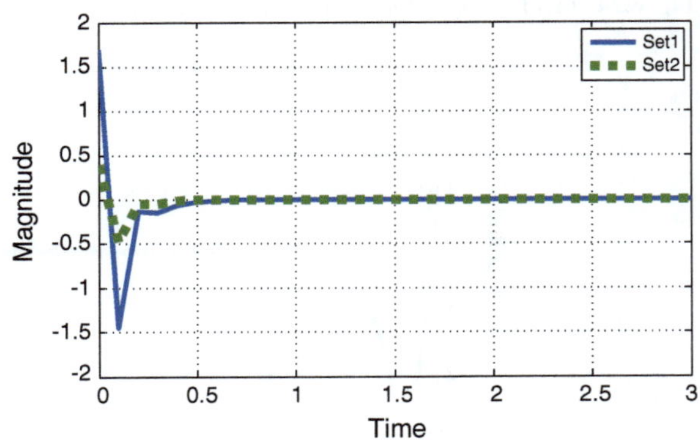

Fig. 6.28 Inverter output current along q-axis

An NCS methodology for the stabilization of a MG system with two sets of microalternator and PV systems is presented in this paper. The microalternator and PV system are modeled in detail. The combination of both systems connected to the main grid and a load is also modeled. The NCS based on this model is subjected to

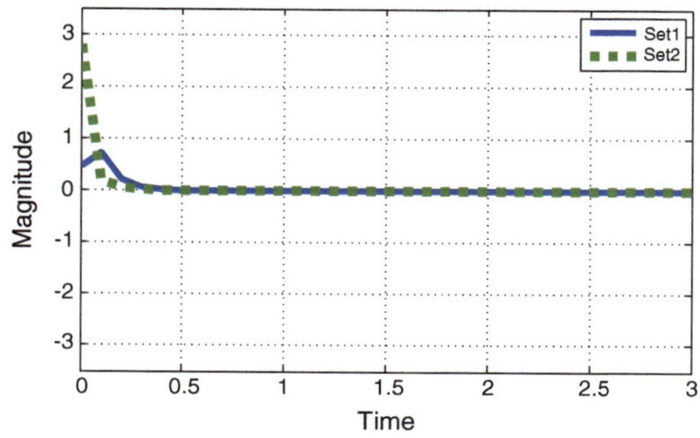

Fig. 6.29 Coupling line current of the filter along d-axis

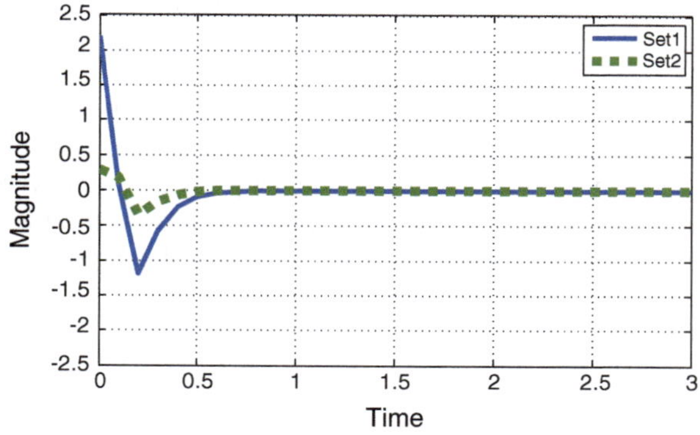

Fig. 6.30 Coupling line current of the filter along q-axis

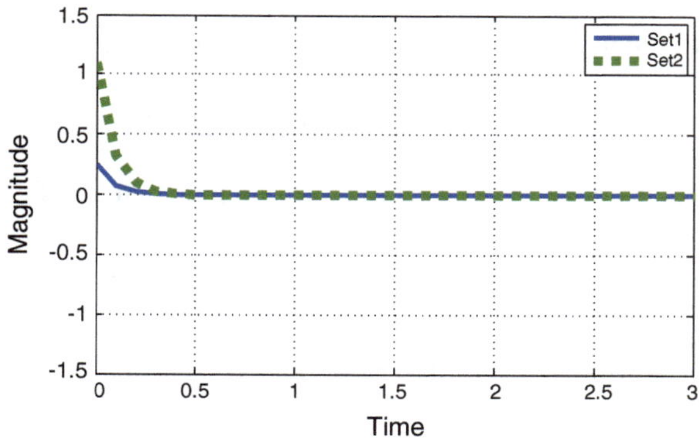

Fig. 6.31 Capacitor voltage of the filter along d-axis

delays and nonstationary packet dropouts. An improved output feedback controller is proposed, which stabilizes the system in the presence of the aforementioned communication infractions (Figs. 6.30, 6.31 and 6.32). Further the simulated results of both sets of microalternator–PV systems are included which demonstrate the effectiveness

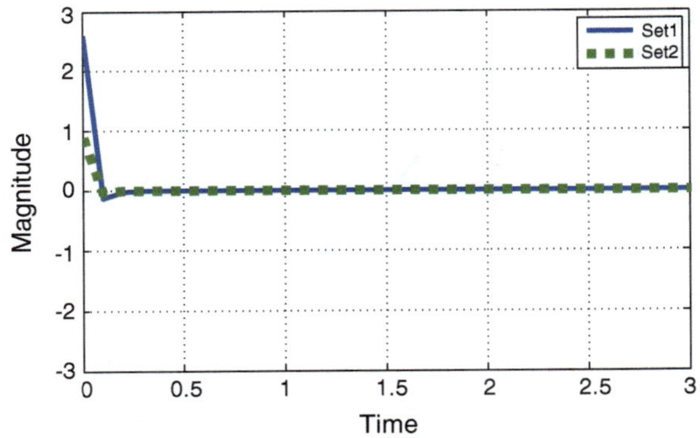

Fig. 6.32 Capacitor voltage of the filter along q-axis

of the controller and attempts to solve a major issue in MGs which is stability of its constituent systems.

Further research can include adopting an output feedback controller in the NCS, and incorporating a Markov process to describe the packet loss procedure. Also additional subsystems can be added to the SoS which may consist of generation units and loads.

6.9 Suggested Problems

Problem 6.1 In the contemporary literature, potential renewable and clean energy sources are interfaced to the grid via pulse-width modulated voltage-sourced-inverters (PWM-VSIs). One candidate of this trend is a multilevel control structure of a current-controlled DG interface employs a power sharing control loop, voltage control loop, and current control loop depicted in Fig. 6.33. The multilevel control structure is widely accepted in DG applications due to the positive benefits of using current-controlled VSIs and its inherent ability to cope with multilevel control and communication standards in power electronic converters.

The objective is to adopt Park's dq-frame to build an appropriate model and hence investigate the dynamic behavior of this model under various operating conditions.

Problem 6.2 Figure 6.34 shows the single line diagram of a sample MG. It is desired to examine the system robustness with respect to voltage disturbances due to the transience of the network-side current and power angle disturbance due to the transition from a grid-frequency-following control to a self-generated-frequency control.

Problem 6.3 A block diagram of the power circuit and the control functions of an inverter-based DG unit in a MG is presented in Fig. 6.35. It is known that due to

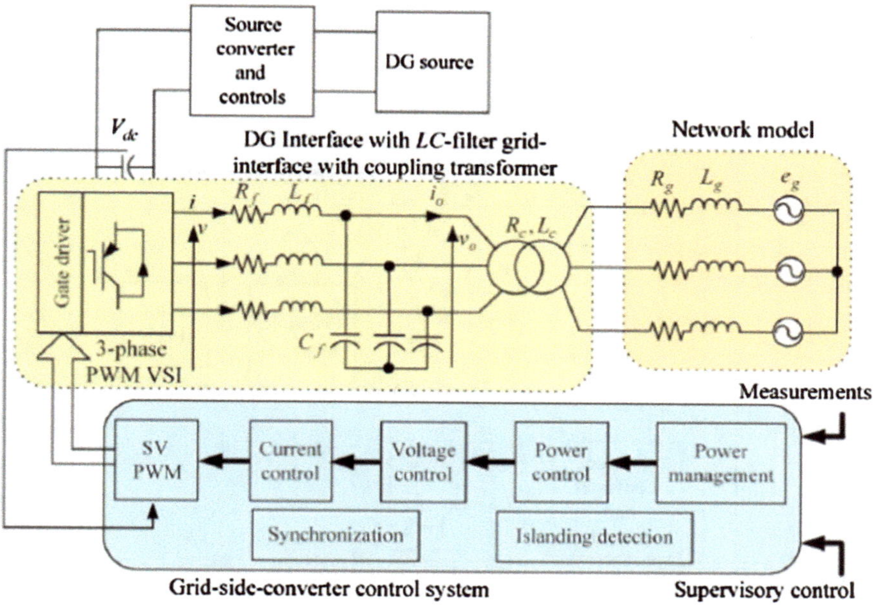

Fig. 6.33 A multilevel control structure of a current-controlled DG interface

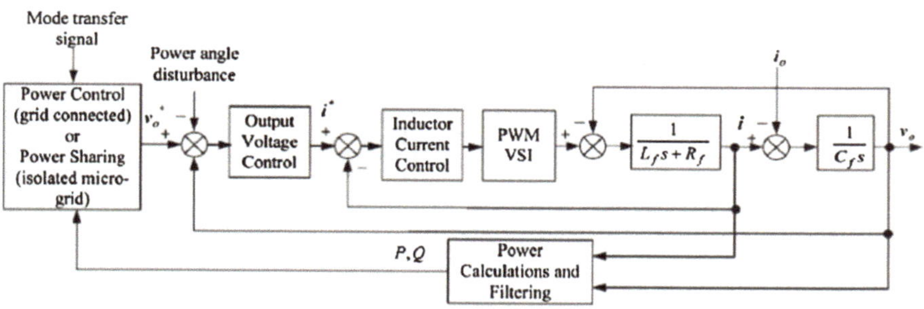

Fig. 6.34 Single line diagram of a sample microgrid

the time– and frequency–scale separations between the power and voltage control loops, voltage and power angle disturbances are weekly coupled. The objective here is to explore the capability of rejecting voltage disturbances by a relatively high bandwidth voltage control loop. In a similar way, it is required to examine the power angle disturbances rejection using the outer power sharing loop.

Problem 6.4 An inverter parallel system which consists of inverter modules and Control Area Network (CAN) bus is shown in Fig. 6.36. Using this architecture, it is desired to investigate the networked control application in power electronics to

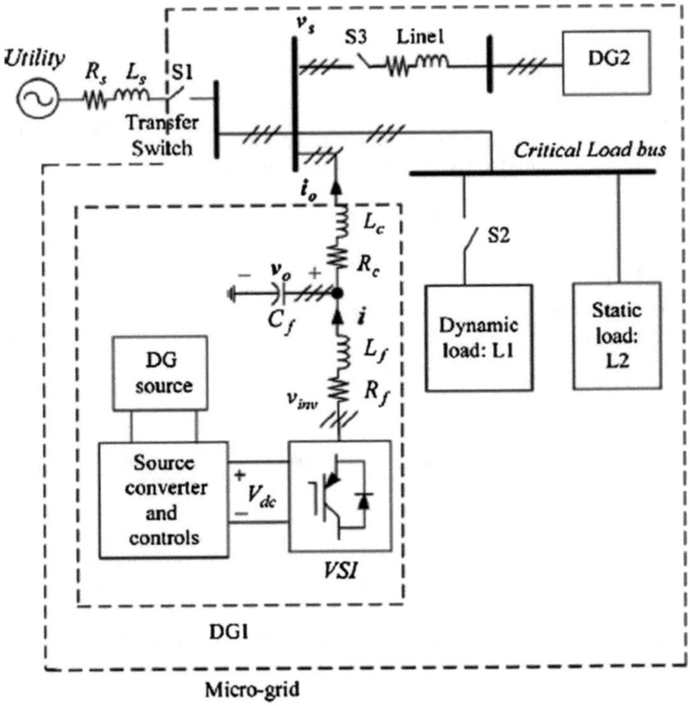

Fig. 6.35 Block diagram of an inverter-based DG unit in a microgrid

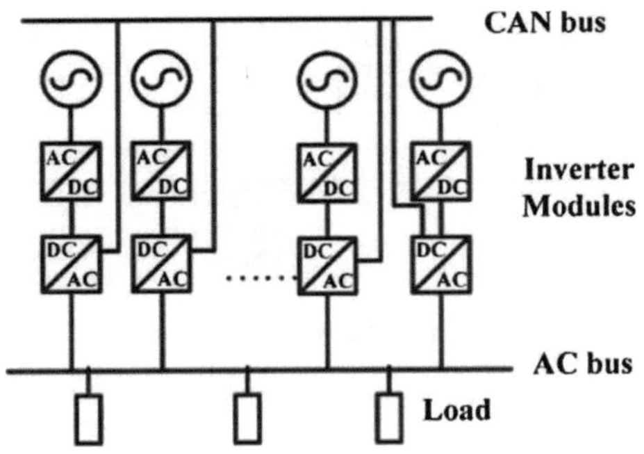

Fig. 6.36 Architecture of inverter parallel system with networked control

improve the load sharing performance in which CAN bus is essentially employed to transfer control signals and realize information sharing among inverters.

Problem 6.5 In Fig. 6.37, the equivalent circuit of inverters connected to an AC bus is presented. The purpose here is to develop a detailed mathematical model expressing the behavior of the active power and reactive power components, the frequency and

Fig. 6.37 Equivalent circuit of j inverters connected to a common AC bus

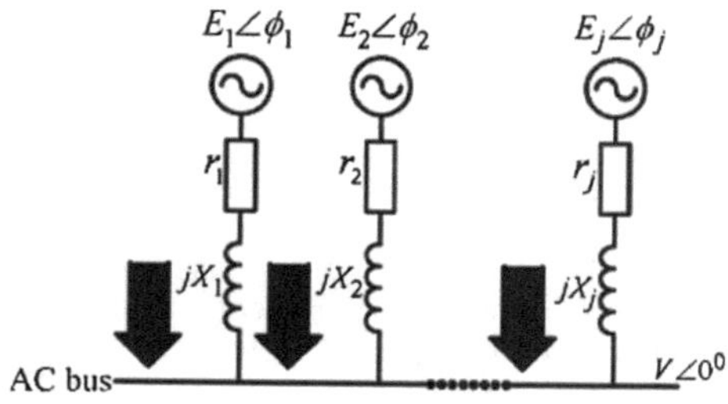

the amplitude of the inverter output voltage reference and identify the transmission signals in the NCS. Then examine the dynamics of the developed model.

References

1. Arulampalam A, Barnes M, Engler A, Goodwin A, Jenkins N (2004) Control of power electronic interfaces in distributed generation microgrids. Int J Electron 91:503–523
2. Ashourian MH, Zin MAA, Mokhtar AS, Mirazimi SJ, Muda Z (2011) Controlling and modeling power-electronic interface DERs in islanding mode operation microgrid. In: Proceedings of IEEE symposium on industrial electronics and applications (ISIEA), pp 161–166
3. Babazadeh M, Karimi H (2011) Robust decentralized control for islanded operation of a microgrid. In: Proceedings of IEEE power and energy society general meeting, pp 1–8
4. European Research Project Microgrids. http://Microgrids.power.ece.ntua.gr
5. European Research Project More Microgrids. http://Microgrids.power.ece.ntua.gr
6. Georgakis D, Papathanassiou S (2004) Operation of a prototype microgrid system based on micro-sources equipped with fast acting power electronics interfaces. In: Proceedings of IEEE 35th annual power electronics specialists conference, vol 4, pp 2521–2526
7. Haruni AMO, Gargoom A, Haque ME, Negnevitsky M (2010) Dynamic operation and control of a hybrid wind-diesel stand alone power systems. In: Proceedings of the twenty-fifth annual IEEE applied power electronics conference and exposition (APEC), pp 162–169
8. Hatziargyriou N, Asano H, Iravani R, Marnay C (2007) Microgrids: an overview of ongoing research, development and demonstration projects. IEEE Power Energy Mag 5:78–94
9. Henshaw M, Siemieniuch C, Sinclair M, Henson S, Barot V, Jamshidi MO, Delaurentis D, Ncube C, Lim SL, Dogan H (2013) Systems of systems engineering: a research imperative. In: Proceedings of IEEE international conference on system science and engineering, Hungary, pp 389–394
10. Hussain MT (2012) Modeling and control of a microgrid including photovoltaic and wind generation. M.Sc. thesis, King Fahd University of Petroleum and Minerals, Saudi Arabia
11. Jamshidi MO (2009) Control of system of systems. In: Proceedings of 7th IEEE international conference on industrial informatics, pp 1–16
12. Jamshidi MO (2008) System of systems engineering: principles and applications. Taylor Francis CRC Publishers, Boca Raton
13. Jamshidi MO (2008) System of systems engineering—new challenges for the 21'st century. IEEE Aerosp Electron Syst Mag 23:4–19
14. Jayalakshmi NS, Gaonkar DN (2011) Performance study of isolated hybrid power system with multiple generation and energy storage units. In: Proceedings of international conference on power and energy systems (ICPS), pp 1–5
15. Jones E, Fitzer C, Barnes M (2006) Investigation of microgrids. In: Proceedings of the 3rd IET international conference on power electronics, machines and drives, pp 510–514

16. Kasal GK, Singh B (2011) Voltage and frequency controllers for an asynchronous generator-based isolated wind energy conversion system. IEEE Trans Energy Convers 26(2):402–416
17. Katiraei F, Iravani MR (2006) Power management strategies for a microgrid with multiple distributed generation units. IEEE Trans Power Syst 21:1821–1831
18. Lasseter RH (2007) Microgrids and distributed generation. J Energy Eng, Am Soc Civ Eng 133(3):144–149
19. Lee DJ, Wang L (2008) Small-signal stability analysis of an autonomous hybrid renewable energy power generation/energy storage system part I: time-domain simulations. IEEE Trans Energy Convers 23(1):311–320
20. Lopes JAP, Moreira CL, Madureira AG (2005) Defining control strategies for analyzing microgrids islanded operation. In: Proceedings of IEEE Russia power technology, pp 1–7
21. Luan X, Shi P, Liu F (2011) Stabilization of networked control systems with random delays. IEEE Trans Ind Electron 58(9):4323–4330
22. Mahamad N, Hadzer CM, Masri S (2004) Application of LC filter in harmonics reduction. In: Proceedings of national power and energy conference, Malaysia
23. Mahmoud MS, Xia Y (2009) Robust stability and stabilization of a class of nonlinear switched discrete-time systems with time-varying delays. J Optim Theory Appl 143:329–355
24. Mahmoud MS (2010) Switched time-delay systems. Springer, New York
25. Mahmoud MS, Selim SZ, Shi P, Baig MH (2012) New results on networked control systems with nonstationary packet dropouts. IET Control Theory Appl 6(15):2442–2452
26. Mahmoud MS, Hussain SA, Abido MA (2014) Modeling and control of microgrid: an overview. J Frankl Inst 351(5):2822–2859
27. Maier MW (1998) Architecting principles for systems-of-systems. Syst Eng 1(4):267–284
28. Moradi R, Karimi H, Karimi-Ghartemani M (2010) Robust decentralized control for islanded operation of two radially connected DG systems. In: Proceedings of IEEE international symposium on industrial electronics (ISIE), pp 2272–2277
29. Piagi P, Lasseter RH (2006) Autonomous control of microgrids. In: Proceedings of IEEE power engineering society general meeting (PES)
30. Prodanovic M, Green TC (2006) High-quality power generation through distributed control of a power park microgrid. IEEE Trans Ind Electron 53(5):1471–1482
31. Quaschning V, Hanitsch R (1996) Numerical simulation of current-voltage characteristics of photovoltaic systems with shaded solar cells. Sol Energy 56:513–520
32. Ray PK, Mohanty SR, Kishor N (2010) Dynamic modeling and control of renewable energy based hybrid system for large band wind speed variation. In: Proceedings of IEEE PES innovative smart grid technologies conference Europe (ISGT Europe), pp 1–10
33. Shabestary SMA, Saeedmanesh M, Rahimi-Kian A, Jalalabadi E (2012) Real-time frequency and voltage control of an islanded mode microgrid. In: Proceedings of the 2nd Iranian conference on smart grids (ICSG), pp 1–6
34. Vasquez JC, Guerrero JM, Miret J, Castilla M, De Vicuna LG (2010) Hierarchical control of intelligent microgrids. IEEE Ind Electron Mag 4:23–29
35. Xiao Z, Wu J, Jenkins N (2010) An overview of microgrid control. Intell Autom Soft Comput 16(2):199–212
36. Yang FW, Wang ZD, Hung YS, Gani M (2006) H_∞ control for networked systems with random communication delays. IEEE Trans Autom Control 51(3):511–518
37. Yu YN (1983) Electric power system dynamics. Academic Press, New York
38. Zhang W, Branicky MS, Phillips SM (2001) Stability of networked control systems. IEEE Control Syst Mag 21(1):84–99
39. Zhang X, Liu J, Liu T, Zhou L (2010) A novel power distribution strategy for parallel inverters in islanded mode microgrid. In: Proceedings of the twenty-fifth annual IEEE applied power electronics conference and exposition (APEC), pp 2116–2120

Chapter 7
Decentralized Voltage Control Methods

7.1 Introduction

The concept of distributed generation (DG) has gained significant acceptance during the last decade [1] due to the (1) high cost of energy, (2) environment concerns, and (3) major advances in DG technologies. Most types of DG units utilize power electronic converters and, in particular, voltage-sourced converter (VSC) units, as the interface media with the utility grid [34, 76]. For a grid-connected DG unit, the interface VSC is conventionally controlled as a current-controlled VSC (CC-VSC) [66]. Thus, direct-quadrature current components of the VSC are used to provide control of instantaneous real- and reactive power exchange (PX) between the VSC and the grid. The VSC synchronization is usually carried by a phase-locked loop (PLL) system, based on voltage signals at the point of common coupling (PCC). The main features of the current-controlled strategy are as follows:

- It provides (almost) independent control of real- and reactive power components.
- It inherently limits the VSC current during abnormal conditions (e.g., grid faults).

Proliferation of Digital Output (DO) units in distribution systems and an increase in the DG depth of penetration [42, 48] indicate that a DG unit is best utilized, if it is capable of:

- operation in a grid-connected mode;
- operation in an islanded or autonomous mode;
- transition between grid-connected and autonomous modes.

In an autonomous (islanded) mode where the frequency and voltage are not dictated by the grid, the conventional current-controlled strategy for the interface VSC is not necessarily the optimum choice. The main reason is that it may result in voltage and/or frequency excursions that lead to either unacceptable operating conditions or instability. To address the problem, the following options can be considered.

M.S. Mahmoud and F.M. AL-Sunni, *Control and Optimization of Distributed Generation Systems*, Power Systems, DOI 10.1007/978-3-319-16910-1_7

- Voltage-controlled VSC (VC-VSC) strategy [6]. Based on this strategy, the VSC provides control over the voltage and frequency at PCC of the DG unit. Drawbacks of this strategy are:
  (1) coupling between instantaneous real- and reactive power components, and
  (2) no inherent current-limit capability for the converter during grid faults.
- Hierarchical (VC-VSC and CC-VSC) strategy. This strategy needs provisions for frequency and voltage supervisory control signals to control the PCC frequency and voltage (in an autonomous mode) through the conventional CC-VSC strategy. However, care must be taken to prevent
  (1) its adverse impact on voltage and/or frequency in the grid-connected mode,
  (2) dynamic interactions among adjacent DO units and synchronous machines (SMs) in an autonomous multi-DG grid, and
  (3) slow dynamic response.
- Hybrid (CC-VSC and VC-VSC) strategy. This approach is based on the transition between CC-VSC and VC-VSC control strategies depending on the mode of operation (i.e., grid-connected or autonomous mode). The main challenge in this approach is a smooth transition between the control strategies which also necessitates fast detection of the change of operating mode.

This section introduces an improved VC-VSC strategy applicable to both grid-connected and autonomous modes of operation of a DG unit. The proposed strategy enables the DG unit:

- to control its exchanged instantaneous real- and reactive power components in a grid-connected mode;
- to control voltage and frequency in an autonomous mode.

Other features of the proposed VC-VSC strategy are that it provides

1. current-limit capability for the VSC during transients,
2. an inherent islanding detection approach [28] without nondetection zone (NDZ),
3. smooth transition capability from a grid-connected mode to an autonomous mode, and
4. ease of compliance with ride-through capability requirements in a grid-connected mode [38].

7.2 Control Strategy I

To describe the proposed control strategy, the study system of Fig. 7.1 is considered. The system is composed of a DO unit and a parallel RLC load, and is required to operate either in a grid-connected or an autonomous (islanded) mode when switch 31 is closed or open, respectively. The DO unit includes a primary energy source and a three-phase VSC. The DO unit is interfaced to the PCC through a series filter and

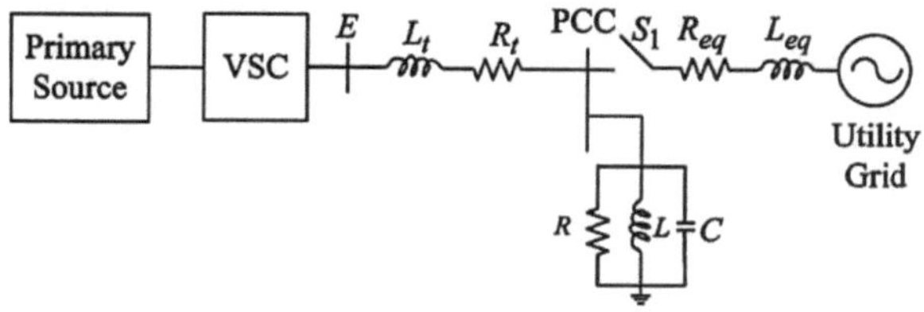

Fig. 7.1 Single-line diagram of the study system

a transformer which are represented by L_t and H_t (Fig. 7.1). The local RLC load is connected to the PCC. The main utility grid is represented by an ideal three-phase voltage source and per-phase R_{eq} and L_{eq} components. To describe the concept of the proposed VSC control, and without the loss of generality, we neglect the dynamics of the DO primary source.

7.2.1 Frequency Control

The proposed VSC frequency control is analogous to that of a SM (i.e., the VSC instantaneous terminal voltage E is to resemble the internal voltage of an SM [46, 57]). The internal voltage of an SM is aligned with the rotor quadrature-axis and rotates with the rotor speed. Due to the rotor relatively high inertia, the frequency of the space vector of the SM internal voltage cannot have abrupt changes, and exhibits low-frequency variations subsequent to the SM dynamics. Thus, we assume a virtual inertia constant H for the VSC unit of Fig. 7.1, such that speed ω of the voltage space vector of the VSC terminal is governed by

$$2H\frac{d\omega}{dt} = P_m - P_{out} \tag{7.1}$$

where P_{out} is the VSC real output power at the PCC and P_m is the VSC input power defined by the droop characteristic

$$P_m - P_{ref} = \frac{1}{D_p}(\omega_{ref} - \omega_{pcJ}). \tag{7.2}$$

ω_{pcc} is the measured frequency at PCC, D_p is the droop constant, and ω_{ref} and P_{ref} are reference values of the converter output frequency and power at the PCC. Considering that the rotating shaft system of an SM is opposed by a viscous damping, (7.1) can be augmented with the damping term $K_D(\omega - \omega_{pcc})$, thus

$$2H\frac{d\omega}{dt} = P_m - P_{out} - K_D(\omega - \omega_{pcc}) \tag{7.3}$$

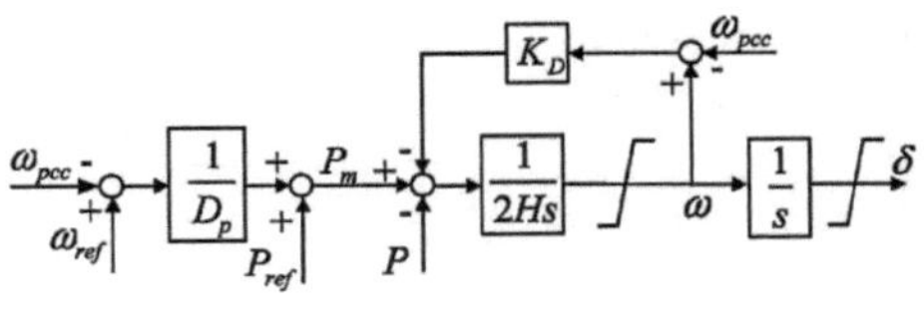

Fig. 7.2 Block diagram of the proposed frequency control of the VSC-based DG unit of Fig. 7.1

where K_D is the damping constant. The relationship between the VSC terminal voltage frequency and position δ is

$$\frac{d\delta}{dt} = \omega. \tag{7.4}$$

Angle δ is required for the VSC modulation scheme. The frequency control approach, based on (7.2)–(7.4), is represented by the block diagram of Fig. 7.2.

When the system of Fig. 7.1 operates in a grid-connected mode (i.e., S_1 is closed), ω_{pcc} is the same as the frequency of the main grid and is fairly constant. Thus, the virtual governor system of Fig. 7.2 should set the reference for the output real power and have no impact on the frequency dynamics. In the grid-connected mode, the damping block ensures that the VSC frequency is the same as that of the main grid.

Under an autonomous operating mode of the system of Fig. 7.1, ω_{PCC} is the same as the VSC frequency. Thus, the damping block Fig. 7.2 has no impact on the system dynamics. The frequency dynamics are determined by the droop characteristic. In the case that the autonomous system includes more than one DO unit, the load disturbance can be shared among the DO units based on appropriate droop characteristics.

7.2.2 Voltage Control

The proposed voltage controller for the DG unit of Fig. 7.1 is shown in Fig. 7.3. The VSC terminal voltage E is determined by signals E_1 and E_2. E_l is determined by the voltage droop block

$$E_1 = E_s - D_q Q \tag{7.5}$$

where Q is the measured output reactive power of the VSC at PCC, D_q is a voltage droop constant, and E_s is the reference value of the VSC terminal voltage. Signal E_2 is the output of the reactive power controller and given by

$$E_2 = \frac{K_Q(Q_{ref} - Q)}{s} \tag{7.6}$$

where K_Q is the integrator gain, and Q_{ref} is the reactive power reference value of the VSC at PCC. Based on Fig. 7.3, the VSC output voltage E is

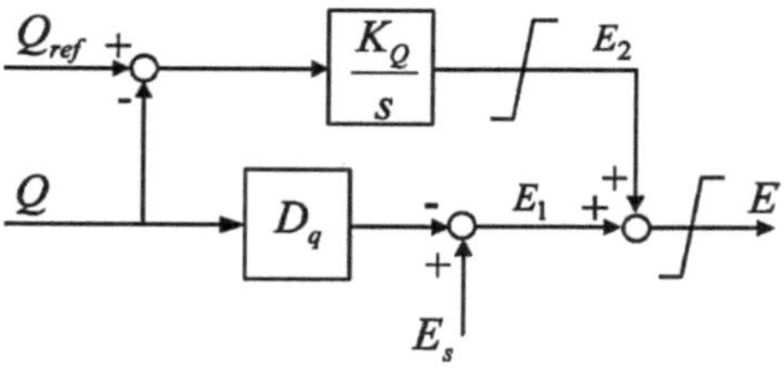

Fig. 7.3 Block diagram of voltage control of the VSC-based DG unit of Fig. 7.1

$$E = E_1 + E_2 = E_s - D_q Q + \frac{K_Q(Q_{ref} - Q)}{s}. \tag{7.7}$$

The voltage control objective of a DG unit depends on its operating mode (i.e., grid-connected or autonomous mode). In a grid-connected mode, voltage control is to set the output reactive power of the VSC at PCC. For example, the reference reactive power is set to zero. This objective can be achieved by the control of Fig. 7.3. However, in an autonomous mode, the DG unit has to supply the load reactive power demand without any reactive power contribution from the main grid. Thus, the reactive power controller (7.6) of the grid-connected mode is no longer applicable and can drive the system unstable. Therefore, E_2 of Fig. 7.3 is set to zero by imposing $K_Q = 0$ during an autonomous mode and only E_1 remains effective. Therefore, the voltage control strategy should change during the transition process from the grid-connected mode to the autonomous mode. This change takes place after islanding is confirmed based on the inherent islanding detection capability of the proposed controller as described in Sect. 7.2.4.

7.2.3 Overcurrent Limiters

A drawback of controlling the VSC of Fig. 7.1 as a voltage-controlled VSC is that the grid faults and disturbances may subject it to overcurrents. To limit overcurrents, two dynamic limiters for δ of Fig. 7.2 and E of Fig. 7.3, are introduced. By means of a PLL system, phase angle δ_{pcc} and amplitude V_{pcc} of the PCC voltage space vector are extracted. The settings of limiters for δ and E are

$$\delta_{pcc} - \Delta\delta \le \delta \le \delta_{pec} + \Delta\delta \tag{7.8}$$

$$V_{pcc} - \Delta E \le E \le V_{pcc} + \Delta E \tag{7.9}$$

where $\Delta\delta$ and ΔE are maximum allowable deviations of δ and E from δ_{pcc} and V_{pcc}, respectively. $\Delta\delta$ and ΔE are defined based on parameters L_t and R_t, and the maximum allowable current of VSC. Limiters for ω of Fig. 7.2 and E_2 of Fig. 7.3 affect the overcurrent indirectly. However, the appropriate selection of these limits can also improve dynamic performance of VSC. The settings for limiters of ω and

E_2 are

$$\omega_{ref} - \Delta\omega \leq \omega \leq \omega_{ref} + \Delta\omega \tag{7.10}$$

$$-\Delta E_2 \leq E_2 \leq \Delta E_2 \tag{7.11}$$

where $\left[\omega_{ref}\ \Delta\omega,\ \omega_{ref} \mid \Delta\omega\right]$ is the allowable frequency range of the VSC. ΔE_2 is determined based on parameters L_t and R_t, and the allowable reactive power range of the VSC.

7.2.4 Islanding Detection Approach

Reactive power control (i.e., upper path of Fig. 7.3), which drives the system to an unstable condition during an autonomous mode of operation, also can be exploited for islanding detection. The detection time may exceed and not satisfy the standard requirements [38] if a PX between the DO and the grid, prior to the islanding instant, is small. To decrease the detection time, a positive feedback with transfer function $G(s)$, based on the instantaneous PCC voltage, is added to the controller of Fig. 7.3 as shown in Fig. 7.4. Transfer function $G(s)$ is given as

$$G(s) = \frac{K_v s}{(1 + T_1 s)(1 + T_2 s)} \tag{7.12}$$

where $1/T_1$ and $1/T_2$ specify upper and lower corner frequencies. $G(s)$ should

(1) ensure that the feedback process is only effective during transients and does not impact the DC steady-state error signal, and
(2) provide attenuation for high-frequency noise.

For example, if a two-second islanding detection is required [38], a 1–10 Hz frequency band for $G(s)$ is selected, and gain K_v is adjusted to achieve the desired islanding detection speed.

In a grid-connected mode of operation,

(1) the PCC voltage is dominantly dictated by the grid, and
(2) a relatively small reactive-power excursion, imposed by the positive feedback control of Fig. 7.4, is also absorbed by the grid and does not result in instability.

To ensure ride-through capability of a grid-connected DG unit, the islanding process must be delayed based on the requirements for "maximum time before cessation of operation" of the adopted standard (e.g., Table 7.1 corresponding to UL1741 [38]). If the PCC voltage violates the specified limits (e.g., those of Table 7.1), then islanding is confirmed. Otherwise, the islanding detection signal is ignored and the unit continues operation. Figures 7.2 and 7.4 show that inputs to the proposed VSC control system are: (1) instantaneous real- and reactive power components of the

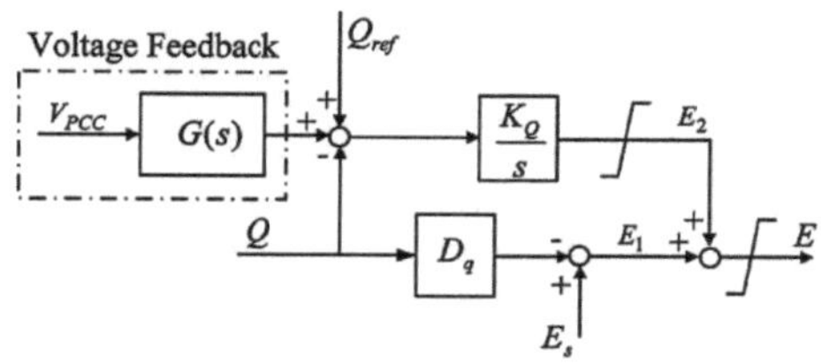

Fig. 7.4 Block diagram of voltage control with positive feedback of the VSC-based DG unit of Fig. 7.1

Table 7.1 Voltage limits based on UL1741

Utility voltage, V (pu)	Maximum time before cessation of operation (s)
$V < 0.5$	0.16
$0.5 \leq V \leq 0.88$	2
$1.1 < V \leq 1.2$	1
V > 1.2	0.16

VSC at PCC, (2) voltage magnitude and phase angle at PCC, and (3) electrical frequency at the PCe. The voltage phase angle and the frequency signal are obtained from the PCC voltage by means of a PLL system. Other signals are obtained from instantaneous PCC voltage and the VSC current. Thus, the control system of Figs. 7.2 and 7.4 is readily implemented.

7.3 Small-Signal Dynamic Analysis

To optimize parameters and demonstrate performance of the proposed controls, a set of small-signal dynamic analysis is performed on the system of Fig. 7.1. The studies include both autonomous (islanded) and grid-connected modes of operation. A small-signal dynamic model of the grid-connected system is developed in the grid dq frame such that the space vector of the grid voltage source is aligned with the direct-axis. Also a small-signal dynamic model of the autonomous system is developed in the VSC dq frame such that the space vector of the VSC terminal voltage is aligned with the direct-axis. The system parameters are given in Table 7.2.

Table 7.3 shows the system eigenvalues corresponding to the (1) grid-connected mode, (2) autonomous mode with reactive power (Q) control, and (3) autonomous mode without reactive power (Q) control. The DG unit supplies I.O/O.O-p.u. real/reactive power in the grid-connected mode. When isolated from the main grid, the DG output power depends on the local RLC load demand (LD). Column 5 of Table 7.3 identifies the dominant states associated with each eigenvalue based on the participation factors. The mode associated with eigenpair (5, 6) is dominantly associated with the RLC load.

Table 7.2 System parameters of Fig. 7.1

Parameters	Physical values	Per unit
Rated capacity	100 kVA	–
Rated voltage (ll,rms)	360 V	–
Rated frequency	60 Hz	–
L_t	0.35 mH	0.1
R_t	0.026 n	0.02
L	2.76 mH	0.8
C	2.55 mF	1.25
R	1.3 Ω	1
L_{eq}	0.345 mH	0.1
R_{eq}	0.013 Ω	0.01
H	0.5 s	–
D_p	–	0.05
D_q	–	0.05
K_v	–	20
K_Q	–	10
T_1	0.159 s	–
T_2	0.0168	–
K_v	–	5

Table 7.3 System eigenvalues: GCM = Grid-connected mode, AM = Autonomous mode

No.	Eigenvalues		
	GCM	AM (with Q control)	AM (no Q control)
1, 2	$-181.3 \pm 1949.4i$	$-183.0 \pm 1571.6i$	$-170.5 \pm 1542.4i$
3, 4	$-170.7 \pm 1150.2i$	$-182.9 \pm 710.5i$	$-156.2 \pm 708.9i$
5, 6	$-3.5 \pm 378.2i$	$-1.2 \pm 385.9i$	$-2.9 \pm 387.7i$
7, 8	$-30.5 \pm 417.2i$	–	–
9	$-9.3 \pm 43.4i$	-19.8	-20.4
10		–	–
11	$-9.8 \pm 9.0i$	0.4	–
12		26.0	–
13	-90	-97.5	–

7.3.1 Dynamics of Grid-Connected Mode

Figure 7.5 shows traces of eigenpair (9, 10) corresponding to the VSC terminal frequency as functions of parameters of the frequency controller of Fig. 7.2. As expected, variations of H do not significantly affect the damping of this mode, however, they noticeably change the mode frequency. However, increasing K_D noticeably increases

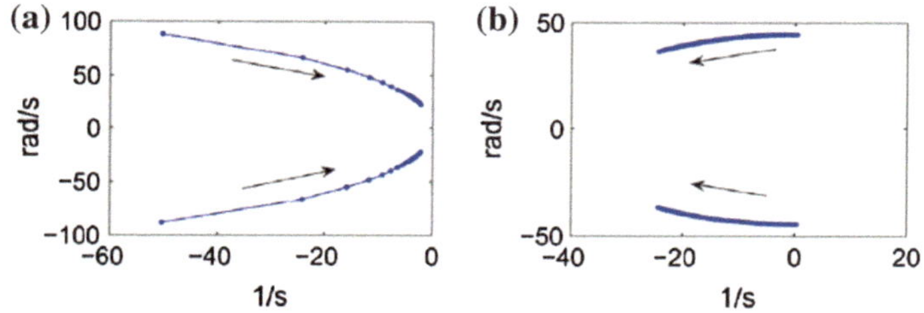

Fig. 7.5 Trace of eigenpair (9, 10) when **a** H changes from 0.1 to 2 s; **b** K_D changes from 1 to 50

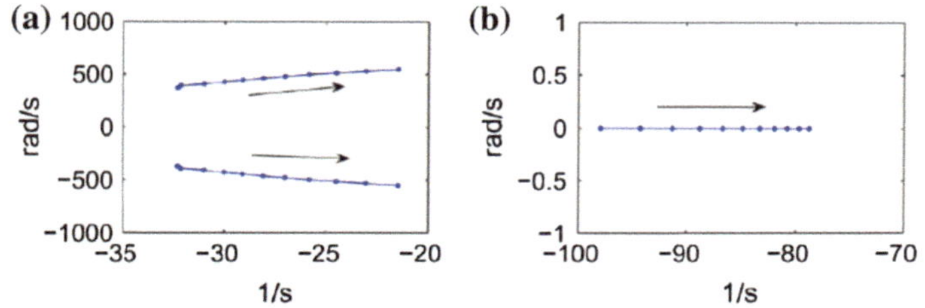

Fig. 7.6 Trace of **a** eigenpair (7, 8) and **b** eigenvalue 13 when D_q changes from 0 to 0.2

the mode damping, while it has insignificant effect on the mode frequency. Thus, H and K_D can be exploited to effectively adjust behavior of the frequency mode (Fig. 7.5).

Variations of D_q of Fig. 7.4 primarily affect eigenpair (7, 8) and eigenvalue 13, Fig. 6. As D q increases, eigenpair (7, 8) and eigenvalue 13 move to the right-hand side. This indicates that a large D_q deteriorates the system behavior in terms of stability. The reason is that the voltage droop characteristic is designed for the autonomous mode and is neither intended nor beneficial for the grid-connected mode.

Variations of K_Q of Fig. 7.4 also affect eigenpair (7, 8) and eigenvalue 13, Fig. 7.7. As K_Q increases, eigenvalue 13 moves to the left-hand side, but the damping of eigenpair (7, 8) decreases.

Figure 7.8a shows a trace of eigenpair (11, 12) when K_v is changed from 1 to 25. For small values of K_v, between 1 to 4, eigenvalues 11 and 12 are real and located in the left-hand plane. However, as K_v increases beyond 4, eigenvalues 11 and 12 form a conjugate pair and move to the right-hand side, and when K_v exceeds 15, they constitute an unstable oscillatory mode.

Figure 7.8b shows impact of the grid stiffness on eigenpair (11, 12) when X_{eq} is changed from 0.1 to 0.5 p.u. ($X_{eq}/R_{eq} = 5$) and $K_v = 5$. Figure 7.8b indicates that for a stiff grid (i.e., $SCR \geq 4$), the oscillatory mode associated with eigenpair (11, 12) is well damped, however, under nonstiff grid conditions (i.e., $SCR < 4$), the mode can be highly unstable. This instability is due to the reactive power controller of Fig. 7.4 through the positive feedback signal. Under nonstiff grid conditions, the reactive power disturbance through the positive feedback is not fully absorbed by the grid and results in the system instability.

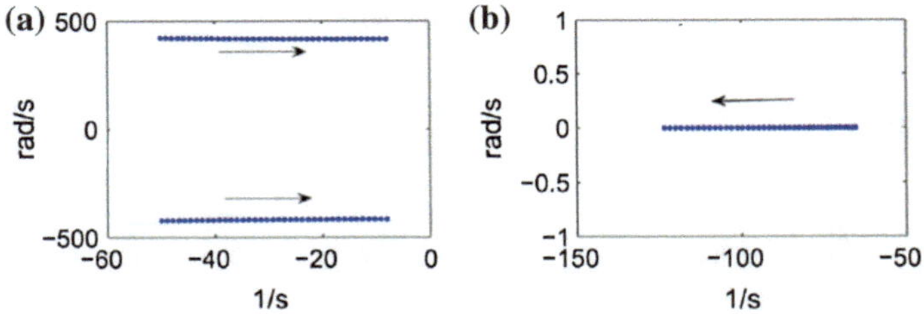

Fig. 7.7 Impact of increasing K_Q from 1 to 20 on the traces of **a** eigenpair (7, 8) and **b** eigenvalue 13

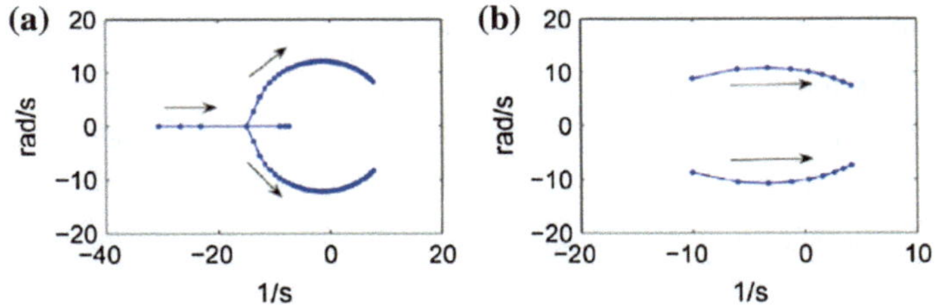

Fig. 7.8 Trace of eigenpair (11, 12) **a** when K_v changes from 1 to 25; **b** when X_{eq} changes from 0.1 to 0.5 p.u., with $X_{eq}/R_{eq} = 5$

Figure 7.9 shows the effect of the X_{eq}/R_{eq} ratio on eigenpairs (9, 10) and (7, 8) when the magnitude of network impedance $Z_{eq} = \sqrt{X_{eq}^2 + R_{eq}^2}$ is kept constant at 0.2 p.u. Figure 7.9 illustrates that as the ratio X_{eq}/R_{eq} increases from 1 to 20 (1) damping of the oscillatory mode associated with eigenpair (7, 8) decreases, but the mode always remains stable, (2) however, damping of the mode associated with eigenpair (9, 10) increases, and the mode becomes unstable for $X_{eq}/R_{eq} < 1$. The reason for instability is that the controller of Fig. 7.2 is designed based on the assumption of decoupled P and Q, while P and Q are tightly coupled at low values of X_{eq}/R_{eq}. Damping of the mode, even when X_{eq}/R_{eq} is relatively small, can be enhanced based on proper selection of K_D (Fig. 7.2). The effect of K_D on eigenpair (9, 10) is shown in Fig. 7.5b.

Eigenanalysis results also reveal that the DG operating point, level of local load, and the amount of power transfer to the grid have an insignificant impact on the system eigenvalues.

7.3.2 Dynamics of Autonomous Mode-Case 1

In an autonomous mode of operation, when S_1 of Fig. 7.1 is open, (1) eigenpair (7, 8) does not exist, and (2) corresponding to the complex conjugate eigenpairs (9, 10) and (11, 12), three real eigenvalues, as shown in the third column of Table 7.3 exist.

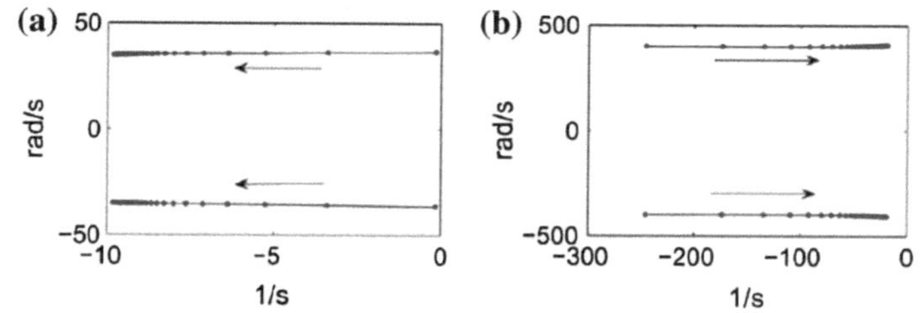

Fig. 7.9 Trace of eigenpairs *a* (9, 10) and **b** (7, 8) when X_{eq}/R_{eq} ratio changes from 1 to 20, with constant $Z_{eq} = 0.2$ p.u.

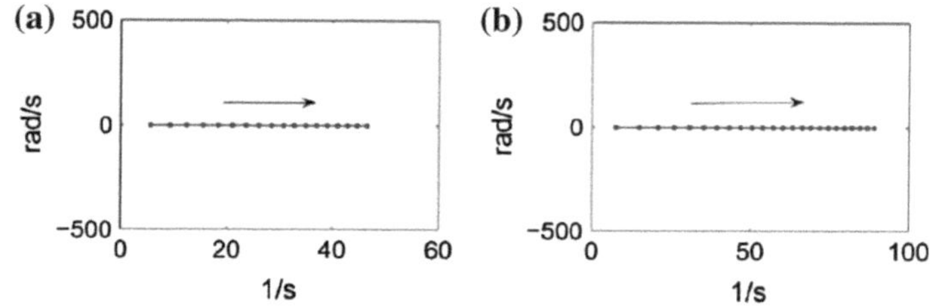

Fig. 7.10 Trace of eigenvalue 12 when **a** K_Q changes from 1 to 20 ($K_v = 5$); **b** K_v changes from 1 to 25 ($K_Q = 10$)

Figure 7.10a shows that eigenvalue 12, associated with the reactive power controller of Fig. 7.4, is monotonically unstable, and the instability is more pronounced as K_Q increases. Figures 7.7a and 7.10a indicate that K_Q can be selected such that (1) in a grid-connected mode, eigenpair (7, 8) remains well damped, and (2) in an autonomous mode, eigenvalue 12 exhibits instability as a means for islanding detection.

Figure 7.10b shows the effect of feedforward gain K_v of Fig. 7.4 on eigenvalue 12. The comparison of Figs. 7.8b and 7.10b shows that K_v can be selected such that at different values of grid stiffness, eigenpair (11, 12) remains stable under the grid-connected mode, and eigenvalue 12 exhibits instability, for islanding detection, during the autonomous mode.

In conformity with the anti-islanding test conditions [38], impacts of load resonant frequency $f = 1/(2\pi\sqrt{LC})$ and load quality factor $q = R\sqrt{C/L}$ on the system eigenvalues are also investigated. The eigenanalysis results show that changing (1) the load resonant frequency within the range of 59.5–60.5 Hz while the output real power from the DG unit is zero, and (2) the quality factor q within the range of 1 and 3, introduces no significant impacts on the system eigenvalues, including unstable eigenvalue 12. Since unstable eigenvalue 12 determines the dynamics of the PCC voltage deviations during the islanding detection process, the load resonant frequency and quality factor have no detrimental impacts on islanding detection. This indicates that the islanding detection approach does not have any NDZ in terms of load parameters.

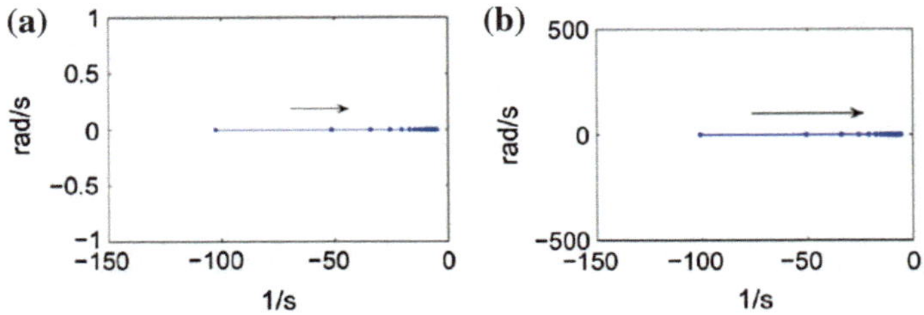

Fig. 7.11 Trace of eigenvalue 9 when **a** H changes from 0.1 to 2 s; **b** D_p changes from 0.01 to 0.2

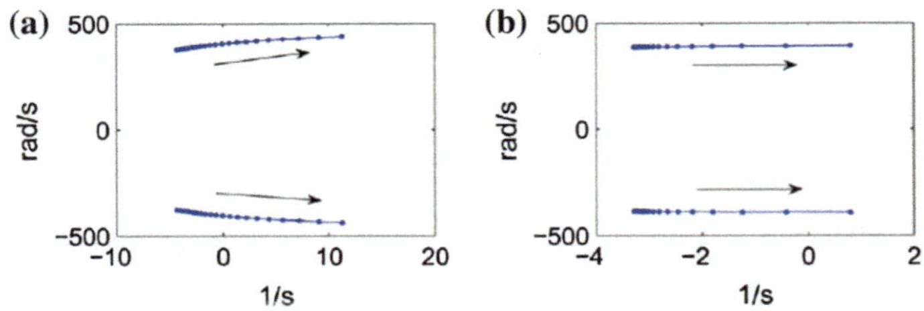

Fig. 7.12 Trace of eigenpair (5, 6) when **a** D_q changes from 0 to 0.2; **b** C changes from 0.5 to 2 p.u. with L = 0.8 p.u.

7.3.3 Dynamics of Autonomous Mode-Case 2

If the system of Fig. 7.1 is expected to remain operational in an autonomous mode and supply the local load, the reactive power controller, Fig. 7.4, which drives the system to instability as part of the islanding detection strategy, must be disabled ($K_Q = 0$) subsequent to an islanding event. Disabling the reactive power controller eliminates eigenvalues 11, 12, and 13, Column 4 of Table 7.3. Corresponding to eigenpair (9, 10) during the grid-connected mode, only eigenvalue 9 exists during an autonomous mode. Figure 7.11 shows impacts of control parameters H and D_p on eigenvalue 9. Figure 7.11a indicates that increasing H noticeably moves eigenvalue 9 to the right-hand side. Figure 7.11b shows that as D_p increases, eigenvalue 9 also moves to the right-hand side. This indicates that a large frequency droop has a detrimental effect on the system stability in the autonomous mode.

Figure 7.12a shows the effect of changing D_q from 0 to 0.2 on the mode associated with eigenpair (5, 6). Figure 7.12a indicates that large values of D_q result in an unstable oscillatory mode. Comparison of Figs. 7.6b and 7.12a indicates that a small D_q (e.g., 0.02) is acceptable.

Figure 7.12b shows the impact of local reactive LD on eigenpair (5, 6) when capacitor C is changed from 0.5 to 2.0 p.u., while L remains constant. The variation of C changes the load from inductive to capacitive load. As the capacitance increases, the oscillatory mode associated with eigenpair (5, 6) moves to the right-half plane. For $C > 1.9$ p.u., the mode becomes unstable. This indicates that for large capacitive loads, voltage droop characteristics should not be used. This is another reason for selecting a small D_q, especially under capacitive load conditions. Eigenanalysis results also reveal that the local real LD does not have a significant impact on the system eigenvalues.

7.4 Time-Domain Simulation Results

Eigenanalysis of Sect. 7.3 provides a systematic approach to select parameters of the VSC controllers of Figs. 7.2 and 7.4. To validate the system performance based on the selected parameters and to illustrate behavior of the system and the controllers due to large-signal transients (e.g., faults and islanding events), a set of time-domain simulation studies are conducted on the system of Fig. 7.1. The EMTDC/PSCAD software package is used as the simulation environment.

7.4.1 Grid-Connected Mode

Figure 7.13 shows responses of the controllers of Figs. 7.2 and 7.4 under a grid-connected mode of operation. Initially, the system is under a steady-state condition and the DO unit delivers 0.8/0.0 p.u. real/reactive power at PCC. The converter is subjected to step changes in reference values of real/reactive power, at time instants $t = 3.0$ s and $t = 4.0$ s, from 0.8/0.0 to 1.0/0.2 p.u., respectively. Figure 7.13a, b shows variations in real- and reactive power at the PCC, and illustrates that the proposed controllers of Figs. 7.2 and 7.4 properly regulate P and Q to the desired values. Figure 7.13a, b also shows that P and Q controllers exhibit some degree of coupling which is the inherent nature of the voltage-controlled strategy.

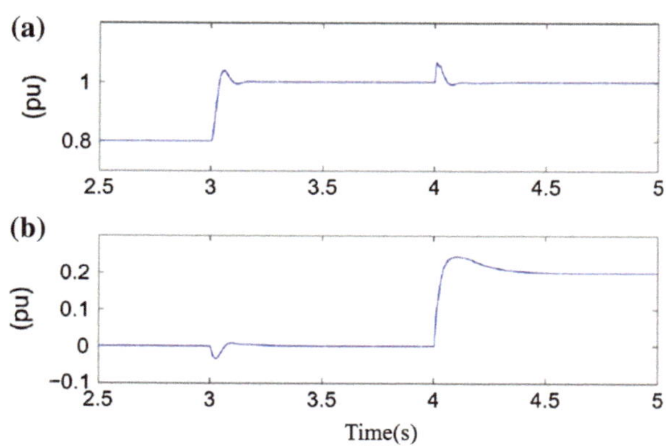

Fig. 7.13 Response of the DG unit in a grid-connected mode to step changes in references of real and reactive power. **a** DG unit real power at PCC. **b** DG unit reactive power at the PCC

7.4.2 Ride-Through Capability of DG Unit

Figure 7.14 shows the ride-through capability of the system of Fig. 7.1, when subjected to a three-phase line-to-ground fault at the PCC. The fault is temporary and cleared five cycles after inception. Figure 7.14a, b shows the PCC rms and instantaneous voltage (Phase A). The PCC voltage during the fault is below 0.5 p.u., however, since the fault period is less than 0.16 s, the voltage drop is within the ride-through capability of the DG unit [38] and the fault should not result in islanding or control change (i.e., $K_Q = 0$) of the DG unit.

Figure 7.14c, d shows the converter output rms and instantaneous current. Figure 7.14c, d indicates that the imposed control limits of Figs. 7.2 and 7.4 confine the converter current within 0 and 2 p.u. Without the limits, a maximum current of about 8 p.u. is encountered.

Figure 7.14e, f shows instantaneous real- and reactive power components of the DG unit at the PCC. Figure 7.14f shows that the instantaneous reactive power undergoes a relatively large variation and slow dynamics after the fault is cleared. This is due to the PCC voltage feedback of Fig. 7.4 which is used to facilitate islanding detection. The reactive power transients settle down within 0.6 s which indicate that the PCC voltage feedback does not adversely affect the system transients, even under a three-phase ground fault conditions at PCC.

Figure 7.15 shows the ride-through capability of the system of Fig. 7.1, due to a voltage drop (0.2 p.u.) for the period of 1.0 s at the grid source. Initially, the system is under a steady-state condition and the DG unit delivers 0.8/0.0 p.u. real/reactive power at PCC. The grid bus voltage drops 0.2 p.u., from 1.0 p.u. to 0.8 p.u., at $t = 3.0$ s. The abnormal condition lasts for one second from $t = 3.0$ s to $t = 4.0$ s.

Figure 7.15a, b shows the PCC voltage. Figure 7.15a shows that the PCC voltage drops below the permissible value of 0.88 p.u. as specified in Table 7.1. However, since the period of voltage drop is less than 2 s, the voltage abnormality is considered within the ride-through capability of the DO unit (Table 7.1) and should not result in islanding or control change (i.e., $K_Q = 0$) of the DG unit [38]. Figure 7.15c, d shows that the peak transient current of the converter is limited to 1.8 p.u.

Figure 7.15e shows that during the time interval of 3.0–4.0 s, power output of the VSC settles down at a lower reference (i.e., 0.64 p.u.). This new reference is calculated by multiplying the original real power reference of Fig. 7.2 by the per unit value of PCC voltage when the PCC voltage drops to (and below) 0.88 p.u. The power reference is intentionally reduced during this time interval, however, the converter current reference is kept at the level prior to voltage drop at $t = 3.0$ s (Fig. 7.15c). Motivation for reducing the power reference is to avoid overcurrent and prevent the overcurrent backup protection to be activated, in case that the period of voltage drop lasts too long (e.g., beyond 1.8 s). Figure 7.15f also shows that the reactive power undergoes transients immediately after a voltage change at $t = 3.0$ or $t = 4.0$ due to the PCC voltage feedback of Fig. 7.4. However, the transients settle down within 0.6 s.

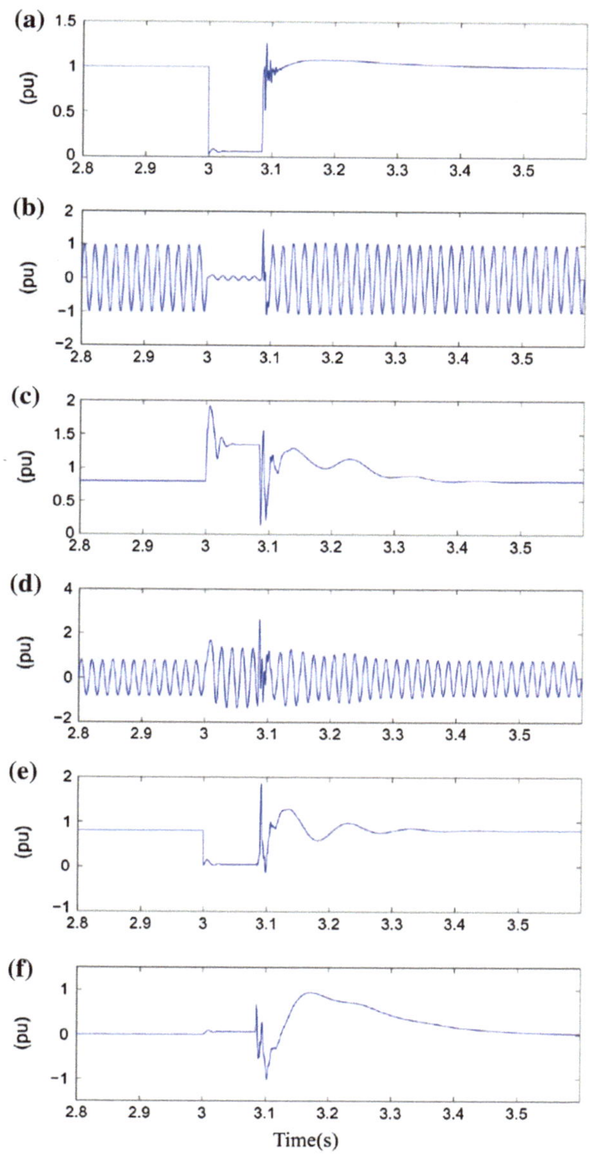

Fig. 7.14 Ride-through transients of the DG unit in a grid-connected mode due to a 5-cycle, three-phase, line-to-ground fault at PCC. **a** PCC rms voltage. **b** PCC instantaneous voltage (phase A). **c** Converter rms current. **d** Converter instantaneous current (phase A). **e** DG unit real power at PCC. **f** DG unit reactive power at PCC

7.4.3 Transition Capability from Grid-Connected to Islanded Mode

Figure 7.16 shows the capability of the system of Fig. 7.1 in terms of transition from a grid-connected mode to an islanded mode, due to a preplanned islanding scenario.

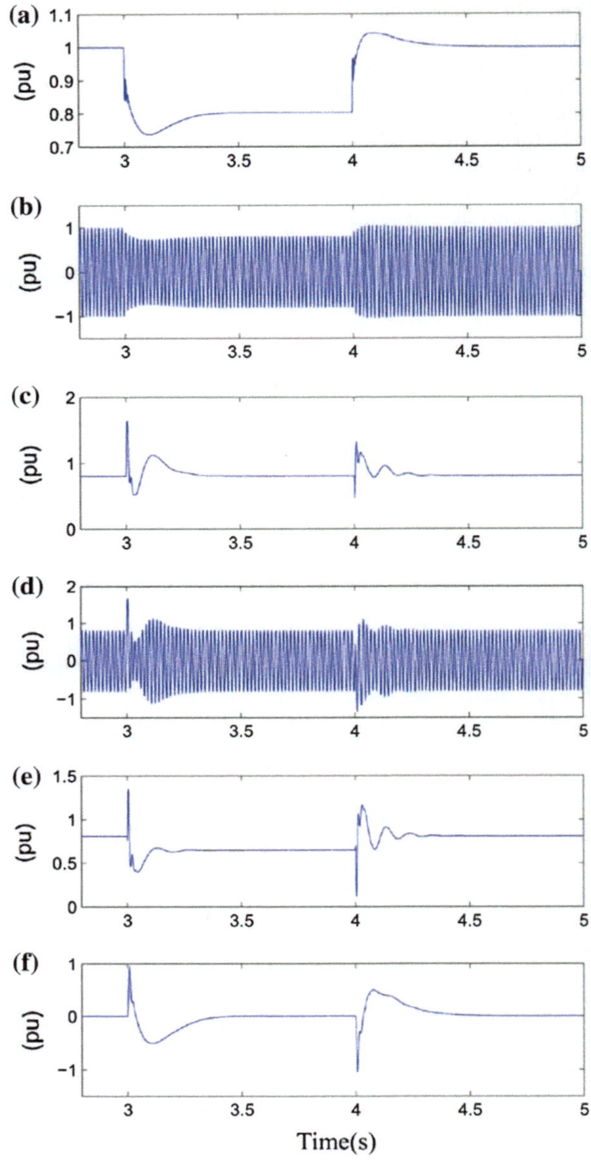

Fig. 7.15 Ride-through transients of the DG unit in a grid-connected mode due to a 1-s voltage drop (0.2 p.u.) of the grid source. **a** PCC rms voltage. **b** PCC instantaneous voltage (phase A). **c** Converter rms current. **d** Converter instantaneous current (phase A). **e** DG unit real power at PCC. **f** DG unit reactive power at PCC

Initially, the DG unit delivers 0.8/0.0 p.u. real/reactive power to the load at 1.0 p.u. PCC voltage. The RLC load is adjusted such that real- and reactive PX with the grid is almost zero. At $t = 3$ s, a preplanned islanding event is imposed by opening switch S_1.

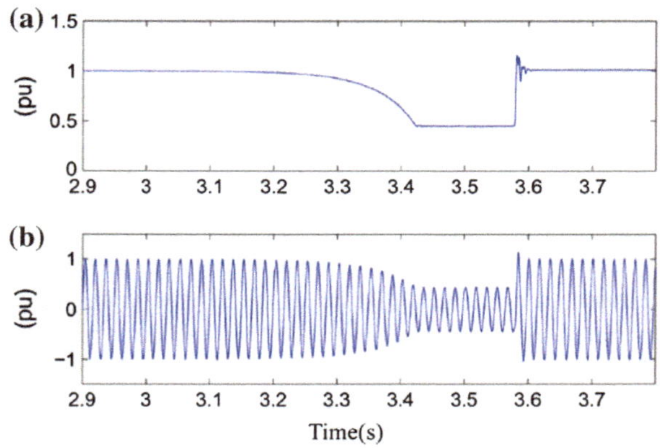

Fig. 7.16 Transients of the VSC terminal voltage during a preplanned islanding scenario of zero real- and reactive PX with the grid. **a** PCC rms voltage. **b** PCC Instantaneous voltage (phase A)

Since PX between the PCC and the grid prior to the islanding instant is insignificant, changes in the PCC voltage immediately after the islanding event ($t = 3.0$ s) are also insignificant (Fig. 7.16a). However, due to the destabilizing effect of the VSC reactive power controller of Fig. 7.4, the PCC voltage gradually deviates from the initial value ($t = 3.0$ s). Change in the voltage is used as the islanding detection signal. Based on Table 7.1:

- islanding detection process is initiated at $t = 3.3$ s corresponding to the PCC voltage of 0.88 p.u. and islanding detection is confirmed after 2 s delay;
- islanding detection process is initiated at $t = 3.42$ s, corresponding to the PCC voltage of 0.5 p.u. and islanding detection is confirmed after a 0.16-s delay.

Rapid PCC voltage drop from 0.88 ($t = 3.3$ s) to 0.5 p.u. ($t = 3.42$ s), Fig. 7.16a, indicates that for the given scenario, 0.5 p.u. PCC voltage initiates islanding detection process and the islanding detection is confirmed after 0.16 s delay (i.e., at $t = 3.58$ s). Subsequent to the islanding detection confirmation, the reactive power control path of the controller of Fig. 7.4 is disabled by setting $K_Q = 0$, and the PCC voltage is restored to its normal value of 1.0 p.u., and the DG unit continues its operation in an autonomous mode. The total islanding detection time is 0.58 s.

Based on the voltage control scheme of Fig. 7.4, the PCC voltage should drop to and remain at zero for the time interval between 3.42 and 3.58 s (Fig. 7.16a). However, to improve dynamic performance of the autonomous system, after islanding detection initiation (e.g., at $t = 3.42$ s corresponding to the PCC voltage of 0.5 p.u.), when the PCC voltage drops to 0.48 p.u., it is maintained at 0.48 p.u. by disabling the positive feedback of Fig. 7.4 (by setting $K_v = 0$). When the positive feedback is disabled, the PCC voltage experiences a very slow rate of drop as shown in the time interval of 3.42–3.58 s of Fig. 7.16. The rate of voltage drop depends on the amount of reactive PX between the PCC and the grid prior to the islanding event at $t = 3.0$ s. Without extending the islanding detection time, this voltage maintenance successfully prevents the PCC voltage from dropping to and remaining at zero for almost 0.16 s, which is an extreme operating condition for the system.

Other islanding scenarios, corresponding to different levels of real- and reactive PX with the grid before the islanding event, were also studied. The studies show that the condition of zero real- and reactive PX is the worst condition in terms of the required islanding detection time (0.58 s) and other islanding cases require less than 0.58 s to be detected. The proposed islanding detection method is robust since it (1) has no NDZ, and (2) is able to distinguish between the loss of the main grid and the grid disturbances.

Figure 7.17 shows the transition capability of the system of Fig. 7.1, from a grid-connected mode to an islanded mode, due to an accidental islanding event which is initiated by a fault. Initially, the DG unit delivers 0.8/0.0 p.u. real/reactive power to the load at 1.0 p.u. PCC voltage. Real- and reactive PX with the grid prior to the fault is almost zero. The grid is subjected to a three-phase to ground fault at the grid-side of S_1 at $t = 3$ s. Islanding occurs 5 cycles after the fault inception by opening S_1.

Figure 7.17a, b shows the PCC rms and instantaneous voltages. Islanding detection process is initiated at $t = 3.0$ s since the PCC voltage immediately drops below 0.5 p.u. (Table 7.1). The islanding detection confirmation time is after another 0.16 s or at $t = 3.16$ s. At $t = 3.083$ sS_1 opens, the fault is cleared and the grid continues its operation as an autonomous system. Immediately after opening S_1, the PCC voltage

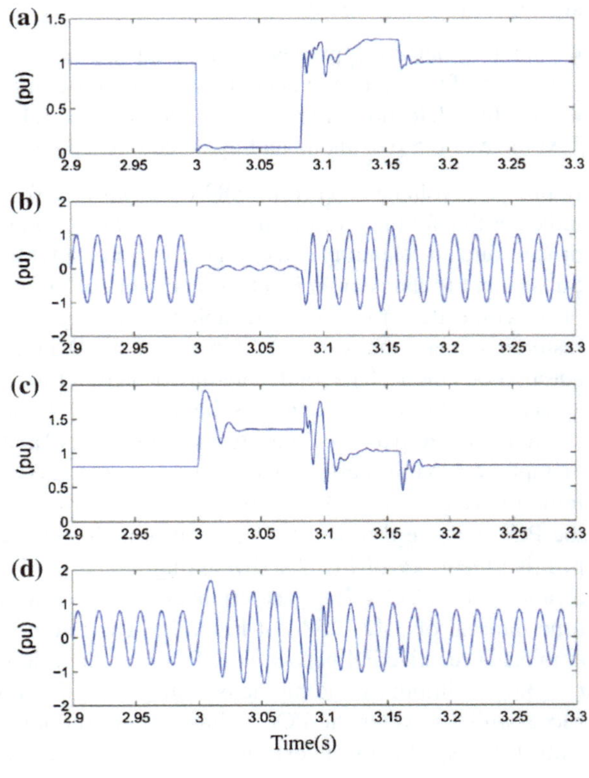

Fig. 7.17 Transients of the DG unit due to a three-phase fault at the grid-side of S, (t = 3.0 s). Islanding occurs after five cycles (t = 3.308 s). **a** PCC rms voltage. **b** PCC instantaneous voltage. **c** Converter rms current. **d** Converter instantaneous current

returns almost to the prefault value since PX with the grid prior to the fault was insignificant. Then the PCC voltage begins to increase due to the positive feedback and the destabilizing voltage effect of the reactive power control of Fig. 7.4. It rises to 1.2 p.u. at $t = 3.13$ s (Fig. 7.17a), which can initiate another islanding detection process and the corresponding islanding detection confirmation after 0.16-s delay (Table 7.1) (i.e., at $t = 3.29$ s). However, the initial islanding detection, which was activated at $t = 3.0$ s, is confirmed at $t = 3.16$ s, and disables the reactive power path of the controller of Fig. 7.4 ($K_Q = 0$). Therefore, the PCC voltage is restored to a value within the permissible range and the autonomous operation of the system continues. Figure 7.17c, d shows the converter rms and instantaneous currents corresponding to phase A.

Similar to the case study of Fig. 7.16a, the PCC voltage deviation of the case study of Fig. 7.17 is also limited for the period of $t = 3.13$ s to $t = 3.16$ s, by setting $K_v = 0$ (Fig. 7.4) when the PCC voltage intends to increase beyond 1.2 p.u. (i.e., at 1.22 p.u.).

7.4.4 Autonomous Mode

Figure 7.18 shows the transient response of the system of Fig. 7.1 to load changes during an autonomous mode of operation. Initially, the system is under a steady-state condition and the DG unit delivers 0.82/0.0 p.u. real/reactive power at 1.01 p.u. PCC voltage to the load. The system is subjected to two consecutive load changes. The load changes are imposed by changing R and L of the load (Fig. 7.1) from 1.25 p.u. and 0.8 p.u. to 1.0 p.u. and 0.74 p.u. (corresponding to step changes of 0.2 p.u. and 0.1 p.u. in real- and reactive power) at $t = 3.0$ s and $t = 3.1$ s, respectively.

Figure 7.18a, b shows variations in real- and reactive power at the PCC, and indicates that the transients settle down within 0.05 s subsequent to the imposed disturbances. The PCC frequency is estimated by a PLL system and shown in Fig. 7.18c, and indicates that the system steady-state frequency drops from 1 p.u. to 0.99 p.u. due to the real power increase, and does not noticeably change due to the reactive power change.

Figure 7.18d shows that the steady-state PCC rms voltage drops 0.01 p.u., from 1.01 p.u. to 1.0 p.u., after real power increase, and drops 0.015 p.u., from 1.0 p.u. to 0.985 p.u., due to the reactive power change. Figure 7.18c, d indicates that the real power change affects both the system frequency and the PCC voltage, while the reactive power change mainly affects the PCC voltage. Figure 7.18c shows that there is a steady-state frequency drop after real power increase. This is a consequence of the ω-P droop characteristic of the VSC. Similarly, steady-state PCC voltage drops after reactive power increase (Fig. 7.18d) due to the V-Q droop characteristic of the VSC.

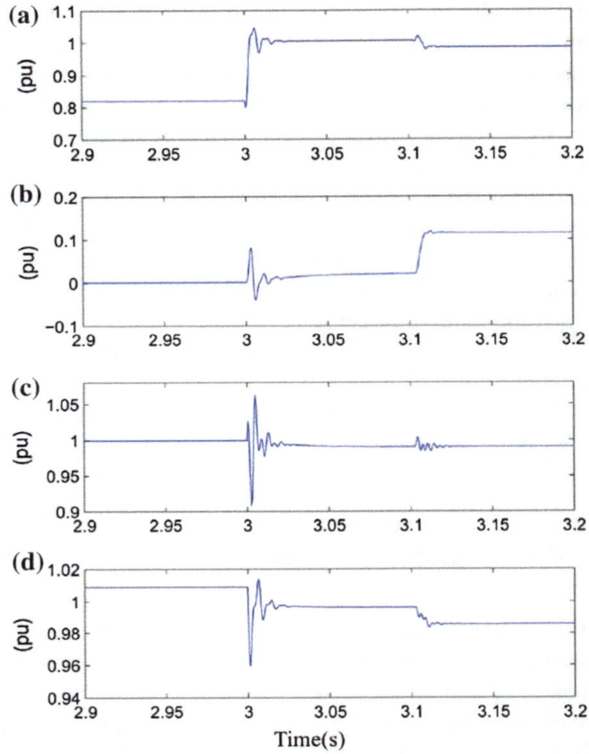

Fig. 7.18 Transient response of the DG unit to load changes ($\Delta P \approx 0.2\,p.u.$, $\Delta Q \approx 0.1\,p.u.$) during the autonomous operation. **a** DG unit real power at PCC. **b** DG unit reactive power at PCC. **c** Estimated frequency at the PCC. **d** PCC rms voltage

7.5 Robust Control Strategy for Multi-Microgrids

This section presents fundamental concepts of a central power management system (PMS) and a decentralized, robust control strategy for autonomous mode of operation of a microgrid (MG) that includes multiple distributed energy resource (DER) units. The DER units are interfaced to the utility grid through VSCs. The frequency of each DER unit is specified by its independent internal oscillator and all oscillators are synchronized by a common time-reference signal received from a global positioning system. The PMS specifies the voltage set-points for the local controllers (LOCs). A linear, time-invariant, multivariable, robust, decentralized, servomechanism control system is designed to track the set-points. Each control agent guarantees fast tracking, zero steady-state error, and robust performance despite uncertainties of the MG parameter, topology, and the operating point.

7.5.1 Introduction

Technical and economical viability of the DER technologies for distribution voltage class applications have resulted in the emergence of the MG concept [4, 48]. Impacts of the DER units on the host MG and their control, protection, and management requirements for successful operation of the MG have been extensively reported in the technical literature [26, 39, 42, 45, 72]. However, the anticipated high depth of penetration of DER units in the MG necessitates a systematic and comprehensive approach to their integration. This sterns from the need to (1) enable the MG to operate in the grid-connected mode, the islanded mode, and the virtual power plant (VPP) mode; (2) respond to the external commands (e.g., market signals); (3) accommodate the MG inherent unbalanced conditions, uncertainties in parameters and topology, and frequent load/generation changes; and (4) make provisions for demand-side integration.

The existing/reported DER control strategies of a MG are as follows:

- Droop-based methods [3, 25, 41, 42, 44, 50, 56, 60, 63, 65, 70]: The main advantage of the droop-based approach is that it obviates the need for communication and operates based on local measurements. However, it presents several limitations [5, 31, 32]: (1) poor transient performance; (2) lack of robustness due to inability to account for load dynamics; and (3) inherent lack of black-start capability.
- Centralized control methods [30, 64]: This approach relies on fast and high-bandwidth communication, and the communication failure can lead to the system collapse.
- Master/slave control method [55]: This method is flexible in terms of connection and disconnection of DER units; however, it requires a dominant DER unit for satisfactory operation.
- Robust servomechanism control method [40]: Although this method is robust to MG parametric uncertainties, it is not readily applicable to multiple DER units.

This section presents a PMS and a control strategy for an islanded multi-DER MG to overcome the drawbacks of the existing approaches. Based on the proposed strategy:

1. the PMS specifies voltage set-points for each voltage-controlled bus, based on a fast power-flow analysis. In this spirit, the PMS precisely controls power flow of the system and achieves a prescribed load sharing among the DER units;
2. local voltage controllers (LVCs) provide tracking of the voltage set-points and rapidly reject disturbances. It is to be noted that LVCs are highly robust to parametric, topological, and unmodeled uncertainties of the MG;
3. an open-loop frequency control and synchronization scheme maintains system frequency. Note that frequency of the system is fixed and cannot deviate due to transients;

4. LVCs are implemented in a decentralized manner which obviates the need for a high-bandwidth communication medium to feed the system's information to a central authority and makes it scalable for larger number of DER units. Further, they enable the system to sudden connection/disconnection of DER units.

7.5.2 System Description

A schematic diagram of a typical radial distribution feeder that is adopted as the study MG system is illustrated in Fig. 7.19a. The MG includes three dispatchable DER units with the voltage rating of 0.6 kV and power ratings of 1.6, 1.2, and 0.8 MVA. It also includes three local loads, and two 13.8-kV distribution line segments. Each DER unit is represented by a 1.5-kV DC voltage source, a VSC, and a series RL filter, and is interfaced to the grid through a 0.6-kV/13.8-kV step-up transformer (with the same power rating as the corresponding DER unit) at its point of coupling (PC) bus. The main utility grid is represented by an AC voltage source behind series R and L elements. The MG can be operated in the grid-connected or the islanded mode based on the status of circuit breaker (CB). The system parameters are given in Table 7.4.

7.5.3 Power Management

A schematic diagram of the proposed control strategy is shown in Fig. 7.19b, which is composed of:

1. The power management system;
2. The LOC of each DER unit; and
3. The MG frequency control and synchronization scheme.

The instantaneous real/reactive power of DER units and loads are communicated to the PMS through a low-bandwidth communication system. Based on the total real/reactive power demand of the MG and a prescribed load sharing strategy, the PMS determines $P_{DG,i}$ and $Q_{DG,i}$, power set-points for DER_i, and determines the voltage set-points (magnitude and angle) for the PC buses (Fig. 7.19a). The set-points are then transmitted to DER LCs. Each LC measures the voltage of its corresponding PC bus and provides voltage tracking based on the received reference set point. Owing to its robust design, each LC also rejects disturbances due to system variations.

The LCs are formulated in the dq frame of reference, and there is a need for (1) a reference angle for $abc(dq)$ to $dq(abc)$ transformation for each LC and (2) a global synchronization mechanism among all LCs. The reference angle for each LC

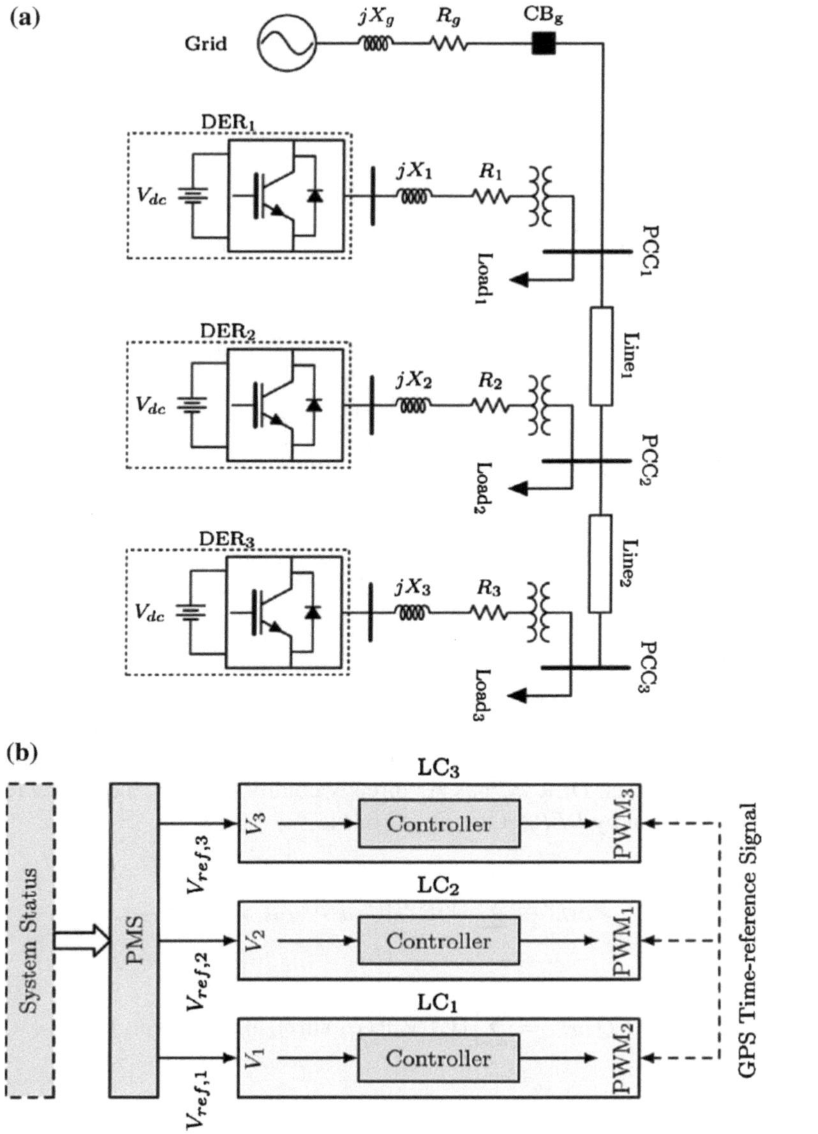

Fig. 7.19 **a** Schematic diagram of the studied microgrid system. **b** Block diagram of the microgrid power management and control system

is generated by an internal oscillator at each DER unit and the three angle signals are synchronized based on a common time-reference signal provided by a GPS, Fig. 7.19b. These entities are further described in Sects. II-B1, II-B2, and II-B3.

Table 7.4 Parameters of the microgrid system

Base values					
$S_{base} = 1.6\,\text{MVA}$		$V_{base,low} = 0.6\,\text{KV}$		$V_{base,high} = 13.8\,\text{KV}$	
Transformers					
0.6/13.8 KV		Δ/Y_g		$X_T = 8\,\%$	
Load parameters					
Load 1		Load 2		Load 3	
R_1	350 Ω	R_2	375 Ω	R_3	400 Ω
X_{LI}	41.8 Ω	X_{L2}	37.7 Ω	X_{L3}	45.2 Ω
X_{CI}	44.2 Ω	X_{C2}	40.8 Ω	X_{C3}	48.2 Ω
R_{11}	2 Ω	R_{l2}	2 Ω	R_{l3}	2 Ω
Line parameters					
R 0.34 Ω/km			$Line_1$ 5 km		
X 0.31 Ω/km			$Line_2$ 10 km		
Filter parameters (based on DERi ratings)					
$X_f = 15\,\%$			Quality factor 50		
Grid parameters					
X_g 2.3 Ω			R_g 2 Ω		

1. *Power management system:* The main function of the PMS is to provide load sharing among DER units based on either a cost function associated with each DER unit or a market signal. In the proposed method, voltage angle δ and magnitude $|V|$ of each DER PC bus are directly controlled; thus, the active/ reactive power injection by DER_i at the ith PC bus is

$$P_{DG,i} = \sum_{k=1}^{N} |V_i||V_k|(G_{ik}\cos\delta_{ik} + B_{ik}\sin\delta_{ik}) \tag{7.13}$$

$$Q_{DG,i} = \sum_{k=1}^{N} |V_i||V_k|(G_{ik}\sin\delta_{ik} + B_{ik}\cos\delta_{ik}). \tag{7.14}$$

Equations (7.13) and (7.14) indicate that power flow of the system is determined based on the voltage magnitude and angle of PC_l, PC_2, and PC_3 (Fig. 7.19a). The power management process should be performed frequently to maintain the prespecified load sharing scheme among the DER units as the MG operating point changes (e.g., due to load changes). The time interval between set-point updates depends on the rate of MG operating point variation.

2. *Frequency Control and Synchronization:* The MG frequency is controlled in an open-loop manner. The LC of each DER unit includes a crystal oscillator which generates $\theta(t) = \int_0^t \omega_0 d\tau$, where $\omega_0 = 2\pi f_0$ and f_0 is the nominal power

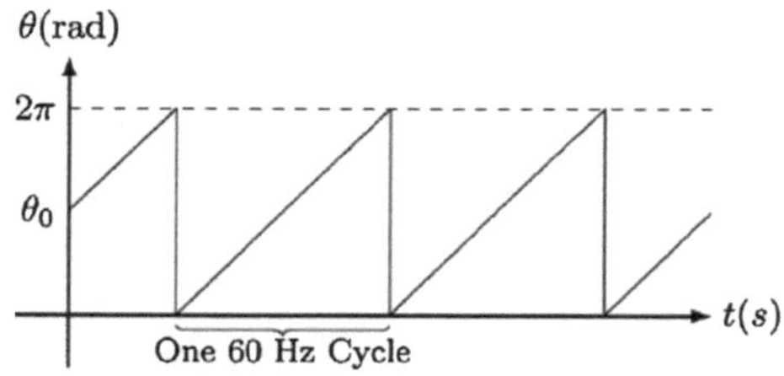

Fig. 7.20 Angle waveform generated by a crystal oscillator

frequency of the MG. Figure 7.20 illustrates the angle waveform deduced from the oscillator of LC_i which is used for the $abc(dq)$ to $dq(abc)$ transformation of the DER_i mathematical model.

Based on the proposed control strategy, all DER units are synchronized by a global synchronization signal that is communicated to the crystal oscillators of DER units (Fig. 7.19b) through a GPS [37]. The global synchronization signal is communicated at relatively large time intervals (e.g. once per second) to I) prevent drift among local oscillators and (2) to initialize incoming DER units. Crystal oscillators with high accuracies (e.g., an error of 2×10^{-6} to 2×10^{-11} seconds per year) and relatively low costs are currently available [71]. All LCs can be synchronized with a high degree of reliability of a common timing source of the GPS radio clock (e.g., with a theoretical accuracy of higher than 1μ s [62]). Although there are 6–10 satellites visible to each area at all times, one can rely on the accuracy of crystal oscillators in case of the unavailability of the synchronizing signal.

3. *Local controllers:* The LC of each DER unit tracks the set-points specified by the PMS and rejects disturbances. Each LC measures the voltage of the corresponding PC bus, and transforms the three-phase voltage to the dq frame based on the angle signal $\theta(t)$ generated by its internal oscillator and synchronized with the time-reference signal received from a GPS. The voltage magnitude and angle set-points received from the PMS $|V_{ref,i}|\angle\delta_i$ are also transformed to the dq frame to generate dq-based reference values (i.e., $V_{d,ref,i} = |V_{ref,i}|\cos\delta_i$ and $V_{q,ref,i} = |V_{ref,i}|\sin\delta_i$). The measured and reference values are provided to the LC to determine the dq voltage components of the terminal of the corresponding DER unit (i.e., $V_{id,i}$ and $V_{iq,i}$) and generate the terminal voltage $v_{t,abc,i,HV}$ at the high-voltage side of the transformer. $v_{t,abc,i,HV}$ is divided by the turn ratio of the transformer and shifted by 30° to obtain $v_{t,abc,i,LV}$ corresponding to the low-voltage side of the transformer. $v_{t,abc,i,LV}$ is then fed to the pulse-width modulation (PWM) signal generator of the interface VSC of the DER unit. It should be noted that although the realistic and theoretical turn ratio and phase shift of the transformer are slightly different, the controller can compensate for the mismatch. Figure 7.21 illustrates a block diagram of LC_i where the measured and reference voltages are transformed to the dq frame of reference. After performing the control action, the outputs are transformed to the abc frame to generate the PWM switching signals of the ith interface VSC. Section 7.6 describes the design

Fig. 7.21 Block diagram of the local control agent

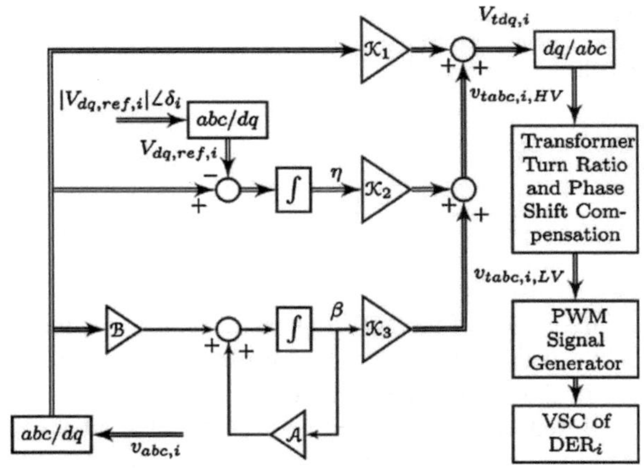

steps of the controller based on the mathematical model of the system. It should be noted that although the controllers are designed based on RLC load models, other types of loads (e.g., motor loads) can be handled by the proposed controllers due to their robustness.

The control strategy is developed based on a decentralized robust servomechanism approach to devise a decentralized controller so that outputs of the system asymptotically track constant reference inputs independent of (1) constant disturbances which the MG is subjected to and (2) variations in the plant parameters and gains of the control system [11]. The robustness and the decentralized nature of the controller are highly desirable for a MG since:

- a centralized controller which requires all inputs to be communicated to a common center is uneconomical due to the complexity and cost of the required high-bandwidth communication system;
- a robust controller overcomes the uncertainty issues of the plant structure and/or parameters.

The conditions for the existence of a centralized system solution based on the servomechanism problem [18] differ from those of a decentralized system [11, 12]. In this case, a plant often has a number of control agents and each control agent has a number of local control inputs and outputs, and a separate and distinct controller is applied to each control agent. In particular, there exists a solution to the problem of stabilizing a plant based on a decentralized control (DEC) system if and only if the plant has no unstable decentralized fixed modes (DFMs) and that certain rank conditions of the plant data hold true [73].

The notion of robustness leads to the problem of constructing a controller for a plant such that the resultant closed-loop system satisfies a given robustness constraint (e.g., having a comparable so-called "real stability radius" for both open-loop and closed-loop systems [14, 49]). This can be done by a decentralized controller design based on the approach of [11–13, 19], subject to a robustness constraint. The constraints depend on the problem under consideration. The principles of application of this DEC scheme to conventional power systems are provided in [21, 22, 24, 67, 68].

In this section, the DEC strategy is applied to a MG system which is generally more susceptible to oscillations caused by transients compared with the conventional power system. To ensure robust performance of the controller, the decentralized controller optimization is carried out subject to a specified robustness constraint [49]. To solve the robust servomechanism problem and to achieve further improvements for the MG performance, the normal optimal control performance index $J = \int_0^{\infty}(x^{'}Qx + u^{'}Ru)d\tau$, which is used for obtaining optimal controllers to reject impulse disturbances, is replaced with the performance index $J = \int_0^{\infty}(e^{'}e+\varepsilon\dot{u}^{'}u)d\tau$, where $\varepsilon > 0$ is a small weighting parameter. The resulting optimal controller is now an optimal servomechanism controller [9]. The micro grid model, existence conditions of the controller, design procedure, and the properties of the closed-loop system (e.g., robustness, gain margin, and tolerance to time delay and unmodeled high-frequency effects) are discussed in the following sections.

7.5.4 Mathematical Model

The proposed DEC is developed based on a linearized model of the MG of Fig. 7.19a in a synchronously rotating dq frame. Figure 7.22 shows a one-line diagram of the MG. The controller is designed based on the fundamental frequency component of the system of Fig. 7.22. Each DER unit is represented by a three-phase controlled voltage source and a series RL branch. Each load is modeled by an equivalent parallel RLC network (Fig. 7.22). Each distribution line is represented by lumped series RL elements.

The MG of Fig. 7.22 is virtually partitioned into three subsystems. The mathematical model of Subsystem1 in the abc frame is

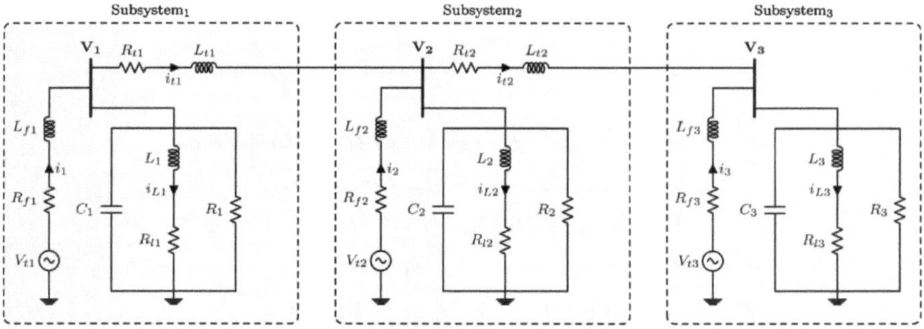

Fig. 7.22 Single-line diagram of the MG of Fig. 7.19 used to derive state-space equations

$$\begin{cases} i_{1,abe} = i_{tl,abc} + C_1 \frac{dv_{1,abc}}{dt} + i_{L1,abc} + \frac{v_{1,abc}}{R_1} \\ v_{tl,abc} = L_{f1} \frac{di_{1,abc}}{dt} + R_{f1} i_{1,abc} + v_{1,abc} \\ v_{l,abc} = L_1 \frac{di_{L_1,abc}}{dt} + R_{l1} i_{L1,abc} \\ v_{l,abc} = L_{t1} \frac{di_{t1,abc}}{dt} + R_{t1} i_{t1,abc} + v_{2,abc} \end{cases} \tag{7.15}$$

where x_{abc} is a 3×1 vector. Assuming a three-wire system, (7.15) is transformed to the synchronously rotating dq frame of reference, as described in Sect. 7.5.3 by [43]

$$f_{dq} = \frac{2}{3} \begin{pmatrix} \cos\theta & \cos(\theta - \frac{2}{3}\pi) & \cos(\theta - \frac{4}{3}\pi) \\ -\sin\theta & -\sin(\theta - \frac{2}{3}\pi) & -\sin(\theta - \frac{4}{3}\pi) \\ \frac{1}{\sqrt{2}} & \frac{1}{\sqrt{2}} & \frac{1}{\sqrt{2}} \end{pmatrix} f_{abc} \tag{7.16}$$

where $\theta(t)$ is the phase angle generated by the crystal oscillator internal to DER_1. Based on (7.15) and (7.16), the mathematical model of *Subsystem*$_1$ in the dq frame is

$$\begin{aligned} \frac{dV_{1,dq}}{dt} &= \frac{1}{C_1} I_{1,dq} - \frac{1}{C_1} I_{tl,dq} - \frac{1}{C_1} I_{L1,dq} - \frac{1}{R_1 C_1} V_{1,dq} - j\omega V_{1,dq}, \\ \frac{dI_{1,dq}}{dt} &= \frac{1}{L_{f1}} V_{t1,dq} - \frac{R_{f1}}{L_{f1}} I_{l,dq} - \frac{1}{L_{f1}} V_{1,dq} - j\omega I_{1,dq}, \\ \frac{dI_{L1,dq}}{dt} &= \frac{1}{L_1} V_{1,dq} - \frac{R_{t1}}{L_1} I_{Ll,dq} - j\omega I_{L1,dq}, \\ \frac{dI_{t1,dq}}{dt} &= \frac{1}{L_{t1}} V_{1,dq} - \frac{R_{t1}}{L_{t1}} I_{t1,dq} - \frac{1}{L_{t1}} V_{2,dq} - j\omega I_{t1,dq} \end{aligned} \tag{7.17}$$

Similarly, the dq frame-based models of *Subsystem*$_2$ and *Subsystem*$_3$, are also developed and used to construct the state-space model of the overall system

$$\begin{aligned} \dot{x} &= Ax + Bu, \\ y &= Cx \end{aligned} \tag{7.18}$$

where

$$\begin{aligned} x^T &= \big(V_{1,d}, V_{1,q}, I_{1,d}, I_{1,q}, I_{L1,d}, I_{L1,q}, I_{tl,d}, I_{tl,q}, \\ &\quad V_{2,d}, V_{2,q}, 1_{2,d}, 1_{2,q}, I_{L2,d}, I_{L2,q}, I_{t2,d}, I_{t2,q}, \\ &\quad V_{3,d}, V_{3,q}, 1_{3,d}, 1_{3,q}, I_{L3,d}, I_{L3,q}\big), \\ u^T &= \big(v_{tl,d}, v_{tl,q}, v_{t2,d}, v_{t2,q}, v_{t3,d}, v_{t3,q}\big), \\ y^T &= \big(V_{1,d}, V_{1,q}, V_{2,d}, V_{2,q}, V_{3,d}, V_{3,q}\big). \end{aligned}$$

$A \subset R^{22\times 22}$, $B \subset R^{22\times 6}$, and $C \subset R^{6\times 22}$ are the state matrices as given in the Appendix. The system (7.18) can alternatively be written as

$$\begin{aligned} \dot{x} &= Ax + B_1u_l + B_2u_2 + B_3u_3, \\ y_l &= C_1x, \\ y_2 &= C_2x, \\ y_3 &= C_3x \end{aligned} \tag{7.19}$$

where

$$\begin{aligned} y_i &= (V_{d,i}, V_{q,i}), \quad i = 1, 2, 3, \\ u_i &= (v_{td,i}, v_{tq,i}), \quad i = 1, 2, 3 \end{aligned}$$

and a decentralized controller

$$U_i(s) = C_i(s)E_i(s), \quad i = 1, 2, 3 \tag{7.20}$$

is to be found, where $E_i(s)$ denotes the system error, $U_i(s)$ denotes the input, and the controller transfer function in (7.20) is restricted to being a proper transfer function $i = 1, 2, 3$.

The **A**-matrix of (7.18), $A \in R^{22\times 22}$, is A = blockdiag$\{A_1, A_2, A_3\}$, where A_1, A_2, and A_3 are defined at the top of the previous page. Other entries of A are zero except for the following: $A_{7,9} = A_{8,10} = -l/L_{t1}$, $A_{9,7} = A_{10,8} = 1/C_2$, $A_{15,17} = A_{16,18} = -1/L_{t2}$, and $A_{17,15} = A_{18,16} = l/C_3$. Nonzero entries of $B_{22\times 6}$ are $B_{a,l} = B_{4,2} = 1/L_{f1}$, $B_{11,a} = B_{12,4} = 1/L_{f2}$, and $B_{19,5} = B_{21,6} = 1/L_{f3}$. Nonzero entries of $C_{6\times 22}$, that is, $C_{1,1}$, $C_{2,2}$, $C_{9,3}$, $C_{17,5}$ and $C_{18,6}$ are unity.

7.6 Control Strategy II

This section elaborates on the properties of the open-loop system, and characterizes its robust properties. Then, a proposed robust decentralized servomechanism controller is designed with a real stability radius constraint imposed [49]. Finally, a robust stability analysis and performance evaluation of the closed-loop system is provided. The decentralized controller will be found so that the real stability radius of the final closed-loop system is approximately the same as the open-loop system (i.e., the closed-loop system should have a robustness index which is not worse than, say, 50 % of the robustness index of the open-loop system).

An extended LTI model of the MG based on (7.18) is given by

$$\begin{aligned} \dot{x} &= Ax + Bu + Ew, \\ Y &= Cx + Fw, \\ e &= y - y_{ref} \end{aligned} \tag{7.21}$$

where $x \in R^n$ is the state, $u \in R^m$ is the input, $y \in R^r$ is the output, $w \in R^n$ is an unmeasurable constant disturbance, $y_{ref} \in R^r$ is the desired constant set-point for the system, and $e \in R^r$ is the error in the system.

We impose the condition that the system must contain $v = 3$ control agents, each corresponding to one of the three virtual subsystems of Fig. 7.22, and we rewrite (7.21) as

$$\begin{aligned} \dot{x} &= Ax + \sum_{i=l}^{v} B_i u_i + Ew, \\ y_i &= C_i x + F_i w, \quad i = 1, 2, \ldots, v \\ e_i &= y_i - y_{ref}^i \quad i = 1, 2, \ldots, v \end{aligned} \tag{7.22}$$

where u_i and y_i are the inputs and outputs of control agent i, and e_i is the error in control agent i, $i = 1, 2, \ldots, v$. The open-loop eigenvalues and transmission zeros [23] of (7.21), corresponding to the system of Fig. 7.22, are given in Table 7.5, which indicates that the system is stable and minimum phase.

Table 7.5 Open-loop plant eigenvalues and transmission zeros

Eigenvalues	Transmission zeros
$-570.79 \pm 4074.8i$	$-1116.71 \pm 377.0i$
$-570.79 \pm 3320.8i$	$-1127.93 \pm 377.0i$
$-547.12 \pm 2519.7i$	$-18.85 \pm 377.0i$
$-547.12 \pm 1765.7i$	$-18.85 \pm 377.0i$
$-29.33 \pm 946.1i$	$-18.85 \pm 377.0i$
$-29.33 \pm 192.1i$	
$-84.85 \pm 377.0i$	
$-39.85 \pm 377.0i$	
$-14.07 \pm 377.0i$	
$-12.66 \pm 377.0i$	
$-11.98 \pm 377.0i$	

7.6.1 Design Requirements

A decentralized controller for the plant (7.22) includes the following desired features:

1. The closed-loop system is asymptotically stable.
2. Steady-state asymptotic tracking and disturbance regulation occurs for (1) all constant set-points $y_{ref}^1, y_{ref}^2, y_{ref}^3$ and (2) all constant disturbances w (i.e., $\lim_{t\to\infty} e_i(t) = 0, i = 1, 2, 3$) for all constant disturbances and set-points.
3. The controller is robust. That is, condition (2) should hold for any perturbations of the plant model (7.22), including dynamic perturbations which do not destabilize the perturbed closed-loop system.
4. The controller should be "fast" with smooth nonpeaking transients (e.g., it should respond to constant set-points and constant disturbance changes within about three cycles of 60 Hz).
5. Low interaction should occur among the output channels of the v control agents, and among the outputs contained in each of the v control agents, for tracking and regulation [20].
6. It is required that the aforementioned conditions should be satisfied for as wide as possible a range of the load parameters R, L, and C of each of the control agents.

These conditions will be achieved by a decentralized controller based on the solution of the DRSP [11, 12].

7.6.2 Existence Conditions

The following existence conditions for a solution to the DRSP, so that conditions (1), (2), and (3) from before all hold, are obtained from [11]. Given plant (7.22), let

$$C_m := \left[C_1^{*T}, C_2^{*T}, \ldots, C_v^{*T}\right]^T \tag{7.23}$$

where

$$C_1^* := \begin{bmatrix} C_1 & 0 & 0 & \ldots & 0 \\ 0 & I_{r1} & 0 & \ldots & 0 \end{bmatrix}$$

$$C_2^* := \begin{bmatrix} C_2 & 0 & 0 & \ldots & 0 \\ 0 & 0 & I_{r2} & \ldots & 0 \end{bmatrix}$$

$$\vdots$$

$$C_v^* := \begin{bmatrix} C_v & 0 & 0 & \ldots & 0 \\ 0 & 0 & 0 & \ldots & I_{rv} \end{bmatrix}$$

and let $B := [B_1, B_2, \ldots, B_v]$. and $C := \left[C_1^T, C_2^T, \ldots, C_v^T\right]^T$.

Theorem 7.1 [11] *Given the system (7.22), then there exists a solution to the DRSP such that conditions (1), (2), and (3) all hold, if and only if the following conditions are all satisfied:*

1. *The system (7.22) has no unstable DFMs.*
2. $m_i \geq \tau_i$, *where* m_i *is the number of outputs of the* ith *control agent and* τ_i *is the number of outputs of* $Subsystem_i$, $i = 1, 2, \ldots, v$.
3. *The system* $\{C_m, \begin{bmatrix} A & 0 \\ C & 0 \end{bmatrix}, B\}$ *has no DFM = 0.*

Remark 7.1 If $m_i = \tau_i, i = 1, 2, \ldots, v$, the condition (3) becomes

$$rank \begin{bmatrix} A & B \\ C & 0 \end{bmatrix} = n + \tau_1 + \tau_2 + \cdots + \tau_v.$$

For the MG of Fig. 7.22, it can be verified that the existence conditions of Theorem 7.1 are all satisfied. In particular, (1) plant (7.22) has no DFMs; (2) it has $m, = \tau_i = 2, i = 1, 2, 3$; and (3) the rank of the matrix $\begin{bmatrix} A & B \\ C & 0 \end{bmatrix}$. given by (7.18), is equal to $n + \tau_1 + \tau_2 + \tau_3 = 22 + 2 + 2 + 2 = 28$.

Remark 7.2 In the plant model (7.21), the three load parameters $R, L,$ and C of each subsystem (Fig. 7.22) can vary and result in structural uncertainty in the plant's nominal model. It is also observed in (7.21) that the load parameters affect neither the output gain matrix C nor the input control matrix B. Therefore, only matrix A of (7.21) is affected by changing load parameters. We use this observation to design a controller with the desirable robustness properties.

The following analysis shows that condition (3) of Theorem 7.1 always holds true for the study MG regardless of its numerical values. The determinant of the matrix $\begin{bmatrix} A & B \\ C & 0 \end{bmatrix}$ is given by ρ_1/ρ_2 where

$$\begin{aligned} \rho_1 &= \left(R_{l1}^2 + \omega^2 L_1^2\right)\left(R_{l2}^2 + \omega^2 L_2^2\right)\left(R_{l3}^2 + w^2 L_3^2\right) \\ &\quad \times \left(R_{t1}^2 + \omega^2 L_{t1}^2)(R_{t2}^2 + \omega^2 L_{t2}^2\right), \\ \rho_2 &= \left(C_1 L_1 L_{f1} L_{t1} C_2 L_2 L_{f2} L_{t2} C_3 L_3 L_{f3}\right)^2. \end{aligned}$$

Clearly, the determinant is always nonzero which implies rank $\begin{bmatrix} A & B \\ C & 0 \end{bmatrix} = 28$ for any numerical value of MG parameters and, thus, the third condition of Theorem 7.1 always holds.

7.6.3 Real Stability Radius Constraints

To evaluate the robustness of a control scheme, the following definition is used [49].

Given a real $n \times n$ matrix A which is asymptotically stable, assume that A is subject to a real perturbation $A \rightarrow A + M\Delta N$, where M and N are given real matrices, and Δ is a real matrix of uncertain parameters. Then, it is desired to find $rstab > 0$, so that (1) $A + M\Delta N$ is asymptotically stable for all real perturbations Δ with the property that $\|\Delta\| < rstab$ and (2) there exists a perturbation Δ^* with the property that $\|\Delta^*\| = rstab$, so that $A + M\Delta^* N$ is unstable. In this case, rstab is called the real stability radius of $\{A, M, N\}$.

7.6.4 Controller Design Procedure

Given the plant (7.22) with $v = 3$, to solve the DRSP, it is necessary [11] that the decentralized controller should include the decentralized servo-compensator

$$\dot{\eta}_i = 0\eta_i + (y_i - y^i_{ref}), \quad i = 1, 2, 3 \tag{7.24}$$

where $\eta_i \in R^2, i = 1, 2, 3$, together with a decentralized stabilizing compensator, which will be assumed to have the structure

$$\begin{aligned} \dot{\beta} &= A\beta + By, \\ u &= K_1 y + K_2 \eta + K_3 \beta \end{aligned} \tag{7.25}$$

where

$$A = \begin{bmatrix} A_1 & 0 & d0 \\ 0 & A_2 & 0 \\ 0 & 0 & A_3 \end{bmatrix}, B = \begin{bmatrix} B_1 & 0 & 0 \\ 0 & B_2 & 0 \\ 0 & 0 & B_3 \end{bmatrix},$$

$$K_1 = \begin{bmatrix} K_1^1 & 0 & 0 \\ 0 & K_1^2 & 0 \\ 0 & 0 & K_1^3 \end{bmatrix}, K_2 = \begin{bmatrix} K_2^1 & 0 & 0 \\ 0 & K_2^2 & 0 \\ 0 & 0 & K_2^3 \end{bmatrix},$$

$$K_3 = \begin{bmatrix} K_3^1 & 0 & 0 \\ 0 & K_3^2 & 0 \\ 0 & 0 & K_3^3 \end{bmatrix}$$

so that the controlled closed-loop system is described by

$$\begin{bmatrix} \dot{x} \\ \dot{\eta} \\ \dot{\beta} \end{bmatrix} = \begin{bmatrix} A + BK_1C & BK_2 & BK_3 \\ C & 0 & 0 \\ BC & 0 & A \end{bmatrix} \begin{bmatrix} x \\ \eta \\ \beta \end{bmatrix} + \begin{bmatrix} 0 \\ -I \\ 0 \end{bmatrix} y_{ref} + \begin{bmatrix} BK_1F + E \\ F \\ BF \end{bmatrix} w,$$

$$y = \begin{bmatrix} C & 0 & 0 \end{bmatrix} \begin{bmatrix} x \\ \eta \\ \beta \end{bmatrix} + Fw. \tag{7.26}$$

In this case, the controller parameters (7.25) are obtained by applying the optimal controller design method [16, 17] to minimize the expected value of the performance index $\int_0^{\infty}(e^{'}e + \varepsilon\dot{u}^{'}\dot{u})d\tau$ given by

$$J = trace(\Gamma) \tag{7.27}$$

where $\Gamma > 0$ is obtained by solving the Lyapunov matrix equation corresponding to

$$\int_0^{\infty}(e^{'}e + \varepsilon\dot{u}^{'}\dot{u})d\tau = \begin{bmatrix} \dot{x}(0) \\ e(0) \\ \dot{\beta}(0) \end{bmatrix}^{'} \Gamma \begin{bmatrix} \dot{x}(0) \\ e(0) \\ \dot{\beta}(0) \end{bmatrix} \tag{7.28}$$

where $\varepsilon = 10^{-6}$, subject to the conditions that

1. the resultant closed-loop system (7.26) is asymptotically stable;
2. the real stability radius of the closed-loop system $\{A_{close}, \begin{bmatrix} A \\ 0 \\ 0 \end{bmatrix}, [I, 0, 0]\}$ should be greater than or equal to $1/2x$ the real stability radius of the open-loop system $\{A, A, I\}$, where

$$A_{close} = \begin{bmatrix} A + BK_1C & BK_2 & BK_3 \\ C & 0 & 0 \\ BC & 0 & A \end{bmatrix}.$$

Substituting for e and $\dot{u}$ from (7.26) in (7.28) provides a closed-form expression for (7.28)

$$A_{close}^T\Gamma + \Gamma A_{close} = -\left(\begin{bmatrix} 0 & I & 0 \end{bmatrix}^{'}\begin{bmatrix} 0 & I & 0 \end{bmatrix} + \varepsilon K^{*T}K^{*}\right) \tag{7.29}$$

where $K^* = [K_1C, K_2, K_3]$ [47]. The control design problem becomes, therefore, a constrained parameter optimization problem whose solution yields the control parameters A, B, K_1, K_2, and K_3.

The reason to impose the robust constraint above is that the load parameters $R_i, L_i, C_i, i = 1, 2, 3$, may vary, but since they affect only the A matrix and not the B_1, B_2, B_3, C_1, C_2, and C_3 matrices of (7.22), then it is desired to minimize the effect of load parameter perturbations by keeping the closed-loop system robustness index, as measured by the real stability radius, as "close" as possible to the open-loop robustness index. In this case, the constraint imposed is such that the closed-loop robustness index is not "worse" than $1/2x$ the open-loop robustness index.

7.6.5 A Decentralized Controller Scheme

On carrying out the optimization of the controller (7.25) as measured by the performance index (7.27), the following decentralized controller is obtained for the MG of Fig. 7.22

$$\begin{aligned} u &= K_1 y + K_2 \eta + K_3 \beta, \\ \dot{\eta} &= y - y_{ref}, \\ \dot{\beta} &= A\beta + By \end{aligned} \tag{7.30}$$

whose variables are defined in the equation at the bottom of the next page.

7.6.6 Properties of the Closed-Loop System

A transfer function representation of the control agents of the decentralized controller (7.30) is given by

$$\begin{aligned}
\begin{bmatrix} U_1(s) \\ U_2(s) \end{bmatrix} &= \frac{1}{s}\begin{bmatrix} -1047 & -649.5 \\ -834.4 & -1103.3 \end{bmatrix}\begin{bmatrix} E_1(s) \\ E_2(s) \end{bmatrix} \\
&+ \frac{1}{s+5313}\begin{bmatrix} -15.83s - 8.4E4 & 3.37s + 1.8E4 \\ -7.97s - 4.1E4 & -110.1s - 7.5E4 \end{bmatrix}\begin{bmatrix} Y_1(s) \\ Y_2(s) \end{bmatrix} \\
\begin{bmatrix} U_3(s) \\ U_4(s) \end{bmatrix} &= \frac{1}{s}\begin{bmatrix} -495.6 & 969.7 \\ -1072 & -838.6 \end{bmatrix}\begin{bmatrix} E_3(s) \\ E_4(s) \end{bmatrix} \\
&+ \frac{1}{s+7702}\begin{bmatrix} -5.11s - 4.3E4 & 104.8s + 5.4E4 \\ -10.38s - 7.5E4 & -132.4s - 9.8E4 \end{bmatrix}\begin{bmatrix} Y_3(s) \\ Y_4(s) \end{bmatrix} \\
\begin{bmatrix} U_5(s) \\ U_6(s) \end{bmatrix} &= \frac{1}{s}\begin{bmatrix} -635.9 & 793 \\ -1002 & -1039 \end{bmatrix}\begin{bmatrix} E_5(s) \\ E_6(s) \end{bmatrix} \\
&+ \frac{1}{s+6913}\begin{bmatrix} -5.51s - 4.0E4 & 6.8s + 2.6E4 \\ -15.39s - 4.84E4 & -118.9s - 8.3E4 \end{bmatrix}\begin{bmatrix} Y_5(s) \\ Y_6(s) \end{bmatrix}
\end{aligned}$$

Table 7.6 Closed-loop eigenvalues using decentralized control

$-1373.4 \pm 5144.2i$
$-857.42 \pm 4033.4i$
$-1440.00 \pm 3666i$
$-1847.00 \pm 2055.9i$
$-519.53 \pm 2529.7i$
$-311.67 \pm 1648.9i$
$-284.21 \pm 431.66i$
$-134.73 \pm 425.55i$
$-18.391 \pm 377.14i$
$-18.133 \pm 377.52i$
$-17.594 \pm 377.75i$
$-91.223 \pm 29.715i$
$-31.809 \pm 5.3436i$
$-60.465 \pm 4.0087i$
-5463.8
-4539.8
-829.98

where $U_i(s)$, $E_i(s)$, $Y_i(s)$, and $i = 1, 2, \ldots, 6$ are the outputs and errors of controllers, and plant's outputs, respectively. Poles of the decentralized controller are 0 (repeated 6 times), -7702, -6913, and -5313. The closed-loop eigenvalues of the plant, using the decentralized controller (7.30), are listed in Table 7.6.

The closed-loop system of plant (7.22) and controller (7.30) is highly robust with respect to changes in the load parameters $R_i, L_i, G_i, i = 1, 2, 3$ of the MG. In particular assume that in the open-loop system (7.21) the matrix A is subject to uncertainty (e.g., due to changes in the load parameters and perturbations in the elements of the matrix A). Also assume that $A \to A(I + \Delta)$, where Δ is an unknown real perturbation matrix, and where A is an asymptotically stable matrix. The largest perturbations Δ for which the system remains stable are determined from the real stability radius of A [49].

In this case, the open-loop perturbed plant $dx/dt = A(I + \Delta)x$ remains asymptotically stable for all real Δ matrix perturbations with $\|\Delta\| < r_{stab}$ if and only if $r_{stab} < r^*_{stab}$, where $r^*_{stab} = 1.495 \times 10^{-3}$. This implies that the open-loop plant $dx/dt = Ax$ remains asymptotically stable for any real perturbation of size 0.15 % or less.

Consider now the closed-loop system obtained by applying the controller (7.30) to (7.22), and assume that the elements of the controller are fixed and that the B_i and $C_i, i = 1, 2, \ldots, v$ elements of the plant are fixed, but that due to load changes in $R_i, L_i, C_i, i = 1, 2, 3$ (and other perturbations), the elements of A are allowed to perturb. We also assume in the closed-loop plant (7.26) that $A \to A(I + \Delta)$, where Δ is an unknown perturbation matrix. In this case, we obtain from [49] that the perturbed

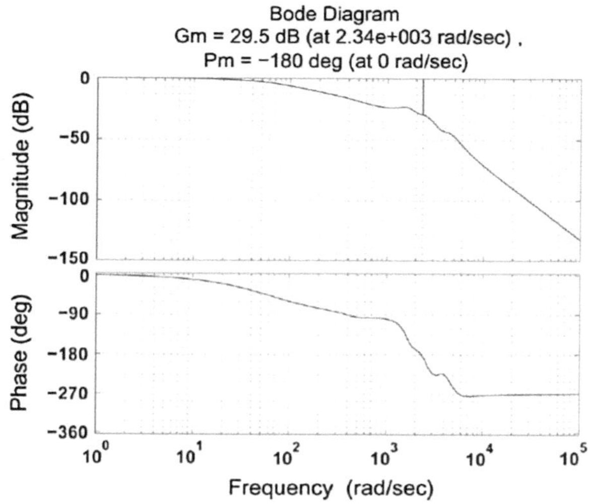

Fig. 7.23 Bode plot of the closed-loop system

closed-loop system remains asymptotically stable for all real Δ matrix perturbations with $\|\Delta\| < r_{stab}$ if and only if $r_{stab} < r^*_{stab}$, where $r^*_{stab} = 7.752 \times 10^{-2}$.

Thus, the stability robustness index of the closed-loop system $r^*_{stab} = 7.752 \times 10^{-2}$ with respect to perturbations of the A matrix is larger than the open-loop system given by $r^*_{stab} = 1.495 \times 10^{-3}$, and so there is no deterioration in robustness of the closed-loop system compared to the open-loop system with respect to perturbations of the elements of A. A bode plot of the closed-loop system is illustrated in Fig. 7.23. Figure 7.23 shows that the closed-loop transfer function from u_1 to y_1 of the first control agent has a gain margin of 29.5 dB at the frequency of 372 Hz (as denoted by a dotted line) and a phase margin of $-180°$ which clearly highlights the robust stability of the closed-loop system.

$$u = \begin{bmatrix} u_1 \\ u_2 \\ u_3 \end{bmatrix}, \ y = \begin{bmatrix} y_1 \\ y_2 \\ y_3 \end{bmatrix},$$

$$K_1 = \begin{bmatrix} -15.83 & 3.37 & 0 & 0 & 0 & 0 \\ -7.97 & -110.14 & 0 & 0 & 0 & 0 \\ 0 & 0 & -5.11 & 104.77 & 0 & 0 \\ 0 & 0 & -10.38 & -132.44 & 0 & 0 \\ 0 & 0 & 0 & 0 & -5.51 & 6.81 \\ 0 & 0 & 0 & 0 & -15.39 & -118.91 \end{bmatrix},$$

$$K_2 = \begin{bmatrix} -1047 & 649 & 0 & 0 & 0 & 0 \\ -834 & -1103 & 0 & 0 & 0 & 0 \\ 0 & 0 & -496 & 970 & 0 & 0 \\ 0 & 0 & -1072 & -839 & 0 & 0 \\ 0 & 0 & 0 & 0 & -636 & 793 \\ 0 & 0 & 0 & 0 & -1002 & -1039 \end{bmatrix},$$

$$K_3 = \begin{bmatrix} 0.18 & 0 & 0 \\ 881.47 & 0 & 0 \\ 0 & -3668.6 & 0 \\ 0 & 4495.3 & 0 \\ 0 & 0 & -1639.3 \\ 0 & 0 & 57906 \end{bmatrix},$$

$$B = \begin{bmatrix} 1 & 579.02 & 0 & 0 & 0 & 0 \\ 0 & 0 & 1 & 205.21 & 0 & 0 \\ 0 & 0 & 0 & 0 & 1 & 12.754 \end{bmatrix}, \quad A \begin{bmatrix} -5313.5 & 0 & 0 \\ 0 & -7702.2 & 0 \\ 0 & 0 & -6913.2 \end{bmatrix}$$

$$A_1 = \begin{bmatrix} \frac{-1}{R_1C_1} & -\omega & \frac{1}{C_1} & 0 & \frac{-1}{C_1} & 0 & \frac{-1}{C_1} & 0 \\ -\omega & \frac{-1}{R_1C_1} & 0 & \frac{1}{C_1} & 0 & \frac{-1}{C_1} & 0 & \frac{-1}{C_1} \\ \frac{-1}{L_{f1}} & 0 & \frac{-R_{f1}}{L_{f1}} & \omega & 0 & 0 & 0 & 0 \\ 0 & \frac{-1}{L_{f1}} & -\omega & \frac{-R_{f1}}{L_{f1}} & 0 & 0 & 0 & 0 \\ \frac{1}{L_1} & 0 & 0 & 0 & \frac{-R_{t1}}{L_1} & -\omega & 0 & 0 \\ 0 & \frac{1}{L_1} & 0 & 0 & -\omega & \frac{-R_{t1}}{L_1} & 0 & 0 \\ \frac{1}{L_{t1}} & 0 & 0 & 0 & 0 & 0 & \frac{-R_{t1}}{L_{t1}} & -\omega \\ 0 & \frac{1}{L_{t1}} & 0 & 0 & 0 & 0 & -\omega & \frac{-R_{t1}}{L_{t1}} \end{bmatrix},$$

$$A_2 = \begin{bmatrix} \frac{-1}{R_2C_2} & -\omega & \frac{1}{C_2} & 0 & \frac{-1}{C_2} & 0 & \frac{-1}{C_2} & 0 \\ -\omega & \frac{-1}{R_2C_2} & 0 & \frac{1}{C_2} & 0 & \frac{-1}{C_2} & 0 & \frac{-1}{C_2} \\ \frac{-1}{L_{f2}} & 0 & \frac{-R_{f2}}{L_{f2}} & \omega & 0 & 0 & 0 & 0 \\ 0 & \frac{-1}{L_{f2}} & -\omega & \frac{-R_{f2}}{L_{f2}} & 0 & 0 & 0 & 0 \\ \frac{1}{L_2} & 0 & 0 & 0 & \frac{-R_{t2}}{L_2} & -\omega & 0 & 0 \\ 0 & \frac{1}{L_2} & 0 & 0 & -\omega & \frac{-R_{t2}}{L_2} & 0 & 0 \\ \frac{1}{L_{t2}} & 0 & 0 & 0 & 0 & 0 & \frac{-R_{t2}}{L_{t2}} & -\omega \\ 0 & \frac{1}{L_{t2}} & 0 & 0 & 0 & 0 & -\omega & \frac{-R_{t2}}{L_{t2}} \end{bmatrix},$$

$$A_3 = \begin{bmatrix} \frac{-1}{R_3C_3} & -\omega & \frac{1}{C_3} & 0 & \frac{-1}{C_3} & 0 \\ -\omega & \frac{-1}{R_3C_3} & 0 & \frac{1}{C_3} & 0 & \frac{-1}{C_3} \\ \frac{-1}{L_{f3}} & 0 & \frac{-R_{f3}}{L_{f3}} & \omega & 0 & 0 \\ 0 & \frac{-1}{L_{f3}} & -\omega & \frac{-R_{f3}}{L_{f3}} & 0 & 0 \\ \frac{1}{L_3} & 0 & 0 & 0 & \frac{-R_{t3}}{L_3} & -\omega \\ 0 & \frac{1}{L_3} & 0 & 0 & -\omega & \frac{-R_{t3}}{L_3} \end{bmatrix}.$$

7.6.7 Other Robustness Measures

1. **Input and Output Gain Margins**: The following definition which is an extension of the classical Single-Input Single-Output (SISO) gain margin to multivariable systems [10] is provided.
Definition: Given a plant $\dot{x} = Ax + Bu, Y = Cx$ controlled by a controller $\dot{\eta} = A_c\eta + B_c y, u = C_c\eta$, assume that the closed-loop system is asymptotically stable. Let $y = Cx$ be replaced by a $y = (1 + K_y)Cx$, where K_y is a constant gain matrix; then if the closed-loop system remains stable $\forall K_y$ with $\|K_y\| < l_y$, the system has an output gain margin of $(1 + l_y)$. Also let Bu be replaced by $B(l + K_u)u$; then if the closed-loop system remains stable VKu with $\|K_u\| < l_u$, the system is said to have an input gain margin of of $(1 + l_u)$. The gain margin can be calculated from the real stability radius of the system, and the closed-loop system (7.22) and (7.30) has output and input gain margins of 1.38 and 1.28, respectively, which are quite satisfactory [15].
2. **Input Time Delay Tolerance**: Often a system may have time delays which are ignored in the modeling of a system, and it is important that the controlled system be robust to such unmodeled effects. The following definition is used to describe such a robust property [36].
Definition: Given a plant $\dot{x} = Ax + Bu, y = Cx$ controlled by the controller $\dot{\eta} = A_c\eta + B_c y, u = C_c\eta$ assume that the closed-loop system is asymptotically stable, and let $u(t)$ be replaced by $u(t - \beta)$, corresponding to a time delay of β s. Then, if there exists $\underline{\beta} > 0$ such that the closed-loop system remains stable $\forall\beta \in [0, \underline{\beta}]$, the closed-loop system has an input time-delay tolerance of $\underline{\beta}$ [36]. In this case, the closed-loop system (7.22) and (7.30) has an input time-delay tolerance of $1.58 \times 10 - 4$ s, which is acceptable considering the switching frequency of the VSCs.
3. **Tolerance to Unmodeled High-Frequency Effects**: It is of interest to determine the extent that the closed-loop system is tolerant to unmodeled high-frequency effects in the plant model (7.21). This is captured by including the term ΔD in the model of the plant as follows:

$$\begin{aligned} \dot{x} &= Ax + Bu, \\ y &= Cx + \Delta Du, \end{aligned} \tag{7.31}$$

where (C, A, B) are given in (7.21), and ΔD is a real perturbation matrix. In this case, on applying the same controller (7.30) to plant (7.31), we obtain that the resultant closed-loop system remains asymptotically stable $\forall\|\Delta D\| < rstab$ if and only if $rstab < rstab^*$, where $rstab^* = 4.737 \times lO^3$ [49]. This implies that the perturbed closed-loop system remains asymptotically stable for all perturbations in the gain matrix ΔD of (7.21) with the property that $(\|\Delta D\|)/(\|CA^{-1}B\|) \times 100\,\% \leq 0.490\,\%$ (where $\|CA^{-1}B\| = 0.9658$), which satisfactorily meets the requirements.

7.7 Decentralized Inverter Control

Decentralized inverter control is essential in DG microgrids for low deployment/ operation cost and high reliability. However, decentralized inverter control suffers from a limited system stability mainly because of the lack of communications among different inverters. We investigate stability enhancement of the droop-based decentralized inverter control in MGs in this section. Specifically, we propose a power sharing-based control strategy which incorporates the information of the total real and reactive power generation of all DG units. The information is acquired by a wireless network (such as a WiFi, ZigBee, and/or cellular communication network) in a decentralized manner. Based on the desired power sharing of each DG unit and the acquired information of total generation, additional control terms are added to the traditional droop controller. We evaluate the performance of the proposed control strategy based on small-signal stability analysis. As timely communication may not be established for a MG with low-cost wireless communication devices, two kinds of analytical models are developed with respect to negligible and nonnegligible communication delays, respectively. Extensive numerical results are presented to demonstrate the system stability under the proposed control strategy with respect to different communication delays.

Although DG has a great potential for economic and environmental benefits, how to establish efficient and reliable control over a large number of DG units is one of the primal problems to be solved in the near future. One promising solution to this problem is the MG, which interconnects a group of DG units and loads at a distribution voltage level in a local area such as a university or a residential community [8, 52]. MGs are established based on localized control and can operate in either a grid-connected mode or an islanded mode, which significantly reduces the complexity of DG unit control. In order to avoid high capital expenditure and low reliability in MG operation, DEC is indispensable [27].

In the sequel, we investigate one of the critical control problems in MGs, i.e., the droop-based decentralized inverter control. The main objective is to achieve efficient real and reactive power sharing while maintaining the MG frequency and voltage [59]. Featured by a DEC strategy without a communication network infrastructure, the traditional droop control strategy emulates the droop characteristics of synchronous generators based on local estimates on real and reactive power generation. However, in a typical MG application where the ratio of line resistance to line reactance is high, the coupling between real and reactive power control deteriorates the system stability [64]. Several methods can be used to improve the stability of traditional droop control in MGs. For instance, an adaptive transient droop function can be added to damp the oscillatory modes of the power sharing controller [59], while the transient droop gains can be adaptively scheduled based on a small-signal analysis with respect to the loading trajectory of each DG unit. The virtual impedance method can be used to decouple the real and reactive power control [33, 65], especially for the transformer coupled DG units which already have significant output inductance [50]. In order to avoid the complexity in impedance design, the virtual frequency and voltage frame-based droop control strategy can be used [51]. The frequency and voltage are transformed to a virtual frame for a completely decoupled relationship between real- and reactive power.

In comparison with the existing approaches which improve the traditional droop control strategy of individual inverters, we address the stability enhancement of droop control by establishing coordination among the inverters based on a wireless network. The objective of this research is to introduce limited communication overhead to improve the stability while maintaining the decentralized structure of the traditional droop control strategy. A power sharing-based control strategy is proposed based on the information provided by a wireless network, i.e., the total real- and reactive power generation of all DG units. The information acquisition is completed in a decentralized manner in accordance with the decentralized inverter control. By evaluating the differences between the desired and actual real- and reactive power sharing of each DG unit, additional control terms are incorporated in the traditional droop controller to improve the system stability. Specifically, a proportional term is incorporated in the frequency droop control to damp the oscillation in real power sharing without affecting the original droop gain. On the other hand, an integral term is incorporated in the voltage droop control to correct the inaccurate reactive power sharing based on the traditional droop control. As timely communication may not be established for a MG with a low-cost wireless network infrastructure, the communication delay may adversely affect the performance of MG control [8, 53].

7.7.1 System Model

Consider an islanded MG with parallel-connected inverters, as shown in Fig. 7.24. In the MG, there are G inverters (corresponding to the DG units), H loads, N nodes (or connection points), and L power transmission lines. Each line in the MG interconnects a pair of two nodes, while each node is connected to an inverter and/or a load. Without loss of generality, we order the elements in the MG such that the sets of inverters, loads, nodes, and lines are given by $\{1, 2, \ldots, G\}$, $\{1, 2, \ldots, H\}$, $\{1, 2, \ldots, N\}$, and $1, 2, \ldots, L$, respectively. A wireless network is established to provide wireless communication links among the inverters. Each inverter is equipped with a wireless communication device which can communicate with a certain set of inverters. The wireless network is considered to be strongly connected, i.e., there exists a communication (possibly multihop) path between any pair of two inverters.

The droop-based decentralized inverter control strategy is considered [59, 64]. As shown in Fig. 7.25, the controller of each individual inverter consists of three parts, i.e., the power controller, voltage controller, and current controller. In this work, we consider the stability of power controller, while the voltage and current controllers are based on traditional proportional-integral (PI) controllers. The D-Q reference frame transformation is considered, where the d-axis and q-axis of the reference frame of each inverter are rotating at the reference frequency [69]. Based on the traditional

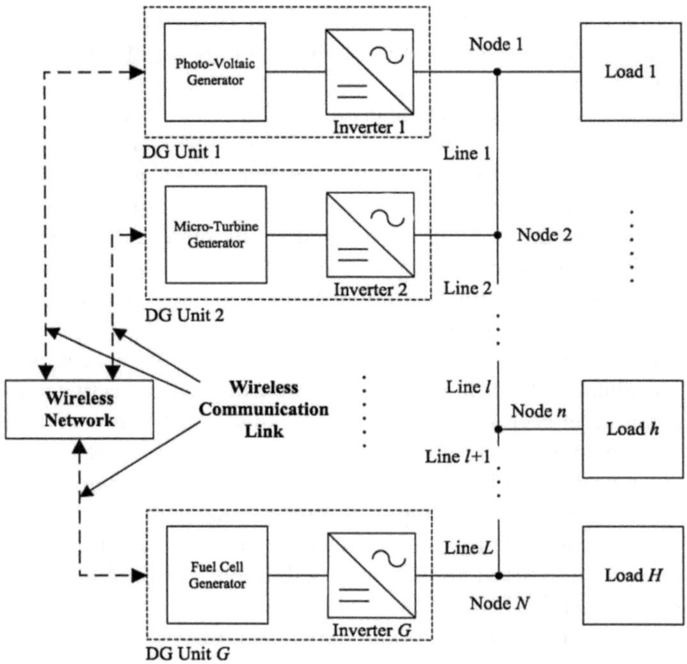

Fig. 7.24 The block diagram of an inverter controller

droop control strategy, the reference frequency and output d-axis voltage of inverter $g(g \in \{1, 2, \ldots, G\})$ are determined by the power controller and are given by

$$W_{R-dr}^{(g)} = \omega_0 - K_{P-dr}^{(g)} P^{(g)} \tag{7.32}$$

$$V_{R-dr,o}^{(g)} = v_{0,od} - K_{Q-dr}^{(g)} Q^{(g)} \tag{7.33}$$

$$v_{R-dr,oq}^{(g)} = 0 \tag{7.34}$$

where ω_o and $v_{O,od}$ are the nominal set-points of frequency and d-axis output voltage, respectively, while $K_{P-dr}^{(g)}$ and $K_{Q-dr}^{(g)}$ are the frequency and voltage droop gains with respect to certain ranges of frequency and voltage magnitude, $[\omega_{\min}^{(g)}, \omega_{\max}^{(g)}]$ and $[v_{\min,od}^{(g)}, v_{\max,od}^{(g)}]$, respectively, given by

$$K_{P-dr}^{(g)} = \frac{\omega_{\max}^{(g)} - \omega_{\min}^{(g)}}{P_{\max}^{(g)}} \tag{7.35}$$

$$K_{Q-dr}^{(g)} = \frac{v_{\max,od}^{(g)} - v_{\min,od}^{(g)}}{Q_{\max}^{(g)}} \tag{7.36}$$

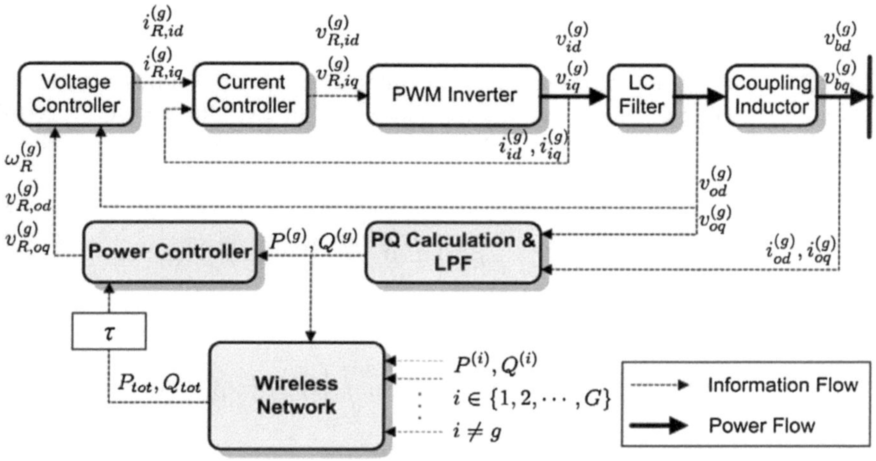

Fig. 7.25 The configuration of a microgrid with wireless communications among inverters

In (7.32)–(7.34), we use the subscript "$-dr$" to denote the values determined based on the traditional droop control strategy. Since no communication network is available for traditional droop control, $\omega_{R-dr}^{(g)}$, $v_{R-dr,od}^{(g)}$, and $v_{R-dr,oq}^{(g)}$ are equal to $\omega_R^{(g)}$, $v_{R,od}^{(g)}$, and $v_{R-oq}^{(g)}$ in Fig. 7.25.

The inputs of the power controller, i.e., the real power ($p^{(g)}$) and reactive power ($Q^{(g)}$), are calculated based on the low-pass filtered (with cutoff frequency we) values of $p^{(g)}$ and $q^{(g)}$, respectively, given by

$$p^{(g)} = v_{od}^{(g)} i_{od}^{(g)} + v_{oq}^{(g)} i_{oq}^{(g)} \tag{7.37}$$

$$q^{(g)} = v_{od}^{(g)} i_{oq}^{(g)} - v_{oq}^{(g)} i_{od}^{(g)}. \tag{7.38}$$

The output $\omega_R^{(g)}$ of the power controller is used by the PWM inverter as reference frequency, while the reference voltage of the PWM inverter needs to be regulated by the following voltage and current controllers.

Based on the references provided by the power controller, $v_{R,od}^{(g)}$ and $v_{R,oq}^{(g)}$, the output of the voltage controller is given

$$\begin{aligned} i_{R,id}^{(g)} = K_{pv}\left[v_{R,od}^{(g)} - v_{od}^{(g)}\right] + K_{iv}\int\left[v_{R,od}^{(g)} - v_{od}^{(g)}\right]dt \\ - \omega_0 C_f v_{oq}^{(g)} + F i_{od}^{(g)} \end{aligned} \tag{7.39}$$

$$i_{R,iq}^{(g)} = K_{pv}\left[v_{R,oq}^{(g)} - v_{oq}^{(g)}\right] + K_{iv}\int\left[v_{R,oq}^{(g)} - v_{oq}^{(g)}\right]dt$$

$$- \omega_0 C_f v_{od}^{(g)} + F i_{oq}^{(g)} \tag{7.40}$$

where K_{pv} and K_{iv} are the proportional and integral gains of the voltage controller, respectively, while C_f and F are the capacitance of the LC filter and feedforward gain, respectively. On the other hand, based on the references provided by the voltage controller, $i_{R,id}^{(g)}$ and $i_{R,iq}^{(g)}$, the output of the current controller is given by

$$v_{R,id}^{(g)} = K_{pi} \left[i_{R,id}^{(g)} - i_{id}^{(g)} \right] + K_{ii} \int \left[i_{R,id}^{(g)} - i_{id}^{(g)} \right] dt$$

$$- \omega_0 L_f I_{Iq}^{(g)} \tag{7.41}$$

$$v_{R,iq}^{(g)} = K_{pi} \left[i_{R,iq}^{(g)} - i_{iq}^{(g)} \right] + K_{ii} \int \left[i_{R,iq}^{(g)} - i_{iq}^{(g)} \right] dt$$

$$- \omega_0 L_f v_{oq}^{(g)} \tag{7.42}$$

where K_{pi} and K_{ii} are the proportional and integral gains of the current controller, respectively, while L_f is the inductance of the LC filter.

The rationale behind the traditional droop control strategy is that the MG frequency and voltage can be considered as two virtual communication links among the inverters. However, this kind of virtual communication is effective only when the transmission lines are mainly inductive, based on which the control of real and reactive power generation can be decoupled [51]. For a typical MG configuration where the line resistance is significant as compared with line inductance, the real and reactive power control is coupled such that the traditional droop control [with respect to (7.32)–(7.34)] deteriorates the system stability.

7.7.2 Power Sharing Control Strategy

In order to improve the system stability under decentralized inverter control, we propose a power sharing-based control strategy which utilizes the information with respect to the total real power generation ($P_{tot} = \sum_{i=1}^{G} P^{(i)}$) and reactive power generation ($Q_{tot} = \sum_{i=1}^{G} Q^{(i)}$) of all DG units, as shown in Fig. 7.25. The gray control blocks correspond to the modified blocks as compared with the traditional droop control strategy. Auxiliary control terms are added to the traditional droop controller, which correspond to the differences between desired and actual sharing of real and reactive power of each DG unit The information is acquired by the wireless network in a decentralized manner.

7.7.3 Controller Design

According to the power sharing based control strategy, the reference frequency and voltages of inverter g are, respectively, given by

$$\omega_R^{(g)} = \omega_{R-dr}^{(g)} + K_{P-sh}^{(g)} \left[\alpha_P^{(g)} P_{tot} - p^{(g)}\right] \quad (7.43)$$

$$v_{R,od}^{(g)} = v_{R-dr,od}^{(g)} + K_{Q-sh}^{(g)} \int \left[\alpha_Q^{(g)} Q_{tot} - Q^{(g)}\right] dt \quad (7.44)$$

$$v_{R,oq}^{(g)} = 0 \quad (7.45)$$

where $K_{P-sh}^{(g)}$ and $K_{Q-sh}^{(g)}$ are the control gains with respect to the real and reactive power sharing, respectively, while $\alpha_P^{(g)}$ and $\alpha_Q^{(g)}$ are the desired sharing of real and reactive power by DG unit g, respectively. Obviously, we have $\sum_{g=1}^{G} \alpha_P^{(g)} = 1$ and $\sum_{g=1}^{G} \alpha_Q^{(g)} = l.$

The rationale behind the controller is to use the actual sharing of real and reactive power to fine tune the reference frequency and voltage of each inverter. With a larger difference, the frequency (or voltage) reference of an individual inverter is adjusted more significantly, and vice versa. The adjustment is stopped when the desired sharing is achieved. Note that an integral tenn is used for the voltage reference since the reactive power sharing of a tradition droop controller is inaccurate for MG applications [35]. It is worth mentioning that our proposed control strategy reduces to the traditional droop control strategy by letting $K_{P-sh}^{(g)} = 0$ and $K_{Q-sh}^{(g)} = O.$

7.7.4 Decentralized Information Acquisition

The information acquisition is carried out based on the wireless network in a decentralized manner. Here, we consider a multiagent coordination-based decentralized information acquisition scheme which can be applied to wireless networks such as WiFi, ZigBee, and cellular communication networks [53, 54]. The communication is fully decentralized such that each inverter only needs to communicate with its direct neighbors to calculate the total generation of all DG units in the MG (i.e., P_{tot} and Q_{tot}). The convergence of the information acquisition is guaranteed based on the average consensus theory [7, 61, 74]. We omit the details of the communication protocol design here because of the space limitation. However, in achieving convergence among the multiagent coordination results with certain accuracy, we need to take into account that a communication delay τ may exist as shown in Fig. 7.20, which corresponds to the period between the information (i.e., $P^{(i)}$ and $Q^{(i)}$ for $i \in \{1, 2, \ldots, G\}$) measurement and the control decision making.

The delay τ depends on the wireless network. If a long-range, low-delay wireless network such as a cellular communication network is available, the communication delay is negligible. The reference frequency and voltage provided by the proposed control strategy are given by (7.43)–(7.45). On the other hand, for most of the MGs where operation cost is one of the major concerns, relying on a high-performance network infrastructure is impractical [59]. Alternatively, low-cost wireless communication devices such as WiFi and ZigBee devices can be used to establish the network infrastructure [27]. For these types of low-cost wireless networks, the communication delay is nonnegligible ($\tau > 0$) in obtaining accurate information acquired by each inverter. The reference frequency and voltage are given by

$$\omega_R^{(g)} = K_{P-sh}^{(g)} \left[\alpha_P^{(g)} P_{tot}(t-\tau) - P^{(g)}(t-\tau) \right] + \omega_{R-dr}^{(g)}(t) \tag{7.46}$$

$$v_{R,od}^{(g)} = K_{Q-sh}^{(g)} \int \left[\alpha_Q^{(g)} Q_{tot}(t-\tau) - Q^{(g)}(t-\tau) \right] dt + v_{R-dr,od}^{(g)}(t) \tag{7.47}$$

$$v_{R,oq}^{(g)} = 0 \tag{7.48}$$

where $\omega_{R-dr}^{(g)}(t)$ and $V_{R-dr,od}^{(g)}(t)$ are given by (7.32) and (7.33), respectively, at time t. Since the communication delay may adversely affect the system stability under our proposed control strategy, we will establish analytical models in the following two sections for performance evaluation with respect to negligible and nonnegligible communication delays, respectively.

Remark 7.3 The following observations stand out:

1. An extension of the multiagent coordination scheme to wireline (or power line) communication networks is straightforward [29, 75].
2. The clock synchronization in MGs can also be done in a decentralized manner [7].
3. The delay can be potentially reduced by selectively introducing additional long-range communication links such as cellular communication links, at an extra monetary cost [53].

7.7.5 Stability Analysis Without Communication Delay

In this section, we present a small-signal model for the performance evaluation of our proposed control strategy. The small-signal model is obtained by linearizing the system state equations at certain operation point. The model can effectively characterize the system stability subject to small disturbances such as load changes (in contrast to large disturbances such as fault and loss of generation) [47], and is widely used to evaluate the stability of decentralized inverter control. In comparison with the small-signal models of traditional droop control strategy, new system states are included in the power controller model since an auxiliary integral term is used for voltage reference calculation. Moreover, in order to capture the effect of introducing a wireless network, the coupling relation among the inverters is established.

In the following analysis, we first establish the model of individual inverter by incorporating new system states and power sharing information. The model is further decomposed into two components in the analysis of combined inverters and MG, corresponding to the traditional droop control term (with new system states) and power sharing control term, respectively. The purpose of the decomposition is to separate the model based on the information obtained from the wireless network from that based on the local information of each inverter, and to facilitate the stability analysis for nonnegligible communication delay, to be discussed in Sect. 7.5.

7.7.6 Model of Individual Inverter

In this subsection, the small-signal model for a single inverter is devised. Our focus is on a model for the power controller, while for the models of voltage controller, current controller, LC 3 the delay can be potentially reduced by selectively introducing additional long-range communication links such as cellular communication links, at an extra monetary cost [53], filter, coupling inductance, and output on common reference frame follow the models of traditional droop controller [59, 64].

1. Power controller: Our proposed power controller incorporates the total real and reactive power generation information of all DG units (i.e., P_{tot} and Q_{tot}). The small-signal state equations of the power controller are given by

$$\begin{aligned}\Delta\dot{\delta}^{(g)} &= \left[-K_{P-d\tau}^{(g)} + K_{P-sh}^{(g)}\alpha_P^{(g)} - K_{P-sh}^{(g)}\right]\Delta P^{(g)} \\ &\quad + \sum_{i=1,i\neq g}^{G}\left[K_{P-sh}^{(g)}\alpha_P^{(g)}\Delta P^{(i)}\right] - \Delta\omega_{com} \qquad (7.49)\end{aligned}$$

$$\begin{aligned}\Delta\dot{P}^{(g)} &= -\omega_c\Delta P^{(g)} + w_c I_{od}^{(g)}\Delta v_{od}^{(g)} + w_c I_{oq}^{(g)}\Delta v_{oq}^{(g)} \\ &\quad + w_c V_{od}^{(g)}\Delta i_{od}^{(g)} + w_c V_{oq}^{(g)}\Delta i_{oq}^{(g)} \qquad (7.50)\end{aligned}$$

$$\begin{aligned}\Delta\dot{Q}^{(g)} &= -\omega_c\Delta Q^{(g)} + w_c I_{oq}^{(g)}\Delta v_{od}^{(g)} - w_c I_{od}^{(g)}\Delta v_{oq}^{(g)} \\ &\quad - w_c V_{oq}^{(g)}\Delta i_{od}^{(g)} + w_c V_{od}^{(g)}\Delta i_{oq}^{(g)} \qquad (7.51)\end{aligned}$$

where $I_{od}^{(g)}$, $I_{oq}^{(g)}$, $V_{od}^{(g)}$, and $V_{oq}^{(g)}$ are the initial values of the inverter output current and voltage, respectively.

Denote $S_Q^{(g)}$ as integrator state of the power controller, which is a new system state we introduced for each inverter in comparison with the model of traditional droop controller. Then, the small-signal model of the integrator is given by

$$\Delta\dot{S}_Q^{(g)} = \alpha_Q^{(g)}\sum_{i=1}^{G}\Delta Q^{(i)} - \Delta Q^{(g)}$$

$$= \left[\alpha_Q^{(g)} - 1\right] \Delta Q^{(g)} + \alpha_Q^{(g)} \sum_{i=1, i \neq 1}^{G} \Delta Q^{(i)}. \tag{7.52}$$

The small-signal model with respect to the output reference voltage of the power controller is given by

$$\Delta v_{R,od}^{(g)} = -K_{Q-dr}^{(g)} \Delta Q^{(g)} + K_{Q-sh}^{(g)} \Delta S_Q^{(g)} \tag{7.53}$$

$$\Delta v_{R,oq}^{(g)} = 0. \tag{7.54}$$

For the inverter, inverter g, whose reference frame is selected as the common reference frame, there is one more small-signal model with respect to the common reference frequency. Since $\Delta\omega_{com} = \Delta\omega_R^{(g)}$ for the specific inverter, we have

$$\begin{aligned}\Delta\omega_{com} &= \left[-K_{P-dr}^{(g)} + K_{P-sh}^{(g)} \alpha_P^{(g)} - K_{P-sh}^{(g)}\right] \Delta P^{(g)} \\ &+ \sum_{i=1, i \neq 1}^{G} \left[K_{P-sh}^{(g)} \alpha_P^{(g)} \Delta P^{(i)}\right].\end{aligned} \tag{7.55}$$

2. Voltage Controller, Current Controller, LC Filter, Coupling Inductance, and Output on Common Reference Frame: Since traditional PI controllers are used for both voltage and current controllers additional states $[S_{V,d}^{(g)} S_{V,q}^{(g)}]$ and $[S_{I,d}^{(g)} S_{I,d}^{(g)}]$ are defined for the integrators of the voltage and current controllers, respectively. Moreover, in order to connect each inverter to the system, the output current and voltage should be converted with respect to the common reference frame. Let the output current of inverter g on the common reference frame be $i_{oDQ}^{(g)} = [i_{oD}^{(g)} i_{oQ}^{(g)}]^T$, and denote the bus voltage of node n on the common reference frame as $v_{oDQ}^{(n)} = [v_{bD}^{(n)} v_{bQ}^{(n)}]^T$. The small-signal models are given in [64].
3. Small-Signal Model of Individual Inverter: For inverter g, denote the state space of the small-signal model as $\Delta x^{(g)}$, which includes the state spaces of the power controller, voltage controller, current controller, LCL filter, and coupling inductance, given by a 14-element vector

$$\begin{aligned}\Delta x^{(g)} = \Big[&\Delta\delta^{(g)} \Delta P^{(g)} \Delta Q^{(g)} \Delta S_Q^{(g)} \Delta S_{V,d}^{(g)} \Delta S_{V,q}^{(g)} \Delta S_{I,d}^{(g)} \\ &\Delta S_{I,q}^{(g)} \Delta i_{id}^{(g)} \Delta i_{iq}^{(g)} \Delta v_{od}^{(g)} \Delta v_{oq}^{(g)} \Delta i_{od}^{(g)} \Delta i_{oq}^{(g)}\Big]^T.\end{aligned} \tag{7.56}$$

In addition to state $\Delta x^{(g)}$, the small-signal model of inverter g has inputs $\Delta P^{(i)} (i \in \{1, 2, \ldots, G\}, i \neq g)$, $\Delta Q^{(i)} (i \in \{1, 2, \ldots, G\}, i \neq g)$, $\Delta v_{bDQ}^{(n)} = [\Delta v_{bD}^{(n)}, \Delta v_{bQ}^{(n)}]^T$, and $\Delta\omega_{com}$. Then, the small-signal model of inverter g is given by

$$\Delta \dot{x}^{(g)} = A_1^{(g)} \Delta x^{(g)} + B_1^{(g)} \Delta v_{bDQ}^{(n)} + B_{com}^{(g)} \Delta \omega_{com} + \sum_{i=1}^{G} B_{P,i}^{(g)} \Delta x^{(i)} + \sum_{i=1}^{G} B_{Q,i}^{(g)} \Delta x^{(i)} \tag{7.57}$$

where matrix $A_1^{(g)}$ is defined with respect to the state variables, given by

$$A_1^{(g)} = \begin{bmatrix} A_{11}^{(g)} & A_{12}^{(g)} \end{bmatrix}_{14 \times 14} \tag{7.58}$$

and $A_{11}^{(g)}$ and $A_{12}^{(g)}$ are given at the bottom of the page. The matrices $B_1^{(g)}$, $B_{com}^{(g)}$, $B_{P,i}^{(g)}$ and $B_{Q,i}^{(g)}$. are defined with respect to the inputs. We have

$$B_1^{(g)} = \begin{bmatrix} 0 & 0 \\ \vdots & \vdots \\ -\frac{\cos(\delta_0^{(g)})}{L_c} & -\frac{\sin(\delta_0^{(g)})}{L_c} \\ \frac{\sin(\delta_0^{(g)})}{L_c} & -\frac{\cos(\delta_0^{(g)})}{L_c} \end{bmatrix}_{14 \times 2} \tag{7.59}$$

$$B_{com}^{(g)} = \begin{bmatrix} -1 0 \dots 0 \end{bmatrix}_{1 \times 14}^{T}. \tag{7.60}$$

$$A_{11}^{(g)} = \begin{bmatrix} 0 & -K_{P-sr}^{(g)} & 0 & 0 & 0 & 0 \\ 0 & -\omega_c & 0 & 0 & 0 & 0 \\ 0 & 0 & -\omega_c & 0 & 0 & 0 \\ 0 & 0 & 0 & 0 & 0 & 0 \\ 0 & 0 & -K_{Q-dr}^{(g)} & K_{Q-sh}^{(g)} & 0 & 0 \\ 0 & 0 & -K_{pv} K_{Q-dr}^{(g)} & K_{pv} K_{Q-sh}^{(g)} & K_{iv} & 0 \\ 0 & 0 & 0 & 0 & 0 & K_{iv} \\ 0 & -K_{P-dr}^{(g)} I_{iq}^{(g)} & \frac{K_{Q-dr}^{(g)} K_{pp} K_{pi}}{L_f} & \frac{K_{Q-sh}^{(g)} K_{pv} K_{pi}}{L_f} & \frac{K_{pi} K_{iv}}{L_f} & 0 \\ 0 & K_{p-dr}^{(g)} I_{id}^{(g)} & 0 & 0 & 0 & \frac{K_{pi} K_{iv}}{L_f} \\ 0 & -K_{P-dr}^{(g)} V_{oq}^{(g)} & 0 & 0 & 0 & 0 \\ 0 & K_{P-dr}^{(g)} V_{od}^{(g)} & 0 & 0 & 0 & 0 \\ -\frac{-V_{bD}^{(n)} \sin(\delta_0^{(g)}) + V_{bQ}^{(n)} \cos(\delta_0^{(g)})}{L^o} & -K_{P-dr}^{(g)} I_{oq}^{(g)} & 0 & 0 & 0 & 0 \\ -\frac{-V_{bD}^{(n)} \cos(\delta_0^{(g)}) - V_{bQ}^{(n)} \sin(\delta_0^{(g)})}{L^o} & K_{P-dr}^{(g)} K_{od}^{(g)} & 0 & 0 & 0 & 0 \end{bmatrix}$$

$$
A_{12}^{(g)} = \begin{bmatrix}
0 & 0 & 0 & 0 & 0 & 0 & 0 & 0 \\
0 & 0 & 0 & 0 & \omega_c I_{od}^{(g)} & \omega_c I_{oq}^{(g)} & \omega_c V_{od}^{(g)} & \omega_c V_{oq}^{(g)} \\
0 & 0 & 0 & 0 & \omega_c I_{oq}^{(g)} & -\omega_c I_{od}^{(g)} & -\omega_c V_{oq}^{(g)} & \omega_c V_{od}^{(g)} \\
0 & 0 & 0 & 0 & 0 & 0 & 0 & 0 \\
0 & 0 & 0 & 0 & -1 & 0 & 0 & 0 \\
0 & 0 & 0 & 0 & 0 & -1 & 0 & 0 \\
0 & 0 & -1 & 0 & -K_{pv} & -\omega_0 C_f & F & 0 \\
\frac{K_{ii}}{L_f} & 0 & 0 & -1 & \omega_0 C_f & -K_{pv} & 0 & F \\
-\frac{R_f+K_{pi}}{L_f} & \omega_{sys}-\omega_0 & -\frac{1+K_{pi}K_{pv}}{L_f} & \frac{-\omega_0 C_f K_{pi}}{L_f} & \frac{F K_{pi}}{L_f} & 0 & & \\
0 & \frac{K_{ii}}{L_f} & \omega_0-\omega_{sys} & -\frac{R_f+K_{pi}}{L_f} & \frac{\omega_0 C_f K_{pi}}{L_f} & -\frac{1+K_{pi}K_{pv}}{L_f} & 0 & \frac{F K_{pi}}{L_f} \\
0 & 0 & \frac{1}{C_f} & 0 & 0 & \omega_{sys} & -\frac{1}{C_f} & 0 \\
0 & 0 & 0 & \frac{1}{C_f} & -\omega_{sys} & 0 & 0 & -\frac{1}{C_f} \\
0 & 0 & 0 & 0 & \frac{1}{L_c} & 0 & -\frac{R_c}{L_c} & \omega_{sys} \\
0 & 0 & 0 & 0 & 0 & \frac{1}{L_c} & -\omega_{sys} & -\frac{R_c}{L_c}
\end{bmatrix}_{14\times 8}
$$

Letting $B_{Q,i}^{(g)} = [b_{Q,i}^{(g)}(j,k)]_{14\times 14}$ we have

$$
b_{Q,i=g}^{(g)}(j,k) = \begin{cases} \alpha_Q^{(g)} - 1, & \textit{if } j = 4 \textit{ and } k = 3 \\ 0, & \textit{otherwise} \end{cases} \tag{7.61}
$$

$$
b_{Q,i\neq g}^{(g)}(j,k) = \begin{cases} \alpha_Q^{(g)}, & \textit{if } j = 4 \textit{ and } k = 3 \\ 0, & \textit{otherwise} \end{cases} \tag{7.62}
$$

Denote $B_{P,i}^{(g)}$ as $[b_{P,i}^{(g)}(j,k)]_{14\times 14}$. We have

$$
b_{P,i=g}^{(g)}(j,k) = \begin{cases}
K_{P-sh}^{(g)}\alpha_P^{(g)} - K_{P-sh}^{(g)}, & \textit{if } J = l, k = 2 \\
I_{iq}^{(g)}\left[K_{P-sh}^{(g)}\alpha_P^{(g)} - K_{P-sh}^{(g)}\right], & \textit{if } J = 9, k = 2 \\
-I_{id}^{(g)}\left[K_{P-sh}^{(g)}\alpha_P^{(g)} - K_{P-sh}^{(g)}\right], & \textit{if } J = lO, k = 2 \\
V_{oq}^{(g)}\left[K_{P-sh}^{(g)}\alpha_P^{(g)} - K_{P-sh}^{(g)}\right], & \textit{if } j = 11, k = 2 \\
V_{od}^{(g)}\left[K_{P-sh}^{(g)}\alpha_P^{(g)} - K_{P-sh}^{(g)}\right], & \textit{if } j = 12, k = 2 \\
I_{oq}^{(g)}\left[K_{P-sh}^{(g)}\alpha_P^{(g)} - K_{P-sh}^{(g)}\right], & \textit{if } J = 13, k = 2 \\
-I_{od}^{(g)}\left[K_{P-sh}^{(g)}\alpha_P^{(g)} - K_{P-sh}^{(g)}\right], & \textit{if } J = l4, k = 2 \\
0, & \textit{otherwise}
\end{cases} \tag{7.63}
$$

$$b_{P,i\neq g}^{(g)}(j,k)=\begin{cases}K_{P-sh}^{(g)}\alpha_P^{(g)}, & if\, J=l, k=2\\ I_{iq}^{(g)}K_{P-sh}^{(g)}\alpha_P^{(g)}, & if\ J=9, k=2\\ -I_{id}^{(g)}K_{P-sh}^{(g)}\alpha_P^{(g)}, & if\ J=lO, k=2\\ V_{oq}^{(g)}K_{P-sh}^{(g)}\alpha_P^{(g)}, & if\ j=11, k=2\\ -V_{od}^{(g)}K_{P-sh}^{(g)}\alpha_P^{(g)}, & if\ j=12, k=2\\ I_{oq}^{(g)}[K_{P-sh}^{(g)}\alpha_P^{(g)}], & if\ J=13, k=2\\ -I_{od}^{(g)}K_{P-sh}^{(g)}\alpha_P^{(g)}, & if\ J=l4, k=2\\ 0, & otherwise\end{cases} \tag{7.64}$$

$$\Delta i_{oDQ}^{(g)}=\left[C_1^{(g)}\right]^T\Delta x^{(g)}$$

$$C_1^{(g)}=\begin{bmatrix}\left[-I_{od}^{(g)}\sin\left(\delta_0^{(g)}\right)-I_{oq}^{(g)}\cos\left(\delta_0^{(g)}\right)\right] & \left[-I_{od}^{(g)}\sin\left(\delta_0^{(g)}\right)-I_{oq}^{(g)}\cos\left(\delta_0^{(g)}\right)\right]\\ 0 & 0\\ \vdots & \vdots\\ 0 & 0\\ \cos\left(\delta_0^{(g)}\right) & -\sin\left(\delta_0^{(g)}\right)\\ \sin\left(\delta_0^{(g)}\right) & \cos\left(\delta_0^{(g)}\right)\end{bmatrix}_{14\times 2}. \tag{7.65}$$

By converting the output current of inverter g to the common reference frame, we have (7.65) at the bottom of the page, without loss of generality, select the reference frame of inverter 1 as the common reference frame, taking account of (7.55), the small-signal model of the common reference frame is given by

$$\Delta\omega_{com}=C_{com,0}\Delta x^{(1)}+\sum_{i=1}^{G}C_{com,i}\Delta x^{(i)} \tag{7.66}$$

$$C_{com,i}=\begin{cases}-K_{P-dr}^{(1)}e_2^T, & if\ \ i=0\\ \left[K_{P-sh}^{(1)}\alpha_P^{(1)}-K_{P-sh}^{(1)}\right]e_2^T, & if\ \ i=1\\ K_{P-sh}^{(1)}\alpha_P^{(1)}e_2^T, & otherwise\end{cases} \tag{7.67}$$

where e_g is the gth unit vector (of size 14), with the gth element being 1 and all other elements being zero.

7.7.7 Combined Inverter Model

Since the information of total generation (P_{tot} and Q_{tot}) is used by each individual inverter, the correlation among different inverters should be considered in the combined inverter model. The states of all inverters are given by

$$\Delta x = \left[\, \Delta x^{(1)}\ \Delta x^{(2)} \ldots \Delta x^{(G)} \,\right]^T . \tag{7.68}$$

Denote the states of all bus voltage and generator output current as Δv_{bDQ} and Δi_{oDQ}, respectively, given by

$$\begin{aligned} \Delta v_{bDQ} &= \left[\, \Delta v_{bDQ}^{(1)}\ \Delta v_{bDQ}^{(2)} \cdots \Delta v_{bDQ}^{(N)} \,\right]^T \\ \Delta i_{oDQ} &= \left[\, \Delta i_{oDQ}^{(1)}\ \Delta i_{oDQ}^{(2)} \cdots \Delta i_{oDQ}^{(G)} \,\right]^T . \end{aligned} \tag{7.69}$$

Then, the small-signal model of all inverters is given by

$$\Delta \dot{x} = A_1 \Delta x + B_1 M_1^{'} \Delta v_{bDQ} \tag{7.70}$$

$$\Delta i_{oDQ} = C_1 \Delta x \tag{7.71}$$

$$\Delta \omega_{com} = C_{com} \Delta x . \tag{7.72}$$

In (7.70), matrix A_1, can be decomposed into two components A_{1-dr} and A_{1-sh} ($A_1 = A_{1-dr} + A_{1-sh}$), corresponding to the traditional droop control term and power sharing-based control term, respectively, where

$$A_{1-dr} = \begin{bmatrix} \left[A^{(1)} + B_{com}^{(1)} C_{com,0}\right] & 0 & \ldots & 0 \\ \left[B_{com,0}^{(2)}\right] & A^{(2)} & \ldots & 0 \\ \vdots & \vdots & \ddots & \vdots \\ \left[B_{com}^{(G)} C_{com,0}\right] & 0 & \ldots & A^{(G)} \end{bmatrix} \tag{7.73}$$

and the value of A_{1-sh} is given at the bottom of this page. Similarly, C_{com} in (7.72) can be decomposed into two components C_{com-dr} and C_{com-sh}($C_{com} = C_{com-dr} + C_{com-sh}$). Based on (7.66) and (7.67), we have

$$C_{com-sh} = \left[\, C_{com,0}\ 0 \ldots 0 \,\right]_{1 \times 14G} \tag{7.74}$$

$$C_{com-dr} = \left[\, C_{com,1}\ C_{com,2} \cdots C_{com,G} \,\right]_{1 \times 14G} . \tag{7.75}$$

In (7.70), $M_1^{'} = [m_1^{'}(i,j)]_{2G \times 2N}$ is a matrix which maps the nodes to the inverters. If inverter g is connected to node n, we have $m_1^{'}(2g-1, 2-1) = m_1^{'}(2g, 2n) = 1$, and 0 otherwise. The matrices B_1, and C_1, are given by

$$B_1 = \begin{bmatrix} B_1^{(1)} & 0 & \dots & 0 \\ 0 & B_1^{(2)} & \dots & 0 \\ \vdots & \vdots & \ddots & \vdots \\ 0 & 0 & \dots & B_1^{(G)} \end{bmatrix} \tag{7.76}$$

$$C_1 = \begin{bmatrix} \left[C_1^{(1)}\right]^T & 0 & \dots & 0 \\ 0 & \left[C_1^{(2)}\right]^T & \dots & 0 \\ \vdots & \vdots & \ddots & \vdots \\ 0 & 0 & \dots & \left[C_1^{(G)}\right]^T \end{bmatrix}. \tag{7.77}$$

7.7.8 Network Model

The network model specifies how the lines and loads are connected to the nodes in the MG. We consider the standard model in this work [64]. Denote the state variable for all lines by

$$\Delta i_{2DQ} = \left[\Delta i_{2DQ}^{(1)} \; \Delta i_{2DQ}^{(2)} \; \dots \; \Delta i_{2DQ}^{(L)} \right]^T \tag{7.78}$$

where $\Delta i_{2DQ}^{(l)} = [\Delta i_{2D}^{(l)} \; \Delta i_{2Q}^{(l)}]^T$ is the state variable of line l. Then, the small-signal model of the network is given by

$$\Delta \dot{i}_{2DQ} = A_2 \Delta i_{2DQ} + B_{2v} \Delta v_{bDQ} + B_{2\omega} \Delta \omega_{com} \tag{7.79}$$

where the matrices A_2, B_{2v}, and $B_{2\omega}$ are defined for the output current, bus voltage, and common reference frame, respectively, given at the bottom of the next page. The matrix $B_{2v}^{(1)} = [b_{2v}^{(1)}(i, j)]_{2\times 2N}$ which is incorporated in B_{2v}, is given by

$$b_{2v}^{(1)}(i, j) = \begin{cases} \frac{1}{L_{lines}^{(l)}}, & \textit{if } (i, j) = (1, 2n_1, -1) \textit{ or } (2, 2n_1) \\ -1/L_{lines}^{(l)}, & \textit{if } (i, j) = (1, 2n_2 - 1) \textit{ or } (2, 2n_2) \\ 0, & \textit{otherwise} \end{cases} \tag{7.80}$$

where n_1, and n_2 are indices (in $\{1, 2, \dots, N\}$) of the two nodes which are connected by line l, respectively.

Similarly, we can define the load model. Denote the states of the (RL) loads in the MG as

$$\Delta i_{3DQ} = \left[\Delta i_{3DQ}^{(1)} \quad \Delta i_{3DQ}^{(2)} \quad \dots \quad \Delta i_{3DQ}^{(H)} \right]^T \tag{7.81}$$

where $\Delta i_3^{(h)} = [\Delta i_{3D}^{(h)} \ \Delta i_{3Q}^{(h)}]^T$ is the state of the small-signal model of load h. Then, the small-signal model of all loads in the MG is given by

$$\Delta \dot{i}_{3DQ} = A_3 \Delta i_{3DQ} + B_{3v} \Delta v_{bDQ} + B_{3\omega} \Delta \omega_{com} \tag{7.82}$$

where matrices A_3, B_{3v}, and $B_{3\omega}$ are given at the bottom of the next page. The matrix $B_{3v}^{(h)} = [b_{3v}^{(h)}(i,j)]_{2\times 2N}$ in B_{3v} is given by

$$b_{3v}^{(h)}(i,j) = \begin{cases} \frac{1}{L_{load}^{(h)}}, & \textit{if } (i,j) = (1, 2n-1) \textit{ or } (2, 2n) \\ 0, & \textit{otherwise} \end{cases} \tag{7.83}$$

where n is the index of the node connecting to load h.

Let R_N be a sufficiently large (virtual) resistance which connects each node to the ground. Then, the relations among the node voltages can be represented by the currents of the inverters, loads, and transmission lines, given by

$$\Delta v_{bDQ} = R_N (M_1 i_{oDQ} + M_2 i_{2DQ} + M_3 i_{3DQ} \tag{7.84}$$

where $M_1 = [m_1(i,j)]_{2N\times 2G}$ is a matrix which maps the inverters to the nodes. If inverter g is connected to node n, we have $m_1(2n-1, 2g-1) = m_1(2n, 2g) = 1$ and 0 otherwise. Similarly, denoting $M_2 = [m_2(i,j)]_{2N\times 2L}$ and $M_3 = [m_3(i,j)]_{2N\times 2H}$ for the lines and loads, we have $m_2(2n-1, 2l-1) = m_2(2n, 2l) = 1$ and $m_2(2n-1, 2l-1) = m_2(2n, 2l) = -1$ if line l is entering and leaving the node n, respectively, while $m_3(2n-1, 2h-1) = m_3(2n, 2h) = -1$ if the load h is connected to node n, and 0 otherwise.

$$A_{1-sh} = \begin{bmatrix} A_{11} & A_{12} & \dots & A_{13} \\ A_{21} & A_{22} & \dots & A_{23} \\ \vdots & \vdots & \ddots & \vdots \\ A_{31} & A_{32} & \dots & A_{33} \end{bmatrix}$$

$$A_{11} = \left[B_{com}^{(1)} C_{com,1} + B_{P,1}^{(1)} + B_{Q,1}^{(1)} \right], \quad A_{12} = \left[B_{com}^{(1)} C_{com,2} + B_{P,2}^{(1)} + B_{Q,2}^{(1)} \right]$$

$$A_{13} = \left[B_{com}^{(1)} C_{com,G} + B_{P,G}^{(1)} + B_{Q,G}^{(1)} \right], \quad A_{21} = \left[B_{com}^{(2)} C_{com,1} + B_{P,1}^{(2)} + B_{Q,1}^{(2)} \right]$$

$$A_{22} = \left[B_{com}^{(2)} C_{com,2} + B_{P,2}^{(2)} + B_{Q,2}^{(2)} \right], \quad A_{23} = \left[B_{com}^{(2)} C_{com,G} + B_{P,G}^{(2)} + B_{Q,G}^{(2)} \right]$$

$$A_{31} = \left[B_{com}^{(G)} C_{com,1} + B_{P,1}^{(G)} + B_{Q,1}^{(G)} \right], \quad A_{32} = \left[B_{com}^{(G)} C_{com,2} + B_{P,2}^{(G)} + B_{Q,2}^{(G)} \right]$$

$$A_{33} = \left[B_{com}^{(G)} C_{com,G} + B_{P,G}^{(G)} + B_{Q,G}^{(G)} \right]$$

7.7.9 Microgrid Model

The small-signal model of the entire MG is given by

$$\begin{bmatrix} \Delta\dot{x} \\ \Delta i_{2DQ} \\ \Delta i_{3DQ} \end{bmatrix} = A_{grid} \begin{bmatrix} \Delta x \\ \Delta i_{2DQ} \\ \Delta i_{3DQ} \end{bmatrix} \tag{7.85}$$

where A_{grid} can be decomposed into two components $A_{grid-dr}$ and $A_{grid-sh}$ ($A_{grid} = A_{grid-dr} + A_{grid-sh}$) with respect to the traditional droop control term and power sharing based control term, respectively, given at the bottom of the page, while R_N is a $2N \times 2N$ diagonal matrix with all diagonal elements equal to R_N. Obviously, the size of A_{grid} is $(14G + 2L + 2H) \times (14G + 2L + 2H)$.

7.7.10 System Stability Evaluation

The system stability is reflected by the eigenvalues of matrix A_{gria}, which are determined by the characteristic equation det $\Psi(\lambda) = 0$. The characteristic matrix $\Psi(A)$ is given by

$$\Psi(\lambda) = \lambda I_0 - A_{grid} \tag{7.86}$$

where I_0 is a $(14G+2L+2H) \times (14G+2L+2H)$ identity matrix. The eigenvalues are often referred to as modes, which reveal different frequency components in the system and the corresponding damping. If the real part of a mode is larger than zero, the system becomes unstable, and vice versa. The modes with smaller real parts are better damped in comparison to the modes with larger real parts.

7.7.11 Stability Analysis with Communication Delay

With communication delay τ, the small-signal model of the MG is given by

$$\begin{bmatrix} \Delta\dot{x}(t) \\ \Delta i_{2DQ}(t) \\ \Delta i_{3DQ}(t) \end{bmatrix} = A_{grid-dr} \begin{bmatrix} \Delta x(t) \\ \Delta i_{2DQ}(t) \\ \Delta i_{3DQ}(t) \end{bmatrix} + A_{grid-sh} \begin{bmatrix} \Delta x(t-\tau) \\ \Delta i_{2DQ}(t-\tau) \\ \Delta i_{3DQ}(t-\tau) \end{bmatrix}. \tag{7.87}$$

Note that the matrices $A_{grid-dr}$ and $A_{grid-sh}$ are readily obtained based on the decomposed model in the previous section. The system stability with respect to different modes is evaluated by solving the characteristic equation $\det \Omega(\lambda, \tau) = 0$ based on the characteristic matrix, given by

$$\Omega(\lambda, \tau) = \lambda I_0 - A_{grid-dr} - A_{grid-sh} e^{-\lambda\tau}. \tag{7.88}$$

It is worth mentioning that, as the communication delay τ approaches zero, we have the following property with respect to the characteristic matrix (7.88):

$$\begin{aligned}\lim_{\tau\to 0} \Omega(\lambda, \tau) &= \lim_{\tau\to 0} [\lambda I_0 - A_{grid-dr} - A_{grid-sh} e^{-\lambda\tau}] \\ &= \lambda I_0 - A_{grid-dr} - A_{grid-sh} \\ &= \lambda I_0 - A_{grid} \\ &= \Psi(\lambda).\end{aligned} \tag{7.89}$$

In other words, the stability analysis reduces to the negligible communication delay case as $\tau \to 0$. However, in comparison with the negligible communication delay case, the characteristic equation for the nonnegligible communication delay case has infinite number of roots since it is transcendental. Therefore, only a reduced set of roots can be obtained based on approximation. We adopt the method proposed in [58] based on the calculation of the eigenvalues of the following matrix

$$\Phi = \left[\frac{\hat{\Theta} \otimes I_0}{A_{grid-sh}\ 0\ \dots\ 0\ A_{grid-dr}} \right] \tag{7.90}$$

where $\otimes$ represents the Kronecker product. Matrix $\hat{\Theta}$ consists of the first M rows of matrix $\Theta = -2D_M/\tau$, where $D_M = [d_M(i, j)]_{(M+1)\times(M+1)}$ is the Chebyshev's differentiation matrix of order M, given by

$$d_M(i, j) = \begin{cases} \frac{d_i(-1)^{i+j}}{d_j(\beta_{i-1}-\beta_{j-1})}, & if\ i \neq j \\ -\frac{1}{2}\frac{\beta_{i-1}}{1-\beta_{i-1}^2}, & if\ i = j \neq 1, M+1 \neq j \\ \frac{2N^2+1}{6}, & if\ i = j = 1 \\ -\frac{2N^2+1}{6}, & if\ i = j = M+1 \end{cases} \tag{7.91}$$

with $d_1 = d_{M+1} = 2$ and $d_2 = \cdots = d_M = 1$, and $\beta_k (k \in 0, l, \dots, M)$ given by

$$\beta_k = \cos\left(\frac{k\pi}{M}\right). \tag{7.92}$$

The value of M corresponds to the number of nodes for Chebyshev's discretization which determines the computational precision and complexity.

$$A_2 = \begin{bmatrix} A_2^{(1)} & 0 & \dots & 0 \\ 0 & A_2^{(2)} & \dots & 0 \\ \vdots & \vdots & \ddots & \vdots \\ 0 & 0 & \dots & A_2^{(L)} \end{bmatrix} \quad B_{2v} = \begin{bmatrix} B_{2v}^{(1)} \\ B_{2v}^{(2)} \\ \vdots \\ B_{2v}^{(L)} \end{bmatrix} \quad B_{2\omega} = \begin{bmatrix} B_{2\omega}^{(1)} \\ B_{2\omega}^{(2)} \\ \vdots \\ B_{2\omega}^{(L)} \end{bmatrix}$$

$$A_2^{(l)} = \begin{bmatrix} -R_{line}^{(l)}/L_{line}^{(l)} & \omega_{sys} \\ -\omega_{sys} & -R_{line}^{(l)}/L_{line}^{(l)} \end{bmatrix} A_{2\omega}^{(l)} = \begin{bmatrix} I_{2Q}^{(l)} \\ -I_{2D}^{(l)} \end{bmatrix}$$

$$A_3 = \begin{bmatrix} A_3^{(1)} & 0 & \dots & 0 \\ 0 & A_3^{(2)} & \dots & 0 \\ \vdots & \vdots & \ddots & \vdots \\ 0 & 0 & \dots & A_3^{(D)} \end{bmatrix} \quad B_{3v} = \begin{bmatrix} B_{3v}^{(1)} \\ B_{3v}^{(2)} \\ \vdots \\ B_{3v}^{(D)} \end{bmatrix} \quad B_{3\omega} = \begin{bmatrix} B_{3\omega}^{(1)} \\ B_{3\omega}^{(2)} \\ \vdots \\ B_{3\omega}^{(D)} \end{bmatrix}$$

$$A_3^{(h)} = \begin{bmatrix} -R_{load}^{(h)}/L_{load}^{(h)} & \omega_{sys} \\ -\omega_{sys} & -R_{load}^{(h)}/L_{load}^{(h)} \end{bmatrix} B_{3\omega}^{(h)} = \begin{bmatrix} I_{3Q}^{(h)} \\ -I_{3D}^{(h)} \end{bmatrix}$$

$$A_{grid-dr} = \begin{bmatrix} [A_{1-dr} + B_1 M_1' R_N M_1 C_1] & [B_1 M_1' R_N M_2] & [B_1 M_1' R_N M_3] \\ \begin{bmatrix} B_{2v} R_N M_1 C_1 + B_{2\omega} C_{com-dr} \end{bmatrix} & [A_2 + B_{2v} R_N M_2] & [B_{2v} R_N M_3] \\ \begin{bmatrix} B_{3v} R_N M_1 C_1 + B_{3\omega} C_{com-dr} \end{bmatrix} & [B_{3v} R_N M_2] & [A_3 + B_{3v} R_N M_3] \end{bmatrix}$$

$$A_{grid-sh} = \begin{bmatrix} [A_{1-sh}] & 0 \dots 0 \\ \begin{bmatrix} B_{2\omega} C_{com-sh} \\ B_{3\omega} C_{com-sh} \end{bmatrix} & \begin{matrix} \vdots & \ddots & \vdots \\ 0 & \dots & 0 \end{matrix} \end{bmatrix}$$

7.7.12 Simulation Results

We use the droop control test system proposed in [64] as an example for performance evaluation. The MG consists of three generators, two lines, and two loads. The three generators have the same capacities and configurations. The desired sharing of the real and reactive power among the three generators are the same, i.e., $\alpha_P^{(g)} = \alpha_Q^{(g)} = 1/3$ for $g \in \{1, 2, 3\}$. With traditional droop control strategy (or equivalently, the real power sharing gain $K_{P-sh}^{(g)} = 0$ and the reactive power sharing gain $K_{Q-sh}^{(g)} = 0$), the trace of the low-frequency modes which are sensitive to frequency and voltage droop gains is shown in Fig. 3. We can see that the low-frequency modes move toward two different directions as the frequency droop gain ($K_{P-dr}^{(g)}$) increases from 1×10^{-5} to 1.2×10^{-4}. Although some modes move toward the stable region with

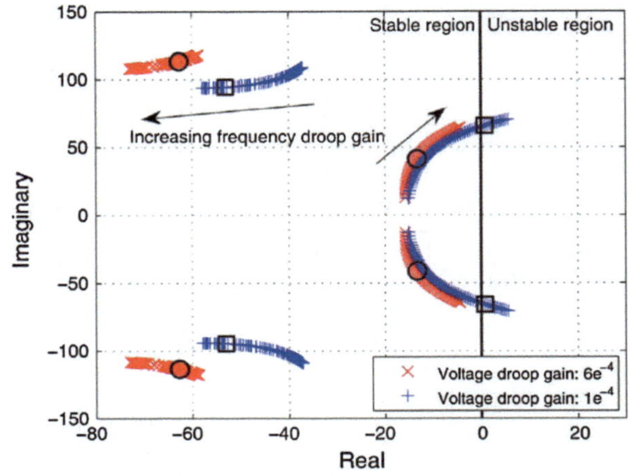

Fig. 7.26 Trace of low-frequency modes versus frequency and voltage droop gains for traditional droop control strategy

negative real parts, the other modes which are close to the imaginary axis move significantly toward the unstable region with positive real parts. On the other hand, the low-frequency modes move slightly toward the stable region as the voltage droop gain ($K^{(g)}_{Q-dr}$) increases. Although increasing the voltage droop gain can potentially improve MG stability, the low-frequency modes are generally less sensitive to the voltage droop gain than that to the frequency droop gain [64], which makes the stability enhancement less effective. Moreover, a change in the voltage droop gain may affect the sharing of reactive power among the DG units, which is not desired in MG operation. On the other hand, the information exchange based on a wireless network can potentially improve the MG stability. In the following performance evaluation, we use the modes marked by squares and circles in Fig. 7.26 (which correspond to frequency droop gains 1×10^{-4} and 5×10^{-5}, respectively) as reference points.

Figure 7.27 demonstrates the impact of real power sharing gain. The reactive power sharing gain is zero. The frequency and voltage droop gains of the reference points with respect to the traditional droop control strategy are given by 5×10^{-5} and 6×10^{-4}, respectively, while the corresponding modes are marked by circles in Fig. 7.26. Similar to the change in frequency droop gain, the low-frequency modes move toward two different directions as the real power sharing gain increases (from 0 to 6.7×10^{-5}). Since some of the modes which are close to the imaginary axis move slightly toward the unstable region, the stability improvement by adjusting the real power sharing gain is not obvious, mainly because the real power sharing based on traditional droop control is already accurate in MG operation [51].

Next, we fix the real power sharing gain to zero (the corresponding modes are marked by squares in Fig. 7.26) and investigate the impact of reactive power sharing gain. Note that the original MG is unstable since the modes marked by squares in Fig. 7.26 are in the unstable region. The impact of reactive power sharing gain is shown in Fig. 7.28. We observe that the low-frequency modes which are in the

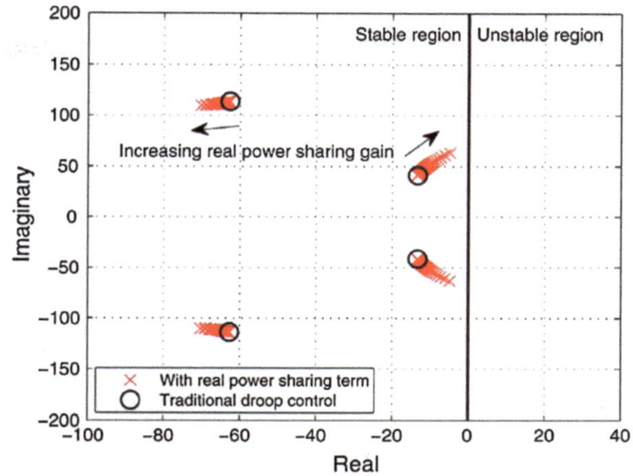

Fig. 7.27 Trace of low-frequency modes versus real power sharing gain

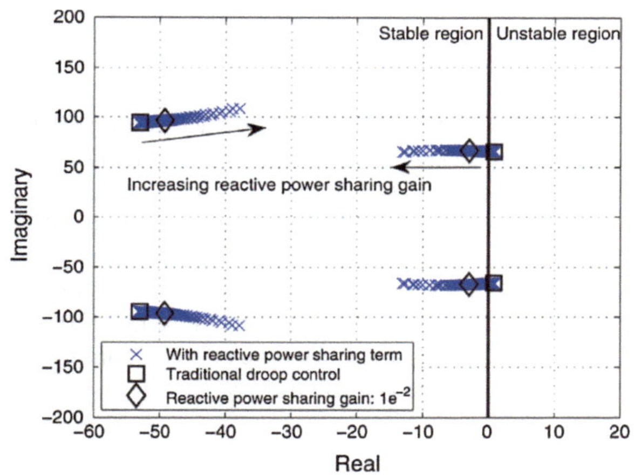

Fig. 7.28 Trace of low-frequency modes versus reactive power sharing gain

unstable region move toward the stable region as the reactive power sharing gain increases (from 0 to 0.1) such that the MG can be stabilized. The reason is that, the information provided by the wireless network can correct the reactive power sharing error based on an additional integral control term [35]. According to traditional droop control strategy, the inverters at different locations of MG share different amount of reactive power as the load changes significantly, due to the fact that the line impedance is not negligible. This is also known as one of the major limitations of traditional droop control in MG [64]. On the other hand, by sharing the reactive power among the DG units evenly, the temporary overloading of DG units can be avoided and the MG stability can be improved based on our proposed control strategy.

From Figs. 7.27 and 7.28, we see that the proposed control strategy can improve system stability (especially by adjusting the reactive power sharing gain) given that the communication delay is negligible. To investigate how the communication delay can affect the system stability, the trace of the modes when communication delay increases from 0 to 50 ms is shown in Fig. 7.29. We choose $M = 2$ for the

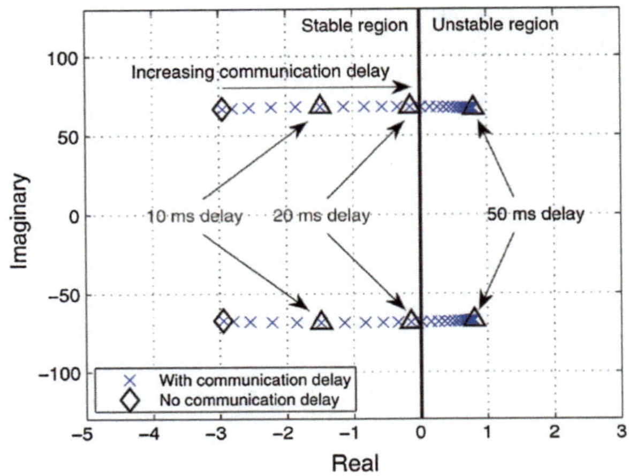

Fig. 7.29 Impact of communication delay on the locations of different modes

Chebyshev's discretization [58]. The frequency droop gain, voltage droop gain, real power sharing gain, and reactive power sharing gain are set to 1×10^{-4}, 1×10^{-4}, 0, and 1×10^{-2}, respectively. The reference points with respect to the performance without communication delay are marked by diamonds in Figs. 7.28 and 7.29, respectively. We only show the low-frequency modes which are close to the imaginary axis since these modes dominate the system stability as the communication delay changes. We can see that the low-frequency modes move toward the unstable region as the communication delay increases. The system becomes unstable again when the communication delay is large (i.e., more than 20 ms for the MG under consideration). How to mitigate the negative effect of communication delay to further improve system stability needs further study.

7.8 Conclusions

This chapter presents a new control strategy for a VSC interfaced DO unit. The strategy controls the interface VSC as a VC-VSC. The VC-VSC concept provides control over frequency and voltage through the VSC at its PCC and thus permits operation of the DO unit in both grid-connected and autonomous modes. The other features of the VC-VSC are:

- it provides ride-through capability for the DO unit under a grid-connected mode;
- it provides smooth transition capability between grid-connected and autonomous (islanded) modes subsequent to planned and accidental islanding events;
- it inherently includes an islanding detection capability without NDZ;
- it can provide current-limit capability for the converter during grid faults.

The chapter also introduces a study system to demonstrate characteristics of the proposed VC-VSC strategy. An eigenanalysis is performed on the system to investigate dynamic stability of the system and select the proposed VC-VSC controller parameters. Time-domain simulation studies, in the PSCADI/ EMTDC environment, are also carried out to

1. validate the eigenanalysis results and
2. evaluate transient performance of the VC-VSC strategy (e.g., ride-through capability, transition from grid-connected to islanded mode, and impact of load variations). The study results confirm merits of the proposed VSC control strategy.

Next, a power management and control strategy is presented for an autonomous, multi-DER MG. The envisioned strategy provides: (1) power management of the overall MG; (2) open-loop frequency control and synchronization; and (3) a local, DEC for each DER unit. The PMS, based on classical power-flow analysis, determines the terminal voltage set-points for DER units. The frequency of the system is controlled in an open-loop manner by utilizing an internal crystal oscillator for each DER unit that also generates the angle waveform required for **dq/abc(abc/dq)** transformations. Synchronization of DER units is achieved by exploiting a GPS-based time-reference signal. The local control of each DER unit, which is the main focus of this paper, is designed based on a new multivariable decentralized robust servomechanism approach which utilizes a linear state-space model of the MG. Various attributes of the controller (i.e., the existence conditions, gain margins, robustness, and tolerance to delays and high-frequency effects) are analytically discussed, and the design procedures are outlined.

Finally, we investigate the stability enhancement of decentralized inverter control in DG MGs. A wireless network is established to acquire the information of total real and reactive power generation of all DG units. A power sharing-based control strategy is proposed, which introduces additional terms to the traditional droop control strategy to capture the differences between the desired and actual real and reactive power generation. An analytical model is developed to evaluate the small-signal stability of the MG under our proposed control strategy. Numerical results indicate that, our proposed control strategy can improve the system stability when the communication delay is small.

Suggested future work includes the communication delay reduction based on communication protocol design, and the voltage stability improvement while maintaining an accurate reactive power sharing among the DG units. Moreover, how to improve MG stability under packet losses in multiagent coordination over an unreliable wireless channel [2] is critical but needs further investigation.

7.9 Suggested Problems

Problem 7.1 A typical MG consisting of of four DGs is shown in Fig. 7.30, where the lines between buses are modeled as series RL branches. In Table 7.7, the electrical system parameters are presented. In this table, K_{PV}, K_{IV}, K_{PC}, and K_{IC} are the parameters of the voltage and current controllers and the voltage and current controller parameters are adopted from [3].

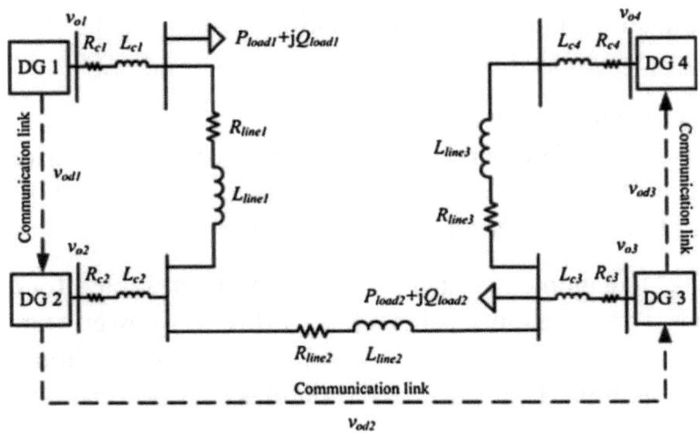

Fig. 7.30 Microgrid system with communication structure

Table 7.7 Electrical microgrid system parameters

DGs	*DG* 1	*DG* 2	*DG* 3	*DG* 4
Rating	33 KVA	33 KVA	25 KVA	25 KVA
m_P	9.4×10^{-5}	9.4×10^{-5}	12.5×10^{-5}	12.5×10^{-5}
n_Q	1.3×10^{-3}	1.3×10^{-3}	1.5×10^{-3}	1.5×10^{-3}
R_c	0.03 Ω	0.03 Ω	0.03 Ω	0.03 Ω
L_c	0.35 mH	0.35 mH	0.35 mH	0.35 mH
K_{PV}	0.1	0.1	0.05	0.05
K_{IV}	420	420	390	390
K_{PC}	15	15	10.5	10.5
K_{IC}	20,000	20,000	16,000	16,000
Lines	*Line* 1	*Line* 2	*Line* 3	
Parameters	$R_{line\,1} = 0.23\,\Omega$	$R_{line\,2} = 0.35\,\Omega$	$R_{line\,3} = 0.23\,\Omega$	
	$L_{line\,1} = 0.318$ mH	$L_{line\,2} = 1.847$ mH	$L_{line\,3} = 0.318$ mH	
Loads	*Load* 1 *Per Phase*		*Load* 2 *Per Phase*	
Parameters	$P_{load\,1} = 12$ kW		$P_{load\,2} = 15.3$ kW	
	$Q_{load\,1} = 12\ KVAr$		$Q_{load\,2} = 7.6\ KVAr$	

Conduct complete simulation studies to investigate the voltage and frequency restoration of the closed-loop MG system. Explore the impact of control gains variation on the transient response.

Problem 7.2 Consider a two-DG MG depicted in Fig. 7.31. It is desired is study the performance of this system and demonstrate the results through a simulation in MATLAB®/SimPowerSystems. The performance is to be evaluated in terms of voltage tracking and perturbations in the load parameters (Table 7.8).

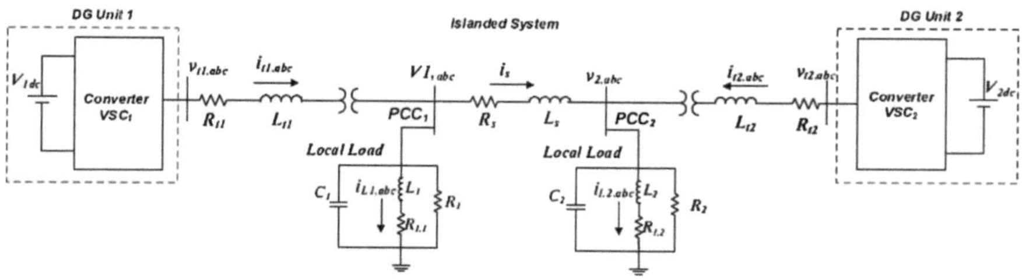

Fig. 7.31 Islanded system consisting of two DG units in radial configuration

Table 7.8 Electrical microgrid system parameters

Quantity	*Value*	*Per unit (pu)*	*Designation*
R_t	1.5 m Ω	0.01	*Resistance of VSC filter*
L_t	300 μH	0.785	*Inductance of VSC filter*
VSC	2.5 MW	1	*Rated power*
V_{base}	600 V	1	*Terminal voltage (line–line)*
f_{sw}	1985 Hz	–	*PWM carrier frequency*
R	76 Ω	1	*Load nominal resistance*
L	111.9 mH	0.554	*Load nominal inductance*
C	62.86 μF	1.805	*Load nominal capacitance*
$Q = R\sqrt{\frac{C}{L}}$	1.8	–	*Load quality factor*
$f_{res} = \frac{1}{2\pi\sqrt{LC}}$	60 Hz	–	*Load resonant frequency*
$q_\ell = \frac{L\Omega_0}{R_\ell}$	120	–	*Inductor quality factor*

Problem 7.3 Coordinated control methods of generation of microsources feeding cluster of loads within a confined neighborhood are of contemporary interests. In Fig. 7.32, a MG scheme is presented that outlines the distributed nature of the power network (PN). The goal of control design is to achieve an optimal compromise between redundancy and (primarily dynamic) performance. It is therefore desired to develop a DEC method to attain this goal.

Problem 7.4 In line of **Problem 7.3**, consider the MG scheme in Fig. 7.33 and investigate a decentralized coordinated control scheme that exploits the modular implementation of a three-phase inverter and attempt to improve redundancy and fault tolerance, reduce footprint space and cost (volume is higher), and increase efficiency.

Problem 7.5 In this problem, we consider the design of a control-communication network for a three-phase inverter system. In Fig. 7.34, a block diagram of a homogeneous PN consisting of parallel-connected inverters that could either operate in

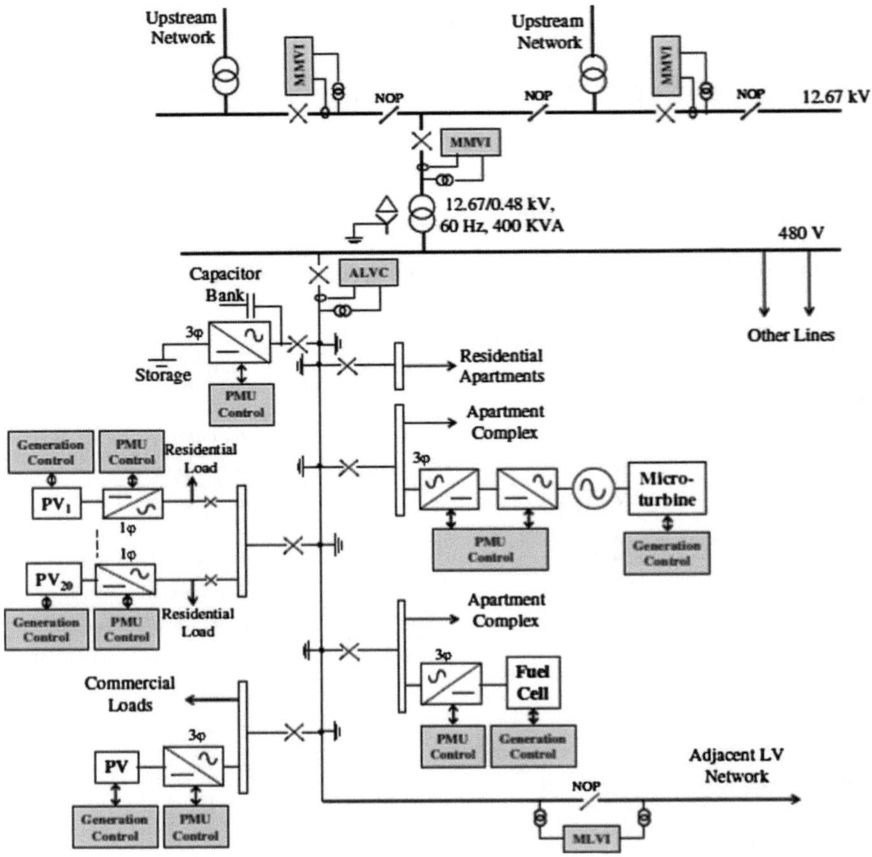

Fig. 7.32 A two-microgrid architecture

stand-alone mode or in MG connected modes of operation. The objective is to effectively regulate the load voltage and ensure equal power sharing among the inverter modules using a distributed control over a wireless communication network (WCN). It is desired to investigate the operation of the homogeneous network under three possible WCN operating scenarios:

- Nominal network (NN) where all the default communication links are functional,
- Routed network (RN), which corresponds to the state where communication among the communication nodes can be ensured by routing the information through the links that are functional and

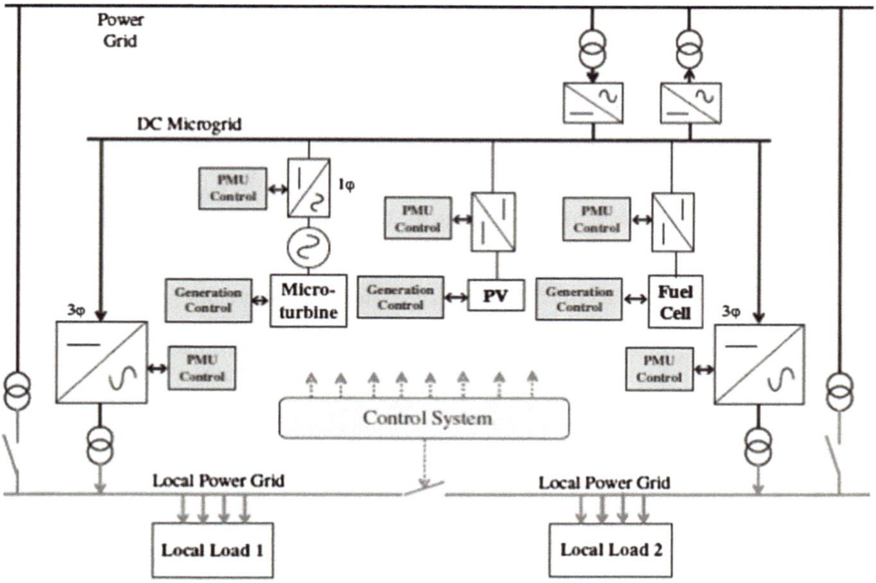

Fig. 7.33 Schematics illustrating a two-coupled microgrid

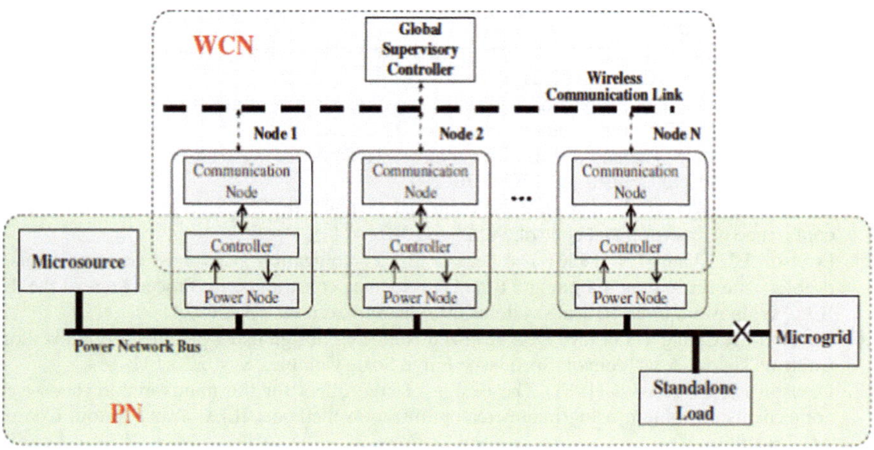

Fig. 7.34 Power network with distributed control

- Clustered network (CN), which corresponds to the state when one or more of the communication nodes cannot communicate with the rest of the WCN and are isolated as a different cluster.

References

1. Ackermann T, Anderson G, Soder L (2001) Distributed generation: a definition. Electr Power Syst Res 57:195–204
2. Aysal TC, Yildiz ME, Sarwate AD, Scaglione A (2009) Broadcast gossip algorithms for consensus. IEEE Trans Signal Process 57(7):2748–2761
3. Barklund E, Pogaku N, Prodanovic M, Hernandez-Aramburo C, Green TC (2008) Energy management in autonomous microgrid using stability—constrained droop control of inverters. IEEE Trans Power Electron 23(5):2346–2352
4. Carrasco JM, Franquelo LG, Bialasiewicz JT, Galvan E, Guisado RCP, Prats MAM, Leon JI, Moreno-Alfonso N (2006) Power electronic systems for the grid integration of renewable energy sources: a survey. IEEE Trans Ind Electron 53(4):1002–1016
5. Chandorkar MC, Divan DM (1996) Decentralized operation of distributed ups systems. In: Proceedings of the international power electronic, drives and energy systems for industrial growth conference, vol 1, pp 565–571
6. Chandorkar MC, Divan DM, Adapa R (1993) Control of parallel connected inverters in standalone AC supply systems. In: IEEE Transactions on Industry Applications, vol 29, no. 1, pt. 1, pp 136–143
7. Choi BJ, Liang H, Shen X, Zhuang W (2012) DCS: Distributed asynchronous clock synchronization in delay tolerant networks. IEEE Trans Parallel Distrib 23(3):491–504
8. Chowdhury S, Chowdhury SP, Crossley P (2009) Microgrids and active distribution networks. Institution of Engineering and Technology, London
9. Davison DE, Davison EJ (2002) Optimal transient response shaping of the servomechanism problem. J Optim Theory Appl 115(3):491–515
10. Davison E, Copeland B (1985) Gain margin and time lag tolerance constraints applied to the stabilization problem and robust servomechanism problem. IEEE Trans Autom Control AC-30(3):229–239
11. Davison EJ (1976) The robust decentralized control of a general servomechanism problem. IEEE Trans Autom Control AC-21(1):14–24
12. Davison EJ (1977) The robust decentralized servomechanism problem with extra stabilizing control agents. IEEE Trans Autom Control AC-22(2):256–258
13. Davison EJ (1978) Decentralized robust control of unknown systems using tuning regulators. IEEE Trans Autom Control AC-23(2):276–289
14. Davison EJ (1986) Robust control for industrial systems. In: Proceedings of the 25th IEEE conference on decision and control, vol 25, pp 367–374
15. Davison EJ, Chang TN (1986) The decentralized stabilization and robust servomechanism problem subject to gain margin and time lag tolerance constraints. In: Proceedings of the 4th IFAC symposium on large scale systems, Zurich, Switzerland, pp 236–241
16. Davison EJ, Chang TN (1986) Decentralized controller design using parameter optimization. Control: Theory Adv Technol, Spec Issue Large Scale Complex Syst 2(2):131–154
17. Davison EJ, Ferguson I (1981) The design of controllers for the multivariable robust servomechanism problem using parameter optimization methods. IEEE Trans Autom Control AC-26(1):93–110
18. Davison EJ, Goldenberg A (1975) The robust control of a general servomechanism problem: the servocompensator. Automatica 11(5):461–471
19. Davison EJ, Ozguner U (1982) Synthesis of the decentralized robust servomechanism problem using local models. IEEE Trans Autom Control AC-27(3):583–600
20. Davison EJ, Scherzinger B (1987) Perfect control of the robust servomechanism problem. IEEE Trans Autom Control AC-32(8):689–702
21. Davison EJ, Tripathi N (1978) The optimal decentralized control of a large power system: load and frequency control. IEEE Trans Autom Control AC-23(2):312–325
22. Davison EJ, Tripathi NK (1979) Decentralized tuning regulators: an application to solve the load and frequency control problem for a large power system. In: Proceedings of the 18th IEEE

conference on decision and control symposium on adaptive processes, Fort Lauderdale, vol 18, pp 879–885
23. Davison EJ, Wang S (1974) Properties and calculation of transmission zeros of linear multivariable systems. Automatica 10:643–658
24. Davison EJ, Rau NS, Palmay FV (1973) The optimal decentralized control of a power system consisting of a number of interconnected synchronous machines. Int J Control 18(6):1313–1328
25. Diaz G, Gonzalez-Moran C, Gomez-Aleixandre J, Diez A (2010) Scheduling of droop coefficients for frequency and voltage regulation in isolated microgrids. IEEE Trans Power Syst 25(1):489–496
26. Doyle MT (2002) Reviewing the impacts of distributed generation on distribution system protection. Proc IEEE Power Eng Soc Summer Meet 1:103–105
27. Farhangi H (2010) The path of the smart grid. IEEE Power Energy Mag 8(I):18–28
28. Funabashi T, Koyanagi K, Yokoyama R (2003) A review of islanding detection methods for distributed resources. Presented at the power tech conference
29. Galli S, Scaglione A, Wang Z (2011) For the grid and through the grid: the role of power line communications in the smart grid. Proc IEEE 99(6):998–1027
30. Guerrero JM, Hang L, Uceda J (2008) Control of distributed uninterruptible power supply systems. IEEE Trans Ind Electron 55(8):2845–2859
31. Guerrero JM, de Vicuna LG, Matas J, Castilla M, Miret J (2004) A wireless controller to enhance dynamic performance of parallel inverters in distributed generation systems. IEEE Trans Power Electron 19(5):1205–1213
32. Guerrero JM, de Vicuna LG, Matas J, Castilla M, Miret J (2005) Output impedance design of parallel-connected ups inverters with wireless load-sharing control. IEEE Trans Ind Electron 52(4):1126–1135
33. Guerrero JM, Matas J, Vicuna LG, Castilla M, Miret J (2007) Decentralized control for parallel operation of distributed generation inverters using resistive output impedance. IEEE Trans Ind Electron 54(2):994–1004
34. Hatziargyriou ND, Meliopoulos APS (2002) Distributed energy sources: technical challenges. Proc IEEE Power Eng Soc Winter Meet 2:1017–1022
35. He J, Li YW (2011) An accurate reactive power sharing control strategy for DG units in a micro grid. In: Proceedings of the PICPE ECCE, pp 551–556, May–June 2011
36. Hu G, Davison EJ (2003) Real stability radius of LTI time-delay systems. Syst Control Lett 50(3):209–219
37. IEEE Standard for Synchro-phasors for Power Systems, IEEE Standard C37.118-2005, 2006, (Revision of IEEE Standard 1344-1995)
38. Inverters, Converters, and Controllers for use in independent power systems, UL Std. 1741 (2002)
39. Jenkins PCDKN, Allan R, Strbac G. Strbac (2000) Embedded generation. Institution of Electrical Engineers, London
40. Karimi H, Davison EJ, Iravani R (2010) Multivariable servomechanism controller for autonomous operation of a distributed generation unit: design and performance evaluation. IEEE Trans Power Syst 25(2):853–865
41. Katiraei F, Iravani MR (2006) Power management strategies for a microgrid with multiple distributed generation units. IEEE Trans Power Syst 21(4):1821–1831
42. Katiraei F, Iravani MR, Lehn PW (2005) Micro-grid autonomous operation during and subsequent to islanding process. IEEE Trans Power Deliv 20(1):248–257
43. Krause P, Wasynczuk O, Sudhoff S (2002) Analysis of electric machinery and drive systems, 2nd edn. Wiley-IEEE Press, Hoboken
44. Krishnamurthy S, Lasseter RH (2009) Consortium for Electric Reliability Technology Solutions (CERTS), Control of wound field synchronous machine gen sets for operation in a CERTS microgrid, Madison, WI, DE-FC02-06CH11350, March 2009
45. Kroposki B, Lasseter R, Ise T, Morozumi S, Papatlianassiou S, Hatziargyriou N (2008) Making microgrids work. IEEE Power Energy Mag 6(3):40–53
46. Kundur P (1994) Power system stability and control. McGrawHill, New York

47. Kundur P, Paserba J, Ajjarapu V, Andersson G, Bose A, Canizares C, Hatziargyriou N, Hill D, Stankovic A, Taylor C, Van Cutsem T, Vittal V (2004) Definition and classification of power system stability IEEE/CIGRE joint task force on stability terms and definitions. IEEE Trans Power Syst 19(3):1387–1401
48. Lasseter RH, Paigi P (2004) Microgrid: a conceptual solution. In: Proceedings of the IEEE 35th annual power electronics specialists on conference, pp 4285–4290, June 2004
49. Li Q, Bernhardsson B, Rantzer A, Davison EJ, Young PM, Doyle JC (1995) A formula for computation of the real stability radius. Automatica 31(6):879–890
50. Li YW, Kao C-N (2009) An accurate power control strategy for power-electronics-interfaced distributed generation units operating in a low-voltage multibus microgrid. IEEE Trans Power Electron 24(12):2977–2988
51. Li Y, Li Y (2011) Power management of inverter interfaced autonomous microgrid based on virtual frequency-voltage frame. IEEE Trans Smart Grid 2(I):30–40
52. Liang H, Choi BJ, Zhuang W, Shen X (2012) Decentralized inverter control in micro grids based on power sharing information through Wireless communications. In: Proceedings of the IEEE GLOBECOM
53. Liang H, Choi BJ, Abdrabou A, Zhuang W, Shen X (2012) Decentralized economic dispatch in micro grids via heterogeneous wireless networks. IEEE J Sel Areas Commun 30(6):1061–1074
54. Liang H, Abdrabou A, Choi BJ, Zhuang W, Shen X, Awad ASA (2012) Multiagent coordination in micro grids via wireless networks. IEEE Wirel Commun 19(3):14–22
55. Lopes JAP, Moreira CL, Madureira AG (2005) Defining control strategies for analyzing microgrids islanded operation. In: Proceedings of the IEEE Russia power tech, pp. 1–7
56. Marwali MN, Keyhani A (2004) Control of distributed generation systems—part I: voltages and currents control. IEEE Trans Power Electron 19(6):1541–1550
57. Marwali MN, Jung J-W, Keyhani A (2004) Control of distributed generation systems-part II: load sharing control. IEEE Trans Power Electron 19(6):1551–1561
58. Milano F, Anghel M (2012) Impact of time delays on power system stability. IEEE Trans Circuits Syst I, Regul Pap 59(3):889–900
59. Mohamed Y, EI-Saadany EF (2008) Adaptive decentralized droop controller to preserve power sharing stability of paralleled inverters in distributed generation microgrids. IEEE Trans Power Electron 23(6):2806–2816
60. Nikkhajoei H, Lasseter RH (2009) Distributed generation interface to the CERTS microgrid. IEEE Trans Power Deliv 24(3):1598–1608
61. Olfati-Saber R, Fax J, Murray R (2007) Consensus and cooperation in networked multi-node systems. Proc IEEE 95:215–233
62. Phadke A, Thorp J, Pai M (eds) (2008) Synchronized phasor measurements and their applications, 1st edn. Springer, New York
63. Piagi P, Lasseter RH (2006) Autonomous control of microgrids. Presented at the IEEE power engineering society, General meeting, Montreal, Canada
64. Pogaku N, Prodanovic M, Green TC (2007) Modeling, analysis and testing of autonomous operation of an inverter-based microgrid. IEEE Trans Power Electron 22(2):613–625
65. Sao CK, Lehn PW (2008) Control and power management of converter fed microgrids. IEEE Trans Power Syst 23(3):1088–1098
66. Schauder C, Mehta H (1993) Vector analysis and control of advanced static VAR compensators. Proc Inst Electr Eng C 140(4):299–306
67. Tripathi NK, Davison EJ (1979) The automatic generation control of a multiarea interconnected system using reduced order models. In: Proceedings of the IFAC symposium on computer applications in large scale power systems, New Delhi, India, 16–18 August, pp 117–126
68. Tripathi NK, Davison EJ (1979) Automatic generation control of a large multi-area interconnected power system by optimal decentralized control. In: Titli A, Singh M (eds) Large scale systems engineering applications. North-Holland, Amsterdam, pp 472–795
69. Undrill JM (1968) Dynamic stability calculations for an arbitrary number of interconnected synchronous machines. IEEE Trans Power App Syst PAS–87(3):835–845

70. Vasquez JC, Guerrero JM, Luna A, Rodriguez P, Teodorescu R (2009) Adaptive droop control applied to voltage-source inverters operating in grid-connected and islanded modes. IEEE Trans Ind Electron 56(10):4088–4096
71. Vig J (2000) Quartz crystal resonators and oscillators for frequency control and timing applications: a tutorial, U.S. Army Communications—Electronics Command, Fort Monmouth, NJ, Technical report, SLCET-TR-88-1 (Rev. 8.4.2)
72. Walling RA, Saint R, Dugan RC, Burke J, Kojovic LA (2008) Summary of distributed resources impact on power delivery systems. IEEE Trans Power Deliv 23(3):1636–1644
73. Wang S-H, Davison EJ (1973) On the stabilization of decentralized control systems. IEEE Trans Autom Control AC-18(5):473–478
74. Xiao L, Boyd S (2003) Fast linear iterations for distributed averaging. Proc IEEE CDC 5:4997–5002
75. Xu Y, Liu W (2011) Novel multiagent based load restoration algorithm for microgrids. IEEE Trans Smart Grid 2(1):152–161
76. Yao J, Popovic D (2004) Stability of a MV distribution network with electronically interfaced distributed generation. In: Proceedings of the IEEE mediterranean electro-technical conference, vol 3, pp 975–978

Chapter 8
Advanced Control Approaches

8.1 Introduction

In the literature, most of the results on the control of wind and solar systems have focused on wind or solar systems. Specifically, there is a significant body of literature dealing with control of wind-based energy generation systems [6, 8, 11, 19–24, 86], while significant contribution has been made to the control of solar-based energy generation systems [32, 35–40].

A supervisory predictive control method was proposed in [70] for short-term optimal management and operation of wind–solar energy generation systems in which the supervisory control system was designed via model predictive control (MPC) to take into account the optimal allocation of generation assignment between the two subsystems. A computationally efficient supervisory control was designed in [69], we designed a computationally efficient supervisory control system for long-term optimal management and operation of an integrated wind–solar energy generation and a reverse osmosis (RO) water desalination system, based on a two timescale decomposition of the integrated system model.

In the following section, a distributed control framework presented is for electrical grid integrated with distributed renewable energy generation systems in order to enable the development of the so-called "smart electrical grid."

8.2 Distributed Control Architecture

The proposed distributed control architecture for electrical grid integrated with distributed renewable energy generation systems and distributed loads is shown in Fig. 8.1. In this architecture, the electrical grid is considered to be decomposed into n control areas. Each control area typically consists of numerous generators and loads. The different control areas are interconnected through bidirectional power lines and electrical power can flow between the different control areas bidirectionally via the

M.S. Mahmoud and F.M. AL-Sunni, *Control and Optimization of Distributed Generation Systems*, Power Systems, DOI 10.1007/978-3-319-16910-1_8

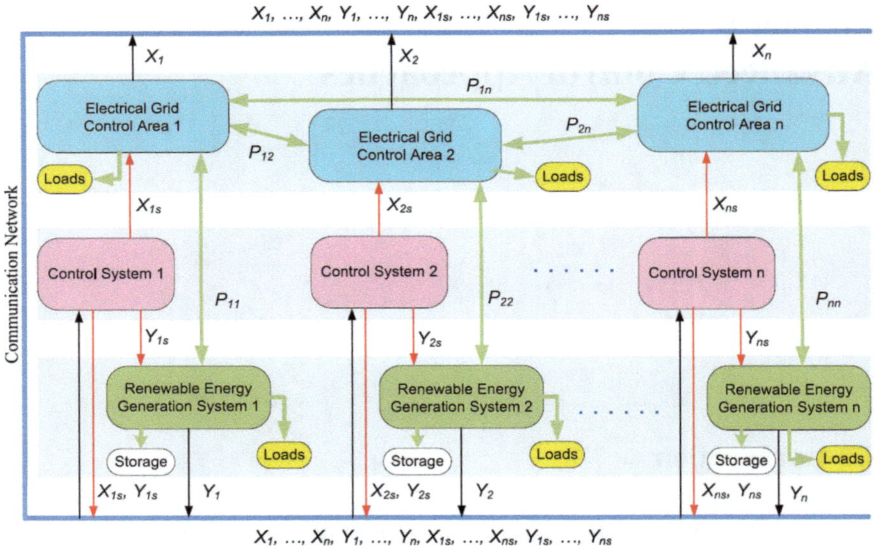

Fig. 8.1 Integrated electrical grid structure

power lines. For example, in Fig. 8.1, electrical power can flow from control area 1 to the other control areas (e.g., control areas 2 to n) and be transmitted to control area 1 from other control areas.

The symbol $P_{ij}(i = 1, \ldots, n$ and $j = 1, \ldots, n)$ is used to denote the power transmitted between control areas i and j. When the energy generated in a control area is not enough to satisfy the total power demands in that control area, the energy-deficient control area may obtain additional energy from other control areas or its associated renewable energy generation system. When the energy generation in a control area is sufficient and there is a surplus of energy, this area may transmit energy to other control areas or its associated loads or storage systems.

To simplify the description, in Fig. 8.1, we assume that all the renewable energy generation systems connected to a control area can be lumped into one equivalent renewable energy generation system which is also connected with the control area using bidirectional power lines. A renewable energy generation system first uses its generated energy to satisfy the loads connected to it and also sends, if the generation capacity permits, extra energy produced to its associated control area. If the energy generated by a renewable energy generation system is not sufficient to satisfy its load demands (LDs), it may also get energy from its associated control area. In Fig. 8.1, the symbol $P_{ii}(i = 1, \ldots, n)$ is used to denote the power transmitted between control area i and the renewable energy generation system i.

It is desired to design a control system for each control area and its associated renewable energy generation system. The control system calculates the operating set-points for the control area and the renewable energy generation system. There is also a real-time communication network integrated in the overall system. The control areas, renewable energy generation systems, and the control systems communicate via the real-time communication network at specific sampling time instants. In Fig. 8.1, the symbols X_i and $Y_i (i = 1, \ldots, n)$ are used to indicate the state of control area i and of the renewable energy generation system i; the symbols X_{is} and $Y_{is} (i = 1, \ldots, n)$ are used to indicate the operating set-points calculated by control system i for the control area i and renewable energy generation system i. At a sampling time t_k, control areas and renewable energy generation systems broadcast their state information at t_k (i.e., $X_1(t_k), \ldots, X_n(t_k), Y_1(t_k), \ldots, Y_n(t_k)$) to the communication network, and the control systems also broadcast their last control actions (e.g., control system i broadcasts $X_{is}(t_{k-1}), Y_{is}(t_{k-1})$) to the communication network and receive all the state and control actions information at t_k. Based on the state information and control actions of other control systems, control system i calculates its own control actions (i.e., $X_{is}(t_k)$ and $Y_{is}(t_k)$).

In the course of design, distributed MPC is particularly suited because of its ability to account for the control actions of other control systems and its ability to take into account optimization considerations and constraints in the calculation of control actions in real time.

The successful development of the proposed distributed control architecture needs to address many key challenging issues including [44–52]:

1. the predictive control of different types of renewables-based energy generation systems; for example, integrated wind and solar energy generation systems,
2. the coordination of a renewable-based energy generation system with the electrical grid and loads, and
3. the cooperation between different control systems.

8.2.1 Integrated Wind/Solar/RO System Modeling

In this section, we introduce the integrated wind/solar energy generation system which is connected to an RO water desalination system and the electrical grid. A schematic of the integrated system is shown in Fig. 8.2.

There is a wind generation subsystem, a solar generation subsystem, and a lead–acid battery bank which is used to overcome periods of scarce generation as well as store energy in the energy generation system. There is a windmill, a multipolar permanent magnet synchronous generator (PMSG), a rectifier, and a DC/DC converter. The DC/DC converter is used to control the operating point of the wind generation subsystem. The mathematical description of the wind subsystem written in a rotor reference frame is as follows:

$$\ddot{i}_q = -\frac{r_s}{L} i_q - \omega_e i_d + \frac{\omega_e \phi_{sr}}{L} - \frac{\pi v_b i_q u_w}{3\sqrt{3} L \sqrt{i_q^2 + i_q^2}}$$

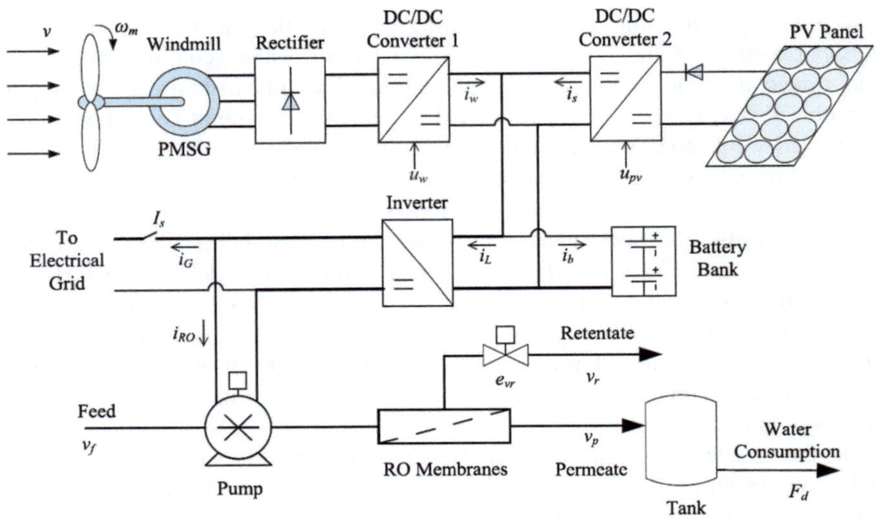

Fig. 8.2 Integrated wind/solar energy generation and water desalination system

$$\dot{i}_d = -\frac{r_s}{L}i_d - \omega_e i_q - \frac{\pi v_b i_d u_w}{3\sqrt{3}L\sqrt{i_q^2 + i_q^2}}$$

$$\dot{\omega}_e = \frac{P}{2J}\left(T_t - \frac{3}{2}\frac{P}{2}\phi_{sr} i_q\right) \tag{8.1}$$

where i_q and i_d are the quadrature current and the direct current (DC) in the rotor reference frame, respectively; r_s and L are the per phase resistance and inductance of the stator windings, respectively; ω_e is the electrical angular speed, ϕ_{sr} is the flux linked by the stator windings; v_b is the voltage on the battery bank terminals, u_w is the control signal (duty cycle of the DC/DC converter (DC/DC converter 1 in Fig. 8.2)), P is the PMSG number of poles, J is the inertial of the rotating parts and T_t is the wind turbine (WT) torque. The WT torque can be written as

$$T_t = \frac{1}{2}C_t(\lambda)\rho A R v^2 \tag{8.2}$$

where ρ is the air density, A is the turbine-swept area, R is the turbine radius, v is the wind speed, and $C_t(\lambda)$ is a nonlinear torque coefficient which depends on the tip speed ratio ($\lambda = R\omega_m/v$ with $\omega_m = 2\omega_e/P$ being the angular shaft speed).

Based on (8.1), we can express the power generated by the wind subsystem and injected into the DC bus as follows:

$$P_w = i_w v_b \tag{8.3}$$

where $i_w = (\pi/(2/\sqrt{3}))\sqrt{i_q^2 + i_d^2}u_w$ denotes the current which is injected to the DC bus by the wind subsystem.

In the solar generation subsystem, there is a photovoltaic (PV) panel array and a half-bridge buck DC/DC converter (DC/DC converter 2 in Fig. 8.2). In this subsystem, similar to the wind subsystem, the converter is used to control the operating point of the PV panels. The mathematical description of the solar subsystem is as follows:

$$
\begin{aligned}
\dot{v}_{PV} &= \frac{i_{PV}}{C} - \frac{i_s}{C}u_{PV} \\
\dot{i}_s &= -\frac{v_b}{L_c} + \frac{v_{PV}}{L_c}u_{PV} \\
i_{PV} &= n_p I_{ph} - n_p I_{rs}\left(\exp\left(\frac{q(v_{PV} + i_{PV}R_s)}{n_s A_c KT}\right) - 1\right)
\end{aligned}
\tag{8.4}
$$

where v_{PV} is the voltage level on the PV panel array terminals, is is the current injected on the DC bus, C and L_c are electrical parameters of the buck converter, u_{PV} is the control signal (duty cycle), i_{PV} is the current generated by the PV array, R_s is the series resistance in the PV circuit, n_s is the number of PV cells connected in series, n_p is the number of series strings in parallel, K is the Boltzman constant, A_c is the cell deviation from the ideal pn junction characteristic, I_{ph} is the photocurrent, and I_{rs} is the reverse saturation current. The power injected by the PV solar module into the DC bus can be computed by

$$P_s = i_s v_b. \tag{8.5}$$

Note that this power indirectly depends on the control signal u_{PV}.

The lead–acid battery bank may be modeled as a voltage source E_b connected in series with a resistance R_b and a capacitance C_b. Based on this simple model, the DC bus voltage expression can be written as follows:

$$v_b = E_b + v_c + i_b R_b, \tag{8.6}$$

where i_b is the current across the battery bank, v_c is the voltage in capacitor C_b and its dynamics is as follows:

$$\dot{v}_c = \frac{1}{C_b} i_b. \tag{8.7}$$

The quantity of electricity stored in the battery bank can be calculated as follows:

$$Q_c = C_b v_c. \tag{8.8}$$

Based on (8.8), the state of charge (SOC) of the battery bank, s_b, can be calculated as follows:

$$s_b = \frac{Q_c}{Q_c^{\max}} = \frac{v_c}{v_c^{\max}} \tag{8.9}$$

where $Q_c^{\max}$ is the maximum capacity of the capacitor corresponding to the maximum voltage $v_c^{\max}$ that can be tolerated by the capacitor. We also introduce the concept of depth of discharge (DOD) of the battery bank and denote it as d_b, which is calculated as follows:

$$d_b = 1 - s_b. \tag{8.10}$$

The integrated system has two operating modes and hence a binary variable, I_s, is used to indicate the operating mode of the integrated system:

- When $I_s = 1$, the integrated system is connected to the electrical grid; and
- When $I_s = 0$, the integrated system works in stand-alone mode.

An energy balance can be written, assuming an ideal voltage inverter, as follows:

$$P_w + P_s = P_{RO} + I_s P_G + i_b v_b, \tag{8.11}$$

where P_{RO} denotes the total power demand from the water desalination system and the term $I_s P_G$ is used to denote the power sent to the electrical grid. Note that when i_b is positive, the battery bank is being charged; and when i_b is negative, the battery bank supplies energy to the RO water desalination system or to the electrical grid. The energy balance of (8.11) can be also expressed as:

$$i_w + i_s = i_{RO} + I_s i_G + i_b, \tag{8.12}$$

where i_{RO} and i_G are the currents injected to the RO water desalination system and the electrical grid, respectively. From (8.12), we obtain the current across the battery bank as follows:

$$i_b = i_w + i_s - i_{RO} - I_s i_G. \tag{8.13}$$

8.2.2 Water Desalination System Description

There is a high-pressure pump, a membrane module, and a water storage tank in the RO water desalination system. Salt water is fed into the system through the pump, which is equipped with a variable frequency drive. The feed water is pressurized to a desired pressure in the pump and then is sent to the membrane module where it is separated into a low-salinity product (or permeate) stream, and a high-salinity brine (or retentate) stream. The permeate stream enters the storage tank which provides desalinated water to satisfy the water consumption demand. Note that the difference between the water drawn from (water consumption) and the water fed into (water production) the storage tank is accumulated in the storage tank.

In the RO system, the pressure downstream of the actuated valve and at the permeate outlet is assumed to be equal to the atmospheric pressure. The RO system model is based on a mass balance taken around the entire system and an energy balance taken around the actuated retentate valve

$$\frac{dv_r}{dt} = \frac{P_{sys}A_p}{\rho_w V} - \frac{1}{2}\frac{A_p e_{vr} v_r^2}{V} \tag{8.14}$$

where v_r is the retentate stream velocity, P_{sys} is the feed pressure, A_p is the pipe cross-sectional area, V is the system volume, ρ_w is the fluid density, and e_{vr} is the retentate valve resistance. The system pressure P_{sys} can be calculated as follows:

$$P_{sys} = \frac{\rho_w A_p}{A_m K_m}(v_f - v_r) + \Delta\pi. \tag{8.15}$$

where A_m is the membrane area, K_m is the membrane overall mass transfer coefficient, v_f is the feed stream velocity, and $\Delta\pi$ is the difference in osmotic pressure between the feed side of the membrane and the permeate side and can be computed as:

$$\Delta\pi = \delta C_{eff}(T_w + 273)$$

where C_{eff} is the effective average concentration at the membrane surface, δ is a constant relating effective concentration to osmotic pressure and T_w is the water temperature in degrees Celsius.

It is noted that the three velocities v_f, v_r and v_p (the permeate stream velocity) satisfy an overall steady-state mass balance as follows (since all pipe cross-sectional areas are assumed to be the same):

$$0 = v_f - v_r - v_p. \tag{8.16}$$

The valve resistance value (e_{vr}) can be used as the manipulated input.

The RO system is operated at energy optimal water recovery, Y_{opt}, which requires that v_p/v_f is adjusted in real time [106, 107]. Based on the Bernoulli equation and ignoring the water elevation change, we can obtain the power needed for the water desalination system as follows:

$$PRO = \frac{1}{\eta}\left(P_{sys}\frac{F_p}{Y_{opt}} + \frac{1}{2}\frac{F_p^3}{Y_{opt}^3 A_p^2}\rho w\right), \quad 0 < \eta < 1 \tag{8.17}$$

where η is the overall power efficiency of the pump of the RO desalination system and $F_p = A_p v_p$ is the permeate flow rate, which is used to satisfy the water consumption and storage demands. If we denote the water consumption demand as F_d and water storage demand as F_s, then we obtain the following equation from a steady-state mass balance:

$$0 = F_p - F_d - F_s. \tag{8.18}$$

Note that the water storage demand Fs can take positive or negative values.

Based on (8.16) and (8.18), the dynamics of the water level in the storage tank, h_l, can be obtained as follows:

$$\dot{h}_l = \frac{F_s}{A_s} = \frac{A_p}{A_s}(v_f - v_r) - \frac{F_d}{A_s} \tag{8.19}$$

where A_s is the cross-sectional area of the water storage tank. Similarly, we define the state of storage (SOS), s_t, for the storage tank as follows:

$$s_t = \frac{h_l}{h_l^{\max}}$$

where $h_l^{\max}$ is the maximum water level in the storage tank.

The dynamics of the integrated wind/solar/RO system can be written in the following compact form:

$$\begin{aligned} \dot{x} &= f(x) + g(x)u \\ h(x) &= 0 \end{aligned} \tag{8.20}$$

where $x = [i_q \; i_d \; \omega_e \; v_{PV} \; i_s \; v_c \; v_r \; h_l]^T$, $u = [u_w \; u_{PV} \; e_{vr}]^T$, and f, g, h are nonlinear vector functions whose explicit forms are omitted for brevity.

The control objectives are:

- For the wind subsystem, the objective is to force the wind subsystem to track the operating trajectory, which is the desired power generation (power reference), P_w^{ref}.
- For the solar subsystem, the objective is to force the solar subsystem to track the operating trajectory, which is the desired power generation, P_s^{ref}.
- For the RO water desalination subsystem, the objective is to regulate the retentate valve resistance to track the reference retentate flow velocity, v_r^{ref}.

8.2.3 Short-Term Supervisory Predictive Control

In what follows, we focus on the stand-alone operating mode of the integrated system ($I_s = 0$) and design a supervisory control system to account for short-term optimal operation of the integrated system. We assume that the storage demand of the storage tank $F_s = 0$ to simplify the description of the controller formulation.

8.2.4 Supervisory Control System Design I

Following [56–58, 67–74], the supervisory MPC optimizes the power references P_w^{ref} and P_s^{ref} (operating points) of the wind and solar subsystems, respectively, but also takes into account the dynamics of the RO system. Note that in this design, i_b and v_r are not considered as decision variables in the formulation of the supervisory MPC because i_b is determined by the current balance of (8.12) in stand-alone operating mode and v_r^{ref} is determined by the water demand F_d. The primary control objective of the supervisory control system is to coordinate the wind and solar subsystems as well as the battery bank to provide enough energy to the RO system to satisfy the power demand of the scheduled water production. In addition, we try to reduce battery short-term charge–discharge cycles which can be caused by the variability of the renewable energy resources or the LD. We operate the wind subsystem as the primary generation system and only activate the solar subsystem when the wind subsystem alone cannot satisfy the power demand.

In the following, we first design the cost function used in the MPC to take into account the control objectives and then formulate the MPC based on the cost function. The supervisory control system is designed via MPC because it can take into account optimality considerations and handle state and input constraints. In this stand-alone operating mode, the operation strategy of the integrated energy generation system is as follows:

- When the wind subsystem can generate enough energy to satisfy the total power demand, only the wind subsystem is activated and operated to track the power demand;
- When the wind subsystem alone cannot generate enough energy to satisfy the total power demand, the solar subsystem is also activated to provide extra energy to satisfy the power demand;
- When the two subsystems are not sufficient to complement the generation to satisfy the total power demand, the battery bank discharges to provide extra power to satisfy the load requirements. However, when the power limit that can be provided by the battery is surpassed, the load must be disconnected to recharge the battery and avoid damages.

We consider the case where the future water demand is known, that is $F_d(t)$ is known. We note that this assumption is a reasonable one given that there has been extensive work in short-term forecasting for weather conditions and municipal water demand [21]. The main implementation element of supervisory predictive control is that the supervisory controller is evaluated at discrete time instants $t_k = t_0 + k\Delta$, $k = 0, 1, \ldots$, with t_0 being the initial time and Δ being the sampling time, and the optimal future power references, P_w^{ref} and P_s^{ref}, for a time period (prediction horizon) are obtained and only the first part (move) of the optimal reference trajectories are sent to the local control systems and implemented on the two units. In order to design this controller, first, a proper number of prediction steps, N, and a sampling time, Δ, are chosen. Before going to the formulation of the supervisory MPC, we design the cost function used in the MPC to take into account the control objectives. Specifically, the proposed design of the cost function is as follows:

$$J_s(t_k) = \int_{t_k}^{t_{k+N}} \alpha(P_{RO}(\tau) - P_w^{ref}(\tau) - P_s^{ref}(\tau))^2 d\tau + \int_{t_k}^{t_{k+N}} \beta P_s^{ref}(\tau)^2 d\tau$$
$$+ \int_{t_k}^{t_{k+N-1}} \zeta(P_b(\tau + \Delta) - P_b(\tau))^2 d\tau \tag{8.21}$$

where α, β and ζ are positive weighting factors on different terms.

Remark 8.1 The first term in the cost function penalizes the difference between the power generated by the wind–solar system and the total power demand, which drives the wind and solar subsystems to satisfy the total demand to the maximum extent. Because there are infinite combinations of P_w^{ref} and P_s^{ref} that can minimize the first term, in order to allow only one solution to the optimization problem and to operate the wind subsystem as the primary generation system, we also put a small penalty on the reference power of the solar subsystem, P_s^{ref}. This term guarantees that the solar generation subsystem is only activated when the wind subsystem alone cannot satisfy the power demand. The third term in the cost function penalizes the change of the power provided by the battery bank to the load to reduce battery short-term charge–discharge cycles. Note that, if one wants to operate the solar system as the primary generation subsystem, the second term in the cost function can be modified to penalize the power reference of the wind subsystem. Note also that, other considerations (for example, charge of the battery bank) can be also taken into account in the design of the cost function by adding additional terms or modifying existing terms.

The distributed MPC design for the supervisory control system is as follows:

$$\min_{P_w^{ref}, P_s^{ref} \in S(\Delta)} \quad J_s(t_k) \tag{8.22}$$
$$s.t.\ P_w^{ref}(\tau) \le \min_{\tau}\{P_w^{\max}(\tau)\}, \quad \tau \in [t_{k+j}, t_{k+j+1}) \tag{8.23}$$
$$P_s^{ref}(\tau) \le \min_{\tau}\{P_{PV,\max}(\tau)\}, \quad \tau \in [t_{k+j}, t_{k+j+1}) \tag{8.24}$$
$$P_w^{ref}(t_{k+j+1}) - P_w^{ref}(t_{k+j}) \le dP_{w,\max} \tag{8.25}$$
$$P_s^{ref}(t_{k+j+1}) - P_s^{ref}(t_{k+j}) \le dP_{s,\max} \tag{8.26}$$
$$\dot{\tilde{x}}(\tau) = f(\tilde{x}(\tau)) + g(\tilde{x}(\tau))u(\tau) \tag{8.27}$$
$$h(\tilde{x}) = 0 \tag{8.28}$$
$$\tilde{x}(0) = x(t_k) \tag{8.29}$$

where $j = 0, \ldots, N-1$, $J_s(t_k)$ is the cost function to be minimized at time t_k, $P_w^{\max}(\tau)$ and $P_s^{\max}(\tau)$ are the maximum powers that can be generated by the wind and solar subsystems at time τ, respectively; $dP_{w,\max}$ and $dP_{s,\max}$ are the maximum allowable increasing values of P_w^{ref} and P_s^{ref} in two consecutive power references, $\tilde{x}$ is the predicted future state trajectory of the integrated system and $x(t_k)$ is the state measurement obtained at time t_k. We denote the optimal solution to the optimization problem of (8.22)–(8.29) as $P_{w,s}^{ref,*}(\tau|t_k)$ and $P_{s,s}^{ref,*}(\tau|t_k)$.

The power references of the wind and solar subsystems generated by the supervisory predictive controller of (8.22)–(8.29) are defined as follows:

$$\begin{aligned} P_w^{ref}(t) &= P_{w,s}^{ref,*}(t|t_k), \quad \forall t \in [t_k, t_{k+1}), \\ P_s^{ref}(t) &= P_{s,s}^{ref,*}(t|t_k), \quad \forall t \in [t_k, t_{k+1}). \end{aligned} \tag{8.30}$$

In the optimization problem of (8.22)–(8.29), the constraints of (8.23) and (8.24) require that the computed power references should be smaller than the minimum of the maximum available within each sampling interval, which means the power references should be achievable for the wind and solar subsystems. The constraints of (8.25) and (8.26) impose limits on the rate of change of the two power references. Note that the future maximum available powers for the wind and the solar subsystems are estimated using the information of future weather conditions forecast. The constraints of (8.27) and (8.28) are the system model of (8.20) with initial condition defined by (8.29). The constraints of (8.23)–(8.29) are inspired by results on the design of Lyapunov-based MPC systems [13, 64].

8.2.5 Simulation Results I

In this subsection, we carry out simulations to demonstrate the effectiveness and applicability of the proposed supervisory MPC of (8.22)–(8.29). The sampling time and the prediction horizon of the MPC are chosen to be $\Delta = 1$ s and $N = 2$. Note that the choice of the prediction horizon is based on the fast dynamics of the wind–solar energy generation system and the RO system, the uncertainty associated with wind speed and is also made to achieve a balance between the evaluation time of the optimization problem of the supervisory MPC of (8.22)–(8.29) and of the desired closed-loop performance. The maximum increasing values of the two power references are chosen to be $dP_{w,\max} = 1000\ W$ and $dP_{s,\max} = 500\ W$, respectively. The RO water desalination system is operated at a recovery rate Y = 0.8 and the overall power efficiency is assumed to be $\eta = 0.7$. The weighting factors in the cost function are chosen to be $\alpha = 1$, $\beta = 0.01$ and $\zeta = 0.4$.

We first carry out simulations under varying environmental conditions without disturbances. The time evolution of wind speed, PV panel temperature and isolation are shown in Fig. 8.3a.

We consider time-varying water demand $F_d(t)$ with step changes as shown in Fig. 8.4a (dashed line). This water demand is reflected as a load current i_L with transient processes on the side of the energy generation system as shown in Fig. 8.4b, which are due to the dynamic properties of the RO system.

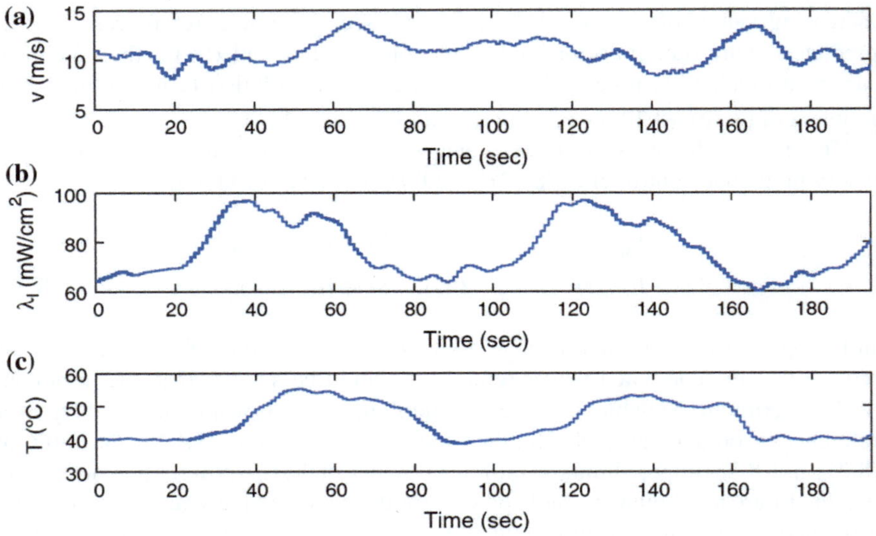

Fig. 8.3 Environmental conditions. **a** Wind speed v; **b** isolation λ_l; and **c** PV panel temperature T

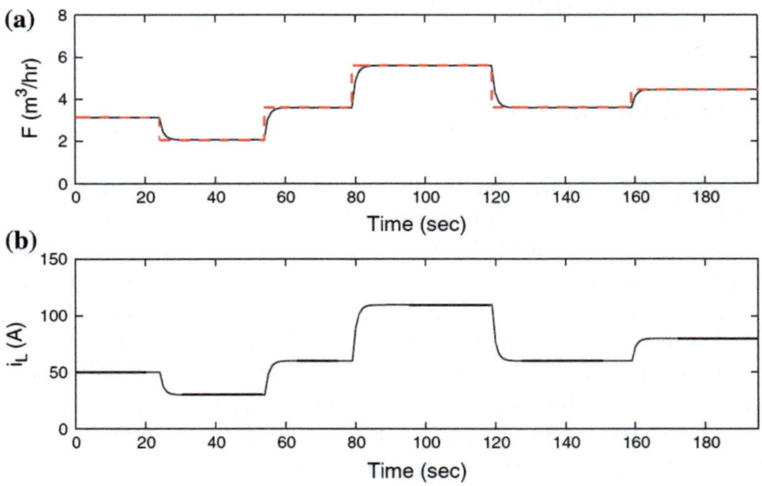

Fig. 8.4 **a** Time-varying water demand (*dashed line*) and actual water production (*solid line*) and **b** corresponding load current i_L

The simulation results are shown in Fig. 8.5. It can be seen from Fig. 8.5a that the wind/solar/battery powers coordinate their behavior to meet the power demand of the RO system. Time evolution of output power and maximum available power from the wind subsystem and the solar subsystem are plotted in Fig. 8.5b, c. When

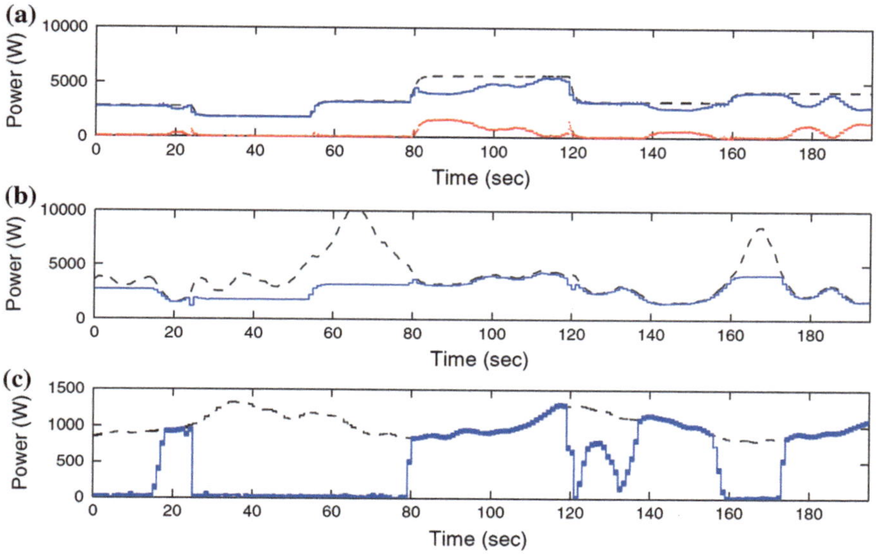

Fig. 8.5 **a** Generated power $P_w + P_s$ (*solid line*), total power demand P_T (*dashed line*), and power provided by battery bank P_b (*dotted line*); **b** Power generated by wind subsystem P_w (*solid line*) and maximum wind generation $P_{w,\max}$ (*dashed line*); and **c** Power generated by solar subsystem P_s (*solid line*) and maximum solar generation $P_{s,\max}$ (*dashed line*)

sufficient energy supply can be extracted from the two subsystems such as during 0.7, 25.8, 125.37, and 155.73 s, the battery is being recharged. In other periods, LD is relatively high and the weather condition, which determines the maximum available generation capacity of the two subsystems, cannot permit sufficient energy supply. Thus, the supervisory controller drives the wind/solar subsystems to their instant maximum capacity and calls the battery bank for shortage compensation. Note that because of the dynamics of the RO desalination system, when there is a step change in the water demand, the RO desalination system takes a short time to track the water demand [see Fig. 8.4a (solid line)].

We have also carried out simulations to evaluate the robustness of the proposed supervisory control system of (8.22)–(8.29) subject to disturbances in wind speed and insulation; specifically, 10 % variation in the wind speed and 5 % variation in the insulation and additional high-frequency disturbances to simulate real forecasting inaccuracy. The profiles of the wind speed and insulation are shown in Figs. 8.6a, b. The simulation results are shown in Figs. 8.7 and 8.8. From these figures, we can see that the proposed supervisory control system of (8.22)–(8.29) operates in a robust fashion to coordinate the wind and solar subsystems as well as the battery bank to meet the total power demand of the desired water production.

We note that the simulations in this section were carried out in a computer with a Core 2 Duo 1.6 GHz processor. The MPC optimization problems were solved by

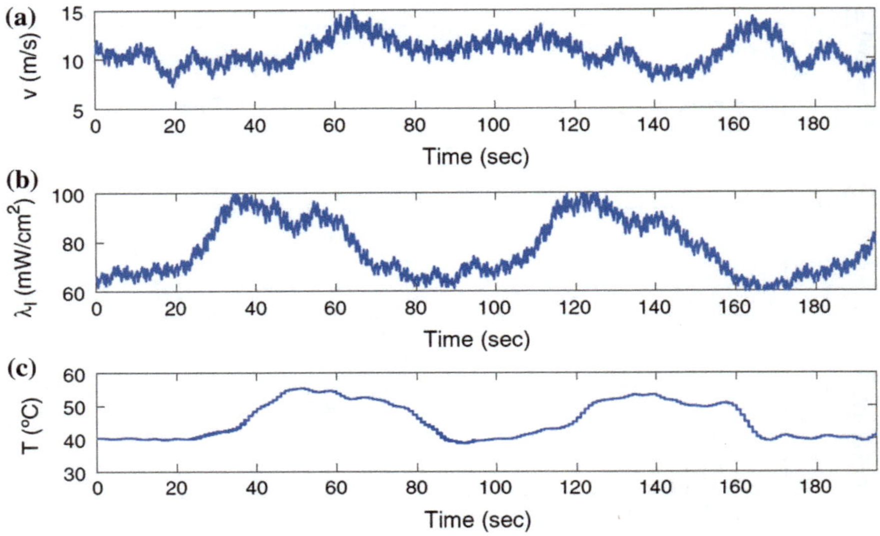

Fig. 8.6 Environmental conditions. **a** Wind speed v; **b** isolation λ_l; and **c** PV panel temperature T

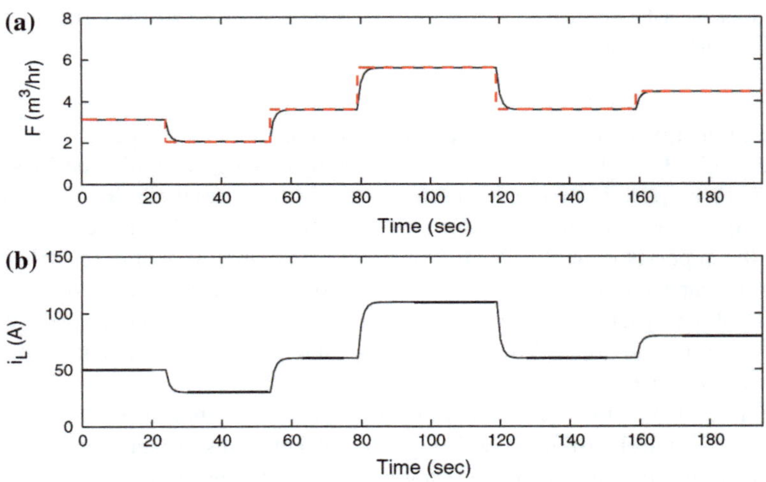

Fig. 8.7 Time-varying water demand Fd and corresponding load current i_L

MATLAB® built-in optimization solver *f-mincon* and the average evaluation time was a little more than one minute. We also note that this evaluation time can be significantly reduced by code optimization, improved processor speed, and model reduction.

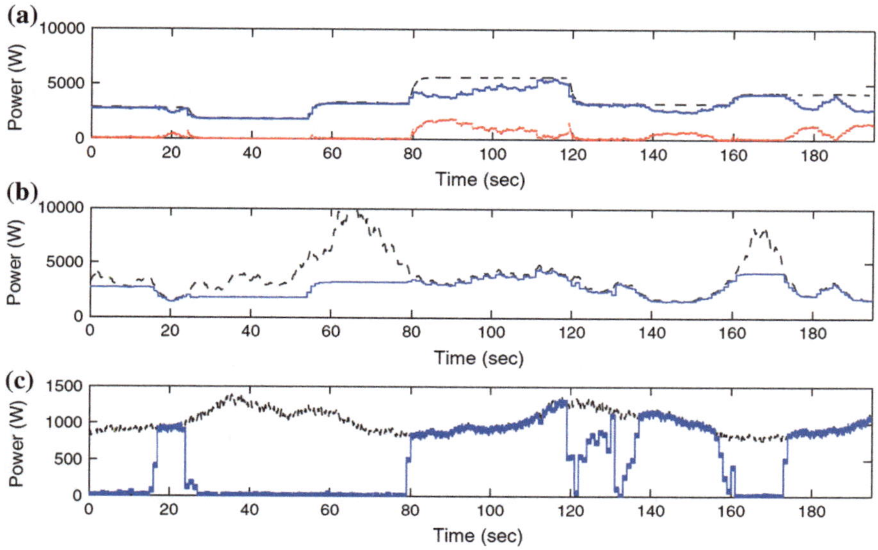

Fig. 8.8 **a** Generated power $P_w + P_s$ (*solid line*), total power demand P_T (*dashed line*), and power provided by battery bank P_b (*dotted line*); **b** Power generated by wind subsystem P_w (*solid line*) and maximum wind generation $P_{w,\max}$ (*dashed line*); and **c** Power generated by solar subsystem P_s (*solid line*) and maximum solar generation $P_{s,\max}$ (*dashed line*)

8.2.6 Integration for Long-Term Operation

Now turning to the electrical grid-connected operating mode of the integrated system ($I_s = 1$), it is desired to design a supervisory predictive control system to accomplish long-term optimal operation of the integrated system.

- The primary control objective is to regulate the integrated system to produce enough desalinated water to satisfy the total water consumption, F_d and storage demand, F_s.
- The secondary objective is to take into account optimality considerations on system operation; for example, battery maintenance and time-varying electric power pricing. The supervisory control system determines the reference retentate flow rate $\left(v_r^{ref}\right)$ for the RO water desalination subsystem and the current sent to the electrical grid $\left(i_G^{ref}\right)$.

In this case, the wind and solar subsystems operate at their maximum power generation points.

8.2.7 Supervisory Control System Design II

In this supervisory MPC design, we explicitly take into account the following battery bank maintenance considerations [54, 75, 82, 85–99]:

1. Small charge/discharge currents are favorable;
2. The charge current should be constrained under a certain upper bound (we set the upper bound of the charge currents based on a simple taper charging approach [54];
3. The DOD of the battery bank should not exceed $d_b^{\max}$ in order to protect the battery bank, and
4. The battery should be charged if extra generated power is available in addition to satisfying water production power demand.

We also take into account the time-varying electric power pricing and trying to achieve economically optimal operation by sending (selling) energy to the electrical grid at high electric power price and obtaining (buying) energy from the electrical grid at low electric power price. In order to send more energy to (or obtain less energy from) the electrical grid, we operate the RO water desalination subsystem at the energy optimal water recovery Y_{opt} so that the energy consumption per unit water produced is minimized [107].

In addition, we assume that there is a preferred SOS, s_t^{opt}, of the storage tank which is a balance between the capacities of the tank to supply unexpected water consumption demand and to store extra water production. We consider the case where the future water consumption demand of the RO subsystem is known; that is, $F_d(t)$, is known. We also assume that future hourly weather conditions (i.e., wind speed, isolation, and PV cell temperature) forecast information is available.

The cost function of the proposed supervisory controller is designed as follows:

$$
\begin{aligned}
J_g(t_k) = \gamma \int_{t_k}^{t_{k+N}} d_b(\tau)d\tau + \xi \int_{t_k}^{t_{k+N}} i_b(\tau)^2 d\tau + \zeta \int_{t_k}^{t_{k+N}} p(\tau)P_G(\tau)d\tau \\
+\varepsilon \int_{t_k}^{t_{k+N}} |s_t(\tau) - s_t^{opt}|d\tau + \theta \frac{\int_{t_k}^{t_{k+N}} PRO(\tau)d\tau}{\int_{t_k}^{t_{k+N}} F_p(\tau)d\tau}
\end{aligned}
\tag{8.31}
$$

where γ, ξ, ζ, ε, and θ are positive weighting factors, $p(\tau)$ denotes the time-varying electric power price, and $P_G(P_G = -i_G v_b)$ is the power drawn from the electrical grid by the integrated system. In this cost function, the first term implies that the battery should be charged if the battery is not fully charged; the second term takes into account that small charge currents are preferred; the third term is used to account for the economic optimization consideration by selling/buying power to/from the electrical grid; the fourth term is used to make sure that the water level in the storage tank is maintained around the optimal water level; and the fifth term penalizes the power consumption per unit of permeate water produced.

For this case, the proposed MPC design for the supervisory control system at time t_k is as follows:

$$\min_{i_G^{ref}, v_r^{ref} \in S(\Delta)} \quad J_g(t_k) \tag{8.32}$$

$$s.t.\ P_{RO}(\tau) - P_w(\tau) - P_s(\tau) + i_G^{ref}(\tau)v_b(\tau) + i_b v_b = 0 \tag{8.33}$$

$$F_p^{\min} \leq F_p(\tau) \leq F_p^{\max} \tag{8.34}$$

$$0 \leq d_b(\tau) \leq d_b^{\max} \tag{8.35}$$

$$s_t^{\min} \leq s_t(\tau) \leq s_t^{\max} \tag{8.36}$$

$$i_b(\tau) \leq i_b^{\max}(s_b(\tau)) \tag{8.37}$$

$$\dot{\tilde{x}}(\tau) = f(\tilde{x}(\tau)) + g(\tilde{x}(\tau))u(\tau) \tag{8.38}$$

$$h(\tilde{x}) = 0 \tag{8.39}$$

$$\tilde{x}(0) = x(t_k) \tag{8.40}$$

We denote the optimal solution to the optimization problem of Eq. (8.32)–(8.40) as $i_G^{ref,*}(\tau|t_k)$ and $v_r^{ref,*}(\tau|t_k)$. The references of battery charge/discharge current and of the RO retentate flow rate sent to the local controllers (LOCs) by the supervisory controller of (8.22)–(8.29) are defined as follows:

$$\begin{aligned} i_G^{ref}(t) &= i_G^{ref,*}(t|t_k), \quad \forall t \in [t_k, t_{k+1}), \\ v_r^{ref}(t) &= v_r^{ref,*}(t|t_k), \quad \forall t \in [t_k, t_{k+1}). \end{aligned} \tag{8.41}$$

In the optimization problem of (8.32)–(8.40), the constraint of (8.33) is an energy balance for the integrated system. The constraint of (8.34) imposes upper and lower bounds ($F_p^{\max}$ and $F_p^{\min}$, respectively) on the permeate flow rate F_p, which are used to guarantee the equipment safety of the membrane module in the RO water desalination subsystem. The constraint of (8.35) requires that the DOD of the battery bank should not exceed $d_b^{\max}$. The constraint of (8.36) imposes upper and lower bounds on the water level in the storage tank. The constraint of (8.37) places an upper bound on the charge current of the battery bank and this upper bound is a function of the current DOD of the battery bank. The constraints of Eqs. (8.38) and (8.39) are the system model of (8.20) with initial condition defined by (8.40).

Note that the dynamics of the integrated system exhibit a two timescale behavior. Specifically, the dynamics of the states, $i_q, i_d, w_e, w_{PV}, i_s$ and v_r, are relatively fast (in the order of seconds); and the dynamics of the states, v_c and h_l, are relatively slow (in the order of minutes). In order to achieve long-term system optimal operation, we take advantage of this two timescale property in the design of the supervisory MPC design of (8.32)–(8.40) to significantly reduce the evaluation time of the MPC. Specifically, in the evaluation of the MPC of (8.32)–(8.40), only the slow system dynamics is taken into account and the fast system states that are (explicitly or implicitly) used in the MPC are estimated by the computed future operating trajectories (decision variables of the MPC). For example, in the calculation of future $F_p(\tau)$, the fast state v_r is assumed to be equal to v_r^{ref}.

8.2.8 Simulation Results II

We carry out a closed-loop simulation to demonstrate the applicability and effectiveness of the supervisory MPC of (8.32)–(8.40) in achieving efficient integration of the wind/solar RO system with the electrical grid. The prediction horizon and the sampling time of the MPC are chosen to be $N = 24$ and $\Delta = 1$ h taking into account that the water demand (for example, of a community) usually presents periodic fluctuations with a period of 24 h. The weighting factors in the cost function are chosen to be $\gamma = 0.001$, $\xi = 1.8 \times 10^{-7}$, $\zeta = 1 \times 10^{-6}$, $\varepsilon = 0.01$ and $\theta = 2 \times 10^{-7}$. The overall RO system pump power efficiency is assumed to be $\eta = 0.7$, the upper bound on db is $d_b^{\max} = 0.8$, the lower and upper bounds on F_p are $F_p^{\min} = 0.1814\,\text{m}^3/\text{h}$ and $F_p^{\max} = 3.9918\,\text{m}^3/\text{h}$, respectively, and the lower and upper bounds on st are $s_t^{\min} = 0$ and $s_t^{\max} = 1$, respectively.

Specifically, simulations were carried out for 1 day (24 h) starting at 8 a.m. We assume that hourly weather forecast, hourly water demand, and hourly grid electricity price are available to the supervisory MPC for the entire 24-h period. A 2-day forecast information of weather conditions, water demand, and grid electricity price, used in our simulations, is shown in Fig. 8.9. We also introduce up to 15, 10, and 10 % deviations and additional high-frequency disturbances to simulate the forecasting inaccuracy for wind speed, PV cell temperature, and isolation, respectively, and also introduce a low-frequency disturbance in the water demand. The actual weather conditions and water demand along with real-time electricity price variations are shown in Fig. 8.10. Figure 8.11 displays the time evolutions of power generation by the wind and solar subsystems and power consumption by the RO subsystem. Figure 8.10a, b shows that in this scenario the supervisory MPC forces the wind/solar subsystems to track the maximum power generation points all the time. Figure 8.12a shows the profile of power trading of the system with the grid. We denote the power purchased from the grid as positive and the power sold to the grid as negative. It can be seen that the supervisory MPC controller outputs optimal grid power trading references at each sampling time taking into account one day ahead variations of electricity price, states of battery bank and water tank as well as wind/solar power generation capabilities.

Note that the periods when grid power is fluctuating indicates that part of battery current is diverted to the grid at that moment so that the upper battery current limit cannot be exceeded. Consequently, profiting from power trading is achieved while system performance is optimized. The cumulative revenue of power trading is shown in Fig. 8.12b.

In addition to efficient grid integration, an advantage of our supervisory MPC design is that it is able to schedule water production to be smooth (nearly uniform

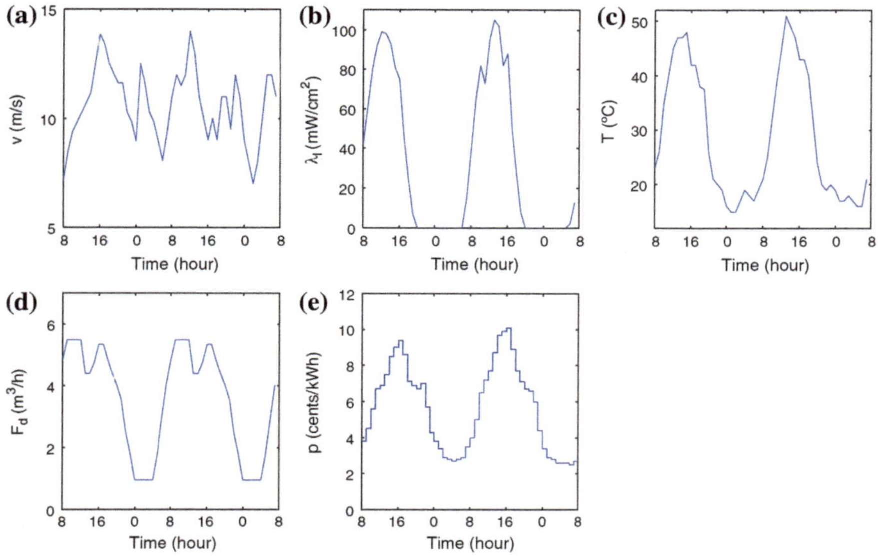

Fig. 8.9 Forecast of weather conditions, water demand, and grid electricity price. **a** Wind speed v; **b** isolation λ_l; **c** PV panel temperature T; **d** water demand F_d; and **e** grid electricity price p

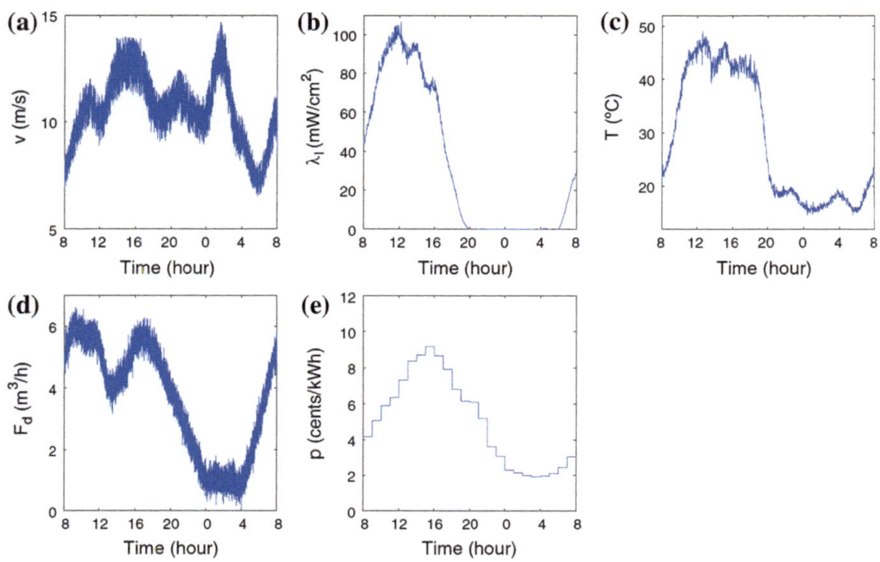

Fig. 8.10 Weather conditions, water demand, and grid electricity price including realistic variations. **a** Wind speed v; **b** insulation λ_l; **c** PV panel temperature T; **d** water demand F_d; and **e** grid electricity price p

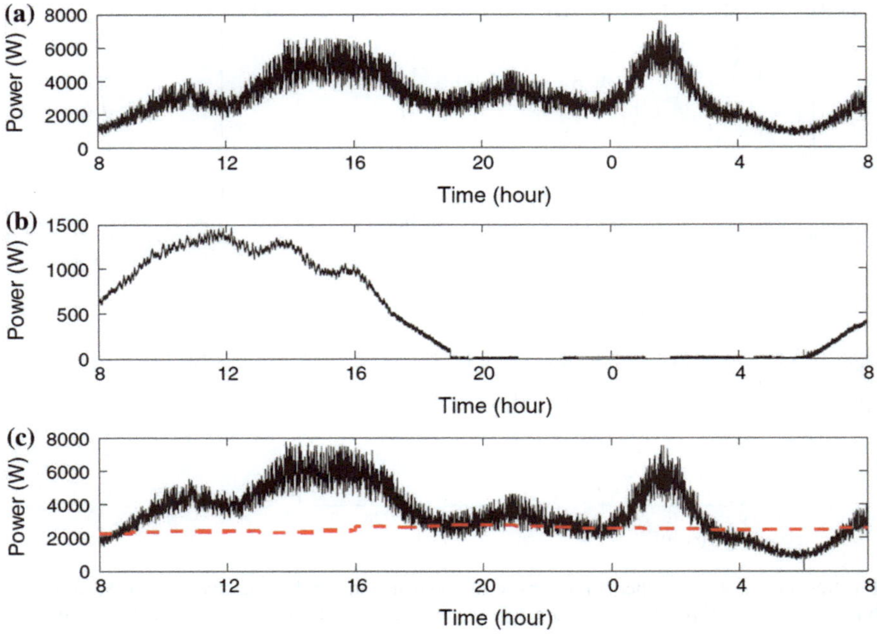

Fig. 8.11 Power generation and consumption. **a** Power generated by the wind subsystem P_w; **b** Power generated by the solar subsystem P_s; **c** Generated power $P_w + P_s$ (*solid line*) and total power demand P_{RO} (*dashed line*)

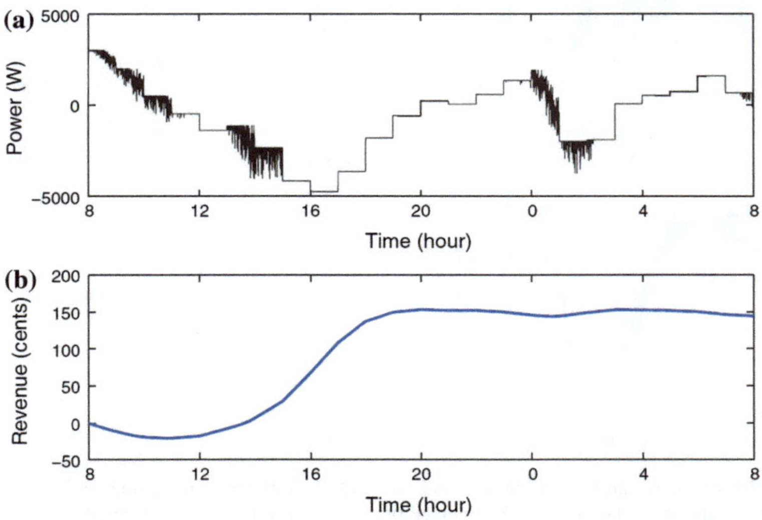

Fig. 8.12 **a** Electrical grid power trading profile P_G and **b** Cumulative revenue of power trading with the grid

with respect to time) by taking into account future conditions/demand variations and by coordinating the subsystems. Given a total amount of water demand, uniformly producing water is not only beneficial in terms of equipment maintenance, but it also consumes a smaller amount of energy as well. In fact, deviation from the optimal operating point as specified in Fig. 8.13a results in additional energy consumption. It can be seen from Fig. 8.13a that water production is relatively smooth despite significantly varying water demand and weather conditions during a day. This is largely due to the optimized utilization of capacities of the battery bank and of the water tank as well as integration with the electrical grid, which jointly act as buffers against external fluctuations. The states of charge of the battery bank and of storage of the water tank are shown in Fig. 8.13b, displaying, as expected, periodic variations.

It is noted that the simulations in this section were carried out in a computer with a Core 2 Quad 2.66 GHz processor. The MPC optimization problems were solved using the open source interior point optimizer I_{popt} and the average evaluation time was less than 0.5 s, which is sufficiently fast compared with the sampling interval of 1 h. We note that the reduction in the evaluation time compared to the one in Sect. 8.2.5 is primarily due to the use of a reduced-order process model in the optimization problem.

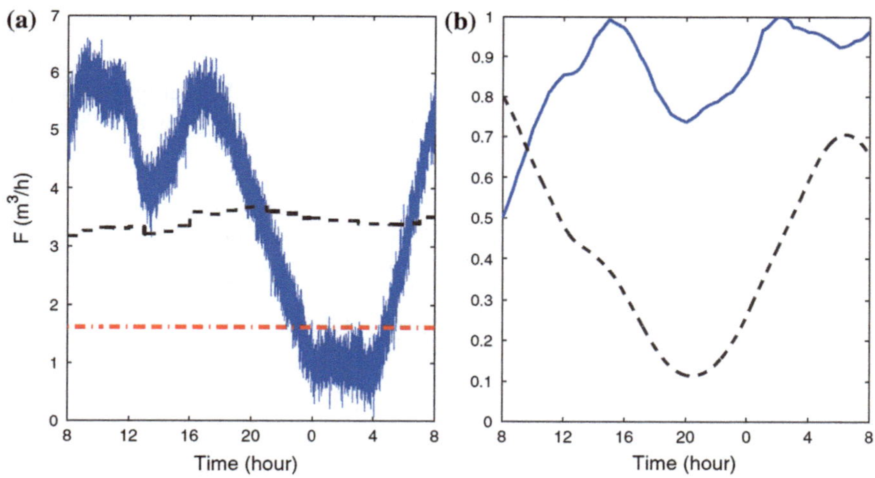

Fig. 8.13 **a** Rate of water demand F_d (*solid line*), permeate flow rate F_p (*dashed line*) and the energy-efficient production flow rate F_s (*dash-dotted line*); **b** Battery state of charge s_b (*solid line*) and water tank state of storage s_t (*dashed line*)

8.2.9 Distributed Energy Systems

A schematic diagram of an inverter-based distributed energy is depicted Fig. 8.14. It clearly shows the primary power source (e.g., PV panels), the voltage source converter (VSC), and the power, voltage, and current control loops.

The power controller provides the voltage references v^*_{odi} and v^*_{oqi} for the voltage controller, and the operating frequency ω_i for the VSC. Note that nonlinear dynamics of each DG in a microgrid (MG) are formulated on its own $d - q$ (direct-quadratic) reference frame. The reference frame of MG is considered as the common reference frame and the dynamics of other DGs are transformed to the common reference frame. The angular frequency of this common reference frame is denoted by ω_{com}.

The nonlinear dynamics of the ith DG, shown in Fig. 8.14, can be written as

$$\begin{cases} \dot{x}_i = f_i(x_i) + k_i(x_i)D_i + g_i(u_i) \\ y_i = h_i(x_i) + d_i u_i \end{cases} \tag{8.42}$$

The term D_i is considered as a known disturbance. Detailed expressions for

$$f_i(x_i),\, g_i(x_i),\, h_i(x_i),\, d_i(x_i),\, k_i(x_i)$$

are adopted from the nonlinear model presented in [68].

In what follows, the control levels are prescribed. The primary control is usually implemented as a CC at each DG by the droop technique. We learned before that droop technique prescribes a desired relation between the frequency and the active

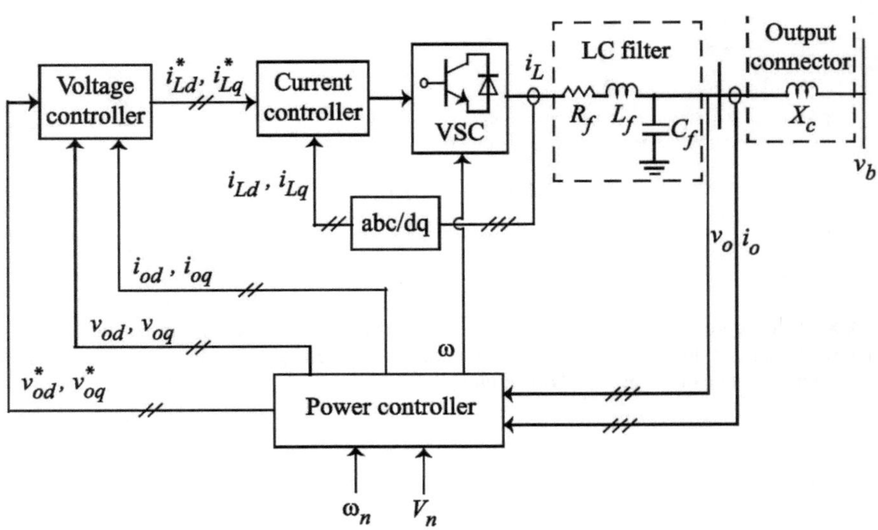

Fig. 8.14 The block diagram of an inverter-based DG

power, and between the voltage amplitude and the reactive power. In this regard, the primary frequency control is

$$\omega_i = \omega_{ni} - m_{P_i} P_i \tag{8.43}$$

where ω_{ni} is the primary frequency control reference and m_{P_i} is the frequency-active power droop coefficient.

The secondary frequency control selects ω_{ni} such that the angular frequency of each DG synchronizes to its nominal value, i.e., $\omega_i \rightarrow \omega_{ref}$. After applying the primary control, the DG output powers satisfy the following equality

$$m_{P1} P_1 = \cdots = m_{PN} P_N. \tag{8.44}$$

Since the active power droop coefficients m_{Pi} are chosen based on the active power rating of DGs, $P_{\max i}$, then (8.44) is equivalent to

$$\frac{P_1}{P_{\max 1}} = \cdots = \frac{P_N}{P_{\max N}} \tag{8.45}$$

The secondary frequency control must also satisfy (8.44) or (8.45). For the secondary frequency control, the outputs and inputs are $y_i = \omega_i$ and $u_i = \omega_{ni}$, respectively.

Remark 8.2 Conventionally, the secondary frequency control is implemented by using a centralized controller for the whole MG having the proportional plus integral (PI) structure. In a centralized control structure, the central controller (CC) communicates with all DGs in the MG through a star communication network. A centralized control structure deteriorates the system reliability.

Following [16], the output active power of each DG can be written as

$$P_i = \frac{|v_{oi}||v_{bi}|}{X_{ci}} \sin(\delta_i) \equiv h_i \sin(\delta_i), \tag{8.46}$$

where δ_i is the angle of the DG reference frame with respect to the common reference frame. v_{oi}, v_{bi}, and X_{ci} are shown in Fig. 8.14. The term h_i can be assumed to be constant since the amplitude of v_{oi} and v_{bi} change slightly around the nominal voltage [16]. Since $_{ci}$ is typically small, δ_i is small, and hence, $\sin(\delta_i)$ is approximately equal to δ_i [16]. Considering these assumptions and differentiating (8.46) yields

$$\dot{P}_i = h_i(\omega_i - \omega_{com}) \tag{8.47}$$

It must be noted that (8.47) provides a direct relationship between the differentiated output power of DGs and their angular frequency with respect to the angular frequency of MG. The global form of (8.47) can be written as

$$\dot{P} = h(\omega - \underline{\omega}_{com}), \tag{8.48}$$

where $h = diag\{h_i\}$ and $\underline{\omega}_{com} = 1_N \otimes \omega_{com}$.

8.2.10 Distributed Frequency Control

The distributed cooperative frequency control is designed to synchronize the frequency of DGs, ω_i in (8.43), to the reference frequency, ω_{ref}, while sharing the active power among DGs based on their power ratings as stated in (8.44).

The nonlinear dynamics of the ith DG in (8.42) are considered. Differentiating the frequency-droop characteristic in (8.43) yields

$$\dot{\omega}_{ni} = \dot{\omega}_i + m_{Pi}\dot{P}_i = u_i, \tag{8.49}$$

where u_i is an auxiliary control to be designed. Note that (8.49) is a dynamic system for computing the control input ω_{ni} from u_i, see Fig. 8.15. The auxiliary control should be designed such that DG frequencies synchronize to the reference frequency ω_{ref}, and (8.44) is satisfied. According to (8.49), the secondary frequency control of a MG including N DGs is transformed to a synchronization problem for a first-order and linear multiagent system

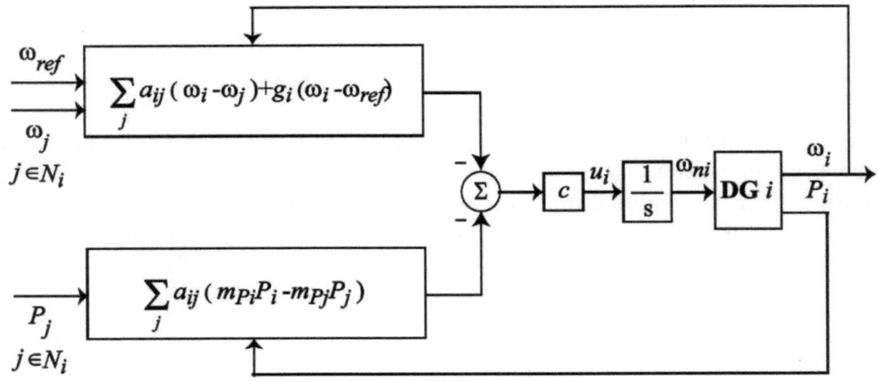

Fig. 8.15 Distributed secondary frequency control diagram

$$\begin{aligned}\dot{\omega}_1 + m_{P1}\dot{P}_1 &= u_1\\ \dot{\omega}_2 + m_{P2}\dot{P}_2 &= u_2\\ &\vdots\\ \dot{\omega}_N + m_{PN}\dot{P}_N &= u_N\end{aligned} \tag{8.50}$$

To achieve synchronization, it is assumed that DGs can communicate with each other through the prescribed communication digraph g. The auxiliary controls u_i are chosen based on each DGs own information, and the information of its neighbors in the communication digraph as

$$u_i = -c\left(\sum_{j\in N_i} a_{ij}(\omega_i - \omega_j) + g_i(\omega_i - \omega_{ref}) + \sum_{j\in N_i} a_{ij}(m_{Pi}P_i - m_{Pj}P_j)\right), \tag{8.51}$$

where $c \in R$ is the control gain. It is assumed that the pinning gain $g_i \geq 0$ is nonzero for only one DG that has the reference frequency ω_{ref}. The global control input u is written as

$$u = -c((L+G)(\omega - \underline{\omega}_{ref}) + Lm_P P), \tag{8.52}$$

where $\omega = [\omega_1\ \omega_2\ \ldots\ \omega_N]^T$, $\underline{\omega}_{ref} = 1_N \otimes \omega_{ref}$, with 1_N the vector of ones with the length of N, $m_P = diag\{m_{Pi}\}$, and $P = [P_1\ P_2\ \ldots\ P_N]^T$. The Kronecker product is $\otimes$.$G \in R^{N\times N}$ is a diagonal matrix with diagonal entries equal to the pinning gains g_i. The global form of dynamics in (8.50) can be written as

$$\dot{\omega} = m_P\dot{P} = -c((L+G)(\omega - \underline{\omega}_{ref}) + Lm_P P). \tag{8.53}$$

The term $(L+G)(\omega - \underline{\omega}_{ref})$ is defined as the global neighborhood tracking error e. The term $\omega - \underline{\omega}_{ref}$ is defined as the global disagreement vector, δ. With reference to the Appendix, the following result is recalled:

Lemma 8.1 *Zero is a simple eigenvalue of L if and only if the directed graph has a spanning tree. Moreover,* $L1_N = 0$*, with* 1_N *being the vector of ones with the length of N.*

Lemma 8.2 *Let the digraph g have a spanning tree and* $g_i \neq 0$ *for at least one root node. Then,* $L + G$ *is a nonsingular M-matrix. Additionally*

$$\|\delta\| \leq \|e\|/\sigma_{\min}(L+G), \tag{8.54}$$

where $\sigma_{\min}(L+G)$ *is the minimum singular value of* $L+G$*, and* $e = 0$ *if and only if* $\delta = 0$.

In the following, it is assumed that the DG for which $g_i \neq 0$ is labeled as DG1. Theorem 8.1 is the main result.

Theorem 8.1 *Let the digraph g have a spanning tree and $g_i \neq 0$ for only one DG placed as a root node of digraph g. Let the auxiliary control u_i be chosen as in (8.51). Then, the DG frequencies ω_i in (8.43) synchronize to ω_{ref}, and the active power among DGs is shared based on their power ratings satisfying (8.45).*

Proof In the steady state, the left sides of (8.53) and (8.48) are equal to zero. Setting the left side of (8.48) equal to zero yields

$$\omega = \underline{\omega}_{com}. \tag{8.55}$$

Note that (8.55) shows that all the DG frequencies synchronize to the MG frequency in steady state. Therefore, according to Lemma 8.1

$$L\omega = 0. \tag{8.56}$$

Setting the left side of (8.53) equal to zero, and considering (8.56) yields

$$Lm_P P + G(\omega - \underline{\omega}_{ref}) = 0. \tag{8.57}$$

The commensurate form of (8.57) can be written as

$$\begin{bmatrix} \sum_{j=1:N} a_{1j} & -a_{12} & \cdots & -a_{1N} \\ -a_{21} & \sum_{j=1:N} a_{2j} & \cdots & -a_{2N} \\ \vdots & \vdots & \ddots & \vdots \\ -a_{N1} & -a_{N2} & \cdots & \sum_{j=1:N} a_{Nj} \end{bmatrix} \begin{bmatrix} m_{P1}P_1 \\ m_{P2}P_2 \\ \vdots \\ m_{PN}P_N \end{bmatrix} + \begin{bmatrix} g_1(\omega_1 - \omega_{ref}) \\ 0 \\ \vdots \\ 0 \end{bmatrix} = 0, \tag{8.58}$$

that equivalently yields:

$$\begin{aligned} &a_{12}(m_{P1}P_1 - m_{P2}P_2) + \cdots + a_{1N}(m_{P1}P_1 - m_{PN}P_N) \\ &+g_1(\omega_1 - \omega_{ref}) = 0, \end{aligned} \tag{8.59}$$

$$(\overline{L} + \overline{G}) \left(\begin{bmatrix} m_{P2}P_2 \\ m_{P3}P_3 \\ \vdots \\ m_{PN}P_N \end{bmatrix} - \begin{bmatrix} m_{P1}P_1 \\ m_{P1}P_1 \\ \vdots \\ m_{P1}P_1 \end{bmatrix} \right) = 0 \tag{8.60}$$

where

$$\overline{L} = \begin{bmatrix} \sum_{j=1:N} a_{2j} & -a_{23} & \dots & -a_{2N} \\ -a_{32} & \sum_{j=1:N} a_{3j} & \dots & -a_{3N} \\ \vdots & \vdots & \ddots & \vdots \\ -a_{N2} & -a_{N3} & \dots & \sum_{j=1:N} a_{Nj} \end{bmatrix} \tag{8.61}$$

$$\overline{G} = \begin{bmatrix} -a_{21} & 0 & \dots & 0 \\ 0 & -a_{31} & \dots & 0 \\ \vdots & \vdots & \ddots & \vdots \\ 0 & 0 & \dots & -a_{N1} \end{bmatrix} \tag{8.62}$$

We observe that (8.60) shows that the set $\{m_{P1}P_1, m_{P2}P_2, \dots, m_{PN}P_N\}$ can be considered on a communication digraph with $m_{P1}P_1$ as the leader node and $m_{P2}P_2$ as the root node. All nodes have access to the leader $m_{P1}P_1$ through the matrix G in (8.62). Since the original digraph g has a spanning tree with $m_{P1}P_1$ as the root node, at least one of the diagonal terms in G is nonzero. Therefore, exploiting Lemma 8.2 shows that all $m_{Pi}P_i$ synchronize to a common value in the steady state which satisfies (8.44), or, equivalently, (8.45). Additionally, according to (8.59), having all $m_{Pi}P_i$ synchronized to a common value shows that ω_1 synchronizes to ω_{ref} and hence, according to (8.55), all DG frequencies synchronize to ω_{ref}. This completes the proof.

The block diagram of the secondary frequency control based on the distributed cooperative control is shown in Fig. 8.15. As seen in this figure, the control input ω_{ni} is written as

$$\omega_{ni} = \int u_i dt. \tag{8.63}$$

8.2.11 Simulation Results III

The MG shown in Fig. 8.16a is used to verify the effectiveness of the proposed secondary control. This MG consists of four DGs. The lines between buses are modeled as series RL branches. The specifications of the DGs, lines, and loads are summarized in Table 8.1. In this table, K_{PV}, K_{IV}, K_{PC} , and K_{IC} are the parameters of the voltage and current controllers in Fig. 8.14. The voltage and current controller parameters are adopted from [68]. The simulation results are extracted by modeling the dynamical equations of MG in MATLAB.

It is assumed that DGs communicate with each other through the communication digraph depicted in Fig. 8.16b. This communication topology is chosen based on the

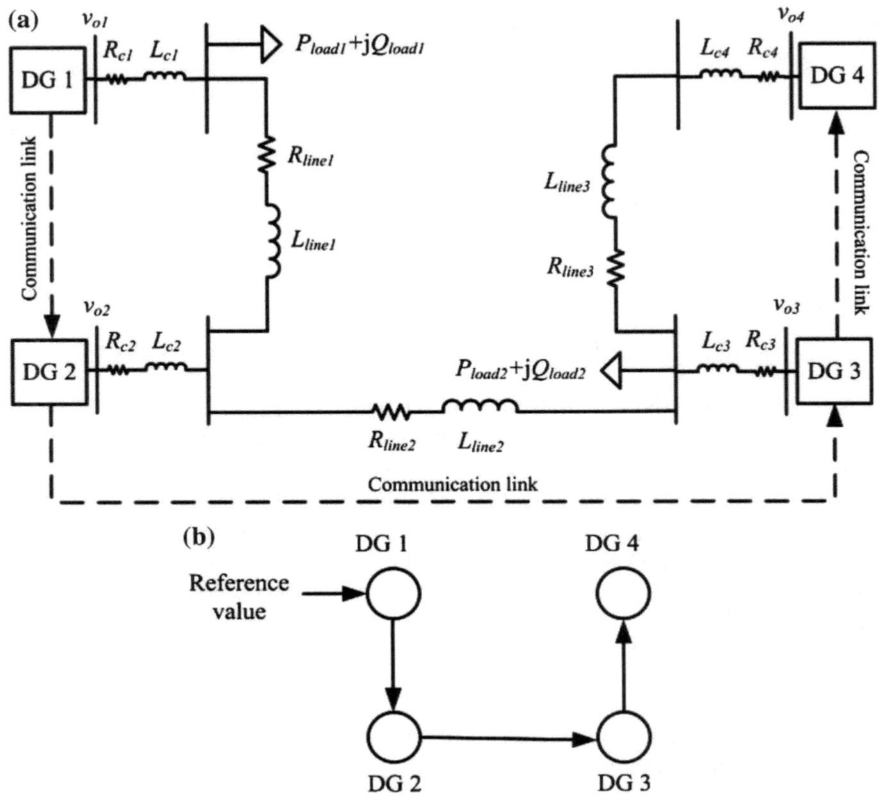

Fig. 8.16 **a** The microgrid test system; **b** The communication digraph

geographical location of DGs. The associated adjacency matrix of the digraph in Fig. 8.17a is

$$A = \begin{bmatrix} 0\,0\,0\,0 \\ 1\,0\,0\,0 \\ 0\,1\,0\,0 \\ 0\,0\,1\,0 \end{bmatrix}. \tag{8.64}$$

DG1 is the only DG connected to the leader node with the pinning gain of $g_1 = 1$. The reference value for the MG angular frequency ω_{ref} is set as 314.16 rad/s (The nominal frequency of the MG is 50 Hz.). The control gain c is set to 400.

It is assumed that the MG is islanded form the main grid at $t = 0$. Figure 8.17 shows frequencies and output powers of DGs before and after applying the secondary frequency control. As seen in Fig. 8.17a, once the primary control is applied, DG operating frequencies all go to a common value that is the operating frequency of

Table 8.1 Specs of the microgrid test system

DGs	*DG1 and 2 (45 kVA rating)*		*DG3 and 4 (34 kVA rating)*			
	m_P	9.43×10^{-5}	m_P	12.5×10^{-5}		
	n_Q	1.34×10^{-3}	n_Q	1.49×10^{-3}		
	R_c	0.03 Ω	R_c	0.03 Ω		
	L_c	0.35 mH	L_c	0.35 mH		
	R_f	0.1 Ω	R_f	0.1 Ω		
	L_f	1.35 mH	L_f	1.35 mH		
	C_f	50 μF	C_f	50 μF		
	K_{PV}	0.1	K_{PV}	0.05		
	K_{IV}	420	K_{IV}	390		
	K_{PC}	15	K_{PC}	10.5		
	K_{IC}	20,000	K_{IC}	16,000		
Lines	Line 1		Line 2		Line 3	
	R_{l1}	0.23 Ω	R_{l2}	0.35 Ω	R_{13}	0.23 Ω
	L_{l1}	318 μH	L_{l2}	1847 μH	L_{l3}	318 μH
Loads	Load 1		Load 2			
	P_{L1} (per phase)	12 kW	P_{L2} (per phase)	15.3 kW		
	Q_{L1} (per phase)	12 kVAr	Q_{L2} (per phase)	7.6 kVAr		

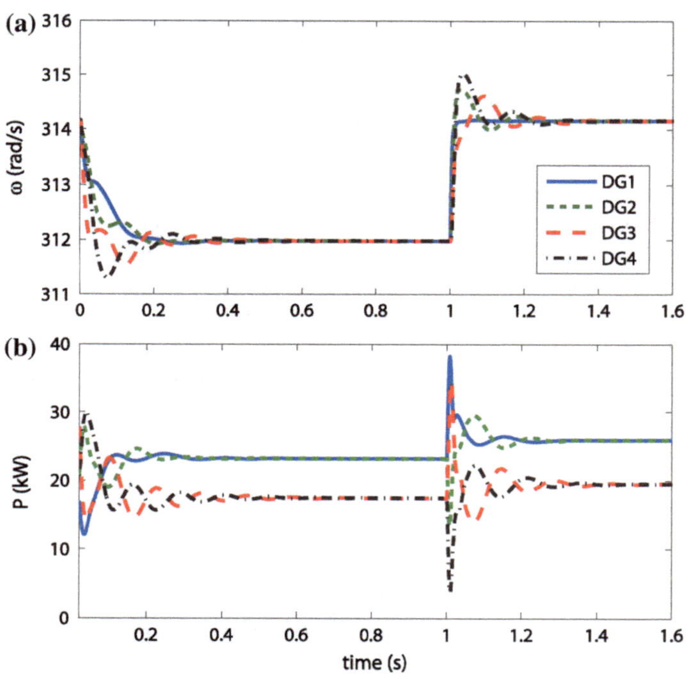

Fig. 8.17 The secondary frequency control with $\omega_{ref} = 314.16$ rad/s : **a** DG angular frequencies; **b** DG output powers

MG. However, the secondary frequency control returns the operating frequency of MG to its nominal value after 0.3 s. Figure 8.17b shows that the DG output powers all satisfy (8.44) and (8.45), and are set according to the power rating of DGs.

8.3 Multilevel Control of Droop-Controlled Microgrids

Islanded MGs have been used in applications like avionics, automotive, marine, or rural areas. The interfaces between the prime movers and the MGs are often based on power electronic converters acting as voltage sources (voltage source inverters (VSIs) in case of AC MGs). These power electronic converters are connected in parallel through the MG. In order to avoid circulating currents among the converters without using any critical communication between them, the droop-control method is often applied (Fig. 8.18).

The conventional droop method is not suitable when the paralleled system must share nonlinear loads because the control units should take into account harmonic currents and, at the same time, to balance active and reactive power. Thus, harmonic current-sharing techniques have been proposed to avoid the circulating distortion power when sharing nonlinear loads. All of them consist in distorting the voltage to enhance the harmonic current-sharing accuracy, resulting in a tradeoff.

By using the droop method, the power sharing is affected by the output impedance of the units and the line impedances. Hence, those virtual output impedance loops can solve this problem. In this sense, the output impedance can be seen as another control variable. However, an important disadvantage of the droop method is its load-dependent frequency deviation, which implies a phase deviation between the output voltage frequency of the uninterrupted power supply (UPS) system and the input voltage provided by the utility mains. This fact can lead to a loss of synchronization since the bypass switch must connect the utility line directly to the MG bus.

Consequently, this method, in its original version, can be only applied to islanded MGs [77]. Hence, this technique is not directly applicable to line interactive MGs since the transition between islanded and grid-connected modes will be difficult due to loss of synchronization. In addition, the inherent tradeoff of this method between

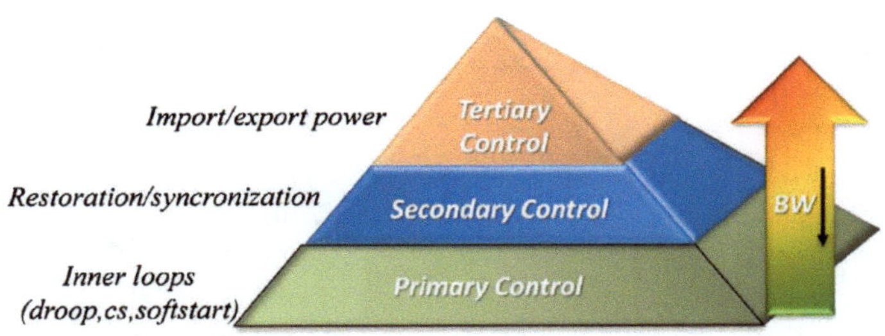

Fig. 8.18 Multilevel control levels of an MG

frequency and amplitude regulation in front of active- and reactive power sharing accuracy cannot be avoided in islanded mode [28]. Significant efforts have been done to improve the droop-control method to avoid the aforementioned frequency deviation.

In case of paralleling DC power converters, the droop method consists of subtracting a proportional part of the output current from the output voltage reference of each module. Thus, a virtual output resistance can be implemented through this control loop. This loop, also called adaptive voltage position, has been applied to improve the transient response of the voltage-regulation modules in low-voltage (LV) high-current applications. In addition, the droop method has also an inherent tradeoff between the voltage regulation and the current sharing between the converters.

In what follows, an external control loop, named secondary control, has been considered to restore the nominal values of the voltage inside the MG. Further, additional tertiary control can be used to bidirectionally control the power flow when the MG is connected to a stiff power source or the mains (in case of AC MGs) [28]. Multilevel control applied to power dispatching in AC power systems is well known, and it has been used extensively for decades. Nowadays, these concepts are starting to be applied to wind-power parks and were proposed for isolated PV systems. However, with the rise of power electronic-based MGs, which are able to operate both in grid-connected and in island mode, multilevel control and energy-management systems are necessary. Some authors proposed secondary [63] and tertiary controllers [98]. The main problem to be solved in such works is the frequency control of the system. However, voltage stability and synchronization issues are also important to achieve enough flexibility to operate in both modes. Only a few works conceived the MG as a whole problem, taking into account the different control levels.

8.3.1 A Generalized Multilevel Structure

The objectives of the grid code ISA-95 are to provide consistent terminology that is a foundation for supplier and manufacturer communications, to provide consistent information models, and to provide consistent operation models which are the foundation for clarifying application functionality and how information is to be used. In this standard, a multilevel control structure is proposed with the following [31]:

- *Level* 5: Enterprise. The enterprise level comprises the superior management policies of a commercial entity. This level has operational and developmental responsibility for the entire enterprise, including all of its plants and their respective production lines.
- *Level* 4: Campus/Plant. The campus or plant level comprises superior management policies of a branch or operational division of an enterprise, usually including the elements of the enterprise financial section that are directly associated with that business entity.

- *Level* 3: Building/Production. The building or production level comprises the management and control policies required to administer the states and behaviors of a building and its environmental and production systems.
  Level 2: Area/Line. The area or production line level comprises the management and control policies required to administer states and behaviors of a specific area or production line.
- *Level* 1: Unit/Cell. The unit or cell level comprises the management and control policies required to govern the states and behaviors of a unit of automation or a manufacturing cell.
  Level 0: Device. The device level comprises the set of field devices that sense and provide actuation of physical processes within the environmental and production systems.

Each level has the duty of a command level and provides supervisory control over lower level systems. In this sense, it is necessary to ensure that the command and reference signals from one level to the lower levels will have low impact in the stability and robustness performance. Thus, the bandwidth must be decreased with an increase in the control level. In order to adapt ISA-95 to the control of an MG, zero to three levels can be adopted as follows [28]:

1. Level 3 (*tertiary control*): This energy-production level controls the power flow between the MG and the grid.
2. Level 2 (*secondary control*): Ensures that the electrical levels into the MG are within the required values. In addition, it can include a synchronization control loop to seamlessly connect or disconnect the MG to or from the distribution system (see Fig. 8.19).
3. Level 1 (*primary control*): The droop-control method is often used in this level to emulate physical behaviors that makes the system stable and more damped. It can include a virtual impedance control loop to emulate physical output impedance.
4. Level 0 (*inner control loops*): Regulation issues of each module are integrated in this level. Current and voltage, feedback and feedforward, and linear and nonlinear control loops can be performed to regulate the output voltage and to control the current while maintaining the system stable.

AC MGs should be able to operate both in grid-connected and islanded modes [76]. The bypass switch is responsible for connecting the MG to the grid. This bypass switch is designed to meet grid-interconnection standards.

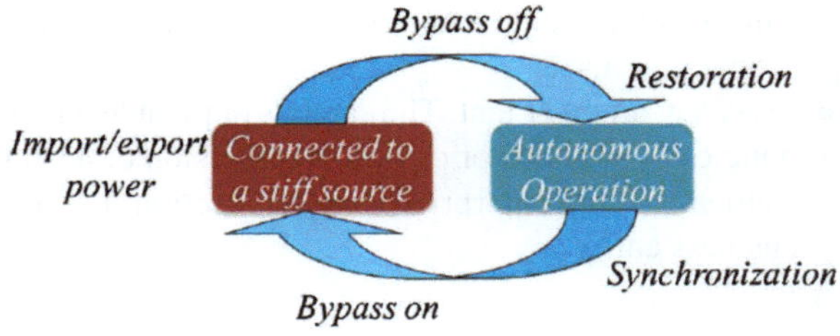

Fig. 8.19 Stiff source connection and autonomous operations

In grid-connected mode, the MG operates according to IEEE 1547–2003. The transition to islanded mode is done by intentional or unintentional events, e.g., grid failures. Thus, proper islanding detection algorithms must be implemented. In islanded mode, the MG must supply the required active and reactive powers as well as provide frequency stability and operate within the specified voltage ranges. Reconnection of the MG to the grid will be done when the grid voltage is within acceptable limits and the phasing is correct. Active synchronization is required to match the voltage, frequency, and phase angle of the MG.

8.3.2 Multilevel Control of AC Microgrids

Nowadays, it is necessary to conceive flexible MGs that are able to operate in both grid-connected and islanded modes. Thus, the study of topologies, architectures, planning, and configurations of MGs are necessary. This section deals with the multilevel control of AC MGs consisting of the same three control levels as presented in the previous section. The Union for the Coordination of Transmission of Electricity (UCTE), Continental Europe, has defined a multilevel control for large power systems, as shown in Fig. 8.20. Such a kind of systems is supposed to operate over large synchronous machines (SMs) with high inertias and inductive networks. However, in power electronic-based MGs, there are no inertias, and the nature of the networks is mainly resistive. Consequently, there are important differences between both systems that we have to take into account when designing their control schemes. This three-level control is organized as follows.

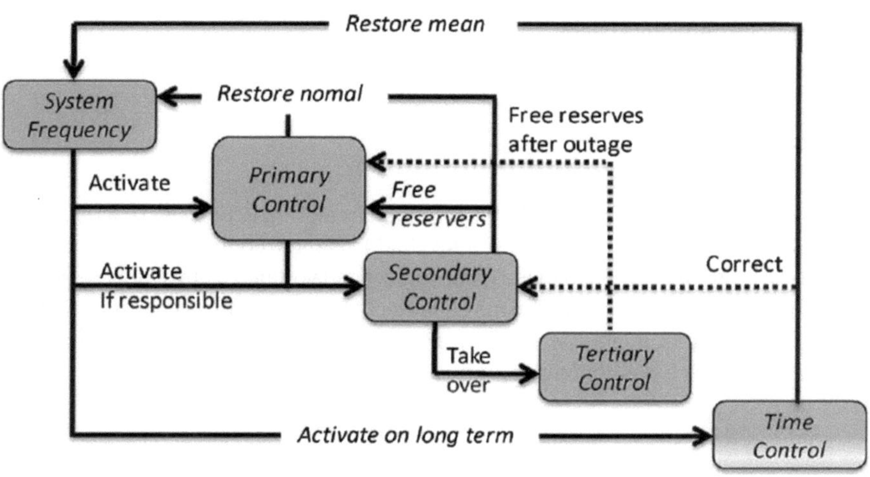

Fig. 8.20 Frame for the multilevel control of a power system as defined by UCTE

1. The primary control deals with the inner control of the DG units, adding virtual inertias and controlling their output impedances.
2. The secondary control is conceived to restore the frequency and amplitude deviations produced inside the MG by the virtual inertias and output virtual impedances.
3. The tertiary control regulates the power flow between the grid and the MG at the point of common coupling (PCC).

8.3.3 Inner Control Loops

The use of intelligent power interfaces between the electrical generation sources and the MG is mandatory. These interfaces have a final stage consisting of DC/AC inverters, which can be classified as current-source inverters (CSIs), which consist of an inner current loop and a phase-locked loop (PLL) to continuously stay synchronized with the grid, and VSIs, consisting of an inner current loop and an external voltage loop. In order to inject current to the grid, CSIs are commonly used, while in island or autonomous operation, VSIs are needed to keep the voltage stable. When these inverters are required to operate in grid-connected mode, they often change their behavior from voltage to current sources. Nevertheless, to achieve flexible MG, i.e., to be able to operate in both grid-connected and islanded modes, VSIs are required to control the exported or imported power to the mains grid and to stabilize the MG [28].

VSIs and CSIs can operate together in an MG. The VSIs are often connected to energy-storage devices, fixing the frequency and voltage inside the MG. The CSIs are often connected to a PV or small WTs that require maximum power point tracking (MPPT) algorithms, although those DG inverters could also work as VSIs if necessary. Thus, we can have a number of VSIs and CSIs, or only VSIs, connected in parallel, forming an MG.

8.3.4 Primary Control

When connecting two or more VSIs in parallel, circulating active and reactive power can appear (see Fig. 8.21). This control level adjusts the frequency and amplitude of the voltage reference provided to the inner current and voltage control loops. The main idea of this control level is to mimic the behavior of a synchronous generator (SYGN), which reduces the frequency when the active power increases. This principle can be integrated in VSIs by using the well-known P/Q droop method [105]

$$\omega = \omega^* - G_P(s)\dot{(}P - P^*) \tag{8.65}$$

$$E = E^* - G_Q(s)\dot{(}Q - Q^*) \tag{8.66}$$

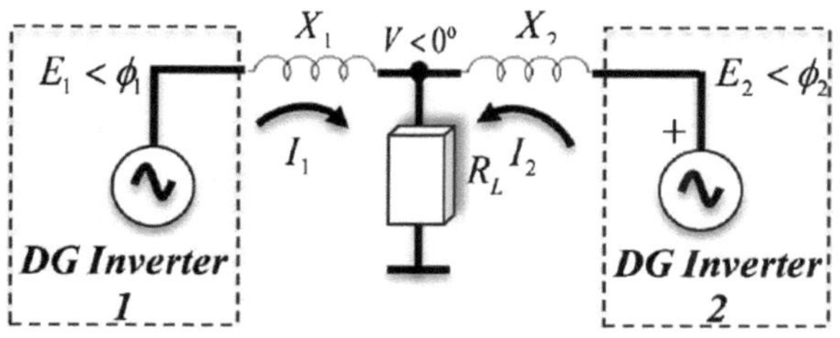

Fig. 8.21 Equivalent circuit of two parallel-connected inverters

with ω and E being the frequency and amplitude of the output voltage reference, ω^* and E^* being their references, P and Q as the active and reactive power, P^* and Q^* as their references, and $G_P(s)$ and $G_Q(s)$ as their corresponding transfer functions (typically proportional droop terms as shown in Fig. 8.22, i.e., $G_P(s) = m$, and $G_Q(s) = n$). Note that the use of pure integrators is not allowed when the MG is in islanded mode since the total load will not coincide with the total injected power, but they can be useful in grid-connected mode to have a good accuracy of the injected active and reactive power [28]. Nevertheless, this control objective will be achieved by the tertiary control level.

The design of $G_P(s)$ and $G_Q(s)$ compensators can be done by using different control-synthesis techniques such that the DC gain of such compensators (named m and n) provide for the static $\Delta P/\Delta\omega$ and $\Delta Q/\Delta V$ deviations to keep the system synchronized and within the voltage stability limits. One design approach of those parameters is as follows:

$$m = \Delta\omega/P_{\max} \tag{8.67}$$

$$n = \Delta V/2Q_{\max} \tag{8.68}$$

with $\Delta\omega$ and ΔV being the maximum frequency and voltage allowed and $P_{\max}$ and $Q_{\max}$ are the maximum active and reactive power delivered by the inverter.

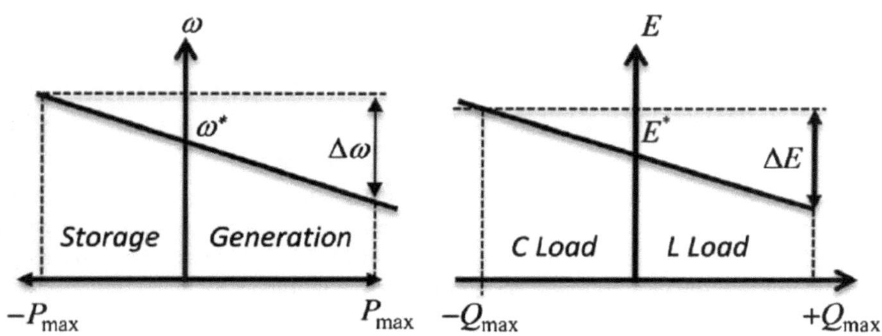

Fig. 8.22 P/Q droop functions

8.3.5 Secondary Control

Now, in order to compensate for the frequency and amplitude deviations, a secondary control is proposed. It basically ensures that the frequency and voltage deviations are regulated toward zero after every change of load or generation inside the MG. The frequency and amplitude levels in the MG, ω_{MG} and E_{MG}, are sensed and compared with the references ω^*_{MG} and E^*_{MG}; the errors processed through the compensators ($\delta\omega$ and δE) are sent to all the units to restore the output voltage frequency and amplitude. Practically, the secondary control should correct the frequency deviation within allowable limit, e.g., ± 0.1 Hz in Nordel (North of Europe) or ± 0.2 Hz in UCTE (Continental Europe). It is defined as

$$\delta P = -\beta \dot{G} - \frac{1}{T_r} \int G dt \tag{8.69}$$

where δP is the output set-point of the secondary controller, β is the gain of the proportional controller, T_r is the time constant of the secondary controller, and G is the area control error (ACE), which is normally calculated in about 5- to 10-s intervals by computers in the dispatch center as

$$G = P_{meas} - P_{sched} + K_{ri}(f_{meas} - f_0) \tag{8.70}$$

with P_{meas} being the sum of the instantaneous measured active power transferred at the PCC, P_{prog} being the resulting exchange program, K_{ri} being the proportional factor of the control area set on the secondary controller, and $f_{meas} - f_0$ being the difference between the instantaneous measured system frequency and the set-point frequency. From (8.69), note that the control action δP is increased by the integral formula if the deviation of ACE remains constant (PI- type controller). This controller is also called load-frequency control or automatic gain controller.

In case of an AC MG, the frequency and amplitude restoration controllers G_ω and G_E can be obtained similarly as follows [28]:

$$\delta\omega = k_{p\omega}(\omega^*_{MG} - \omega_{MG}) + k_{i\omega} \int (\omega^*_{MG} - \omega_{MG}) dt + \Delta\omega_S \tag{8.71}$$

$$\delta E = k_{pE}(E^*_{MG} - E_{MG}) + k_{iE} \int (E^*_{MG} - E_{MG}) dt \tag{8.72}$$

where $k_{p\omega}$, $k_{i\omega}$, k_{pE}, and k_{iE} are the control parameters of the secondary control compensator and $\Delta\omega_S$ is a synchronization term which remains equal to zero when the grid is not present. In this case, $\delta\omega$ and δE must be limited in order not to exceed the maximum allowed frequency and amplitude deviations.

8.3.6 Tertiary Control

When the MG is operating in grid-connected mode, the power flow can be controlled by adjusting the frequency (changing the phase in steady state) and amplitude of the voltage inside the MG [63]. By measuring the P/Q through the static bypass switch, P_G and Q_G can be compared with the desired P_G^* and Q_G^*. The control laws PI_P and PI_Q can be expressed as in the following [28]:

$$\omega_{MG}^* = k_p P(P_G^* - P_G) + k_{iP} \int (P_G^* - P_G)dt \tag{8.73}$$

$$E_{MG}^* = k_p Q(Q_G^* - Q_G) + k_{iQ} \int (Q_G^* - Q_G)dt \tag{8.74}$$

with k_pP, k_iP, k_pP, and k_{iQ} being the control parameters of the tertiary control compensator. Here, ω_{MG}^* and E_{MG}^* are also saturated in case of being outside of the allowed limits. These variables are inner generated in island mode ($\omega_{MG}^* = \omega_i^*$ and $E_{MG}^* = E_{MG}^*$) by the secondary control. When the grid is present, the synchronization process can start, and ω_{MG}^* and E_{MG}^* can be equal to those measured in the grid. Thus, the frequency and amplitude references of the MG will be the frequency and amplitude of the mains grid. After the synchronization, these signals can be given by the tertiary control (8.73) and (8.74). Notice that, depending on the sign of P_G^* and Q_G^*, the active and reactive power flows can be exported or imported independently.

Note that by making k_iP and k_iQ equal to zero, the tertiary control will act as a primary control of the MG, thus allowing the interconnection of multiple MGs, forming a cluster.

This control loop can also be used to improve the power quality (PQ) at the PCC. In order to achieve voltage-dip ride through, the MG must inject reactive power to the grid, thus achieving inner voltage stability. Islanding detection is also necessary to disconnect the MG from the grid and disconnect both the tertiary control references as well as the integral terms of the reactive power PI controllers, to avoid voltage instabilities.

Experimental results were done by using an MG laboratory with a DG generator working as VSI. The MG was able to operate in islanded mode as well as in grid-connected mode.

After the MG is completely synchronized, it can be connected to the grid and command P and Q. Since the MG is based on VSIs, if there is some non-planned grid disconnection, the MG can remain working as an island.

Figure 8.23 shows the experimental waveforms of the current P and Q injected to the grid. At 1.2 s, the MG was connected to the grid, and the tertiary control started. At this point, we changed the P^* and Q^* as follows. First, we started with $P^* = 1$ kW and $Q^* = 0$ var, injecting real power to the grid and achieving unity power factor. At $t = 4.8$ s, we changed the Q^* from 0 to -500 var, and hence, the MG was acting

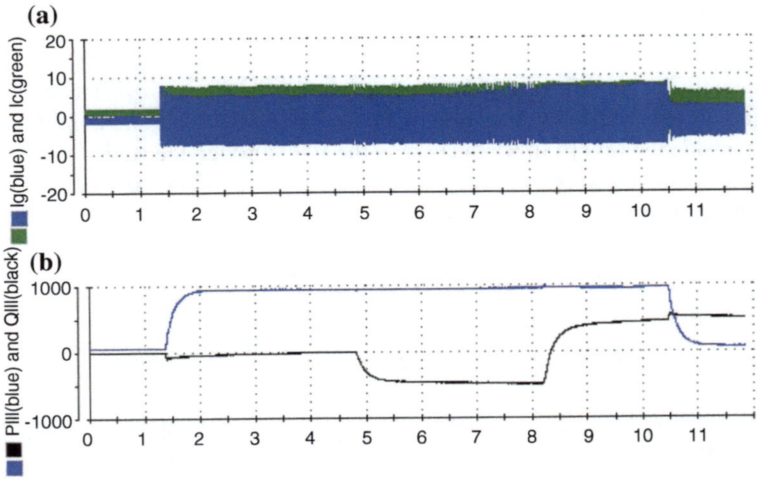

Fig. 8.23 Experimental results. **a** Currents at the PCC and at the MG load. **b** Active and reactive power injected to the grid

like a capacitor. Then, at 9.2 s, we suddenly changed Q^* from -500 to $+500$ var; this time, the MG was acting like an inductor. Finally, at $t = 10.5$ s, P^* was fixed to zero without changing the reactive power.

8.3.7 Simulation Results IV

Computer simulation results from a three-inverter MG are now presented. The inverters consisted of a full bridge with an LC filter, rated at 5 kVA. The LOC consisted of current and voltage loops, the P and Q calculations, and the droop control with a virtual output impedance of 50 μH. In this example, the AC MG multilevel control bandwidth of level 0 is 5 kHz for the voltage control loop and 20 kHz for the current control loop. The bandwidth for levels 1 and 2 are 30 Hz and 3 Hz, respectively. The selected control parameters are listed in Table 8.2.

Figure 8.24a, b shows, respectively, the active power and reactive sharing dynamics of the MG system. First, the three-inverter system connected to the grid was started up. The grid reference was fixed from 0 to 1 kW at $t = 2.5$ s by the tertiary control, while the three inverters gave 650 W each. At $t = 2.5$ s, a preplanned islanding scenario occurred, and the grid was disconnected from the MG, operating in autonomous (islanded) operation. Afterward, at $t = 5$ s, one inverter (DG#1) was disconnected suddenly from the grid, and inverters DG#2 and DG#3 provided the power to the local loads. At $t = 7.5$ s, inverter DG#2 was disconnected; thus, inverter DG#3 is supplying all the required power. Notice the flexible operation of the MG, which was able to operate both in grid-connected and in islanded modes.

Table 8.2 AC MG control-system parameters

Parameter	Symbol	Value	Units
Power stage			
Grid voltage	V_g	311	V
Grid frequency	f	50	Hz
Grid inductance	L_g	Ie-3	H
Grid resistance	R_g	1	Ω
Loss resistance inverter I	R_{liossl}	0.1	Ω
Loss resistance inverter II	R_{loss2}	0.11	Ω
Loss resistance inverter III	R_{loss3}	0.09	Ω
Inverter I inductance	L_I	50e-3	H
Inverter II inductance	L_2	55e-3	H
Inverter III inductance	L_3	45e-3	H
Inverter 1 resistance	r_l	0.1	Ω
Inverter II inductance	r_2	0.15	Ω
Inverter III inductance	r_3	0.05	Ω
Load	R_L	25	Ω
Primary control			
Derivative frequency droop	m_d	0.0001	W/rd
Proportional frequency droop	m_p	0.0015	Ws/rd
Proportional amplitude droop	n_p	0.001	VAr/V
Secondary control			
Proportional frequency droop	k_{pw}	1	Ws/rd
Integral frequency droop	k_{iw}	10	W/rd
Proportional amplitude droop	k_{pE}	1	VAr/V
Integral amplitude droop	k_{iE}	100	VAr's/V
Tertiary control			
Proportional phase term	k_{pP}	le-5	Ws/rd
Integral phase term	k_{iP}	0.1	W/rd
Proportional amplitude term	k_{pQ}	1	W/rd's
Integral amplitude term	k_{iQ}	100	VAr's/Y

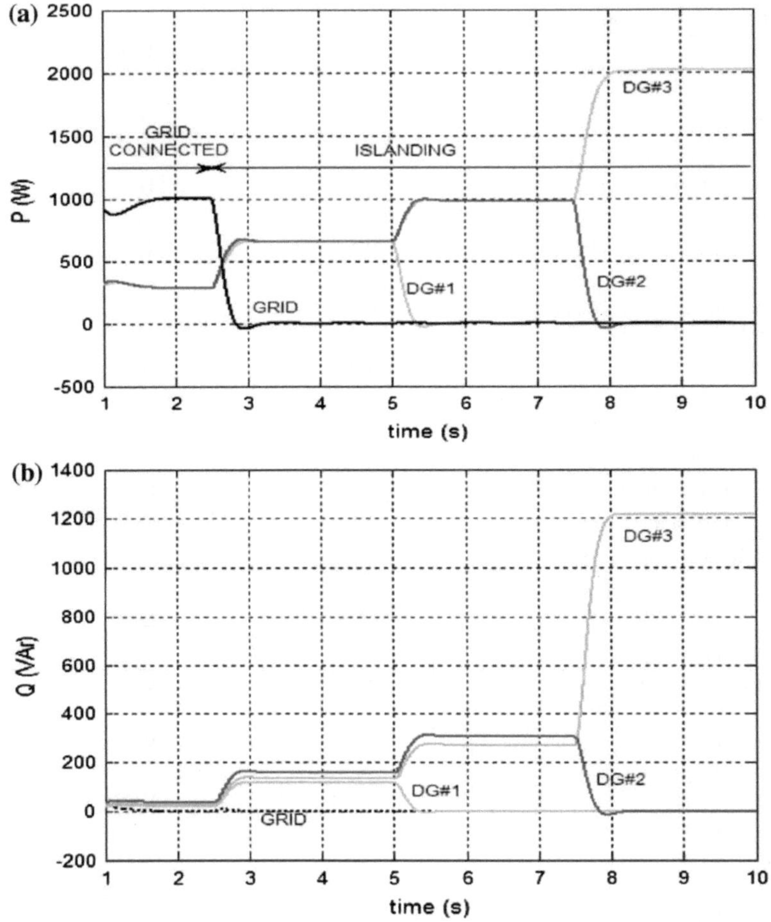

Fig. 8.24 Transient response of a two-inverter AC MG. **a** Active power and **b** reactive power

During the islanded mode, the primary control produces frequency and amplitude deviations, which can be compensated by the secondary control loops. Figure 8.25a, b shows, respectively, the frequency- and amplitude restoration action done by the secondary control. Note that these control loops avoid the inherent steady-state error produced by the primary control (detailed waveforms are shown in Fig. 8.26).

Figure 8.27 shows the transient response of the frequency and amplitude of the AC MG for a non-planned islanded scenario that occurred at $t = 2.5$ s. After 1 s, the islanding operation is detected, the tertiary control loops are disabled, and the frequency and amplitude references for the secondary control are self-generated.

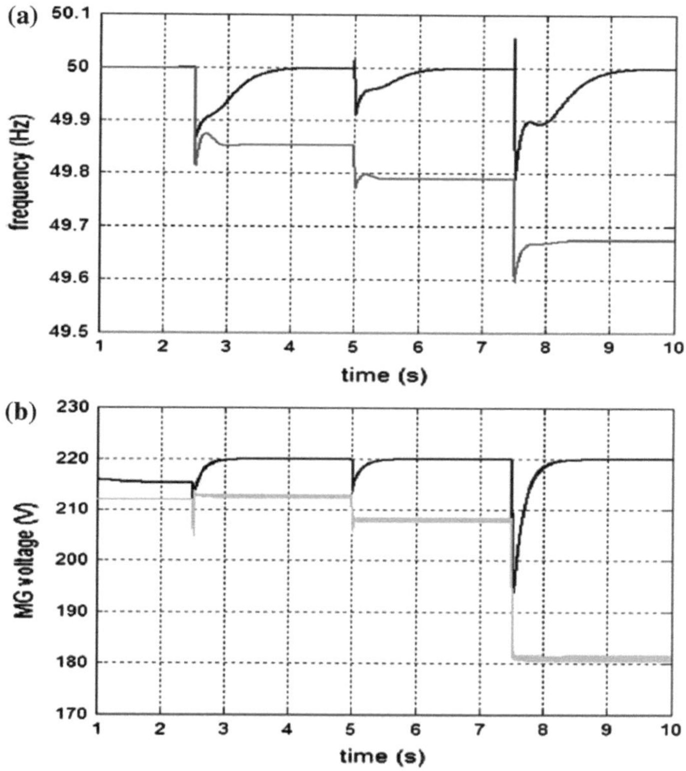

Fig. 8.25 **a** Frequency and **b** voltage rms transient response of the AC MG (*gray line*) without and (*black line*) with the secondary control

Fig. 8.26 Detail of the frequency and amplitude restorations

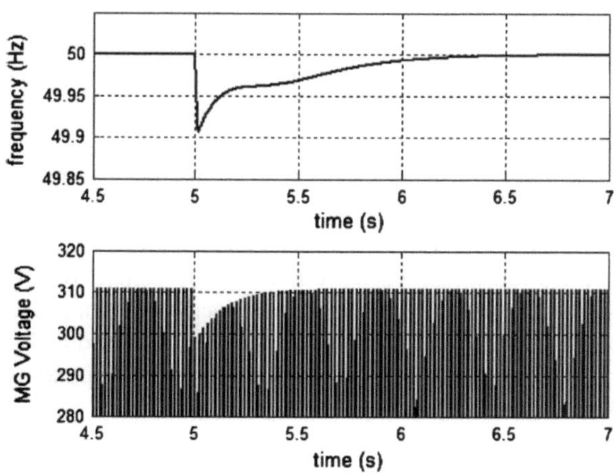

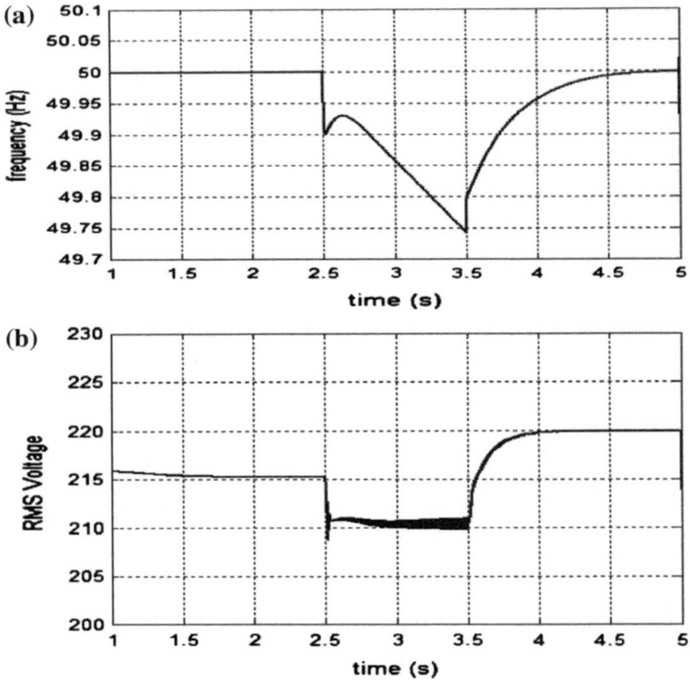

Fig. 8.27 Islanding detection for a non-planned islanded scenario. Transient response of the **a** frequency and **b** amplitude

Notice that the small transient frequency deviation can be used to detect that the AC MG is operating in islanded mode. This transient barely affects the performance of the MG system.

Figure 8.28 shows the active- and reactive power flow from the grid to the MG. In this case, the three inverters remained connected to the grid, and the tertiary control changed the reference of the active power while keeping the reactive power equal to zero in steady state.

Finally, Fig. 8.29 shows the P dynamics during different scenarios: At $t = 0$ s, the MG is in islanded mode; after the synchronization process, at $t = 5$ s, the MG is connected to the grid; and at $t = 10$ s, the reference of P is changed from zero to 1 kW. Figure 8.30 shows a detail of the voltage difference between the MG and the grid during the synchronization process, showing the action of the synchronization loop. Consequently, the operation during island and grid-connected modes, as well as the transitions between modes and its corresponding synchronization process, has been performed successfully.

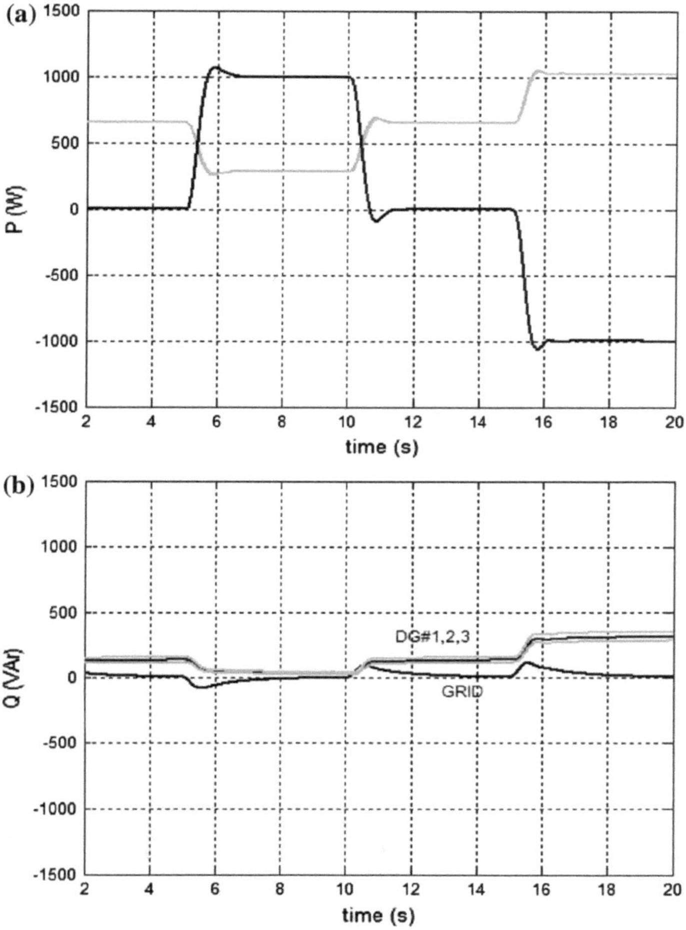

Fig. 8.28 Tertiary control: **a** active power and **b** reactive power

8.4 Multilevel Control of DC Microgrids

This section deals with the multilevel control of DC MGs consisting of the following three control levels, similarly conceived as with the AC MG. In this case, the control structure is easier to perform as follows. The primary control consists of a resistive virtual output-impedance loop, integrating the soft-start approach. The secondary control is based on an external common controller to restore the voltage deviation inside the DC MG. The tertiary control regulates the current flow from/to an external stiff DC source, which can be, for instance, a DC distribution system or DC/AC converter connected to the grid or an AC MG part.

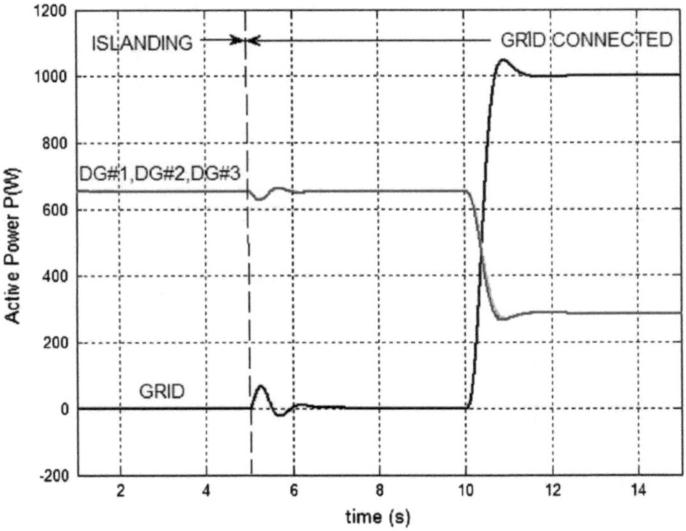

Fig. 8.29 Transfer process from islanding to grid-connected mode

Fig. 8.30 Voltage difference between the grid and the MG during the synchronization process

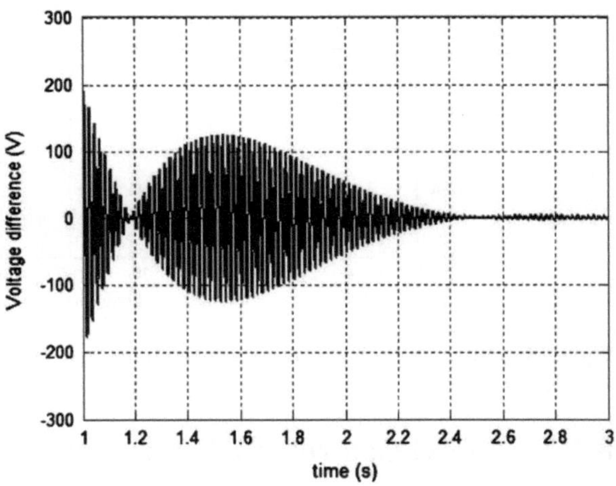

8.4.1 Primary Control

Figure 8.31 shows the equivalent circuit of two DC power supplies connected in parallel sharing a common load through resistive output impedances. If there is some voltage difference, this will circulate a current between both DC sources. In order to reduce the circulating current, we can program virtual output impedances by using a primary control. This control level adjusts the voltage reference provided

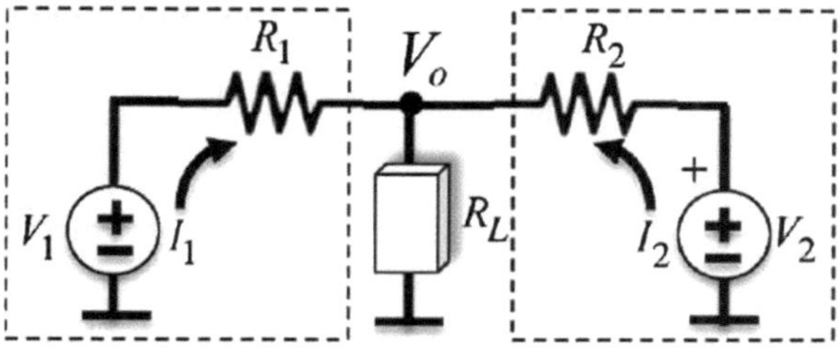

Fig. 8.31 Equivalent circuit of two parallel-connected DC power supplies

to the inner current and voltage control loops (level 0). It includes the virtual output impedance loop in which the output voltage can be expressed as [41]:

$$v_o^* = v_{ref} - R_D i_o \tag{8.75}$$

where i_o is the output current, R_D is the virtual output impedance, and v_{ref} is the output voltage reference at no load. Assuming that ε_v is the maximum allowed voltage deviation, R_D and v_{ref} must be designed as follows:

$$v_{ref} = v_n - \varepsilon_v/2 \tag{8.76}$$

$$R_D = \varepsilon_v/i_{\max} \tag{8.77}$$

with v_n being the nominal output voltage and $i_{\max}$ as the maximum output current.

This control loop provides resistive output impedance to the power converters to compensate for the difference between the voltage references $\Delta v_o^* = v_{o1}^* - v_{o2}^*$. Thus, the current sharing between the two converters $\Delta i_o = i_{o1} - i_{o2}$ (see Fig. 8.28) can be expressed as follows:

$$\Delta i_o = \Delta v_o^*/R_D. \tag{8.78}$$

This control loop not only allows parallel operation of the converters, but also improves the dynamic performance of the output voltage. However, it has the inherent load-dependent voltage deviation.

8.4.2 Secondary Control

Now to solve the problem of the voltage deviation, a secondary control is proposed. The voltage level in the MG v MG is sensed and compared with the voltage reference v_{MG}^*, and the error processed through a compensator is sent to all the units δv_o to restore the output voltage, see Fig. 8.32. The controller can be expressed as follows:

$$\delta v_o = k_p(v_{MG}^* - v_{MG}) + k_i \int (v_{MG}^* - v_{MG})dt \tag{8.79}$$

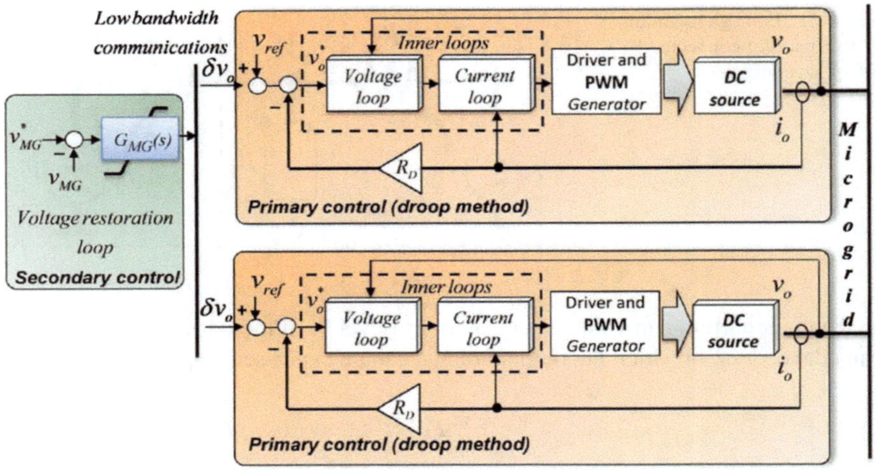

Fig. 8.32 Primary and secondary controls of a DC MG

k_p and k_i being the control parameters of the secondary control compensator. Notice that δv_o must be limited in order not to exceed the maximum voltage deviation. Finally, (8.75) becomes

$$v_o^* = v_{ref} + \delta v_o - R_D i_o. \tag{8.80}$$

In order to connect the MG to a DC stiff source, first, we have to measure the voltage of this stiff source, and that will be the reference of the secondary control loop. After the transient, the MG can be connected to the DC stiff source through the static bypass switch. At that moment, the MG does not have any exchange of power with the external DC source.

8.4.3 Tertiary Control

Once the MG is connected to the DC source, the power flow can be controlled by changing the voltage inside the MG. As can be seen in Fig. 8.33, by measuring the current (or the power) through the static bypass switch i_G, it can be compared with the desired positive or negative current i_G^* (or power), depending whether we want to import or export energy.

The controller can be expressed as follows:

$$\delta v_o = k_p^{'}(i_G^* - i_G) + k_i^{'} \int (i_G^* - i_G)dt \tag{8.81}$$

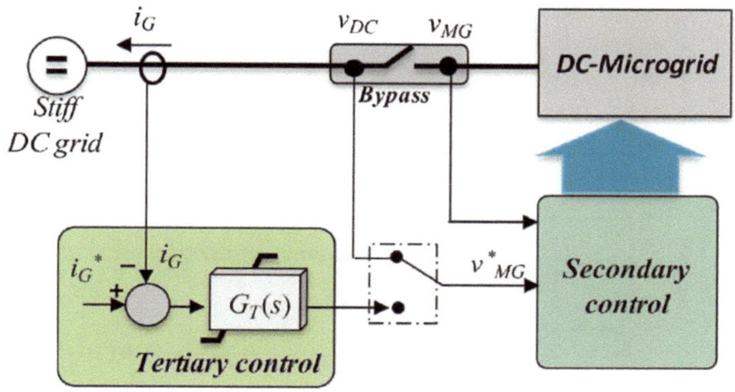

Fig. 8.33 Tertiary control and synchronization loop of a DC MG

where $k_p^{'}$ and $k_i^{'}$ are the control parameters of the tertiary control compensator. Here, δv_o is also saturated in case of being outside the limits of $\pm\varepsilon_v$. Notice that, depending on the sign of I_G^*, the power flow can be exported ($I_G^* > 0$) or imported ($I_G^* < 0$).

8.4.4 Simulation Results V

For simulation purposes, the virtual impedance R_D is set at 2 Ω while the proportional and integral terms are selected as $k_p = 1,\ k_i = 5000\,\mathrm{s}^{-1},\ k_{p'} = 10V/A,\ k_{i'} = 3000\ V/As$. Simulation results of two DC/DC buck converters connected in parallel, forming a DC-MG, are presented in order to show the feasibility of the proposed multilevel control. The MG voltage was selected at 400 V, the control was implemented by using voltage and current loops, and the virtual output impedance was rated at 2 Ω.

Figure 8.34 shows a set of waveforms derived from the use of the different levels of the proposed control scheme. In all of them, at 8 ms, the primary control was activated, and at 15 ms, there is a sudden step of the load. Figure 8.34a shows the voltage droop caused by the primary control. Notice that an increase of the current produces higher voltage droop.

Figure 8.34b shows the effect of the secondary control, which compensates for the voltage deviations caused by the primary control. Finally, Fig. 8.34c shows the action of the tertiary control, which regulates the current flow to/from a stiff DC source connected to the MG. Note that at 17 ms, a reference of 20 A is imposed by this controller, producing a voltage deviation inside the voltage reference to inject the desired current to the DC stiff grid (Fig. 8.33).

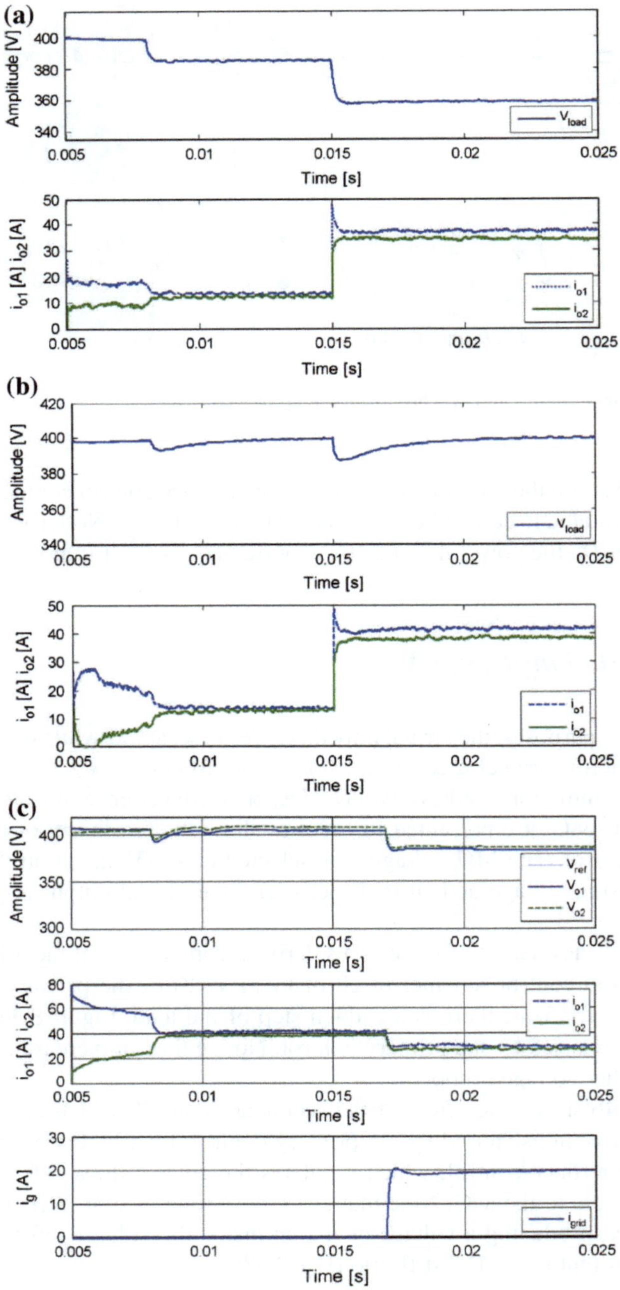

Fig. 8.34 Behavior of the voltage and currents of the DC MG: **a** primary, **b** secondary, and **c** tertiary control

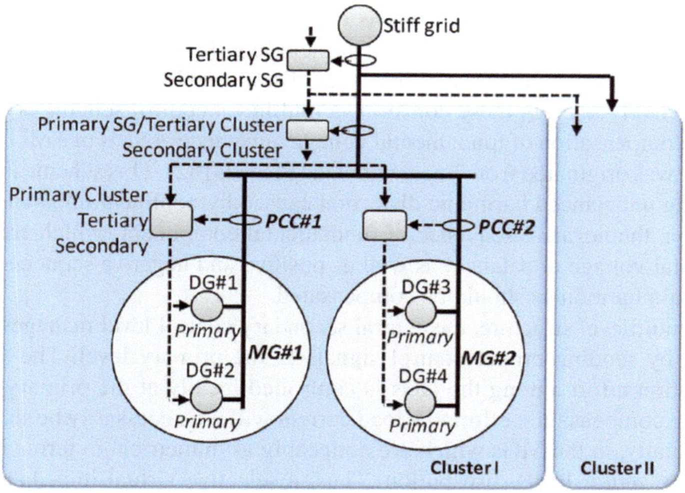

Fig. 8.35 Multiple MGs forming an SG configuration

8.5 Enhanced Compensation Technique

Some recent approaches are being developed to control the DG interface converter aiming to compensate PQ problems. A single-phase DG which injects harmonic current to compensate voltage harmonics is presented. However, in the case of sever harmonic distortion, a large amount of the interface converter capacity is used for compensation and it may interfere with the power supply by the DG. Harmonic compensation approaches of [34, 67, 89] are essentially based on making the DG units of a power distribution system emulate a resistance at harmonic frequencies. A method for compensation of voltage harmonics in an islanded MG was presented in [48] based on the resistance emulation and applies a harmonic power droop characteristic in order to share the compensation effort among DGs.

The approach presented in [10] is based on controlling each DG unit of a MG as a negative sequence conductance to compensate voltage unbalance. The conductance reference is determined by applying a droop characteristic which uses negative sequence reactive power to provide the compensation effort sharing. The control system of [10] is implemented in d_q (synchronous) reference frame while [81] addresses the voltage unbalance compensation using $\alpha\beta$ (stationary) frame control.

The control method presented in [52] is based on using a two-inverter interface converter (one connected in shunt and the other in series with the grid) in order to control power flow and also to compensate the voltage unbalance. This two-inverter structure can be unattractive considering the cost and volume of the DG interface converter. In addition, it should be noted that the methods presented in [10, 34, 48, 67, 81, 89] are designed to enhance voltage quality at the DG terminal while the PQ at the "Sensitive Load Bus (SLB)" is an important concern in MGs.

Furthermore, if the voltage distortion is compensated locally, it may be amplified in the other buses of the electrical system including the SLB. This phenomenon is called whack-a-mole in the case of harmonic distortion [101–103].

To effectively address these concerns, a multilevel control scheme is developed for direct compensation of fundamental voltage unbalance at SLB of a MG where the unbalance was originated from linear unbalanced loads [82]. This scheme is extended considering unbalanced harmonic distortion caused by nonlinear unbalanced loads. In this case, the negative sequence of fundamental component (which creates SLB fundamental voltage unbalance) as well as positive and negative sequences of SLB voltage main harmonics should be compensated.

In the multilevel structure, the central secondary control level manages the compensation by sending proper control signals to the primary level. The sharing of compensation effort among the DGs is controlled locally at the primary level. By sharing the compensation effort, the load current will not necessarily be shared properly, especially, in the MGs which are noticeably asymmetrical in terms of the line impedances and/or loads distribution. Thus, a selective virtual impedance loop is proposed for each DG unit to improve the load sharing.

8.5.1 Microgrid Multilevel Control Scheme

Figure 8.36 shows application of the proposed multilevel control scheme to a general MG which includes a number of electronically interfaced DGs connected to the

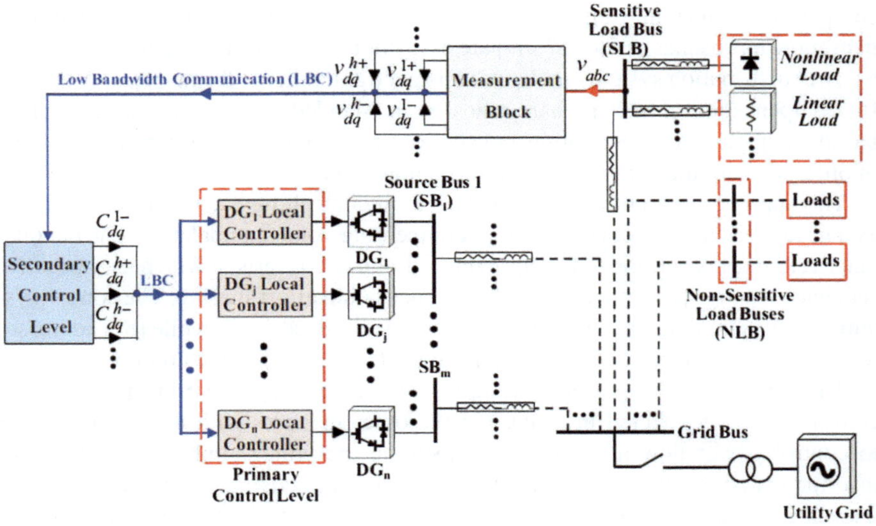

Fig. 8.36 Multilevel control scheme for a general microgrid

Source Buses (SB). Each DG unit may consist of power generators and/or energy-storage systems. Furthermore, some dispersed loads including balanced/unbalanced linear and nonlinear loads are connected to the Load Buses including SLB and Non-Sensitive Load Buses (NLB). The multilevel scheme consists of two control levels. The primary level comprises DG local controllers and the secondary level is a CC which sends proper reference signals to each of the DGs in order to reduce the voltage unbalance and harmonic distortion at the microgrid SLB to the required level.

The secondary controller can be far from DGs and SLB. Thus, as shown in Fig. 8.36, SLB voltage information is sent to this controller by means of low bandwidth communication (LBC). Low bandwidth is applied to avoid dependence on the availability of a high bandwidth which may endanger the system reliability. In order to ensure LBC adequacy, positive and negative sequences of SLB voltage fundamental and main harmonic components are extracted in d_q frame and the resultant DC values are transmitted to the secondary controller. In Fig. 8.36, superscripts "+", "−", "1" and "h" represent positive sequence, negative sequence, fundamental component, and *h*th harmonic component, respectively. For instance, v_{dq}^{h+} is the positive sequence of *h*th harmonic voltage in d_q frame. The details of voltage components extraction are depicted in "Measurement Block" of Fig. 8.37. As seen, in order to extract v_{dq}^{1+}, v_{dq}^{1-}, v_{dq}^{h+}, and v_{dq}^{h-}, at first, the measured three-phase voltage of SLB (v_{abc}) is transformed to d_q reference frames rotating at ω, $-\omega$, $h\omega$ and $-h\omega$, respectively. ω is the system angular frequency estimated by a PLL [42]. Afterwords, three second-order 5 Hz low pass filters (LPF) are applied. The second-order filters are used since the first-order ones were not able to provide acceptable performance.

On the other hand, as shown in Figs. 8.36 and 8.37, compensation references for fundamental component unbalance and *h*th harmonic positive and negative sequences

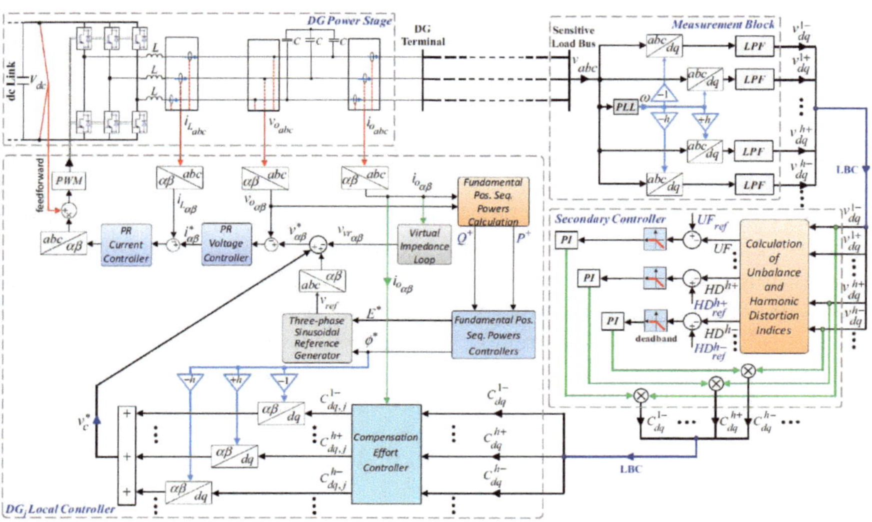

Fig. 8.37 Detailed block diagram of the control system

(C_{dq}^{1-}, C_{dq}^{h+}, and C_{dq}^{h-}, respectively) which are also in d_q frame are generated by the secondary controller and sent to the DGs local controllers using LBC. As shown in Fig. 8.37, these references are fed to Compensation Effort Controller block of each DG local controller and the outputs ($C_{dq,j}^{1-}$, $C_{dq,j}^{h+}$ and, $C_{dq,j}^{h-}$ as the compensation references for DG_j) are transformed to $\alpha\beta$ frame, added together and injected as a reference for the DG voltage controller. The rotation angles of transformations are set to $-\phi^*, h\phi^*$, and $-h\phi^*$ for the compensation references of voltage unbalance and positive and negative sequences of hth voltage harmonic, respectively. ϕ^* is the DG voltage reference phase angle generated by the active power controller [82].

It should be noted that the harmonic and also unbalance variations in the practical grids are usually slow [49]. Thus, it is not necessary to provide a very fast control action. This fact confirms the sufficiency of LBC for the proposed compensation approach. On the other hand, low communication bandwidth can be provided at a relatively low cost.

The total time needed to transmit the measured and control signals between control levels and to perform the required control actions depends on the data size, bandwidth, and delay of communication and also the processing time of the measurement and control devices [8]. Figure 8.36 is depicted assuming one SLB in the MG. However, in the case of multiple SLBs, the MG can be divided to some control regions that each has a dedicated secondary controller and includes one SLB. Each secondary controller manages the voltage quality enhancement at the corresponding SLB by sending control signal to the DG units of the region. It is also noteworthy that the proposed control approach is applicable for voltage quality enhancement in both grid-connected and islanded modes of MG operation. It is only necessary to make the necessary changes to the power controllers of DG units depending on the operation mode.

8.5.2 DG Local Control

The structure of each DG power stage and LOC is shown in Fig. 8.37. As it can be seen, a feedforward loop may be included to consider small variations of DC link voltage (V_{dc}). The local control of DGs is performed in $\alpha\beta$ reference frame. As shown in "DG_j local controller" block, the reference of the DG output voltage in $\alpha\beta$ frame ($v_{\alpha\beta}^*$) is provided by power controllers, virtual impedance loop, and compensation effort controller. Then, according to $v_{\alpha\beta}^*$ and the instantaneous measured output voltage ($v_{0_{\alpha\beta}}$), the reference current ($i_{\alpha\beta}^*$) is generated. On the other hand, LC filter inductor current is measured, transformed to $\alpha\beta$ frame ($i_{L_{\alpha\beta}}$), and controlled by the current controller to provide voltage reference for pulse-width modulator (PWM).

8.5.3 Fundamental Positive Sequence Powers Controllers

Control of the active and reactive powers is performed assuming a mainly inductive MG. The power controllers determine the reference values of DGs output voltage phase angle and amplitude (ϕ^* and E^*, respectively) considering the operation mode of MG, i.e., islanded or grid-connected. Design of the power controllers is sufficiently studied in the literature, [12, 13] and its references.

8.5.4 Voltage and Current Controllers

The following proportional-resonant (PR) voltage and current controllers are applied in this paper.

$$G_V(s) = k_{pV} + \sum_{k=1,3,5,7} \frac{2k_{rVk}\omega_{cV}s}{s^2 + 2\omega_{cV}s + (k.\omega_0)^2} \tag{8.82}$$

$$G_I(s) = k_{pI} + \sum_{k=1,3,5,7} \frac{2k_{rIk}.\omega_{cI}.s}{s^2 + 2\omega_{cI}s + (k\omega_0)^2} \tag{8.83}$$

where k_{pV} (k_{pI}) and k_{rVk}(k_{rIk}) are the proportional and kth harmonic (including fundamental component as the first harmonic) resonant coefficients of the voltage (current) controller, respectively. ω_{cV} and ω_{cI} represent voltage and current controllers cutoff frequencies, respectively.

In order to design these controllers, the closed-loop transfer function of local control system is extracted [81]. Bode diagram of this transfer function using the power stage and LOC parameters (Tables 8.3 and 8.4) is depicted in Fig. 8.38. As seen, the gain and phase angle at the resonant frequencies are fixed at unity and zero, respectively. Thus, proper tracking of the voltage reference is ensured. The output impedance of the interface inverter can be expressed as $Z_o(s) = Z_o'(s) + Z_{vr}(s)$ where $Z_{vr}(s)$ represents the virtual impedance and $Z_0'(s)$ is the inverter output impedance without addition of the virtual impedance [81]. Figure 8.39 shows $Z_o'(s)$ magnitude plot of DG units. It can be observed that the magnitude is approximately zero at fundamental and 3rd, 5th, and 7th harmonic frequencies.

8.5.5 Virtual Impedance Loop

The block diagram of the virtual impedance is depicted in Fig. 8.40 where R_{vr}^{1+}, R_{vr}^{1-} and R_{vr}^{h} represent the virtual resistance for fundamental positive sequence, fundamental negative sequence, and hth harmonic (both positive and negative sequences) components of DG output current, respectively. L_{vr} and ω_0 are respectively the virtual

Table 8.3 Power stage parameters

Dc link voltage	LC filter inductance	LC filter capacitance	Nonlinear loads tie lines
V_{dc} (V)	L (mH)	$C(\mu F)$	$Z(\Omega, mH)$
650	1.8	25	0.1, 1.8
Distribution lines	Nonlinear loads		Linear load
$Z_1, Z_2, Z_3(\Omega, mH)$	$C_{NL}/R_{NL1}, R_{NL2}/L_{NL}(\mu F)/(\Omega)/(mH)$		$Z_L(\Omega, mH)$
0.1, 1.8	235/50, 200/0.084		50, 20

Table 8.4 DGS local controller parameters

Power controllers $(DG1/DG2)$ [82]					
m_D (rad/W)		m_P (rad/W.s)		n_P (V/VAr)	
$0.6 \times 10^{-5}/1.2 \times 10^{-5}$		$0.6 \times 10^{-4}/1.2 \times 10^{-4}$		$1.4 \times 10^{-1}/2.8 \times 10^{-1}$	
Virtual impedance $(DG1/DG2)$					
$R_{vr}^{1+}(\Omega)$	L_{vr} (mH)	$R_{vr}^{1-}(\Omega)$	$R_{vr}^{3}(\Omega)$	$R_{vr}^{+}(\Omega)$	$R_{vr}^{7}(\Omega)$
0.3/0.6	2.5/5	1.5/3	2/4	4/8	4/8
Voltage/current controller					
k_{pV}/k_{pI}	k_{rV1}/k_{rI1}	k_{rV3}/k_{rI3}	k_{rV5}/k_{rI5}	k_{rV7}/k_{rI7}	ω_{cV}/ω_{cI} (rad/s)
1/5	100/1000	200/400	50/100	100/100	2/2

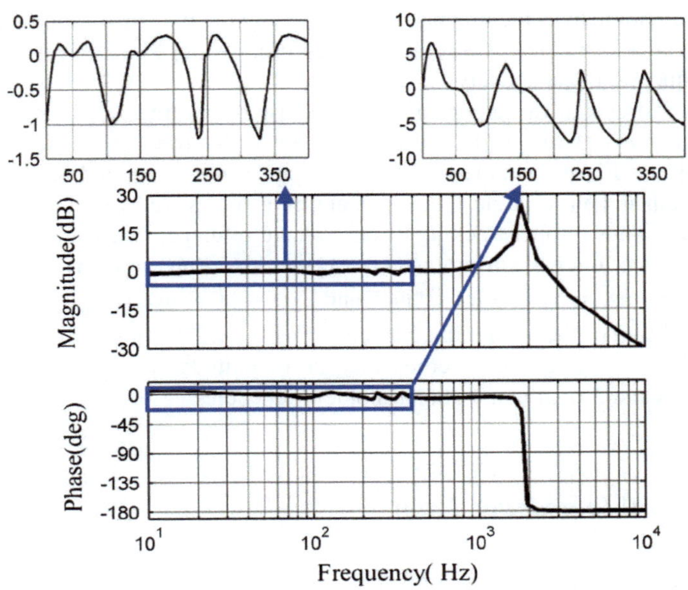

Fig. 8.38 Bode diagram of closed-loop transfer function

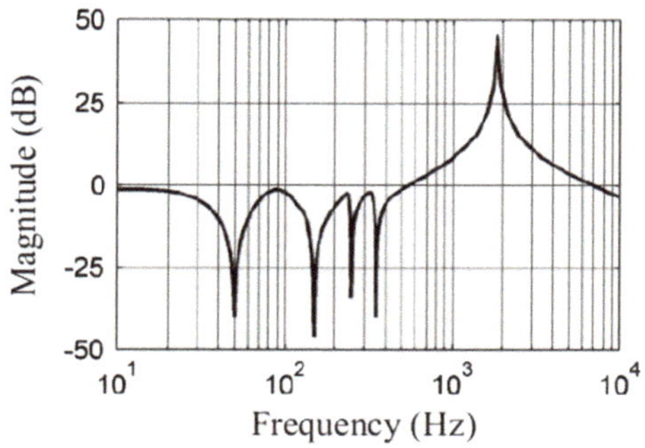

Fig. 8.39 Magnitude of $Z_0'(s)$

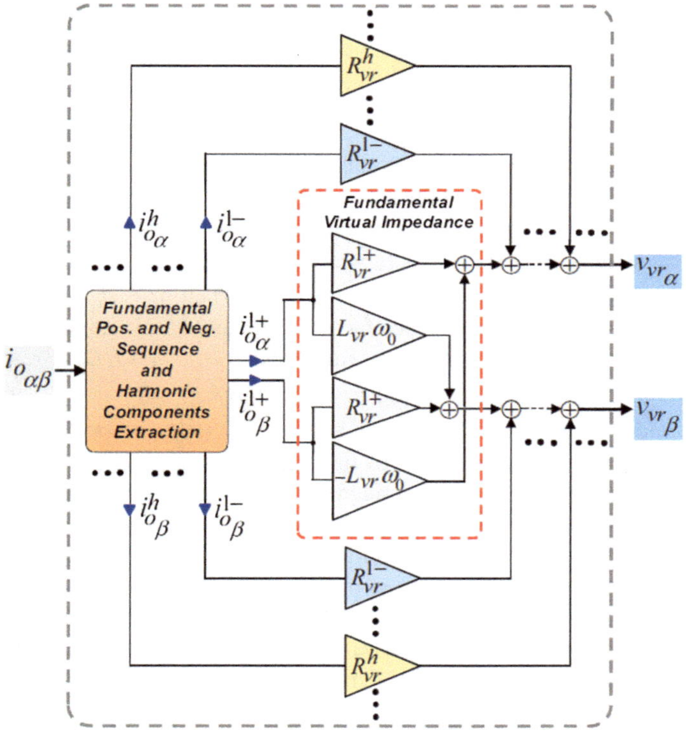

Fig. 8.40 Block diagram of selective virtual impedance

inductance against fundamental positive sequence current and the rated frequency. In order to provide proper control of fundamental positive sequence powers, the MG is made more inductive by including L_{vr}. However, a small R_{vr}^{1+} is added to damp the system oscillations [26].

The basic structure of the fundamental frequency virtual impedance has been proposed in [33]. Moreover, it is demonstrated in [5, 11, 26] that the virtual impedance can improve the sharing of nonlinear (harmonic) load among parallel converters. Hence, the basic structure is extended by including virtual resistances for the fundamental negative sequence (R_{vr}^{1-}) and the main harmonic components (R_{vr}^{h}, $h = 3, 5$ and 7) of the DG output current in order to improve the sharing of these current components. Output current components are extracted according to [73, 74].

The sharing improvement is achieved at the expense of distorting DGs output voltage as a result of voltage drop on the virtual resistances. Thus, for selection of virtual resistance values, a tradeoff should be considered between the amount of output voltage distortion and sharing accuracy. Furthermore, considering double rating of $DG1$ in the studied MG, its virtual impedances at fundamental and harmonic frequencies are set at half value to improve the current sharing (see Table 8.4).

8.5.6 Compensation Effort Controller

The compensation effort controller manages the sharing of compensation workload among the MG DGs. The block diagram of DG_j effort controller is shown in Fig. 8.41. As seen, DG unit output current in $\alpha\beta$ frame ($i_{o_{\alpha\beta}}$) is fed to this controller and positive and negative sequences of its α-axis fundamental component ($i_{o_\alpha}^{1+}$ and $i_{o_\alpha}^{1-}$.) and hth harmonic component ($i_{o_\alpha}^{h+}$ and $i_{o_\alpha}^{h-}$) are extracted. Then, $i_{o_\alpha}^{1+}$, $i_{o_\alpha}^{1-}$, $i_{o_\alpha}^{h+}$ and $i_{o_\alpha}^{h-}$ are applied to calculate current unbalance factor (UF_I) and harmonic distortion indices of hth harmonic positive and negative sequences (HD_I^{h+} and HD_I^{h-}, respectively). UF_I, HD_I^{h+} and HD_I^{h-} are calculated as the ratio of $i_{o_\alpha}^{1-}$, $i_{o_\alpha}^{h+}$ and $i_{o_\alpha}^{h-}$ rms values ($I_{o_\alpha}^{1-}$, $I_{o_\alpha}^{h+}$ and $I_{o_\alpha}^{h-}$, respectively) to rms value of $i_{o_\alpha}^{1+}(i_{o_\alpha}^{1+})$, respectively. Note that using β-components for calculation of unbalance and harmonic distortion indices leads to the same results because the magnitude of α- and β-components is equal for both positive and negative sequences of fundamental and harmonic components.

Finally, the references for compensation of fundamental unbalance and hth harmonic positive and negative sequences by DG_j ($C_{dq,j}^{1-}$, $C_{dq,j}^{h+}$and $C_{dq,j}^{h-}$, respectively) are calculated as shown in Fig. 8.41 where $S_{0,j}$ is the rated capacity of DG_j and subscript "max" represent the maximum value. By multiplying the ratio of DG_j rated capacity ($S_{0,j}$) to the total capacity of the microgrid DGs ($\sum_{l=1}^{n} S_{0,l}$), compensation effort of each DG will be proportional to its rated capacity. UF_I, HD_I^{h+} and HD_I^{h-} can be considered as the indices of compensation effort because as shown in the simulation results, compensation of SLB voltage unbalance and hth harmonic positive and negative sequences is achieved through injecting corresponding current components by the DGs. Thus, the terms ($UF_{I,\max} - UF_I$), ($HD_{I,\max}^{h+} - HD_I^{h+}$) and ($HD_{I,\max}^{h-} - HD_I^{h-}$) in Fig. 8.41 contribute toward sharing of compensation effort. In fact, increase of each component compensation effort leads to the increase of corresponding index. Consequently, ($UF_{I,\max} - UF_I$), ($HD_{I,\max}^{h+} - HD_I^{h+}$), or

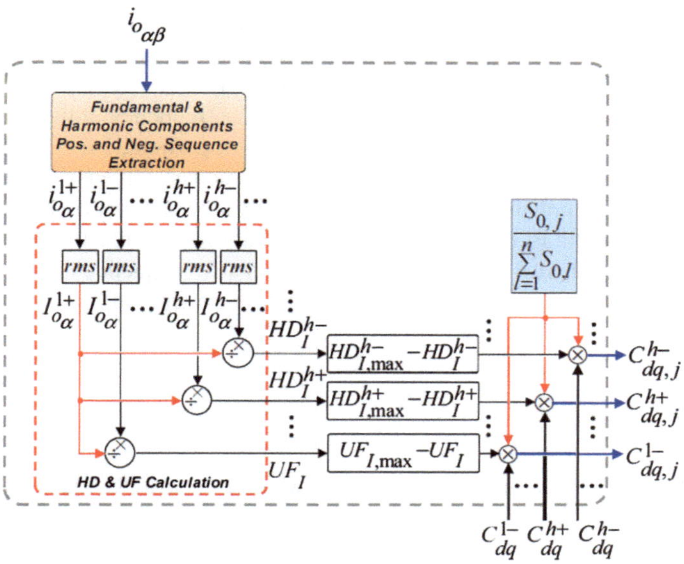

Fig. 8.41 Block diagram of compensation effort controller

$(HD_{I,\max}^{h-} - HD_I^{h-})$ decrease and it leads to compensation effort decrease. So, inherent negative feedback exists in the effort controller. It is assumed that the maximum values of unbalance factor and harmonic distortion indices are unity. This assumption is valid for most of the practical cases; however, larger constants can be used as the maximum values.

8.5.7 Secondary Controller

The block diagram of the secondary controller is also shown in Fig. 8.37. As seen, d_q components of SLB voltage fundamental positive and negative sequences (v_{dq}^{1+} and v_{dq}^{1-}) and hth harmonic positive and negative sequences (v_{dq}^{h+} and v_{dq}^{h-}) are used to calculate voltage unbalance factor (UF) and hth harmonic positive and negative sequence distortion indices (HD^{h+} and HD^{h-}, respectively). Calculation block is similar to "HD & UF Calculation" block of Fig. 8.41. Then, UF , HD^{h+}, and HD^{h-} are compared with the reference values (UF_{ref}, HD_{ref}^{h+}, and HD_{ref}^{h-}, respectively) and the errors are fed to PI controllers. Afterwords, the outputs of these controllers are multiplied by v_{dq}^{1-}, v_{dq}^{h+}, and v_{dq}^{h-} to generate C_{dq}^{1-}, C_{dq}^{h+}, and v_{dq}^{h-}, respectively. If the unbalance factor or any of the harmonic distortion indices are less than the reference value, the respective dead-band block prevents the increase of the distortion by the PI controller.

It is well known that with the increase of proportional coefficient of PI controllers, the response time is reduced, but, the control system becomes more prone to instability. On the other hand, in order to minimize the effect of PI controllers phase lag on the compensation performance, the corner angular frequency of these controllers which can be calculated as the ratio of integral to proportional coefficients, should be set at one decade or more below the frequency of undercompensation component [49]. Harmonic and unbalance variations are usually slow; thus, it is not necessary to apply high bandwidth PI controllers. Here, secondary level comprises PI controllers for compensation of SLB voltage fundamental negative sequence and 3rd, 5th and 7th harmonic components. The parameters of PI controllers are listed in Table 8.5.

8.5.8 Simulation Results VI

Figure 8.42 shows the simulation test system which is a two-DG islanded MG comprising two SB, one SLB and one NLB. A diode rectifier and a star-connected linear load are connected to SLB. It is assumed that one phase of nonlinear load is disconnected to create unbalanced voltage distortion. Furthermore, a balanced nonlinear load is connected to NLB. Switching frequency of the DGs inverters is 10 kHz. The test system parameters are listed in Table 8.3. Note that in this Table, the impedances of linear load and lines are presented in terms of resistance (Ω) and inductance (mH). Simulations are performed using MATLAB/Simulink®. Three simulation steps are considered:

- **Step 1** ($0 \leq t < 2$ s):
  DGs operate only with fundamental positive sequence virtual impedance and secondary control is not acting.
- **Step 2** ($2 \leq t < 4$ s):
  Virtual resistances for fundamental negative sequence and harmonic components are added.
- **Step 3** ($4 \leq t < 7$ s):
  Secondary control is activated. The reference values of unbalance factor and harmonic distortion indices are 0.2 %.

Table 8.5 Secondary PI controller parameters

Fundamental negative sequence		3rd Harmonic positive and negative sequences	
Proportional	Integral	Proportional	Integral
3	90	1.25	110
5th Harmonic positive and negative sequences		7th Harmonic positive and negative sequences	
Proportional	Integral	Proportional	Integral
1	150	0.95	200

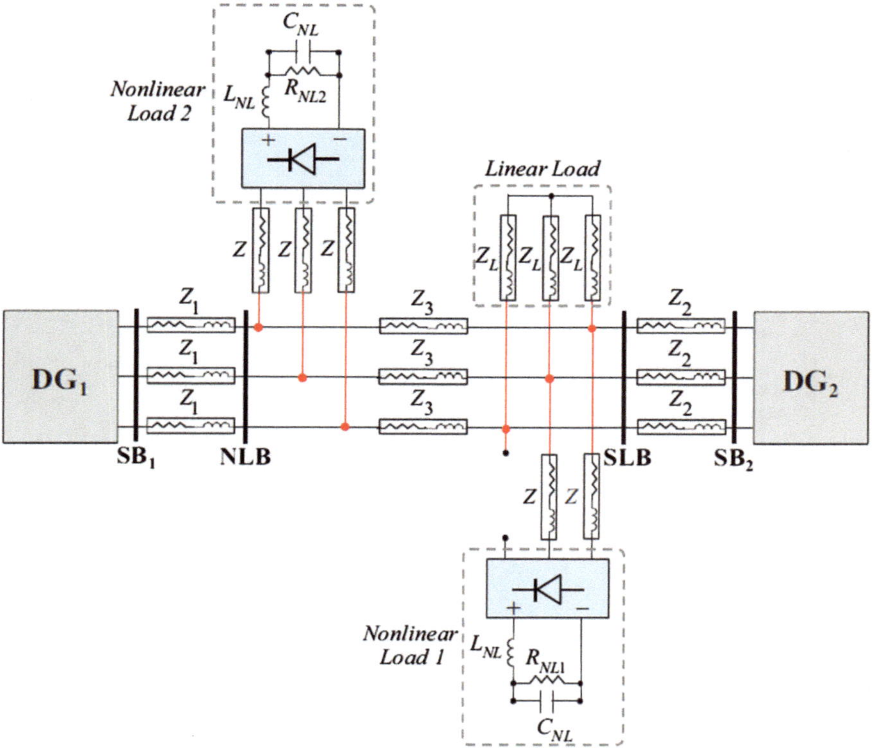

Fig. 8.42 Test system for simulation studies

As seen in Table 8.6, before activating the virtual resistances for fundamental negative sequence and harmonic components, DGs output voltages are approximately free of distortion. This fact can also be observed in Fig. 8.43 as the low values of UF and HD^{3-} before $t = 2$ s. It demonstrates the effectiveness of LOCs in tracking the voltage reference. But, as shown in Table 8.6 and Fig. 8.43, SLB voltage is distorted noticeably due to voltage drops on distribution lines. It should be noted that in order to avoid excessive length, simulation results of other distortion indices are not included.

Table 8.7 shows negative sequence single-phase waveforms at fundamental and 3rd harmonic frequencies as well as three-phase waveforms of DGs output current in different simulation steps. Considering double rating of $DG1$, it can be noticed from Table 8.7 that the load current is not properly shared in the first simulation step. In fact, all components of the load current except fundamental positive sequence one are shared according to the test system topology and before adding R_{vr}^{1-}, R_{vr}^{3}, R_{vr}^{5} and R_{vr}^{7}, $DG2$ will supply larger portions of the fundamental negative sequence and the harmonic currents. As mentioned before, fundamental positive sequence component of the load is shared by using droop controllers. Figure 8.44 demonstrates the

Table 8.6 Some voltage waveforms

	Step1	Step 2	Step 3
SB_1	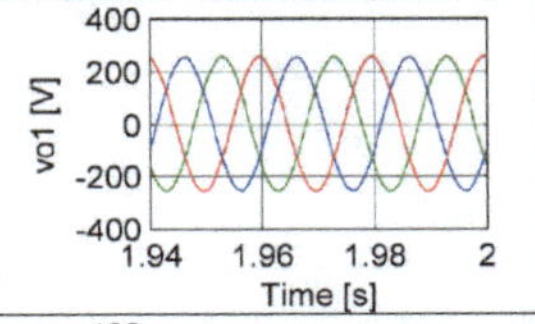	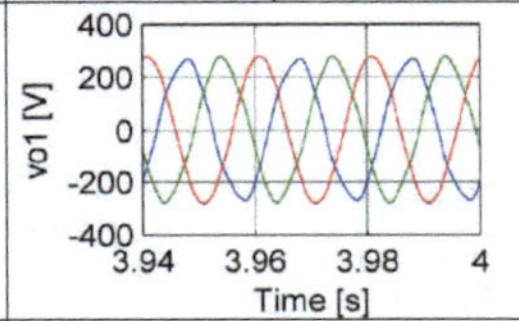	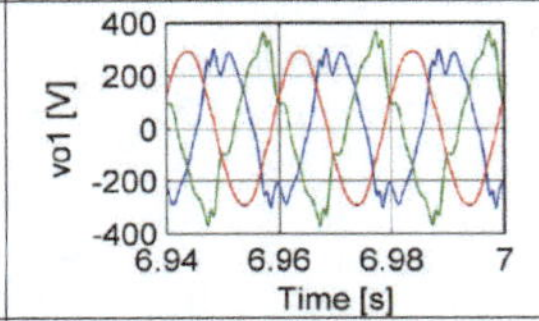
SB_2	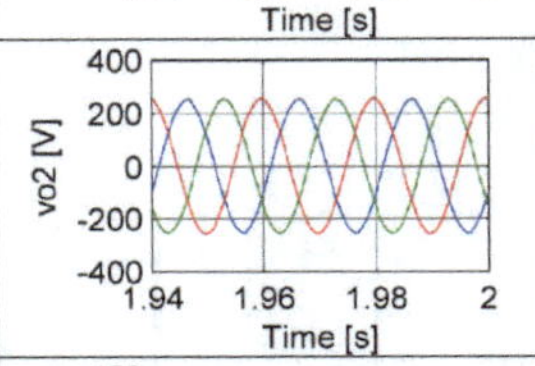	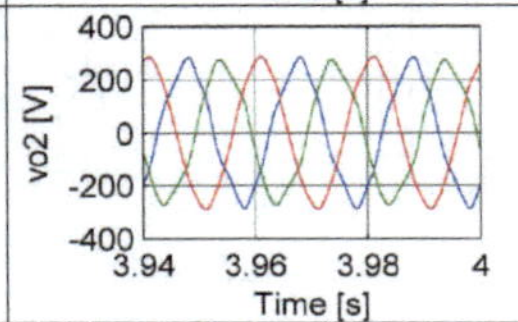	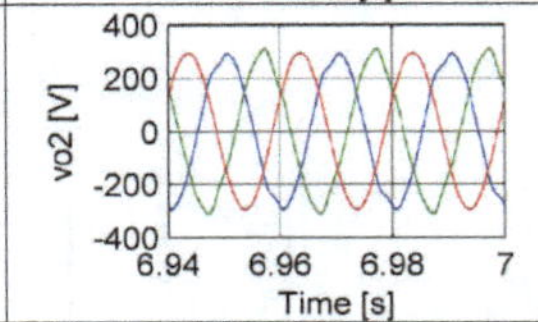
SLB	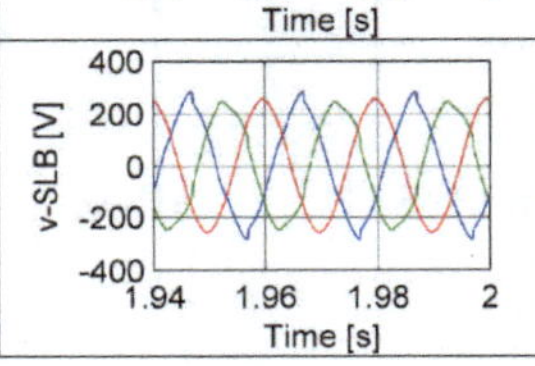	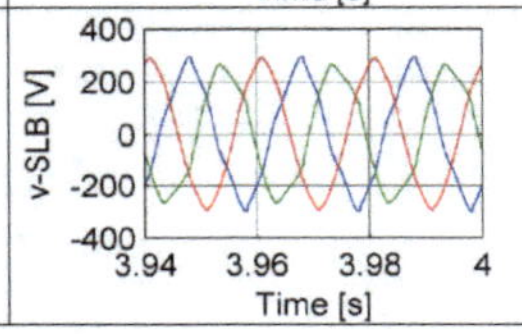	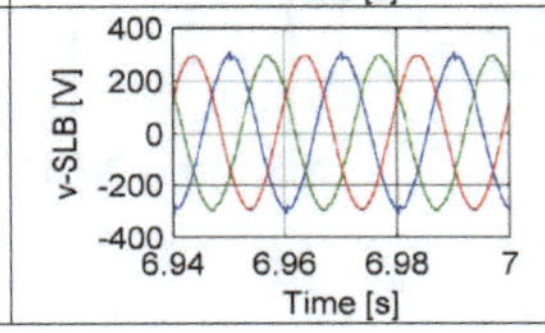

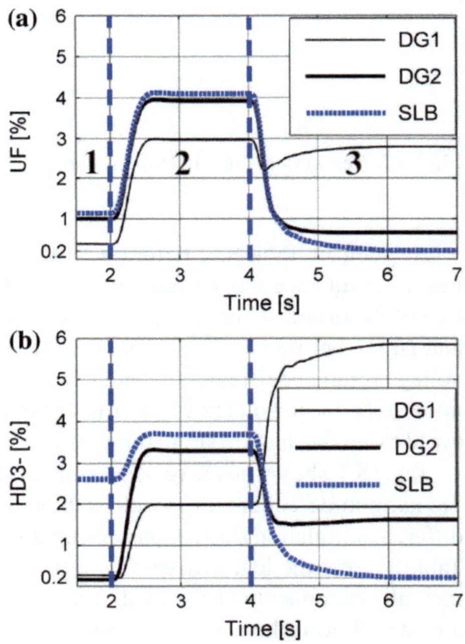

Fig. 8.43 **a** DGs and SLB voltage unbalance factor, **b** DGs and SLB voltage distortion index for 3rd harmonic negative sequence

Table 8.7 DGS output current waveforms

	Step1	Step 2	Step 3
fundamental negative sequence (DG1: solid DG2: dashed)	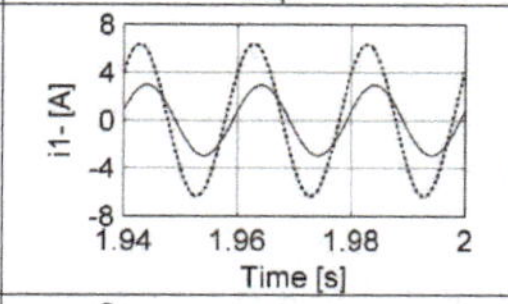	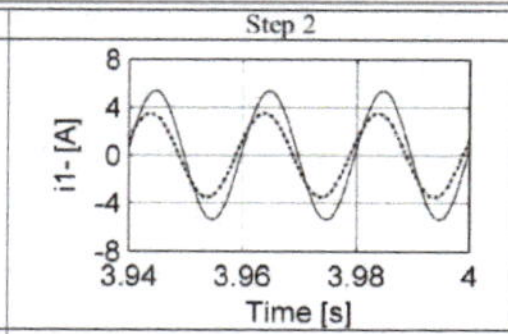	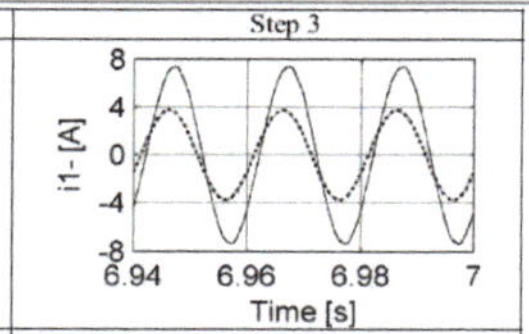
3rd harmonic negative sequence (DG1: solid DG2: dashed)	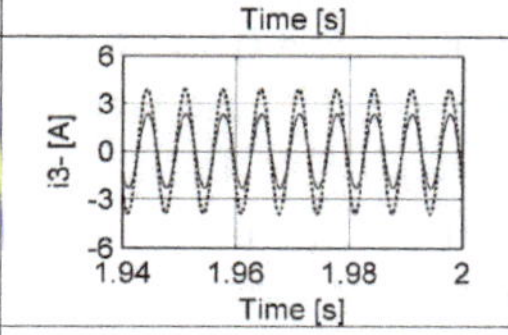	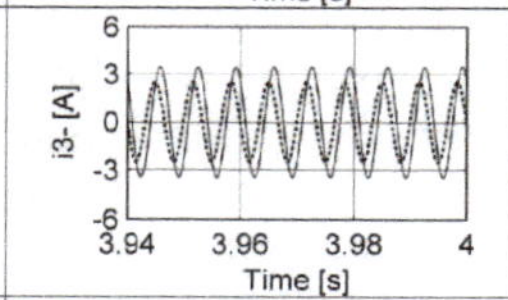	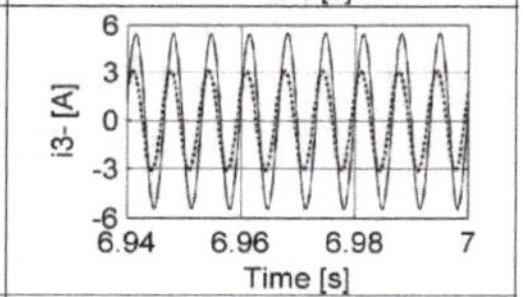
total output current (DG1: solid DG2: dashed)	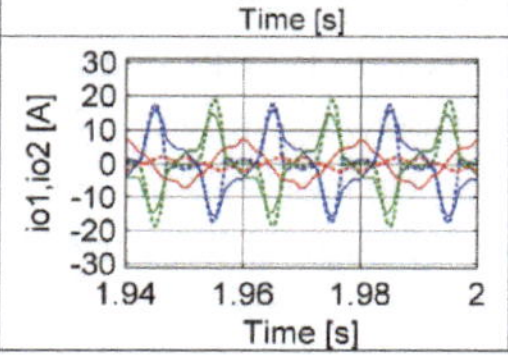	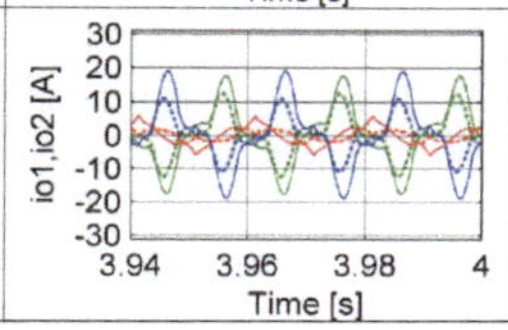	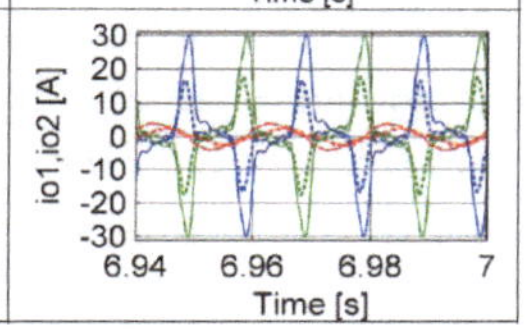

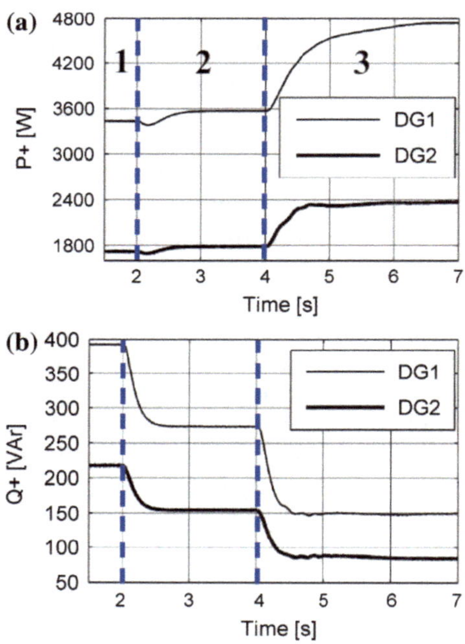

Fig. 8.44 DGs positive sequence output powers: **a** P+, **b** Q+

proper sharing of P^+ and Q^+ between DGs throughout the under-study interval. It demonstrates the effectiveness of the droop controllers.

In simulation Step 2, virtual resistances for fundamental negative sequence and harmonic components are added. As seen in Table 8.7, the current sharing is improved noticeably; however, still is not in proportion to the DGs-rated powers. The sharing improvement is achieved at the expense of voltage distortion increase at DGs terminals and consequently at SLB, as can be observed in Table 8.6 and Fig. 8.43. On the other hand, it can be seen in Fig. 8.44 that the addition of these virtual resistances leads to the change of fundamental positive sequence powers. In fact, due to nonlinear nature of the diode rectifiers, fundamental positive sequence component cannot be considered completely decoupled from the other components.

In the last simulation step, selective compensation of SLB voltage main harmonics and fundamental negative sequence component is activated at $t = 4$ s. As seen in Fig. 8.43, UF and HD^{3-} track the reference values properly. The other harmonic components which are not shown in this paper show the same behavior. Consequently, SLB voltage quality is significantly improved as seen in Table 8.6.

As observed in Table 8.7, fundamental negative sequence and 3rd harmonic negative sequence of DGs output current increase to provide compensation. The same behavior has been achieved for other harmonic components. The increase of these current components leads to the increase of DGs output current as shown in Table 8.7. However, due to double capacity, the increase of $DG1$ current is significantly higher. In addition, it can be observed that the current sharing is noticeably improved after compensation activation. This fact reveals the effectiveness of the proposed compensation effort controller and virtual impedance loop.

It can be observed also in Table 8.6 that the compensation is achieved by the increase of $DG1$ output voltage distortion. Note that the impedance of the distribution line between SB_1 and SLB is relatively high; also, the fundamental negative sequence and harmonic components of the load which are supplied by this DG are approximately twice of the amounts of $DG2$. Thus, in order to compensate the voltage drops on the lines and the virtual resistances and provide approximately sinusoidal voltage at SLB, $DG1$ output voltage becomes noticeably distorted. On the other hand, due to low value of the line impedance between SB_2 and SLB and also lower load portion of $DG2$, the distortions of SLB and $DG2$ voltages change with a similar behavior. Moreover, it can be seen in Fig. 8.44 that active and reactive powers change as a result of compensation. As mentioned before, it is originated from coupling between fundamental positive sequence and other components.

8.6 Distributed Cooperative Control

MGs, as the building blocks of an intelligent electric network, the so-called smart gird, are small-scale power systems that facilitate the effective integration of distributed generators (DGs) [4, 19]. Proper control of MGs is a prerequisite for stable and economically efficient operations of smart grids (SGs) [29, 30]. MGs can operate in both grid-connected and islanded operating modes.

In normal operation, the MG is connected to the main grid. In the event of a disturbance, the MG disconnects from the main grid and enters the islanded operation. Once a MG is islanded, the so-called primary control maintains the voltage and frequency stability [17, 78]. However, even in the presence of primary control, voltage and frequency can still deviate from their nominal values. To restore the voltage and frequency of DGs to their nominal values, the so-called secondary control is also required [3, 29, 58].

Conventional secondary controls of MGs assume a centralized control structure that requires a complex communication network [3, 29, 61], in some cases, with two-way communication links. In turn, this can adversely affect the system reliability. Alternatively, distributed cooperative control structures, with sparse communication network, are suitable alternatives for the secondary control of MGs. Distributed cooperative control is recently introduced in power systems [103], to regulate the output power of multiple PV generators.

Networked multiagent systems have received increasing attention due to their flexibility and computational efficiency. These systems are inspired by the natural phenomena such as swarming in insects, flocking in birds, thermodynamics laws, and synchronization and phase transitions in physical and chemical systems. In these phenomena, the coordination and synchronization process necessitates that each agent exchange information with other agents according to some restricted communication protocols [60].

In this section, distributed cooperative control of multiagent systems is adopted to implement the secondary control of MGs. The term "distributed" means that the controller requires a communication network by which each agent only receives the information of its neighboring agents. The term cooperative means that, in contrast to the competitive control, all agents act as one group toward a common synchronization goal and follow cooperative decisions [60]. Distributed cooperative control of multiagent systems is mainly categorized into the regulator synchronization problem and the tracking synchronization problem. In regulator synchronization problem, also called leaderless consensus, all agents synchronize to a common value that is not prescribed or controllable. In tracking synchronization problem, all agents synchronize to a leader node that acts as a command generator [53]. Neighboring agents can communicate with each other. The leader is only connected to a small portion of the agents.

Distributed cooperative control for multiagent systems with nonlinear or nonidentical dynamics has been recently introduced in the literature [15]. Considering DGs in a MG as agents in a networked multiagent system, the secondary control design resembles a tracking synchronization problem. The dynamics of DGs in MGs are nonlinear and nonidentical; input–output feedback linearization is used to transform the nonlinear heterogeneous dynamics of DGs to linear dynamics. Thus, the secondary voltage control is transformed to a second-order tracking synchronization problem. The Lyapunov technique is then adopted to derive fully distributed control protocols for each DG.

Our objective hereafter is to design the secondary voltage control of a MG system. The salient features of the proposed control methodology are given here.

- The secondary voltage control of MGs is implemented using the concept of distributed cooperative control of multiagent systems.
- Input–output feedback linearization is used to solve the tracking synchronization problem for nonlinear and heterogeneous multiagent systems.
- The proposed secondary voltage control obviates the requirement for a CC and requires only a sparse communication structure with one-way communication links which is cheaper and can be more reliable.
- Desired response speeds can be obtained by tuning the control parameters.

8.6.1 Microgrid Control Levels

A MG is able to operate in both grid-connected and islanded modes. The voltage and frequency of the MG in the grid-connected mode are dictated by the main grid [3, 29]. The MG can switch to the islanded mode due to the preplanned scheduling or unplanned disturbances. Subsequent to the islanding process, the primary control [18, 29] maintains the voltage and frequency stability of the MG. Primary control avoids voltage and frequency instability by keeping these values in prespecified ranges. However, it might not return the MG to the normal operating conditions, and an additional control level is required to restore the voltage and frequency. This functionality is provided by the secondary control, which compensates for the voltage and frequency deviations caused by the primary control [3, 29]. The secondary control operates with a longer time frame than primary control. This facilitates the decoupled operation and design of primary and secondary control levels.

Similar to the previous sections, primary control is usually implemented as a LOC at each DG. This control level always exists and takes action in the event of disturbances. Coordinated control of the primary LOCs can be achieved by the active and reactive power droop techniques [3, 29]. Droop technique prescribes a desired relation between the frequency and active power *P*, and between the voltage amplitude and reactive power *Q*. The frequency and voltage droop characteristics for the ith DG are given by

$$w_i = w_{ni} - m_{P_i} P_i \tag{8.84}$$

$$V^*_{0,magi} = V_{ni} - n_{Q_i} Q_i \tag{8.85}$$

where $V^*_{0,magi}$ is the reference value for the output voltage magnitude that is provided for the internal voltage control loop of DG, w_i is the angular frequency of the DG dictated by the primary control, P_i and Q_i are the measured active and reactive power at the DGs terminal, m_{P_i} and n_{Q_i} are the droop coefficients, and V_{ni} and w_{ni} are the primary control references [3, 29]. The droop coefficients are selected based on the active and reactive power ratings of each DG.

The secondary control sets the references for the primary control V_{ni} and w_{ni} in (8.84) so as to regulate the frequency and voltage amplitude to their prescribed nominal values. Conventionally, the secondary control is implemented for each DG using a centralized controller having the PI structure [3, 29]. This secondary control is centralized and requires a star communication structure. In a star communication structure, it is necessary to have a communication link between all DGs and the CC. Due to the centralized structure of this controller, this control scheme can potentially be unreliable. Alternatively, a distributed cooperative control structure is proposed in this paper.

8.6.2 Large-Signal Inverter-Based Model

The proposed secondary voltage control is designed based on the large-signal nonlinear dynamical model of the DG. The block diagram of an inverter-based DG is shown in Fig. 8.45. It contains an inverter bridge, connected to a primary DC power source (e.g., PV panels or fuel cells). The control loops, including the power, voltage, and current controllers, adjust the output voltage and frequency of the inverter bridge [45, 68]. Given the relatively high switching frequency of the inverter bridge, the switching artifacts can be safely neglected via average-value modeling. As stated in [68], DC bus dynamics can be safely neglected, assuming an ideal source from the DG side.

It should be noted that the nonlinear dynamics of each DG are formulated in its own $d-q$ (direct-quadrature) reference frame. It is assumed that the reference frame

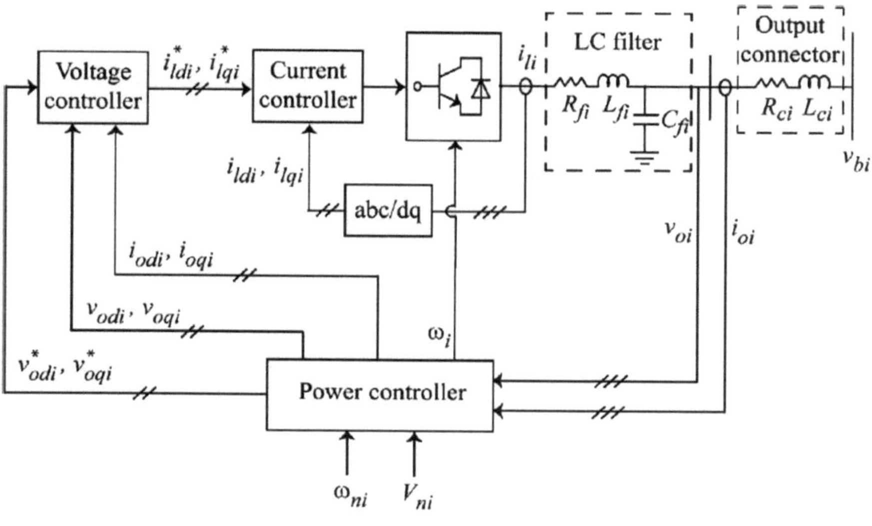

Fig. 8.45 Block diagram of an inverter-based DG

of the ith DG is rotating at the frequency of w_i. The reference frame of one DG is considered as the common reference frame with the rotating frequency of w_{com}. The angle of the ith DG reference frame, with respect to the common reference frame, is denoted as δ_i and satisfies the following differential equation:

$$\dot{\delta}_i = w_i - w_{com}. \tag{8.86}$$

Although different angular frequencies are considered for reference frames, all of the reference frames rotate synchronously at a common angular frequency due to the presence of the frequency-droop characteristic in (8.84).

The power controller block, shown in Fig. 8.46, contains the droop technique in (8.84) and provides the voltage references v^*_{odi} and v^*_{oqi} for the voltage controller, as well as the operating frequency w_i for the inverter bridge. Two LPFs with the cutoff frequency of w_{ci} are used to extract the fundamental component of the output active and reactive powers, denoted as P_i and Q_i, respectively. The differential equations of the power controller can be written as

$$\dot{P}_i = -w_{ci}P_i + w_{ci}(v_{odi}i_{odi} + v_{oqi}i_{oqi}), \tag{8.87}$$
$$\dot{Q}_i = -w_{ci}Q_i + w_{ci}(v_{oqi}i_{odi} - v_{odi}i_{oqi}) \tag{8.88}$$

where v_{odi}, v_{oqi}, i_{odi}, and i_{oqi} are the direct and quadrature components of v_{oi} and i_{oi} in Fig. 8.45. As seen in Fig. 8.46, the primary voltage control strategy for each DG aligns the output voltage magnitude on the d-axis of the corresponding reference frame. Therefore

$$\begin{cases} v^*_{odi} = V_{ni} - n_{Q_i}Q_i \\ v^*_{oqi} = 0. \end{cases} \tag{8.89}$$

The block diagram of the voltage controller is shown in Fig. 8.47, [45, 59]. The differential algebraic equations of the voltage controller are written as

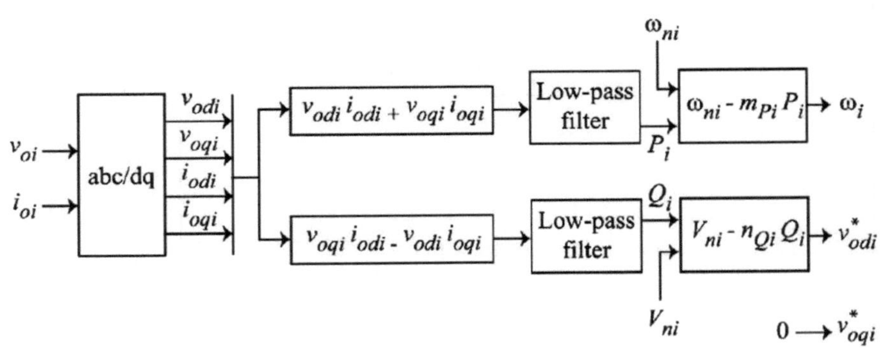

Fig. 8.46 Block diagram of the power controller

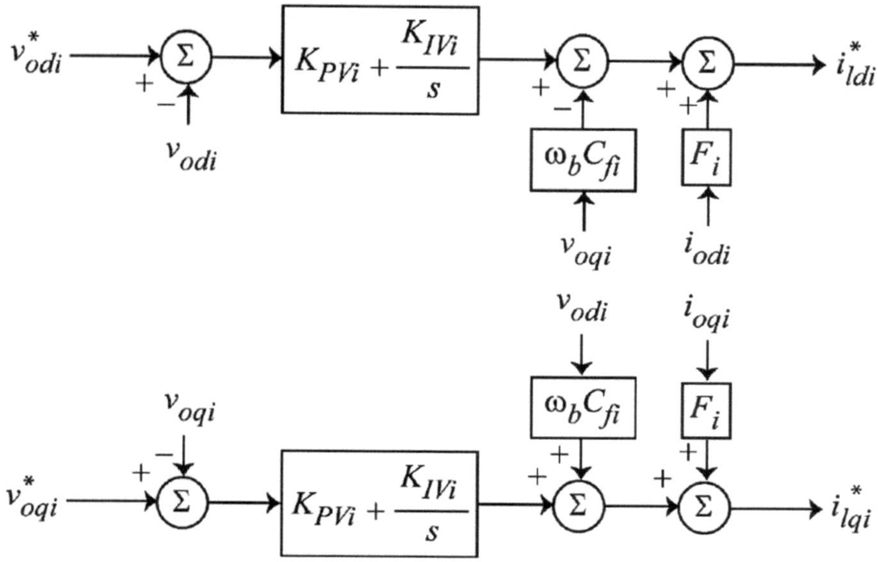

Fig. 8.47 Block diagram of the voltage controller

$$\dot{\phi}_{di} = v_{odi}^* - v_{odi} \tag{8.90}$$

$$\dot{\phi}_{qi} = v_{oqi}^* - v_{oqi} \tag{8.91}$$

$$i_{ldi}^* = F_i i_{odi} - w_b C_{fi} v_{oqi} + K_{PVi}(v_{odi}^* - v_{odi}) + K_{IVi}\phi_{di} \tag{8.92}$$

$$i_{lqi}^* = F_i i_{oqi} + w_b C_{fi} v_{odi} + K_{PVi}(v_{oqi}^* - v_{oqi}) + K_{IVi}\phi_{qi} \tag{8.93}$$

where ϕ_{di} and ϕ_{qi} are the auxiliary state variables defined for PI controllers in Fig. 8.47. w_b is the nominal angular frequency. Other parameters are shown in Figs. 8.45 and 8.47.

The block diagram of the current controller is shown in Fig. 8.48, [45, 59]. The differential algebraic equations of the current controller are written as

$$\dot{\gamma}_{di} = i_{ldi}^* - i_{ldi} \tag{8.94}$$

$$\dot{\gamma}_{qi} = i_{lqi}^* - i_{lqi} \tag{8.95}$$

$$v_{ldi}^* = -w_b L_{fi} i_{lqi} + K_{PCi}(i_{ldi}^* - i_{ldi}) + K_{ICi}\gamma_{di} \tag{8.96}$$

$$v_{lqi}^* = w_b L_{fi} i_{ldi} + K_{PCi}(i_{lqi}^* - i_{lqi}) + K_{ICi}\gamma_{qi} \tag{8.97}$$

where γ_{di} and γ_{qi} are the auxiliary state variables defined for the PI controllers in Fig. 8.48 and i_{ldi} and i_{lqi} are the direct and quadrature components of i_{l_i} in Fig. 8.45. Other parameters are shown in Figs. 8.45 and 8.48.

The differential equations for the output LC filter and output connector are as follows:

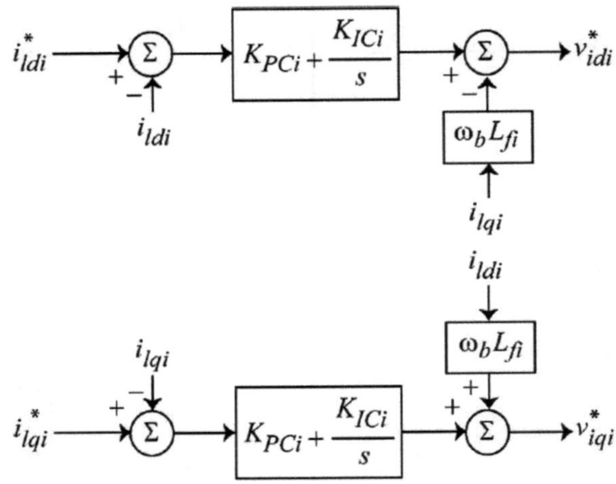

Fig. 8.48 Block diagram of the current controller

$$\dot{i}_{ldi} = -\frac{R_{fi}}{L_{fi}} i_{ldi} + w_i i_{lqi} + \frac{1}{L_{fi}} v_{idi} - \frac{1}{L_{fi}} v_{odi} \tag{8.98}$$

$$\dot{i}_{lqi} = -\frac{R_{fi}}{L_{fi}} i_{lqi} - w_i i_{ldi} + \frac{1}{L_{fi}} v_{iqi} - \frac{1}{L_{fi}} v_{oqi} \tag{8.99}$$

$$\dot{v}_{odi} = w_i v_{oqi} + \frac{1}{C_{fi}} i_{ldi} - \frac{1}{C_{fi}} i_{odi} \tag{8.100}$$

$$\dot{v}_{oqi} = -w_i v_{odi} + \frac{1}{C_{fi}} i_{lqi} - \frac{1}{C_{fi}} i_{oqi} \tag{8.101}$$

$$\dot{i}_{odi} = -\frac{R_{ci}}{L_{ci}} i_{odi} + w_i i_{oqi} + \frac{1}{L_{ci}} v_{odi} - \frac{1}{L_{ci}} v_{bdi} \tag{8.102}$$

$$\dot{i}_{oqi} = -\frac{R_{ci}}{L_{ci}} i_{oqi} + w_i i_{oqi} - w_i i_{odi} + \frac{1}{L_{ci}} v_{oqi} - \frac{1}{L_{ci}} v_{bqi}. \tag{8.103}$$

Equations (8.86)–(8.103) form the large-signal dynamical model of the ith DG. The large-signal dynamical model can be written in a compact form as

$$\begin{cases} \dot{x}_i = f_i(x_i) + K_i(x_i)D_i + g_i(x_i)u_i \\ y_i = h_i(x_i) \end{cases} \tag{8.104}$$

where the state vector is

$$x_i = [\delta_i P_i Q_i \phi_{di} \phi_{qi} \gamma_{di} \gamma_{qi} i_{ldi} i_{lqi} v_{odi} v_{oqi} i_{odi} i_{oqi}]^T. \tag{8.105}$$

The term $D_i = [w_{com} v_{bdi} v_{bqi}]^T$ is considered as a known disturbance. The detailed expressions for $f_i(x_i)$, $g_i(x_i)$, and $K\gamma_i(x_i)$ can be extracted from (8.86) to (8.103).

The secondary voltage control selects V_{ni} in (8.84) such that the terminal voltage amplitude of each DG approaches its nominal value, i.e., $v_{0,magi} \rightarrow v_{ref}$. Since the amplitude of the DG output voltage is

$$v_{0,magi} = \sqrt{v_{odi}^2 + v_{oqi}^2} \tag{8.106}$$

the synchronization of the voltage amplitude can be achieved by choosing the control input V_{ni} such that $v_{odi} \rightarrow v_{ref}$. Therefore, for the secondary voltage control, the output and control input are set to $y_i = v_{odi}$ and $u_i = V_{ni}$, respectively.

8.6.3 Cooperative Secondary Voltage Control

From a dynamical viewpoint, a MG resembles a nonlinear and heterogeneous multiagent system, where each DG is an agent. The secondary control of MGs is a tracking synchronization problem, where all DGs try to synchronize their terminal voltage amplitude to prespecified reference values. For this purpose, each DG needs to communicate with its neighbors only. The required communication network can be modeled by a communication graph. For further details about graph theory, the reader is referred to the Appendix.

8.6.4 Feedback Linearization and Tracking Synchronization

As discussed in the foregoing section, the secondary voltage control chooses appropriate control inputs V_{ni} in to (8.84) to synchronize the voltage magnitudes of DGs $v_{0,magi}$ to the reference voltage v_{ref}. The synchronization of the voltage magnitudes of DGs $v_{0,magi}$ is equivalent to synchronizing the direct term of output voltages v_{odi}. Therefore, the secondary voltage control should choose u_i in (8.104) such that $y_i \rightarrow y_0, \forall i$, where $y_0 \equiv v_{ref}$.

Since the dynamics of DGs in a MG are nonlinear and might not be all identical, input–output feedback linearization can be used to facilitate the secondary voltage control design. In input–output feedback linearization, a direct relationship between the dynamics of the output y_i (or equivalently v_{odi}) and the control input u_i (or equivalently V_{ni}) is generated by repetitively differentiating y_i with respect to time.

For the dynamics of the ith DG in (8.104), the direct relationship between the y_i and u_i is generated after the second derivative of the output y_i as

$$\ddot{y}_i = L_{F_i}^2 h_i + L_{g_i} L_{F_i} h_i u_i \tag{8.107}$$

$$F_i(x_i) = f_i(x_i) + K_i(x_i) D_i. \tag{8.108}$$

$L_{F_i} h_i$ is the Lie derivative [85] of h_i with respect to F_i and is defined by $L_{F_i} h_i = \nabla h_i F_i = (\partial(h_i)/\partial x_i) F_i . L_{F_i}^2 h_i$ is defined by $L_{F_i}^2 h_i = L_{F_i}(L_{F_i} h_i) = (\partial(L_{F_i} h_i)/\partial x_i) F_i$.

An auxiliary control v_i is defined as

$$v_i = L_{f_i}^2 h_i + L_{g_i} L_{F_i} h_i u_i. \tag{8.109}$$

Equations (8.107) and (8.109) result in the second-order linear system

$$\ddot{y}_i = v_i, \forall i. \tag{8.110}$$

By choosing appropriate v_i, the synchronization for y_i is provided. The control input u_i is implemented by v_i as

$$u_i (L_{g_i} L_{F_i} h_i)^{-1} (-L_{F_i}^2 h_i + v_i). \tag{8.111}$$

In the following, the procedure for designing appropriate v_i is elaborated. First, (8.110) and the first derivative of y_i are written as

$$\begin{cases} \dot{y}_i \equiv y_{i,1} \\ \dot{y}_{i,1} = v_i, \end{cases} \tag{8.112}$$

or equivalently

$$\dot{\mathbf{y}}_i = A \mathbf{y}_i + B v_i, \forall i \tag{8.113}$$

where $\mathbf{y}_i = [y_i \ y_{i,1}]^T$, $B = [0 \ 1]^T$, and $A = \begin{bmatrix} 0 & 1 \\ 0 & 0 \end{bmatrix}$.

Using input–output feedback linearization, the nonlinear dynamics of each DG in (8.104) are transformed to (8.113) and a set of internal dynamics. The commensurate reformulated dynamics of the reference generator can be expressed as

$$\dot{\mathbf{y}}_0 = A \mathbf{y}_0 \tag{8.114}$$

where $\mathbf{y}_0 = [y_0 \ \dot{y}_0]^T$. It should be noted that, since $y_0 = v_{ref}$ is constant, $\dot{y}_0 = 0$.

It is assumed that DGs can communicate with each other through a communication network described by the digraph G_r. Based on the digraph G_r, the ith DG may need to transmit $\mathbf{y}_i$ in (8.113) through the communication network. It is assumed that only one DG has the access to the reference $\mathbf{y}_0$ in (8.114) by a weight factor known as the pinning gain g_i. The secondary voltage control problem is to find a distributed v_i in (8.111) such that $y_i \to y_0, \forall i$. To solve this problem, the cooperative team objectives are expressed in terms of the local neighborhood tracking error

$$e_i = \sum_{j \in N_i} a_{ij}(y_i - y_j) + g_i(y_i - y_0) \tag{8.115}$$

where a_{ij} denotes the elements of the communication digraph adjacency matrix. The pinning gain g_i is nonzero for one DG.

For a MG including N DGs, the global error vector for graph G_r is written from (8.115) as

$$e = ((L + G) \otimes I_2)(Y - Y_0) \equiv ((L + G) \otimes I_2)\delta \tag{8.116}$$

where $Y = \left[y_1^T \; y_2^T \; \ldots \; y_N^T\right]^T$, $e = \left[e_1^T \; e_2^T \; \ldots \; e_N^T\right]^T$, $Y_0 = 1_N y_0$ (1_N is the vector of ones with the length of N.), $G = diag\{g_i\}$, I_2 is the identity matrix with two rows and two columns, and δ is the global disagreement vector. The Kronecker product is shown as $\otimes$. Y can be written as

$$\dot{Y} = (I_N \otimes A)Y + (I_N \otimes B)v \tag{8.117}$$

where $v = \left[v_1 \; v_2 \; \ldots \; v_N\right]^T$ is the global auxiliary control vector. $\dot{Y}_0$ can be written as

$$\dot{Y} = (I_N \otimes A)Y_0. \tag{8.118}$$

The following definitions and lemmas are required for designing the auxiliary controls v_i.

Definition 8.1 (A, B) are stabilizable if there exists a matrix S such that all eigenvalues of $A - BS$ have a strictly negative real part.

Definition 8.2 A matrix is Hurwitz if all of its eigenvalues have a strictly negative real part.

Definition 8.3 A symmetric matrix P is positive definite if $x^T Px$ is positive for all nonzero column vector x, and $x^T Px$ is zero only for $x = 0$.

Lemma 8.3 ([104]) *Let (A, B) be stabilizable. Let the digraph G_r have a spanning tree and $g_i \neq 0$ for one DG placed on a root node of the digraph G_r. Let $\lambda_i (i \in \{1, 2, \ldots, N\})$ be the eigenvalues of $L + G$. The matrix*

$$H = I_N \otimes A - c(L + G) \otimes BK \tag{8.119}$$

with $c \in R$ and $k \in R^{1\times 2}$ is Hurwitz if and only if all of the matrices $A - c\lambda_i BK$, $\forall i \in \{1, 2, \ldots, N\}$ are Hurwitz.

Lemma 8.4 ([104]) *Let (A, B) be stabilizable and matrices $Q = Q^T$ and $R = R^T$ be positive definite. Let feedback gain K be chosen as*

$$K = R^{-1}B^T P_1 \tag{8.120}$$

where P_1 is the unique positive definite solution of the control algebraic Riccati equation (ARE)

$$A^R P_1 + P_1 A + Q - P_1 B R^{-1} B^T P_1 = 0. \tag{8.121}$$

Then, all of the matrices $A - c\lambda_i BK$, $\forall i \in \{1, 2, \ldots, N\}$, are Hurwitz if $c \geq (1/2\lambda_{\min})$, where $\lambda_{\min} = \min_{i\in N} Re(\lambda_i)(Re(\lambda_i)$ denotes the real part of $\lambda_i)$.

Theorem 8.2 *Let the digraph G_r have a spanning tree and $g_i \neq 0$ for one DG placed on a root node of the digraph G_r. It is assumed that the internal dynamics of each DG are asymptotically stable. Let the auxiliary control v_i in (8.111) be*

$$v_i = -cKe_i \tag{8.122}$$

where $c \in R$ is the coupling gain and $K \in R^{1\times 2}$ is the feedback control vector. Then, all y_i in (8.113) synchronize to y_0 in (8.114) and, hence, the direct term of DG output voltages v_{odi} synchronizes to v_{ref}, if K is chosen as in (8.120) and

$$c \geq \frac{1}{2\lambda_{\min}} \tag{8.123}$$

where $\lambda_{\min} = \min_{i\in N} Re(\lambda_i)$.

Proof Consider the Lyapunov function candidate

$$V + \frac{1}{2}\delta^T P_2 \delta \quad P_2 = P_2^T, P_2 > 0 \tag{8.124}$$

where δ is the global disagreement vector in (8.116). Then

$$\dot{V} = \delta^T P_2 \dot{\delta} = \delta^T P_2(\dot{Y} - \dot{Y}_0) = \delta^T P_2((I_N \otimes A)\delta + (I_N \otimes B)v) \tag{8.125}$$

The global auxiliary control v can be written as

$$v = -c(I_N \otimes K)((L + G) \otimes I_2)\delta. \tag{8.126}$$

Placing (8.126) into (8.125) yields

$$\dot{V} = \delta^T P_2(I_N \otimes A - c(L + G) \otimes BK)\delta \equiv \delta^T P_2 H \delta. \tag{8.127}$$

From Lemmas 8.3 and 8.4, H is Hurwitz. Therefore, given any positive real number β, the positive definite matrix P_2 can be chosen such that the following Lyapunov equation holds

$$P_2H + H^T P_2 = -\beta I_{2N}. \tag{8.128}$$

Placing (8.128) in (8.127) yields

$$\begin{aligned}\dot{V} &= \delta^T P_2 H\delta = \frac{1}{2}\delta^T (P_2H + H^T P_2)\delta \\ &= -\frac{\beta}{2}\delta^T I_{2N}\delta.\end{aligned} \tag{8.129}$$

Note (8.129) shows that $\dot{V} \leq 0$. Therefore, the global disagreement vector δ, (8.110), and (8.122) are asymptotically stable and all y_i in (8.113) synchronize to y_0 in (8.114). Hence, the direct term of DG output voltages v_{odi} synchronizes to v_{ref}. If the internal dynamics are asymptotically stable, then they are all bounded. This completes the proof.

The block diagram of secondary voltage control based on distributed cooperative control is shown in Fig. 8.49. As seen, the control input V_{ni} is implemented using (8.111). Each DG has a $\dot{v}_{odi}$ calculator block based on (8.100).

Choosing the coupling gain c and the feedback control vector K based on (8.120) and (8.123) ensures the asymptotic stability of the controller. Moreover, these controller parameters can adjust the response speed of the secondary voltage control.

8.6.5 Required Sparse Communication Topology

The proposed secondary voltage control must be supported by a local communication network that provides its required information flows. This communication graph should be designed to reduce transmission delays and the required information flows between components. Long communication links are not desired. For the MGs with a small geographical span, the communication network can be implemented by Control Area Network (CAN) Bus and PROFIBUS communication protocols [2]. It should be noted that communication links contain an intrinsic delay. However, in this paper, the communication link delays are assumed to be zero. Since the timescale of the secondary control is sufficiently large, the aforementioned assumption is valid and the communication link delays do not significantly affect the system performance.

According to the results of the theorem in Sect. 8.6.4, the communication topology should be a graph containing a spanning tree in which the secondary control of each DG only requires information about that DG and its immediate neighbors in the communication graph. Therefore, the communication requirements for implementing the proposed control are rather mild. Given the physical structure of the MG, it is not difficult to select a graph with a spanning tree that connects all the DGs in an

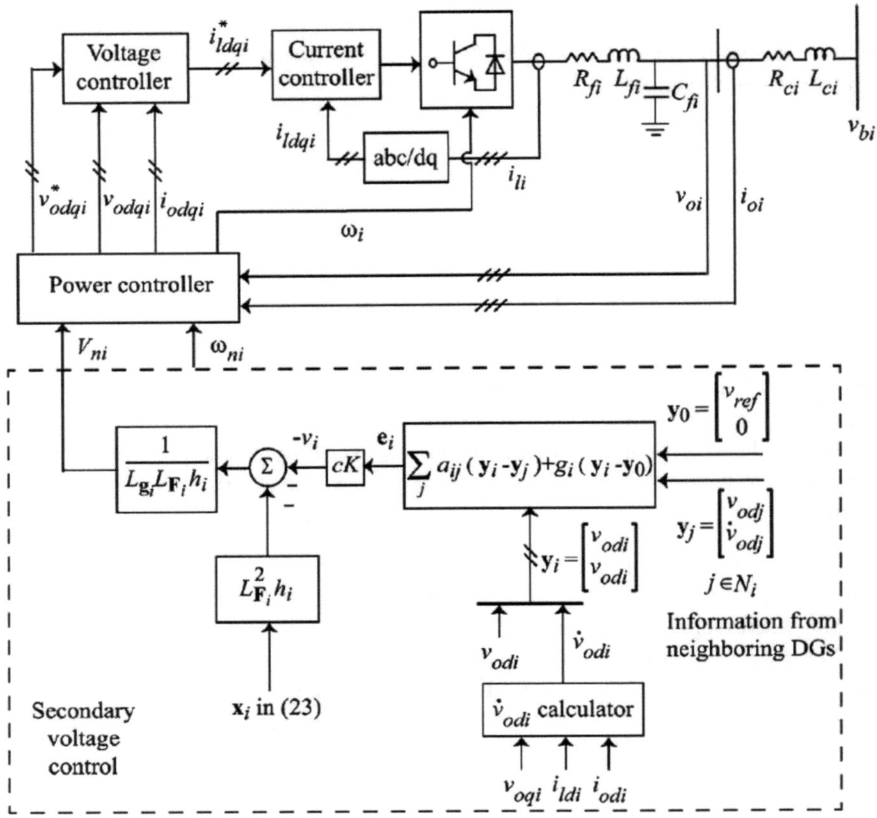

Fig. 8.49 Block diagram of the proposed secondary voltage control

optimal fashion. Such optimal connecting graphs can be designed using operations research or assignment problem solutions. The optimization criteria can include minimal lengths of the communication links, maximal use of existing communication links, and minimal number of links, and so on.

8.6.6 Simulation Results VII

The effectiveness of the proposed secondary voltage control is verified by simulating an islanded MG in MATLAB. Figure 8.50 illustrates the single-line diagram of the MG test system. This MG consists of four DGs. The lines between buses are modeled as series RL branches. The specifications of the DGs, lines, and loads are summarized in Table 8.8.

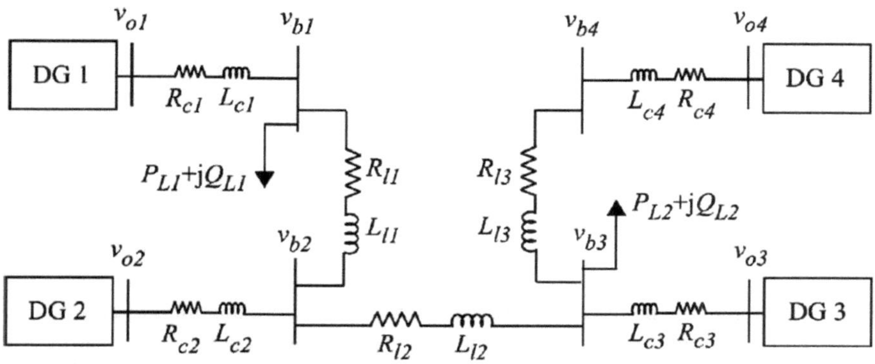

Fig. 8.50 Single-line diagram of the microgrid test system

Table 8.8 Specification of the microgrid test system

DGs	DG1 & 2 (45 kVA rating)		DG3 & 4 (34 kVA rating)			
	m_P	9.42×10^{-5}	m_P	12.43×10^{-5}		
	n_Q	1.34×10^{-3}	n_Q	1.52×10^{-3}		
	R_c	0.03 Ω	R_c	0.03 Ω		
	L_c	0.35 mH	L_c	0.35 mH		
	R_f	0.1 Ω	R_f	0.1 Ω		
	L_f	1.35 mH	L_f	1.35 mH		
	C_f	50 μF	C_f	50 μF		
	K_{PV}	0.1	KPV	0.05		
	K_{IV}	420	KIV	390		
	K_{PC}	15	KPC	10.5		
	K_{IC}	20,000	KIC	16,000		
Lines	Line 1		Line2		Line 3	
	R_{l1}	0.23 Ω	R_{l2}	0.35 Ω	R_{l3}	0.23 Ω
	L_{l1}	318 μH	L_{l2}	1847 μH	L_{l3}	318 μH
Loads	Load 1		Load 2			
	P_{L1} (per phase)	12 kW	P_{L2} (per phase)	15.3 kW		
	Q_{L1} (per phase)	12 kVAr	Q_{L2} (per phase)	7.6 kVAr		

It is assumed that DGs communicate with each other through the communication digraph depicted in Fig. 8.51. This communication topology is chosen based on the geographical location of DGs. The associated adjacency matrix of the digraph in Fig. 8.51 is

$$A_G = \begin{bmatrix} 0\,0\,0\,0 \\ 1\,0\,0\,0 \\ 0\,1\,0\,0 \\ 1\,0\,0\,0 \end{bmatrix}. \tag{8.130}$$

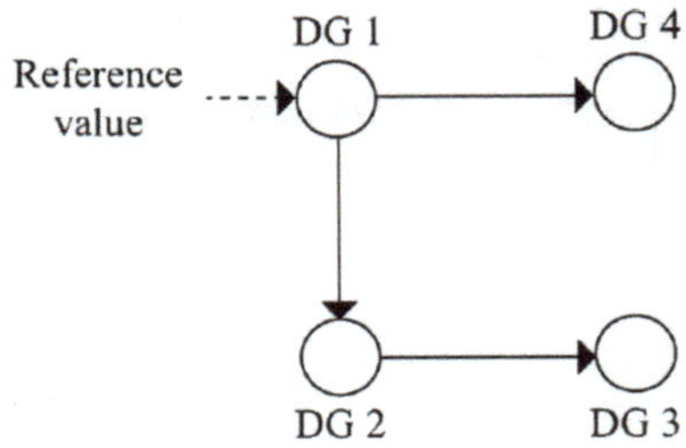

Fig. 8.51 Topology of the communication digraph

DG1 is the only DG that is connected to the leader node with the pinning gain $g_1 = 1$.

In the following, first, the effectiveness of the proposed secondary voltage control is shown for three different reference voltage values. Then, the effects of the ARE parameters on the transient response of the controller are studied.

Now, the coupling gain in (8.122) is $c = 4$ which satisfies (8.123). The solution of the ARE in (8.121) is used to calculate the feedback control vector K in (8.122). In (8.121), the ARE parameters are chosen as $Q = \begin{bmatrix} 50000 & 0 \\ 0 & 1 \end{bmatrix}$ and $R = 0.01$. The resulting feedback control vector is $K = [2236\ 67.6]$.

In the first case, namely Case A, the microgird is islanded from the main grid at $t = 0$, while the secondary control is active. Figure 8.52a–c shows the DG terminal

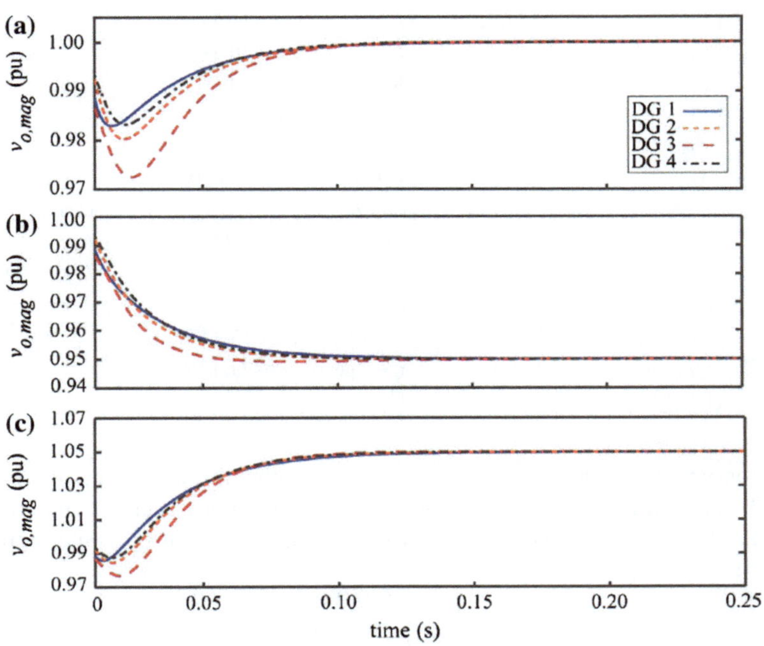

Fig. 8.52 DG output voltage magnitudes for Case A: **a** when $v_{ref} = 1$ p.u., **b** when $v_{ref} = 0.95$ p.u., and **c** when $v_{ref} = 1.05$ p.u.

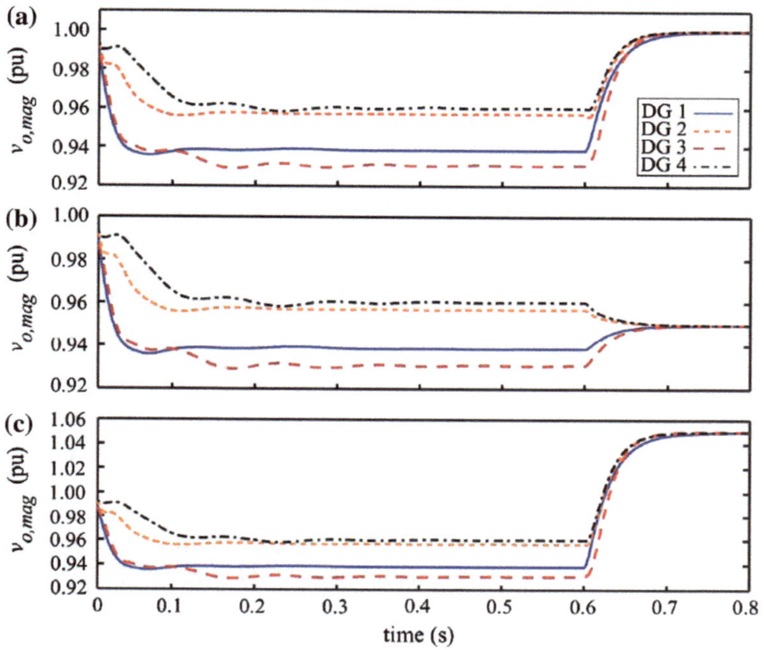

Fig. 8.53 DG output voltage magnitudes for Case B: **a** when $v_{ref} = 1$ p.u., **b** when $v_{ref} = 0.95$ p.u., and **c** when $v_{ref} = 1.05$ p.u.

voltage amplitudes when the reference voltage value is set to 1, 0.95, and 1.05 p.u., respectively. As seen in Fig. 8.52, the secondary control returns all DG terminal voltage amplitudes to the prespecified reference values after 0.1 s.

It should be noted that the secondary control level always exists as a supervisory control level and take actions in the event of disturbances. However, to highlight the effectiveness of the proposed secondary control, a new case study namely Case B, is considered. It is assumed that the MG is islanded from the main grid at $t = 0$, and the secondary control is applied at $t = 0.6$ s. Figure 8.53a–c shows the simulation results when the reference voltage value is set to 1, 0.95, and 1.05 p.u., respectively. As seen in Fig. 8.53, while the primary control keeps the voltage amplitudes stable, the secondary control returns all terminal voltage amplitudes to the prespecified reference values after 0.1 s.

8.7 Conclusions

In order to enable the development of the "smart electrical grid," a conceptual distributed control framework is presented for the electrical grid integrated with distributed renewable energy generation systems. The different elements and their interactions in the distributed control framework are first introduced. With focus on a specific wind/solar energy generation system connected to a RO water desalination system and the electrical grid, a supervisory predictive controller is designed via MPC to operate the integrated system taking into account short-term and long-term optimal maintenance and operation considerations.

Next, the secondary voltage and frequency control of MGs are designed based on the distributed cooperative control of multiagent systems. The MG is considered as a multiagent system with DGs as its agents. DGs can communicate with each other through a communication network modeled by a digraph. Input–output feedback linearization is used to transform the nonlinear dynamics of DG to linear dynamics. Feedback linearization converts the secondary voltage and frequency controls to first-order tracking synchronization problems. The control inputs are designed such that each DG only requires its own information and the information of its neighbors on the communication digraph. The proposed MG secondary control requires a sparse communication structure with one-way communication links and is more reliable than centralized secondary controls.

The next section has presented a general approach of multilevel control for MGs. The multilevel control stems from ISA-95. A three-level control is applied to AC and DC MGs. On the one hand, the control of AC MG mimics a large-scale power system AC grid, pointing out the similarities between both systems. On the other hand, the multilevel control of DC MGs presents novel features that can be useful in distributed power systems applications, such as telecommunication DC voltage networks, among others.

Consequently, flexible MGs are obtained which can be used fully for AC or DC interconnection with an AC or DC distribution system, controlling the power flow from the MG to these systems. In addition, these MGs are able to operate in both island or stiff source-connected modes, as well as to achieve a seamless transfer from one mode to the other.

By using the proposed approach, as shown Fig. 8.35, a multi-MG cluster can be performed, constituting an SG. In this sense, the tertiary control could provide high-level inertias to interconnect more MGs, thus acting as the primary control of the cluster. In this sense, MGs will behave like a voltage source with high inertias. Thus, a superior control level could send all the references to each cluster of MGs to restore the frequency and amplitude, i.e., the secondary control of the cluster level. Finally, the tertiary cluster control can fix the active and reactive power to be provided by this cluster or act like a primary control to interconnect more clusters. As a result, we could scale the hierarchy of control as necessary.

Using this approach, the system turns more flexible and expandable; and consequently, it could integrate more and more MGs without changing the local multilevel control system associated to each MG.

We have reviewed the current status of MG decentralized control (DEC) and methods to analyze and assess MG stability. We have also considered the issue of in-situ decentralized testing of MG controllers. Time-synchronization techniques such as the GPS timing signal and the PTP are very likely to play a significant role in both MG control and controller testing. Similarly, advances in numerical techniques that assess conventional power system stability are also likely to play a role in MGs as well.

The future trends in multilevel control for MGs are essentially related to energy-management systems, giving references from and to the tertiary control in order to optimize the efficiency of the MG. Another important issue will be the clusters of MGs, which are expected to be developed in the near future by interconnecting intelligent MGs. Each microgrid will have a number of energy services, such as active/reactive power demand/generation, storage capability, and so on, which could be of mutual interest among MGs. Thus, multiagents could negotiate the interchange of energy between MGs or microgrid clusters. With multiagents and multilevel control being a clear trend of research in MGs, technologies like communication systems are becoming important to make these applications feasible.

A multilevel control structure consisting of primary and secondary levels is proposed for MGs. The secondary level controls selective compensation of SLB voltage fundamental negative sequence and positive and negative sequences of main harmonics by sending proper control signals to the primary level. A new method for sharing of harmonic compensation effort is presented. Moreover, a selective virtual impedance scheme is proposed to improve load sharing among the microgrid DGs. The control system design is discussed in detail. Simulation results show that the SLB voltage quality is enhanced significantly by using the proposed compensation method while the load current is shared properly.

Finally, more industrial applications will push the research in this area after the recent final approval of the IEEE Standard 1547.4, which allows MGs to operate in island under certain conditions [37]. This standard constitutes a clear breakthrough toward new codes and industrial equipment that will be needed for extra functionalities required by the MG operations.

In this paper, the concept of distributed cooperative control of multiagent systems is adopted to implement the secondary voltage control of MGs. Input–output feedback linearization is used to transform the nonlinear dynamics of DGs to linear dynamics. Feedback linearization converts the secondary voltage control to a second-order tracker synchronization problem. The controller for each DG is fully distributed. Each DG only requires its own information and the information of some neighbors. The proposed MG secondary control requires a sparse communication network with one-way communication links and is more reliable than centralized secondary controls. It is shown that the controller parameters can effectively tune the controller synchronization speed.

8.8 Suggested Problems

Problem 8.1 It is desired to test the performance of secondary control approach in droop-controlled MGs. For this purpose, a computer simulation setup using MATLAB Simulink environment is constructed as shown in Fig. 8.54. Use the data in Table 8.9 where all the parameters are the same for both DG units.

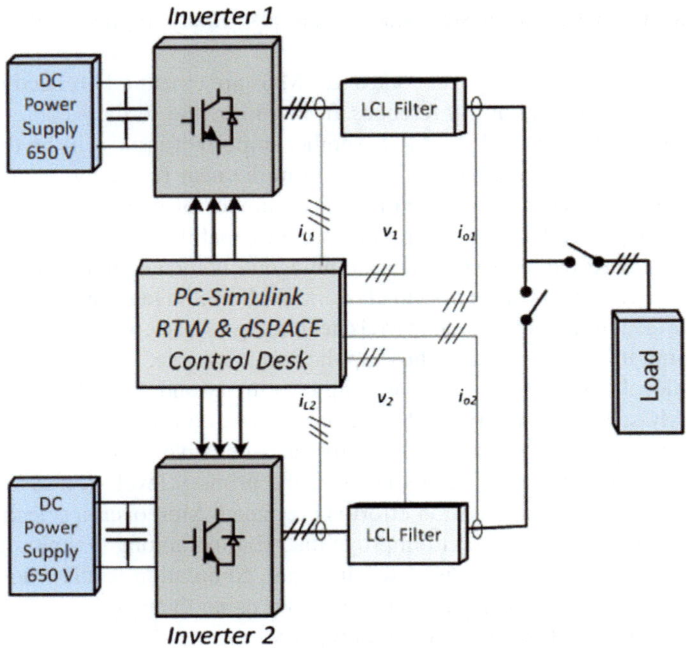

Fig. 8.54 Schematic setup

In the simulation setup, two DG inverters are forming an islanded MG, including an electrical setup with the two Danfoss 2.2 kW inverters, the dSpace1103 control board, LCL filters, and measurement LEM sensors. A diode rectifier is used as nonlinear load loaded by a capacitor, and a 200 Ω linear load, as well as an RL load with a 15 mH inductance connected to the common point. The switching frequency was 10 kHz.

For prescribed distributed voltage and frequency controls, generate the performance results to restore frequency and voltage variations as well as active and reactive power sharing is thought.

Consider the four scenarios for both stages

- *Scenario 1* ($0 < t < 5$ s): DG units operate without load and secondary control is not enabled.
- *Scenario 2* ($5 < t < 10$ s): A linear load 400 Ω is added to the common point.
- *Scenario 3* ($10 < t < 20$ s): The linear load is changed to 200 Ω.
- *Scenario 4* ($20 < t < 40$ s): Distributed secondary control (DSC) finally is activated.

Comment on the ensuing simulation results.

Table 8.9 Electrical and control system parameters

Type	Parameters		Value
	Symbol	Quantity	
Electrical setup	V_{dc}	DC voltage	650
	V_{MG}	MG voltage	311 V
	F	MG frequency	50 Hz
	C	Filter capacitance	25 μF
	L	Filter inductance	1.8 mH
	L_o	Output impedance	1.8 mH
	R_L	Resistance load	200 Ω/400 Ω
	L_L	Inductance load	15 mH
Inner loops	k_{pI}	Current integral term	0.7
	k_{iI}	Current integral term	100
	k_{pv}	Voltage integral term	0.35
	k_{iV}	Voltage integral term	400
Droop control	k_{pP}	Active power droop coefficient	0.0000 1 Ws/rd
	k_{iP}	Frequency integral term	0.0008 Ws/rd
	k_{pQ}	Reactive power droop coefficient	0.16 VAr/V
	R_v	Virtual resistance	1 Ω
	L_v	Virtual inductance	4 mH
Secondary control	k_{pF}	Frequency proportional term	1
	k_{iF}	Frequency integral term	2 s^{-1}
	k_{pE}	Amplitude proportional term	1
	k_{iE}	Amplitude integral term	2 s^{-1}
	k_{pQ}	Reactive power proportional term	0.0001 VAr/V
	k_{iQ}	Reactive power integral term	0.2 VAr/Vs

Problem 8.2 Consider the simulation setup in Fig. 8.54 with the objective of examining the effects of DSC on frequency and reactive power control, for a certain amount of communication latency delay. Display the results and compared it with the conventional centralized secondary control.

Problem 8.3 Considering that the primary control can be used to balance the energy between the DG units and energy-storage elements including batteries. Depending on the SoC of the batteries, the contribution of active power can be adjusted according to the availability of energy from each DG unit. Deevlop a frequency-droop function that reflects the level of charge of batteries.

Problem 8.4 In the conventional droop method used by large power systems, the output impedance of SYGNs, and the line impedance, are considered mainly inductive. By using power electronics, the output impedance will depend on the control strategy used by the inner control loops. In LV applications, the line impedance is nearly pure resistive. Show that one can modify the control droops (8.65) and (8.66) to reflect the impedance angle θ into the form

$$\omega = \omega^* - G_P(s)[(P - P^*)\sin\theta - (Q - Q^*)\cos\theta] \tag{8.131}$$
$$E = E^* - G_Q(s)[(P - P^*)\cos\theta + (Q - Q^*)\sin\theta] \tag{8.132}$$

By including the virtual output impedance loop, develop an expression that the expresses output voltage using appropriate voltage references to ensure inductive behavior at the line frequency.

Problem 8.5 A key feature of MGs with distributed energy sources (DES) is that the sources are dispersed over a wide area. These sources are interconnected to each other and to loads by a distribution network. Furthermore, the distributed MG may be connected to the main power grid at some point as well. Figure 8.55a shows a distributed MG structure connected to the main grid. The figure also shows the MG line impedances ($Z_{01}, Z_{12}, \ldots, Z_{n-1,n}$). The source is connected to the MG distribution network by an inverter interface through a filter, e.g., an LCL filter, shown in Fig. 8.55b.

It is desired to classify the distributed generation (DG) architectures into appropriate classes with respect to the interconnecting impedances Z_{01} shown in Fig. 8.55a. Discuss the impact on the voltage magnitude and phase. Identify the possible types of the grid.

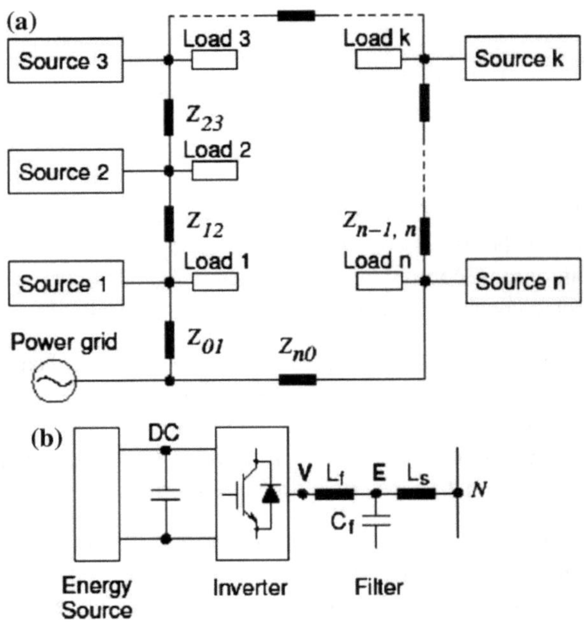

Fig. 8.55 Microgrid with distributed sources and loads

Problem 8.6 It is known that stability is a critical issue in a MG in which the source power electronic interfaces (PEIs) are controlled in a decentralized way. Each interface is controlled based only on local measurement; and therefore, it is significant to analyze how the individual control systems interact to ensure overall stability. Reflect this fundamental om the MG with distributed sources and loads shown in Fig. 8.55a. Using small-signal analysis, develop a computational method to determining the MG stability scalable to large systems.

Problem 8.7 Consider the MG with distributed sources and loads shown in Fig. 8.55a. Address in detail the main approaches to examine the effect of the interconnecting line impedance on droop-based control. Discuss the impact of droop gains on MG stability margin.

Problem 8.8 It is desired to investigate the multilevel control of MGs coupled with large power system. For this purpose, consider a three control level structure depicted in Fig. 8.56. It is supposed to operate over large SMs with high inertias and inductive networks. However, as it is well known that power electronics based MGs, there are no inertias, and the nature of the networks is mainly resistive. In turn, there are important differences between both systems that need to be taken into account when designing their control schemes.

The objective is to identify the task/function of each control level, specify an necessary addition to guarantee uniform performance and hence develop a math model of each level.

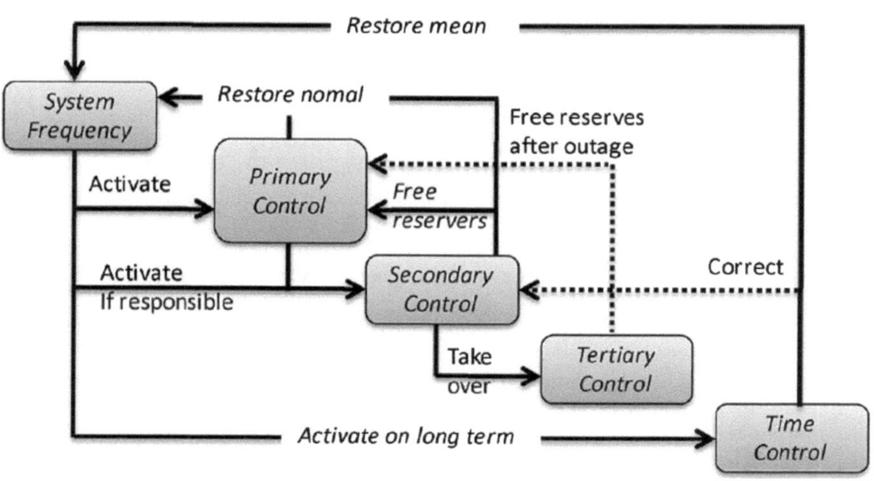

Fig. 8.56 Frame for the multilevel control of a power system, defined by UCTE

References

1. Ahuja RK, Magnanti TL, Orlin JB (1993) Network flows: theory, algorithms, applications. Prentice-Hall, Englewood Cliffs
2. Bassi E, Benzi F, Lusetti L, Buja GS (1995) Communication protocols for electrical drives. In: Proceedings of 21st international conference on industrial electronics, pp 706–711
3. Bidram A, Davoudi A (2012) Multilevel structure of microgrids control system. IEEE Trans Smart Grid 3(12):1963–1976
4. Bidram A, Hamedani-golshan ME, Davoudi A (2012) Capacitor design considering first swing stability of distributed generations. IEEE Trans Power Syst 27(4):1941–1948
5. Borup U, Blaabjerg F, Enjeti PN (2001) Sharing of nonlinear load in parallel-connected three-phase converters. IEEE Trans Ind Appl 37(6):1817–1823
6. Bryan J, Duke R, Round S (2004) Decentralized generator scheduling in a nanogrid using dc bus signaling. Proc IEEE Power Eng Soc Gen Meet 1:977–982
7. Burkard R, Dell'Amico M, Martello S (2009) Assignment problems. SIAM, Philadelphia
8. Buse DP, Wu QH (2004) Mobile agents for remote control of distributed systems. IEEE Trans Ind Electron 51(6):1142–1149
9. Camacho EF, Berenguel M (1997) Robust adaptive model predictive control of a solar plant with bounded uncertainties. Int J Adapt Control Signal Process 11:311–325
10. Cheng PT, Chen C, Lee TL, Kuo SY (2009) A cooperative imbalance compensation method for distributed-generation interface converters. IEEE Trans Ind Appl 45(2):805–815
11. Chiang SJ, Yen CY, Chang KT (2001) A multimodule parallelable series-connected PWM voltage regulator. IEEE Trans Ind Electron 48(3):506–516
12. Chinchilla M, Arnaltes S, Burgos JC (2006) Control of permanent-magnet generators applied to variable-speed wind-energy systems connected to the grid. IEEE Trans Energy Convers 21:130–135
13. Christofides PD, El-Farra NH (2005) Control of nonlinear and hybrid process systems: designs for uncertainty constraints and time-delays. Springer, Berlin
14. Dang DQ, Wang Y, Cai W (2008) Nonlinear model predictive control (NMPC) of fixed pitch variable speed wind turbine. In: Proceedings of 2008 IEEE international conference on sustainable energy technologies, Singapore, pp 29–33
15. Das A, Lewis FL (2010) Distributed adaptive control for synchronization of unknown nonlinear networked systems. Automatica 46:2014–2021
16. Delghavi MB, Yazdani A (2012) A unified control strategy for electronically interfaced distributed energy resources. IEEE Trans Power Deliv 27:803–812
17. Diaz G, Gonzalez-Moran C, Gomez-Aleixandre J, Diez A (2010) Scheduling of droop coefficients for frequency and voltage regulation in isolated microgrids. IEEE Trans Power Syst 25(1):489–496
18. Divshali PH, Alimardani A, Hosseinian SH, Abedi M (2012) Decentralized cooperative control strategy of microsources for stabilizing autonomous VSC-based microgrids. IEEE Trans Power Syst 27(4):1949–1959
19. Fahimi B, Kwasinski A, Davoudi A, Balog RS, Kiani M (2011) Charge it. IEEE Power Energy Mag 9:54–64
20. Fax J, Murray R (2004) Information flow and cooperative control of vehicle formations. IEEE Trans Autom Control 49(9):1465–1476
21. Ghiassi M, Zimbra DK, Saidane H (2008) Urban water demand forecasting with a dynamic artificial neural network model. J Water Resour Plan Manag 134:138–146
22. Guerrero JM, de Vicuna LG, Matas J, Castilla M, Miret J (2004) A wireless controller to enhance dynamic performance of parallel inverters in distributed generation systems. IEEE Trans Power Electron 19(5):1205–1213
23. Guerrero JM, Garcia de Vicuna L, Matas J, Castilla M, Miret J (2005) Output impedance design of parallel-connected UPS inverters with wireless load-sharing control. IEEE Trans Ind Electron 52(4):1126–1135

24. Guerrero JM, Matas J, de Vicuna LG, Castilla M, Miret J (2006) Wireless-control strategy for parallel operation of distributed-generation inverters. IEEE Trans Ind Electron 53(5):1461–1470
25. Guerrero JM, Garcia de Vicuna L, Uceda J (2007) Uninterruptible power supply systems provide protection. IEEE Ind Electron Mag 1(1):28–38
26. Guerrero JM, Matas J, Garcia de Vicuna L, Castilla M, Miret J (2007) Decentralized control for parallel operation of distributed generation inverters using resistive output impedance. IEEE Trans Ind Electron 54(2):994–1004
27. Guerrero JM, Hang L, Uceda J (2008) Control of distributed uninterruptible power supply systems. IEEE Trans Ind Electron 55(8):2845–2859
28. Guerrero JM, Vasquez JC, Matas J, Castilla M, de Vicuna LG (2009) Control strategy for flexible microgrid based on parallel line-interactive UPS systems. IEEE Trans Ind Electron 56(3):726–736
29. Guerrero JM, Vasquez JC, Matas J, Castilla M, Vicuna LGD, Castilla M (2011) Multilevel control of droop-controlled AC and DC microgrids a general approach toward standardization. IEEE Trans Ind Electron 58(1):158–172
30. Guerrero JM, Chandorkar M, Lee TL, Loh PC (2013) Advanced control architectures for intelligent microgrids part I: decentralized and multilevel control. IEEE Trans Ind Electron 60(4):1254–1262
31. Hamilton SL, Gunther EW, Drummond RV, Widergren SE (2006) Interoperability a key element for the grid and DER of the future. In: Proceedings of IEEE PES transmission and distribution, pp 927–931
32. Hamrouni N, Jraidi M, Cherif A (2008) New control strategy for 2-stage grid-connected photovoltaic power system. Renew Energy 33:2212–2221
33. He J, Li YW (2010) Analysis and design of interfacing inverter output virtual impedance in a low voltage microgrid. In: Proceedings of energy conversion congress and exposition (ECCE), pp 2857–2864
34. He J, Li YW, Munir MS (2012) A flexible harmonic control approach through voltage controlled DG-grid interfacing converters. IEEE Trans Ind Electron 59(1):444–455
35. Hua CC, Liao KA, Lin JR (2002) Parallel operation of inverters for distributed photovoltaic power supply system. In: Proceedings of IEEE PESC conference, pp 1979–1983
36. Hui Q, Haddad W (2008) Distributed nonlinear control algorithms for network consensus. Automatica 42:2375–2381
37. IEEE Guide for Design, Integration, and Operation of Distributed Resource Island Systems with Electric Power Systems, IEEE Std 1547.4.2011
38. IEEE Guide for Design, Operation, and Integration of Distributed Resource Island Systems with Electric Power Systems, IEEE Standard 1547.4-2011 (2011)
39. Jadbabaie A, Lin J, Morse AS (2003) Coordination of groups of mobile autonomous agents using nearest neighbor rules. IEEE Trans Autom Control 48(6):988–1001
40. Johansen TA, Storaa C (2002) Energy-based control of a distributed solar collector field. Automatica 38:1191–1199
41. Kakigano H, Miura Y, Ise T, Uchida R (2007) DC voltage control of the DC micro-grid for super high quality distribution. IEEE Trans Ind Appl 127(8):890–897
42. Karimi-Ghartemani M, Boon-Teck O, Bakhshai A (2011) Application of enhanced phase-locked loop system to the computation of synchrophasors. IEEE Trans Power Deliv 26(1): 22–32
43. Katiraei F, Iravani MR (2005) Power management strategies for a microgrid with multiple distributed generation units. IEEE Trans Power Syst 21(4):1821–1831
44. Katiraei F, Iravani MR, Lehn PW (2005) Microgrid autonomous operation during and subsequent to islanding process. IEEE Trans Power Del 20(1):248–257
45. Keyhani A, Marwali MN, Dai M (2010) Integration of green and renewable energy in electric power systems. Wiley, Hoboken
46. Lasseter RH (2002) Microgrid. In: Proceedings of IEEE power engineering society winter meeting, New York, USA, vol 1, pp 305–308

47. Lasseter RH, Akhil A, Marnay C, Stevens J, Dagle J, Guttromson R, Meliopoulous AS, Yinger R, Eto J (2002) White paper on integration of distributed energy resources. The CERTS microgrid concept. In: Proceedings of consortium for electricity reliability technology solutions, pp 1–27, April 2002
48. Lee TL, Cheng PT (2007) Design of a new cooperative harmonic filtering strategy for distributed generation interface converters in an islanding network. IEEE Trans Power Electron 22(5):1919–1927
49. Lee TL, Li JC, Cheng PT (2009) Discrete frequency tuning active filter for power system harmonics. IEEE Trans Power Electron 24(5):1209–1217
50. Lee C-T, Chuang C-C, Chu C-C, Cheng P-T (2009) Control strategies for distributed energy resources interface converters in low voltage microgrid. In: Proceedings of IEEE ECCE, pp 2022–2029
51. Li X, Wang X, Chen G (2004) Pinning a complex dynamical network to its equilibrium. IEEE Trans Circuits Syst I, Reg Papers 51(10):2074–2087
52. Li Y, Vilathgamuwa DM, Loh PC (2005) Microgrid power quality enhancement using a three-phase four-wire grid-interfacing compensator. IEEE Trans Ind Appl 41(6):1707–1719
53. Li Z, Duan Z, Chen G, Huang L (2010) Consensus of multi-agent systems and synchronization of complex networks: a unified viewpoint. IEEE Trans Circuits Syst I, Reg Papers 57(1):213–224
54. Linden D, Reddy TB (eds) (2002) Handbook of batteries, 3rd edn. McGraw-Hill, New York
55. Liu J, Muffoz de la Peffa D, Christofides PD (2009) Distributed model predictive control of nonlinear process systems. AIChE J 55:1171–1184
56. Liu J, Chen X, Muffoz de la Peffa D, Christofides PD (2010) Sequential and iterative architectures for distributed model predictive control of nonlinear process systems. AIChE J 56:2137–2149
57. Lopes JAP, Moreira CL, Madureira AG (2006) Defining control strategies for microgrids islanded operation. IEEE Trans Power Syst 21(2):916–924
58. Marinescu B, Bourles H (1999) Robust predictive control for the flexible coordinated secondary voltage control of large scale power system. IEEE Trans Power Syst 14(4):1262–1268
59. Marwali MN, Keyhani A (2004) Control of distributed generation systems part I: voltage and currents control. IEEE Trans Power Electron 19(6):1541–1550
60. McArthur SDJ, Davidson EM, Catterson VM, Dimeas AL, Hatziargyriou ND, Ponci F, Funabashi T (2007) Multi-agent systems for power engineering applications-part I: concepts, approaches, technical challenges. IEEE Trans Power Syst 22(4):1743–1752
61. Mehrizi-Sani A, Iravani MR (2010) Potential-function based control of a microgrid in islanded and grid-connected models. IEEE Trans Power Syst 25(4):1883–1891
62. Mehrizi-sani A, Iravani R (2012) Online set-point adjustment for trajectory shaping in microgrid applications. IEEE Trans Power Syst 27(1):216–223
63. Mehrizi-Sanir A, Iravani R (2009) Secondary control for microgrids using potential functions: modeling issues. In: Proceedings of CYGRE, pp 182.1–182.9
64. Mhaskar P, El-Farra NH, Christofides PD (2006) Stabilization of nonlinear systems with state and control constraints using Lyapunov-based predictive control. Syst Control Lett 55:650–659
65. National Institute of Standards and Technology (2010) NISF framework and roadmap for smart grid interoperability standards
66. Novak P, Ekelund T, Jovik Y, Schmidtbauer B (1995) Modeling and control of variable-speed wind-turbine drive system dynamics. IEEE Control Syst Mag 15:28–37
67. Pogaku N, Green TC (2006) Harmonic mitigation throughout a distribution system: a distributed-generator-based solution. IEE Proc Gener Trans Distrib 153(3):350–358
68. Pogaku N, Prodanovic M, Green TC (2007) Modeling, analysis and testing of autonomous operation of an inverter-based microgrid. IEEE Trans Power Electron 22(2):613–625
69. Qi W, Liu J, Christofides PD (2011) Supervisory predictive control for long-term scheduling of an integrated wind/solar energy generation and water desalination system. IEEE Trans Control Syst Technol. doi:10.1109/TCST.2011.2119318

70. Qi W, Liu J, Chen X, Christofides PD (2011) Supervisory predictive control of standalone wind-solar energy generation systems. IEEE Trans Control Syst Technol 19:199–207
71. Qu Z (2009) Cooperative control of dynamical systems: applications to autonomous vehicles. Springer, New York
72. Ren W, Beard RW (2008) Distributed consensus in multi-vehicle cooperative control. Springer, Berlin
73. Rodriguez P, Timbus AV, Teodorescu R, Liserre M, Blaabjerg F (2007) Flexible active power control of distributed power generation systems during grid faults. IEEE Trans Ind Electron 54(5):2583–2592
74. Rodriguez P, Luna A, Candela I, Mujal R, Teodorescu R, Blaabjerg F (2011) Multi-resonant frequency-locked loop for grid synchronization of power converters under distorted grid conditions. IEEE Trans Ind Electron 58(1):127–138
75. Ruddell AJ, Dutton AG, Wenzl H, Ropeter C, Sauer DU, Merten J, Orfanogiannis C, Twidell JW, Vezin P (2002) Analysis of battery current microcycles in autonomous renewable energy systems. J Power Sources 112:531–546
76. Salam AA, Mohamed A, Hannan MA (2008) Technical challenges on microgrids. ARPN J Eng Appl Sci 3(6):64–69
77. Salomonsson D, Soder L, Sannino A (2007) An adaptive control system for a dc microgrid for data centers. In: 42nd IEEE IAS annual meeting conference record, pp 2414–2421
78. Sao CK, Lehn W (2008) Control and power management of converter fed microgrids. IEEE Trans Power Syst 23(3):1088–1098
79. Savaghebi M, Jalilian A, Vasquez J, Guerrero J (2012) Secondary control scheme for voltage unbalance compensation in an islanded droop-controlled microgrid. IEEE Trans Smart Grid 3(6):797–807
80. Savaghebi M, Jalilian A, Vasquez J, Guerrero J (2012) Secondary control for voltage quality enhancement in microgrids. IEEE Trans Smart Grid 3(12):1893–1902
81. Savaghebi M, Jalilian A, Vasquez JC, Guerrero JM (2012) Secondary control scheme for voltage unbalance compensation in an islanded droop-controlled microgrid. In: IEEE trans, smart grid, Early Access January 2012
82. Savaghebi M, Jalilian A, Vasquez JC, Guerrero JM (2012) Secondary control scheme for voltage unbalance compensation in an islanded droop-controlled microgrid. In: IEEE trans, smart grid, Early Access February 2012
83. Schnberger J, Duke R, Round SD (2006) DC-bus signaling: a distributed control strategy for a hybrid renewable nanogrid. IEEE Trans Ind Elecron 53(5):1453–1460
84. Simoes MG, Bose BK, Spiegel RJ (1997) Fuzzy logic based intelligent control of a variable speed cage machine wind generation system. IEEE Trans Power Electron 12:87–95
85. Slotine JJE, Li W (2009) Applied nonlinear control. Prentice-Hall, Upper Saddle River
86. Spee R, Enslin JH (1995) Novel control strategies for variable-speed doubly fed wind power generation systems. Renew Energy 6:907–915
87. Sun X, Lee Y-S, Xu D (2003) Modeling, analysis, and implementation of parallel multi-inverter systems with instantaneous average-current sharing scheme. IEEE Trans Power Electron 18(3):844–856
88. Suter G, Werner TG (2009) The distributed control center in a smart grid. In: Proceedings of CIRED, pp 1–4
89. Takeshita T, Matsui N (2003) Current waveform control of PWM converter system for harmonic suppression on distribution system. IEEE Trans Ind Electron 50(6):1134–1139
90. Tan K, Islam S (2004) Optimum control strategies in energy conversion of PMSG wind turbine system without mechanical sensors. IEEE Trans Energy Convers 19:392–399
91. Thringer T (2002) Grid-friendly connecting of constant-speed wind turbines using external resistors. IEEE Trans Energy Convers 17(4):537–542
92. Thiringer T, Linders J (1993) Control by variable rotor speed of fixed-pitch wind turbine operating in speed range. IEEE Trans Energy Convers 8:520–526
93. Uhlen K, Foss BA, Gjosaeter OB (1994) Robust control and analysis of a wind-diesel hybrid power plant. IEEE Trans Energy Convers 9:701–708

94. U.S. Department of Energy (2008) The smart grid: an introduction, Technical report
95. Valenciaga F, Puleston PF (2005) Supervisor control for a stand-alone hybrid generation system using wind and photovoltaic energy. IEEE Trans Energy Convers 20:398–405
96. Valenciaga F, Puleston PF (2008) High-order sliding control for a wind energy conversion system based on a permanent magnet synchronous generator. IEEE Trans Energy Convers 23:860–867
97. Valenciaga F, Puleston PF, Mantz RJ, Battaiotto PE (2000) An adaptive feedback linearization strategy for variable speed wind energy conversion systems. Int J Energy Res 24:151–161
98. Vanthournout K, De Brabandere K, Haesen E, Van den Keybus J, Deconinck G, Belmans R (2005) Agora: distributed tertiary control of distributed resources. In: Proceedings 15th power systems computation conference, Liege, Belgium, 22–25 Aug 2005
99. Venkat AN, Hiskens IA, Rawlings JB, Wright SJ (2008) Distributed MPC strategies with application to power system automatic generation control. IEEE Trans Control Syst Technol 16:192–1206
100. Viczel P (2007) Power electronic converters in dc microgrid. In: Proceedings IEEE 5th International CPE, pp 1–6
101. Villeneuve PL (2004) Concerns generated by islanding. IEEE Power Energy Mag 2(3):49–53
102. Wada K, Fujita H, Akagi H (2002) Considerations of a shunt active filter based on voltage detection for installation on a long distribution feeder. IEEE Trans Ind Appl 38(4):1123–1130
103. Xin H, Qu Z, Seuss J, Maknouninejad A (2011) A self-organizing strategy for power flow control of photovoltaic generators in a distribution network. IEEE Trans Power Syst 26(3):1462–1473
104. Zhang H, Lewis FL, Das A (2011) Optimal design for synchronization of cooperative systems: state feedback, observer, output feedback. IEEE Trans Autom Control 56(8):1948–1952
105. Zhong Q-C, Weiss G (2009) Static synchronous generators for distributed generation and renewable energy. In: Proceedings of IEEE PES/PSCE, pp 1–6
106. Zhu A, Christofides PD, Cohen Y (2009) Effect of thermodynamic restriction on energy cost optimization of RO membrane water desalination. Ind Eng Chem Res 48(2009):6010–6021
107. Zhu A, Rahardianto A, Christofides PD, Cohen Y (2010) Reverse osmosis desalination with high permeability membranes dost optimization and research needs. Desalin Water Treat 15:256–266

Chapter 9
Real-Time Implementation

9.1 Neural-Network-Based Secondary Control

In this section, a neural network (NN)-based distributed secondary control (DSC) to regulate the output voltage and frequency of a smart autonomous microgrid (MG) system is presented. Generally, secondary controller is implemented in a centralized manner using constant gain proportional-plus-integral (PI) controller which may perform well under certain operating conditions only. Also the failure of centralized controller can lead to instability of the whole system. On the other hand, the control technique proposed in this paper is a distributed one based on *NN* which increases the robustness of the system. A well-trained NN supplies the controller with suitable gains according to each operating point. Before training the NN, evolutionary optimization technique, *differential evolution* (DE), is employed to obtain the optimal gains of controller at each operating load condition which forms the training set for NN. Simulation results show that the proposed controller damps the oscillations caused by load changes and restores the output voltage and frequency of the system to their nominal values. The controller also ensures proper load sharing. The performance of the controller is also compared with fixed-gain controller.

9.1.1 Introduction

A networked MG is termed as '*smart microgrid*' [2, 6]. Figure 9.1 shows its simple architecture. It will have high penetration of DG units which when integrated alone raises number of issues [3]. It also makes use of renewable energy resources making it cost effective and environment friendly. The most vital aspect of smart MG is *distributed/decentralized* control using communication network. In other words, it will employ networked control system (NCS) so that we can have a network of DG units exchanging information. Control is the key point here which will be implemented

M.S. Mahmoud and F.M. AL-Sunni, *Control and Optimization of Distributed Generation Systems*, Power Systems, DOI 10.1007/978-3-319-16910-1_9

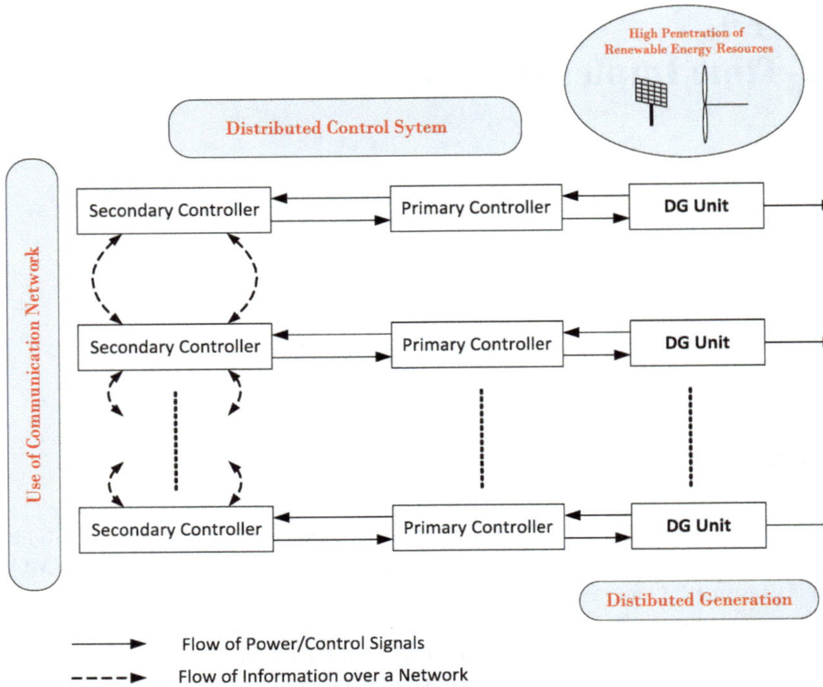

Fig. 9.1 Architecture of smart autonomous microgrid

in a distributed fashion contrary to centralized control in several conventional techniques which can be seen in the literature [4, 5]. This will ensure stability of system, power balancing, proper load sharing, and voltage and frequency regulation.

One of the widely used and crucial control technique is the "multilevel control" [8, 9]. There are three main control levels, each taking care of particular responsibilities. Primary Control level ensures the proper load sharing between the converters. Secondary control removes any steady state error introduced by primary control. Tertiary Control deals with global responsibilities like energy transfer to and from the grid. Tertiary control comes into picture only when the MG is connected to the utility grid. Tertiary control is a decentralized control (DEC) level responsible for global optimization of MG. Since autonomous operation of MG does not involve tertiary control level, we will be only focusing on the primary and secondary control levels of an MG.

The primary control makes use of the droop-based control techniques for its operation. But due to various reasons discussed in next sections, primary level alone is not sufficient for the stable operation of overall MG. To global controllability, a secondary control level is often used. This concept is already used from a long time in the large electrical power systems [10] and is recently adopted in the MG concept.

Secondary control strategies using NCS have been proposed in literature. A pseudo-decentralized control architecture is proposed in [11] which can be used for the optimization of wireless communication network (WCN) with the help of a global supervisory control and local controllers (LOCs). In [12], NCS strategy was applied to a parallel inverter system to achieve superior load sharing and good robustness. Investigation of centralized secondary controller in an MG with primary voltage droops is carried out in [13]. This controller regulates the voltage at pilot points in the MG. In [14], a networked controlled parallel multi-inverter system is proposed to achieve precise load sharing among each module, a centralized controller is used here along with the LOCs.

Most of the works in the literature are based on the *centralized* secondary control, where all the DG in the MG are supervised by a common secondary control. This controller is often termed as microgrid central controller (MGCC). It makes use of communication channel for both sensing the measurements and to send the control signal. MGCC is relatively slow in functioning [15, 16, 17]. If there is any failure in MGCC, the whole system is at risk of going unstable. Depending on only one central control unit for the proper operation is a big drawback in itself. Any fault in controller/communication network can lead to the system collapse.

A distributed secondary controller based on averaging algorithms is proposed in [18]. The controller, which is also termed as distributed averaging proportional integral (*DAPI* controller), regulates the system frequency under time-varying loads. Recently, a new method of implementing secondary control in a distributed fashion using the NCS approach is proposed in [19]. This concept has proved to be better as both the primary and secondary controls are implemented in a distributed way, resulting in individual secondary control for each DG unit.

But the proposed controller in [19] is based on fixed PI gains which may perform well under some operating conditions only but not all. The gains of secondary controller were randomly tuned and lacked a proper defined procedure. Consequently, improper tuning of controller results in the bad adaptation to varying operating conditions. Moreover, PI controllers are not robust enough to accommodate the variations in the load. It is desired to have an intelligent PI-type controller, which when load changes can self-tune its controller gain [20, 21].

A distributed secondary controller based on NN is proposed which can operate over a wide range of operating points. Using DE, the optimized gains of secondary controller are obtained and serve as training pattern for the NN. The voltage and frequency regulation, load sharing performance of the controller are demonstrated using MATLAB®/Simulink® simulations. Performance comparison between the proposed controller and fixed-gain controller is also performed. The simulation results show that the proposed secondary control ensures stable operation of the system under varying loads.

9.1.2 An Autonomous Microgrid

As we learned before, autonomous mode operation of MG is also known as *islanded* MG. This can be caused by two reasons. One is due to any network fault or some failure in the utility grid and another is due to performance of maintenance at planned intervals. An electrical switch will disconnect the MG from main utility grid and result in the autonomous operation of MG [26]. In this work, MG involving only inverter-interfaced DG units is considered. In [23], it is mentioned that the prime mover can be replaced with a *DC* source without loss of generality because they both essentially serve the same purpose. This simplification allows us to study the behavior of inverter-based generators without actually using a prime mover.

Figure 9.2 shows a general structure of an inverter-based MG in autonomous mode. The inverters are classified as voltage source inverters (VSIs) and current source inverter (CSI). Due to their ride through capability and improved power quality (PQ), VSI is mostly used [24]. Load is connected through a LC filter and coupling inductance. The power, voltage, and current controllers constitute the primary control level of any individual DG unit. Small signal modeling of each of the part of MG can be carried out by following the procedure carried out in [26, 36].

As mentioned in [36], d-q reference frame was used to formulate the nonlinear dynamics of DG units. The reference frame of one DG is considered as the common reference frame with frequency ω_{com}. The angle δ between an individual reference frame and common reference frame, satisfies the following equation:

$$\dot{\delta} = \omega - \omega_{com}$$

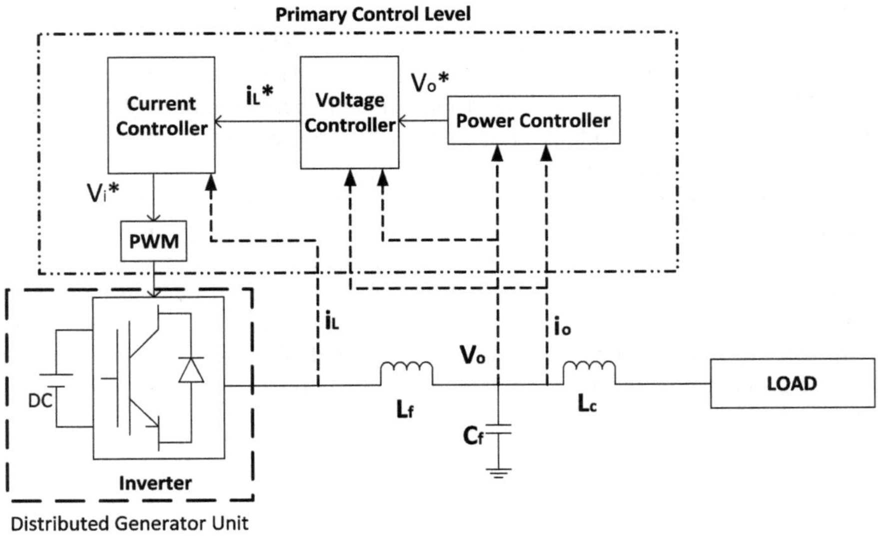

Fig. 9.2 MG in autonomous mode with primary control level

9.1.3 Primary Control

The control technique used at this level is known as *droop*-based control [10]. This type of control makes use of local measurements and does not need any communication medium for its operation. Droop control is a decentralized strategy which ensures proper load sharing. Its main purpose is to share active and reactive powers among DG units at the same time maintaining the output levels of voltage and frequency within limits. In droop technique, there is a desired relationship between the active power P and angular frequency ω and between reactive power Q and voltage V as given below:

$$\omega = \omega_n - m_p P, \quad V = V_n - n_q Q$$

where V_n and ω_n are the nominal values of output voltage and angular frequency, respectively. P and Q are the real and reactive powers, respectively. m_p and n_q are the real and reactive power droop gains, respectively. The frequency ω is set according to the droop gain m_p and output voltage V is set as per droop gain n_q (Fig. 9.3).

Therefore, the output frequency/voltage is decreased when there is an increase in the load real/reactive power and vice versa. The $P\omega$ and QV droop characteristics are shown in Fig. 3.22.

The primary control level can be divided into three different parts namely power, voltage, and current controller. The power controller, shown in Fig. 9.4, sets the inverter output voltage magnitude and frequency with the help of "Droop" characteristics. Basically, it mimics the operation of a synchronous generator (SYGN) which will change the frequency of the output voltage if any change in load is sensed. First, the instantaneous powers are calculated using output voltages and currents, by filtering these instantaneous values by a low-pass filter (LPF) we get the average real and reactive powers. These average values are passed through their respective droop gains to obtain the angular frequency and voltage [29]. The control strategy is

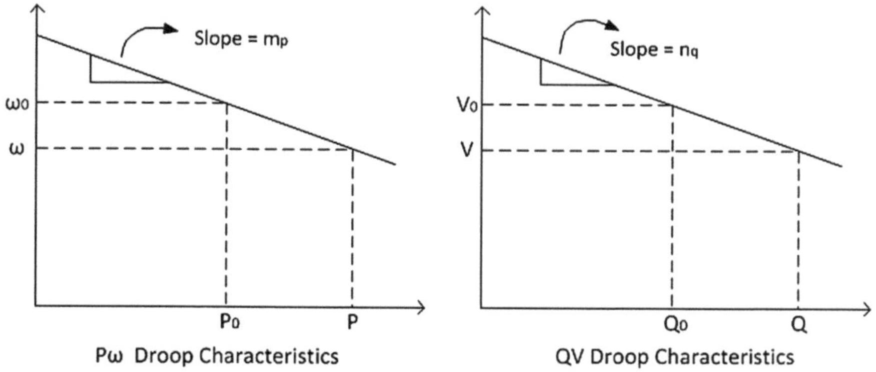

Fig. 9.3 Droop characteristics

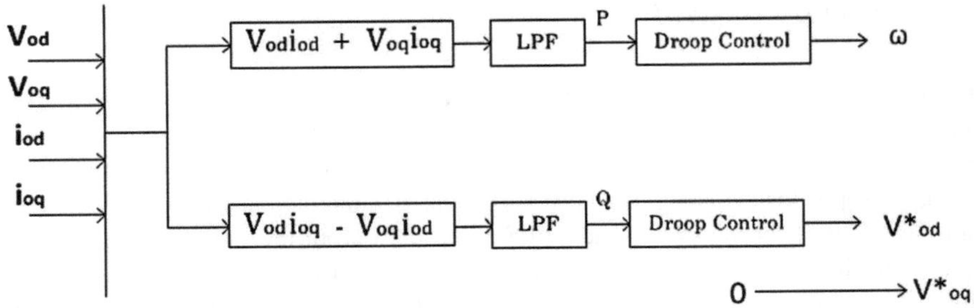

Fig. 9.4 Block diagram of power controller

chosen such that the output voltage magnitude reference is aligned to the d-axis of the inverter reference frame and q-axis reference is set to zero.

The block diagram of voltage controller is shown in Fig. 9.5, a PI controller is used to achieve the output voltage control. The corresponding state equations are given by

$$\dot{\phi}_d = v^*_{od} - v_{od}, \quad \dot{\phi}_q = v^*_{oq} - v_{oq}$$

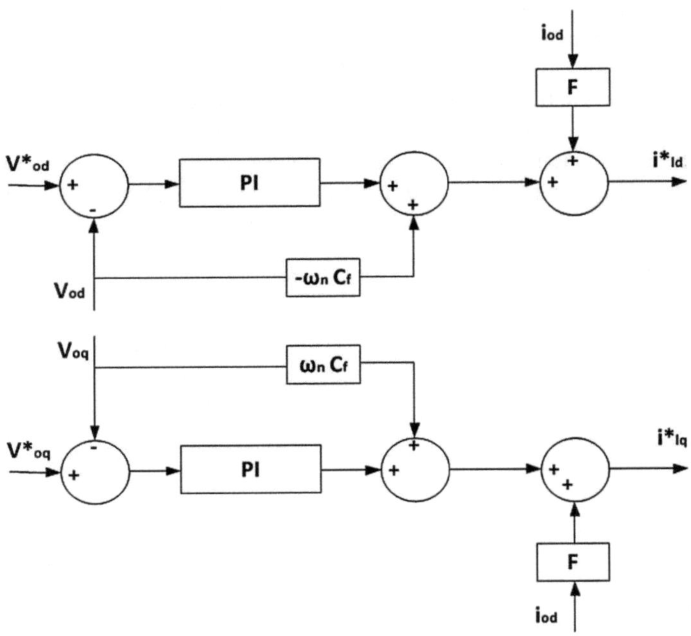

Fig. 9.5 Block diagram of voltage controller

where ϕ_d and ϕ_q are the d-q axis state variables of voltage controller (integrator states), respectively.

$$i_{ld}^* = Fi_{od} - \omega_n C_f v_{oq} + K_{pv}(v_{od}^* - v_{od}) + K_{iv}\phi_d$$
$$i_{lq}^* = Fi_{oq} + \omega_n C_f v_{od} + K_{pv}(v_{oq}^* - v_{oq}) + K_{iv}\phi_q$$

The block diagram of current controller is shown in Fig. 9.6, a PI controller is used to achieve the output filter inductor current. The corresponding state equations are given by

$$\dot{\gamma}_d = i_{ld}^* - i_{ld}, \quad \dot{\gamma}_q = i_{lq}^* - i_{lq}$$

where γ_d and γ_q are the d-q axis state variables of current controller (integrator states), respectively.

$$v_{id}^* = -\omega_n L_f i_{lq} + K_{pc}(i_{ld}^* - i_{ld}) + K_{ic}\gamma_d$$
$$v_{iq}^* = \omega_n L_f i_{ld} + K_{pc}(i_{lq}^* - i_{lq}) + K_{ic}\gamma_q$$

The main purpose of voltage and current controllers is to reject the high-frequency disturbances and damp the output filter to avoid any resonance with the external network. The PI controller provides zero steady state error and stabilizes the system. As it can be seen in figures, additional feedforward gain and decoupling terms are also used. These PI controllers make use of the local measurements to perform the control action. Detailed analysis on modeling and behavior of this system can be found in [36].

Due to virtual impedance and virtual inertias within the primary control, deviations are produced inside the MG which can be observed during its transient behavior.

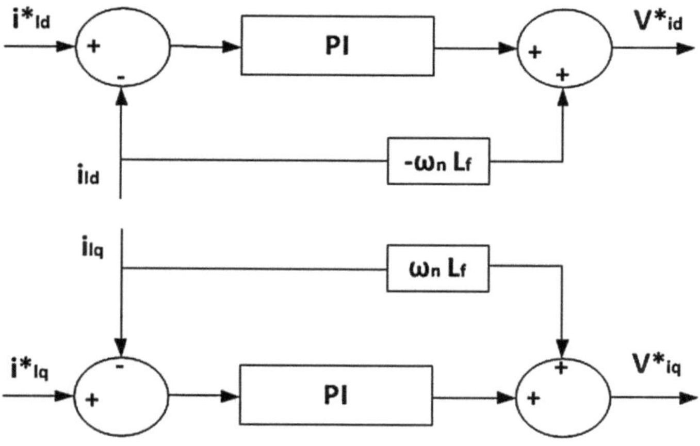

Fig. 9.6 Block diagram of current controller

These deviations can make the output voltage and frequency levels go out of the specified range and hence can cause the destabilization of the system. The main advantages of primary control are that it is fast, does not need any communication medium, and make use of local measurements.

Primary control is a tradeoff between voltage regulation and power sharing. Good sharing of power is achieved at an expense of error in output voltage and vice versa. Poor transient performance, lack of robustness, and steady state error are its main drawback. Therefore, a secondary control level is deployed which brings back the output voltage frequency within the allowable limits.

9.1.4 Distributed Secondary Control

The block diagram of distributed secondary controller is shown in Fig. 9.7. This controller ensures zero steady state error and regulates the deviations produced in output frequency and voltage due to load change toward zero. The control law at the secondary level is given as follows:

$$\delta\omega = K_{p\omega}(\omega_n - \omega_{avg}) + K_{i\omega}\int(\omega_n - \omega_{avg})dt \tag{9.1}$$

$$\delta V = K_{pv}(V_n - V_{avg}) + K_{iv}\int(V_n - V_{avg})dt \tag{9.2}$$

where $K_{p\omega}$, $K_{i\omega}$ are the PI controller parameters for frequency control, K_{pv}, K_{iv} are the PI controller parameters for voltage control. ω_n and V_n are the frequency and voltage set-points. ω_{avg} and V_{avg} are the average values of frequency and voltage, respectively. $\delta\omega$ and δV are the control signals produced by secondary controller.

At each sample time, each DG unit measures their respective output voltage and frequency and sends this information to other DG units by means of a communication network. DG units average the frequency/voltage measurements and compare it with the reference values to produce an error signal. The secondary control then processes this error signal to produce control signals. Because the deviations are produced by the *droop* control, these control signals are sent to the primary control level to remove the steady state error. The output voltage and frequency are restored to their nominal values as follows:

$$\omega = \omega_n - m_p P + \delta\omega, \quad V = V_n - n_q Q + \delta V$$

$\delta\omega$ and δV are the control signals received by primary control from the secondary control to restore the output frequency and voltage, respectively, to their nominal values.

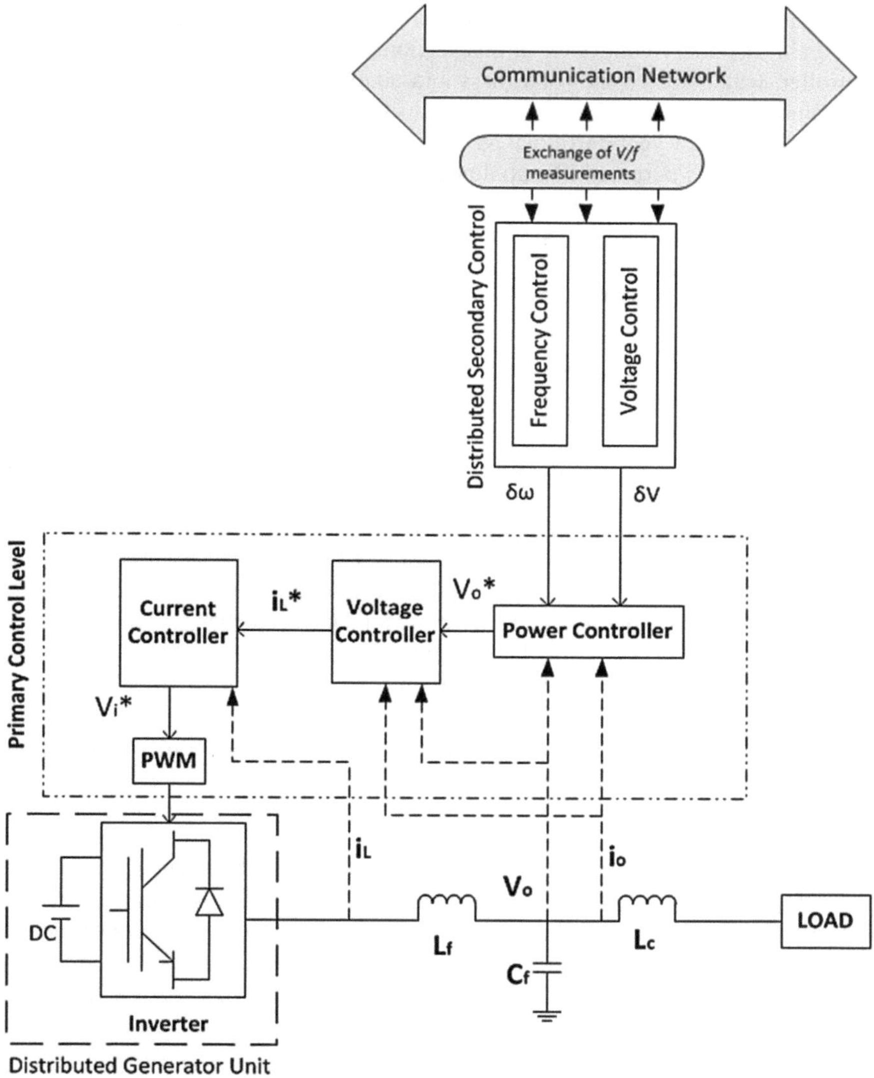

Fig. 9.7 Distributed secondary controller

9.1.5 Neural-Network-Based Distributed Secondary Control

The controller discussed in the above section is based on the fixed-gain PI scheme. Under certain operating points or conditions, this fixed-gain scheme may work fine but its performance degrades at other operating conditions. Also suitable PI gains are obtained using time-consuming trail-and-error methods. Poor tuning of gains deteriorates the system performance. To increase the robustness and adaptability of fixed-gain secondary controller, in this section we propose a NN-based secondary controller, which solves the robustness and adaptability problem of PI controller maintaining its simplicity, reliability, and feasibility.

Over the past few years, artificial NNs are being widely used in the field of control system for various purposes like nonlinear modeling, tuning controller parameters, system identification, etc. [30]. NN consists of *neurons* which are simple computational units. A neuron is a building block of NN and it resembles information processing model of the human brain. The structure of a neuron is shown in Fig. 9.8. Any kth neuron can be defined mathematically as [31]

$$v_k = \sum_{j=1}^{p} w_{kj}x_j + w_{ko}, \quad y_k = f(v_k)$$

where $x_1, x_2 \ldots x_p$ denotes input signals, $w_{k1}, w_{k2} \ldots w_{kp}$ denotes the synaptic weights of kth neuron, w_{k0} is the bias, v_k denotes the linear combiner output, $f(.)$ is the activation function, and y_k denotes the output of the neuron.

Figure 9.9 illustrates the block diagram of NN-based secondary controller. This controller can self-tune the PI gains as per various operating conditions. A trained artificial NN provides optimal gains to the secondary controller whenever the load changes, i.e., input to the NN is the load value and its output are the corresponding PI gains. The secondary controller then produces a control signal as per the control law given by expression in (9.1) and (9.2). The control signals produced are sent to the primary control level of the respective DG unit for compensating the errors. This way the proposed secondary controller dynamically regulates the output voltage and frequency for time-varying load.

Before using the NN for self-tuning, it has to be trained offline by learning (or training) process. Training is effective only if the network output matches the desired output for each training pattern. For this purpose, a training set is required, which is a set of input and desired output data. It is very important to have a proper training set which can otherwise affect the accuracy of NN [35]. Therefore, *evolutionary* computational technique known as DE is used to obtain a proper training set. For each

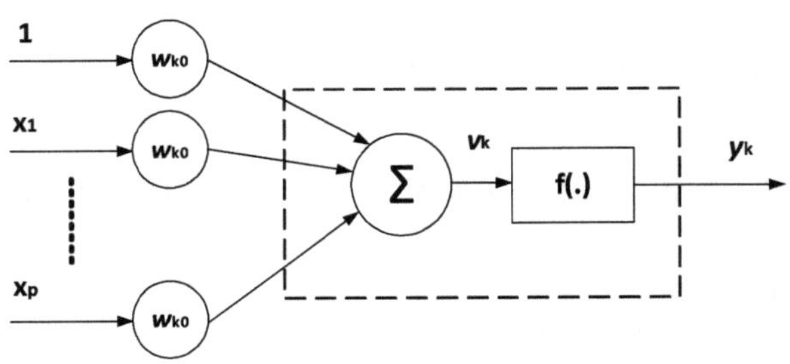

Fig. 9.8 Artificial neuron model

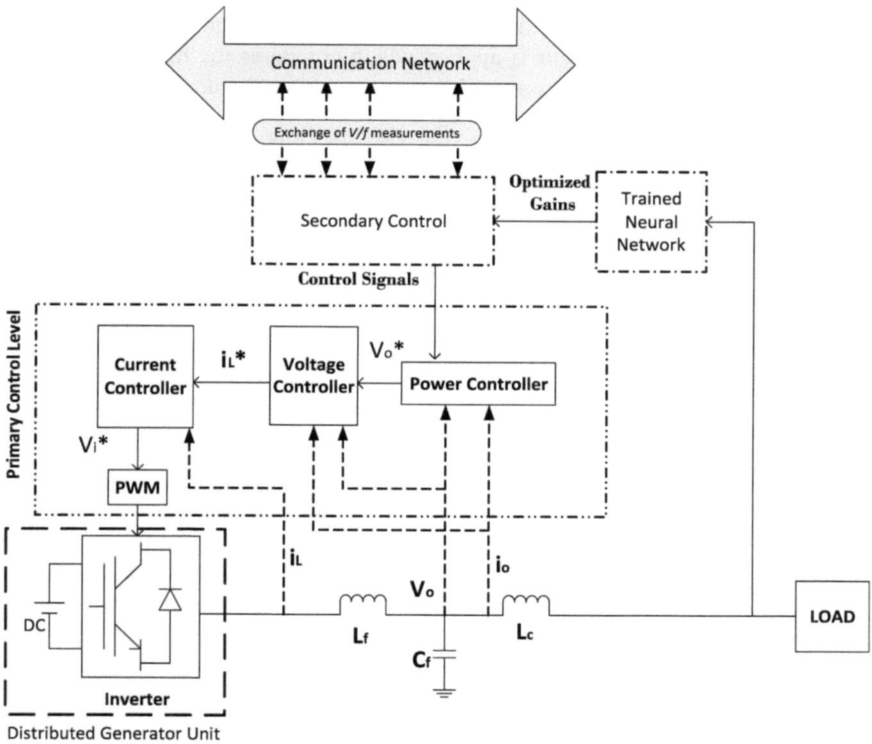

Fig. 9.9 Neural-network-based distributed secondary control

load value (operating point), *DE* is employed to perform the optimization process and to provide the optimized values of PI gains which will give the proper regulation of output voltage and frequency.

9.1.6 Differential Evolution

Most of the problems relating to engineering science cannot be solved using analytical methods, especially global optimization problems are solved using *evolutionary* algorithms [38]. These algorithms are used to find the near optimal solution for wide range of problems. DE is one such novel evolutionary algorithm which uses simple population-based stochastic search for optimizing functions with real value parameters [39].

DE produces a new vector by adding perturbation of two vectors to a third vector. This process is the main differential and is termed as *mutation*. The new vector produced is combined with predefined parameters in accordance with a set of rules. This process is called as *crossover*. This operation is performed to enhance the searching process. Thereafter, an operator is applied which compares the fitness function of two competing vectors to determine who can survive for the next generation. This process is known as *selection* process [40, 41].

The objective function (or performance index) used is the integral of time multiply squared error (ITSE) defined as follows:

$$J_{ITSE} = \int te^2(t)dt$$

where e is the error which is equal to $(\omega_n - \omega_{avg})$ for frequency control and $(V_n - V_{avg})$ for voltage control. The optimization problem is defined as

$$\min[\max(J_{ITSE})]$$

DE algorithm is coded in MATLAB/Simulink and implemented online in order to minimize the integral error and to obtain the optimal PI gains of the secondary controller. DE specifications and the final optimized gains are shown in Fig. 9.11. With the help of a flowchart, the DE process of is explained in detail in Fig. 9.10.

The fitness versus number of iterations graph for frequency control corresponding to above optimization details is shown in Fig. 9.12, where the fitness corresponds to ITSE. It can be seen that the fitness is gradually reduced which in turn reduces the steady state error.

To obtain the optimal PI gains for one operating condition, approximately 1 h was required. To reduce the collection time of training set, only 36 operating points were considered. From 100 to 7500 W, the above optimization process was repeated for 36 different load values with an interval of approximately 200 W. For each load value, optimal PI gains were obtained for both frequency and voltage control. The load values and their corresponding optimized controller gains form the *training set* for NN.

9.1.7 NN Training

NN resembles the adaptive control since they learn from the set of example data rather than having to be programmed in a conventional way [33] and so a set of data called *training set* is required to train the NN and adjust its synaptic weights and thresholds. The training data were obtained from the DE optimization algorithm explained in previous section.

To design and train an artificial NN, the NN *toolbox* [34] available in MATLAB/Simulink is used. The command "*nntool*" opens the network/data manager window, which allows import, create, use, and export NNs and data. The structure of NN used is of feedforward type as shown in Fig. 9.13. It consists of two hidden

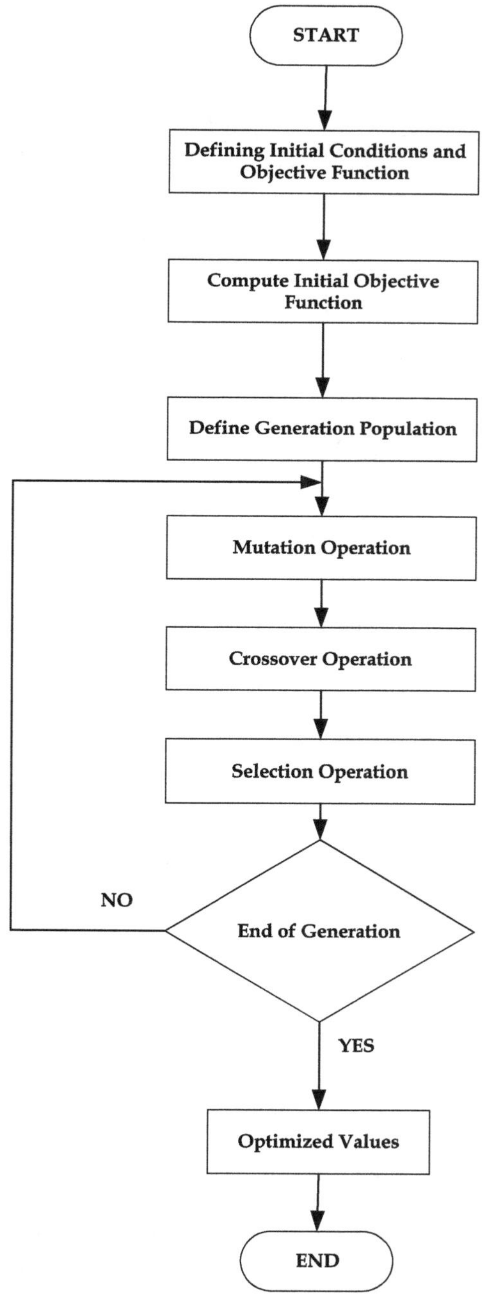

Fig. 9.10 Flow-chart of differential evolution

Specifications	No. of Iterations	Initial Population	Mutation Factor	Crossover Factor	ODE Solver
Value	100	20	0.5	0.6	ODE45

Optimized Values / Load Value	Frequency Control		Voltage Control	
	Proportional Gain	Integral Gain	Proportional Gain	Integral Gain
4500 Watts	7.4072	10.8121	10.5367	10.7784

Fig. 9.11 Optimization details

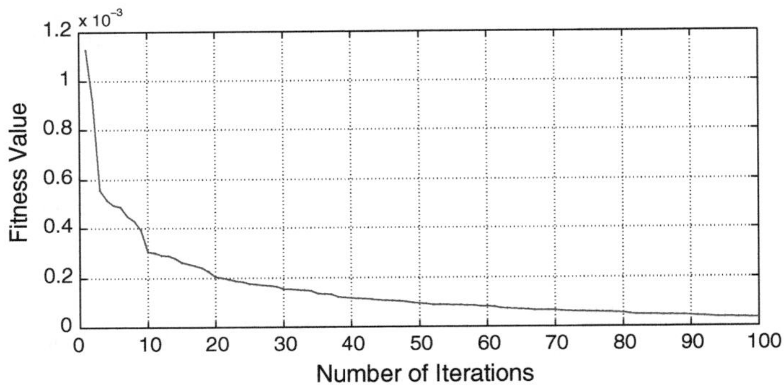

Fig. 9.12 Fitness versus iteration curve

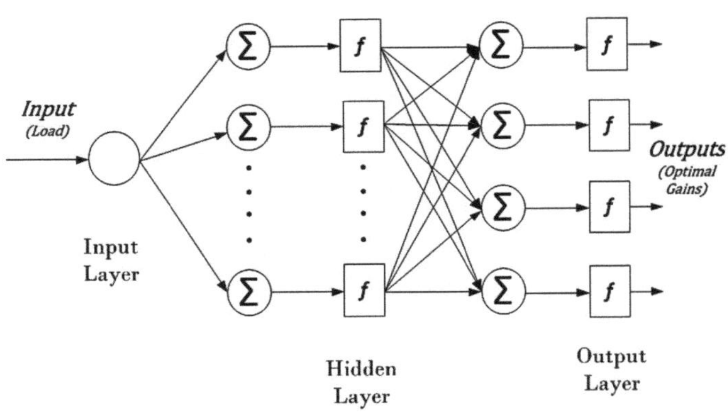

Fig. 9.13 Structure of feedforward NN

layer with 1 input node, 10 Hidden nodes and 4 output nodes. As can be seen, the flow of signal is unidirectional, i.e., output of each neuron is connected to the input of a neuron in the next layer. Depending on the activity level at the input of a neuron, the activation function defines its output [32].

Table 9.1 Neural network training details

S.no.	Network property
1	Adaption learning function: improved gradient descent
2	Performance function: mean squared normalized error
3	Transfer function: hyperbolic tangent sigmoid transfer function

To train the NN Levenberg–Marquardt backpropagation [36] algorithm was used. It is a type of back propagation algorithm [37] mostly used for approximation of function, mode identification and classification, data compression, and so on. The other details of NN training are tabulated in Table 9.1. The NN inputs are the load values R_L and the outputs generated by the NN are the optimal gains of secondary controller corresponding to that load.

9.1.8 Simulation Results

The simulations were performed in MATLAB/Simulink environment. A nonlinear model of the multiple DG units is designed using *SimPowerSystems* Library. Figure 9.14 shows an autonomous MG system developed in the Simulink. There are a total of three DG units connected to a three-phase load by means of line impedance given by $R_{l1} = 0.23\,\Omega$, $L_{l1} = 31.8\,\mu\text{H}$, $R_{l2} = 0.35\,\Omega$, $L_{l2} = 184.7\,\mu\text{H}$ and $R_{l3} = 0.18\,\Omega$, $L_{l3} = 0.0022$. The other parameters of the system and their values are given in Table 9.2.

Initially, the MG system is operated under no load conditions with only primary control enabled. After 5 s, a load of 4.5 kW is applied to the system. The response of output frequency and voltage from no load to sudden application of load is shown in Figs. 9.15 and 9.16. Frequency and voltage deviations can be seen after t = 5 s when load is realized on the system. DG3 which is closely located to the load responds quickly and took the major part of the transient when compared to other DG units.

To regulate the output voltage and frequency to their nominal values and to eliminate the steady state error secondary control is enabled. To demonstrate the effectiveness of proposed controller, a comparative analysis between the fixed-gain secondary control and NN-based secondary control is performed.

Figures 9.17, 9.18 and 9.19 shows the comparison between the fixed-gain DSC and NN-based DSC. Regulation of output voltage, frequency, and load sharing capability of two controllers is depicted in these figures, respectively. By observing the

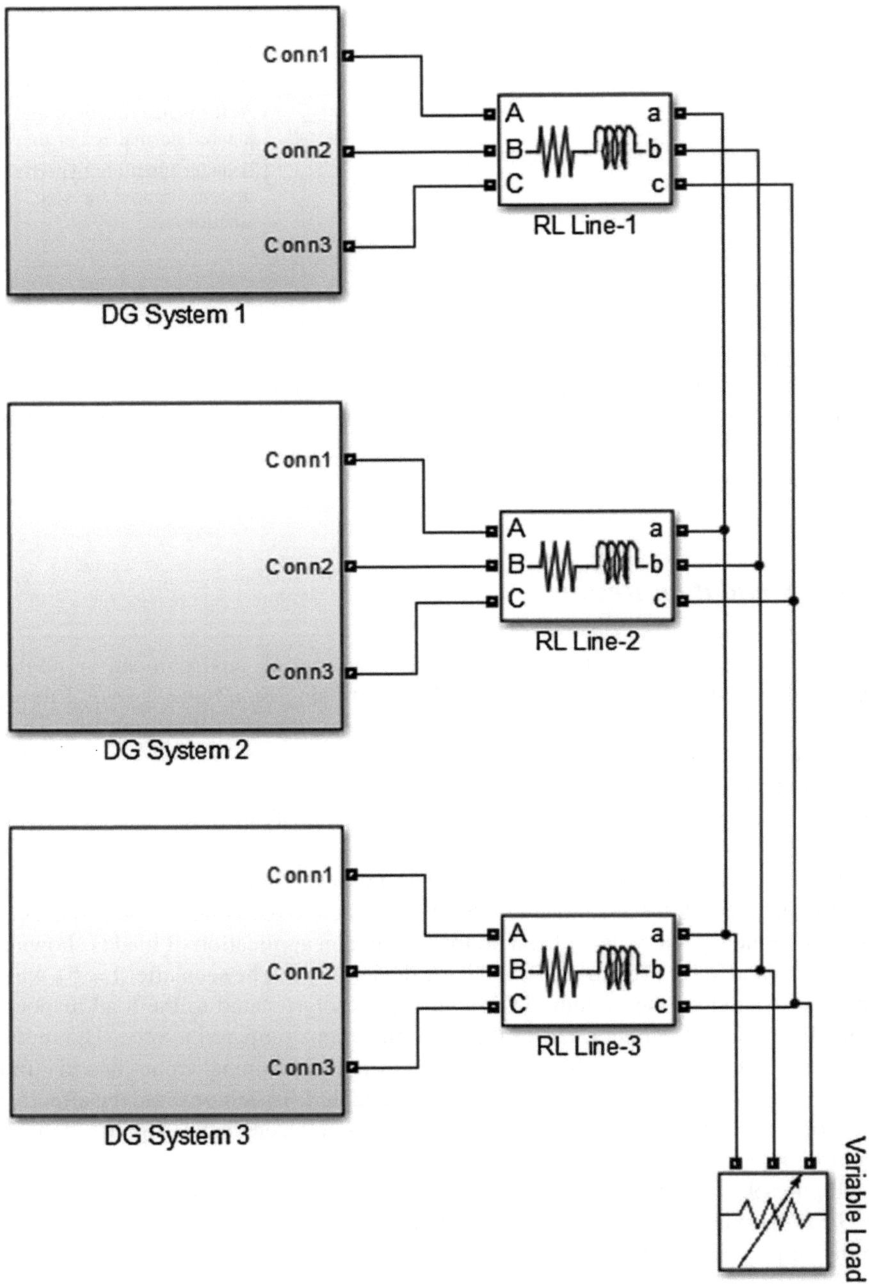

Fig. 9.14 Simulink model of three DG system

Table 9.2 System parameters

Symbol	Quantity	Value
L_f	Filter inductance	1.35 mH
r_f	Filter resistance	0.1 Ω
C_f	Filter capacitance	50 μF
L_c	Coupling inductance	0.35 mH
r_c	Coupling resistance	0.03 Ω
V_n	Nominal voltage	381 V
ω_n	Nominal frequency	314 rad/s
ω_c	Cutoff frequency of LPF	31.4 rad/s
f_s	Switching frequency	8 kHz
m_p	Real power droop gain	9.4×10^{-5}
n_q	Reactive power droop gain	1.3×10^{-3}
K_{pv}	Proportional gain of voltage controller	0.05
K_{iv}	Integral gain of voltage controller	390
K_{pc}	Proportional gain of current controller	10.5
K_{ic}	Integral gain of current controller	16,000
F	Feedforward gain of voltage controller	0.75

comparison results, it can be seen that proposed controller is superior than the conventional one in responding to the load changes, its response is quicker than the conventional one.

To demonstrate the robustness and adaptability of the proposed controller under varying loads, the system is subjected to variable load with respect to time. The load is changed after every 20 s. Figures 9.20 and 9.21 show the output voltage and frequency response of all the 3 DG units under time-varying load. At each instant when there is a sudden change in load, transients can be observed. These transients are slowly regulated toward zero by the controller so that output voltage and frequency are slowly restored to their nominal values. Therefore, proper regulation of output voltage and

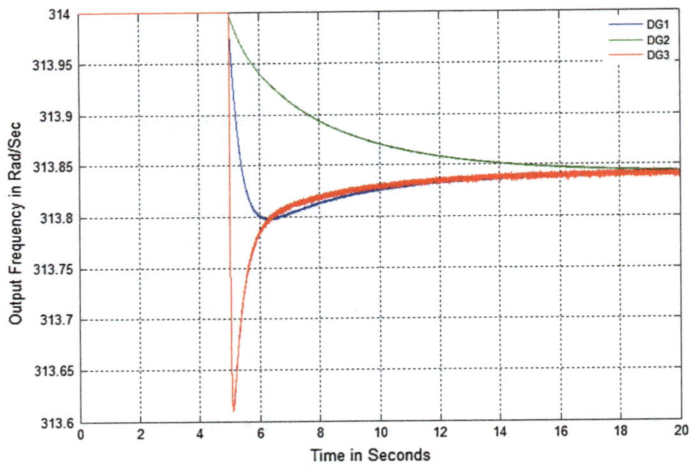

Fig. 9.15 Frequency response under sudden application of load

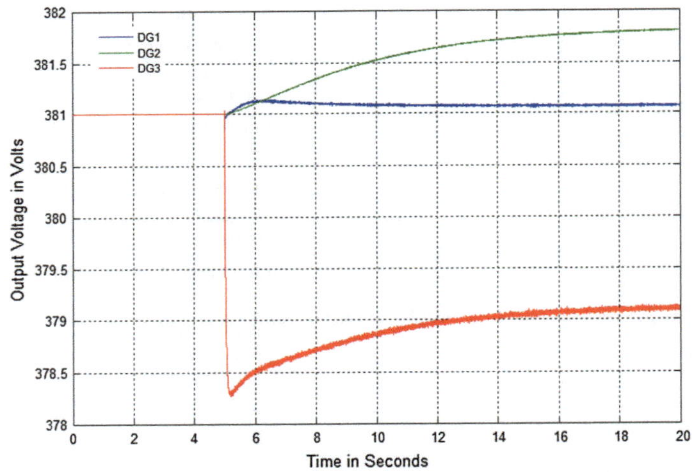

Fig. 9.16 Voltage response under sudden application of load

frequency is achieved. Figure 9.22 shows the load sharing response among the DG units. As can be seen, DG1 and DG2 reacted slow to the load change compared to DG3, which shared the major part after every change in load. After sometime, all the DG units are seen sharing equal amount of load. Therefore, proper load balancing is also achieved.

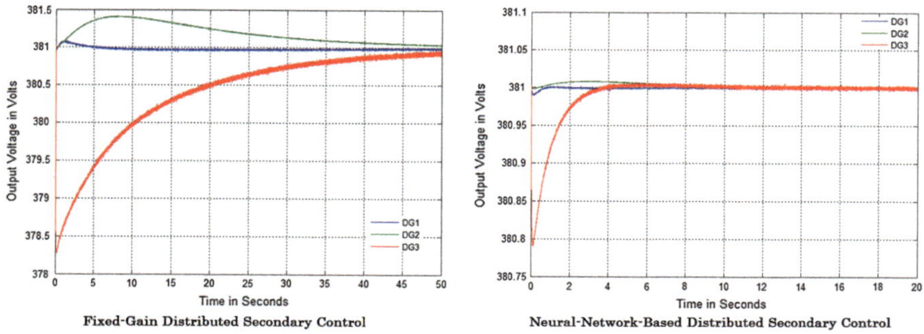

Fig. 9.17 Performance comparison for voltage regulation

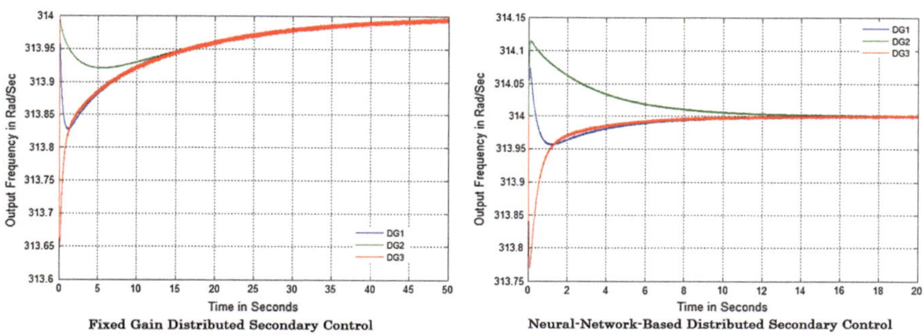

Fig. 9.18 Performance comparison for frequency regulation

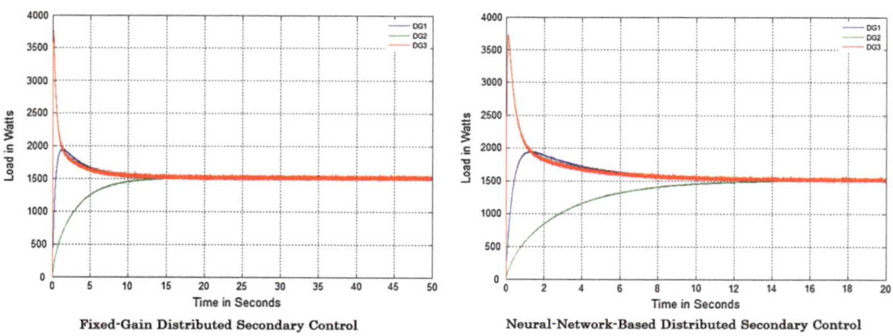

Fig. 9.19 Performance comparison for load sharing

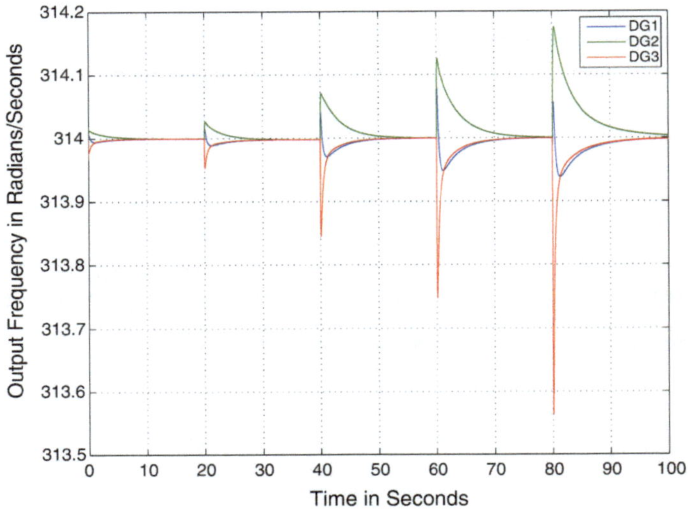

Fig. 9.20 Output frequency under varying load

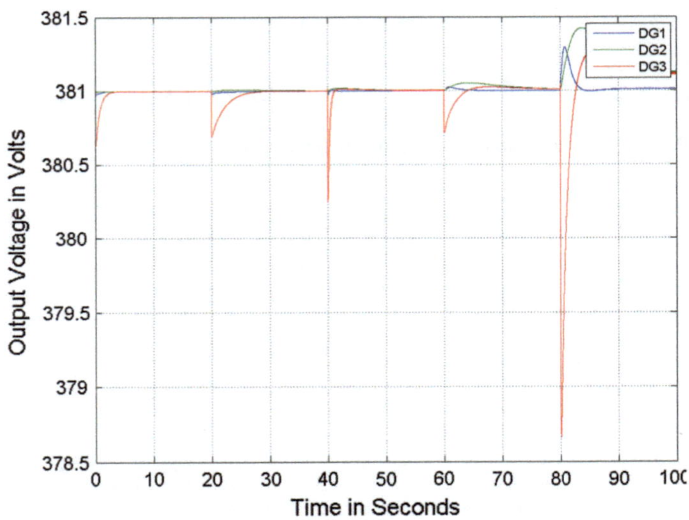

Fig. 9.21 Output voltage under varying load

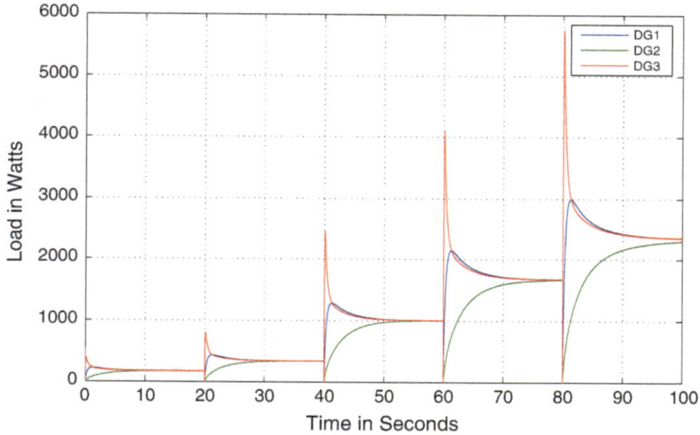

Fig. 9.22 Load sharing among the DG units

9.2 Optimal Control for Autonomous Microgrid

9.2.1 Introduction

Nowadays, with the fast growing concern over greenhouse gas (GHG) emissions and other environmental issues, DG sources such as wind, photovoltaic (PV), micro turbines, and fuel cells are being connected rapidly to the electricity network [3, 9, 18, 43]. Most DG units have been interfaced using power electronics to facilitate the operation and control of DGs [6]. Connecting multiple customers with multiple DG sources and storage units is defined as MG [26]. MG operates in autonomous mode as well as grid-connected mode [4, 15, 22, 35, 47]. In the grid-connected mode, the MG is used to improve the dynamic response of the utility. For example, a grid-connected PV is connected at the end of transmission line to improve the supply PQ and to stabilize line voltage [3]. On the other side, autonomous MG is used to support the loads which are far away from center stations to eliminate the cost of the long transmission lines [35]. Control and dynamics of DGs are the most important MG aspects as several control strategies have presented to control the inverter-based DGs in both mode [17, 19, 31, 38, 46]. In the autonomous mode, current, voltage, and power controllers are generally used to control the DG. The voltage and current PI controllers are implemented to damp the output filter and reject high-frequency disturbances to avoid resonance with the external network [35]. The droop power controller is utilized to share the powers between DGs. The behavior of the inverter of the DG is similar to the synchronous machine (SM) behavior where the voltage magnitude depends on reactive power, while the power angle depends mainly on real power. So the voltage magnitude of the MG can be obtained by adjusting active power while the frequency can be adjusted using the reactive power. The stability is affected by power sharing coefficients and PI controller parameters [22]. In general, selection of the power sharing coefficients and controller parameters carefully will promote the system performance against disturbances and load changes. Recently, this problem has been solved using trial and error methods where the researchers tried to study the effect of changing power sharing coefficients and the controller parameters on the MG stability [28, 29]. Heuristic techniques such as genetic algorithm and particle swarm optimization (PSO) have been used to solve the trial problems [34]. PSO has been considered as one of the favorable optimization technique due to its robustness, computational efficiency, and simplicity. Generally, PSO emulated the organisms' behavior such as bird flocking and fish schooling. It merges social psychology principles in socio-cognition human agents and evolutionary computations.

For dynamic analysis of the power system, using offline simulation usually is not enough especially if the system includes power inverter devices since the operation of these inverters depends mainly on a high switching frequency. The real-time digital simulator (RTDS) is a well-established real-time simulation tool for prototyping and hardware-in-loop testing. It also provides accurate, fast, reliable, and cost effective study of complex power systems [37]. Optimal structures, unbalance problems, and control strategies of the MGs containing multienergy generators have been studied using MG test beds built in the laboratory [37]. In particular, RTDS has been used for real-time tests for DGs and its controllers of MGs [13, 23, 24]. A new unified controller for use with each DG system in the MG was proposed and implemented on RTDS [23].

In this section, analysis and controller design of the autonomous MG on RTDS have been investigated. The dynamic model of an autonomous MG containing VSI, power, current, and voltage controllers, coupling inductance, LC filter, lines, and loads has been developed and implemented on RTDS small time-step environment to accommodate high switching frequency. With the aim of stability enhancement of the autonomous MG, a nonlinear time domain simulation-based objective function is employed and the controller design is formulated as an optimization problem where PSO is used to solve this design problem. The controller parameters and the power sharing coefficients are optimized to enhance the stability of the MG. The MG performance with the proposed optimal controllers has been examined under different disturbances using the nonlinear time domain simulations. The performance of the considered autonomous MG has been verified using RTDS.

9.2.2 Autonomous Microgrid Controller

The autonomous MG with its controllers is shown in Fig. 9.23. The power, voltage, and current controllers are used to control the output powers of the DG units as well as to maintain the voltage and frequency of this autonomous MG system [13].

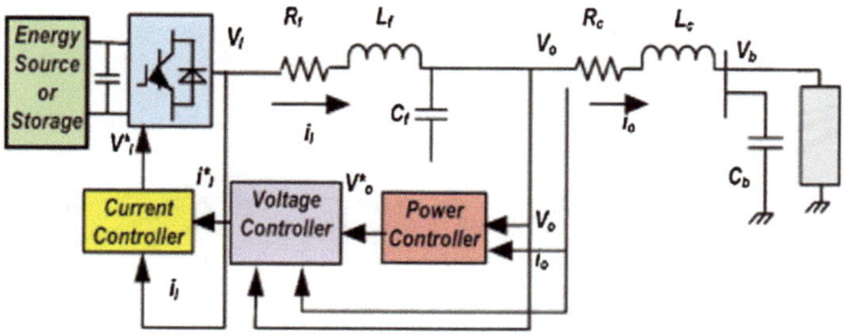

Fig. 9.23 Autonomous microgrid controllers

9.2.3 Power Controller

In the autonomous MG, the DG units are controlled to share the load demands (LDs) between them. The conventional droop method shown in Fig. 9.24 is a common control method used to share the loads between these DGs. The power sharing concept is shown in Fig. 9.25. It emulates the SYGN where the active power depends on the power angle, while the reactive power mostly depends on the output voltage amplitude. First, the measured output current and voltage are used to determine the instantaneous measured active and reactive powers. Then these measured powers are filtered through LPF to obtain the real power "P_c" and reactive power "Q_c". Finally, the frequency corresponding to the active power and the d-component of the output voltage reference v^*_{od} corresponding to the reactive power can be determined as follows [17]:

$$\omega = \omega_n - m_p P_c, \quad \dot{\theta} = \omega \tag{9.3}$$

$$v^*_{od} = V_n - n_q Q_c, \quad v^*_{oq} = 0 \tag{9.4}$$

where ω_n and V_n are the nominal values of DG angular frequency and DG voltage magnitude, P_c and Q_c are the measured real and reactive powers after the low-pass filtering, m_p and n_q are the droop gains or the real and reactive power droop slopes. The d-component of the output voltage is taken as a reference for the inverter

Fig. 9.24 Power controller

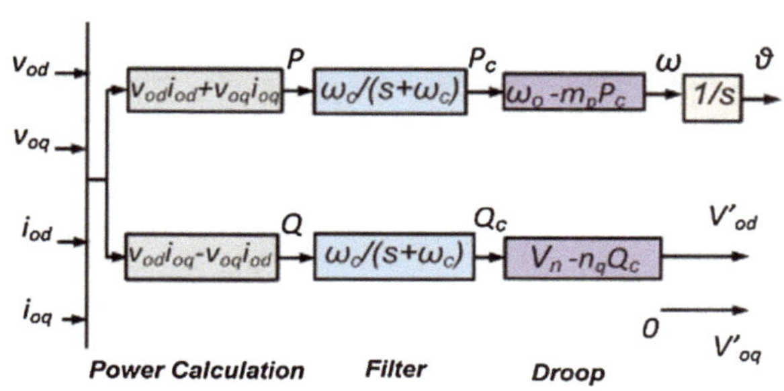

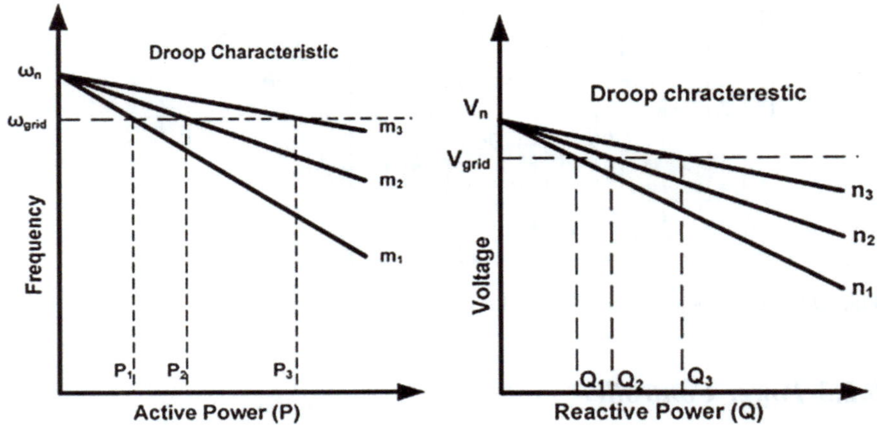

Fig. 9.25 Droop characteristics

reference frame while the q-component is set to zero. The conventional droop method depends mainly on the power calculation, filter characteristics, the droop coefficients, and the output impedance [14]. These parameters make the transient response very limited then the inverter dynamics cannot be independently controlled.

9.2.4 Voltage Controller

The PI voltage controller shown in Fig. 9.26 is implemented to control the output voltage. The state equations of the voltage controller are written as [17]:

$$\phi_d^* = v_{od}^* - v_{od}, \quad \phi_q^* = v_{oq}^* - v_{oq} \tag{9.5}$$

$$\begin{aligned} i_{ld}^* &= Fi_{od} - \omega_n C_f v_{oq} + K_{pv}(v_{od}^* - v_{od}) + K_{iv}\phi_d \\ i_{lq}^* &= Fi_{oq} + w_n C_f v_{od} + K_{pv}(v_{oq}^* - v_{oq}) + K_{iv}\phi_q \end{aligned} \tag{9.6}$$

where F is the voltage controller feedforward gain.

Fig. 9.26 Voltage controller

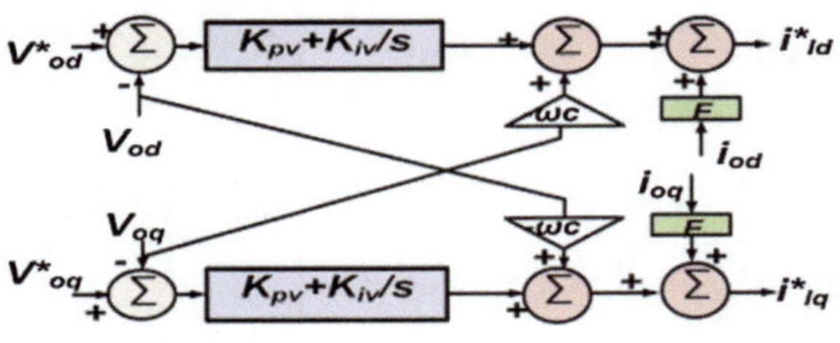

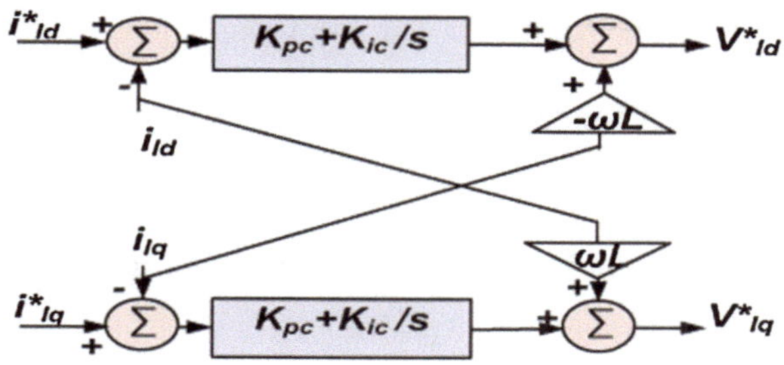

Fig. 9.27 Current controller

9.2.5 Current Controller

Figure 9.27 shows the current controller. The output filter inductor current is controlled by a standard PI controller. The state equations of the current controller along with the algebraic equations are:

$$\gamma_d^* = i_{ld}^* - i_{ld}, \quad \gamma_q^* = i_{lq}^* - i_{lq} \tag{9.7}$$

$$v_{id}^* = -\omega_n L_f i_{lq} + K_{pc}(i_{ld}^* - i_{ld}) + K_{ic}\gamma_d$$
$$v_{iq}^* = \omega_n L_f i_{lq} + K_{pc}(i_{lq}^* - i_{lq}) + K_{ic}\gamma_q \tag{9.8}$$

9.2.6 Coupling Inductance and Filter

The state equations of the coupling inductance and LC filter is given as follows:

$$\dot{i}_{ld} = -\frac{r_f}{L_f} i_{ld} + \omega i_{lq} + \frac{1}{L_f}(v_{id} - v_{od}) \tag{9.9}$$

$$\dot{i}_{lq} = -\frac{r_f}{L_f} i_{lq} - \omega i_{ld} + \frac{1}{L_f}(v_{iq} - v_{oq}) \tag{9.10}$$

$$\dot{v}_{od} = \omega v_{oq} + \frac{1}{C_f}(i_{ld} - i_{od}) \tag{9.11}$$

$$\dot{v}_{oq} = -\omega v_{od} + \frac{1}{C_f}(i_{lq} - i_{oq}) \tag{9.12}$$

$$\dot{i}_{od} = -\frac{r_c}{L_c} i_{od} + \omega i_{oq} + \frac{1}{L_c}(v_{od} - v_{bd}) \tag{9.13}$$

$$\dot{i}_{oq} = -\frac{r_c}{L_c} i_{oq} - \omega i_{od} + \frac{1}{L_c}(v_{oq} - v_{bq}) \tag{9.14}$$

It is assumed that the required output voltage of the inverter is same as the input voltage of the inverter.

9.2.7 Lines Model

On a common reference frame, the line current state equations of the ith line connected between nodes j and k can be written as:

$$\dot{i}_{lineDi} = -\frac{r_{linei}}{L_{linei}} i_{lineDi} + \omega i_{lineQi} + \frac{1}{L_{linei}}(v_{bDj} - v_{bDk}) \tag{9.15}$$

$$\dot{i}_{lineQi} = -\frac{r_{linei}}{L_{linei}} i_{lineQi} - \omega i_{lineQi} + \frac{1}{L_{linei}}(v_{bQj} - v_{bQk}) \tag{9.16}$$

9.2.8 Load Model

Considering general RL load model, the state equations of the load current and load voltages of the ith node are written as:

$$\dot{i}_{loadDi} = -\frac{R_{loadi}}{L_{loadi}} i_{loadDi} + \omega i_{loadQi} + \frac{1}{L_{loadi}} v_{bDi} \tag{9.17}$$

$$\dot{i}_{loadQi} = -\frac{R_{loadi}}{L_{loadi}} i_{loadQi} - \omega i_{loadDi} + \frac{1}{L_{loadi}} v_{bQi} \tag{9.18}$$

$$\dot{v}_{bDi} = \omega v_{bQi} + \frac{1}{C_j}(1_{oDi} - i_{loadDi} \pm i_{lineDi,j}) \tag{9.19}$$

$$\dot{v}_{bQi} = -\omega v_{bDi} + \frac{1}{C_f}(i_{oQi} - i_{loadQi} \pm i_{lineQi,j}) \tag{9.20}$$

9.3 Problem Formulation

To enhance the MG stability, nonlinear time model is presented. To achieve MG stability, the power sharing and voltage and current controller parameters are the main effecting parameters. The design problem is formulated as an optimization problem where PSO is proposed to obtain the optimized parameters. PSO is a population-based stochastic optimization technique [34]. It is inspired by social behavior of bird flocking or fish schooling. The inertia, the personal best, and the group best are the parameters that affect the particles motion. For the next time step, the direction of the movement corresponding to each particle is determined by the interaction of these components. The inertia is defined as the tendency of the particles to move in the original movement direction. The best estimation of the particle objective function is called personal best. The best position of the group is the best swarm solution in any given time step. The steps and advantages of the PSO are summarized in [1]. The proposed PSO-based approach is implemented. Nonlinear time domain simulation-based objective function is given as follows:

$$J = \int_{t=0}^{t_{sim}} (P_m - P_{ref})^2 dt \tag{9.21}$$

The objective to minimize the objective function "J" with the following constraints:

$$k_{pv}^{min} \leq k_{pv} \leq k_{pv}^{max} \tag{9.22}$$

$$k_{pv}^{max} \leq k_{iv} \leq k_{iv}^{max} \tag{9.23}$$

$$k_{pc}^{min} \leq k_{pc} \leq k_{pc}^{max} \tag{9.24}$$

$$k_{ic}^{min} \leq k_{ic} \leq k_{ic}^{max} \tag{9.25}$$

$$m_p^{min} \leq m_p \leq m_p^{max} \tag{9.26}$$

$$n_q^{min} \leq n_q \leq n_q^{max} \tag{9.27}$$

where k_{pv}, k_{iv}, k_{pc}, k_{ic}, m_p, and n_q are the optimized parameters.

9.3.1 Results and Discussions

The autonomous MG considered in this work contains three inverter-based DGs (10 kVA) connected with two loads through coupling inductances, filters, and two transmission lines. The system parameters and optimal parameters are given in Tables 9.3 and 9.4. The MATLAB nonlinear model code of the autonomous MG was developed to investigate the MG stability. First, the impact of the optimal power

Table 9.3 System parameters of the autonomous microgrid

Parameter	Value	Parameter	Value
f_s	8 kHz	V_n	381 V
L_f	1.35 mH	L_c	0.35 mH
C_f	50e-6 F	C_b	50e-6 F
r_f	0.1 Ω	r_c	0.03 Ω
ω_n	314.1 rad/s	ω_c	31.416 rad/s
$r_1 + jx_1$	(0.23 + j0.1) Ω	$r_2 + jx_2$	(0.35 + j0.58) Ω

Table 9.4 Optimal parameters

	PI voltage controller		PI current controller		Power sharing coefficients	
	k_{pv}	k_{iv}	k_{pc}	k_{ic}	m_p	n_q
DG1	0.531	1.555	39.082	11.311	7.37e?5	0.0013
DG2	0.6145	1.911	23.848	20.104	9.4e?05	0.0013
DG3	0.4598	0.840	31.461	22.682	0.0005	0.0027

sharing and optimized controller parameters on the system stability is assessed due to different disturbances such as over load, fault and step change disturbances. Second, the performance of the proposed controller is verified by implementing the autonomous MG in RTDS as shown in Fig. 9.28.

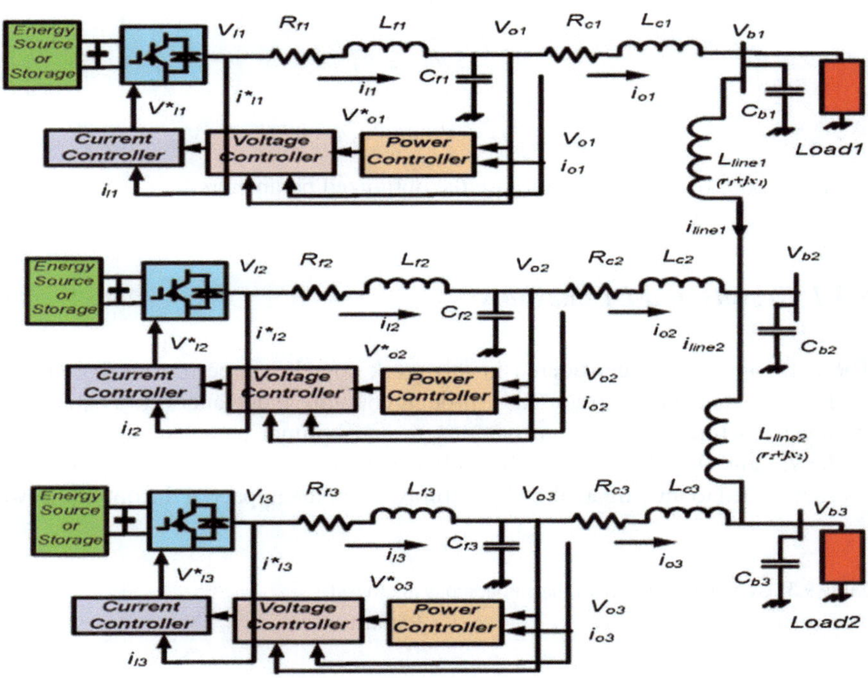

Fig. 9.28 Three inverter-based DGs autonomous microgrid

9.3.2 Nonlinear Time Domain Simulation

Nonlinear time domain-based objective function is proposed to minimize the error in the calculated active power as well as the error in the calculated reactive power. First, the performance of the MG with the proposed controllers and optimal settings under different disturbances has been investigated through the nonlinear time domain simulations. Figures 9.29, 9.30, 9.31, 9.32, 9.33 and 9.34 show the system response at overload disturbance. The results show the satisfactory damping characteristics of the MG considered. The responses show that the damping characteristics are greatly enhanced and the system performance in terms of overshoots and settling time is improved significantly. The output active and reactive power responses of the three DGs are given in Figs. 9.29 and 9.30 while DQ output current responses of the three DGs are presented in Figs. 9.31 and 9.32. Figures 9.33 and 9.34 depict the output d-axis and q-axis voltage responses of all the three inverters for this disturbance.

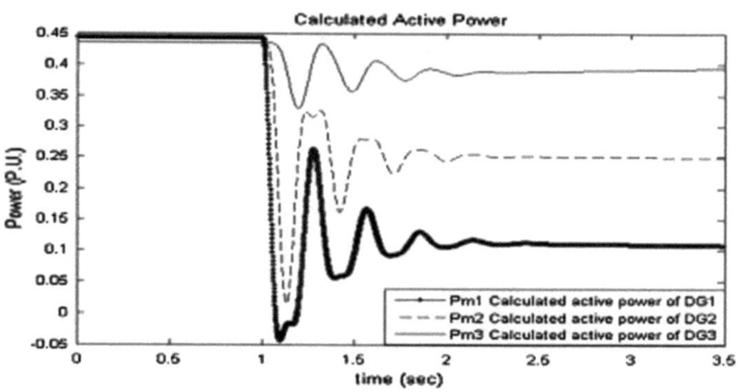

Fig. 9.29 Output active power response of the three DGs at over load disturbance

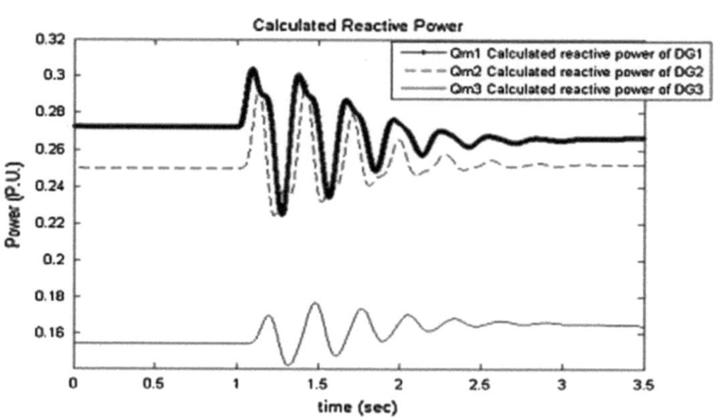

Fig. 9.30 Output reactive power response

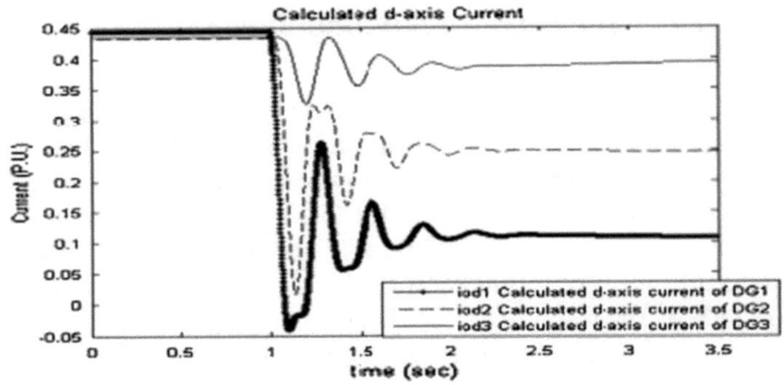

Fig. 9.31 *D*-axis output current response

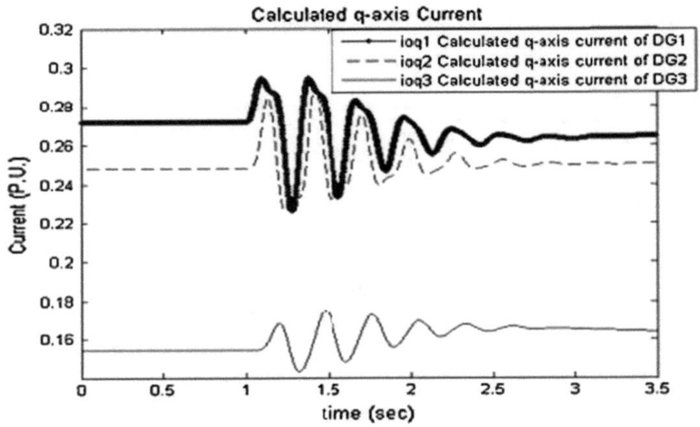

Fig. 9.32 *Q*-axis output current response

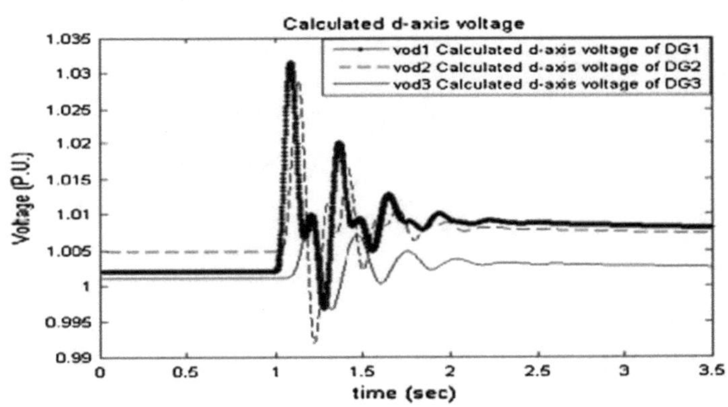

Fig. 9.33 *D*-axis output voltage response

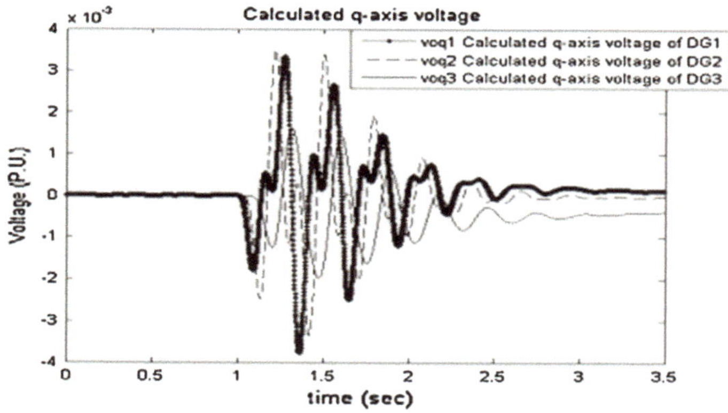

Fig. 9.34 *Q*-axis output voltage response

Second, the proposed controllers with the optimal settings under fault disturbance have been investigated through the nonlinear time domain simulations. The measured as well as the reference of the d-axis and q-axis currents are given in Figs. 9.35, 9.36, 9.37, 9.38, 9.39 and 9.40 at fault disturbance. It could be seen from results that the controller capability during the fault is satisfactory so that the measured currents are tracking the reference currents without significant overshoot and delay time. Finally, a step change of reference voltage has been applied to assure the controller capability with the optimal settings in terms of overshoot and delay time. Reference voltage of the DG1 has been stepped up and stepped down. The d-axis measured and the reference voltages of the DG1 are shown in Figs. 9.41 and 9.42 under this disturbance. The measured voltages are following the reference voltages without significant overshoots and delay time.

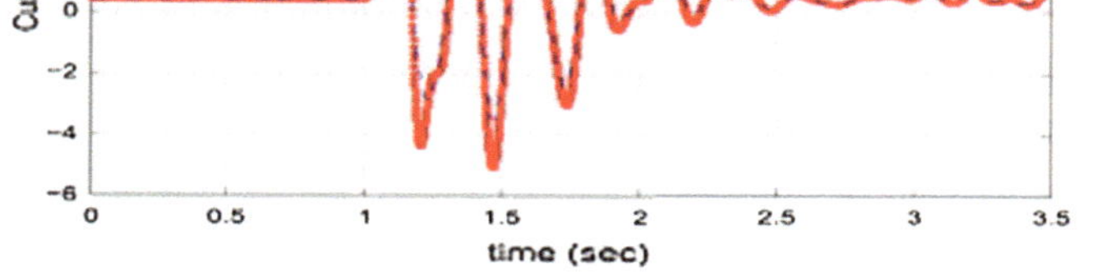

Fig. 9.35 Measured data and reference *d*-axis inductor currents of the DG1 when fault disturbance occurred at load 1

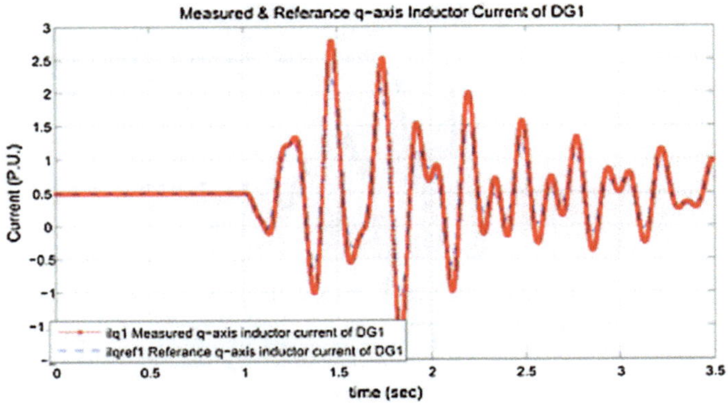

Fig. 9.36 Measured data and reference q-axis inductor currents of the DG1

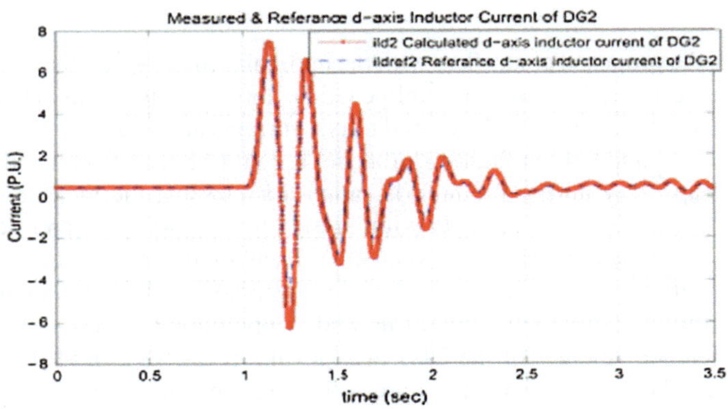

Fig. 9.37 Measured data and reference d-axis inductor currents of the DG2

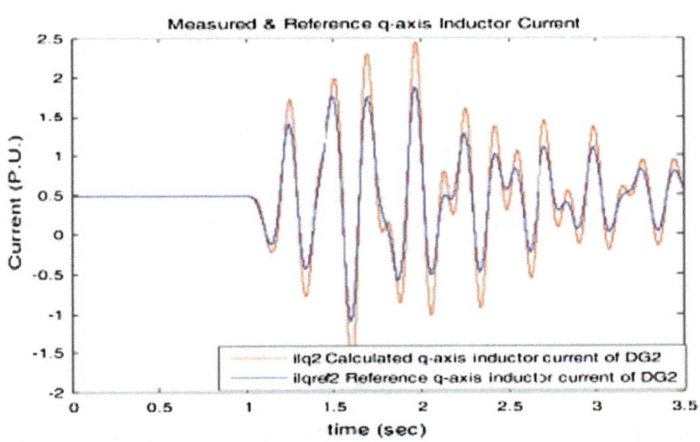

Fig. 9.38 Measured data and reference q-axis inductor currents of the DG2

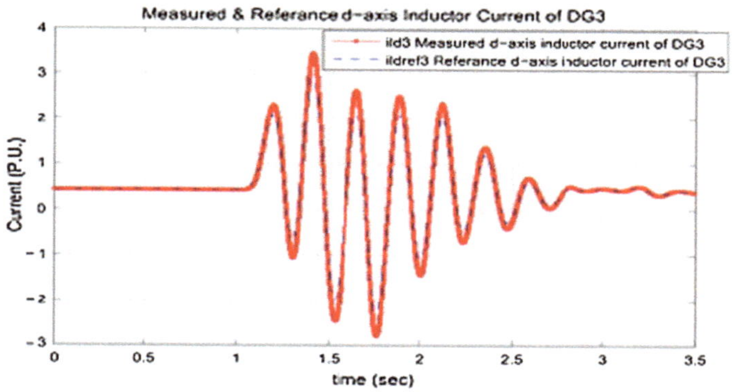

Fig. 9.39 Measured data and reference *d*-axis inductor currents of the DG3

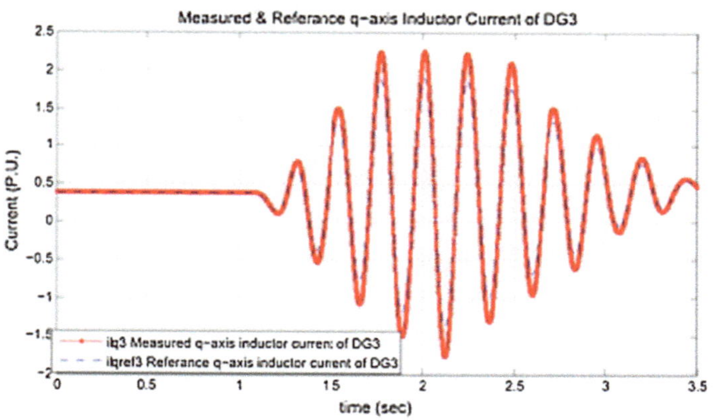

Fig. 9.40 Measured data and reference *q*-axis inductor currents of the DG3

9.3.3 Experimental Results

The autonomous MG including power, current, and voltage controllers has been developed and implemented in RTDS as shown in Fig. 9.43. The performance of the proposed controllers has been tested extensively. The power controller of the MG is shown in Fig. 9.44 while the current and voltage controllers are shown in Fig. 9.45. The MG presented in Sect. 9.2 is analyzed using the RTDS as well. The setup of the MG including the interfacing inverters is simulated using the RTDS (Figs. 9.46 and 9.47). The optimal parameters of the controller obtained by PSO are used to assess the controller performance of the autonomous MG in real time. The

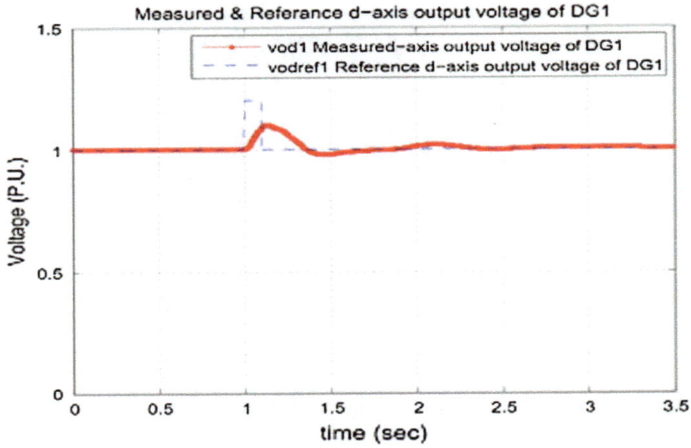

Fig. 9.41 Measured data and reference *d*-axis output voltage of the DG1 with a step-up change in the reference voltage of DG1

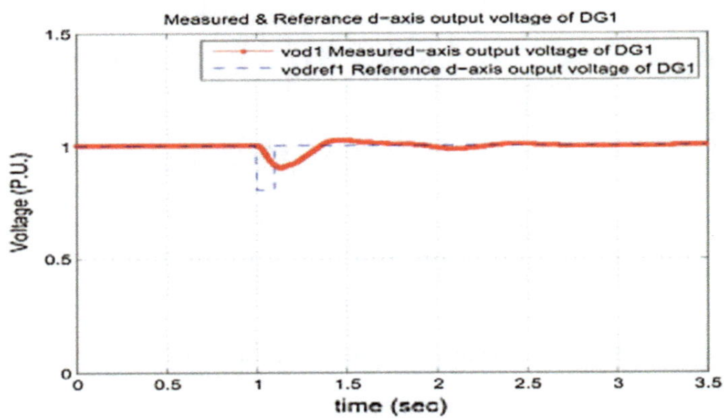

Fig. 9.42 Measured data and reference *d*-axis output voltage of the DG1 with a step-down change in the reference voltage of DG1

effectiveness of the proposed approach for stability enhancement is investigated. Step change of the reference voltage of DG1 has been applied. Figure 9.49 shows the output voltage of the DG1 and its reference due to step-up change while Fig. 25 shows the output voltage of the DG1 and its references due to step-down change. The output and reference currents of the three DGs are shown in Figs. 9.48 and 9.49 when a short circuit disturbance has been applied at both load 1 and 2. The results show the controller effectiveness due to these disturbances. As shown in the RTDS

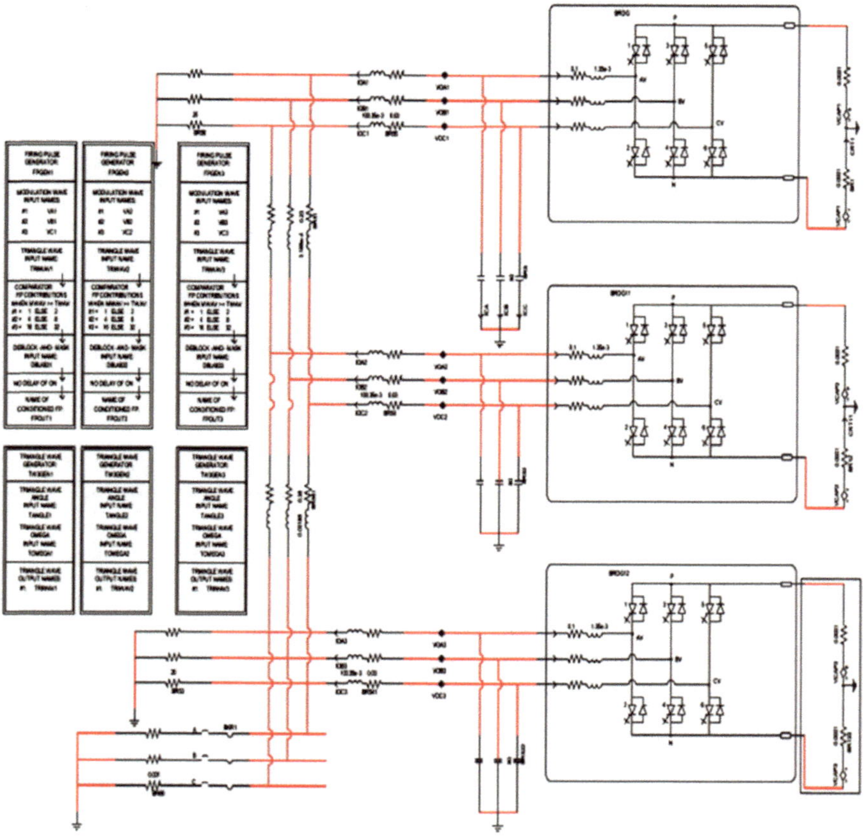

Fig. 9.43 Autonomous microgrid in RTDs

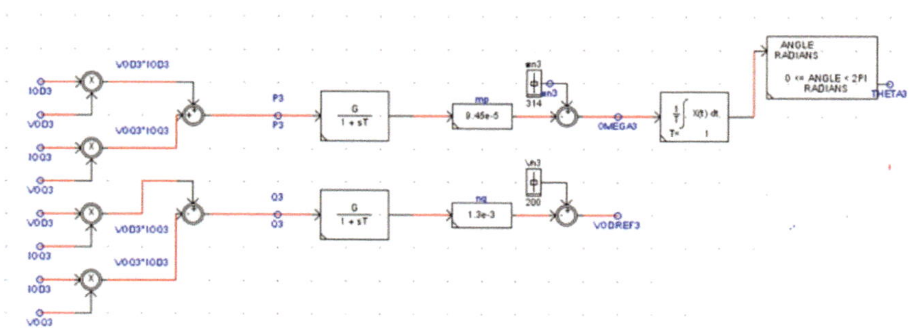

Fig. 9.44 Power controller in RTDs

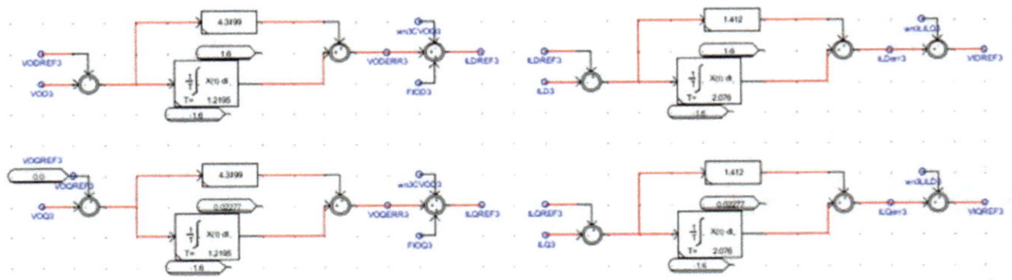

Fig. 9.45 Voltage and current controllers in RTDs

Fig. 9.46 Output voltages of the DG1 and its reference with a step-up change in the reference voltage of DG1

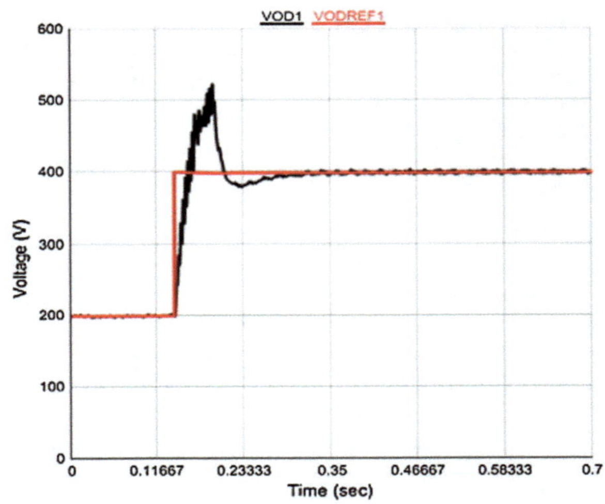

Fig. 9.47 Output voltage of the DG2 and its reference with a step-down change in the reference voltage of DG2

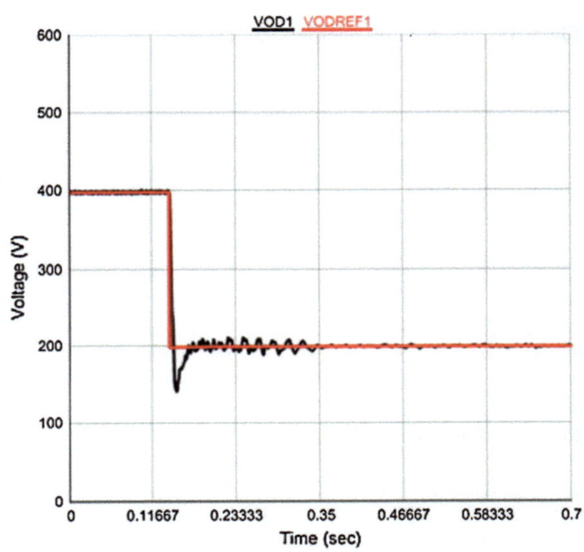

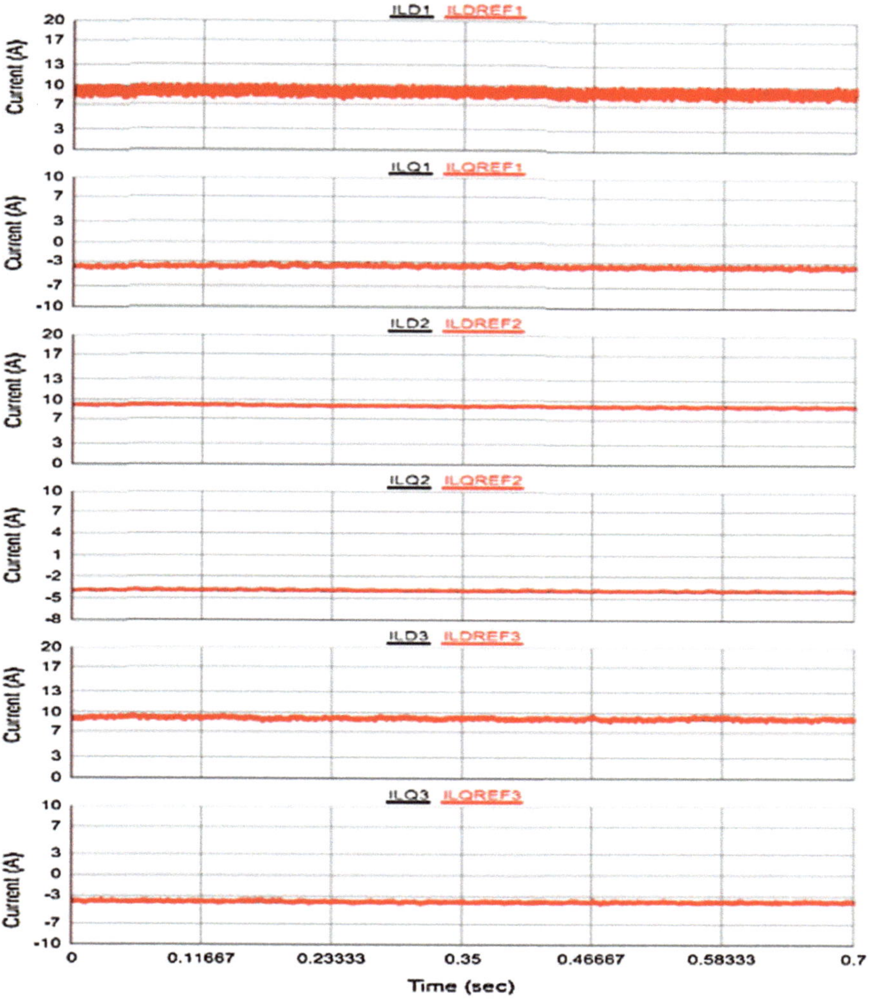

Fig. 9.48 Output and reference currents of the three DGs with a short circuit at load 1

results, the capability of the controller to track the reference during the step change is satisfactory without significant overshoot and delay time. It was observed that the oscillations in the experimental results need to be reduced by tuning the filter elements. Having been out of the main scope of this work, the harmonic current filtering design should be considered in future work [45].

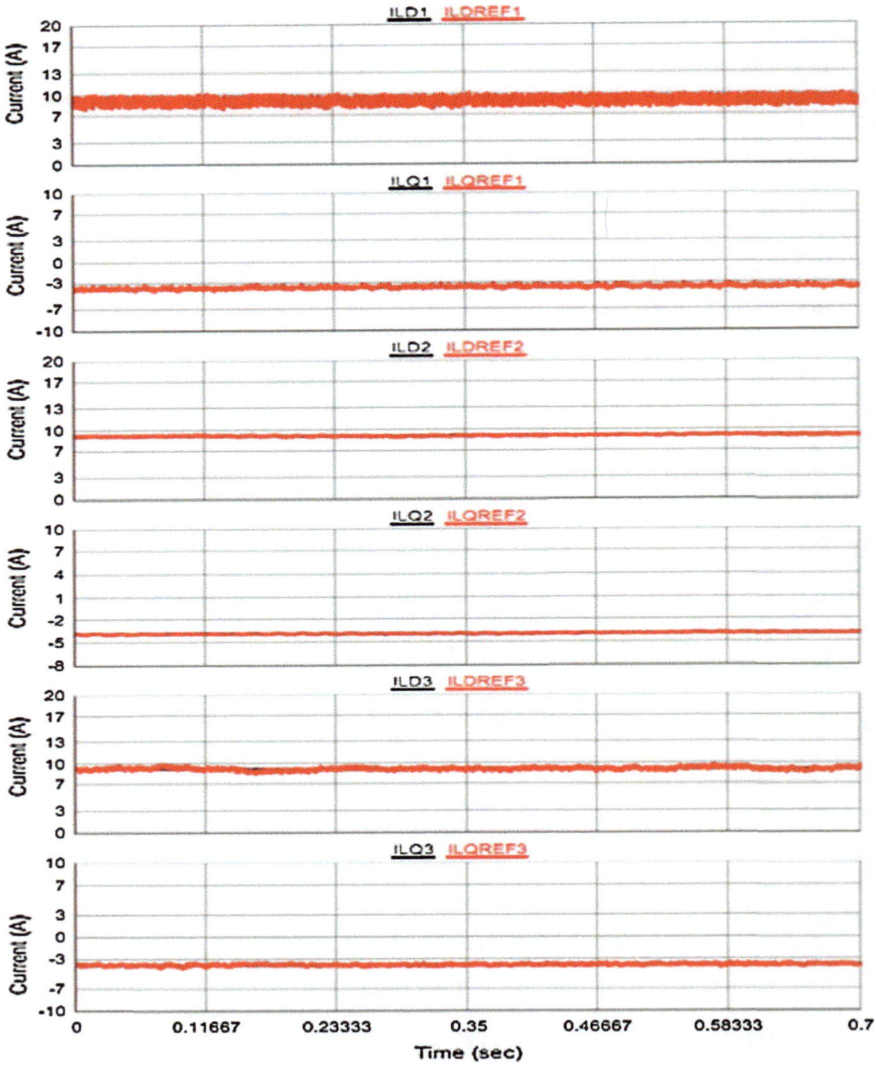

Fig. 9.49 Output and reference currents of the three DGs with a short circuit at load 2

9.4 Distributed Control for Autonomous Microgrid

9.4.1 Introduction

The concept of MG is well known from the time it was originally introduced a decade ago [21]. Control of autonomous MG is a key aspect which has been the focus of researchers over the past few years. In our work we propose a multilevel control scheme for the autonomous MG system wherein both the primary and secondary controllers are implemented in distributed way resulting in individual secondary control for each generating unit. Secondary controller is based on averaging algorithms and uses the concept of network control for its operation.

The RTDS is specially designed for real-time power system transient electromagnetic simulation. RTDS facilitates accurate, reliable, and cost effective study of power systems. The RTDS computations are fast because of distributed parallel processing. It also supports hardware-in-the-loop simulations. RTDS allows investigating effects of disturbances on power system equipment and transmission networks so as to prevent outages or complete failure of the system. It can perform analytical studies of a system and can also be used to educate or train operators, engineers, and students and has improved ability of the simulation accuracy [10].

In this paper, we use the capability of RTDS for implementation of distributed control for an autonomous MG system. Specifically, we develop an accurate model of autonomous MG system and its controller in real-time environment. In Sect. 9.2, detailed description of RTDS hardware and software used to carry out this study is discussed. Section 9.3 presents a brief overview of MG distributed control. Section 9.4 provides a detailed description of RTDS implementation of distributed control of MG. The results and discussion is carried out in Sect. 9.5 and finally conclusion is given in Sect. 9.6.

9.4.2 Real-Time Digital Simulator

The RTDS was developed by Manitoba HVDC Research Center in 1980s. It is a fully digital, real-time power system simulator used to conduct close-loop testing of protection, power electronics and control equipment [25, 39]. RTDS generally simulates power systems in real-time environment with a time-step size of the order 50 s with the help of parallel operating digital signal processors. Power electronic devices with time step as small as 1.4–2.5 s, i.e., with higher switching frequency can also be simulated with sufficient accuracy. Additionally, it has capability of incorporating real devices in closed loop simulation environment. RTDS works in continuous sustained real time, which means equations representing any power system or network can be solved fast enough to simultaneously produce the output conditions. As the solver is real time, it can be connected to power system components for tuning purposes [7]. The RTDS equipment used for our study is basically a combination of advanced computer hardware and comprehensive software [12] which are discussed in following sections.

9.4.3 Description of RTDS Hardware

RTDS is designed to solve the electromagnetic transient simulation algorithm developed by Dr. Herman Dommel [11] and have a customized parallel processing architecture. It is made up of standard 19" rack as shown in Fig. 9.50. Each rack has a provision for 18 processing element (PE) cards, 1 inter-rack communication (IRC), and 1 workstation interface card (WIC) [40]. The communication and processor cards are linked through a common backplane which facilitates exchange of information.

The PE cards are equipped with a processor responsible for calculating the overall network behavior. The processor capacity is up to 13 millions floating point operations per second. The PE card has interface to connect external signals. Two analog channels can be selected to monitor variables being computed on that card and these channels can be scaled online to suit the external device. In general, PE cards perform two different types of calculation namely nodal analysis (network solution) and the auxiliary components. The former solves the branch currents and node voltages based on network impedance and contribution of the auxiliary components. The later is nothing but transmission lines, networks, transformers, etc. which provides admittance matrix overlays and current injections to the network solution. The overall network response is simulated by the combined solution of network solution and auxiliary components.

Two different types of PE cards are used by current RTDS: Triple processor card (3PC) and giga processor card (GPC). In our work, we are using GPC which is

Fig. 9.50 Standard RTDS racks

Fig. 9.51 A giga processor card

recently developed and contains two RISC processors running at 1 GHz. A typical GPC processor is shown in Fig. 9.51. The processors can communicate directly at any time using shared memory without accessing the backplane. GPC has strong computational power enabling simultaneous calculation of more than one component model.

RTDS can also carry out the testing of physical protection and control equipment and thus input–output (I/O) channels were a primary design consideration. The PE cards and I/O can directly communicate with each other without a central communication link. The I/O developed for GPC is known as GT-I/O, shown in Fig. 9.50, which include analog and digital input–output with 16-bit data converters. They connect to GPC through 2 GHz fiber optic links completely isolated from the simulator. The IRC card comes into picture when multiple racks are connected with each other. Its function is to transfer the data generated on one rack to another rack and share the information among racks. The information transfer takes place at a speed of 500 MHz. WIC is not involved in the real-time simulation rather its purpose is to download a case prior to its run. WIC is the communication between RTDS and the host computer network over an Ethernet communication link during online run. The

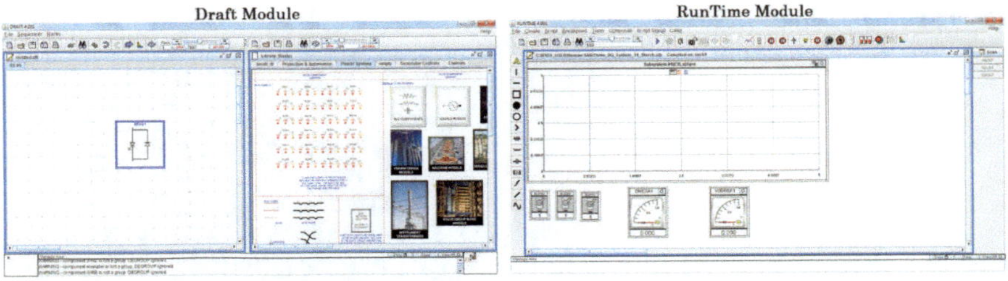

Fig. 9.52 Screenshots of draft and runtime modules

rate of data transfer is 10 MHz. It also synchronizes the calculations and coordinates between the PE cards to ensure their proper operation.

9.4.4 Description of RTDS Software

The software of RTDS is an organized three-level hierarchy structure with a low-level operating system, a mid-level compiler and communication, and a graphical user interface at the top level. The user works only with the highest level user interface.

In this work, a high-level graphical user interface known as RSCAD is used to construct, run, operate the simulation circuits, and also to record/document the results. This user interface is installed in the host workstation. There are two modules namely *draft* and *runtime* in RSCAD as shown in Fig. 9.52.

The *draft* module contains component selection library on the right side containing icons representing various system elements which have been coded for real-time simulation. Any component can be dragged on to the left side where a blank area is provided to assemble the system model. With the help of a data menu, the parameters of any particular component can be entered. After drafting any network or system the next step is to compile it by which a simulation code is generated. Once after compilation, *runtime* module can be used to run.

The *runtime* module with the help of a WIC and Ethernet, communicates back and forth with the simulator. This bidirectional communication allows in downloading the simulations and as well as running them on the screen. This module has variety of options from plotting a response to changing a parameter online or can even switch ON/OFF any particular variable. The plots are of high resolution displaying every time step recorded by the WIC. Slow moving signals such as RMS voltage of a bus can be monitored on a continuous basis to allow observation of transient behavior.

9.4.5 Distributed Control Scheme

MG is a complex network consisting of DG, communication medium, transmission lines and controllers, etc. In this study, we are interested in the autonomous operation

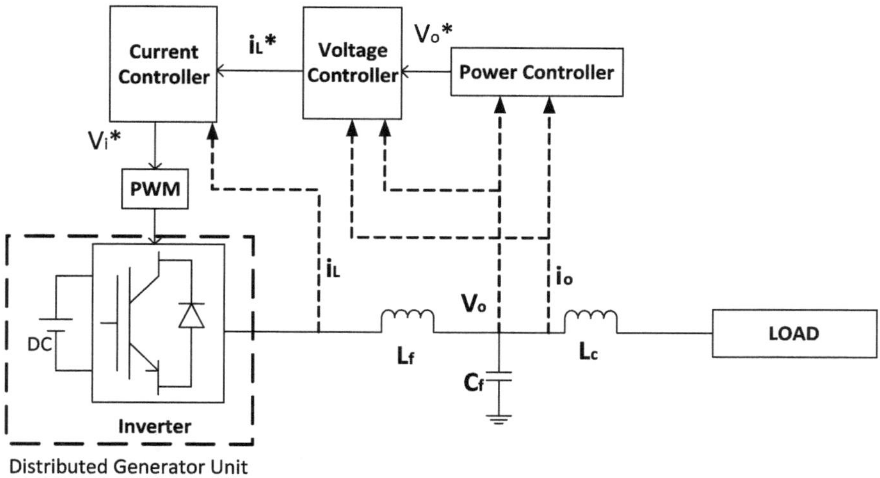

Fig. 9.53 Block diagram of DG supplying a load

mode of MG, wherein an MG is responsible for supplying a load with its own control structure. The main task of MG in this mode is to ensure that voltage and frequency supplied are within the specified limits and also to take care that the load is equally shared among its generating units.

An inverter-based MG is adopted for this study. The block diagram is shown in Fig. 9.53. The model consists of DG units supplying a common load via LC filter and coupling inductance. It consists of *droop*-based power controller which sets the output voltage and frequency of MG [5, 8]. Voltage and current controllers reject the high-frequency disturbances and damp the output filter to avoid any resonance with the external network. Further details regarding this inverter-based MG model can be found in [33].

Control of islanded MG is a key aspect which aims toward the stable operation of MG and has been the focus of researchers over the past few years. An overview of the different methods of modeling and control is reported in [27]. One of the widely used control technique is the "MultiLevel Control" of MG [16, 32, 44] consisting of primary and secondary levels of control. Both primary and secondary controls are essential to obtain overall stability of the system.

In this section, multilevel distributed control of MG using NCS approach is implemented in real-time environment using RTDS. In this control technique, both primary and secondary controls are implemented in distributed fashion resulting in individual secondary control for each generating unit. Primary control makes use of local measurements and does not need any communication medium for its operation. It ensures proper load sharing at the same time maintaining the output levels of voltage and frequency within limits. Primary control alone is not sufficient for the stable operation of overall MG; to achieve global controllability, a secondary control is deployed.

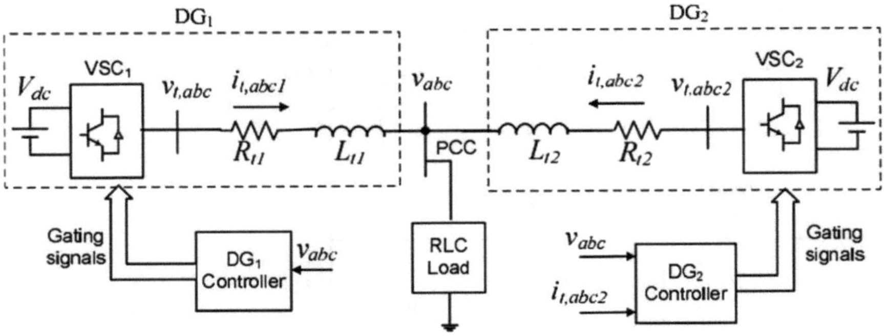

Fig. 9.54 Distributed secondary control for individual DG unit

In what follows, distributed PI controller is used at the secondary level of control. This controller removes steady state error in output voltage and frequency produced by the primary control. Secondary control also regulates the deviations produced due to load changes to zero. It is based on averaging algorithm, where the secondary controller collects the measurements (voltage, frequency) of all DG units via communication system, averages them, and produces a suitable control signal to be send to primary control level [42]. Figure 9.54 shows the overall control block diagram of distributed control of islanded MG. The control law at the secondary level is given by following equations:

$$\delta\omega = K_{p\omega}(\omega_n - \omega_{avg}) + K_{i\omega}\int(\omega_n - \omega_{avg})dt \tag{9.28}$$

$$\delta V = K_{pv}(V_n - V_{avg}) + K_{iv}\int(V_n - V_{avg})dt \tag{9.29}$$

where $K_{p\omega}$, $K_{i\omega}$ are the PI controller parameters for frequency control, K_{pv}, K_{iv} are the PI controller parameters for voltage control. ω_n and V_n are the frequency and voltage set-points. ω_{avg} and V_{avg} are the average values of frequency and voltage, respectively. $\delta\omega$ and δV are the control signals produced by secondary controller.

9.4.6 RTDS Implementation

In this section, the real-time implementation of autonomous MG model and its distributed control is explained in detail. Figure 9.55 shows the laboratory setup of equipments. The studies were carried out using one standard RTDS rack developed by RTDS technologies [30] and a work-station installed with RSCAD is used to develop the complete simulation test bed of MG system. During the simulations power system, control system, and voltage-sourced converter (VSC) small time-step components from the RTDS model library [41] are used.

Fig. 9.55 RTDS/RSCAD setup in the laboratory

Inverter-based MG can be conveniently constructed using the VSC small time-step modeling library. All the VSC components are assembled in VSC bridge box. Figure 9.56 shows the blue VSC bridge box icon, detailed circuit inside the box, and configuration table for inverter bridge. As can be seen, the MG system consists of three distributed generating units supplying a common load through necessary filter and coupling circuits. The three-phase inductor currents, output voltage, and output currents denoted as I_L, V_O, and I_O, respectively, are monitored.

This study involves the use of power electronic device inverters which requires pulse-width modulation (PWM) for its operation. A triangular wave generator along with firing pulse generator is used for this purpose. Figure 9.57 illustrates these blocks for DG1 unit with their respective configuration tables.

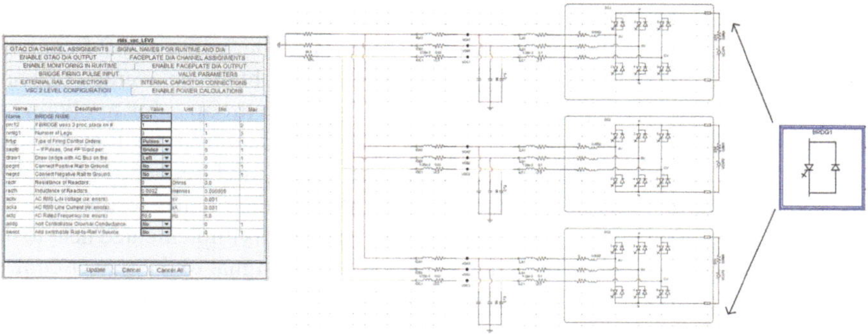

Fig. 9.56 RTDS equivalent model for autonomous MG system

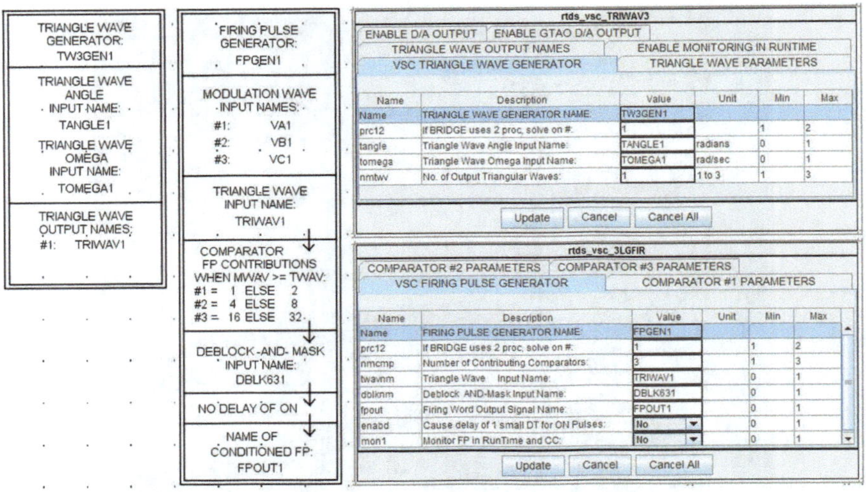

Fig. 9.57 Triangular wave generator and firing pulse blocks

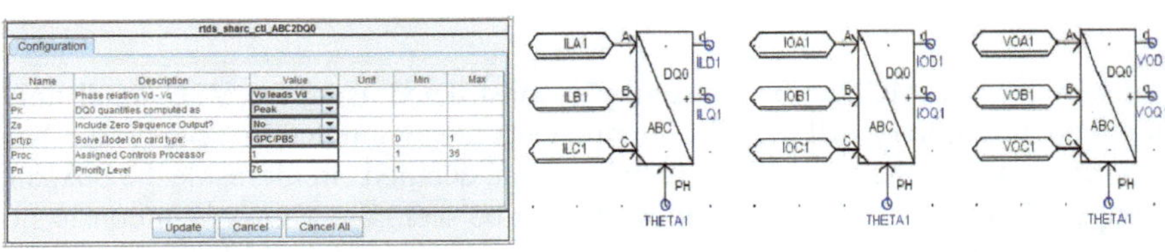

Fig. 9.58 Block for ABC to DQ0 transformation

To perform analysis, three-phase AC voltages and currents are reduced into DC quantities using *ABC-DQ0* transformation. Figure 9.58 shows the block which performs this transformation for DG1. The angle *THETA* used here is obtained from the power controller circuit discussed ahead.

The power controller model for DG1 is implemented as shown in Fig. 9.59, it is based on droop action. Instantaneous powers are calculated using output voltages and currents and then passed through a low-pass butterworth filter with a cutoff frequency of 5 Hz for filtering the harmonics at this level. We then obtain the average real and reactive powers which are passed through their respective droop gains to obtain the angular frequency and voltage, respectively [2]. As is evident from the figure, active power (*P*), reactive power (*Q*), output voltage (*VODREF*) and frequency (*OMEGA*) are monitored variables at this level. Implementation of distributed secondary controller is also explained in this figure. Secondary controller is implemented near to the primary controller since the controller sends the control signal to the power control to regulate the output voltage and frequency values as per the control law explained in Eqs. (9.1) and (9.2).

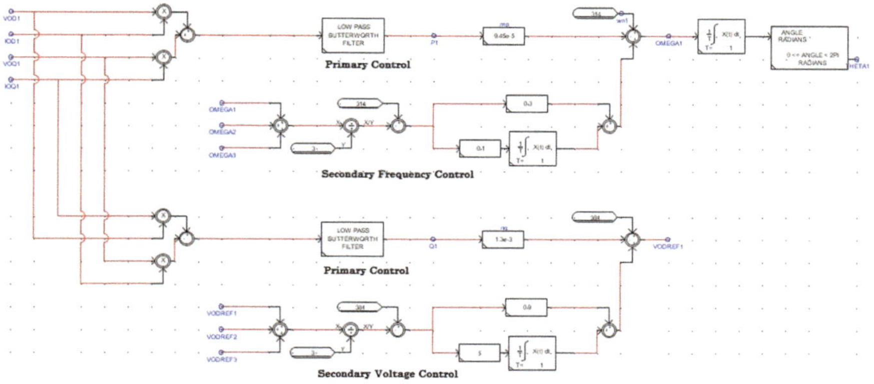

Fig. 9.59 RTDS equivalent model for power controller

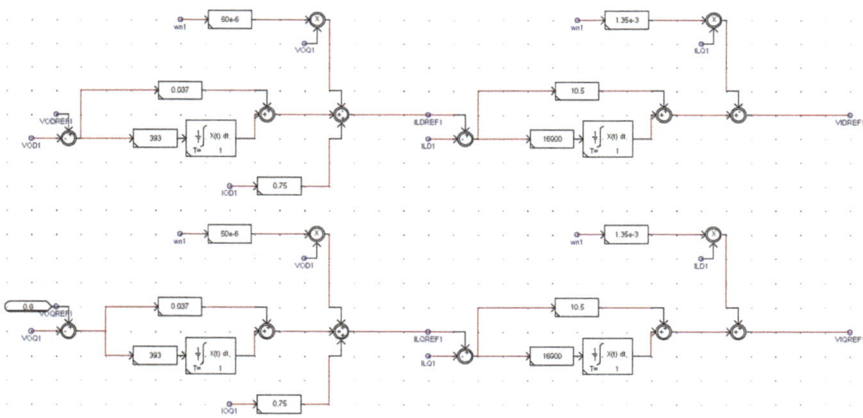

Fig. 9.60 RTDS equivalent model for voltage and current controllers

The voltage controller and current controllers for DG1 consisting of PI controls are implemented as shown in Fig. 9.60. These controllers reject high-frequency disturbances and damp the output filter to avoid any resonance with the external network. These PI controllers make use of the local measurements to perform the control action, an additional feedforward gain and decoupling terms are also used. Details on these controllers can be found in [20, 36].

The output of current controller is transformed into three-phase voltage signals using *DQ0-ABC* transformation. Such block used for DG1 is shown in Fig. 9.61. These voltage signals serve as the modulation waves for generating the firing pulses.

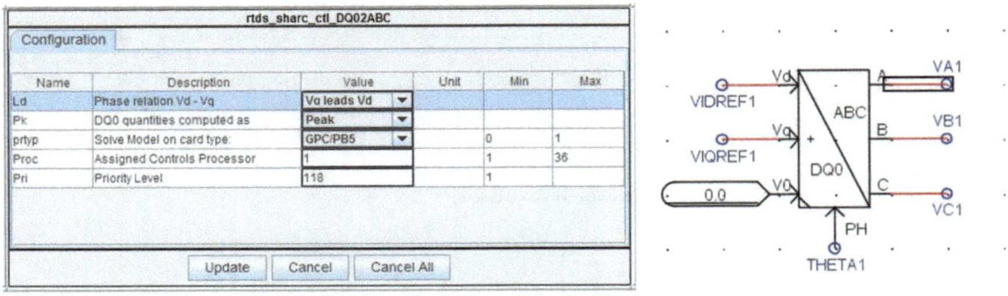

Fig. 9.61 Block for DQ0 to ABC transformation

9.4.7 Simulation Results

In this section, in order to confirm the real-time performance of the proposed distributed secondary controller, MG system described in the Sect. 9.4 is simulated in RTDS. Simulation results of the RTDS are compared with the results of the same system simulated in MATLAB environment. Response of the controller and its load sharing capability under fault disturbance is also shown.

The specifications of the MG are described in Table 9.5. The primary controller parameters for RTDS and MATLAB are given in Table 9.6. The secondary controller parameters for RTDS and MATLAB are given in Table 9.7.

Figures 9.62, 9.63 and 9.64 shows the waveforms of three-phase inductor currents, output voltage and currents of the inverter, respectively, when supplying a load of 5 kW. It can be seen that the inductor currents are balanced set with near sinusoidal nature. The output voltages and currents are also balanced set with sinusoidal nature and negligible amount of switching frequency components.

Table 9.5 System parameters

Parameter	Value
Filter inductance (L_f)	1.35 mH
Filter resistance (r_f)	0.1 Ω
Filter capacitance (C_f)	50 μF
Coupling inductance (L_c)	0.35 mH
Coupling resistance (r_c)	0.03 Ω
Nominal voltage (V_n)	381 V
Nominal frequency (ω_n)	314 rad/s
Cutoff frequency of LPF (ω_c)	31.4 rad/s or 5 Hz

Table 9.6 Primary control parameters

Parameter	Value
Real power droop gain (m_p)	9.4×10^{-5}
Reactive power droop gain (n_q)	1.3×10^{-3}
Proportional gain of voltage controller (K_{pv})	0.037
Integral gain of voltage controller (K_{iv})	393
Proportional gain of current controller (K_{pc})	10.5
Integral gain of current controller (K_{ic})	16,000
Feedforward gain of voltage controller (F)	0.75

Table 9.7 Secondary control parameters

Parameter	RTDS	MATLAB
Proportional gain for secondary frequency control	0.5	4.3656
Integral gain for secondary frequency control	0.1	9.1206
Proportional gain for secondary voltage control	0.9	7.3677
Integral gain for secondary voltage control	0.5	3.6765

9.4.8 Comparison of RTDS and MATLAB Results

The same system shown in Fig. 9.56, is also implemented in MATLAB using Simulink/Sim power systems. Comparative study between RTDS and MATLAB simulations are presented here using the results for power sharing, voltage, and frequency regulation. Initially, the system is operated under no load condition and at t = 4 s, a load of 3 KW is suddenly realized on the system. From the RTDS curve, it looks as if the load is applied before 4 s. This is because the load is manually switched and output recording of RTDS starts to time after refreshing while the simulation is running in real time.

Figure 9.65 illustrates the load sharing among the DG units. From t = 0 to t = 4 s, there is no active power shared as the load is zero on the system. It can be seen from RTDS response that under no load condition, a very small amount of active power

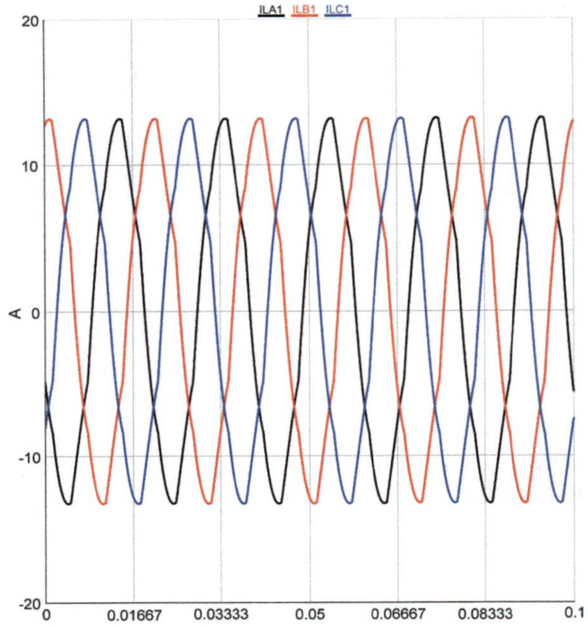

Fig. 9.62 Three-phase inductor current

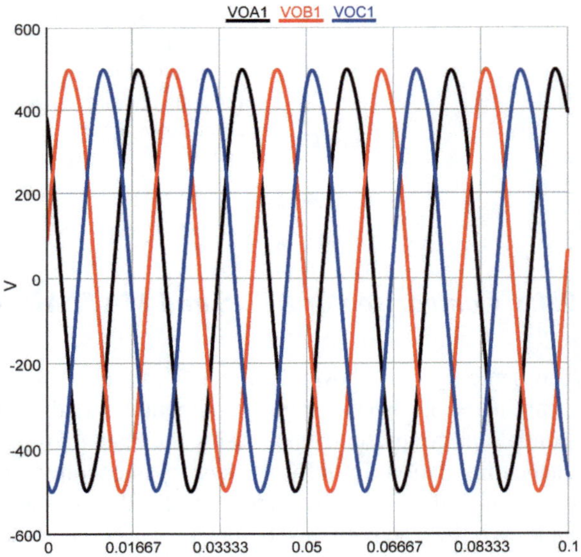

Fig. 9.63 Three-phase output voltage

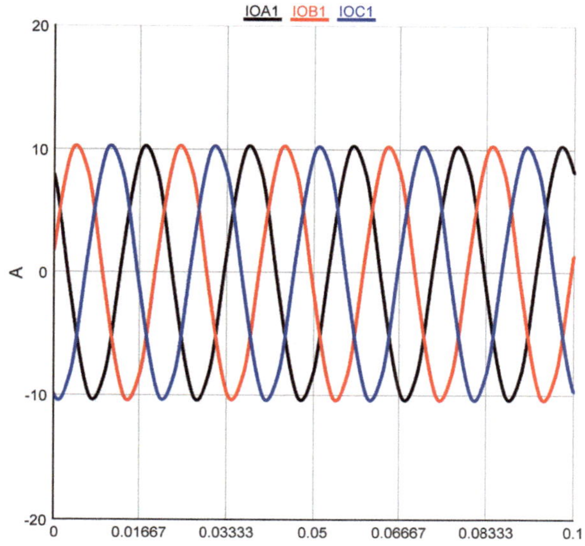

Fig. 9.64 Three-phase output current

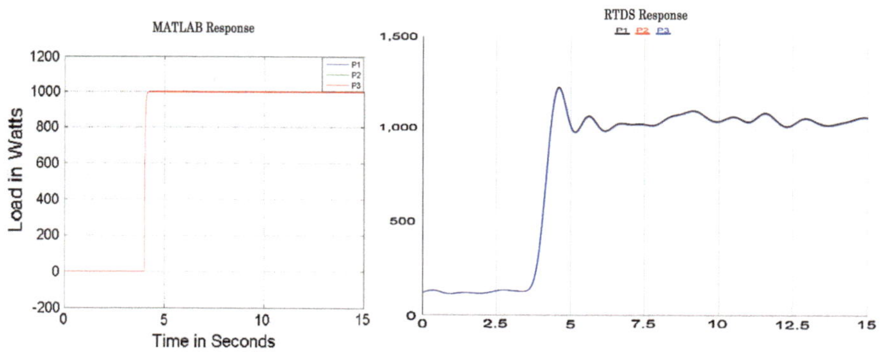

Fig. 9.65 Load sharing response of MATLAB and RTDS

is shared which is because of the presence of active resistance in the circuit. After $t = 4$ s, each DG unit shares equal load of 1 kW which sums up to 3 kW, as there are 3 DG units, which is the total load on the system. It is clear from comparison that both the results are almost same.

Figures 9.66 and 9.67 illustrate the regulation of output frequency and output voltage, respectively, under load change. Due to sudden load change, transients can be seen at $t = 4$ s which are at once regulated toward zero by the controller. The controller operates as expected and successfully brings back the deviated output values to their nominal set-points or nominal values. The set-points for output frequency is 314 rad/s and for output voltage is 381 V. From the above results, it can be seen that the steady

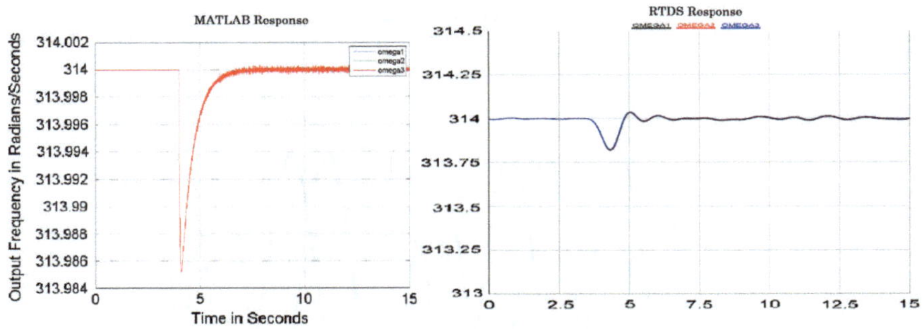

Fig. 9.66 Output frequency response of MATLAB and RTDS

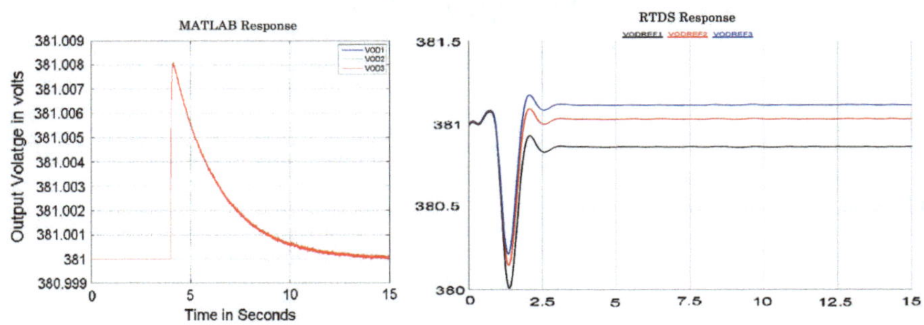

Fig. 9.67 Output voltage response of MATLAB and RTDS

state and transient responses of RTDS and MATLAB agree well and are consistent with each other.

9.4.9 Load Sharing During Faults

In this section, load sharing capability of the controller during faults is demonstrated. For better understanding, we simplify the system by considering only two DG units. VSC valve breaker is used to apply the fault. This fault is applied manually using a switch and will completely isolate the DG unit from the system. Initially, a load of 5 kW is applied on the system. The DG units are seen sharing the equal load of 2.5 kW which is illustrated by Fig. 9.68. If due to some fault condition, DG1 shuts down then the total load has to be supplied by DG2. This condition is illustrated by Fig. 9.69, where a fault on DG1 is applied and isolated from the network. As can be seen from the figure, the controller acts quickly and the total load of 5 kW is supplied by DG2 alone whereas power delivered by DG1 is now reduced to zero.

Similarly is the case when a similar fault is applied on DG2 and the total load is delivered by DG1 alone which is demonstrated in Fig. 9.70. Figure 9.71 illustrates

Fig. 9.68 Load shared by DG units at 5 kW load

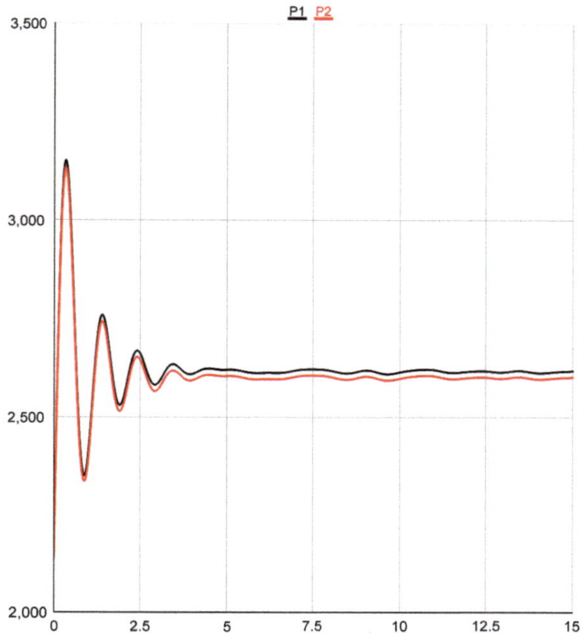

Fig. 9.69 Load sharing when the fault is on DG1

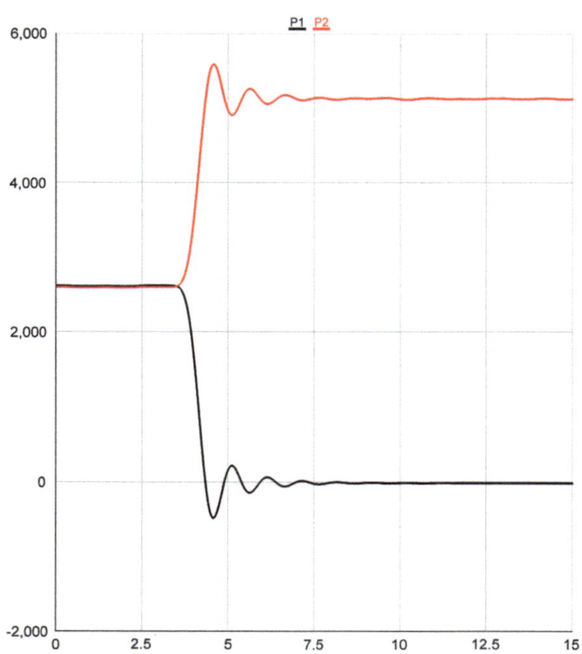

Fig. 9.70 Load sharing when the fault is on DG2

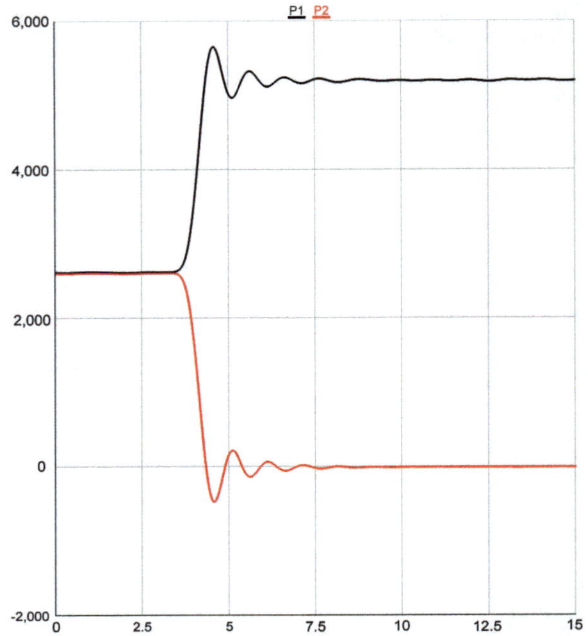

Fig. 9.71 Load sharing when the fault on DG2 is removed

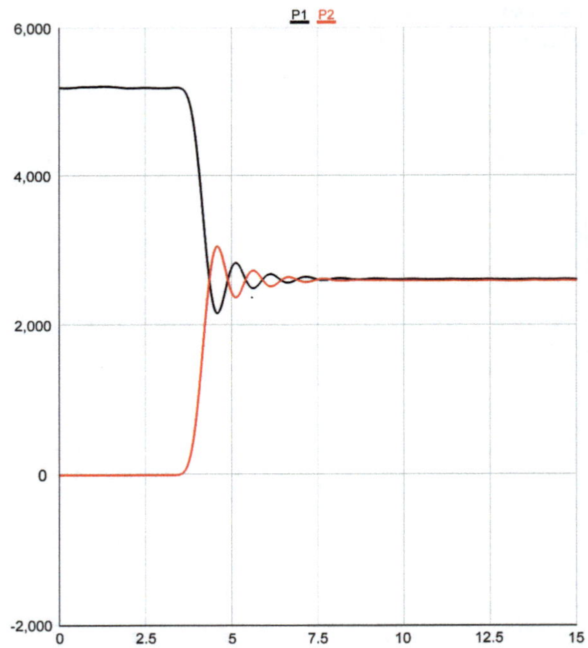

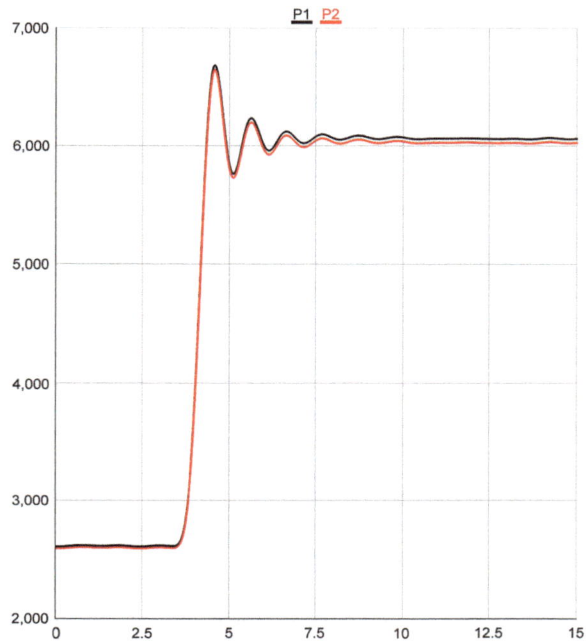

Fig. 9.72 Load shared when the load is increased from 5 to 12 kW

the condition when the fault on DG2 is removed, as can be seen both the DGs start to share the same load.

Finally, an additional load of 7 kW is added to the system, due to this addition a step increase in power delivered by both the DGs is observed as can be seen in Fig. 9.72. As the total load on system now is 12 kW, both DG units are now seen sharing equal load of 6 kW. Therefore, the controller shares the load evenly among generating units even after fault conditions.

9.5 Experimental Verification of Inverter-Based Microgrid

In this section, experimental verification of inverter-based MG will be carried out. A 220 V (per phase RMS), 50 Hz prototype MG was built to test and verify the model results. As shown in Fig. 3.53 the test system consists of three inverters of equal rating (10 kVA) with two load banks, one at each bus 1 and bus 3. These inverters are controlled to share the real and reactive powers over the lines 1 and 2. System parameters are given in Table 9.8.

Network is resistance dominated as is the case in low-voltage (LV) distribution systems. DG1 and DG2 are located relatively close together compared to DG3. In this test system only resistive loads were used to verify the model. A resistive load

Table 9.8 Test system parameters

Inverter parameters (10 kVA rating)			
Parameter	Value	Parameter	Value
f_s	8 kHz	m_p	9.4e-5
L_f	1.35 mH	n_q	1.3e-3
Cf	50 μF	K_{pv}	0.05
r_f	0.1Ω	K_{iv}	390
L_c	0.35 mH	K_{pc}	10.5
r_{Lc}	0.03 Ω	K_{ic}	16e3
W_c	31.41	F	0.75
Network and load parameters (see Fig. 9.10)			

of 5.8 kW (=25 Ω per phase) at bus 1 and 7.3 kW (=20 Ω per phase) at bus 2 is considered as an initial operating point.

In the test system analyzed in this paper all the inverters are equally rated. Hence, the droop gains of all the inverters were chosen to be equal so that they equally share the fundamental power. The nominal frequency droop was 0.3 % at the maximum real power output, whereas the nominal voltage droop was 2 % at the maximum reactive power output. In a practical scenario, there can be various types of sources of DG and the droop gains and sharing ratios may be dictated by the economic interest of the system operators. The choice of droop gains in such cases is further discussed in [129]. The objective of this paper is to investigate the stability of the system for the chosen values of droop gains. The model discussed in the previous section is general and it allows the users to investigate the system stability for any chosen combination of the system parameters.

In the design of the inverter output filter, the main criteria was to achieve attenuation by a factor of 100 in switching frequency ripple in the output voltage. This requires a resonant frequency of 10× less than the switching frequency, which is a common rule of thumb. The filter inductor value was chosen to have low ripple content in the inductor current and, when combined with the capacitor value, gives the required resonant frequency. A large coupling inductance results in a poor bus voltage regulation. Hence, in this application the coupling inductance was chosen to provide a reasonable coupling impedance between the inverter output and the connection bus with a good bus voltage regulation.

The proportional and integral gains of the voltage controller, shown in Table 3.1, were chosen using classical pole-zero and bode techniques to yield a bandwidth of 400 Hz for the voltage controller. The current feedforward gain F was chosen to yield a low output impedance and hence improve the disturbance rejection of the inverter system. The current controller was designed for 1.6-kHz bandwidth with good rejection of high-frequency disturbance. Although, the control was implemented in the discrete time domain, the equivalent continuous domain gains are provided in Table 9.9 for construction of the model.

Table 9.9 Initial Conditions

Par	Value	Par	Value
V_{od}	[380.8 381.8 380.4]	V_{oq}	[0 0 0]
I_{od}	[11.4 11.4 11.4]	I_{oq}	[0.4 − 1.45 1.25]
l_{id}	[11.4 11.4 11.4]	I_{lq}	[−5.5 − 7.3 − 4.6]
V_{bd}	[379.5 380.5 379]	V_{bq}	[−6 − 6 − 5]
W_O	[314]	δ_0	[0 1.9e−3 −0.0113]
I_{line1d}	[−3.8]	I_{line1q}	[0.4]
I_{line2d}	[7.6]	I_{line2q}	[−1.3]

9.5.1 Modeling Results

Recall that a complete model of the test system was obtained using the procedure outlined in Chap. 3. Initial conditions of the system are given in Table 9.9. These steady state operating point conditions were obtained from a MATLAB/Simulink time-step simulation of the system. However, it is possible to use a more general load-flow solution as is often done in conventional power system modeling to obtain initial steady state conditions [10]. The value chosen for virtual resistor (VIR) was 1000 Ω.

Figure 3.54 shows the complete eigenvalues of the system for the initial conditions given in Table 9.9. It can be seen that a large range of frequency components exist and that these fall into three different clusters. Using (3.76), participation values of the different states in these eigenvalues can be obtained. This analysis shows that the high-frequency modes in cluster "3" are sensitive to the state variables of LCL filter block of inverters and the line currents. The modes in cluster "2" are largely sensitive to the static variables of voltage controller, current controller, and output LC filter. The low-frequency dominant modes shown in cluster "1" are largely sensitive to the state variables of the power controller.

Figure 3.55 shows the trajectory of the two pairs of complex conjugate dominant low-frequency eigenvalues (part of cluster I) as a function of the real power droop gain m_p (the same value used for all the three inverters). The eigenvalues marked with λ_{1-2} are largely sensitive to the state variables of real power part of the power controllers of inverters 1 and 2, as given in Table 9.10. Similarly, eigenvalue marked as λ_{1-3} are highly sensitive to the state variables of real power part of the power controllers of inverters 1 and 3. It is therefore apparent that the modes λ_{1-2} and λ_{1-3} represent the dynamics of real power sharing of the DGs. However, these modes are also sensitive to the reactive power. This is a consequence of the coupling of real and reactive powers in the network due to the presence of highly resistive lines.

Figure 3.55 shows that as m_p is increased, modes λ_{1-2} and λ_{1-3} move toward unstable region making the system more oscillatory and eventually leading to instability. It is to be noted that large droop gain is necessary to improve the transient response of DGs, whereas a LPF with low cutoff frequency is needed to achieve good

Table 9.10 Sensitivity of low frequency

Sensitivity of λ_{1-2}		Sensitivity of λ_{1-3}	
State	Participation	State	Participation
P_1	0.15	P_1	0.12
Q_1	0.05	Q_1	0.06
P_2	0.3	P_3	0.32
Q_2	0.03	Q_3	0.03
δ_2	0.5	δ_3	0.57
Remaining states $\leq$ 0.005			

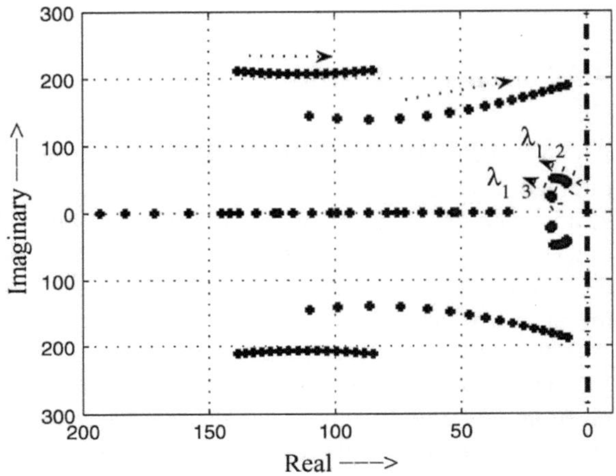

Fig. 9.73 Trace of low-frequency modes as function of reactive power droop gain: $3.17e^{-4}(0.5\,\%$ droop) $\leq n_q \leq 4.8e^{-3}$ (8 % droop)

attenuation of high-frequency distortion components in the measured power and to avoid any interaction with inner current controllers. Also, from Table 9.10 it can be observed that the dominant mode λ_{1-2} is highly sensitive to the states of the power controller of inverter 2. Hence, in this system, inverter 2 is the most critical element from the point of view of system stability. However, the low-frequency dominant modes are less sensitive to the reactive power droop gain compared to the active power droop gain, as shown in Fig. 9.73.

9.5.2 Experimental Results

First, to verify the low frequency modes within the model, a disturbance in load current i_{load1} was arranged. This requires the addition of a controlled current source in parallel to R_{load1} shown in Fig. 10 and the addition of a disturbance term to (3.74). The disturbance was chosen to be a step change of 3.8-kW real power. The experimental system was excited with the same 3.8-kW step change in load at bus 1. A second set of tests was used to examine the low-frequency modes under a severe step change in *RL* load connected at bus 1. A third test was used to examine the high-frequency modes. Due to the presence of significant damping, a large disturbance in the load was needed to capture the high-frequency modes. A step change of 27 kW (from no load) at bus 1 was considered.

Figures 9.74, 9.75, 9.76, 9.77, 9.78 and 9.79 show the response of state variables P, Q, V_{od}, and i_{ld} of all the three inverters obtained from the model and experiment. It is to be noted that the waveforms corresponding to the experimental results are actually the internal variables of inverters that were captured by using onboard D/A

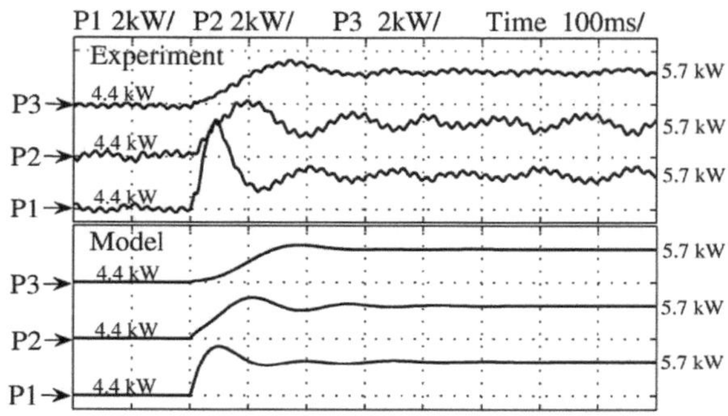

Fig. 9.74 Active power (filtered) response of microsources with 3.8 kW of step change in load power at bus 1

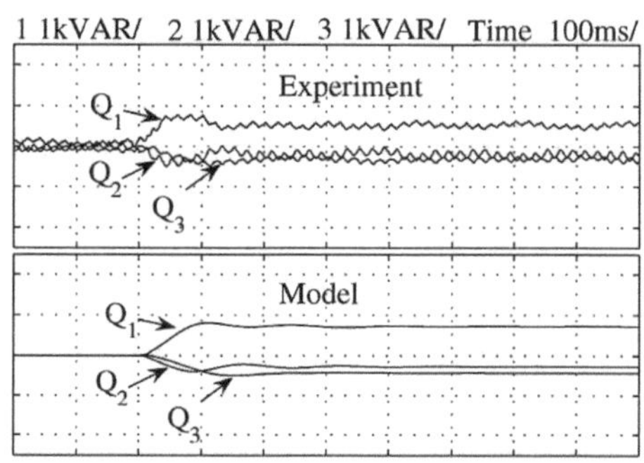

Fig. 9.75 Reactive power exchange between the microsources with 3.8 kW of step change in load power at bus 1 (Initial values: $Q1 = 0$, $Q2 = -200$, $Q3 = +200$; Final values: $Q1 = +600$, $Q2 = -300$, $Q3 = -200$)

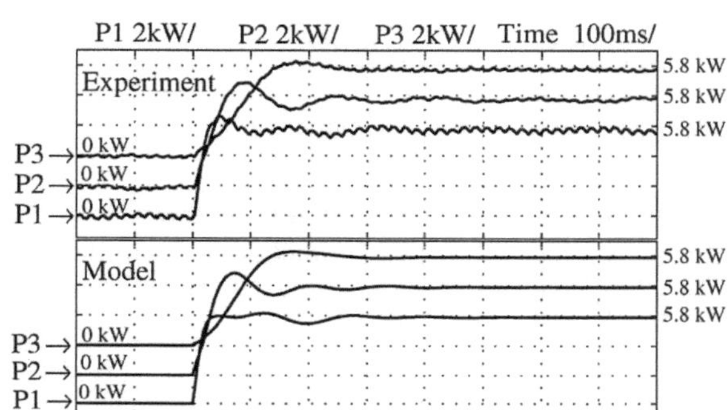

Fig. 9.76 Active power (filtered) response of microsources with 16.8 kW and 12 kVAR RL load step change at bus 1

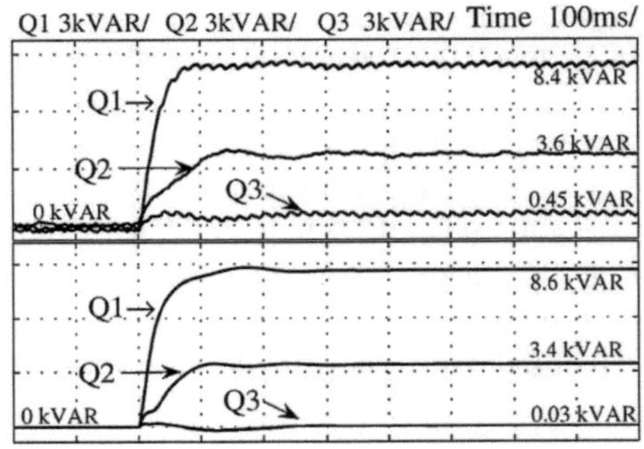

Fig. 9.77 Reactive power (filtered) response of microsources with 16.8 kW and 12 kVAR RL load step change at bus I

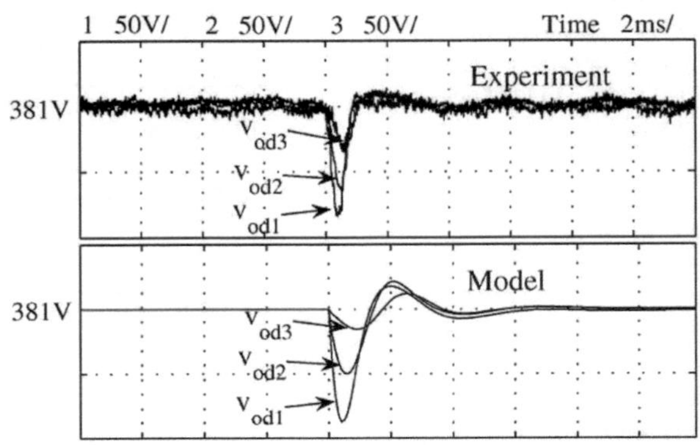

Fig. 9.78 Output voltage (d-axis) response with 27 kW of step change in load power at bus 1

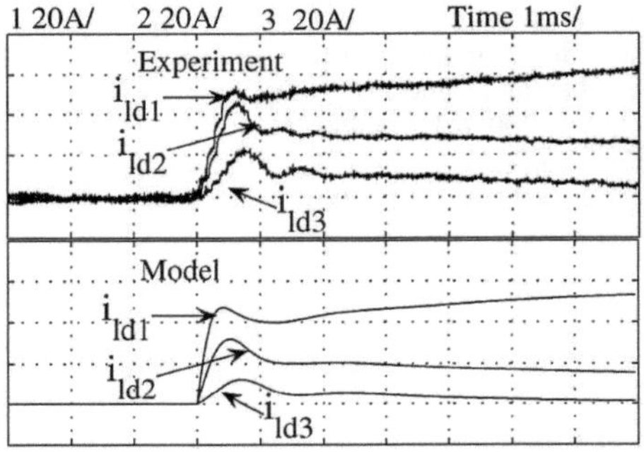

Fig. 9.79 Inductor current (d-axis) response with 27 kW of step change in load power at bus 1

converters. Also, all these figures depict only the variation in the signal from their initial point (relative change).

Figure 9.74 shows the Digital Output (DO) fundamental output power response for a 3.8-kW step change in load 1, for both the model and experimental system. The dominant, poorly damped low-frequency modes λ_{1-2} (marked with a circle in Fig. 3.55) of frequency 7.2 Hz can be clearly observed in the fundamental power. Although, a slight difference in the magnitude exists, the response obtained from the model matches with the response obtained from the practical test system. Due to a slight unequal DC offset in the measured output phase currents a small 50-Hz component was observed in the output power in the experimental case.

In Fig. 9.74 it can be seen that DG1, which is nearest to the changed load, took the major part of the transient whereas DG2 and DG3 have responded more slowly, depending on the effective impedance seen from the load point. Hence, during large changes in the load, closely located DGs may be overloaded and can be tripped out due to the limited overload capacity of the inverters.

Figure 9.75 shows the fundamental reactive power sharing. It can be seen that a considerable amount of reactive power was exchanged between the inverters even though the step was in the real power. This was because of the presence of significant resistance in the lines. This effect can be reduced by increasing the voltage magnitude droop but this will be at the expense of voltage quality. This is one of the major limitations of conventional droop control applied in LV grids. Again, it can be observed that the experimental results closely match the model results.

To investigate the low-frequency mode response under severe test load conditions, a test involving a step change of an *RL* load was conducted. In this test, there was initially no load connected to the system and then a load of 16.8 kW and 12 kVAR at bus 1 was switched on. Figures 9.76 and 9.77 show the active and reactive power response of the inverters under such load transient. In both cases the results from the model closely match the experimental results. This indicates that the dominant low-frequency modes are not particularly dependent on the system load level. Also, it can be inferred that the reactive power sharing is rather poor in this case. However, this can be improved by increasing the reactive power droop gains but at the expense of poor bus voltage regulation.

Figure 9.78 depicts the output voltage response of all the three inverters for a 27-kW load change. The high-frequency modes of frequency around 350 Hz in cluster "2" shown in Fig. 3.54 can be observed in the output voltage response. A notch of 20 % for less than 1 ms can be seen in the output voltage of inverter 1 which is closer to the load point. Although, a slight difference exists in the magnitude of response obtained from the linear model compared to the test system, the oscillatory response in both cases matches. Also, the response obtained from the experimental setup is more damped than that of the model. It is to be noted that the output voltage state variable has the maximum participation in these modes. It would be interesting to investigate the possible excitation of these modes under harmonic loads because the lower order significant harmonic frequencies will fall in the range of frequencies of these modes.

Figure 9.79 shows the inductor current response of all the three inverters. In this case, high-frequency modes with a frequency of around 800 Hz in cluster "3" in Fig. 3.54 can be observed in the response. It was observed that these modes are highly sensitive to the system load level which determines their damping.

9.6 Conclusions

A nonlinear model of an autonomous MG is developed. The stability of the MG has been investigated using the nonlinear time domain simulations. The design problem is formulated as an optimization problem. The minimization of the error between the measured power and reference power has been considered as an objective function to enhance the system stability. PSO technique is employed to search for the optimal settings of the optimized parameters. Different disturbances such as step change in the reference voltage, overload conditions, and fault at load have been applied to demonstrate the effectiveness of the proposed design approach. The results assure the effectiveness and robustness of the proposed PSO-based approach. The controller capability to follow the reference power has been tested. RTDS results verified the effectiveness of the proposed controllers under different disturbances and operating conditions for the system considered.

The next section describes the real-time implementation of multilevel distributed control for autonomous MG system using RTDS platform. The effectiveness of the model and controller is demonstrated with the help of experimental results in RTDS compared with the simulated results in MATLAB. Additionally, fault analysis with respect to load sharing aspect of the controller is also displayed. The proposed controller dynamically regulates the output voltage and frequency during load changes to their nominal values. Results verify the controller to be reliable and robust.

References

1. Abido M (2002) Optimal design of power-system stabilizers using particle swarm optimization. IEEE Trans Energy Convers 17(3):406–413
2. Akagi H, Kanazawa Y, Nabae A (1984) Instantaneous reactive power compensators comprising switching devices without energy storage components. IEEE Trans Ind Appl 3:625–630
3. Alajmi N, Ahmed K, Finney S, Williams B (2013) A maximum power point tracking technique for partially shaded photovoltaic systems in microgrids. IEEE Trans Ind Electron 60(4):1596–1606
4. Anderson R, Boulanger A, Powell W, Scott W (2011) Adaptive stochastic control for the smart grid. Proc IEEE 99(6):1098–1115
5. Arulampalam A et al (2004) Control of power electronic interfaces in distributed generation microgrids. Int J Electron 91(9):503–523
6. Basak P, Chowdhury S, Halder S, Chowdhury S (2012) A literature review on integration of distributed energy resources in the perspective of control, protection and stability of microgrid. Renew Sustain Energy Rev 16(8):5545–5556
7. Bo Z, Klimek A, Ren Y, He J (2008) A real time digital simulation system for testing of integrated protection schemes. In: Proceedings of the power system technology and IEEE power India conference, pp 1–5
8. Chandorkar MC, Divan DM, Adapa R (1993) Control of parallel connected inverters in standalone AC supply systems. IEEE Trans Ind Appl 29(1):136–143
9. Coster E, Myrzik J, Kruimer B, Kling W (2011) Integration issues of distributed generation in distribution grids. Proc IEEE 99(1):28–39

10. Deng Y, Foo S, Li H (2009) Real time simulation of power flow control strategies for fuel cell vehicle with energy storage by using real time digital simulator (RTDS). In: Proceedings of the 6th IEEE power electronics and motion control conference (IPEMC), pp 2323–2327
11. Dommel HW (1969) Digital computer solution of electromagnetic transients in single and multiphase. Networks 88(4):388–399
12. Forsyth P, Kuffel R (2007) Utility applications of a RTDS simulator. In: Proceedings of the power engineering conference (IPEC), pp 112–117
13. Forsyth P, Kuffel R (2008) Utility applications of a RTDS simulator. In: Proceedings of the IPEC international power engineering conference, pp 112–117
14. Green T, Prodanovic M (2007) Control of inverter-based microgrids. Electric Power Syst Res 77:1204–1213
15. Guerrero J, Blaabjerg F, Zhelev T, Hemmes K, Monmasson E, Jemei S, Comech M, Granadino R, Frau J (2010) Distributed generation: toward a new energy paradigm. IEEE Ind Electron Mag 4(1):52–64
16. Guerrero JM et al (2011) Hierarchical control of droop-controlled AC and DC microgrids—a general approach toward standardization. IEEE Trans Ind Electron 58(1):158–172
17. Hassan M, Abido M (2011) Optimal design of microgrids in autonomous and grid-connected modes using particle swarm optimization. IEEE Trans Power Electron 99(6):755–769
18. Ilic M (2011) Dynamic monitoring and decision systems for enabling sustainable energy services. Proc IEEE 99(1):58–79
19. Kamel R, Chaouachi A, Nagasaka K (2011) Detailed analysis of microgrid stability during islanding mode under different load conditions. Eng J 3:508–516
20. Kawabata T, Miyashita T, Yamamoto Y (1990) Dead beat control of three phase PWM inverter. IEEE Trans Power Electron 5(1):21–28
21. Lasseter B (2001) Microgrids (distributed power generation). IEEE Power Eng Soc Winter Meet 1:146–149
22. Lasseter RH (2011) Smart distribution: coupled microgrids. Proc IEEE 99(6):1074–1082
23. Li Y, Vilathgamuwa D, Loh P (2004) Design, analysis, and real-time testing of a controller for multibus microgrid system. IEEE Trans Power Electron 19(5):1195–1204
24. Li H, Steurer M, Shi K, Woodruff S, Zhang D (2006) Development of a unified design, test, and research platform for wind energy systems based on hardware-in-the-loop real-time simulation. IEEE Trans Ind Electron 53(4):1144–1151
25. Li Q, Woodruff S, Steurer M (2007) Study of power loss of small time-step VSC model in RTDS. In: Proceedings of the power engineering society general meeting. IEEE Press, pp 1–7
26. Llaria A, Curea O, Jimenez J, Camblong H (2011) Survey on microgrids: unplanned islanding and related inverter control techniques. Int J Renew Energy 36:2052–2061
27. Mahmoud MS, Hussain SA, Abido MA (2014) Modeling and control of microgrid: an overview. J Frankl Inst 351(5):2822–2859
28. Majumder R, Ghosh A, Ledwich G, Zare F (2009) Load sharing and power quality enhanced operation of a distributed microgrid. IET Renew Power Gener 3(2):109–119
29. Majumder R, Chaudhuri B, Ghosh A, Majumder R, Ledwich G, Zare F (2010) Improvement of stability and load sharing in an autonomous microgrid using supplementary droop control loop. IEEE Trans Power Syst 25(2):796–808
30. McLaren PG et al (1992) A real time digital simulator for testing relays. IEEE Trans Power Deliv 7(1):207–213
31. Mehrizi-Sani A, Iravani R (2010) Potential function based control of a microgrid in islanded and grid-connected modes. IEEE Trans Power Syst 25(4):1883–1891
32. Mohamed YR, AA Radwan (2011) Hierarchical control system for robust microgrid operation and seamless mode transfer in active distribution systems. IEEE Trans Smart Grid 2(2):352–362
33. Nagaraju P, Prodanovic M, Green TC (2007) Modeling, analysis and testing of autonomous operation of an inverter-based microgrid. IEEE Trans Power Electron 22(2):613–625
34. Panigrahi B, Abraham A, Das S (eds) (2010) Computational intelligence in power engineering. Springer, Berlin

35. Planas E, Gil-de-Muro A, Andreu J, Kortabarria I, Martnez de Alegra I (2013) General aspects, hierarchical controls and droop methods in microgrids: a review. Renew Sustain Energy Rev 17:147–159
36. Prodanovic M, Green TC (2003) Control and filter design of three-phase inverters for high power quality grid connection. IEEE Trans Power Electron 18(1):373–380
37. Real Time Digital Simulator Power System and Control User Manual (2009) RTDS Technologies
38. Rocabert J, Luna A, Blaabjerg F, Rodriguez P (2012) Control of power converters in AC microgrids. IEEE Trans Power Electron 27(11):4734–4749
39. Rockefeller GD (1969) Fault protection with a digital computer. IEEE Trans Power Appar Syst 4:438–464
40. Rosendahl GK, Wierckx RP, Maguire TL, Woodford DA (1989) A parallel machine for real time power system simulations. In: Proceedings of the IEEE on electrical and computer engineering
41. RSCAD/RTDS Manual (2007) RTDS Technologies Inc
42. Shafiee Q, Vasquez JC, Guerrero JM (2012) Distributed secondary control for islanded microgrids—a networked control systems approach. In: Proceedings of the IEEE 38th annual conference on industrial electronics—IECON'12
43. Tonkoski R, Lopes L, El-Fouly T (2011) Coordinated active power curtailment of grid connected PV inverters for overvoltage prevention. IEEE Trans Sustain Energy 2(2):139–147
44. Vasquez JC et al (2010) Hierarchical control of intelligent microgrids. Ind Electron Mag 4(4):23–29
45. XiongfeI W, Frede B, Zhe C (2013) Autonomous control of inverter-interfaced distributed generation units for harmonic current filtering and resonance damping in an islanded microgrid. IEEE Trans Ind Appl 99. http://dx.doi.org/10.1109/TIA.2013.2268734
46. Xue-song Z, Li-qiang C, You-Jie M (2011) Research on control of microgrid. In: Third international conference on measuring technology and mechatronics automation, proceedings of the IEEE, pp 1129–1132
47. Zhao B, Zhang X, Chen J (2012) Integrated microgrid laboratory system. IEEE Trans Power Syst 27(4):2175–2185

Chapter 10
Appendix

10.1 Important Facts in Linear Algebra

We assume that the reader already has basic training in linear algebra and for a more complete introduction, the reader should resort to a book such as [5]. Some familiarity with the numerical software MATLAB® is also encouraged. In this book, we deal mostly with finite-dimensional linear spaces, which are also often called linear vector spaces. For generality, we consider the linear space to be n-dimensional. A linear space is typically denoted by the letter V (for vector space). Although most of the time we will deal with vectors of real numbers $\Re$, occasionally we will encounter vectors of complex numbers $\mathbb{C}$. As a reminder, for instance, the eigenvalues or eigenvectors of a real matrix could be complex. For simplicity, our review will be conducted for linear spaces over the field of real numbers $\Re$ with the understanding that most definitions and results generalize to the complex case with little change.

10.1.1 Basic Notions

Definition 10.1 (*A linear space or a vector space*) A set (of vectors) V is considered a linear space over the field $\Re$ if its elements, called vectors, are closed under two basic operations: scalar multiplication and vector summation "+". That is, given any two vectors $v_1, v_2 \in V$ and any two scalars $\alpha, \beta \in \Re$, the linear combination $v = \alpha v_1 + \beta v_2$ is also a vector in V. Furthermore, the addition is commutative and associative, it has an identity 0, and each element has an inverse, "$-v$", such that $v + (-v) = 0$. The scalar multiplication respects the structure of $\Re$, that is, $\alpha(\beta)v = (\alpha\beta)v$, $1v = v$ and $0v = 0$. The addition and scalar multiplication are related by the distributive laws: $(\alpha + \beta)v = \alpha v + \beta v$ and $\alpha(v + u) = \alpha v + \alpha u$.

M.S. Mahmoud and F.M. AL-Sunni, *Control and Optimization of Distributed Generation Systems*, Power Systems, DOI 10.1007/978-3-319-16910-1_10

For example, $\Re^n$ is a linear space over the field of real numbers $\Re$. To be consistent, we always use a column to represent a vector:

$$[x_1 x_2 \ldots x_n]^t = \begin{bmatrix} x_1 \\ x_2 \\ \vdots \\ x_n \end{bmatrix} \in \Re^n \tag{10.1}$$

where $[x_1 x_2 \ldots x_n]^t$ means the (row) vector $[x_1 x_2 \ldots x_n]$ is transposed. Given two scalars $\alpha,\ \beta \in \Re$ and two vectors $x = [x_1 x_2 \ldots x_n]^t \in \Re^n$ and $y = [y_1 y_2 \ldots y_n]^t \in \Re^n$, their linear combination is a componentwise summation weighted by α and β:

$$\begin{aligned} \alpha\ x + \beta\ y &= \alpha\,[x_1 x_2 \ldots x_n]^t\ +\ \beta\,[y_1 y_2 \ldots y_n]^t \\ &= [\alpha x_1 + \beta y_1 \alpha x_2 + \beta y_2 \ldots \alpha x_n + \beta y_n]^t \end{aligned} \tag{10.2}$$

We will now provide a brief review of basic notions and frequently used notation associated with a linear vector space V (that is, $\Re^n$).

Definition 10.2 (*Subspace*) A subset W of a linear space is called a subspace if the zero vector) is in W and $w = \alpha w_1 + \beta w_2 \in W$ for all $\alpha, \beta \in \Re$ and $w_1, w_2 \in W$.

Definition 10.3 (*Spanned subspace*) Given a set of vectors $S = \{v_i\}_{i=1}^m$, the subspace spanned by S is the set of all finite linear combinations $\sum_{i=1}^m \alpha_i v_i$ for all $[\alpha_1 \alpha_2 \ldots \alpha_n]^t$. This subspace is usually denoted by Span(S).

For example, the two vectors $v_1 = \begin{bmatrix} 1\ 0\ 0 \end{bmatrix}^t$ and $v_2 = \begin{bmatrix} 1\ 1\ 0 \end{bmatrix}^t$ span a subspace of $\Re^3$ whose vectors are of the general form $v = \begin{bmatrix} x\ y\ 0 \end{bmatrix}^t$.

Definition 10.4 (*Linear independence*) A set of vectors $S = \{v_i\}_{i=1}^m$ is linearly independent if

$$[\alpha_1 v_1 + \alpha_2 v_2 + \cdots \alpha_m v_m] = 0$$

implies

$$\alpha_1 = \alpha_2 = \cdots = \alpha_m = 0$$

On the other hand, a set of vectors $\{v_i\}_{i=1}^m$ is said to be linearly dependent if there exist $[\alpha_1 \alpha_2 \ldots \alpha_n] \in \Re$ not all zero such that

$$[\alpha_1 v_1 + \alpha_2 v_2 + \cdots \alpha_m v_m] = 0$$

Definition 10.5 (*Basis*) A set of vectors $B = \{b_i\}_{i=1}^n$ of a linear space V is said to be a basis if B is a linearly independent set and B spans the entire space V; i.e., $V = span(B)$.

(Properties of a basis). Suppose B and B' are two bases for a linear space V. Then:

1. B and B' contain exactly the same number of linearly independent vectors. This number, say n, is the dimension of the linear space V.
2. Let $B = \{b_i\}_{i=1}^n$ and $B' = \{b'_i\}_{i=1}^n$. Then each basis vector of B can be expressed as a linear combination of those in B'; i.e.,

$$b_j = a_{1j}b'_1 + a_{2j}b'_2 + \cdots + a_{nj}b'_n = \sum_{i=1}^{n} a_{ij}b'_i, \tag{10.3}$$

   for some $a_{ij} \in R, i, j = 1, 2, \ldots, n$.
3. Any vector $v \in V$ can be written as a linear combination of vectors in either of the bases:

$$v = x_1b_1 + x_2b_2 + \cdots + x_nb_n = x_nb_n = x'_1b'_1 + x'_2b'_2 + \cdots + x'_nb'_n, \tag{10.4}$$

   where the coefficients $\{x_i \in R\}_{i=1}^n$ and $\{x'_i \in R\}_{i=1}^n$ are uniquely determined and are called the coordinates of v with respect to each basis.

In particular, if B and B' are two bases for the linear space R^n, we may put the basis vectors as columns of two $n \times n$ matrices and also call them B and B', respectively:

$$B \doteq [b_1, b_2, \ldots, b_n], \quad B' \doteq [b'_1, b'_2, \ldots, b'_n] \in R^{n\times n}. \tag{10.5}$$

Then we can express the relationship between them in the matrix form $B = B'A$ as

$$[b_1, b_2, \ldots, b_n] = [b'_1, b'_2, \ldots, b'_n] \begin{bmatrix} a_{11} & a_{12} & \ldots & a_{1n} \\ a_{21} & a_{22} & \ldots & a_{2n} \\ \vdots & \vdots & \ddots & \vdots \\ a_{n1} & a_{n2} & \ldots & a_{nn} \end{bmatrix}. \tag{10.6}$$

The role of the $n \times n$ matrix is to transform one basis (B') to the other (B). Since such a transformation can go the opposite way, the matrix A must be invertible. So we can also write $B' = BA^{-1}$.

If v is a vector in V, it can be expressed in terms of linear combinations of either basis as

$$v = x_1b_1 + x_2b_2 + \cdots + x_nb_n = x_1^tb_1^t + x_2^tb_2^t + \cdots + x_n^tb_n^t. \tag{10.7}$$

Thus, we have

$$v = [b_1, b_2, \ldots, b_n]\begin{bmatrix} x_1 \\ x_2 \\ \vdots \\ x_n \end{bmatrix} = [b_1^t, b_2^t, \ldots, b_n^t]\begin{bmatrix} a_{11} & a_{12} & \ldots & a_{1n} \\ a_{21} & a_{22} & \ldots & a_{2n} \\ \vdots & \vdots & \ddots & \vdots \\ a_{n1} & a_{n2} & \ldots & a_{nn} \end{bmatrix}\begin{bmatrix} x_1 \\ x_2 \\ \vdots \\ x_n \end{bmatrix}.$$

Since the coordinates of v with respect to B^t are unique, we obtain the following transformation of coordinates of a vector from one basis to the other:

$$\begin{bmatrix} x_1^t \\ x_2^t \\ \vdots \\ x_n^t \end{bmatrix} = \begin{bmatrix} a_{11} & a_{12} & \ldots & a_{1n} \\ a_{21} & a_{22} & \ldots & a_{2n} \\ \vdots & \vdots & \ddots & \vdots \\ a_{n1} & a_{n2} & \ldots & a_{nn} \end{bmatrix}\begin{bmatrix} x_1 \\ x_2 \\ \vdots \\ x_n \end{bmatrix}. \tag{10.8}$$

Let $x = [x_1, x_2, \ldots, x_n]^t \in R^n$ and $x^t = [x_1^t, x_2^t, \ldots, x_n^t]^t \in R^n$ denote the two coordinate vectors. We may summarize in matrix form the relationships between two bases and coordinates with respect to the bases as

$$B^t = BA^{-1}, \quad x^t = Ax. \tag{10.9}$$

Be aware of the difference in transforming bases from transforming coordinates!

10.1.2 Inner Product and Orthogonality

Definition 10.6 (*Inner product*) A function is an inner product[1] if

1. $< u, \alpha v + \beta w >= \alpha < u, v > + \beta < u, w >, \quad R^n \times R^n \to R$
2. $< u, v >=< v, u >$
3. $< v, v >\geq 0$, and $< v, v >= 0 \Leftrightarrow v = 0$.

For each vector v, $\sqrt{< v, v >}$ is called its norm.

The inner product is also called a metric, since it can be used to measure length and angles.

For simplicity, a standard basis is often chosen for the vector space R^n as the set of vectors

$$e_1 = [1, 0, 0, \ldots, 0]^t, \quad e_2 = [0, 1, 0, \ldots, 0]^t, e_n = [0, 0, 0, \ldots, 0, 1]^t. \tag{10.10}$$

The matrix $I = [e_1, e_2, \ldots, e_n]$ with these vectors as columns is exactly the $n \times n$ identity matrix.

[1] In some literature, an inner product is also called a dot product, denoted by $u \cdot v$. However, in this book, we will not use that name.

Definition 10.7 (*Canonical inner product on* R^n) Given any two vectors $x = [x_1, x_2, \ldots, x_n]^t$ and $y = [y_1, y_2, \ldots, y_n]^t$ in R^n, we define the canonical inner product to be

$$< x, y > \doteq x^t y = x_1 y_1 + x_2 y_2 + \cdots + x_n y_n. \tag{10.11}$$

This inner product induces the standard 2-norm, or Euclidean norm, $\| \cdot \|_2$, which measures the length of each vector as

$$\|x\|_2 \doteq \sqrt{x^t x} = \sqrt{x_1^2 + x_2^2 + \cdots + x_n^2}. \tag{10.12}$$

Notice that if we choose another basis B^t related to the above standard basis I as $I = B^t A$, then the coordinates of the vectors x, y related to the new basis are x^t and y^t, respectively, and they relate to x, y by $x^t = Ax$ and $y^t = Ay$. The inner product in terms of the new coordinates becomes

$$< x, y >= x^t y = (A^{-1} x^t)^t (A^{-1} y^t) = (x^t) A^{-t} A^{-1} (y^t). \tag{10.13}$$

We denote this expression of the inner product with respect to the new basis by

$$< x^t, y^t >_{A^{-t}A^{-1}} \doteq (x^t)^t A^{-t} A^{-1} (y^t). \tag{10.14}$$

This is called an induced inner product from the matrix A. Knowing the matrix $A^{-t} A^{-1}$, we can compute the canonical inner product directly using coordinates with respect to the nonstandard basis B^t.

Definition 10.8 (*Orthogonality*) Two vectors x, y are said to be orthogonal if their inner product is zero: $< x, y >= 0$. This is often indicated as $x \perp y$.

10.1.3 Kronecker Product and Stack of Matrices

Definition 10.9 (*Kronecker product of two matrices*) Given two matrices $A \in R^{m \times n}$ and $B \in R^{k \times l}$, their Kronecker product, denoted by $A \otimes B$, is a new matrix

$$A \otimes B \doteq \begin{bmatrix} a_{11} B & a_{12} B & \ldots & a_{1n} B \\ a_{21} B & a_{22} B & \ldots & a_{2n} B \\ \vdots & \vdots & \ddots & \vdots \\ a_{n1} B & a_{n2} B & \ldots & a_{mn} B \end{bmatrix} \in R^{mk \times nl} \tag{10.15}$$

If A and B are two vectors, i.e., $n = l = 1$, the product $A \otimes B$ is also a vector but of dimension mk.

In MATLAB, one can easily compute the Kronecker product by using the command C = kron(A, B).

Definition 10.10 (*Stack of a matrix*) Given an $m \times n$ matrix $A \in R^{m \times n}$ the stack of the matrix A is a vector, denoted by A^s, in R^{mn} obtained by stacking its n column vectors, say $a_1, a_2, \ldots, a_n \in R^m$, in order

$$A^s \doteq \begin{bmatrix} a_1 \\ a_2 \\ \vdots \\ a_n \end{bmatrix} \in R^{mn}. \tag{10.16}$$

As mutually inverse operations, A^s is called A stacked, and A is called A^s unstacked.

The Kronecker product and stack of matrices together allow us to rewrite algebraic equations that involve multiple vectors and matrices in many different but equivalent ways. For instance, the equation

$$u^t A v = 0 \tag{10.17}$$

for two vectors u, v, and a matrix A of proper dimensions can be rewritten as

$$(v \otimes u)^t A^s = 0. \tag{10.18}$$

The second equation is particularly useful when A is the only unknown in the equation.

10.2 Linear Transformations and Matrix Groups

Linear algebra studies the properties of the linear transformations, or linear maps, between different linear spaces. Since such transformations can be represented as matrices, linear algebra to a large extent studies the properties of matrices.

Definition 10.11 (*Linear transformation*) A linear transformation from a linear (vector) space R^n to R^m is defined as a map $L : R^n \to R^m$ such that

- $L(x + y) = L(x) + L(y), \quad \forall x, y \in R^n$;
- $L(\alpha x) = \alpha L(x), \quad \forall x \in R^n, \alpha \in R$.

With respect to the standard bases of R^n and R^m, the map L can be represented by a matrix $A \in R^{m \times n}$ such that

$$L(x) = Ax, \quad \forall x \in \Re^n. \tag{10.19}$$

The ith column of the matrix A is then nothing but the image of the standard basis vector $e_i \in \Re^n$ under the map L; i.e.,

$$A = [L(e_1), L(e_2), \ldots, L(e_n)] \in \Re^{m \times n}.$$

The set of all (real) $m \times n$ matrices is denoted by $M(m, n)$. When viewed as a linear space, $M(m, n)$ can be identified as the space $\Re^{mn}$. When there is little ambiguity, we refer to a linear map L by its matrix representation A. If $n = m$, the set $M(n, n) \doteq M(n)$ forms an algebraic structure called a ring (over the field R). That is, matrices in $M(n)$ are closed under both matrix multiplication and summation: If A, B are two $n \times n$ matrices, so are $C = AB$ and $D = A + B$.

Linear maps or matrices that we encounter in computer vision often have a special algebraic structure called a group.

Definition 10.12 A group is a set G with an operation "o" on the elements of G that:

- is closed: if $g_1, g_2 \in G$, then also $g_1 o g_2 \in G$;
- is associative: $(g_1 o g_2) o g_3 = g_1 o (g_2 o g_3)$, for all $g_1, g_2, g_3 \in G$;
- has a unit element $e : eog = goe = g$, for all $g \in G$;
- is invertible: For every element $g \in G$, there exists an element $g^{-1} \in G$ such that $gog^{-1} = g^{-1} = g^{-1}og = e$.

Definition 10.13 (*The general linear group $GL(n)$*) The set of all $n \times n$ nonsingular (real) matrices with matrix multiplication forms a group. Such a group of matrices is usually called the general linear group and denoted by $GL(n)$.

Definition 10.14 (*Matrix representation of a group*) A group G has a matrix representation or can be realized as a matrix group if there exists an injective map

$$R : G \to GL(n); \quad g \to R(g),$$

Note that a map $f(\cdot)$ is called injective if $f(x) \neq f(y)$ as long as $x \neq y$, which preserves the group structure of G. That is, the inverse and composition of elements in G are preserved by the map in the following way: Such a map is called a group homomorphism in algebra.

$$R(e) = I_{n \times n}, \quad R(goh) = R(g)R(h), \quad \forall g, h \in G. \tag{10.20}$$

Below, we identify a few important subsets of $M(n)$ that have special algebraic structures (as examples of matrix groups) and nice properties.

The group $GL(n)$ itself can be identified as the set of all invertible linear transformations from $\Re^n$ to $\Re^n$ in the sense that for every $A \in GL(n)$, we obtain a linear map

$$L : \Re^n \to \Re^n; \quad x \to Ax. \tag{10.21}$$

Notice that if $A \in GL(n)$, then so is its inverse: $A^{-1} \in GL(n)$. We know that an matrix is invertible if and only if its determinant is nonzero. Therefore, we have

$$\det(A) \neq 0, \quad \forall A \in GL(n). \tag{10.22}$$

The general linear group, when matrices are considered to be known only up to a scalar factor, $GL(n)/R$, is referred to as the projective transformation group, whose elements are called projective matrices or homographies.

Matrices in $GL(n)$ of determinant $+1$ forma subgroup called the special linear group, denoted by $SL(n)$. That is, $\det(A) = +1$ for all $A \in SL(n)$. It is easy to verify that if $A \in SL(n)$, then so is A^{-1}, since $\det A^{-1} = \det A^{-1}$.

Definition 10.15 (*The affine group* $A(n)$) An affine transformation L from $\Re^n$ to $\Re^n$ is defined jointly by a matrix $A \in GL(n)$ and a vector $b \in \Re^n$ such that

$$L : \Re^n \rightarrow \Re^n; \quad x \rightarrow Ax + b. \tag{10.23}$$

The set of all such affine transformations is called the affine group of dimension and is denoted by $A(n)$.

Notice that the map L so defined is not a linear map from $\Re^n$ to $\Re^n$ unless $b = 0$. Nevertheless, we may imbed this map into a space one dimension higher so that we can still represent it by a single matrix. If we identify an element $a \in \Re^n$ with $\begin{bmatrix} x \\ 1 \end{bmatrix} \in \Re^{n+1}$, then L becomes a map from $\Re^{n+1}$ to $\Re^{n+1}$ in the following sense:

Observe that this is the so-called homogeneous representation of x. Notice that this identification does not preserve the vector structure of $\Re^n$.

$$L : \Re^{n+1} \rightarrow \Re^{n+1}; \quad \begin{bmatrix} x \\ 1 \end{bmatrix} \mapsto \begin{bmatrix} A & b \\ 0 & 1 \end{bmatrix} \begin{bmatrix} x \\ 1 \end{bmatrix}. \tag{10.24}$$

Thus, a matrix of the form

$$\begin{bmatrix} A & b \\ 0 & 1 \end{bmatrix} \in \Re^{(n+1 \times n+1)}, \quad A \in GL(n), b \in \Re^n, \tag{10.25}$$

fully describes an affine map, and we call it an affine matrix. This matrix is an element in the general linear group $GL(n+1)$. In this way, $A(n)$ is identified as a subset (and in fact a subgroup) of $GL(n+1)$. The multiplication of two affine matrices in the set $A(n)$ is

$$\begin{bmatrix} A_1 & b_1 \\ 0 & 1 \end{bmatrix} \begin{bmatrix} A_2 & b_2 \\ 0 & 1 \end{bmatrix} = \begin{bmatrix} A_1 A_2 & A_1 b_2 + b_1 \\ 0 & 1 \end{bmatrix} \in \Re^{(n+1 \times n+1)}, \tag{10.26}$$

which is also an affine matrix in $A(n)$ and represents the composition of two affine transformations.

Given $\Re^n$ and its standard inner product structure, $< x, y >= x^t y, \forall x, y \in \Re^n$, let us consider the set of linear transformations (or matrices) that preserve the inner product.

Definition 10.16 (*The orthogonal group* $O(n)$) An $n \times n$ matrix A (representing a linear map from $\Re^n$ to itself) is called orthogonal if it preserves the inner product, that is,

$$< Ax, Ay >=< x, y >, \forall x, y \in \Re^n. \quad (10.27)$$

The set of all $n \times n$ orthogonal matrices forms the orthogonal group of dimension , and it is denoted by $O(n)$.

Obviously, $O(n)$ is a subset (and in fact a subgroup) of $GL(n)$. If R is an orthogonal matrix, we must have $\Re^t R = R\Re^t = I$. Therefore, the orthogonal group $O(n)$ can be characterized as

$$O(n) = \{R \in GL(n) | \Re^t R = I\}. \quad (10.28)$$

The determinant $\det(R)$ of an orthogonal matrix R can be either $+1$ or -1. The subgroup of $O(n)$ with determinant $+1$ is called the special orthogonal group and is denoted by $SO(n)$. That is, for any $R \in SO(n)$, we have $\det(R) = +1$. Equivalently, one may define $SO(n)$ as the intersection $SO(n) = O(n) \cap SL(n)$. In the case $n = 3$, the special orthogonal matrices are exactly the 3×3 rotation matrices.

The affine version of the orthogonal group gives the Euclidean (transformation) group:

Definition 10.17 (*The Euclidean group* $E(n)$) A Euclidean transformation L from $\Re^n$ to $\Re^n$ is defined jointly by a matrix $R \in O(n)$ and a vector $T \in \Re^n$ such that

$$L : \Re^n \to \Re^n; \quad x \mapsto Rx + T. \quad (10.29)$$

The set of all such transformations is called the Euclidean group of dimension and is denoted by $E(n)$.

Obviously, $E(n)$ is a subgroup of $A(n)$. Therefore, it can also be embedded into a space one dimension higher and has a matrix representation

$$\begin{bmatrix} R & T \\ 0 & 1 \end{bmatrix} \in \Re^{(n+1 \times n+1)}, \quad R \in O(n), T \in \Re^n. \quad (10.30)$$

If R further belongs to $SO(n)$, such transformations form the special Euclidean group, which is traditionally denoted by $SE(n)$. When $n = 3$, $SE(3)$, represents the conventional rigid body motion in $\Re^3$, where R is the rotation of a rigid body and T is the translation (with respect to a chosen reference coordinate frame).

Since all the transformation groups introduced so far have natural matrix representations, they are matrix groups. To summarize their relationships, we have

$$SO(n) \subset O(n) \subset GL(n), \quad SE(n) \subset A(n) \subset GL(n+1). \quad (10.31)$$

Since these groups themselves admit a differential structure, they belong to the Lie groups.

10.3 Elements of Graph Theory

In this section, some preliminary knowledge of graph theory [4] is introduced for the analysis throughout the book.

Let $G(V, E, A)$ be a directed graph of order n, where $V = \{s_1, \ldots, s_n\}$ is the set of nodes, $E \subseteq V \times V$ is the set of edges, and $A = [a_{ij}] \in \Re^{n\times n}$ is a weighted adjacency matrix. The node indexes belong to a finite index set $I = \{1, 2, \ldots, n\}$. An edge of G is denoted by $e_{ij} = (s_i, s_j)$, where the first element s_i of the e_{ij} is said to be the tail of the edge and the other s_j to be the head. The adjacency elements associated with the edges are positive, that is, $e_{ij} \in E \Leftrightarrow a_{ij} > 0$. If a directed graph has the property that $a_{ij} = a_{ji}$ for any $i, j \in I$, the directed graph is called undirected. The Laplacian with the directed graph is defined as $L = \Delta - A \in \Re^{n\times n}$, where $\Delta = [\Delta_{ij}]$ is a diagonal matrix with $\Delta_{ii} = \sum_{j=1}^{n} a_{ij}$. An important fact of L is that all the row sums of L are zero and thus 1 is an eigenvector of L associated with the zero eigenvalue. The set of neighbors of node s_i is denoted by $N_i = \{s_j \in V : (s_i, s_j) \in E\}$. A directed path is a sequence of ordered edges of the form $(s_{i1}, s_{i2}), (s_{i2}, s_{i3}), \ldots$, where $s_{ij} \in V$ in a directed graph. A directed graph is said to be strongly connected, if there is a directed path from every node to every other node. Moreover, a directed graph is said to have spanning trees, if there exists a node such that there is a directed path from every other node to this node.

10.3.1 Basic Results

Lemma 10.1 *If the graph G has a spanning tree, then its Laplacian L has the following properties:*

1. *Zero is a simple eigenvalue of L, and $\mathbf{1_n}$ is the corresponding eigenvector, that is, $L\mathbf{1_n} = \mathbf{0}$.*
2. *The rest $n - 1$ eigenvalues all have positive real parts. In particular, if the graph G is undirected, then all these eigenvalues are positive and real.*

Lemma 10.2 *Consider a directed graph G. Let $D \in \Re^{n\times|E|}$ be the 01-matrix with rows and columns indexed by the nodes and edges of G, and $E \in \Re^{|E|\times n}$ be the 01-matrix with rows and columns indexed by the edges and nodes of G, such that*

$$D_{uf} = \begin{cases} 1 \text{ if the node } u \text{ is the tail of the edge } f \\ 0 \text{ otherwise} \end{cases}$$

$$E_{fu} = \begin{cases} 1 \text{ if the node } u \text{ is the head of the edge } f \\ 0 \text{ otherwise} \end{cases}$$

where $|E|$ is the number of the edges. Let $Q = diag\{q_1, q_2, \ldots, q_{|E|}\}$, where $q_p (p = 1, \ldots, |E|)$ is the weight of the pth *edge of G (i.e., the value of the adjacency matrix on the* pth *edge). Then the Laplacian of G can be transformed into $L = DQ(D^T - E)$.*

10.3.2 Laplacian Spectrum of Graphs

This section is a concise review of the relationship between the eigenvalues of a Laplacian matrix and the topology of the associated graph. We list a collection of properties associated with undirected graph Laplacians and adjacency matrices, which will be used in subsequent sections of the paper.

A graph $\mathcal{G}$ is defined as

$$\mathcal{G} = (\mathcal{V}, \mathcal{A}) \tag{10.32}$$

where $\mathcal{V}$ is the set of nodes (or vertices) $\mathcal{V} = \{1, \ldots, N\}$ and $\mathcal{A} \subseteq \mathcal{V} \times \mathcal{V}$ the set of edges (i, j) with $i \in \mathcal{V}$, $j \in \mathcal{V}$. The degree d_j of a graph vertex j is the number of edges which start from j. Let $d_{\max}(\mathcal{G})$ denote the maximum vertex degree of the graph $\mathcal{G}$.

10.3.3 Properties of Adjacency Matrix

We denote $\mathbf{A}(\mathcal{G})$ by the (0, 1) adjacency matrix of the graph $\mathcal{G}$. Let $\mathbf{A}_{ij} \in \Re$ be its i, j element, then $\mathbf{A}_{i,i} = 0, \forall i = 1, \ldots, N$, $\mathbf{A}_{i,j} = 0$ if $(i, j) \notin \mathcal{A}$ and $\mathbf{A}_{i,j} = 1$ if $(i, j) \in \mathcal{A}, \forall i, j = 1, \ldots, N, i \neq j$. We will focus on *undirected* graphs, for which the adjacency matrix is symmetric.

Let $\mathcal{S}(\mathbf{A}(\mathcal{G})) = \{\lambda_1(\mathbf{A}(\mathcal{G})), \ldots, \lambda_N(\mathbf{A}(\mathcal{G}))\}$ be the spectrum of the adjacency matrix associated with an undirected graph $\mathcal{G}$ arranged in nondecreasing semi-order.

(1) *Property 1*: $\lambda_N(\mathbf{A}(\mathcal{G})) \leq d_{\max}(\mathcal{G})$.

This property together with Proposition 1 implies

(2) *Property 2*: $\gamma_i \geq 0, \forall \gamma_i \in \mathcal{S}(d_{\max} I_N - \mathbf{A})$.

We define the Laplacian matrix of a graph $\mathcal{G}$ in the following way:

$$L(\mathcal{G}) = \mathbf{D}(\mathcal{G}) - \mathbf{A}(\mathcal{G}) \tag{10.33}$$

where $\mathbf{D}(\mathcal{G})$ is the diagonal matrix of vertex degrees d_i (also called the valence matrix). Eigenvalues of Laplacian matrices have been widely studied by graph theorists. Their properties are strongly related to the structural properties of their associated graphs. Every Laplacian matrix is a singular matrix. By Gershgorin theorem, the real part of each nonzero eigenvalue of $L(\mathcal{G})$ is strictly positive.

For undirected graphs, $L(\mathcal{G})$ is a symmetric, positive semidefinite matrix, which has only real eigenvalues. Let $\mathcal{S}(L(\mathcal{G})) = \{\lambda_1(L(\mathcal{G})), \ldots, \lambda_N(L(\mathcal{G}))\}$ be the spectrum of the Laplacian matrix L associated with an undirected graph $\mathcal{G}$ arranged in nondecreasing semiorder. Then,

(3) *Property 3*:

1. $\lambda_1(L(\mathscr{G})) = 0$ with corresponding eigenvector of all ones, and $\lambda_2(L(\mathscr{G}))$ iff $\mathscr{G}$ is connected. In fact, the multiplicity of 0 as an eigenvalue of $L(\mathscr{G})$ is equal to the number of connected components of $\mathscr{G}$.
2. The modulus of $\lambda_i(L(\mathscr{G})), i = 1, \ldots, N$ is less then N.

The second smallest Laplacian eigenvalue $\lambda_2(L(\mathscr{G}))$ of graphs is probably the most important information contained in the spectrum of a graph. This eigenvalue, called the algebraic connectivity of the graph, is related to several important graph invariants, and it has been extensively investigated.

Let $L(\mathscr{G})$ be the Laplacian of a graph $\mathscr{G}$ with N vertices and with maximal vertex degree $d_{\max}(\mathscr{G})$. Then properties of $\lambda_2(L(\mathscr{G}))$ include

(4) *Property 4*:

1. $\lambda_2(L(\mathscr{G})) \leq (N/(N-1)) \min\{d(v), v \in \mathscr{V}\}$;
2. $\lambda_2(L(\mathscr{G})) \leq v(\mathscr{G}) \leq \eta(\mathscr{G})$;
3. $\lambda_2(L(\mathscr{G})) \geq 2\eta(\mathscr{G})(1 - \cos(\pi/N))$;
4. $\lambda_2(L(\mathscr{G})) \geq 2(\cos\frac{\pi}{N} - \cos 2\frac{\pi}{N})\eta(\mathscr{G}) - 2\cos\frac{\pi}{N}(1 - \cos\frac{\pi}{N})d_{\max}(\mathscr{G})$

where $v(\mathscr{G})$ is the vertex connectivity of the graph $\mathscr{G}$ (the size of a smallest set of vertices whose removal renders $\mathscr{G}$ disconnected) and $\eta(\mathscr{G})$ is the edge connectivity of the graph $\mathscr{G}$ (the size of a smallest set of edges whose removal renders $\mathscr{G}$ disconnected).

Further relationships between the graph topology and Laplacian eigenvalue locations are discussed in [4] for undirected graphs. Spectral characterization of Laplacian matrices for directed graphs can be found in [2].

A lemma about Laplacian L associated with a balanced digraph G is given hereafter:

Lemma 10.3 *If G is balanced, then there exists a unitary matrix*

$$V = \begin{pmatrix} \frac{1}{\sqrt{n}} & * & \ldots & * \\ \frac{1}{\sqrt{n}} & * & \ldots & * \\ \vdots & \vdots & & \vdots \\ \frac{1}{\sqrt{n}} & * & \ldots & * \end{pmatrix} \in C^{m\times n} \tag{10.34}$$

such that

$$V^*LV = \begin{pmatrix} 0 & \\ & H \end{pmatrix} = \Lambda \in C^{n\times n}, \quad H \in C^{(n-1)\times(n-1)} \tag{10.35}$$

Moreover, if G has a globally reachable node, $H + H^*$ is positive definite.

Proof Let $V = [\zeta_1, \zeta_2, \ldots, \zeta_n]$ be a unitary matrix where $\zeta_i \in C^n (i = 1, \ldots, n)$ are the column vectors of V and

$$\zeta_1 = (1/\sqrt{n})1 = (1/\sqrt{n}, 1/\sqrt{n}, \ldots, 1/\sqrt{n})^T$$

Notice that if G is balanced, it implies that $\zeta_1^* L = 0$. Then we have

$$\begin{aligned} V^* L V &= V^* L[\zeta_1, \zeta_2, \ldots, \zeta_n] \\ &= \begin{pmatrix} \zeta_1^* \\ \zeta_2^* \\ \vdots \\ \zeta_n^* \end{pmatrix} [0_n, L\zeta_2, \ldots, L\zeta_n] \\ &= \begin{pmatrix} 0 & 0_{n-2}^T \\ \bullet & H \end{pmatrix} \end{aligned}$$

Furthermore, if G has a globally reachable node, then $L + L^T$ is positive semidefinite, see **Theorem** 7 in [6]. Hence, $V^*(L + L^T)V$ is also positive semidefinite. Since it is known that "zero" is a simple eigenvalue of L, therefore, $H + H^*$ is positive definite.

10.4 Matrix Algebra

In what follows, we let $A \in \mathbb{C}^{n \times n}$ and denote its conjugate transpose $\bar{A}^t := A^*$. Suppose A has p distinct eigenvalues, $\lambda_1, \ldots, \lambda_p$, with the ith one being of multiplicity m_i. The spec of all eigenvalues of A is called *spectrum*, denoted by $\sigma(A)$. Let the characteristic polynomial of A be

$$\Delta_A(\lambda) := det(\lambda I - A) = \lambda^n + c_{n-1}\lambda^{n-1} + \cdots + + c_0. \quad (10.36)$$

Then we have the following:

1. $det(A) = (-1)^n c_0 = \prod_{i=1}^{p} \lambda_i^{m_i}$. Furthermore,

$$det(A) = det(A^t), \ \ det(AB) = det(A)det(B), \ \ \forall \ A, B \in \mathbb{C}^{n \times n}.$$

2. $Trace(A) = \sum_{i=1}^{p} a_{ii} = \sum_{i=1}^{p} m_i \lambda_i = (-1)^{n-1} c_{n-1}$. Moreover,

$$Trace(A + B) = Trace(A) + Trace(B).$$

3. If λ is an eigenvalue of A, its complex conjugate $\bar{\lambda}$ is an eigenvalue of A^*.
4. If A is real, so are $c_i' s$. Therefore, if λ is an eigenvalue, so is $\bar{\lambda}$.
5. If A is both real and symmetric, $\sigma(A)$ is real.

6. In general, A is called Hermitian if $A = A^*$. For real matrix, this is synonymous to being symmetric. For a Hermitian matrix A, $\sigma(A)$ is real, and it admits a complete set of n orthogonal eigenvectors (even if the eigenvalues are not distinct). Denote the normalized versions of these eigenvectors by $x_, \ldots, x_n$ where x_i corresponds to the eigenvalue λ_i where we allow the possibility for $\lambda_i = \lambda_j$ for $i \neq j$. Then we can rewrite A as:

$$A = \sum_{i=1}^{n} \lambda_i x_i x_i^* .$$

which is known as the *spectral representation* of A.
7. If A is not Hermitian, but semisimple (that is, has a set of n linearly independent eigenvectors x_i's, it still admits a spectral representation, this time of the form

$$A = \sum_{i=1}^{n} \lambda_i x_i y_i^* .$$

where y_i^* is the ith row of M^{-1}, with $M = [x_1, \ldots, x_n]$. This is known as the *eigenvector dyadic expansion* of A.
8. If a Hermitian matrix A has only positive (respectively, nonnegative) eigenvalues, it is called a *positive definite* (respectively, *nonnegative definite*) matrix, and this property is symbolically displayed as $A > 0$ (respectively, $A \geq 0$). A is said to be *negative definite* (respectively, *nonpositive definite*) if $-A > 0$ (respectively, $-A \geq 0$). A positive definite matrix A has the property that $x^* A x > 0$ for all $x \in \mathbb{C}^n$ which is not zero.

10.4.1 Inverse of Block Matrices

Let A be a square matrix of appropriate dimension and partitioned in the form

$$A = \begin{bmatrix} A_1 & A_2 \\ A_3 & A_4 \end{bmatrix} \tag{10.37}$$

where both A_1 and A_4 are square matrices. If A_1 is invertible, then

$$\Delta_1 = A_4 - A_3 A_1^{-1} A_2$$

is called the Schur complement of A_1. Alternatively, if A_4 is invertible, then

$$\Delta_4 = A_1 - A_2 A_4^{-1} A_3$$

is called the Schur complement of A_4.

It is well known that matrix A is invertible if and only if either

$$A_1 \quad \textit{and} \quad \Delta_1 \quad \textit{are invertible},$$

or

$$A_4 \quad \textit{and} \quad \Delta_4 \quad \textit{are invertible}.$$

Specifically, we have the following equivalent expressions:

$$\begin{bmatrix} A_1 & A_2 \\ A_3 & A_4 \end{bmatrix}^{-1} = \begin{bmatrix} \Upsilon_1 & -A_1^{-1} A_2 \Delta_1^{-1} \\ -\Delta_1^{-1} A_3 A_1^{-1} & \Delta_1^{-1} \end{bmatrix} \tag{10.38}$$

or

$$\begin{bmatrix} A_1 & A_2 \\ A_3 & A_4 \end{bmatrix}^{-1} = \begin{bmatrix} \Delta_4^{-1} & -\Delta_4^{-1} A_2 A_4^{-1} \\ -A_4^{-1} A_3 \Delta_4^{-1} & \Upsilon_4 \end{bmatrix} \tag{10.39}$$

where

$$\begin{aligned} \Upsilon_1 &= A_1^{-1} + A_1^{-1} A_2 \Delta_1^{-1} A_3 A_1^{-1} \\ \Upsilon_4 &= A_4^{-1} + A_4^{-1} A_3 \Delta_4^{-1} A_2 A_4^{-1} \end{aligned} \tag{10.40}$$

Important special cases are

$$\begin{bmatrix} A_1 & 0 \\ A_3 & A_4 \end{bmatrix}^{-1} = \begin{bmatrix} A_1^{-1} & 0 \\ -A_4^{-1} A_3 A_1^{-1} & A_4^{-1} \end{bmatrix} \tag{10.41}$$

and

$$\begin{bmatrix} A_1 & A_2 \\ 0 & A_4 \end{bmatrix}^{-1} = \begin{bmatrix} A_1^{-1} & -A_1^{-1} A_2 A_4^{-1} \\ 0 & A_4^{-1} \end{bmatrix} \tag{10.42}$$

10.4.2 Matrix Inversion Lemma

Let $A \in \mathrm{R}^{n \times n}$ and $C \in \mathrm{R}^{m \times m}$ be nonsingular matrices. By using the definition of matrix inverse, it can be easily verified that

$$[A + B\,C\,D]^{-1} = A^{-1} - A^{-1}\,B\,[D\,A^{-1}B + C^{-1}]^{-1}\,DA^{-1} \tag{10.43}$$

10.4.3 Range, Kernel, Rank and Eigenvectors

Let A be a general $m \times n$ matrix that also conveniently represents a linear map from the vector space $\Re^n$ to $\Re^n$.

Definition 10.18 (*Range, span, null space, and kernel*) Define the range or span of A, denoted by range (A) or span (A), to be the subspace of $\Re^m$ such that $y \in range(A)$ if and only if $y = Ax$ for some $x \in \Re^n$. Define the null space of A, denoted by null (A), to be the subspace of $\Re^n$ such that $x \in null(A)$ if and only if $Ax = 0$. When A is viewed as an abstract linear map, $null(A)$ is also referred to as the kernel of the map, denoted by $ker(A)$.

Notice that the range of a matrix A is exactly the span of all its column vectors; the null space of a matrix A is exactly the set of vectors which are orthogonal to all its row vectors (for a definition of orthogonal vectors see Definition 10.9). The notion of range or null space is useful whenever the solution to a linear equation of the form $Ax = b$ is considered. In terms of range and null space, this equation will have a solution if $b \in range(A)$ and will have a unique solution only if $null(A) = \oslash$ (the empty set).

In MATLAB, the null space of a matrix can be computed using the command

$$>> \quad Z = \mathbf{null}(A)$$

Definition 10.19 (*Rank of a matrix*) The rank of a matrix is the dimension of its range:

$$rank(A) \doteq \dim(range(A)). \tag{10.44}$$

(Properties of matrix rank): For an arbitrary $m \times n$ matrix A, its rank has the following properties:

- $rank(A) = n - \dim(null(A))$.
- $0 \leq rank(A) \leq \min\{m, n\}$.
- $rank(A)$ is equal to the maximum number of linearly independent column (or row) vectors of A.
- $rank(A)$ is the highest order of a nonzero minor of A.
- **Sylvester? inequality**: Let B be an $n \times k$ matrix. Then AB is an $m \times k$ matrix and

$$rank(A) + rank(B) - n \leq rank(AB) \leq \min\{rank(A), rank(B)\}. \tag{10.45}$$

- For any nonsingular matrices $C \in \Re^{m \times m}$ and $D \in \Re^{n \times n}$, we have

$$rank(A) = rank(CAD). \tag{10.46}$$

In MATLAB, the rank of a matrix A is just

$$>> \quad \mathbf{rank}(A)$$

It should be noted that a minor of order k is the determinant of a $k \times k$ submatrix of A.

Definition 10.20 (*Orthogonal complement to a subspace*) Given a subspace S of $\Re^n$, we define its orthogonal complement to be the subspace $S^\perp \subseteq \Re^n$ such that $x \in S^\perp$ if and only if $x^t y = 0$ for all $y \in S$. We write $\Re^n = S \oplus S^\perp$.

The notion of orthogonal complement is used in this book to define the? oimage? of an image of a point or a line. Also, with respect to any linear map A from $\Re^n$ to $\Re^m$, the space $\Re^n$ can be decomposed as a direct sum of two subspaces,

$$\Re^n = null(A) \oplus null(A)^\perp,$$

and $\Re^m$ can be decomposed similarly as

$$\Re^m = range(A) \oplus range(A)^\perp.$$

We also have the following not so obvious relationships:

Theorem 10.1 *Let be a linear map from $\Re^n$ to $\Re^m$. Then:*

1. $null(A)^\perp = range(A^t)$,
2. $range(A)^\perp = null(A^t)$,
3. $null(A^t) = null(AA^t)$,
4. $range(A) = range(AA^t)$,

Proof To prove part c: $null(AA^t) = null(A^t)$, we have

1. $AA^t x = 0 \Rightarrow < x, AA^t x >= \|A^t x\|^2 = 0 \Rightarrow A^t x = 0$, hence $null(AA^t) \subseteq null(A^t)$.
2. $A^t x = 0 \Rightarrow AA^t x = 0$; hence $null(AA^t) \supseteq null(A^t)$.

To prove part d, range (AA^t) = range (A) , we first need to prove that $\Re^n$ is a direct sum of range (A^t) and null (A), i.e., part a of the theorem. Part b can then be proved similarly. We prove this by showing that a vector x is in null (A) if and only if it is orthogonal to range (A^t) : $x \in null(A) \Leftrightarrow < Ax, y >= 0, \forall y \Leftrightarrow < x, A^t y >= 0, \forall y$. Hence null (A) is exactly the subspace that is the orthogonal complement to range (A^t) (denoted by range $(A^t)^\perp$). Therefore, $\Re^n$ is a direct sum of range (A^t) and null (A). Now to complete our proof of part d, let $Img_A(S)$ denote the image of a subspace S under the map A. Then we have range $(A) = Img_A(\Re^n) = Img_A(range(A^t)) = range(AA^t)$ (in the second equality we used the fact that $\Re^n$ is a direct sum of range (A^t) and null (A)).

In fact, the same result holds even if the domain of the linear map is replaced by an infinite-dimensional linear space with an inner product (i.e., $\Re^n$ is replaced by a Hilbert space). In that case, this theorem is also known as the *finite-rank operator fundamental lemma* [5]. We will later use this result to prove the singular value decomposition (SVD). But already it implies a result that is extremely useful in the study of multiple view geometry.

Lemma 10.4 (Rank reduction lemma) *Let $A \in \Re^{n\times n}$ be a matrix and let W be a matrix of the form*

$$W = \begin{bmatrix} M & 0 \\ AB & AA^t \end{bmatrix} \in \Re^{(m+n)\times(k+n)} \tag{10.47}$$

for some matrices $M \in \Re^{m\times k}$ and $B \in \Re^{n\times k}$. Then, regardless of what B is, we always have

$$rank(M) = rank(W) - rank(A). \tag{10.48}$$

The proof is easy using the fact $range(AB) \subseteq range(A) = range(AA^t)$ with the second identity from the previous theorem, and we leave the rest of the proof to the reader as an exercise.

A linear map from $\Re^n$ to itself is represented by a square $n \times n$ matrix A. For such a map, we sometimes are interested in subspaces of $\Re^n$ that are invariant under the map. This notion turns out to be closely related to the eigenvectors of the matrix A. More rigorously speaking, a subspace $S \subset \Re^n$ is invariant if $A(S) \subseteq S$.

Definition 10.21 (*Eigenvalues and eigenvectors of a matrix*) Let A be an $n \times n$ complex matrix in $C^{n\times n}$. A nonzero vector $v \in C^n$ is said to be its (right) eigenvector if

$$Av = \lambda v \tag{10.49}$$

for some scalar $\lambda \in C$; λ is called an eigenvalue of A. Similarly, a nonzero row vector $\eta^t \in C^n$ is called a left eigenvector of A if $\eta^t A = \lambda \eta^t$ for some $\lambda \in C$.

Although A will mostly be a real matrix in this book, to talk about its eigenvectors, it is more convenient to think of it as a complex matrix (with all entries that happen to be real).

Unless otherwise stated, an eigenvector by default means a right eigenvector. The set of all eigenvalues of a matrix A is called its spectrum, denoted by $\sigma(A)$. The MATLAB command

$$[V, D] = \mathbf{eig}(A)$$

produces a diagonal matrix D of eigenvalues and a full-rank matrix V whose columns are the corresponding eigenvectors, so that $AV = VD$.

We give the following facts about eigenvalues and eigenvectors of a matrix without a proof.

(Properties of eigenvalues and eigenvectors): Given a matrix $A \in \Re^{n\times n}$, we have:

1. If $Av = \lambda v$, then for the same eigenvalue λ, there also exists a left eigenvector η^t such that $\eta^t A = \lambda \eta^t$ and vice versa. Hence $\sigma(A) = \sigma(A^t)$.
2. The eigenvectors of A associated with different eigenvalues are linearly independent.
3. All its eigenvalues $\sigma(A)$ are the roots of the (characteristic) polynomial equation $\det(\lambda I - A) = 0$. Hence $\det(A)$ is equal to the product of all eigenvalues of A.
4. if $B = PAP^{-1}$ for some nonsingular matrix P , then $\sigma(B) = \sigma(A)$.

5. If A is a real matrix, then $\lambda \in C$ is an eigenvalue implies that its conjugate $\bar{\lambda}$ is also an eigenvalue. Simply put, $\sigma(A) = \bar{\sigma}(A)$ for real matrices.

10.4.4 Symmetric and Skew-Symmetric Matrices

Definition 10.22 (*Symmetric matrix*) A matrix $S \in \Re^{n\times n}$ is called symmetric if $S^t = S$. A symmetric matrix S is called positive (semi-)definite, if $x^t Sx > 0$ (or $x^t Sx \geq 0$) for all $x \in \Re^n$, denoted by $S > 0$ (or $S \geq 0$).

(Properties of symmetric matrices): If S is a real symmetric matrix, then:

1. All eigenvalues of S must be real, i.e., $\sigma(S) \subset R$.
2. Let (λ, v) be an eigenvalue–eigenvector pair. If $\lambda_i \neq \lambda_j$, then $v_i \perp v_j$; i.e., eigenvectors corresponding to distinct eigenvalues are orthogonal.
3. There always exist n orthonormal eigenvectors of S, which form a basis for $\Re^n$.
4. $S > 0 (S \geq 0)$ if $\lambda_i > 0 (\lambda_i \geq 0) \forall i = 1, 2, \dots, n$; i.e., S is positive (semi-)definite if all eigenvalues are positive (nonnegative).
5. $S \geq 0$ and $\lambda_1 \geq \lambda_2 \geq \cdots \geq \lambda_n$; then $\max_{\|x\|_2=1} < x, Sx >= \lambda_1$ and $\min_{\|x\|_2=1} < x, Sx >= \lambda_n$.

From point 3, we see that if $V = [v_1, v_2, \dots, v_n] \in \Re^{n\times n}$ is the matrix of all the eigenvectors, and $\lambda = diag\{\lambda_1, \lambda_2, \dots, \lambda_n\}$ is the diagonal matrix of the corresponding eigenvalues, then we can write

$$S = V \Lambda V^t,$$

where V is an orthogonal matrix. In fact, V can be further chosen to be in $SO(n)$ (that is, of determinant + 1) if n is odd, since $V \Lambda V^t = (-V)\Lambda(-V)^t$ and $\det(-V) = (-1)^n \det(V)$.

Definition 10.23 (*Induced 2-norm of a matrix*) Let $A \in \Re^{m\times n}$. We define the induced 2-norm of A (as a linear map from $\Re^n$ to $\Re^m$) as

$$\|A\|_2 \doteq \max_{\|x\|_2=1} \|A\|_2 = \max_{\|x\|_2=1} \sqrt{< x, A^t Ax >}.$$

Similarly, other induced operator norms on A can be defined starting from different norms on the domain and codomain spaces on which A operates.

Let A be as above. Then $A^t A \in \Re^{n\times n}$ is clearly symmetric and positive semi-definite, so it can be diagonalized by a orthogonal matrix V. The eigenvalues, being nonnegative, can be written as σ_i^2. By ordering the columns of V so that the eigenvalue matrix Λ has decreasing eigenvalues on the diagonal, we see, from point 5 of the preceding fact, that $A^t A = V diag\{\sigma_1^2, \sigma_2^2, \dots, \sigma_n^2\} V^t$ and

$$\|A\|_2 = \sigma_1$$

The induced 2-norm of a matrix $A \in \Re^{m \times n}$ is different from the? -norm? of A viewed as a vector in $\Re^{mn}$. To distinguish them, the latter one is conventionally called the Frobenius norm of A, precisely defined as $\|A\|_f = \sqrt{\sum_{i,j} a_{ij}^2}$. Notice that $\sum_{i,j} a_{ij}^2$ is nothing but the trace of $A^t A$(or AA^t). Thus, we have

$$\|A\|_f = \sqrt{trace(A^t A)} = \sqrt{\sigma_1^2, \sigma_2^2, \ldots, \sigma_n^2}.$$

The inverse problem of retrieving from the symmetric matrix $S = A^t A$ is usually solved by Cholesky factorization. For the given S, its eigenvalues must be nonnegative. Thus, we have $S = V \Lambda V^t = A^t A$ for $A = \Lambda^{(\frac{1}{2})} V^t$, where $\Lambda^{(\frac{1}{2})} = diag\{\sigma_1, \sigma_2, \ldots, \sigma_n\}$ is the? quare root? of the diagonal matrix λ. Since $\Re^t R = I$ for any orthogonal matrix, the solution for A is not unique: RA is also a solution. Cholesky factorization then restricts the solution to be an uppertriangular matrix (exactly what we need for camera calibration in Chap. 6). In MATLAB, the Cholesky factorization is given by the command A = chol(S).

Definition 10.24 (*Skew-symmetric matrix*) A matrix $A \in \Re^{n \times n}$ is called skew-symmetric (or antisymmetric) if $A^t = -A$.

(Properties of a skew-symmetric matrix). If A is a real skew-symmetric matrix, then:

1. All eigenvalues of A are either zero or purely imaginary, i.e., of the form $i\omega$ for $i = \sqrt{-1}$ and some $\omega \in R$.
2. There exists an orthogonal matrix V such that

$$A = V \Lambda V^t, \tag{10.50}$$

where Λ is a block diagonal matrix $\Lambda = diag\{A_i, \ldots, A_m, 0, \ldots, 0\}$, where each A_i is a 2×2 real skew-symmetric matrix of the form

$$A_i = \begin{bmatrix} 0 & a_i \\ -a_i & 0 \end{bmatrix} \in \Re^{2 \times 2}, \quad i = 1, 2, \ldots, m. \tag{10.51}$$

From point 2, we conclude that the rank of any skew-symmetric matrix must be even. A commonly used skew-symmetric matrix in computer vision is associated with a vector $u \in \Re^3$, denoted by

$$\widehat{u} = \begin{bmatrix} 0 & -u_3 & u_2 \\ u_3 & 0 & -u_1 \\ -u_2 & u_1 & 0 \end{bmatrix} \in \Re^{3 \times 3}. \tag{10.52}$$

The reason for such a definition is that $\widehat{u}v$ is equal to the conventional cross product $u \times v$ of two vectors in $\Re^3$. Then we have rank $(\widehat{u}) = 2$ if $u \neq 0$ and the (left and right) null space of $\widehat{u}$ is exactly spanned by the vector u itself. That is, $\widehat{u}u = 0$ and $u^t\widehat{u} = 0$. In other words, columns and rows of the matrix $\widehat{u}$ are always orthogonal to u.

Obviously, $A^t\widehat{u}A$ is also a skew-symmetric matrix. Then $A^t\widehat{uA} = \widehat{v}$ for some $v \in \Re^3$. We want to know what the relationship between v and A, u is.

(Hat operator). If A is a 3×3 matrix of determinant 1, then we have

$$A^t\widehat{u}A = \widehat{A^{-1}u}. \tag{10.53}$$

This is an extremely useful fact, which will be extensively used in our book. For example, this property allows us to push a matrix through a skew-symmetric matrix in the following way: $\widehat{u}A = A^{-t}\widehat{u}A = A^{-t}\widehat{A^{-1}u}$. We leave to the reader as an exercise to think about how this result needs to be modified when the determinant of A is not!, or when A is not even invertible.

10.5 Singular Value Decomposition

The SVD is a useful tool to capture essential features of a matrix (that represents a linear map), such as its rank, range space, null space, and induced norm, as well as to generalize the concept of eigenvalue eigenvector pair to non-square matrices. The computation of the SVD is numerically well conditioned, making it extremely useful for solving many linear algebraic problems such as matrix inversion, calculation of the rank, linear least-squares (LLS) estimate, projections, and fixed-rank approximations.

We give hereafter a complete description and proof for the SVD theorem. Recall first that the inner product defined on $\mathbb{C}^n$ is $< x, \ y > = \ x^*y, \ \forall x, y \in \mathbb{C}^n$. We now introduce the following important lemma:

Lemma 10.5 *Let $A \in \mathbb{C}^{m\times n}$ and A^* be its conjugate transpose. We then always have:*

$$\mathcal{N}(AA^*) = \mathcal{N}(A^*), \quad \mathcal{R}(AA^*) = \mathcal{R}(A).$$

Proof To prove $\mathcal{N}(AA^*) = \mathcal{N}(A^*)$, we have:

1. $AA^*x = \theta \implies < x, AA^*x > = ||A^*x||^2 = 0 \implies A^*x = \theta$. Hence $\mathcal{N}(AA^*) \subseteq \mathcal{N}(A^*)$.
2. $A^*x = \theta \implies AA^*x = \theta$. Hence $\mathcal{N}(AA^*) \supseteq \mathcal{N}(A^*)$.

To prove $\mathscr{R}(AA^*) = \mathscr{R}(A)$, we need to prove that $\mathbb{C}^n$ is a direct sum of $\mathscr{R}(A^*)$ and $\mathscr{N}(A)$. We prove this by showing that a vector x is in $\mathscr{N}(A)$ if and only if it is orthogonal to $\mathscr{R}(A^*)$:

$$x \in \mathscr{N}(A), \implies < A^*x, y >= 0, \ \forall y \in \mathscr{R}(A^*) \implies < x, Ay >= 0, \ \forall y$$

Hence $\mathscr{N}(A)$ is exactly the subspace which is orthogonal supplementary to $\mathscr{R}(A^*)$ (sometimes denoted as $\mathscr{R}(A^*)^{\perp}$). Therefore $\mathbb{C}^n$ is a direct sum of $\mathscr{R}(A^*)$ and $\mathscr{N}(A)$. Let $\Im_A(S)$ denote the image of a subspace S under the map A. Then we have:

$$\mathscr{R}(A) = \Im_A(\mathbb{C}^n) = \Im_A \mathscr{R}(A^*) = \mathscr{R}(AA^*)$$

wherein the second equality we used the fact that $\mathbb{C}^n$ is a direct sum of $\mathscr{R}(A^*)$ and $\mathscr{N}(A)$.

We are now ready to give a complete proof for the SVD Theorem is:

Theorem 10.2 (Singular value decomposition) *Let* $\mathsf{F} = \Re$ *or* $\mathbb{C}$. *Let* $A \in \mathsf{F}^{n\times n}$ *be a matrix of rank* r. *Then there exist matrices* $U \in \mathsf{F}^{m\times m}$ *and* $V \in \mathsf{F}^{n\times n}$, *and* $\Sigma_1 \in \Re^{n\times n}$ *such that:*

1. $V = [V_1 : V_2]$, $V_1 \in \mathsf{F}^{n\times r}$, *satisfies:*
   V is unitary, that is, $V^*V = I_{n\times n}$,
   $\mathscr{R}(V_1) = \mathscr{R}(A^*)$, *the columns of* V_1 *form an orthonormal basis of* $\mathscr{R}(A^*)$,
   $\mathscr{R}(V_2) = \mathscr{N}(A)$, *the columns of* V_2 *form an orthonormal basis of* $\mathscr{N}(A)$,
   The columns of V *form a complete orthonormal basis of eigenvectors of* A^*A.
2. $U = [U_1 : VU_2]$, $U_1 \in \mathsf{F}^{m\times r}$, *satisfies:*
   U is unitary, that is, $U^*U = I_{m\times m}$,
   $\mathscr{R}(U_1) = \mathscr{R}(A^*)$, *the columns of* U_1 *form an orthonormal basis of* $\mathscr{R}(A)$,
   $\mathscr{R}(U_2) = \mathscr{N}(A)$, *the columns of* U_2 *form an orthonormal basis of* $\mathscr{N}(A^*)$,
   The columns of U *form a complete orthonormal basis of eigenvectors of* AA^*.
3. $\Sigma_1 = diag(\sigma_1, \ldots, \sigma_r) \in \Re^{r\times r}$ *such that* $\sigma_1 \geq \sigma_2 \ldots \sigma_r > 0$. $A \in \mathsf{F}^{m\times n}$ *has dyadic expansion*

   $$A = U_1 \, \Sigma_1 \, V_1^*, \implies A = \sum_{i=1}^{r} \sigma_i u_i v_i^*$$

   where u_i, v_i *are the columns of* U_1 *and* V_1, *respectively.*
4. $A \in \mathsf{F}^{m\times n}$ *has an SVD*

   $$A = U \, \Sigma \, V^*, \quad \Sigma = \begin{bmatrix} \Sigma_1 & 0 \\ 0 & 0 \end{bmatrix}$$

Proof

1. $A \in \mathsf{F}^{m\times n}$ has rank r, hence the nonnegative (or, equivalently, positive semi-definite) Hermitian matrix AA^* has rank r according to Lemma 10.5. It has n nonnegative eigenvalues σ_i^2 ordered as

$$\sigma_1^2 \geq \sigma_2^2 \geq \cdots \sigma_r^2 > 0 = \sigma_{r+1}^2 = \cdots = \sigma_n^2$$

to which corresponds a complete orthonormal eigenvector basisn $(\nu)_{i=1}^2$ of AA^*. This family of vectors ($\in \mathsf{F}^m$) form the columns of a unitary $n \times n$ matrix, say, V. From Lemma 10.5, $\mathscr{R}(AA^*) = \mathscr{R}(A^*)$ and $\mathscr{N}(AA^*) = \mathscr{N}(A)$, the properties listed in 1 follow.

2. Define a diagonal matrix $\Sigma_1 = diag(\sigma_1, \sigma_2, \ldots, \sigma_r) \in \Re^{r\times r}$. We then have $AA^*V_1 = V_1\Sigma_1^2$, hence

$$(AV_1\Sigma_1^{-1})^*(AV_1\Sigma_1^{-1}) = I_{r\times r}$$

This defines an $m \times r$ matrix:

$$U_1 = AV_1\Sigma_1^{-1}$$

It follows that $U_1^*U_1 = I_{r\times r}$. Since A^*A and AA^* both have exactly r nonzero eigenvalues, it follows that the columns of U_1 form an orthonormal basis for $\mathscr{R}(AA^*)$ and $\mathscr{R}(A)$. Thus the properties of U_1 listed in 2 hold. Now define an $m \times (m-r)$ matrix U_2 with orthonormal columns which are orthogonal to columns of U_1. Then $U = [U_1 : U_2]$ is clearly an unitary matrix. From the proof of Lemma 10.5, columns of U_2 form an orthonormal basis of $\mathscr{N}(A^*)$ or $\mathscr{N}(AA^*)$. Therefore, columns of U_2 are all the eigenvectors corresponding to the zero eigenvalue. Hence columns of U form a complete orthonormal basis of eigenvectors of AA^*. List 2 is then fully proven.
3. Since $U_1 = AV_1\Sigma_1^{-1}$, we have

$$A = U_1\Sigma_1V_1^*.$$

The dyadic expansion directly follows.
4. The SVD follows because

$$A[V_1 : V_2] = [U_1\Sigma_1 : 0] = [U_1 : U_2]\Sigma \implies A = U\Sigma V^*$$

10.5.1 Geometric Interpretation

Notice that in the **SVD** of a square matrix $A = U\Sigma V^t \in \Re^{n\times n}$, columns of $U = [u_1, u_2, \ldots, u_n]$ and columns of $V = [v_1, v_2, \ldots, v_n]$ form orthonormal bases for

$\Re^n$. The **SVD** essentially states that if A (as a linear map) maps a point x to y, then coordinates of with respect to the basis U are related to coordinates of x with respect to the basis V by the diagonal matrix $\sum$ that scales each coordinate by the corresponding singular value.

Theorem 10.3 *Let $A \in \Re^{n\times n} = U \sum V^t$ be a square matrix. Then A maps the unit sphere $S^{n-1} \doteq \{x \in \Re^n : \|x\|_2 = 1\}$ to an ellipsoid with semi-axes $\sigma_i u_i$, where u_i is the ith column of U.*

Proof Let x, y be such that $Ax = y$. The set $\{u_i\}_{i=1}^n$ is an orthonormal basis for $\Re^n$. With respect to such a basis x has coordinates

$$[\alpha_1, \alpha_2, \dots, \alpha_n]^t = [< v_1, x >, < v_2, x >, \dots, < v_n, x >]^t.$$

That is, $x = \sum_{i=1}^n \alpha_i v_i$. With respect to the basis $\{u_i\}_{i=1}^n$, has coordinates

$$[\beta_1, \beta_2, \dots, \beta_n]^t = [< u_1, y >, < u_2, y >, \dots, < u_n, y >]^t.$$

We also have $y = \sum_{i=1}^n \beta_i u_i = Ax = \sum_{i=1}^n \sigma_i u_i v_i^t x = \sum_{i=1}^n \sigma_i < v_i, x > u_i$. Hence $\sigma_i \alpha_i = \beta_i$. Now $\|x\|_2^2 = \sum_{i=1} \alpha_i^2 = 1, \forall x \in S^{n-1}$, and so we have $\sum_{i=1}^n \beta_i^2/\sigma_i^2 = 1$, which implies that the point y satisfies the equation of an ellipsoid with semi-axes of length σ_i. This is illustrated in Fig. 10.2 for the case $n = 2$.

10.5.2 Example A.1

The SVD of matrix

$$A = \begin{bmatrix} 0 \\ 3 \\ 4 \end{bmatrix}$$

is $A = U\Sigma V^*$ where

$$U = \begin{bmatrix} 0 & -0.6 & -0.8 \\ 0.6 & 0.64 & -0.84 \\ 0.8 & -0.48 & 0.36 \end{bmatrix}, \quad \Sigma = \begin{bmatrix} 5 \\ 0 \\ 0 \end{bmatrix}, \quad V = 1$$

Since matrix A has $rank = 1$, it has a single nonzero singular value. Therefore, the spectral norm of A is $\sigma_M(A) = 5$.

10.5.3 Some Properties of the SVD

Let $A = U\Sigma V^*$ be the SVD of the $n \times m$ matrix A, with singular values $\sigma_1, \sigma_2, \ldots, \sigma_r$, $r = \min(n, m)$. Denote the columns of the $n \times n$ unitary matrix U as $u_i, i = 1, 2, \ldots, n$, and those of the $m \times m$ unitary matrix V as v_i, $i = 1, 2, \ldots, m$. The following statements hold true:

1. For $i = 1, 2, \ldots, \min(n, m)$, the column vector u_i is an eigenvector of AA^* corresponding to the eigenvalue σ_i^2. Any remaining columns are eigenvectors corresponding to the eigenvalue 0.
2. Similarly, for $i = 1, 2, \ldots, \min(n, m)$, the column vector v_i is an eigenvector of A^*A corresponding to the eigenvalue σ_i^2. Any remaining columns are eigenvectors corresponding to the eigenvalue 0.
3. For $i = 1, 2, \ldots, \min(n, m)$, the vectors u_i and v_i satisfy

$$Av_i = \sigma_i u_i, \quad A^*u_i = \sigma_i v_i$$

4. Given a square $n \times n$ matrix A, the following properties hold [3]:

$$\begin{aligned}
\sigma_M(A) &= \max_{x \in \mathbb{C}^n} \frac{||Ax||_2}{||x||_2}, \\
\sigma_m(A) &= \min_{x \in \mathbb{C}^n} \frac{||Ax||_2}{||x||_2}, \\
\sigma_m(A) &\leq |\lambda_i(A)| \leq \sigma_M(A), \quad \textit{where } \lambda_i(A) \textit{ is the ith,} \\
&\quad \textit{eigenvalue of } A, \\
\sigma_m(A) &= \frac{1}{\sigma_M(A^{-1})}, \quad \sigma_M(A) = \frac{1}{\sigma_m(A^{-1})}, \quad \textit{if} A^{-1} \textit{ esists,} \\
\sigma_M(\alpha A) &= |\alpha| \sigma_M(A), \quad \textit{with } \alpha \textit{ any complex number,} \\
\sigma_M(A + B) &\leq \sigma_M(A) + \sigma_M(B), \\
\sigma_M(AB) &\leq \sigma_M(A)\, \sigma_M(B), \\
\sigma_m(A) - \sigma_M(B) &\leq \sigma_m(A + B) \leq \sigma_m(A) + \sigma_M(B), \\
\max(\sigma_M(A), \sigma_M(B)) &\leq \sigma_M([A \;\; B]) \leq \sqrt{2} \max(\sigma_M(A), \sigma_M(B)), \\
\max_{i,j} |A_{ij}| &\leq \sigma_M(A) \leq n \max_{i,j} |A_{ij}|, \quad \textit{with } A_{ij} \textit{ the } (i, j) \textit{ element,} \\
&\quad \textit{of } A, \\
\sum_{i=1}^{n} \sigma_i^2(A) &= Tr(A^*A)
\end{aligned}$$

Problems involving orthogonal projections onto invariant subspaces of $\mathbb{C}$, such as the LLS problem, can be easily solved using **SVD**.

Definition 10.25 (*Generalized (Moore Penrose) inverse*) Given a matrix $A \in \Re^{m\times n}$ of rank r with its SVD $A = U\Sigma V^t$, we then define the generalized inverse of A to be

$$A^{\perp} = V\Sigma^{\perp}U^t, \quad \Sigma^{\perp} = \begin{bmatrix} \Sigma_1^{-1} & 0 \\ 0 & 0 \end{bmatrix}_{n\times m}.$$

The generalized inverse is sometimes also called the pseudoinverse.

In MATLAB, the pseudoinverse of a matrix is computed by the command X = pinv(A). **(Properties of generalized inverse)**.

- $AA^{\perp}A = A, \quad A^{\perp}AA^{\perp} = A^{\perp}.$

The generalized inverse can then be used to solve linear equations in general.

Proposition 10.1 (Least-squares solution of a linear systems) *Consider the problem $Ax = b$ with $A \in \Re^{m\times n}$ of rank $r \le \min(m, n)$. The solution x^* that minimizes $\|Ax - b\|_2$ is given by $x^* = A^{\perp}b$.*

The following two results have something to do with the sensitivity of solving linear equations of the form $Ax = b$.

Proposition 10.2 (Perturbations) *Consider a nonsingular matrix $A \in \Re^{n\times n}$. Let δA be a full-rank perturbation. Then*

- $|\sigma_k(A + \delta A) - \sigma_k(A)| \le \sigma_1(\delta A), \quad \forall k = 1, 2, \ldots, n.$
- $\sigma_n(A\delta A) \ge \sigma_n(A)\sigma_n(\delta A),$
- $\sigma_1(A^{-1}) = \frac{1}{\sigma_n(A)},$

where σ_i denotes the ith *singular value.*

Proposition 10.3 (Condition number) *Consider the problem $Ax = b$, and consider a? erturbed? full-rank problem $(A + \delta A)(x + \delta x) = b$. Since $Ax = b$, then to first-order approximation, $\delta x = -A^{\dagger}\delta Ax$. Hence $\|\delta x\|_2 \le \|A^{\dagger}\|_2\|\delta A\|_2\|x\|_2$, from which*

$$\frac{\|\delta x\|_2}{\|x\|_2} \le \|A^{\dagger}\|_2\|A\|_2\frac{\|\delta A\|_2}{\|A\|_2} \doteq k(A)\frac{\|\delta A\|_2}{\|A\|_2},$$

where $k(A) = \|A^{\dagger}\|_2\|A\|_2$ is called the condition number of A. It is easy to see that $k(A) = \sigma_1/\sigma_n$ if A is invertible.

Last but not the least, one of the most important properties of the **SVD** *is related to a fixed-rank approximation of a given matrix. Given a matrix A of rank r, we want to find a matrix B such that it has fixed rank $p < r$ and the Frobenius norm of the difference $\|A - B\|_f$ is minimal. The solution to this problem is given simply by setting all but the first p singular values to zero*

$$B \doteq U\sum_{(p)}V^t,$$

where $\sum_{(p)}$ *denotes the matrix obtained from* $\sum$ *by setting to zero its elements on the diagonal after the* pth *entry. The matrix B has exactly the same induced 2-norm of, i.e.,* $\sigma_1(A) = \sigma_1(B)$, *and satisfies the requirement on the rank.*

Proposition 10.4 (Fixed-rank approximation) *Let* A, B *be defined as above. Then* $\|A - B\|_f^2 = \sigma_{p+1}^2 + \cdots + \sigma_r^2$. *Furthermore, such a norm is the minimum achievable.*

The proof is an easy exercise that follows directly from the properties of orthogonal projection and the properties of the ***SVD*** *given above.*

After we have gone through all the trouble proving this theorem, you must know that SVD has become a numerical routine available in many computational softwares such as MATLAB. Within MATLAB, to compute the SVD of a given $m \times n$ matrix A, simply use the command

$$>> \quad [U,\ S,\ V] = \mathbf{SVD(A)}$$

which returns matrices $U,\ S,\ V$ satisfying $A = USV^*$ (where S represents Σ as defined above).

10.5.4 The QR Decomposition

A matrix in $GL(n)$ has n independent rows (or columns). A matrix in $O(n)$ has orthonormal rows (or columns). The Gram–Schmidt procedure can be viewed as a map from $GL(n)$ to $O(n)$, for it transforms a nonsingular matrix into an orthogonal one. Call $L_+(n)$ the subset of $GL(n)$ consisting of lower triangular matrices with positive elements along the diagonal. Such matrices form a subgroup of $GL(n)$.

Theorem 10.4 (Gram–Schmidt procedure) *For every* $A \in GL(n)$, *there exists a lower triangular matrix* $L \in \Re^{n\times n}$ *and an orthogonal matrix* $E \in O(n)$ *such that*

$$A = LE. \tag{10.54}$$

Proof Contrary to the convention of the book, for simplicity in this proof all vectors indicate row vectors. That is, if v is an -dimensional row vector, it is of the form: $v = [v_1, v_2, \ldots, v_n] \in \Re^n$. Denote the ith row vector of the given matrix A by a_i for $i = 1, 2, \ldots, n$. The proof consists in constructing L and E iteratively from the row vectors a_i:

$$\begin{array}{llll}
l_1 \doteq a_1 & & \rightarrow e_1 \doteq l_1/\|l_1\|_2, \\
l_2 \doteq a_2 - (a_2, \varepsilon_1)\varepsilon_1 & & \rightarrow e_2 \doteq l_2/\|l_2\|_2, \\
\vdots \quad \vdots \vdots & & \quad \vdots \qquad \vdots \\
l_n \doteq a_n - \sum_{i=1}^{n-1} < a_{i+1}, e_i > e_i & & \rightarrow e_n \doteq l_n/\|l_n\|_2.
\end{array}$$

Then $E = [e_1^t, \dots, e_n^t]^t$, and the matrix L is obtained as

$$L = \begin{bmatrix} \|l_1\|_2 & 0 & \dots & 0 \\ < a_2, e_1 > & \|l_2\|_2 & \dots & 0 \\ \vdots & \ddots & \ddots & \vdots \\ < a_2, e_1 > & \dots & < a_n, e_{n-1} > & \|l_n\|_2 \end{bmatrix}.$$

By construction is orthogonal; i.e. $EE^t = E^t E = I$.

Remark 10.1 The Gram–Schmidt? procedure has the peculiarity of being causal, in the sense that the ith row of the transformed matrix E depends only upon rows with index $J \le i$ of the original matrix A. The choice of the name E for the orthogonal matrix above is not accidental. In fact, we will view the Kalman filter (to be reviewed in the next appendix) as a way to perform a Gram–Schmidt orthonormalization in a special Hilbert space, and the outcome E of the procedure is traditionally called the innovation.

There are a few useful variations to Gram–Schmidt procedure. By transposing $A = LE$, we get $A^t = E^t L^t \doteq QR$. Notice that $R = L^t$ it is an upper triangular matrix. Thus, by applying Gram–Schmidt procedure to the transpose of a matrix, we can also decompose it into the form QR where Q is an orthogonal matrix and R an upper triangular matrix. Such a decomposition is called the QR decomposition. In MATLAB, this can be done by the command $[Q, R] = qr(A)$. Furthermore, by inverting $A^t = E^t L^t$, we get $A^{-t} = L^{-t} E \doteq KE$. Notice that $K = L^{-t}$ is still an upper triangular matrix. Thus, we can also decompose any matrix into the form of an upper triangular matrix followed by an orthogonal There are a few useful variations to Gram–Schmidt procedure. By transposing , we get . Notice that is an upper triangular matrix. Thus, by applying Gram–Schmidt procedure to the transpose of a matrix, we can also decompose it into the form where is an orthogonal one. The latter one is the kind of "QR decomposition" we use in Chap. 6 for camera calibration.

10.6 Useful Formulae

In what follows, some standard formulae that are of common use in the book are presented.

10.6.1 Ackermann's Formula for Eigenvalue Assignment

In [1], a formula is provided to compute the feedback gain matrix K such that a set of eigenvalues of the linear system

$$\dot{x}(t) = Ax(t) + Bu(t)$$

at desired locations. The formula emerges by organizing a three-step procedure of converting the system matrices into controller canonical form, solving for the gain and converting the gain back. It has the form:

$$
\begin{aligned}
K &= [0, \dots, 0, 1]\ P_c \alpha_c(A) \\
P_c &= [B, AB, A^2B, \dots, A^{n-1}B] \\
\alpha_c(A) &= A^n + \alpha_1 A^{n-1} + \alpha_2 A^{n-2} + \dots + \alpha_n I
\end{aligned}
$$

where α_j are the coefficients of the desired characteristic polynomial

$$
\begin{aligned}
\alpha_c(s) &= (s - s_1)\ (s - s_2)\ \cdots\ (s - s_n) \\
&= s^n + \alpha_1 s^{n-1} + \alpha_2 s^{n-2} + \dots + \alpha_n
\end{aligned}
$$

and $s_1, \cdots, s_n$ are the desired eigenvalues.

10.6.2 Parseval Formula

Let $f(t)$ be a continuous-time signal over the time interval $[t_1, t_2]$. Then the *continuous-time signal energy* over the length $L = t_2 - t_1]$ is defined by

$$E_L = \int_{t_1}^{t_2} |f(t)|^2\, dt$$

whereas the *total continuous-time signal energy* is given by

$$E_\infty = \int_{-\infty}^{\infty} |f(t)|^2\, dt \tag{10.55}$$

From Laplace tramsform methods [Gajic], we recall the frequency domain convolution which states that *the Frequency transform of a product of two signals in time is proportional to the convolution of their convolution of their Fourier transforms in the frequency domain*, that is, for two continuous-time signals $x_1(t)$ and $x_2(t)$, we have

$$
\begin{aligned}
\mathscr{F}[x_1(t)x_2(t)] &= \frac{1}{2\pi}\ X_1(j\omega) * X_2(j\omega) \\
&= \frac{1}{2\pi} \int_{-\infty}^{\infty} X_1(j(\omega - \lambda)) * X_2(j\lambda)\, d\lambda
\end{aligned} \tag{10.56}
$$

where $\mathscr{F}[x]$ is the Fourier transform of x and $*$ stands for the convolution operation. Based thereon, we can now establish the relationship between signal energy in the time and frequency domains. Rewriting (10.56) as

$$\mathscr{F}[x_1(t)x_2(t)] = \int_{-\infty}^{\infty} x_1(t)x_2(t)\, e^{-j\omega t}\, dt \tag{10.57}$$

Since (10.57) is valid for any ω, it must be valid for $\omega = 0$. In view of (10.55)–(10.57), we have

$$\int_{-\infty}^{\infty} x_1(t)x_2(t)\, e^{-j\omega t}\, dt = \frac{1}{2\pi}\int_{-\infty}^{\infty} X_1(j(\omega-\lambda)) * X_2(j\lambda)\, d\lambda \tag{10.58}$$

Letting $x_1(t) = x_2(t) = x(t)$ with $x(t)$ being a real function such that $|x(t)|^2 = x^2(t)$, it follows that

$$\begin{aligned} E_\infty = \int_{-\infty}^{\infty} x^2(t)\, dt &= \frac{1}{2\pi}\int_{-\infty}^{\infty} X(j\lambda)X(j\lambda)d\,\lambda \\ = \frac{1}{2\pi}\int_{-\infty}^{\infty} |X(j\lambda)|^2\, d\lambda &= \int_{-\infty}^{\infty} |X(jf)|^2\, df \end{aligned} \tag{10.59}$$

Note that $\lambda = 2\pi f$, is a dummy variable of integration, plays the role of the angular frequency. The quantity $|X(jf)|^2$ is known as the *energy spectrum*. The result established in (10.59) is known as *Parseval Theorem* which has great importance in signal processing and communications.

10.6.3 Frobenius Formula

Let $\lambda_1, \lambda_2, \ldots, \lambda_n$ are the eigenvalues of the $n \times n$ matrix A and let $f(x)$ be a function which is analytic inside a circle in the complex plane that contains all the λ_i. Then $f(\lambda_1), f(\lambda_2), \ldots, f(\lambda_n)$ are the eigenvalues of the matrix function $f(A)$. For example, the state transition matrix $\Phi(t, t_o) = e^{A(t-t_o)}$ has eigenvalues $\gamma_i = e^{\lambda_i(t-t_o)}$. It can be verified that the eigenvectors of A and $\Phi(t, t_o)$ are the same.

10.7 Inequalities

All mathematical inequalities are proved for completeness. They are termed facts due to their high frequency of usage in the analytical developments.

10.7.1 Inequality 1

For any real matrices Σ_1 , Σ_2 and Σ_3 with appropriate dimensions and $\Sigma_3^t\ \Sigma_3\ \leq\ I$, it follows that

$$\Sigma_1 \Sigma_3 \Sigma_2 + \Sigma_2^t \Sigma_3^t \Sigma_1^t \ \leq\ \alpha\ \Sigma_1 \Sigma_1^t + \alpha^{-1}\ \Sigma_2^t \Sigma_2, \quad \forall \alpha > 0$$

Proof This inequality can be proved as follows. Since $\Phi^t \Phi \geq 0$ holds for any matrix Φ, then take Φ as

$$\Phi = [\alpha^{1/2}\ \Sigma_1 - \alpha^{-1/2}\ \Sigma_2]$$

Expansion of $\Phi^t \Phi \geq 0$ gives $\forall \alpha > 0$

$$\alpha\ \Sigma_1 \Sigma_1^t + \alpha^{-1}\ \Sigma_2^t \Sigma_2 -\ \Sigma_1^t \Sigma_2 -\ \Sigma_2^t \Sigma_1 \geq 0$$

which by simple arrangement yields the desired result. ∇∇∇

10.7.2 Inequality 2

Let $\Sigma_1,\ \Sigma_2,\ \Sigma_3$ and $0 < R = R^t$ be real constant matrices of compatible dimensions and $H(t)$ be a real matrix function satisfying $H^t(t)H(t) \leq I$. Then for any $\rho > 0$ satisfying $\rho \Sigma_2^t \Sigma_2 < R$, the following matrix inequality holds:

$$(\Sigma_3 + \Sigma_1 H(t) \Sigma_2) R^{-1} (\Sigma_3^t + \Sigma_2^t H^t(t) \Sigma_1^t) \ \leq\ \rho^{-1} \Sigma_1 \Sigma_1^t + \Sigma_3 \Big(R - \rho \Sigma_2^t \Sigma_2\Big)^{-1} \Sigma_3^t$$

Proof The proof of this inequality proceeds like the previous one by considering that

$$\Phi = [(\rho^{-1}\ \Sigma_2 \Sigma_2^t)^{-1/2} \Sigma_2 R^{-1} \Sigma_3^t - (\rho^{-1}\ \Sigma_2 \Sigma_2^t)^{-1/2} H^t(t) \Sigma_1^t]$$

Recall the following results:

$$\rho \Sigma_2^t \Sigma_2 < R,$$

$$[R - \rho \Sigma_2^t \Sigma_2]^{-1} = [R^{-1} + R^{-1} \Sigma_2^t [\rho^{-1} I\ -\ \Sigma_2 R^{-1} \Sigma_2^t]^{-1} \Sigma_2 R^{-1} \Sigma_2$$

and

$$H^t(t)H(t) \leq I \Longrightarrow H(t)H^t(t) \leq I$$

Expansion of $\Phi^t \Phi\ \geq\ 0$ under the condition $\rho \Sigma_2^t \Sigma_2\ <\ R$ with standard matrix manipulations gives

$$\Sigma_3 R^{-1} \Sigma_2^t H^t(t) \Sigma_1^t + \Sigma_1 H(t) \Sigma_2 R^{-1} \Sigma_3^t + \Sigma_1 H(t) \Sigma_2 \Sigma_2^t H^t(t) \Sigma_1^t \le$$
$$\rho^{-1} \Sigma_1 H(t) H^t(t) \Sigma_1^t + \Sigma_3^t R^{-1} \Sigma_2 [\rho^{-1} I \ \Sigma_2 \Sigma_2^t]^{-1} \Sigma_2 R^{-1} \Sigma_3^t \Longrightarrow$$
$$(\Sigma_3 + \Sigma_1 H(t) \Sigma_2) R^{-1} (\Sigma_3^t + \Sigma_2^t H^t(t) \Sigma_1^t) - \Sigma_3 R^{-1} \Sigma_3^t \le$$
$$\rho^{-1} \Sigma_1 H(t) H^t(t) \Sigma_1^t + \Sigma_3^t R^{-1} \Sigma_2 [\rho^{-1} I - \Sigma_2 \Sigma_2^t]^{-1} \Sigma_2 R^{-1} \Sigma_3^t \Longrightarrow$$
$$(\Sigma_3 + \Sigma_1 H(t) \Sigma_2) R^{-1} (\Sigma_3^t + \Sigma_2^t H^t(t) \Sigma_1^t) \le$$
$$\Sigma_3 [R^{-1} + \Sigma_2 [\rho^{-1} I - \Sigma_2 \Sigma_2^t]^{-1} \Sigma_2 R^{-1}] \Sigma_3^t +$$
$$\rho^{-1} \Sigma_1 H(t) H^t(t) \Sigma_1^t =$$
$$\rho^{-1} \Sigma_1 H(t) H^t(t) \Sigma_1^t + \Sigma_3 \Big(R - \rho \Sigma_2^t \Sigma_2 \Big)^{-1} \Sigma_3^t$$

which completes the proof. ∇∇∇

10.7.3 Inequality 3

For any real vectors β, ρ and any matrix $Q^t = Q > 0$ with appropriate dimensions, it follows that

$$-2\rho^t \beta \le \rho^t \, Q \, \rho + \beta^t \, Q^{-1} \, \beta$$

Proof Starting from the fact that

$$[\rho + Q^{-1}\beta]^t \, Q \, [\rho + Q^{-1}\beta] \ge 0, \quad Q > 0$$

which when expanded and arranged yields the desired result. ∇∇∇

10.7.4 Inequality 4 (Schur Complements)

Given a matrix Ω composed of constant matrices $\Omega_1, \Omega_2, \Omega_3$ where $\Omega_1 = \Omega_1^t$ and $0 < \Omega_2 = \Omega_2^t$ as follows:

$$\Omega = \begin{bmatrix} \Omega_1 & \Omega_3 \\ \Omega_3^t & \Omega_2 \end{bmatrix}$$

We have the following results:

(A) $\Omega \geq 0$ if and only if either

$$\begin{cases} \Omega_2 \geq 0 \\ \Pi = \Upsilon \Omega_2 \\ \Omega_1 - \Upsilon \, \Omega_2 \, \Upsilon^t \geq 0 \end{cases} \tag{10.60}$$

or

$$\begin{cases} \Omega_1 \geq 0 \\ \Pi = \Omega_1 \Lambda \\ \Omega_2 - \Lambda^t \, \Omega_1 \, \Lambda \geq 0 \end{cases} \tag{10.61}$$

hold where Λ, Υ are some matrices of compatible dimensions.
(B) $\Omega > 0$ if and only if either

$$\begin{cases} \Omega_2 > 0 \\ \Omega_1 - \Omega_3 \, \Omega_2^{-1} \, \Omega_3^t > 0 \end{cases}$$

or

$$\begin{cases} \Omega_1 \geq 0 \\ \Omega_2 - \Omega_3^t \, \Omega_1^{-1} \, \Omega_3 > 0 \end{cases}$$

hold where Λ, Υ are some matrices of compatible dimensions.

In this regard, matrix $\Omega_3 \, \Omega_2^{-1} \, \Omega_3^t$ is often called the Schur complement $\Omega_1(\Omega_2)$ in Ω.

Proof (**A**) To prove (10.60), we first note that $\Omega_2 \geq 0$ is necessary. Let $z^t = [z_1^t \quad z_2^t]$ be a vector partitioned in accordance with Ω. Thus we have

$$z^t \, \Omega \, z = z_1^t \Omega_1 z_1 + 2 z_1^t \Omega_3 z_2 + z_2^t \Omega_2 z_2 \tag{10.62}$$

Select z_2 such that $\Omega_2 z_2 = 0$. If $\Omega_3 z_2 \neq 0$, let $z_1 = -\pi \Omega_3 z_2, \; \pi > 0$. Then it follows that

$$z^t \, \Omega \, z = \pi^2 \, z_2^t \Omega_3^t \Omega_1 \Omega_3 z_2 - 2\pi \, z_2^t \Omega_3^t \Omega_3 z_2$$

which is negative for a sufficiently small $\pi > 0$. We thus conclude $\Omega_1 \, z_2 = 0$ which then leads to $\Omega_3 z_2 = 0, \; \forall \, z_2$ and consequently

$$\Omega_3 = \Upsilon \, \Omega_2 \tag{10.63}$$

for some Υ.

Since $\Omega \geq 0$, the quadratic term $z^t \; \Omega \; z$ possesses a minimum over z_2 for any z_1. By differentiating $z^t \; \Omega \; z$ from (10.62) wrt z_2^t, we get

$$\frac{\partial(z^t \; \Omega \; z)}{\partial z_2^t} = 2\Omega_3^t \; z_1 + 2\Omega_2 \; z_2 = 2\Omega_2 \; \Upsilon^t \; z_1 + 2\Omega_2 \; z_2$$

Setting the derivative to zero yields

$$\Omega_2 \; \Upsilon \; z_1 = -\Omega_2 \; z_2 \tag{10.64}$$

Using (10.63) and (10.64) in (10.62), it follows that the minimum of $z^t \; \Omega \; z$ over z_2 for any z_1 is given by

$$\min_{z_2} \; z^t \; \Omega \; z = z_1^t [\Omega_1 \; - \; \Upsilon \; \Omega_2 \; \Upsilon^t] z_1$$

which prove the necessity of $\Omega_1 \; - \; \Upsilon \; \Omega_2 \; \Upsilon^t \geq 0$.

On the other hand, we note that the conditions (10.60) are necessary for $\Omega \geq 0$ and since together they imply that the minimum of $z^t \; \Omega \; z$ over z_2 for any z_1 is nonnegative, they are also sufficient.

Using similar argument, conditions (10.61) can be derived as those of (10.60) by starting with Ω_1.

The proof of **(B)** follows as direct corollary of **(A)**. ∇∇∇

10.7.5 Inequality 5

For any quantities u and v of equal dimensions and for all $\eta_t = i \in \mathscr{S}$, it follows that the following inequality holds:

$$||u \; + \; v||^2 \; \leq \; [1 + \beta^{-1}] \; ||u||^2 \; + \; [1 + \beta]||v||^2 \tag{10.65}$$

for any scalar $\beta > 0, \quad i \in \mathscr{S}$

Proof Since

$$\begin{aligned} &[u \; + \; v]^t \; [u \; + \; v] = \\ &u^t \; u \; + \; v^t \; v \; + \; 2 \; u^t \; v \end{aligned} \tag{10.66}$$

It follows by taking norm of both sides of (10.66) for all $i \in \mathscr{S}$ that

$$||u \; + \; v||^2 \; \leq \; ||u||^2 \; + \; ||v||^2 \; + \; 2 \; ||u^t \; v|| \tag{10.67}$$

We know from the triangle inequality that

$$2\ ||u^t\ v||\ \leq\ \beta^{-1}\ ||u||^2\ +\ \beta\ ||v||^2 \tag{10.68}$$

On substituting (10.68) into (10.67), it yields (10.65). ∇∇∇

10.8 Lemmas

The basic tools and standard results that are utilized in robustness analysis and resilience design in the different chapters are collected hereafter.

Lemma 10.6 *The matrix inequality*

$$-\ \Lambda\ +\ S\ \Omega^{-1}\ S^t\ <\ 0 \tag{10.69}$$

holds for some $0 < \Omega = \Omega^t \in \Re^{n\times n}$, *if and only if*

$$\begin{bmatrix} -\Lambda & S\mathcal{X} \\ \bullet & -\mathcal{X} - \mathcal{X}^t + \mathcal{Z} \end{bmatrix}\ <\ 0 \tag{10.70}$$

holds for some matrices $\mathcal{X} \in \Re^{n\times n}$ *and* $\mathcal{Z} \in \Re^{n\times n}$.

Proof ($\Longrightarrow$) By Schur complements, inequality (10.69) is equivalent to

$$\begin{bmatrix} -\Lambda & S\Omega^{-1} \\ \bullet & -\Omega^{-1} \end{bmatrix}\ <\ 0 \tag{10.71}$$

Setting $\mathcal{X} = \mathcal{X}^t = \mathcal{Z} = \Omega^{-1}$, we readily obtain inequality (10.70).
($\Longleftarrow$) Since the matrix $[I \quad S]$ is of full rank, we obtain

$$\begin{bmatrix} I \\ S^t \end{bmatrix}^t \begin{bmatrix} -\Lambda & S\mathcal{X} \\ \bullet & -\mathcal{X} - \mathcal{X}^t + \mathcal{Z} \end{bmatrix} \begin{bmatrix} I \\ S^t \end{bmatrix}\ <\ 0 \Longleftrightarrow$$
$$-\ \Lambda\ +\ S\ \mathcal{Z}\ S^t\ <\ 0 \Longleftrightarrow -\ \Lambda\ +\ S\ \Omega^{-1}\ S^t\ <\ 0\ ,\quad \mathcal{Z} = \Omega^{-1} \tag{10.72}$$

which completes the proof. ∇∇∇

Lemma 10.7 *The matrix inequality*

$$A\boldsymbol{P} + \boldsymbol{P}A^t + D^t\ \boldsymbol{R}^{-1}\ D\ +\ \boldsymbol{M} <\ 0 \tag{10.73}$$

holds for some $0 < \boldsymbol{P} = \boldsymbol{P}^t \in \Re^{n\times n}$, *if and only if*

$$\begin{bmatrix} A\mathscr{V} + \mathscr{V}^t A^t + \boldsymbol{M} & \boldsymbol{P} + A\mathscr{W} - \mathscr{V} & D^t\boldsymbol{R} \\ \bullet & -\mathscr{W} - \mathscr{W}^t & 0 \\ \bullet & \bullet & -\boldsymbol{R} \end{bmatrix} < 0 \tag{10.74}$$

holds for some $\mathscr{V} \in \Re^{n\times n}$ *and* $\mathscr{W} \in \Re^{n\times n}$.

Proof ($\Longrightarrow$) By Schur complements, inequality (10.73) is equivalent to

$$\begin{bmatrix} AP + PA^t + M & D^tR \\ \bullet & -R \end{bmatrix} < 0 \tag{10.75}$$

Setting $\mathscr{V} = \mathscr{V}^t = \mathrm{P}$, $\mathscr{W} = \mathscr{W}^t = \mathrm{R}$, it follows from **Lemma** (10.6) with Schur complements that there exists $\mathrm{P} > 0$, $\mathscr{V}$, $\mathscr{W}$ such that inequality (10.74) holds.
($\Longleftarrow$) In a similar way, Schur complements to inequality (10.74) imply that:

$$\begin{bmatrix} A\mathscr{V} + \mathscr{V}^t A^t + \mathrm{M} & \mathrm{P} + A\mathscr{W} - \mathscr{V} & D^t\mathrm{R} \\ \bullet & -\mathscr{W} - \mathscr{W}^t & 0 \\ \bullet & \bullet & -\mathrm{R} \end{bmatrix} < 0$$

$$\Longleftrightarrow \begin{bmatrix} I \\ A \end{bmatrix} \begin{bmatrix} A\mathscr{V} + \mathscr{V}^t A^t + \mathrm{M} + D^t\mathrm{R}^{-1}D & \mathrm{P} + A\mathscr{W} - \mathscr{V} \\ \bullet & -\mathscr{W} - \mathscr{W}^t \end{bmatrix} \begin{bmatrix} I \\ A \end{bmatrix}^t < 0$$

$$\Longleftrightarrow AP + PA^t + D^t\ R^{-1}\ D\ +\ M < 0, \quad \mathscr{V} = \mathscr{V}^t \tag{10.76}$$

which completes the proof. ∇∇∇

Lemma 10.8 *Given any* $x \in \Re^n$*:*

$$\max\ \{[x^t\ RH\Delta G\ x]^2 :\ \Delta \in \Re\} = x^t\ RHH^tR\,x\ \ x^t\ G^tG\ x$$

Lemma 10.9 *Given matrices* $0 \le X = X^t \in \Re^{p\times p}$, $Y = Y^t < 0 \in \Re^{p\times p}$, $0 \le Z = Z^t \in \Re^{p\times p}$, *such that*

$$[\xi^t\ Y\ \xi]^2\ -\ 4\,[\xi^t\ X\ \xi\ \ \xi^t\ Z\ \xi]^2\ >\ 0$$

for all $0 \ne \xi \in \Re^p$ *is satisfied. Then there exists a constant* $\alpha > 0$ *such that*

$$\alpha^2\ X\ +\ \alpha\ Y\ +\ Z\ <\ 0$$

Lemma 10.10 *For a given two vectors* $\alpha \in \boldsymbol{R}^n$, $\beta \in \boldsymbol{R}^m$ *and matrix* $\boldsymbol{N} \in \boldsymbol{R}^{n\times m}$ *defined over a prescribed interval* Ω, *it follows for any matrices* $X \in \boldsymbol{R}^{n\times n}$, $Y \in \boldsymbol{R}^{n\times m}$, *and* $Z \in \boldsymbol{R}^{m\times m}$, *the following inequality holds*

$$-\,2\int_{\Omega} \alpha^t(s)\,\boldsymbol{N}\,\beta(s)\,ds\ \le\ \int_{\Omega} \begin{bmatrix} \alpha(s) \\ \beta(s) \end{bmatrix}^t \begin{bmatrix} X & Y - \boldsymbol{N} \\ Y^t - \boldsymbol{N}^t & Z \end{bmatrix} \begin{bmatrix} \alpha(s) \\ \beta(s) \end{bmatrix} ds$$

where

$$\begin{bmatrix} X & Y \\ Y^t & Z \end{bmatrix} \geq 0$$

An algebraic version of **Lemma** 10.10 is stated below

Lemma 10.11 *For a given two vectors* $\alpha \in \boldsymbol{R}^n$, $\beta \in \boldsymbol{R}^m$ *and matrix* $\boldsymbol{N} \in \boldsymbol{R}^{n\times m}$ *defined over a prescribed interval* Ω, *it follows for any matrices* $X \in \boldsymbol{R}^{n\times n}$, $Y \in \boldsymbol{R}^{n\times m}$, *and* $Z \in \boldsymbol{R}^{m\times m}$, *the following inequality holds*

$$\begin{aligned} -2\,\alpha^t\,\boldsymbol{N}\,\beta \;&\leq\; \begin{bmatrix} \alpha \\ \beta \end{bmatrix}^t \begin{bmatrix} X & Y-\boldsymbol{N} \\ Y^t-\boldsymbol{N}^t & Z \end{bmatrix} \begin{bmatrix} \alpha \\ \beta \end{bmatrix} \\ &= \alpha^t X\alpha + \beta^t(Y^t-\boldsymbol{N}^t)\alpha + \alpha^t(Y-\boldsymbol{N})\beta + \beta^t Z\beta \end{aligned}$$

subject to

$$\begin{bmatrix} X & Y \\ Y^t & Z \end{bmatrix} \geq 0$$

Lemma 10.12 *Let* $0 < Y = Y^t$ *and* $M,\ N$ *be given matrices with appropriate dimensions. Then it follows that*

$$Y + M\,\Delta\,N + N^t\,\Delta^t\,M^t \;<\; 0\ ,\quad \forall\ \Delta^t\,\Delta \leq I$$

holds if and only if there exists a scalar $\varepsilon > 0$ *such that*

$$Y + \varepsilon\,M\,M^t + \varepsilon^{-1}\,N^t\,N \;<\; 0$$

In the following lemma, we let $X(z) \in \mathrm{R}^{n\times p}$ be a matrix function of the variable z. A matrix $X_*(z)$ is called the orthogonal complement of $X(z)$ if $X^t(z)\,X_*(z) = 0$ and $X(z)\,X_*(z)$ is nonsingular (of maximum rank).

Lemma 10.13 *Let* $0 < L = L^t$ *and* $X,\ Y$ *be given matrices with appropriate dimensions. Then it follows that the inequality*

$$L(z) + X(z)\,P\,Y(z) + Y^t(z)\,P^t\,X^t(z) \;>\; 0 \tag{10.77}$$

holds for some P *and* $z = z_o$ *if and only if the following inequalities*

$$X_*^t(z)\,L(z)\,X_*(z) \;>\; 0,\quad Y_*^t(z)\,L(z)\,Y_*(z) \;>\; 0 \tag{10.78}$$

hold with $z = z_o$.

It is significant to observe that feasibility of matrix inequality (10.77) with variables P and z is equivalent to the feasibility of (10.78) with variable z and thus the matrix variable P has been eliminated from (10.77) to form (10.78). Using Finsler's lemma, we can express (10.78) in the form

$$L(z) - \beta X(z) X^t(z) > 0, \quad L(z) - \beta Y(z) Y^t(z) > 0 \qquad (10.79)$$

for some $\beta \in$ R.

The following is a statement of the reciprocal projection Lemma

Lemma 10.14 *Let* $P > 0$ *be a given matrix. The following statements are equivalent:*
(i) $M + Z + Z^t < 0$
(ii) the linear matrix inequality problem (LMIP)

$$\begin{bmatrix} M + P - (V + V^t) & V^t + Z^t \\ V + Z & -P \end{bmatrix} < 0$$

is feasible with respect to the general matrix V.

10.9 Linear Matrix Inequalities

It has been shown that a wide variety of problem arising in system and control theory can conveniently reduced to a few standard convex or quasiconvex optimization problems involving linear matrix inequalities (LMIs). The resulting optimization problems can then be solved numerically very efficiently using commerically available interior point methods.

10.9.1 Basics

One of the earliest LMIs arises in Lyapunov theory. It is well known that the differential equation

$$\dot{x}(t) = A\, x(t) \qquad (10.80)$$

has all of its trajectories converge to zero (stable) id and only if there exists a matrix $P > 0$ such that

$$A^t\, P + A\, P < 0 \qquad (10.81)$$

This leads to the LMI formulation of stability, that is , *a linear time-invariant system is asymptotically stable if and only if there exists a matrix* $0 < P = P^t$ *satisfying the LMIs*

$$A^t P + A P < 0, \quad P > 0$$

Given a vector variable $x \in R^n$ and a set of matrices $0 < G_j = G_j^t \in R^{n \times n}, \; j = 0, \ldots, p$, then a basic compact formulation of a LMI is

$$G(x) \stackrel{\Delta}{=} G_0 + \sum_{j=1}^{p} x_j G_j > 0 \tag{10.82}$$

Notice that (10.82) implies that $v^t G(x) v > 0 \; \forall 0 \neq v \in R^n$. More importantly, the set $\{x \,| G(x) > 0$ is convex. Nonlinear (convex) inequalities are converted to LMI form using Schur complements in the sense that

$$\begin{bmatrix} Q(x) & S(x) \\ \bullet & R(x) \end{bmatrix} > 0 \tag{10.83}$$

where $Q(x) = Q^t(x), \; R(x) = R^t(x), \; S(x)$ depend affinely on x, is equivalent to

$$R(x) > 0, \quad Q(x) - S(x) R^{-1}(x) S^t(x) > 0 \tag{10.84}$$

More generally, the constraint

$$Tr[S^t(x) \, P^{-1}(x) \, S(x)] < 1, \quad P(x) > 0$$

where $P(x) = P^t(x) \in R^{n \times n}, \; S(x) \in R^{n \times p}$ depend affinely on x, is handled by introducing a new (slack) matrix variable $Y(x) = Y^t(x) \in\in R^{p \times p}$ and the LMI (in x and Y):

$$TrY < 1, \quad \begin{bmatrix} Y & S(x) \\ \bullet & P(x) \end{bmatrix} > 0 \tag{10.85}$$

Most of the time, our LMI variables are matrices. It should clear from the foregoing discussions that a quadratic matrix inequality in the variable P can be readily expressed as LMI in the same variable.

10.9.2 Some Standard Problems

Here we provide some common convex problems that we encountered throughout the monograph. Given an LMI $G(x) > 0$, the corresponding LMIP is to *find a feasible* $x \equiv x^f$ *such that* $G(x^f) > 0$, or *determine that the LMI is infeasible*.

It is obvious that this is a convex feasibility problem.

The generalized eigenvalue problem (GEVP) is to minimize the maximum generalized eigenvalue of a pair of matrices that depend affinely on a variable, subject to an LMI constraint. GEVP has the general form

$$\begin{aligned} &\textit{minimize } \lambda \\ &\textit{subject to } \lambda B(x) - A(x) > 0, \quad B(x) > 0, \; C(x) > 0 \end{aligned} \tag{10.86}$$

where A, B, C are symmetric matrices that are affine functions of x. Equivalently stated

$$\begin{aligned} &\textit{minimize } \lambda_M[A(x), B(x)] \\ &\textit{subject to } B(x) > 0, \; C(x) > 0 \end{aligned} \tag{10.87}$$

where $\lambda_M[X, Y]$ denotes the largest generalized eigenvalue of the pencil $\lambda Y - X$ with $Y > 0$. This is problem is quasiconvex optimization problem since the constraint is convex and the objective, $\lambda_M[A(x), B(x)]$, is quasiconvex.

The eigenvalue problem (EVP) is to minimize the maximum eigenvalue of a matrix that depend affinely on a variable, subject to an LMI constraint. EVP has the general form

$$\begin{aligned} &\textit{minimize } \lambda \\ &\textit{subject to } \lambda I - A(x) > 0, \quad B(x) > 0 \end{aligned} \tag{10.88}$$

where A, B are symmetric matrices that are affine functions of the optimization variable x. This is problem is convex optimization problem.
EVPs can appear in the equivalent form of minimizing a linear function subject to an LMI, that is,

$$\begin{aligned} &\textit{minimize } c^t x \\ &\textit{subject to } G(x) > 0 \end{aligned} \tag{10.89}$$

where $G(x)$ is an affine function of x. Examples of $G(x)$ include

$$PA + A^t P + C^t C + \gamma^{-1} P B B^t P < 0, \quad P > 0$$

It should be stressed that the standard problems (LMIPs, GEVPs, EVPs) are tractable, from both theoretical and practical viewpoints:
They can be solved in polynomial time.

They can be solved in practice very efficiently using commercial software

10.9.3 The S-procedure

In some design applications, we faced the constraint that some quadratic function be negative whenever some other quadratic function is negative. In such cases, this constraint can be expressed as an LMI in the data variables defining the quadratic functions.

Let $G_o, \ldots, G_p$ be quadratic functions of the variable $\xi \in \mathbb{R}^n$:

$$G_j(\xi) \triangleq \xi^t R_j \xi \; + \; 2u_j^t \xi \; + \; v_j, \quad j = 0, \ldots, p, \quad R_j = R_j^t$$

We consider the following condition on $G_o, \ldots, G_p$:

$$G_o(\xi) \; \leq \; 0 \;\; \forall \xi \quad such \;\; that \quad G_j(\xi) \; \geq \; 0, \quad j = 0, \ldots, p \tag{10.90}$$

It is readily evident that if there exist scalars $\omega_1 \geq 0, \ldots, \omega_p \geq 0$ such that

$$\forall \xi, \qquad G_o(\xi) \; - \; \sum_{j=1}^{p} \; \omega_j \; G_j(\xi) \; \geq \; 0 \tag{10.91}$$

then inequality (10.90) holds. Observe that if the functions $G_o, \ldots, G_p$ are affine, then Farkas lemma state that (10.90) and (10.91) are equivalent. Interestingly enough, inequality (10.91) can written as

$$\begin{bmatrix} R_o & u_o \\ \bullet & v_o \end{bmatrix} - \sum_{j=1}^{p} \omega_j \begin{bmatrix} R_j & u_j \\ \bullet & v_j \end{bmatrix} \; \geq \; 0 \tag{10.92}$$

The above discussions were stated for non-strict inequalities. In case of strict inequality, we let $R_o, \ldots, R_p \in \mathbb{R}^{n \times n}$ be symmetric matrices with the following qualifications:

$$\xi^t R_o \xi \; > \; 0 \;\; \forall \xi \quad such \;\; that \quad \xi^t G_j \xi \; \geq \; 0, \quad j = 0, \ldots, p \tag{10.93}$$

Once again, it is obvious that there exist scalars $\omega_1 \geq 0, \ldots, \omega_p \geq 0$ such that

$$\forall \xi, \qquad G_o(\xi) \; - \; \sum_{j=1}^{p} \; \omega_j \; G_j(\xi) \; > \; 0 \tag{10.94}$$

then inequality (10.93) holds. Observe that (10.94) is an LMI in the variables $R_o, \omega_1, \ldots, \omega_p$.

It should be remarked that the **S**-procedure deals with non-strict inequalities allows the inclusion of constant and linear terms. In the strict version, only quadratic functions can be used.

10.10 Lyapunov Map and Lyapunov Equation

An important type of linear equation that we will encounter in our book is of Lyapunov type[2]: Find a matrix $X \in C^{n \times n}$ that satisfies the equation

$$AX + XB = 0 \tag{10.95}$$

for a given pair of matrices $A, B \in C^{n \times n}$. Although solutions to this type of equation can be difficult in general, simple solutions exist when both A and B have n independent eigenvectors. Suppose $\{u_i \in C^n\}_{i=1}^n$ are the n right eigenvectors of A, and $\{v_j \in C^n\}_{j=1}^n$ are the n left eigenvectors of B; i.e.,

$$Au_i = \lambda_i u_i; \quad v_j^* B = \eta_j v_j^* \tag{10.96}$$

for eigenvalues λ_i, η_j for each i, j. Here v^* means the complex conjugate and transpose of v, since v can be complex. (Lyapunov map). For the above matrix A and B, the n^2 eigenvectors of the Lyapunov map

$$L : X \mapsto AX + XB \tag{10.97}$$

are exactly $X_{ij} = u_i v_j^* \in C^{n \times n}$, and the corresponding eigenvalues are $\lambda_i + \eta_j \in C, i, j = 1, 2, \ldots, n$.

Proof The n^2 matrices $\{X_{ij}\}_{i,j=1}^n$ are linearly independent, and they must be all the eigenvectors of L.

Due to this fact, any matrix X that satisfies the Lyapunov equation $AX + XB = 0$ must be in the subspace spanned by eigenvectors X_{ij} that have zero eigenvalues: $\lambda_i + \eta_j = 0$. In MATLAB, the command X = lyap(A, B, C) solves the more general Lyapunov equation $AX + XB = -C$.

In this book, we often look for solutions X with extra requirements on its structure. For instance, X needs to be real and symmetric (Chap. 6), or X has to be a rotation matrix (Chap. 10). If so, we have only to take the intersection of the space of solutions to the Lyapunov equation with the space of symmetricmatrices or rotation matrices.

[2] It is also called Sylvester equation in some literature.

10.11 Persistence of Excitation and Sufficiently Rich Inputs

We start with the following definition:

Definition 10.26 The vector $\phi \in R^n$ is persistently excited (PE) with level α_0 if it satisfies

$$\int_t^{t+T_0} \phi(\tau)\phi^t(\tau)d\tau \geq \alpha_0 T_0 I \tag{10.98}$$

for some $\alpha_0 > 0$, $T_0 > 0$ and $\forall t \geq 0$.

Since $\phi\phi^t$ is always positive semidefinite, the PE condition requires that its integral over any interval of time of length T_0 is a positive definite matrix.

Definition 10.27 The signal $u \in R$ is called sufficiently rich of order n if it contains at least $\frac{n}{2}$ distinct nonzero frequencies.

For example, $u = \sum_{i=1}^{10} \sin \omega_i t$, where $\omega_i \neq \omega_j$ for $i \neq j$ is sufficiently rich of order 20. A more general definition of sufficiently rich signals and associated properties may be found in [3].

Let us consider the signal vector $\phi \in R^n$ generated as

$$\phi = H(s)u, \tag{10.99}$$

where $u \in R$ and $H(s)$ is a vector whose elements are transfer functions that are strictly proper with stable poles.

Theorem 10.5 *Consider (10.99) and assume that the complex vectors*

$$H(j\omega_1), \ldots, H(j\omega_n)$$

are linearly independent on the complex space

$$C^n \;\; \forall\, \omega_1, \omega_2, \ldots, \omega_n \in R, \quad \omega_i \neq \omega_j, \quad i \neq j$$

Then ϕ is PE ifand only if u is sufficiently rich of order n.

Proof The proof of Theorem 10.5 can be found in [3].

We demonstrate the use of Theorem 10.5 for the Example 8.3, where

$$\phi = H(s)u$$

and

$$H(s) = \begin{bmatrix} 1 \\ -\frac{b}{s+u} \end{bmatrix}.$$

In this case $n = 2$ and

$$H(j\omega_1) = \begin{bmatrix} 1 \\ -\frac{b}{j\omega_1 + a} \end{bmatrix}, \quad H(j\omega_2) = \begin{bmatrix} 1 \\ -\frac{b}{j\omega_2 + a} \end{bmatrix}.$$

We can show that the matrix $[H(j\omega_1), H(j\omega_2)]$ is nonsingular, which implies that $H(j\omega_1), H(j\omega_2)$ are linearly independent for any ω_1, ω_2 different than zero and $\omega_1 \neq \omega_2$.

Let us choose

$$u = \sin \omega_0 t$$

for some $\omega_0 \neq 0$ which is sufficiently rich of order 2. According to Theorem 10.5, this input should guarantee that ϕ is PE for the Example 8.3. Ignoring the transient terms that converge to zero exponentially fast, we can show that at steady state

$$\phi = \begin{bmatrix} \sin \omega_0 t \\ c_0 \sin(\omega_0 t + \varphi_0) \end{bmatrix},$$

where

$$c_0 = \frac{|b|}{\sqrt{\omega_0^2 + a^2}}, \quad \varphi_0 = arg\left(\frac{-b}{j\omega_0 + a}\right).$$

Now

$$\phi\phi^t = \begin{bmatrix} \sin^2 \omega_0 t & c_0 \sin \omega_0 t \sin(\omega_0 t + \varphi_0) \\ c_0 \sin \omega_0 t \sin(\omega_0 t + \varphi_0) & c_0^2 \sin^2(\omega_0 t + \varphi) \end{bmatrix}$$

and

$$\int_t^{t+T_0} \phi(\tau)\phi^t(\tau) d\tau = \begin{bmatrix} a_{11} & a_{12} \\ a_{12} & a_{22} \end{bmatrix},$$

where

$$\begin{aligned}
a_{11} &= \frac{T_0}{2} - \frac{\sin 2\omega_0(t + T_0) - \sin 2\omega_0 t}{4\omega_0} \\
a_{12} &= c_0 \frac{T_0}{2} \cos \varphi_0 + c_0 \frac{\sin \varphi_0}{4\omega_0}(\cos 2\omega_0 t - \cos 2\omega_0(t + T_0)), \\
a_{22} &= c_0^2 \frac{T_0}{2} - c_0^2 \frac{\sin 2(\omega_0(t + T_0) + \varphi_0) - \sin 2(\omega_0 t + \varphi_0)}{4\omega_0}.
\end{aligned}$$

Choosing $T_0 = \frac{\pi}{\omega_0}$ it follows that

$$a_{11} = \frac{T_0}{2}, \quad a_{12} = \frac{T_0 c_0}{2} \cos\varphi_0, \quad a_{22} = \frac{T_0}{2} c_0^2$$

and

$$\int_t^{t+T_0} \phi(\tau)\phi^t(\tau)d\tau = \frac{T_0}{2} \begin{bmatrix} 1 & c_0 \cos\varphi_0 \\ c_0 \cos\varphi_0 & c_0^2 \end{bmatrix},$$

which is a positive definite matrix. We can verify that for $\alpha_0 = \frac{1}{2} \frac{(1-\cos^2\varphi_0)c_0^2}{1+c_0^2} > 0$,

$$\int_t^{t+T_0} \phi(\tau)\phi^t(\tau)d\tau \geq T_0 \alpha_0 I,$$

which implies that ϕ is PE.

Let us consider the plant model

$$y = \frac{b(s^2+4)}{(s+5)^3} u,$$

where b is the only unknown parameter. A suitable parametric model for estimating b is

$$z = \theta^* \phi,$$

where

$$z = y, \quad \theta^* = b, \quad \phi = \frac{s^2+4}{(s+5)^3} u.$$

In this case $\phi \in R$ and $H(s) = \frac{s^2+4}{(s+5)^3}$; i.e., $n = 1$ in Theorem 10.5. Let us use Theorem 10.5 to choose a sufficiently rich signal u that guarantees ϕ to be PE. In this case, according to the linear independence condition of Theorem 10.5 for the case of $n = 1$, we should have

$$|H(j\omega_0)| = \frac{4-\omega_0^2}{(25+\omega_0^2)^{3/2}} \neq 0$$

for any $\omega_0 \neq 0$. This condition is clearly violated for $\omega_0 = 2$, and therefore a sufficiently rich input of order 1 may not guarantee ϕ to be PE. Indeed, the input $u = \sin 2t$ leads to $y = 0$, $\phi = 0$ at steady state, which imply that the output y and regressor ϕ carry no information about the unknown parameter b. For this example $u = \sin\omega_0 t$ will guarantee ϕ to be PE, provided $\omega_0 \neq 2$. Also, $u = \text{constant} \neq 0$

and $u = \sum_{i=1}^{m} \sin \omega_i t$, $m \geq 2$, will guarantee that ϕ is PE. In general, for each two unknown parameters we need at least a single nonzero frequency to guarantee PE, provided of course that $H(s)$ does not lose its linear independence as demonstrated by the above example.

The two-parameter case leads to a differential equation, which has exactly the same form as in the case of an arbitrary number of parameters.

10.12 Notes and References

The analysis presented in this chapter made extensive use of the standard texts including [3–5].

References

1. Ackermann J (1972) Fer Entwurf Linearer Regelungssysteme im Zustandsraum. Regelungstech, Prozess-Datenverarb 7:297–300
2. Angeli D, Bliman PA (2009) Convergence speed of unsteady distributed consensus: decay estimate along the settling spanning-trees. SIAM J Control Optim 48(1):1–32
3. Chiang RY, Safonov MG (1992) User's guide, robust control toolbox. The MathWorks, Natick
4. Godsil C, Royle G (2001) Algebraic graph theory. Springer, New York
5. Noble B, Daniel J (1988) Applied linear algebra, 3rd edn. Prentice Hall, New York
6. Olfati-Saber R, Murray R (2004) Consensus problems in networks of agents with switching topology and time delays. IEEE Trans Autom Control 49(9):1520–1533

Index

A
Ackerman formula, 558
Ancillary services, 15
Autonomous microgrid controller, 488
Autonomous mode, 327

B
Basics of microgrids, 16
Batteries, 72
Boost converter control, 68

C
Central energy manager, 210
Combined heat and power (CHP) plants, 47
Commercial factors, 8
Common coupling point, 210
Components and formation, 92
Consumer benefits, 28
Control loops, 67
Control tasks of distributed systems, 10

D
DC load control, 38
DC microgrids, 35
Decentralized control, 117
Decentralized inverter control, 348
Distributed capacity, 7
Distributed control architecture, 379
Distributed cooperative control, 440, 441
Distributed energy systems, 400
Distributed generation, 3
Distributed resources, 6
Dynamics of autonomous mode, 318, 320
Dynamics of grid-connected mode, 316

E
Eigenvalues and eigenvectors, 548
Electric power dynamic demand, 14
Electrical energy storage, 71
Energy storage, 31
Environmental benefits, 28
Environmental factors, 7

F
Factors of distributed generation growth, 7
Features of the grid, 12
Feedback linearization and tracking synchronization, 447
Flywheels, 73
Frequency control, 311
Frequency synchronization, 332
Frobenius formula, 560
Fuel cells, 53

G
Generator side control, 68
Grid-connected mode, 321
Grid-connected operation, 216
Grid-connected solar systems, 60
Grid-islanded operation, 216
Grid-side controller, 11

H
Horizontal axis turbines, 56

I
Inequalities, 560
Input-side controller, 10
Integrating distributed energy resources, 12

M.S. Mahmoud and F.M. AL-Sunni, *Control and Optimization of Distributed Generation Systems*, Power Systems, DOI 10.1007/978-3-319-16910-1

Internal combustion (IC) engines, 49
Inverter interfaces, 74
Islanded mode, 252
Islanding detection approach, 314

K
Kernel of a matrix, 546

L
Laplacian spectrum, 541
Linear matrix inequalities, 568
Local controllers, 333
Local load control, 15

M
Matrix inverse, 544
Matrix inversion lemma, 545
Mechanical storage systems, 72
Micro cogeneration systems, 48
Microgrid central controller, 211
Microgrid control levels, 442
Microgrid control tasks, 18
Microgrid subsystems, 210
Microsource and load controllers, 212
Microsource control, 23
Microsource controller, 210
Microturbines, 51, 210, 213
Mode-transition capability, 323
Monitoring techniques, 30
Multilevel control, 120

N
National/regulatory factors, 8
Null space, 546

O
Optimal control for autonomous microgrid, 487
Overcurrent limiters, 313

P
Parseval formula, 559
Power balancing principle, 38
Power management system, 332
Power sharing control strategy, 352
Primary power source, 400
Protection requirements, 25
PV sources control, 36

Q
QR decomposition, 557

R
Range of a matrix, 546
Rank of a matrix, 546
Rectifier control, 70
Reliability concepts, 13
Renewable energy generation, 55
Robust energy management, 33

S
Singular value decomposition (SVD), 551
Skew-symmetric matrix, 549, 550
Small-scale hydrogeneration, 57
Small-signal dynamic analysis, 315
Small wind turbine systems, 62
Smart grid, 26
Solar photovoltaic generation, 58
Solar technology basics, 59
SoS control, 116
Spinning reserve, 14
Stirling engines, 49
Storage control, 37
Storage devices, 214
Storage technologies, 70
Supercapacitors, 74
Supply-demand, 12
Symmetric matrices, 549
System of systems (SoS), 116

T
Technology of distributed generation, 3
Types of wind turbine systems, 63

V
Value of distributed generation, 5
Vertical axis turbines, 56
Voltage control, 312
Voltage-controlled voltage-sourced converter (VC-VSC), 310
Voltage controller, 490
Voltage source converter, 400
Voltage source inverters, 74
Voltage-sourced converter, 309

W
Wind power plants, 55
Wind turbine fundamentals, 66

MIX
Papier aus verantwortungsvollen Quellen
Paper from responsible sources
FSC® C105338

If you have any concerns about our products,
you can contact us on
ProductSafety@springernature.com

In case Publisher is established outside the EU,
the EU authorized representative is:
Springer Nature Customer Service Center GmbH
Europaplatz 3, 69115 Heidelberg, Germany

Printed by Libri Plureos GmbH
in Hamburg, Germany